€140
118
Bio

Pathophysiological Aspects of Proteases

Sajal Chakraborti · Naranjan S. Dhalla
Editors

Pathophysiological Aspects of Proteases

Springer

Editors
Sajal Chakraborti
Department of Biochemistry
and Biophysics
University of Kalyani
Kalyani, West Bengal
India

Naranjan S. Dhalla
Institute of Cardiovascular Sciences
University of Manitoba
Manitoba
Canada

ISBN 978-981-10-6140-0 ISBN 978-981-10-6141-7 (eBook)
DOI 10.1007/978-981-10-6141-7

Library of Congress Control Number: 2017948636

Printed on acid-free paper

This Springer imprint is published by Springer Nature
The registered company is Springer Nature Singapore Pte Ltd.
The registered company address is: 152 Beach Road, #21-01/04 Gateway East, Singapore 189721, Singapore

This book on Pathophysiological Aspects of Proteases is dedicated to Prof. K. Muniyappa, who was born on 8 September 1952 at Chickaballapur in the state of Karnataka, India. His individual and group achievements in the field of biochemistry and molecular biology have been spectacular and have secured him unique position in several areas of frontier research.

Prof. K. Muniyappa is a Former Chairman of the Department of Biochemistry at the Indian Institute of Science, Bangalore, and the Founding Coordinator for the National Research Associateship Program (and a similar program for the Northeast region) of the Department of Biotechnology, Govt. of India. He obtained his B.Sc. and M.Sc. degrees from the University of Mysore and doctoral degree from the Indian Institute of Science, Bangalore. He did his postdoctoral research at Yale University School of Medicine, USA. He has served as the American Cancer Society Visiting Professor at the University of Washington, Seattle and Royal Society Scholar at MRC, London. His research interests involve various aspects of chromosome biology in both prokaryotes and eukaryotes. Beyond research laboratory, his interest includes capacity building in higher education and research.

Prof. Muniyappa has received a plethora of national and international prizes and awards. To mention a few: S.S. Bhatnagar Prize (highest national award in Science & Technology), IISc Alumni Award for Excellence in Research, J.C. Bose National Fellowship, Sir M. Visvesvaraya Award by Govt. of Karnataka (highest state award), Kempe Gowda award for persons of eminence by GoK, Golden Jubilee award of the University of Mysore, Prof. Vishwanath Memorial award, American Cancer Society Eleanor Roosevelt Award, Yamagiwa International Cancer award (Switzerland) and the Rockefeller Foundation Career Development Award (USA). He is an elected Fellow of all the major Science Academies in

India, a Member of The Third World Academy of Sciences (TWAS), the Founding Member of the Karnataka State Academy for Science and Technology and Member of the Karnataka Vision Group in Science and Technology. He has been the Chairman/Member of several Task Force, PAC and Award committees of DBT, CSIR, ICMR, DST, UGC, UPSC and TWAS, and faculty/scientist selection committees of several universities and National Research Institutes.

Prof. Muniyappa undoubtedly is a legendary figure in Indian Science. Prof. Muniyappa has excellent ability to inspire and motivate young researchers. We feel honoured to dedicate this book to Prof. Muniyappa and wish him good health and success in his long fruitful activities.

Preface

Thou hast made me endless, such as thy pleasure.
This frail vessel thou emptiest again and again, and fillest it ever with fresher life. This little flute of a reed thou hast carried over hills and dales and hast breathed through it melodies eternally new. At the immortal touch of thy hands my little heart loses its limits in a great joy and gives birth to utterance ineffable. Thy infinite gifts come to me only on these very small hands of mine. Ages pass and still thou pourest and still there is room to fill.

Rabindranath Tagore (Gitanjali: Song of offerings)

This book is a product of a long-standing interest of the editors to handover compilations of chapters on a number of proteases and their pathophysiological consequences to the scientific fraternity.

The existence of proteases has been known for centuries. Their use was noted very early in our everyday life. In early days, the properties of proteases were exploited by man for food processing. Proteases from *Aspergillus oryzae* were used to modify wheat gluten, a component of bread indigestible to many people, which also affects yield of loaf processing. Gastric juice was demonstrated to be full of proteases and known to be responsible for food digestion.

Proteases play important physiological functions including cell division, regulation of functional and structural protein turnover and activation of zymogens, formation and lysis of blood clot, entry of sperm into ovum and fertilization, processing and transport of secretary proteins across membrane, regulation of gene expression and also in infection of pathogens. Based on the catalytic site on substrate, proteases are mainly classified into endoproteases and exoproteases. However, considering the mechanism of catalysis, proteases are classified into six distinct classes: aspartic, glutamic, metallo, cysteine, serine and threonine proteases.

In the recent past, inhibitors of angiotensin converting enzyme (ACE) and HIV proteases have shown substantial therapeutic success in developing drugs. However, due to difficulties of clear understanding of the selectivity of the active site of target protease, discovery of new drugs appears to be challenging. Therefore, more basic research is needed on proteases with reference to their physiological and pathological consequences. In this book, eminent scholars illuminated us with new information, which, we believe, will surely benefit researchers to unravel, at least, some of the yet unknown complexities of proteases and their pathological implications.

This book contains twenty-nine chapters, which are contributed by established investigators from across the globe. We are grateful to all the authors for enthusiasm, energy and precious time that they spent on their respective chapters to materialize this project. We would like to thank Praveenkumar Vijayakumar, Dr. Madhurima Kahali and Vinoth Selvamani (Springer) for their persuasiveness to achieve our goal. We are also thankful to Prof. Sankar Kumar Ghosh, Vice Chancellor, University of Kalyani for his encouragement.

Kalyani, India Sajal Chakraborti
Manitoba, Canada Naranjan S. Dhalla

Contents

About the Editors

Dr. Sajal Chakrabori is a Professor of Biochemistry in the Department of Biochemistry and Biophysics, University of Kalyani, Kalyani, West Bengal, India. He received his Ph.D. in Biochemistry from the University of Calcutta in 1982, and D.Sc. in Biochemistry from the University of Kalyani in 2004. His main areas of interest are regulation of phospholipase A2 and -D, Na+/K+ -ATPase, Ca2+ATPase and NADPH oxidase in pulmonary vascular cells; biochemistry of proteases and anti-proteases; and oxidant and calcium signalling by different stimulants such as angiotensin-1, thromboxan A2, endothelin-1 and Leukotriene D4 in regulating pulmonary vascular tone.

He received a Fulbright visiting fellowship to pursue research at the Department of Medicine, The Johns Hopkins University, USA, from 1987 to 1989, and at the Pulmonary Division, University of Utah Medical School and New York Medical College, New York, USA, from 1989 to mid-1990. He joined the University of Kalyani as a Reader in Biochemistry in 1998. Dr. Chakraborti is actively involved in research and has more than 100 publications in prominent journals to his credit. He has also contributed chapters to books published by Springer and Elsevier. He has edited four books with Springer namely: Proteases in Health and Disease; Role of Proteases in Cellular Dysfunction; Regulation of Membrane Na/K-ATPase; and Regulation of Ca2+ ATPases, V-ATPases and F-ATPases.

Dr. Naranjan S. Dhalla is Distinguished Professor at the University of Manitoba, Winnipeg, Canada. His expertise includes the subcellular and molecular basis of heart function in health and disease. He has been engaged in multidisciplinary research and education for promoting the scientific basis of cardiology, as well as training of professional manpower for combating heart disease for over 45 years.

Part I
Role of Proteases in the Development of Cancer

Role of Proteases in Breast Cancer

Sandra Radenkovic, Kristina Gopcevic, Gordana Konjevic
and Vladimir Jurisic

Abstract

Proteolytic enzymes comprise five classes depending on their catalytic mechanisms: serine, cysteine, aspartic, threonine, and matrix metalloproteinases (MMPs). Proteases play an important role in all stages of cancer progression: cancerogenesis, invasiveness, and metastasis. Level and activation of proteases is associated with progression of breast cancer. Cathepsins are proteases involved in tumor formation, and activity of Cathepsin D (CatD) is considered as an independent tumor marker for breast cancer patients. Processes of apoptosis, cell proliferation, growth, angiogenesis, and metastasis are enhanced by CatD. MMPs represent a family of proteases that are involved in processes of invasion and metastasis through cleavage of basement membrane and remodeling of extracellular matrix. Considering that breast cancer had highly aggressive biology, it has been investigated and found that high activity of MMPs, particularly MMP-2 and MMP-9, is engaged in all stages of tumor progression. As proteases are involved in physiological processes such as immune response and in pathological conditions, it seems that examinations of molecular pathways of breast cancer could define a therapeutic target based on proteases.

S. Radenkovic (✉) · G. Konjevic
Department of Radiation Oncology and Diagnostics, Institute of Oncology and Radiology of Serbia, Pasterova 14, 11000 Belgrade, Serbia
e-mail: stankovics@ncrc.ac.rs; radenkovics6@gmail.com

K. Gopcevic · G. Konjevic
Faculty of Medicine, University of Belgrade, Belgrade, Serbia

V. Jurisic
Faculty of Medical Sciences, University of Kragujevac, Kragujevac, Serbia

S. Chakraborti and N.S. Dhalla (eds.), *Pathophysiological Aspects of Proteases*,
DOI 10.1007/978-981-10-6141-7_1

Keywords
Proteases · Breast cancer · MMPs · Cathepsins

1 Introduction

Activity, structure, and role of proteases and their inhibitors were extensively investigated over past 20 years. Enzymes called proteases are subdivided into four subclasses: serine ($\sim$180), cysteine ($\sim$160), aspartic ($\sim$25), threonine ($\sim$30), and matrix metalloproteinases ($\sim$200). Proteases in established classes cleave peptide bonds through several different mechanisms. Serine proteases, cysteine proteases, and amino-terminal nucleophile hydrolases (threonine proteases) had an enzyme side chain in the active cleft that involves in catalysis (covalent catalysis) [1]. On the other side, aspartic acid proteases and metalloproteinases have peptide-bond hydrolysis without covalent involvement, by creating a highly reactive water molecule [2].

Proteases and their function is essential for physiological processes, as well as pathological processes such as cell cycle progression, cell migration, tissue remodeling, neuronal outgrowth, hemostasis, wound healing [3]. Considering that proteolysis is essential for life, all organisms have proteases [4–6].

Considering that increased expression of proteases is involved in processes of tumor progression and autoimmmune diseases, blocking of their expression is usually therapeutic approach [7, 8]. Proteases are usually used as markers for prognosis and follow-up, and they are involved in process of cell death, such as caspases. Matrix metalloproteinases and cathepsins have an important role in cancer progression, especially in breast cancer [8, 9], but exact function must be investigated furthermore [10].

Proteases are involved in all steps of tumor formation and dissemination: tumor initiation, metastasis, and invasion. Each class of human proteases has specific endogenous inhibitor: Cystatins are inhibitors of cysteine proteases; serpins are inhibitors of serine proteases; and specific tissue inhibitors of metalloproteinases (TIMPs) are inhibitors of metalloproteinases [10, 11]. Thus, progression of malignancy is tightly connected to the activity of both proteases involved in cancer and specific inhibitors of proteases [8].

In complex process of tumor formation are included many specific proteases which make prognosis very difficult. The published investigations have shown that cancer progression is connected with activity of different proteases. Cancer cells express proteases, but they also induce its expression in non-neoplastic neighboring cells favoring tumor expansion [11]. The cells of adjacent tissue have been transformed and enhance all steps of cancer progression [8]. The cancer degradome represents a group of all enzymes include in processes of tumor formation and metastasis [12].

2 Proteases in Breast Cancer

Today, breast cancer represents an important health problem in world cause [13]. However, screening which includes mammography and ultrasound has helped detection of non-palpable lesions [14]. Also, investigation regarding tumor progression and new therapeutics enhance survival of breast cancer patients [8, 14].

Breast cancer is one of the most common cancers in women, and, even after chemotherapy, 50% of the patients die from distant relapse or metastasis [15]. Predictive factors can be used to predict response or lack of response to a particular therapy, and prognostic factors can be useful in making decisions concerning patients who should receive adjuvant therapy [16, 17].

Considering that new targets for therapy, prognosis, and follow-up are crucial for breast cancer management, further investigation regarding pathophysiology of breast cancer is necessary. Novel studies showed that insufficient expression of adhesion molecules, such as cadherins, and shorter overall survival of breast cancer patients [18]. In addition to the cadherins family, other adhesion molecules have been studied aiming at understanding their role in mammary carcinogenesis in order to associate such discoveries with the development of new therapeutic targets. MMPs as well as claudins are involved in invasive pseudopodes of tumor edge and are associated with increased metastatic potential [8].

In this review, we evaluate the role of proteases in breast cancer progression and management.

3 Cathepsins in Breast Cancer

The maintenance of the steady state in the cell requires the precise regulation of protein synthesis as well as the protein degradation. For intracellular protein degradation two systems are responsible: the lysosomal as the main system, and the ubiquitin-proteasomal system. Among the approximately 50 known lysosomal hydrolases, the aspartic, serine, and cysteine proteases are the best studied [10, 12]. In an organism, there is a dynamic equilibrium between cells necessary for growth and development and cell which are not necessary. These unwanted cells are eliminated by apoptosis, programmed cell death, with involvement of caspases. Lysosomal acid hydrolases comprise a group of 14 members of three catalytic mechanisms: cysteine (B, C, F, H, L, K, O, S, V, W, and X), aspartate (D and E), and serine (G) cathepsins [11].

Cathepsins are produced as inactive proenzymes, and the process of autolysis converts them in active forms. Cathepsins usually are exposed at the cell surface. They are able to degrade proteins of ECM in the extracellular space [19].

Cysteine protease Cathepsin B (EC 3.4.22.1) is widely distributed and functions in intracellular protein catabolism, in processing of antigens in the immune response, hormone activation, and bone turnover. The role of propeptide is to drive cathepsin precursor in the endoplasmic reticulum, to enable its hydrolysis. The proregion inhibits cathepsin proteolytic activity that preserves its three-dimensional structure at neutral pH and maintains the precursor in an inactive state until cleaved [20]. Cathepsins are processed to the active forms either autocatalytically or by other proteases and transferred to lysosomes either through endosome maturation or through endosome lysosome fusion in late endosomes. However, it has been shown that breast cancer carcinomas as well as colon, colorectal, cervix, head and neck, liver, lung, melanoma, ovarian, pancreatic, prostate, and thyroid tumors have increased expression of CatB [10, 19, 20].

Cathepsin D (CatD), (EC 3.4.23.5) a soluble aspartyl protease, has a role in cancer development and considered as an independent tumor marker for breast cancer patients. CatD is synthesized in rough endoplasmic reticulum as preprocathepsin D (pCatD), 52 kDa inactive precursor, and after its subsequent conversion by proteolytic processing to its active form, CatD (44 kDa) becomes two-chain molecule of 31 and 14 kDa. After activation CatD is transported to the acidic endosomes, lysosomes and phagosomes [11, 21].

In breast cancer cells, unlike in normal ones, are found high levels of pro-cathepsin D due to both overexpression of the gene and altered processing of the protein. [22]. pCatD (the propeptide of CatD) is a glycoprotein secreted by human breast cancer cell lines which are hormone dependent. Two biological activities of pCatD, mitogenic on estrogen-depleted MCF-7 cells, suggesting that it is an estrogen-regulated autocrine mitogen, and an acidic proteolytic activity on different substrates and on proteoglycans and basement membranes, have been demonstrated in vitro [10]. Both activities—growth promoting and extracellular proteolytic—suggest that pCatD may have prognostic importance in breast cancer [21]. It has been reported that both pCatD and CatD induce proliferation and migration of cancer cells, fibroblasts, and endothelial cells [11, 21]. It has been evidenced pCatD may induce mitogenic response by both proteolytic-dependent and proteolytic-independent mechanisms by the induction of fibroblast and cancer cell proliferation [11]. This study confirmed upregulation of pCatD expression in estrogen receptor-positive breast cancer cell lines treated with estrogen. In an excellent study of Ohri and colleagues has been demonstrated that, as a mitogen, pCatD asts as a protein ligand rather than as an enzyme, and that expression and secretion of pCatD achieved by transfection of pCatD construct in HBL-100 cells increased the metastatic potential of this cell line [21].

CatD is distributed in lysosomes, endosomes, and phagosomes. This enzyme is primarily involved in degrading missfolded or non-functional proteins intracellularly. Up to now, many important roles of CatD in normal physiological and pathological conditions were studied [10]. Physiological functions of CatD are based on its ability to degrade intracellular proteins, and activate and degrade polypeptide hormones and growth factors. Furthermore, CatD takes part in several steps of mitosis, regeneration, activation of leukocytes, and regulation of vassal wall permeability. The proteolytic activity of the enzyme is regulated by various intralysosomal factors including pH, products of metabolism, hormones, growth factors, and specific inhibitors. Hormones, e.g., estrogens and progesterone, can regulate the synthesis of cathepsin D and its receptor mannose 6-phosphate [10, 11, 21].

Both extracellular and intracellular activities of CatD are involved in multiple tumor progression steps [22]. Due to the lysosomal membrane permeabilization, CatD is transferred to the cytosol and through several steps leads to the release of cytochrome c from mitochondria into the cytosol, where it is inhibited by pepstatin A, which partially delay IFN-gamma and Fas/APO-induced apoptosis [8, 11]. Thus, proteolytic activity of CatD is one of the key steps in apoptotic cascade, besides Bcl-2 protein family, TNF, and p53 [23, 24]. CatD accumulation in an estrogen-positive breast cancer cells could be a consequence of higher secretion of its proenzyme [11].

Overexpression of CatD at the mRNA and protein level, demonstrated by various methodological approaches: in situ hybridization, cytosolic immunoassay, Northern and Western Blot, immunohistochemistry has been confirmed. Concentration of CatD was higher 2–50 folds, compared to its concentrations in other cells of normal mammary glands. CatD overexpression is correlated with increased risk of clinical metastasis and short survival in breast cancer patients. In the plasma of patients with metastatic breast carcinoma, the increased serum pCatD levels were detected [21]. These studies suggest that CatD stimulates cancer cell proliferation, fibroblast outgrowth, angiogenesis, and metastasis and is a marker of poor prognosis in breast cancer.

4 Cadherins

Among the mechanisms responsible for initiation and progression of tumor, the control of cellular adhesion and motility is one of the crucial mechanisms. Normal cells are connected with each other and the surrounding stroma by different adhesive molecules.

The beginning phase of metastatic cascade includes the local invasion of the tissue by tumor cells which should separate from each other, by lacking restrictions in homotypic cell adhesion and cell–cell contact inhibition. Endothelial cadherins (E-cadherins), a family of transmembrane glycoproteins, represent the best-studied molecular markers expressed in epithelial and breast cancer cells [8, 25, 26]. They are responsible for the structure of epithelial tissues by providing the connection between endothelial cells and the surrounding stroma. E-cadherin is one of the most important molecules in cell–cell adhesion in epithelial tissues. It is localized on the surfaces of epithelial cells in regions of cell–cell contact known as adherens junctions.

The extracellular domain of E-cadherin mediates calcium-dependent homotypic interaction with E-cadherin molecules on adjacent endothelial cells, while the intracellular domain binds cytosolic catenins and provides a link to the actin cytoskeleton [8, 25].

Therefore, in many carcinomas, the lack of expression of E-cadherin correlates with an invasive and dedifferentiated phenotype. Furthermore, the normal function of E-cadherins depends on their association with catenins. The loss of adhesive E-cadherin molecules and the E-cadherin–catenin complexes leads, morphologically, into impaired structure of the cytoskeleton and allows amoeboid movement and motility, but also deregulates the cell cycle and leads to cell proliferation. The observed decrease in adhesive force facilitates dispersion of carcinoma cells from the primary tumor mass. In addition to promoting passive dissemination of carcinoma cells, loss of E-cadherin function can also promote cell invasiveness [8, 25, 26].

Aside from these, there is yet another mechanism to change E-cadherin function, its proteolytic modification. E-cadherin can be altered by degradation of E-cadherin's extracellular domains by stromelysin-1 (MMP-13), a member of MMP-family that has been closely connected with tumor progression [27, 28]. Constitutive expression of active stromelysin in mammary epithelial cells results in cleavage of E-cadherin and progressive phenotypic changes in vitro, including loss of catenins from cell–cell contacts, downregulation of cytokeratins, and upregulation of vimentin and MMP-9 [8]. These changes result in unfavorable epithelial-to-mesenchymal transition of cellular phenotype. In vivo stromelysin expression promotes breast carcinogenesis that involves certain genomic changes [8, 25, 27]. Thus, a decrease in cell–cell adhesion is associated with malignant conversion, tumor cell invasion, and the beginning of metastasis.

The process of epithelial-to-mesenchymal transition (EMT) is characteristic of most carcinomas and represents the progression toward malignancy, associated with loss of epithelial differentiation and, consequently, a shift toward a mesenchymal phenotype. This process is considered as a crucial event in late stage tumorigenesis and a prerequisite for tumor infiltration and metastasis [8, 29]. EMT provides mesenchymal characteristics for epithelial cells to overcome the physical constraints imposed on them by intercellular junctions and adopt a motile phenotype. The transition from a well-differentiated epithelial phenotype to an unfavorable invasive mesenchymal phenotype may include diverse molecular mechanisms that may independently enhance motility and invasiveness. EMT is characterized by proteolytic loss of E-cadherin or endothelial to dysfunctional neural-cadherin (N-cadherin) "switching" and by increasing motility plays a crucial role during invasion and metastasis of carcinomas [27]. Without tight junctions, provided by E-cadherin, mesenchymal lacks an apical–basolateral polarity and transform to an elongated morphology. This morphology is characterized by front–back asymmetry that facilitates motility and locomotion [8, 27, 28]. At the invasive edge of the mesenchymal cells are situated filopodial extensions, enriched with integrin family receptors that interact with the ECM to enhance motility. Filopodial extensions contain MMPs, particularly MMP-2 and MMP-9, that promote metastasis by digesting components of the basement membranes and ECM [8].

Transforming growth factor β (TGF-β) is crucial protein in inducing EMT of various epithelial cell types. During EMT, many proteins are upregulated: Src kinase, integrin-linked kinase, integrin h-5, and MMP-11, MMP-12, and MMP-14 that induce cytoskeletal remodeling and promote cell motility and invasiveness [8, 27]. The plasticity of malignant carcinoma cells due to diverse molecular mechanisms that enhance EMT also allows incomplete EMT, reversion to an epithelial phenotype, and collective migration. As these mechanisms can manifest in a series of independent and reversible steps, EMT represents just one mechanism in the global metastatic carcinoma development process [8, 25]. Moreover, loss of tumor suppressor gene, p53, allows an upregulation of MMP-2 and downregulation of E-cadherin modulation of Slug and regulation of cancer cell metastasis [29, 30].

The additional loss of heterotypic adhesion of endothelial cells, maintained by β1 and β2 integrins (e.g., VLA-1), as receptors for different components of ECM, BM, a highly specialized layer of the extracellular matrix composed of dense, intertwined fibrils of collagen type IV and glycoproteins, such as laminin and heparan sulfate proteoglycan, as well as similar components in the extracellular matrix (ECM), also allows detachment and invasion [8, 28]. A fundamental difference of neoplastic cells is the loss of anchorage requirement for cell survival and growth. The anchorage-independent growth of tumor cells may result from an uncoupling of cell survival signals transduced from the ECM by attachment, coupled with activation of cell cycle progression, and is associated with neoplastic transformation [8].

5 Proteolytic Pathways

In the process of degradation of surrounding tissue tumor cells use two proteolytic systems. It has been proposed that a cascade of degradative enzymes clears a path at the advancing edges of invasive tumors, through which the cancer cells move [27]. The inactive, pro-enzyme forms of the degradative enzymes (zymogens) are locally secreted and become active by the action of their regulatory enzymes which are produced either by the tumor cells themselves or the surrounding stromal cells (fibroblasts, mast cells). Such zymogens include pro-cathepsins, plasminogens and procollagenases, heparanases and matrix metalloproteinases (MMPs). In the process of invasion, the most important pathways are MMPs and plasminogen activation system [8, 31]. Plasminogen activation is an extracellular proteolytic system, which converts the inactive plasminogen into a trypsin-like serine protease plasmin. Plasmin has very broad substrate specificity and can degrade matrix proteins and can convert other zymogens (i.e., MMPs) to their active forms by proteolysis. Two plasminogen activators have been defined urokinase plasminogen activator (uPA) and tissue-type PA (tPA) [8, 32]. It has been shown that uPA is crucial in cell migration and invasion [8, 33].

Plasminogen activation by uPA occurs at the cell surface where the inactive proenzyme of uPA is bound by its cellular receptor uPAR. Pro-uPA is activated by an unknown mechanism but remains bound to its receptor (uPAR) where it is capable of activating plasminogen. uPA is often secreted, aside from tumor cells, by stromal fibroblasts. In this way, highly focalized centers exist where controlled areas of proteolysis can occur at the leading edge of the invading cell [8, 31]. The inhibition of receptor-bound uPA by plasminogen inhibitors, PAI-1 and PAI-2, regulates the invasive process and influences the "proteolytic activity map." The potential importance of u-PAR for the development of minimal residual disease in solid cancer and its biological relevance for tumor cell dormancy has been described due to its influence on the angiogenic switch [8, 33].

The uPA-plasmin system has a role in the control of gelatinase (MMP-2, MMP-9) activity. MMP-2 and MMP-9 are secreted in the form of inactive zymogens that are activated extracellularly, a fundamental process for the control of their activity. Both gelatinases are associated with the cell surface, binding of uPA and plasminogen to the cell surface results in gelatinase activation without the action of other proteinases, and the inhibition of uPA or plasminogen binding to the cell surface can blocks gelatinase activation [3, 8]. Plasmin in soluble phase degrades both proteinases and proteinase activation [8, 33].

The term "cascade" warrants some revision for some other postulated proteolytic cascades. For example, in proteolytic matrix remodeling, the existence of a cascade, or even of a proteolytic pathway, is questioned by considerable biochemical and genetic evidence. Perhaps the concept of a redundant network of interactions is more appropriate in this case. It is now considered that many of the metalloproteinases once thought to degrade the matrix might in fact be limited processing enzymes, the substrates of which are cytokines and growth factors [3, 34]. It is also unclear whether the plasminogen activator–plasmin–metalloproteinase pathway exists anywhere but in vitro, and the 30-year-old concept that the matrix must be broken down to allow cell migration (in normal or in metastatic instances) is under close scrutiny and revision [34]. Indeed, the possibility that the proteolytic degradation of the extracellular matrix might not be required for migration at all has been investigated [35].

5.1 Matrix Metalloproteinases (MMPs)

MMPs are a family of structurally related zinc-dependent endopeptidases that are engaged, aside from various physiological processes such as embryogenesis, reproduction, uterine involution, angiogenesis, in pathological conditions, such as tumor invasion, by mediating degradation of basement membrane and remodeling of extracellular matrix. As invasiveness and metastasis are important biological characteristics of malignant tumors, these processes are extensively investigated and it has been shown that increased expression and activity of MMPs are among the earliest and most sustained events in tumor progression [8, 16].

In tumor progression, increased expression and activity of matrix metalloproteinases plays a role in angiogenesis, invasion, and metastasis. They are a family of zinc metalloproteinases, secreted as latent proenzymes, which are activated by proteolytic cleavage and are inhibited by the tissue inhibitors of metalloproteinases (TIMPs) [8, 36]. There are 27 structurally related known members, which differ according to their individual substrate specificities, including stromelysin 1 and 2, gelatinase A and B, matrilysin, neutrophil collagenase, and interstitial collagenase. Gene-encoding stromelysin 3 has been identified and is specifically overexpressed in breast carcinomas. The gene was expressed from stromal cells in all of the invasive breast carcinoma tested and only in the direct vicinity of the neoplastic cells of the tumor. The transformed cells secrete a factor that induces expression of stromelysin 3 from the stromal fibroblasts and allows progression of the expanding tumor cell mass through the ECM into surrounding tissue [37].

MMPs have been subdivided into four subclasses based on their substrate specificity and cell localization:

(1) Collagenases (MMP-1 and MMP-8) degrade fibrillar collagen;
(2) Gelatinases (MMP-2 and MMP-9) have a preference for collagen type IV or denaturated collagen, i.e., gelatin;
(3) Stromelysins (MMP-3, MMP-7, and MMP-10) degrade a wide variety of ECM substrates such as proteoglycans, laminin, and fibronectin;
(4) Collagenase-3 (MMP-13) has the widest spectrum of substrates;
(5) Membrane-type MMPs (MT-1 to MT-6-MMP) integral plasma membrane proteins capable of activating MMPs (20).

MMP are secreted as soluble inactive enzymes which are activated by intrinsic Zn^{+} ions, located in the center of the molecule and by extrinsic Ca^{+}. Enzyme activity of MMPs is tightly regulated, and deregulation of MMPs activity can lead to uncontrolled degradation of ECM, a process crucial in tumor metastasis. The most commonly associated MMPs with the processes of metastasis are MMP-2 (gelatinase A) and MMP-9 (gelatinase B) due to their ability to degrade collagen type IV, a major component of vascular basement membrane [8, 31].

Also, MMPs, specifically MMP-2 and MMP-9, are essential for the switch to the angiogenic phenotype during tumor progression. It has been shown that MMP-2 and MMP-9 is expressed in breast carcinomas. Increased MMP expression required for the angiogenic switch and may predict when a non-angiogenic, microscopic, dormant tumor becomes angiogenic and invasive [8, 38].

5.2 Tissue Inhibitors of MMPs (TIMPs)

Tissue inhibitors of MMPs (TIMPs) are tight-binding inhibitors of the active forms of MMPs. There are four of homologous proteins, referred to as the tissue inhibitors of MMPs (TIMP-1 to TIMP-4) [39]. In addition to inhibiting matrix metalloproteinase (MMPs) activity, TIMP-2 has been shown to suppress mitogenesis of both host (endothelial, fibroblast) and tumor cells via receptor tyrosine kinase (RTK) inactivation. Potent anti-tumor activity of TIMP-2 is composed of three separate activities. These are direct inhibition of MMP activity, MMP-independent anti-angiogenic effects, and direct downregulation of tumor cell EGFR signaling [8, 39, 40]. It has been shown that TIMP1, although can inhibit MMP-9, stimulates tumor growth and increases proliferation of normal and tumor cells [41]. TIMP1 stimulates proliferation via MMP-independent and MMP-dependent pathways [8, 16]. Breast cancer cells have on their surface receptors for TIMP1 (CD63). Process of bounding TIMP1 for CD63-integrin β1 complex leads to intracellular signaling and activation of mitogen-activated protein kinases (MAPK) which result in cell proliferation. Growth factors activate MAPK signaling pathway and lead to cell proliferation. Studies have been shown that TIMP1 differently acts with cells and these depend on different expression of receptors for TIMP1 on cell surface. Increased expression of CD63 is connected with degree of malignance. TIMP1 also can activate Ras which also can induce tyrosine-specific phosphorylation and activate MAPK pathway [41]. In addition to the above MMP-independent mode of inducing cell proliferation, TIMP1 also via MMP can induce cell proliferation [42]. It has been shown that TIMP1 by blocking constitutive active MMPs prevents MMP-mediated degradation of newly produced growth factors and growth factors so that they cannot stimulate cell proliferation [36, 37].

In addition to the cell proliferation effects, it has been shown that TIMP1 has anti-apoptotic effect. It has been confirmed that high levels of TIMP1 expression induce constitutive activation of FAK (focal adhesion kinase). FAK is activator of PI (3) kinases which regulate Bcl2, protein important for cell survival and avoiding of apoptosis. Bcl2 and BclXL are mitochondrial membrane proteins which blocks Bax, protein which released cytochrome C from mitochondria [23]. Considering that cytochrome C activates caspases, major mediators of apoptosis, by retaining cytochrome C in mitochondria, block apoptosis. Also, it has been shown that TIMP1 increases expression of Bcl2 and BclXL. Due to this mechanism, there is a decreased sensitivity of tumor cells with high expression of TIMP1 to chemotherapeutics which act by inducing apoptosis [42].

In the beginning, it was considered that TIMP1 has anti-angiogenic effects, but in the time it has been shown that TIMP1 enhances angiogenesis. It has been confirmed that production of endostatins and angiostatins, inhibitors of angiogenesis, is mediated by MMPs. MMPs by degradation of collagens release endostatins and angiostatins from collagens [36]. Due to this mode, TIMP1 by inhibition of MMPs and block production of endostatins and angiostatins enhances tumor angiogenesis. Moreover, it has been shown that breast cancers in rats with high expression of TIMP1 are with higher volume and had better vascularization compared to controls [16, 41]. In addition, it has been shown that TIMP1 directly stimulates vascular endothelial growth factor (VEGF), which helps blood vessels growth. Mechanism by which TIMP1 stimulation of VEGF helps angiogenesis is not clarified.

Several studies have been shown that TIMP1 expression is increased in breast cancer tissue compared to benign or normal breast tissue. In addition, it has been confirmed that RNA and protein expression of TIMP1 are significantly increased in malignant breast tissue compared to benign breast tissue. It has been shown that RNA TIMP1 expression increases with invasive potential of transformed cells. Normal tissue has low expression of TIMP1, cancer in situ has higher expression of TIMP1, and invasive breast cancer has highest expression of TIMP1 [43, 44].

Invasion of tumor cells into ECM is an active process that involves four closely related processes:

1. Separation of tumor cells from each other;
2. Encroachment into local tissues by an active process involving lysis of host cells and degradation of the ECM by proteolysis;
3. Following separation from each other and degradation of the surrounding stroma, tumor cells must then be able to enter the region cleared by the action of the proteolytic enzymes by an active process of cell motility;
4. Tumor cells must also promote their own vascularization by inducing capillary growth into and around the tumor.

Cyclical repetition of these four steps leads to progressive invasion of tumors. Therefore, local invasion by tumor cells allows them to reach a blood or lymph vessel and offers them an undesired opportunity to spread to distant body sites [1, 23, 45, 46].

6 Metastatic Cascade

Although metastasis is the major cause of death in cancer patients [1, 47], it is well established that metastasis formation is an inefficient process. Studies have demonstrated that late events in metastasis formation are largely responsible for the inefficiency of this process. The number of circulating tumor cells and tumor emboli correlates with the size and age of the primary tumor. However, the number of circulating tumor cells does not correlate with the clinical outcome of metastases [1, 8, 10]. The inefficiency of tumor cells in completing the metastatic cascade is the result of the fact that successful formation of metastatic foci consists of several highly complex and interdependent steps. Each step is rate limiting in that failure to complete any of these events completely disrupts metastasis formation. The steps involved in metastasis formation are thought to be similar in all tumors and are characterized as follows:

1. Tumorigenesis. After the initial neoplastic transformation, the tumor cells undergo progressive proliferation that is accompanied by further genetic changes and development of a heterogeneous tumor cell population with varying degrees of metastatic potential. The initial growth of the primary tumor is supported by the surrounding tissue microenvironment, which eventually becomes rate limiting for further growth [1, 8, 10].
2. Angiogenic switch. As the tumor grows and central tumor cells become hypoxic, the tumor initiates recruitment of its own blood supply by a process called the "angiogenic switch" which involves the secretion of various angiogenic factors and the removal or suppression of angiogenesis inhibitors [3, 8, 10].

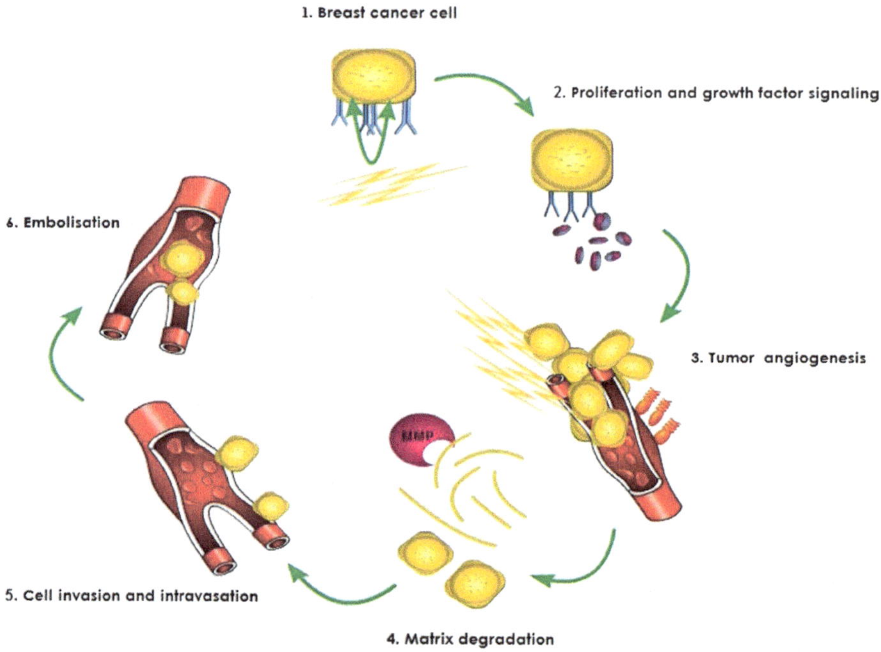

Fig. 1 Role of MMPs in metastatic cascade

3. Clonal dominance of invasive phenotype. Continued genetic alterations in the tumor cell population lead to selection of tumor cell clones with distinct growth advantage and acquisition of an invasive phenotype. Invasive tumor cells downregulate cell–cell adhesion, alter their attachment to the ECM by changing E-cadherin, integrin expression profiles, and by proteolytically altering the matrix accomplish stromal invasion. Tumor cells lose "contact inhibition" that prevents normal cells from continuing to divide when they contact their nearest neighbors. MMPs degrade the components of ECM and basement membrane and allow migration of tumor cells (Fig. 1) [8].
4. Circulating of cancer cells in blood vessels [8, 10].
5. Dormancy of cancer cells at lymph nodes and organs [8, 10].
6. Extravasation and growth at the secondary site. Arrested tumor cells proliferate in response to paracrine growth factors or become dormant. The poor growth of tumor cells after extravasation from the circulation is a major factor contributing to the inefficiency of the metastatic process.
7. Angiogenesis in metastatic foci. Finally, continued growth of the metastatic foci is also dependent on angiogenesis. The development of this neovascular network at the metastatic site enhances the metastatic potential of these cells just as it does for the primary tumor [8, 10].

8. Evasion of immune response. Tumor cell evasion of immune response in metastatic foci includes antigenic modulation and immunosuppression and prevents their eradication (Fig. 1) [1, 8, 10].

Therefore, detachment of single cells or clumps of cells from the tumor mass are associated with lack of cell adhesiveness in tumor populations, regulated by a variety of cell surface molecules such as the cadherins and integrins.

Circulating tumor cells arrests at distant sites and by a repeated process of invasion may colonize a secondary site for growth. In some instances, this organ preference of metastasis can be explained simply in terms of the anatomical relationship of the organ with the site of primary tumor growth (Ewing theory of metastasis) [3, 8]. Some cancers express organ preference that is not associated with a non-specific entrapment process, but rather with specific determinants, which actively promote the growth of the metastatic cells, thereby providing favorable soil (Padgett's theory of metastasis). Fully metastatic behavior is attributable to the expression of only one, or a few, metastasis promoting genes (e.g., CD44) or to the loss of an equally small number of metastasis-suppressing genes (e.g., nm23) [8, 10].

After arriving in a secondary site, metastatic cells begin proliferating, undergo apoptosis, or remain as solitary dormant cells. The process of metastasis, although dangerous, is extremely inefficient with the majority of the cells undergoing apoptosis and thus becoming clinically irrelevant. The cells that begin proliferating and dormant cells are responsible for cancer recurrence [8, 38]. It has been shown that tumor dormancy, a complex and still poorly understood phenomenon observed both in experimental models and in patients, has been connected with insufficient angiogenesis. A defined event, termed "angiogenic switch" characterized by an imbalance between pro- and anti-angiogenic factors, by the influence of two proteolytic systems, uPAR and MMPs, often marks interruption of the dormant state, thus triggering invasive tumor growth [8, 48].

7 Clinical Correlation and Prognosis

There is an urgent need for more efficient prognostic markers that can accurately and reliably identify breast carcinomas with potential to cause recurrence and/or death [17]. There are data that show that markers for survival prognosis could be found among the proteins that participate or regulate the processes associated with cancer progression. Possible candidates include MMPs and their tissue inhibitors, TIMPs [8, 15].

8 MMP-2 and MMP-9 in Tumor Tissue of Breast Cancer Patients

Although there is a need for diagnostic as well as prognostic markers in breast cancer, the studies did not show a new marker who would have an important impact on disease progression [9, 49, 50]. However, molecular examinations have shown that breast cancer is divided into at 4 types of tumors: luminal (ER+, PR+, and HER2−), HER2 overexpressing (ER−, PR−, and HER2+), basal-like (ER−, PR−, HER2−, and CK5/6+/EGFR+), and breast-like (ER−, PR−, and HER2−), with diverse biology and prognosis [9, 36]. It has been shown that patients with basal-like carcinomas had shorter relapse-free survival and worse survival, and the aggressive biology of basal-like carcinomas requires a new therapeutics for treatment [9, 13]. Investigation of activity of gelatinases in tumor tissue and sera of breast cancer patients could help in designing newer more adequate therapeutics.

It has been shown that gelatinase activity in the tissue of breast carcinomas patients can be a diagnostic and staging markers, but also as potential target for new therapies [37]. Radenkovic et al. have been shown that tumor tissue of breast cancer patients had higher latent and active forms of MMP-2 and MMP-9 compared to peritumor tissue (Fig. 2) [9, 50]. Also, it has been confirmed that activity of latent and active forms of MMP-2 and MMP-9 in tissue of breast carcinomas show stage-dependent increase. These results indicate that activity of gelatinases could be used as biomarker of progression of breast cancer patients [9].

Radenkovic et al. show that tumor size is positively associated with latent and active MMP-2 activity [9], while other studies show correlation between MMP-9 and tumor size. Other study investigated detection of MMP-2 and MMP-9 by immunohistochemistry and confirmed positive association of MMP-2 and tumor size [51]. Moreover, it has been shown a positive correlation between gelatinase activity and lymph node positivity of breast cancer patients [9, 52]. The shown association of gelatinase activity and tumors size and lymph node involvement

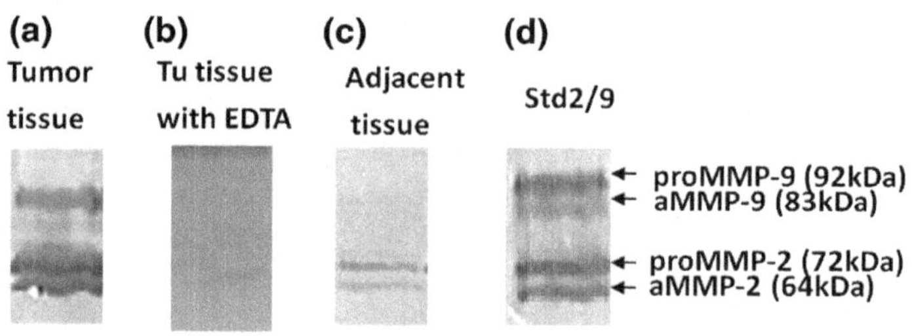

Fig. 2 Electrophoretic traces of SDS-PAGE zymography. **a** The activity of pro and active form of MMP-2 and pMMP-9 in tumor tissue of breast cancer patients. **b** Inhibited MMP-2 and MMP-9 activities in tumor tissue of breast cancer patients with EDTA. **c** The activity of pro and active form of MMP-2 and pMMP-9 in adjacent tissue of breast cancer patient. **d** Standard proteins marker

suggests that MMP-2 and MMP-9 are involves in cancerogenesis as well as in metastasis.

It has been confirmed that estrogen and progesteron receptor tumor positivity is associated with better relapse-free survival and overall survival of breast cancer patients [17, 53]. Radenkovic et al. have shown that hormone receptor tumor negativity is associated with tumor activity of active forms of gelatinases [9]. Since estrogen and progesteron receptor negative tumors are associated with worse outcome, the above-mentioned studies suggest that aggressive nature of hormone-negative tumors is associated with MMP-2 and MMP-9 activity.

Almost one-fifth of patients with breast cancer have tumors with increased expression of the HER2 [13]. The studies have been found that HER2 induces MMP-2 and MMP-9 activity and gene expression, therefore stimulating an invasiveness and migratory phenotype of breast cancer cells [8, 9]. The published studies have shown no difference of gelatinase level between patients with HER2-positive tumors in tumor tissue compared those with HER2 negative tumors [9].

It has been shown that basal-like breast cancers had poor differentiation and worse outcome and usually had haematogenic pathway of dissemination, while luminal A breast cancers are likely to spread into axillary nodes and bones and mostly had better survival [9, 13, 29]. Pro-MMP-2 is involved in the process of angiogenesis. Particularly, proMMP-2 degrades of collagen type IV and, therefore, allows the $\alpha v\beta 3$ integrin bounding of newly formed blood vessels [37]. In these sense, it was found a higher activity of proMMP-2 in tumor tissue of patients with basal-like cancers, indicating that haematogenic spread of basal-like tumors is at least in part regulated by MMP-2.

It has been found that surrounding tissue had MMP-2 and MMP-9 activities produced by stromal cells [36, 37]. It is proposed that tumor cells induce stromal cells via interleukins, growth factors, and interferons in order to produce gelatinases [29, 37]. Also, it has been shown association of higher MMP-2 activity in adjacent tissue and basal-like cancers. [9]. It seems that examinations of adjacent tissue are essential for enlightening of tumor biology.

The above results confirmed that MMP-2 and MMP-9 could be used as prognostic biomarker as well as therapeutic target in breast cancer patients [9, 10].

9 MMP-2 and MMP-9 in Sera of Breast Cancer Patients

It is suggested that MMP-2 and MMP-9 produced in tumor tissue entrance into vessels, and therefore, these enzymes could serve as potential diagnostics and staging biomarkers for breast cancer patients [15, 17, 37].

The published studies have been shown that activities of gelatinases in sera are associated with tumor size and lymph node involvement in breast cancer patients [54]. Stankovic S et al. have shown significant stage-dependent increase of MMP-2 and MMP-9 sera activity (Fig. 3) in breast cancer patients [15], and this indicates

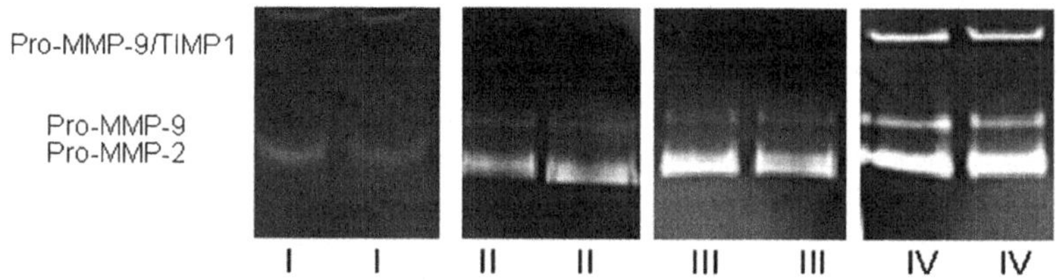

Fig. 3 Representative SDS-PAGE zymography gel showing relative abundance of MMP-2 and MMP-9 in the sera with different clinical stages of breast cancer patients (clinical stage of breast cancer *II*, *III*, and *IV*)

that these enzymes could be used as possible staging and follow-up markers in breast cancer patients. Moreover, it has been found that MMP-9/TIMP-1 complex activity is detected only in metastatic stage of the disease. This result suggests that MMP-9/TIMP-1 complex plays an important role in disease progression [8, 15]. In these sense, MMP-9 and MMP-2 were investigated as prognostic markers for breast cancer patients in many studies. It has been confirmed that plasma MMP-9 levels are inversely associated with RFS and with survival [15, 55]. All these studies suggest importance of MMP-2 and MMP-9 sera activity evaluation in breast cancer patients.

It has been shown that ER induces MMP-9 gene expression and activity and therefore enhances growth of tumor cells and progression [13, 17, 53]. However, some studies have shown a positive association of MMP-9 expression and ER positivity in breast cancer patients, while others studies did not shown any association [15, 54, 56].

HER2 represents an important prognostic marker for breast cancer patients [13, 15]. The published studies found that invasiveness of HER2-positive tumors is enhanced by MMP expression [15, 36, 37]. Moreover, HER2 expressions via stimulation of proteolytic network of enzymes help cancerogenesis, as well tumor progression in breast cancers [15, 55].

Although soluble MMPs have been shown to be produced as zymogen (proMMP-2) in cells, the fate of MMPs and TIMPs after their secretion from producing cells or spatial regulation of their molecular activations is not fully understood [15, 57]. It has been shown that cancer cells as well as tumor infiltrating lymphocytes secrete MMPs, MT!-MMP, and TIMPs and these secretions induce anti-tumor immune response in metastases [15, 57, 58]. It is proposed that MMPs produced by TIL can spread closely to cancer cell in order to activate there [57].

Activity of MMP-2 and MMP-9 in sera of breast cancer patients could be implemented as staging biomarkers. The complex network between lymphocyte and MMPs requires further investigation. The association of the increase in MMP-2 and MMP-9 activity in serum and prognostic and therapeutic parameters, tumor size and axillary lymph node status, suggests a role of MMP-2 and MMP-9 in prognostic stratification of breast cancer patients. This may be helpful not only in the new approaches of "stromal-directed" therapy, but also in avoiding overtreatment with chemotherapy of breast cancer patients [8, 15].

10 Conclusion

Considering that breast cancer had highly aggressive biology, it has been investigated and found that high activity of MMPs, particularly MMP-2 and MMP-9, is engaged in all stages of tumor progression. As proteases are involved in physiological processes such as immune response and in pathological conditions, it seems that examinations of molecular pathways of breast cancer could define a therapeutic target based on proteases.

Acknowledgements This study was supported by the Grant of the Ministry of Science of Serbia No. 175056.

References

1. Drag M, Salvesen GS (2010) Emerging principles in protease-based drug discovery. Nat Drug Discov 9:690–701
2. Rawlings ND, Morton FR, Kok CY et al (2008) MEROPS: the peptidase database. Nucleic Acids Res 36:D320–D325
3. Sabeh F, Shimizu-Hirota R, Weiss SJ (2009) Protease-dependent versus-independent cancer cell invasion programs: three-dimensional amoeboid movement revisited. J Cell Biol 185:11–19
4. Gopcevic KR, Rovcanin BR, Tatic SB et al (2013) Activity of superoxide dismutase, catalase, glutathione peroxidase, and glutathione reductase in different stages of colorectal carcinoma. Dig Dis Sci 58:2646–2652
5. Gopcevic K, Rovcanin B, Kekic D et al (2011) Matrix metalloproteinases and membrane damage markers in sera of patients with acute myocardial infarction. Mol Cell Biochem 350:163–168
6. Hu J, Van den Steen PE, Sang QX, Opdenakker G (2010) Matrix metalloproteinase inhibitors as therapy for inflammatory and vascular diseases. Nat Rev Drug Discov 6:480–498
7. Radenkovic S, Brajovic M, Konjevic G et al (2010) Gelatinases A and B activities in the serum of patients with various coronary artery disease stages. Vojnosanit Pregl 67:825–830
8. Konjevic G, Stankovic S (2007) Matrix metalloproteinases in the process of invasion and metastasis of breast cancer. Archieve Oncol 41:136–140
9. Radenkovic S, Konjevic G, Jurisic V et al (2014) Values of MMP-2 and MMP-9 in tumor tissue of basal-like breast cancer patients. Cell Biochem Biophys 68:143–152
10. Turk B, Turk D, Turk V (2012) Protease signalling: the cutting edge. EMBO J 31:1630–1643
11. Lopez-Otin C, Hunter T (2010) The regulatory crosstalk between kinases and proteases in cancer. Nat Rev Cancer 10:278–292
12. Perez-Silva JG, Espanol Y, Velasco G, Quesada V (2016) The degradome database: expanding roles of mammalian proteases in life and disease. Nucleic Acids Res 44:D351–D355

13. Radenkovic S, Konjevic G, Isakovic A et al (2014) HER2-positive breast cancer patients: correlation between mammographic and pathological findings. Radiat Prot Dosim 162:125–128
14. Radenkovic S, Milosevic Z, Konjevic G et al (2013) Lactate dehydrogenase, catalase, and superoxide dismutase in tumor tissue of breast cancer patients in respect to mammographic findings. Cell Biochem Biophys 66:287–295
15. Stankovic S, Konjevic G, Gopcevic K et al (2010) Activity of MMP-2 and MMP-9 in sera of breast cancer patients. Pathol Res Pract 206:241–247
16. Konjević G (2009) STAT proteins in cancerogenesis and therapy of malignancies. Srp Arh Celok Lek 137:98–105
17. Mansour EG, Ravdin PM, Dressler L (1994) Prognostic factors in early breast carcinoma. Cancer 74:381–400
18. Rossetti C, Reis Bda C, Delgado Pde O et al (2015) Adhesion molecules in breast carcinoma: a challenge to the pathologist. Rev Assoc Med Bra 61:81–85
19. Gocheva V, Joyce JA (2007) Cysteine cathepsins and the cutting edge of cancer invasion. Cell Cycle 6:60–64
20. Johansson AC, Steen H, Ollinger K, Roberg K (2003) Cathepsin d mediates cytochrome C release and caspase activation in human fibroblast apoptosis induced by staurosporine. Cell Death Differ 10:1253–1259
21. Ohri SS, Vashishta A, Proctor M et al (2008) The propeptide of cathepsin D increases proliferation, invasion and metastasis of breast cancer cells. Int J Oncol 32:491–498
22. Foekens JA, Look MP, Bolt-de Vries J et al (1999) Cathepsin-D in primary breast cancer: prognostic evaluation involving 2810 patients. Br J Cancer 79:300–307
23. Jurisic V, Srdic-Rajic T, Konjevic G et al (2011) TNF-α induced apoptosis is accompanied with rapid CD30 and slower CD45 shedding from K-562 cells. J Membr Biol 239:115–122
24. Jurisic V, Kraguljac N, Konjevic G et al (2005) TNF-alpha induced changes in cell membrane antigen expression on K-562 cells associated with increased lactate dehydrogenase (LDH) release. Neoplasma 52:25–31
25. Bracke ME, van Roy FM, Mareel MM (1996) The E-cadherin/catenin complex in invasion and metasta-sis. Curr Top Microbiol Immunol 213:123–161
26. Birchmeier W, Behrens J (1994) Cadherin expression in carcinomas: role in the formation of cell junctions and the prevention of invasiveness. Biochem Biophys Acta 1198:11–26
27. Maeda M, Shintani Y, Wheelock MJ, Johnson KR (2006) Src activation is not necessary for transforming growth factor (TGF)-beta-mediated epithelial to mesenchymal transitions (EMT) in mammary epi-thelial cells. PP1 directly inhibits TGF-beta receptors I and II. J Biol Chem 281:59–68
28. Lochter A, Galosy S, Muschler J (1997) Matrix metalloproteinase stromelysin-1 triggers a cascade of molecular alterations that lead to stable-epithelial-to-mes-enchymal conversion and a premalignant phenotype in mammary epithelial cells. J Cell Biol 139:1861–1872
29. Polyak K, Hahn WC (2006) Roots and stems: stem cells in cancer. Nat Med 12:296–300
30. Lee JM, Dedhar S, Kalluri R, Thompson EW (2006) The epithelial-mesenchymal transition: new insights in signaling, development, and disease. J Cell Biol 172:973–981
31. Hofmann UB, Houben R, Brocker EB, Becker JC (2005) Role of matrix metalloproteinases in melanoma cell invasion. Biochimie 87:307–314
32. Laufs S, Schumacher J, Allgayer H (2006) Urokinase-receptor (u-PAR): an essential player in multiple games of cancer: a review on its role in tumor progression, invasion, metastasis, proliferation/dormancy, clinical outcome and minimal residual disease. Cell Cycle 16:1760–1771
33. Mazzieri R, Masiero L, Zanetta L et al (1997) Control of type IV col-lagenase activity by components of the urokinase-plasmin system: a regulatory mechanism with cell-bound reactants. EMBO J 16:2319–2332

34. Wolf K, Muller R, Borgmann S (2003) Amoeboid shape change and contact guidance: T-lymphocyte crawling through fibrillar collagen is independent of matrix remodeling by MMPs and other proteases. Blood 102:3262–3269
35. McQuibban GA, Gong JH, Tam EM et al (2000) Inflammation dampened by gelatinase a cleavage of monocyte chemoattractant protein-3. Science 289:1202–1206
36. McCawely LJ, Matrisian LM (2000) Matrix metalloproteinases: multifunctional contributors to tumour progression. Mo Med Today 6:149–156
37. Foda HD, Zucker S (2001) Matrix metalloproteinase in cancer invasion, metastasis and angiogenesis. Drug Discovery Today 6:478–482
38. Indraccolo S, Favaro E, Amadori A (2006) Dormant tumors awaken by a short-term angiogenic burst: the spike hypothesis. Cell Cycle 5:1751–1755
39. Yamashita K, Suzuki M, Iwata H et al (1996) Tyrosine phosphorylation is crucial for growth signaling by tissue inhibitors of metalloproteinases (TIMP-1 and TIMP-2). FEBS Lett 396:103–107
40. Jurisic V, Radenkovic S, Konjevic G (2015) The actual role of LDH as tumor marker, biochemical and clinical aspects. Adv Exp Med Biol 867:115–124
41. Woessner JF Jr (1991) Matrix metalloproteinases and their inhibitors in connective tissue remodeling. FASEB J 5:2145–2154
42. Würtz SO, Schrohl AS, Mouridsen H, Brünner N (2008) TIMP-1 as a tumor marker in breast cancer–an update. Acta Oncol 47:580–590
43. Nakopoulou L, Giannopoulou I, Lazaris ACh et al (2003) The favorable prognostic impact of tissue inhibitor of matrix metalloproteinases-1 protein overexpression in breast cancer cells. APMIS 111:1027–1036
44. Chabottaux V, Sounni NE, Pennington CJ et al (2006) Membrane-type 4 matrix metalloproteinase promotes breast cancer growth and metastases. Cancer Res 66:5165–5172
45. Konjevic G, Spuzic I (1992) Evaluation of different effects of sera of breast cancer patients on the activity of natural killer cells. J Clin Lab Immunol 38:83–93
46. Konjevic G, Radenkovic S, Srdic T et al (2011) Association of decreased NK cell activity and IFNγ expression with pSTAT dysregulation in breast cancer patients. J BUON 16:219–226
47. Konjevic G, Jurisic V, Spuzic I (2001) Association of NK cell dysfunction with changes in LDH characteristics of peripheral blood lymphocytes (PBL) in breast cancer patients. Breast Cancer Res Treat 66:255–263
48. Townson JL, Chambers AF (2006) Dormancy of solitary metastatic cells. Cell Cycle 5:1744–1750
49. Colovic M, Todorovic M, Colovic N et al (2014) Appearance of estrogen positive bilateral breast carcinoma with HER2 gene amplification in a patient with aplastic anemia. Pol J Pathol 65:66–69
50. Shah FD, Shukla SN, Shah PM et al (2009) Clinical significance of matrix metalloproteinase 2 and 9 in breast cancer. Indian J Cancer 46:194–202
51. Li HC, Cao DC, Liu Y et al (2004) Prognostic value of matrix metalloproteinases (MMP-2 and MMP-9) in patients with lymph node-negative breast carcinoma. Breast Cancer Res Treat 88:75–85
52. Talvensari-Mattila A, Pakko P, Blanco-Sequeiros G, Turpeenniemi-Hujanen T (2001) Matrix metalloproteinases-2 is associated with the risk for a relapse in postmenopausal patients with node positive breast carcinoma treated with antiestrogen adjuvant therapy. Breast Cancer Res Treat 65:55–61
53. Konjevic G, Brankovic-Magic M, Nikolic-Vukosavljevic D et al (1994) Steroid receptors and NK cell activity in patients with breast cancer. Srp Arh Celok Lek 122:139–142
54. La Rocca G, Pucci-Minafra I, Marrazzo A et al (2004) Zymographic detection and clinical correlations of MMP-2 and MMP-9 in breast cancer sera. Br J Cancer 90:1414–1421

55. Pellikainen JM, Ropponen KM, Kataja VV et al (2004) Expression of matrix metalloproteinase (MMP)-2 and MMP-9 in breast cancer with a special reference to activator protein-2, HER2, and prognosis. Clin Cancer Res 10:7621–7628
56. Hupperets PS, Volovics L, Schouten LJ et al (1997) The prognostic significance of steroid receptor activity in tumor tissues of patients with primary breast cancer. Am J Clin Oncol 20:546–551
57. Koyama S (2004) Enhanced cell surface expression of matrix metalloproteinases and their inhibitors, and tumor-induced host response in progression of human gastric carcinoma. Dig Dis Sci 49:1621–1630
58. Konjevic G, Spuzic I (1993) Stage dependence of NK cell activity and its modulation by interleukin 2 in patients with breast cancer. Neoplasma 40:81–85

The Multifunctional Post-proline Dipeptidyl Peptidase, DPP9, in Mice, Cell Biology and Immunity

Margaret G. Gall and Mark D. Gorrell

Abstract

Dipeptidyl peptidase 9 (DPP9) is a ubiquitous intracellular post-proline protease of the DPP4 (S9b) family of atypical serine proteases. Emerging data support roles for DPP9 in intracellular signalling, particularly in the epidermal growth factor receptor pathway, in immune cells, particularly in macrophages and antigen processing, and in energy metabolism. The focus of this review is the roles of DPP9 in regulating physiological and cellular processes. Such data is derived from a genetically modified mouse strain and from manipulations of cell lines. The mouse strain that lacks DPP9 enzyme activity is homozygous lethal. DPP9 alters behaviours, such as cell adhesion, of cancer cell lines. This review points to the functional importance of DPP9 in immunity, metabolism and cancer.

Keywords

Dipeptidyl peptidase · Mouse models · Neonate development
Fibroblast activation protein

Abbreviations

AMC	Amino methylcoumarin
AMPK	AMP-activated protein kinase
B-CLL	B-cell chronic lymphocytic leukaemia
CAF	Cancer-associated fibroblasts
CXCL	CXC chemokine ligand

M.G. Gall · M.D. Gorrell (✉)
Centenary Institute and Sydney Medical School, University of Sydney, Sydney, NSW 2042, Australia
e-mail: m.gorrell@centenary.usyd.edu.au

S. Chakraborti and N.S. Dhalla (eds.), *Pathophysiological Aspects of Proteases*,
DOI 10.1007/978-981-10-6141-7_2

DEN	Diethylnitrosamine
DPP	Dipeptidyl peptidase
DSS	Dextran sulphate sodium
DTT	Dithiothreitol
ECM	Extracellular matrix
EGF	Epidermal growth factor
EGFR	Epidermal growth factor receptor
EMT	Epithelial–mesenchymal transition
ESFT	Ewing sarcoma family of tumours
FAP	Fibroblast activation protein
FGF	Fibroblast growth factor
HCC	Hepatocellular carcinoma
HFD	High fat diet
HSC	Hepatic stellate cells
IRS	Insulin receptor substrate
IκBα	Nuclear factor of kappa light polypeptide gene enhancer in B-cells inhibitor, alpha
MMP	Matrix metalloproteinase
NEM	N-ethylmaleimide
NPY	Neuropeptide Y
PARP	Poly(ADP-ribose) polymerase
PGC-1α	PPAR coactivator-1α
PBMC	Peripheral blood mononuclear cells
pNA	P-nitroaniline
POP	Prolyl oligopeptidase
PPAR	Peroxisome proliferator-activated receptor
SUMO	Small ubiquitin-like modifier
TAA	Thioacetamide
TAM	Tumour-associated macrophages
TGF	Transforming growth factor
TNF-α	Tumour necrosis factor α
VEGF-A	Vascular endothelial growth factor-A
WT	Wild-type

1 Introduction

1.1 The Dipeptidyl Peptidase 4 (DPP4) Enzyme Family

The DPP4 family of proteases has important roles in regulating physiological and cellular processes. The ubiquitous DPP4 family is the S9b protease subfamily of the prolyl oligopeptidases [1] of atypical serine proteases that contain a conserved catalytic triad of serine, aspartate and histidine [2]. Prolyl endopeptidase (PEP) and the DPP4 family enzymes possess their ability to hydrolyse a post-proline peptide bond. The DPP4 family enzymes can cut two residues from the N terminus of a protein substrate [3–5]. The presence of a proline near the N terminus of a peptide (Fig. 1) generally confers resistance against protease degradation and occurs in many different biologically active peptides, including chemokines, incretins and neuropeptides. Thus, the specialised ability of dipeptidyl peptidases to cleave the post-proline bond is useful to degrade such peptides [6].

The DPP4 protein family consists of six members: four enzymatic members and two non-enzymatic members. DPP4, DPP8, DPP9 and FAP have DPP activity, while FAP is also a post-proline endopeptidase [3–5, 7]. The two non-enzymatic members are DPP6 (DPP-X) and DPP10, also known as dipeptidyl peptidase like (DPL) protein-1 and DPL-2 [8–10] (Fig. 2).

1.1.1 DPP4

DPP4 is the prototypical and the most well-characterised member of the family. It displays enzymatic activity in both the dimeric cell surface membrane-bound form and the soluble circulating form [12, 13]. DPP4 is expressed by epithelial cells in the liver, gut, uterus and kidney, by immune organs, by capillary endothelium [14],

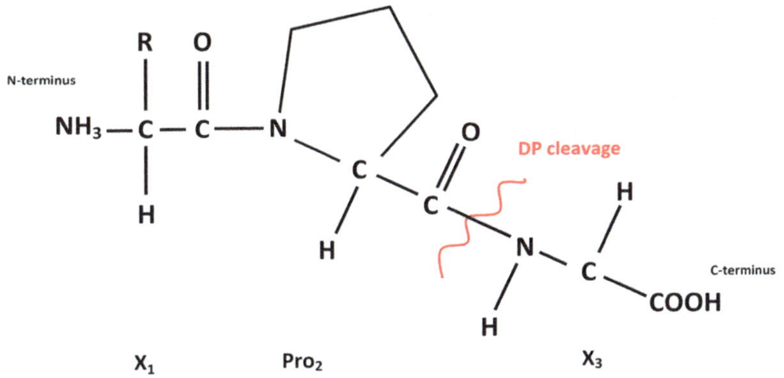

Fig. 1 Cleavage of a post-proline bond. A tri-peptide containing a proline at position two creates a bend in the peptide chain. Proline-containing peptides pose a specific problem for proteases due to structural constraints imposed by the pyrrolidine ring that prevents hydrolysis by most peptidases. Dipeptidyl peptidase cleavage by DPP4, DPP8 and DPP9 occurs C-terminal to the proline residue

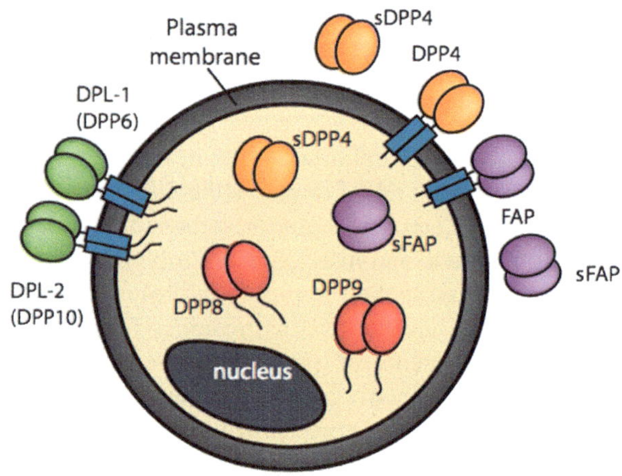

Fig. 2 Cellular localisation of the six members of the DPP4 protein family. DPP4 and FAP are type-II proteins that are membrane-bound cell surface proteases and also have soluble forms. DPP8 and DPP9 are intracellular proteases. DPP6 and DPP10 are membrane-bound, non-enzymatic proteins. Adapted from [11]

and by acinar cells of mucous and salivary glands and pancreas [15]. The soluble form is present in serum, seminal fluid, saliva, kidney, liver and bile [12, 15–18].

Functionally, DPP4 is involved in numerous processes throughout the body, including immunological, haematological, endocrine and metabolic. These include the ability to participate in chemokine inactivation and also to have a role in apoptosis, lymphocyte activation and cell migration, along with the capacity to cleave a wide range of small bioactive proteins [16, 19–22]. Most noted of these is the ability of DPP4 to rapidly inactivate the incretins glucagon-like peptide-1 (GLP-1) and glucose insulinotropic peptide (GIP) [23] leading to the inhibition of DPP4 as a successful type 2 diabetes therapy. DPP4 is also known as CD26 and, as such, has roles in the immune system in T cell activation and proliferation and T helper 1 responses to foreign antigens. Cell surface expression on T cells greatly increases following stimulation with antigen or mitogens [15, 17, 24].

1.1.2 FAP

FAP, which is also known as seprase, has 52% amino acid sequence identity with DPP4. However, FAP possesses both dipeptidyl peptidase and endopeptidase activities, which enables it to cleave a post-proline bond two or more amino acids from the N terminus of a protein (Fig. 3) [7].

While DPP4 is abundant in most tissues, FAP is at very low levels [18] with high-level expression being limited to sites of tissue remodelling and areas of stromal activation, such as in solid tumours, wound healing, tissue damage and inflammation [25, 26], during mouse embryogenesis [27] and in activated hepatic stellate cells in cirrhotic liver [28–30]. FAP has been isolated in a truncated soluble form from human plasma and serum [31, 32]. FAP activity levels in normal and

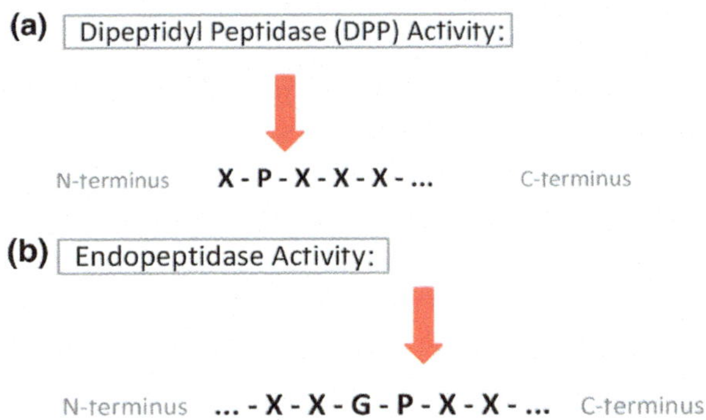

Fig. 3 Enzyme activity of FAP. **a** Dipeptidyl peptidase activity involves the cleavage of two amino acids of the N terminus of a protein following a proline residue. **b** Endopeptidase activity involves the cleavage of a post-proline bond that is more than two amino acids from the N terminus of a protein

diseased tissue from humans, mice and baboons have been measured using the novel FAP-specific substrate, 3144-AMC. In mice, uterus, pancreas, submaxillary gland and skin showed the greatest levels of FAP activity [18].

While FAP has been shown to cleave some DPP4 substrates in vitro [33], until recently there were no known physiological substrates of FAP's dipeptidyl peptidase activity. The first study to characterise the physiological DPP substrate repertoire of endogenous FAP in mammalian plasma showed a potential function of FAP in neuropeptide signalling in liver and liver cancer [34]. In addition to its dipeptidyl peptidase activity, FAP also has endopeptidase activity cleaving α2-antiplasmin [31] and denatured type I collagen [35]. The gelatinase activity of FAP most likely contributes to extracellular matrix degradation.

Because of its pattern of expression by activated stromal fibroblasts in many cancers, FAP has been studied as a potential therapeutic target in tumours. FAP roles in cancer biology have been reviewed previously [36–38]. Strategies for targeting FAP for cancer therapy and the role of FAP seem to be very contextual, so the expression pattern and substrates of FAP require greater definition to better predict the effects of targeting this protease.

1.1.3 DPP6 and DPP10

DPP6 and DPP10 have 53% amino acid similarity with each other and are highly homologous to DPP4 with sequence identities of 33 and 32%, respectively [5, 10]. They both form dimers but lack the catalytic serine and a nearby tryptophan that are essential for enzymatic activity. DPP6 has a wide tissue distribution, but DPP10, in contrast, has expression limited to human brain, adrenal gland and pancreas [5, 10, 39]. The functional effects of these proteins appear to be through binding interactions, and there is evidence of an association with neuronal diseases and asthma [10, 40, 41].

1.1.4 DPP8 and DPP9

DPP8 and its closest relative DPP9 are the most recently discovered members of the DPP4 gene family [4, 5, 42, 43]. As DPP4, DPP8 and DPP9 all have DPP4-like enzymatic activity and ubiquitous expression, the discovery of DPP8 and DPP9 has called for the reinterpretation of previous DPP4 data [4, 42]. This was necessary to determine which functions had been attributed to DPP4 that may instead be accounted for by DPP8 and DPP9.

Unlike DPP4, DPP8 and DPP9 are intracellular proteins and are therefore likely to have different biological roles. As there are currently no substrates or inhibitors that distinguish DPP8 from DPP9, their characteristics are usually outlined together but this review will focus upon DPP9. DPP9 is currently under study in many fields including cell biology, immune-biology and tumour biology [44].

2 Structure of DPP8 and DPP9

DPP8 and *DPP9* are located on human chromosomes 15q22 and 19p13.3 [42, 43] and mouse chromosomes 9 and 17, respectively. In *DPP4* and *FAP*, the sequence adjacent to the active-site serine is encoded by two exons while the homologous region in *DPP8* and *DPP9* is encoded by a single exon, which suggests that the *DPP8* and *DPP9* genes arose at an earlier evolutionary stage and may have arisen by gene duplication [43]. The amino acid homology between human and mouse DPP8 and DPP9 is 95 and 92% identity, respectively [4, 43]. No crystal structures of DPP8 or DPP9 exist, but homology models have been built using the known structures of DPP4, DPP6 and FAP [45–47]. In these models, DPP9 is depicted with an α/β-hydrolase domain containing a Ser-Asp-His catalytic triad and an 8-blade β-propeller domain containing two glutamic acids essential to functionality of the catalytic pocket (Fig. 4).

While structurally similar to DPP4 and FAP, DPP8 and DPP9 are longer at the N terminus but lack a transmembrane domain and are intracellular proteins [4, 42]. Another structural difference between DPP4 and DPP9 is that selective point mutations in the C-terminal loop can inactivate DPP9 while maintaining the dimeric structure [48]. This confirms the importance of the C-terminal loop for DPP8 and DPP9 enzymatic function. Finally, since DPP8 has been shown to have a larger substrate pocket than DPP4 [47], DPP8 and DPP9 might have a larger β-propeller domain and/or may contain a separate element of tertiary structure at the N terminus of the protein.

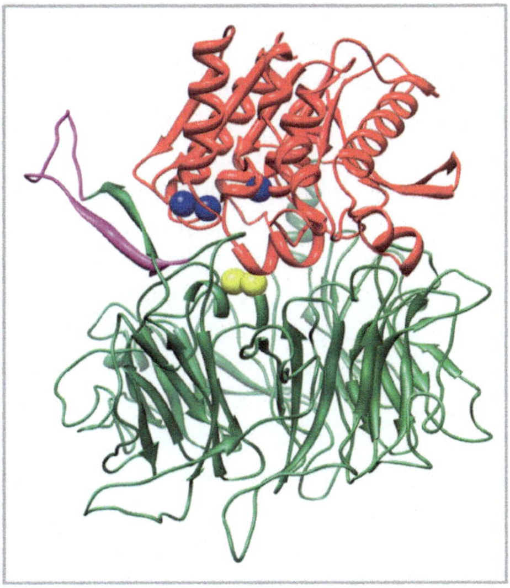

Fig. 4 A model of DPP9 protein structure. DPP9 residues 51–863 are depicted in ribbon representation of the monomer. The α/β-hydrolase domain is in *red* and the 8-blade β-propeller domain in *green*. The catalytic triad (Ser-Asp-His) is shown as *blue* spheres in the hydrolase domain and 2 glutamates essential for catalysis as *yellow* spheres in the propeller domain. An extended arm (^{285}VEVIHVPSPALEERKTDSYR^{304}) in the propeller surface of DPP9 that is critical in SUMO1–DPP9 interaction is coloured *pink*. Modified from [44, 45]

3 Expression Profiles and Tissue Distribution of DPP8 and DPP9

Both DPP8 and DPP9 have isoforms. Human DPP8 has several mRNA transcripts identified: one is abundant in testis, prostate and muscle [5, 42], while another longer transcript variant has intense signals in adult testis [49]. Bioinformatics analysis of NCBI database sequences has identified at least two other isoforms [50].

Two forms of DPP9 cDNA have been cloned, with the short form (DPP9-S) encoding 863 amino acids [4, 43] and the long form (DPP9-L) encoding 971 amino acids [4]. DPP9-S is ubiquitously expressed including high levels in liver, heart and skeletal muscle, while DPP9-L is much less abundant and is predominantly expressed in skeletal muscle. The N-terminal extension of DPP9-L contains a nuclear localisation signal (NLS) allowing it to enter the nucleus [51]. Endogenous nuclear DPP9 has been visualised in mouse fibroblasts [52].

In canine and porcine small intestine, lung, kidney and pancreas, differential relative abundance of DPP8 and DPP9 has been measured by RT-PCR analysis with DPP9 having the greater expression [53]. In adult mice, greater numbers of DPP9 transcripts are in brain, skin, colon and thymus compared with DPP4 [53, 54].

DPP8 mRNA is upregulated in human adult testis compared to foetal testis [49]. DPP8 and DPP9 is in spermatozoids embedded in the epithelium of the seminiferous tubules [55], while the mRNA is in spermatogonia and spermatids [56]. DPP9-S has been purified from bovine testes [57]. Taken together, these studies suggest a role for DPP8 and DPP9 in spermatogenesis.

Using in situ hybridisation, immunohistochemistry and enzyme assays on baboon, mouse and human tissues, the ubiquitous expression and distribution of DPP8 and DPP9 have been visualised in lymphocytes and epithelial cells in the gastrointestinal tract, liver, lung, spleen, lymph node and skin, as well as in pancreatic acinar cells, muscle, adrenal gland, testis and Purkinje cells and the granular layer of the cerebellum [56]. These comprehensive findings concord with related studies [55, 58, 59].

Analyses of mRNA and protein levels in Sprague Dawley rat and cynomolgus monkey showed similar ubiquitous expression and tissue distribution correlated with DPP8/9 enzymatic activity and also detected the DPP8 and DPP9 proteins in some capillary endothelia [14]. DPP8 and DPP9 have been shown to be in rat primary endothelial cells of aortic, endocardial and cardiac microvascular origin [60] with a greater abundance of DPP8 protein over DPP9. However, in human carotid artery endothelial cells, DPP9 has been the only DPP4-like enzyme detected using immunohistochemistry, suggesting a regulated expression of DPPs in endothelial cells.

Both DPP8 and DPP9 occur in tumours [4, 56, 61], hepatocytes [56, 62, 63], macrophages [64] and lymphocytes [56, 63, 65] but are absent from activated stellate cells [56].

Recent reviews provide more detailed information regarding the expression and distribution of DPP8 and DPP9 [19, 22, 44, 66].

4 Enzymatic Activity and Substrate Specificity of DPP8 and DPP9

Although originally thought to be monomers, both DPP8 and DPP9 are, like DPP4, catalytically active as dimers [48, 67, 68]. Both DPP8 and DPP9 have optimum activity at neutral pH of 7.0–8.5 [5, 42, 48], which is consistent with intracellular localisation. DPP8 and DPP9 have reversible changes in enzymatic activity intensity, dependent on the redox state of their cysteine moieties, with activity decreased by oxidation and increased by reduction [45, 69]. DPP9 is allosterically activated by SUMO1 [70]; however, the small peptide region of SUMO1 that interacts with DPP9 is also a non-competitive inhibitor of DPP9 activity with inhibition dependent on the DPP9 arm motif (Fig. 4) [71].

Both enzymes have similar substrate specificity, preferring a proline in the penultimate P1 position and small or basic amino acid residues at the P2 position of synthetic substrates [48, 68, 69]. Similarly, preferred natural substrates have a P1 proline preceded by an alanine in the P2 position [72], but alanine can be in P1

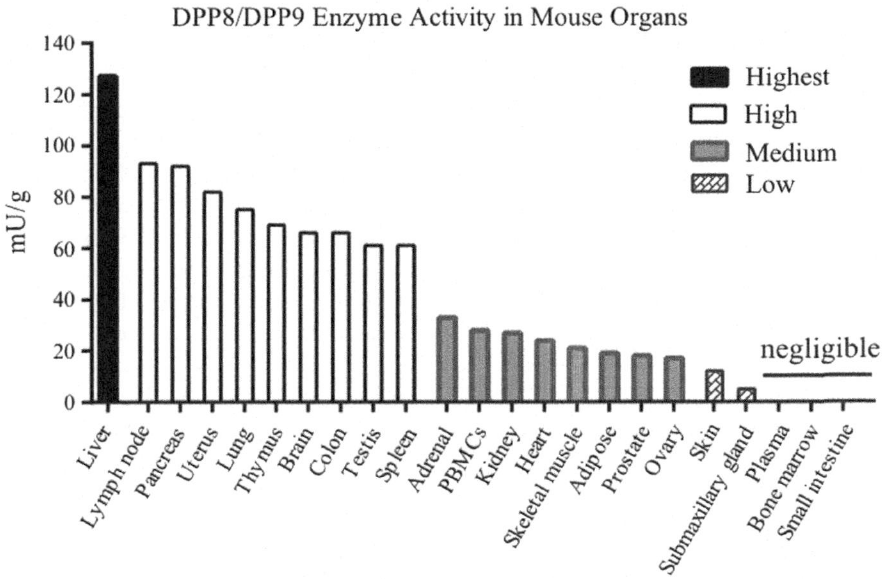

Fig. 5 DPP8/DPP9 enzyme activity in mouse organs. The approximate intensity of DPP8/9 enzyme activity was measured by Yu and colleagues [56] from mouse organs in DPP4-gene knockout mice with/without the addition of the DPP8/9 inhibitor NEM (N-ethylmaleimide). NEM-inhibited activity was subtracted from total DPP activity to estimate the activity derived from DPP8 and DPP9. Modified from [44]

[67, 73]. Ala-Pro- and Gly-Pro-containing fluorogenic and chromogenic substrates have been routinely used to detect DPP8/9 enzymatic activity (Fig. 5), but H-Arg-Pro-pNA is also hydrolysed [4, 56]. We recently showed that DPP9 is much more efficient than DPP8 at cutting after alanine in P1 [73].

DPP8 and DPP9 are able to cleave a number of naturally occurring extracellular peptides and chemokines in vitro but with reduced rates of hydrolysis compared to DPP4 [67, 73, 74]. Whether extracellular substrates of these intracellular DPPs have physiological relevance is unclear. It has previously been shown that neuropeptide Y (NPY) can be cleaved by DPP4, DPP8, DPP9 and FAP [33, 34, 67, 75, 76]. NPY extracted from rat brain can be cleaved in the presence of a DPP4 inhibitor, suggesting that DPP8 and DPP9 can cleave NPY in vivo [76]. Moreover, NPY is a substrate for DPP8 and DPP9 in intact cells, not just cell extracts, and in doing so regulates Ewing sarcoma family tumour (ESFT) cells by reducing NPY-induced cell death [77].

A role of DPP9 in antigen presentation and peptide turnover was first shown by the identification of the antigenic peptide $RU1_{34–42}$ as its first natural substrate. Silencing of DPP8 and DPP9 using siRNA resulted in increased RU1 presentation only in DPP9-silenced cells, which indicated that the $RU1_{34–42}$ antigen is an in vivo substrate of DPP9 [69].

DPP9 substrates have been sought using a degradomic approach and resulted in the identification and validation in vitro of numerous potential substrates, with adenylate kinase 2 and calreticulin confirmed as DPP8 and DPP9 substrates. Those substrates suggested a role for DPP8 and DPP9 in cellular metabolism and homoeostasis [72].

Most recently, two-dimensional differential in-gel electrophoresis (DIGE) was applied to cytoplasmic and nuclear extracts of mouse cells lacking endogenous DPP9 activity (the DPP9-GKI mouse, see below) to identify novel DPP9 substrates [73], with nine being confirmed as substrates by MALDI-TOF or immunoblotting. This study also identified a DPP9-specific consensus site for cleavage that was not recognised by DPP8, suggesting that these two proteases have different in vivo roles. The substrates included two key immune regulators, CXCL10 and IL1RA, and the putative DPP9 substrates S100-A10, SET and NUCB1, which are mediators of immunity and inflammation [73].

5 Non-Enzymatic Functions of DPP8 and DPP9

Early investigations using overexpression in the HEK-293T epithelial cell line showed potential roles for DPP8 and DPP9 in apoptosis, wound healing and cell migration. In vitro overexpression of DPP8 and DPP9 resulted in impaired cell adhesion, migration and monolayer wound healing for DPP9 and impaired migration and wound healing for DPP8, suggesting that those outcomes may result from direct protein–protein interactions or via altered expression of other proteins involved in cell adhesion and migration [78]. Both DPP8 and DPP9 also enhance apoptosis. Using enzyme-negative mutants of each protein, the role of DPP8 and DPP9 in apoptosis appeared to be independent of enzymatic activity in HEK-293T cells.

Recent work showed that apoptosis and cell death were unaffected by DPP9 knock-down or enzyme inhibition, but found that enzyme inhibition or gene silencing of DPP9 in Huh7 cells resulted in less cell mobility and adhesion compared to control cells [79], indicating that DPP9 enzymatic function is important for these processes. Differences between that recent study [79] and the early study [78] may reflect that overexpression data can be cell-type specific [30] or can be misleading.

A non-enzymatic activity that needs further exploration is the association of H-Ras with DPP9 [62]. Whether there is direct binding between these two proteins and the functional significance of this association are unknown.

6 Inhibition of DPP8 and DPP9

Many early DPP4 inhibitors were later found to also inhibit DPP8 and DPP9, raising the question of which functions attributed to DPP4 activity should be attributed to DPP8 and/or DPP9. This has led to the development of many inhibitors that are selective for DPP8 and DPP9 over DPP4 and FAP [80–85], but are not selective for DPP8 or DPP9 alone. While some isoindoline inhibitors were originally thought to inhibit DPP8 alone [82, 86], these were later shown to inhibit both DPP8 and DPP9 [81].

The safety of targeting DPP8/9 activity for therapeutic use has been controversial, with toxic effects reported in rats with the use of the DPP8/9 selective inhibitor, UAMC00132 [80], while a rebuttal study on rats and mice using vildagliptin at high doses that inhibit DPP4, DPP8 and DPP9 showed no toxicity or mortality in animals [87]. Moreover, the compound 1G244 does not cause pathological symptoms when inhibiting DPP8 and DPP9 [81]. With the recent design of two analogues of 1G244 that are 10-fold selective for DPP8 over DPP9 [84], there is good progress towards selectively differentiating between these two enzymes by inhibition of either DPP8 or DPP9, rather than both.

Alternative approaches to chemical inhibitors have been examined for targeting DPP4 and FAP. As the DPPs have non-enzymatic functions, antibodies might block cell surface or extracellular protein-binding interactions, or might remove DPP-expressing cells. Immunotargetting DPP4 is an example [88–90]. A separate approach involved a non-substrate non-apeptide, the HIV-1 Tat protein, the first natural inhibitor of DPP4, as an immunosuppressor [91]. Whether non-apeptides or antibodies can enter cells to act on cytoplasmic or nuclear DPP8 and DPP9 is unknown.

More recently, it was found that, while the small ubiquitin-like protein modifier SUMO1 interacts with DPP9 leading to allosteric activation of the peptide [70], the E67-interacting loop peptide acts as a non-competitive inhibitor [71]. This highlights the potential modulation of enzyme activity by peptides that mimic interaction surfaces. Targeting the non-enzymatic activities of DPP8 and DPP9 as a therapeutic approach would require further identification and analyses of protein-binding partners, such as those required for the localisation of DPP9 to the leading edge of moving cells [79].

7 Biological Functions of DPP8 and DPP9

Knowledge of the physiological functions of DPP8 and DPP9 is emerging with investigations in normal homoeostasis and pathophysiological conditions. Data on enzymatic activity, substrate specificity and cell and tissue distribution of these proteases provide some guidance regarding their biological functions in cell survival, cell biology, disease, inflammation and cell-mediated and humoral immunity.

7.1 DPP8 and DPP9 in Immunity

7.1.1 DPP8 and DPP9 in Innate Immunity

The non-specific defence mechanisms against pathogens include physical barriers, such as skin integrity, blood-borne chemicals and certain immune cells. DPP8 and DPP9 have been shown to have extensive in vivo expression in epithelial cells of the gut and skin and normal immunological tissues and also are expressed by all major lymphocyte populations [42, 56, 63]. DPP8 and DPP9 have also been shown to be upregulated upon B or T cell activation [63]. Several T cell leukaemia cell lines highly express DPP8 and DPP9 mRNA transcripts [4].

DPP8/9 enzyme activity has been found in human lymphocytes, monocytes, Jurkat and U937 cells [65] and also in human and mouse primary leucocytes and B and T cell lines [42, 80, 92]. DPP9 is at low levels in both M1 and M2 macrophages derived from U937 cells but abundant in the macrophage-rich regions of plaques [64, 93]. Using the mouse macrophage cell line J774 and bone marrow-derived monocytes and macrophages, an inhibitor of DPP8 and DPP9 has been shown to lessen mouse M1 macrophage activation [64].

Natural killer (NK) cells play a major role in host rejection of both tumours and virally infected cells in the innate immune system. An early study showed that inhibition of DPP4 enzyme activity results in the suppression of stimulatory cytokines causing a reduction of DNA synthesis and cell cycle progression [94]. However, the inhibitors used in that study were later shown to also inhibit DPP8/9 activity [80]; therefore, those effects were probably DPP8/9 mediated. Further support for that conclusion derives from the finding that the presence or absence of DPP4 on the cell surface of NK cells does not influence the natural cytotoxicity of these cells [95]. In a lung metastasis model, NK cytolytic function against tumour cells was less in DPP4-deficient than in WT rats [96].

7.1.2 DPP8 and DPP9 in Adaptive Immunity

The antigen-specific adaptive immune response is complex and includes formation of peripheral and tissue-resident memory leucocytes that improve the efficiency of future responses, and depends upon antigen presentation and recognition. DPP9 is the dominant protease that degrades the antigenic peptide $RU1_{34-42}$ such that downregulation of DPP9 increases the presentation of this antigen [69].

While mouse splenic B lymphocytes express little DPP4 mRNA, DPP8 and DPP9 mRNA are expressed at greater levels and stimulation with pokeweed mitogen and lipopolysaccharide of mouse splenocytes and Jurkat T- and Raji B-cell lines upregulates both proteins [63]. DPP8 and DPP9 mRNA are downregulated by dithiothreitol treatment and upregulated by mitomycin c treatment of Raji cells. Contrary to this, human Jurkat cells or peripheral blood mononuclear cells (PBMC) stimulated with phytohaemagglutinin do not change DPP8 or DPP9 expression levels [48].

DPP8 and DPP9 have each been detected by in situ hybridisation in lymphocytes in the mantle and paracortical zones of human lymph node and interfollicular cells of baboon spleen, and DPP8/9 activity has been detected in baboon spleen and

Jurkat cells [56]. Activation and proliferation of immune cells appears to be influenced by DPP8 and DPP9 enzymatic activity, and selective DPP8/9 inhibition can reduce cytokine production due to the induction of TGF-β secretion, as well as DNA synthesis and T cell proliferation [80, 97]. This is suggestive of DPP8/9 enzyme activity having important roles in the regulation of immune function.

7.2 DPP8 and DPP9 in Cell Biology and Cell Survival

Both DPP8 and DPP9 are involved in cell-extracellular matrix interactions, but DPP8 does not influence cell adhesion [78]. While DPP8 and DPP9 are very similar intracellular enzymes and both ubiquitously expressed, DPP9 appears to have a more pronounced role in cell biology.

DPP9 mRNA is elevated in cartilage from osteoarthritis patients [98]. A study of normal and keloid-derived skin fibroblasts in vitro showed that inhibitors of DPP8 and DPP9 could suppress fibroblast proliferation and decrease secretion of both the fibrogenic cytokine TGF-β_1 and procollagen type 1, which is important in the treatment of fibrotic skin disorders and keloid scar [99].

In sarcoma cell lines, NPY-driven tumour cell death mediated by the nuclear protein modifying enzyme poly(ADP-ribose) polymerase (PARP-1) and apoptosis-inducing factor can be abolished by the overexpression of DPP8 and DPP9 and enhanced by downregulating these proteases [77]. In contrast, overexpression of DPP9 in epithelial tumour cell lines is anti-proliferative and enhances intrinsic apoptosis [62, 78]. DPP9 overexpression can also cause less epidermal growth factor (EGF)-mediated Akt pathway activation in HepG2 cells but has not been observed in cells stimulated with other growth factors, suggesting that this activation was growth factor dependent [62]. Since experiments in which DPP9 is overexpressed on the one hand and orgene silenced or inhibited on the other has produced differing data, the pro- or anti-apoptotic activity of this enzyme may depend on its in vitro culture environment and/or the cell type in which the experiment is undertaken.

The DPP inhibitor vildagliptin synergistically enhances parthenolide's action in leukaemia, lymphoma and primary human acute myeloid leukaemia cell lines [100]. This synergy was due to the inhibition of DPP8 and DPP9 rather than DPP4, suggesting that DPP8 or DPP9 inhibition might be a chemosensitising agent for leukaemia cells, as has been found for FAP in fibrosarcoma [101].

The use of a wide variety of cell lines and tissues can provide insights into the physiological and pathological roles of DPP8 and DPP9 in cell behaviour and survival.

7.3 DPP8 and DPP9 in Disease and Inflammation

There is increasing evidence for an association of DPP8 and DPP9 with disease pathogenesis (Fig. 6). In a gene expression profile study of human hepatocellular

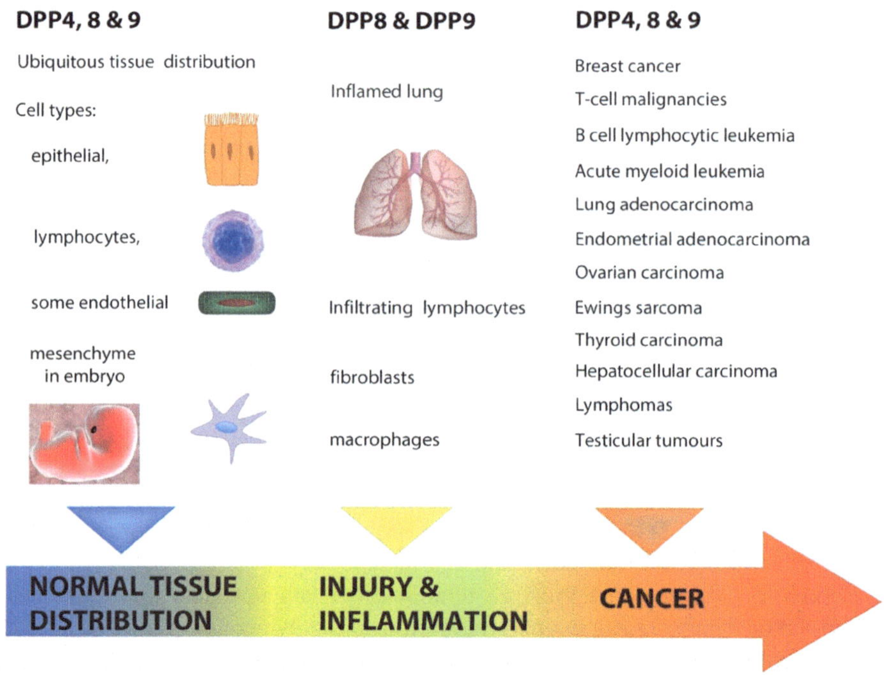

Fig. 6 Summary of studies that have detected mRNA, protein or enzyme activity of DPP4, DPP8 and DPP8 in tissues or cells without disease and increased levels in disease

carcinoma (HCC) tissue, mRNA of DPP9 showed differential expression and upregulation between non-tumour HCC-bearing liver and normal liver [102].

Expression profiling of breast and ovarian carcinoma cell lines, along with 293T and HeLa cell lines, showed ubiquitous but differential expression of DPP8 and DPP9 mRNA and protein across these cell lines. There was poor correlation between mRNA transcript and protein levels for both DPP8 and DPP9, which suggests that DPP8 and DPP9 may be regulated post-transcriptionally in breast and ovarian cancer cell lines [66]. This contrasted with normal mouse tissues, where mRNA and enzyme levels of DPP8 and DPP9 were associated [56].

DPP8 mRNA expression is greater than other DPPs in breast and ovarian cancer cell lines [66], and DPP9 mRNA is abundant in chronic myelogenous leukaemia (K-562) and immortalised cervical cancer (HeLa) cells [43], suggesting that DPP8 and DPP9 may have roles in tumour pathogenesis.

DPP8 and DPP9 mRNA is in tumour infiltrating lymphocytes and DPP9 mRNA is upregulated in human testicular tumours [56]. DPP activity inhibition partially attenuates dextran sulphate sodium (DSS)-induced colitis in mice, and DPP8 mRNA is differentially expressed during colitis development, which is suggestive of role of DPP8 in inflammatory bowel disease [103]. DPP8 mRNA and protein expression are significantly upregulated in B-cell chronic lymphocytic leukaemia (B-CLL) cells compared to normal tonsil B lymphocytes, which suggests that DPP expression may have biological relevance in B-CLL disease states [36].

Meningiomas are inter-cranial tumours derived from the arachnoid cap cells and include a wide variety of subtypes. DPP8/9 enzymatic activity has been detected in all benign meningiomas, with elevated levels in atypical meningiomas that are associated with greater cell proliferation [61]. An aggressive group of human paediatric malignancies, ESFT, are driven by an aberrant transcription factor that upregulates specific target genes including NPY. ESFT cell death is stimulated by exogenous and endogenous NPY; however, this effect is prevented by cleavage of NPY by cell surface DPP4 and intracellular DPP8 and DPP9. Thus, DPPs are survival factors for EFST and may become potential therapeutic targets for these tumours [77].

A pathophysiologically significant role for DPP8 and DPP9 in asthma is suggested by data from a rat asthma model in which DPP8 and DPP9 and DPP8/9 enzyme activity were primarily detected in the bronchial epithelium of the airways, particularly with allergy-like inflammation [59]. DPP8 and DPP9, along with other DPPs, are elevated in human articular cartilage in osteoarthritis patients, which implicates all of these enzymes in the cascades leading to cartilage breakdown and/or collagenolysis that occurs in arthritis [104].

The expression, localisation and activity patterns of DPP8 and DPP9 have been examined in a model of transient and unilateral cerebral ischaemia in rats [105]. That study found that mRNA expression of DPP9 was diminished in the ischaemic region of the brain at 6 h and 3 days after induced ischaemia, whereas DPP8 levels remained the same at all time points. Also, DPP8 was present in activated microglia and macrophages at day 3 post-ischaemia and in astroglial cells at day 7 post-ischaemia [105]. Those data suggest that DPP8 and DPP9 have potential roles in cerebral inflammation.

As a role for DPP4 in atherosclerosis emerged [106], it was important to investigate the expression and role of DPP8 and DPP9. While DPP4 is only present in endothelial cells of plaque, DPP8 and DPP9 are also in macrophages with a significant upregulation of DPP9 in both pro-inflammatory M1 and anti-inflammatory M2 macrophages during monocyte-to-macrophage differentiation [93]. Primary endothelial cells from aortic, endocardial and cardiac microvascular regions of the rat heart contain DPP8/9 enzymatic activity, with DPP8 protein more abundant than DPP9 [60]. The localisation and expression of DPP8 and DPP9 proteins in endothelia have spatial heterogeneity, with DPP8 more highly expressed in the cardiac microvascular endothelium and DPP9 predominant in the human carotid artery endothelial cells. DPP8 and DPP9 are abundant in the macrophage-rich regions of human and mouse atherosclerotic plaques [64, 93]. M1 macrophages have a role in atherogenesis, so those data suggest potential therapeutic prospects in atherosclerosis and/or plaque rupture by inhibition of DPP9.

A genome-wide association study that focused on human pulmonary fibrosis identified *DPP9* as one of several novel susceptibility loci. Although the evidence for greater expression of *DPP9* in lung tissue of pulmonary fibrosis cases compared to controls was slight, this group speculated that DPP9 may be involved in the integrity of lung epithelia via cell-to-cell adhesion [107]. This finding needs further investigation of whether DPP9 has a role in the development of pulmonary fibrosis.

8 DPP4 Family Mouse Phenotypes

8.1 Mouse Models

As the mouse genome is amenable to genetic manipulation, many studies involve the use of gene knockout (GKO) mice as models for investigation of the biological properties of proteins in a physiological and/or pathophysiological context. In GKO mice, the protein of interest is absent from the resultant mouse. In the DPP4 gene family, GKO mice have been created for both DPP4 and FAP. The DPP4-GKO mouse was produced by a targeted inactivation in the *DPP4* gene and resulted in healthy mice [108]. Similarly, the FAP-GKO mouse was constructed through exon deletion and also created fertile mice with no overt developmental or haematological defects or increased cancer susceptibility [109]. It has been speculated that, in the absence of these proteins, other proteins are upregulated and have compensatory roles. To date, no DPP8-GKO mice have been reported.

8.1.1 DPP9 Gene Knock-in (GKI) Mouse

Unlike GKO mice where the entire protein is absent from the mouse, in a GKI mouse the protein of interest is present but inactivated; thus, it is still able to fulfil all functions apart from the one that has been removed. By targeting the DPP9 protein catalytic site with a serine to alanine point mutation, we were able to create a DPP9 gene knock-in mouse (DPP9-GKI) that lacks DPP9 enzymatic activity while maintaining normal DPP9 protein structure and thus the non-enzymatic functions of DPP9 [52]. This DPP9-GKI mouse exhibits neonate lethality, suggesting that DPP9 enzymatic activity is essential for early neonatal survival in mice [52]. However, no morphological defect or cause of death has been identified. Further studies using neonatal liver and gut from DPP9-GKI mice compared to WT have shown that DPP9 enzymatic activity influences the expression of many genes [110].

Taqman PCR arrays and qPCR analyses of neonatal liver revealed differential expression of genes involved in cell growth (epidermal growth factor (EGF), epidermal growth factor receptor (EGFR), vascular endothelial growth factor-A (VEGF-A), innate immunity (tumour necrosis factor (TNF)-α, interleukin-1 β, IκBα (nuclear factor of kappa light polypeptide gene enhancer in B-cells inhibitor, alpha)) and metabolism signalling pathways, such as insulin receptor substrate-2 (INS-2). Neonatal liver and gut showed differential expression of the master transcription coactivator peroxisome proliferator-activated receptor (PPAR)-γ coactivator-1α (PGC-1α) and genes involved in long-chain-fatty-acid uptake and esterification, long-chain fatty acyl-CoA binding, trafficking and transport into mitochondria, lipoprotein metabolism, adipokine transport and gluconeogenesis in the DPP9-GKIhomozygote neonatal mice compared to WT. In addition, immunoblots showed that DPP9 enzyme activity can modulate AMP-activated protein kinase phosphorylation in the Huh7 human hepatoma cell line [110].

The differential expression of genes involved in cell growth, innate immunity, lipid metabolism and gluconeogenesis suggests roles for DPP9 enzymatic activity in the regulation of immune and metabolic pathways in the neonate [110].

9 Conclusion

DPP9 is emerging as an interesting and unique atypical serine protease that has potential roles in tumours, leukaemia, atherosclerosis, inflammation and energy metabolism and in T, B and NK cells and endothelial cells. Numerous studies have been reported now that detail DPP9 expression and its upregulation in brain, joint, liver, lung and gut inflammation. However, a dominant homoeostatic or disease role for DPP9 has not been identified. Despite strong expression in muscle, no function in muscle or heart has arisen. The enzyme activity of DPP9 has been implicated in neonate survival, metabolism and immune system, and in macrophage function, EGF receptor signalling and cell migration.

The impediments to a greater understanding of DPP9 are the lack of any selective enzyme inhibitor, specific enzyme assay or adult GKO or GKI mice.

Acknowledgements This work was supported by National Health and Medical Research Council of Australia Project Grants 512282 and 1113842 (MDG), and a Sydney Medical School Foundation/Francis M. Hooper Scholarship for Medical Research through the University of Sydney (MGG).

References

1. Rawlings ND, Barrett AJ, Finn R (2016) Twenty years of the MEROPS database of proteolytic enzymes, their substrates and inhibitors. Nucleic Acids Res 44:D343–D350
2. Gorrell MD, Yu DMT (2005) Diverse functions in a conserved structure: the dipeptidyl peptidase IV gene family. In: Robinson JW (ed) Trends in protein research. Nova Science Publishers Inc, New York, pp 1–78
3. Abbott CA, Yu DMT, McCaughan GW, Gorrell MD (2000) Post proline peptidases having DP IV like enzyme activity. Adv Exp Med Biol 477:103–109
4. Ajami K, Abbott CA, McCaughan GW, Gorrell MD (2004) Dipeptidyl peptidase 9 has two forms, a broad tissue distribution, cytoplasmic localization and DPIV-like peptidase activity. BBA—Gene Struct Expr 1679:18–28
5. Qi SY, Riviere PJ, Trojnar J, Junien JL, Akinsanya KO (2003) Cloning and characterization of dipeptidyl peptidase 10, a new member of an emerging subgroup of serine proteases. Biochem J 373:179–189
6. Cunningham DF, O'Connor B (1997) Proline specific peptidases. Biochim Biophys Acta 1343:160–186

7. Aertgeerts K, Levin I, Shi L, Snell GP, Jennings A, Prasad GS, Zhang Y et al (2005) Structural and kinetic analysis of the substrate specificity of human fibroblast activation protein α. J Biol Chem 280:19441–19444
8. Kim J, Nadal MS, Clemens AM, Baron M, Jung S-C, Misumi Y, Rudy B et al (2008) Kv4 accessory protein DPPX (DPP6) is a critical regulator of membrane excitability in hippocampal CA1 pyramidal neurons. J Neurophysiol 100:1835–1847
9. Wada K, Yokotani N, Hunter C, Doi K, Wenthold RJ, Shimasaki S (1992) Differential expression of two distinct forms of mRNA encoding members of a dipeptidyl aminopeptidase family. Proc Natl Acad Sci USA 89:197–201
10. Chen T, Ajami K, McCaughan GW, Gai W-P, Gorrell MD, Abbott CA (2005) Molecular characterization of a novel dipeptidyl peptidase like 2 short form (DPL2-s) that is highly expressed in the brain and lacks dipeptidyl peptidase activity. Biochim Biophys Acta 1764:33–43
11. Kirby MS, Yu DMT, O'Connor SP, Gorrell MD (2010) Inhibitor selectivity in the clinical application of dipeptidyl peptidase-4 inhibition. Clin Sci 118:31–41
12. Durinx C, Lambeir AM, Bosmans E, Falmagne JB, Berghmans R, Haemers A, Scharpé S et al (2000) Molecular characterization of dipeptidyl peptidase activity in serum—Soluble CD26/dipeptidyl peptidase IV is responsible for the release of X-Pro dipeptides. Eur J Biochem 267:5608–5613
13. Ogata S, Misumi Y, Tsuji E, Takami N, Oda K, Ikehara Y (1992) Identification of the active site residues in dipeptidyl peptidase IV by affinity labeling and site-directed mutagenesis. Biochemistry 31:2582–2587
14. Harstad EB, Rosenblum JS, Gorrell MD, Achanzar WE, Minimo L, Wu J, Rosini-Marthaler L et al (2013) DPP8 and DPP9 expression in cynomolgus monkey and sprague dawley rat tissues. Regul Pept 186:26–35
15. Gorrell MD, Gysbers V, McCaughan GW (2001) CD26: a multifunctional integral membrane and secreted protein of activated lymphocytes. Scand J Immunol 54:249–264
16. Keane FM, Chowdhury S, Yao T-W, Nadvi NA, Gall MG, Chen Y, Osborne B et al (2012) Targeting dipeptidyl peptidase-4 (DPP-4) and fibroblast activation protein (FAP) for diabetes and cancer therapy. In: Dunn B (ed) Proteinases as drug targets. Royal Society of Chemistry, Cambridge, UK, pp 119–145
17. Gorrell MD, Wickson J, McCaughan GW (1991) Expression of the rat CD26 Antigen (dipeptidyl peptidase IV) on subpopulations of rat lymphocytes. Cell Immunol 134:205–215
18. Keane FM, Yao T-W, Seelk S, Gall MG, Chowdhury S, Poplawski SE, Lai JH et al (2014) Quantitation of fibroblast activation protein (FAP)-specific protease activity in mouse, baboon and human fluids and organs. FEBS Open Bio 4:43–54
19. Yu DMT, Yao T-W, Chowdhury S, Nadvi NA, Osborne B, Church WB, McCaughan GW et al (2010) The dipeptidyl peptidase IV family in cancer and cell biology. FEBS J 277:1126–1144
20. Gorrell MD (2005) Dipeptidyl peptidase IV and related enzymes in cell biology and liver disorders. Clin Sci 108:277–292
21. Tinoco AD, Tagore DM, Saghatelian A (2010) Expanding the dipeptidyl peptidase 4-regulated peptidome via an optimized peptidomics platform. J Am Chem Soc 132:3819–3830
22. Waumans Y, Baerts L, Kehoe K, Lambeir A-M, De Meester I (2015) The dipeptidyl peptidase family, prolyl oligopeptidase and prolyl carboxypeptidase in the immune system and inflammatory disease, including atherosclerosis. Front Immunol 6:387–405
23. Mentlein R, Gallwitz B, Schmidt WE (1993) Dipeptidyl-peptidase IV hydrolyses gastric inhibitory polypeptide, glucagon-like peptide-1(7–36)amide, peptide histidine methionine and is responsible for their degradation in human serum. Eur J Biochem 214:829–835
24. Bengsch B, Seigel B, Flecken T, Wolanski J, Blum HE, Thimme R (2012) Human Th17 cells express high levels of enzymatically active dipeptidyl peptidase IV (CD26). J Immunol 188:5438–5447

25. Garin-Chesa P, Old LJ, Rettig WJ (1990) Cell surface glycoprotein of reactive stromal fibroblasts as a potential antibody target in human epithelial cancers. Proc Natl Acad Sci USA 87:7235–7239
26. Kelly T, Huang Y, Simms AE, Mazur A (2012) Fibroblast activation protein-alpha: a key modulator of the microenvironment in multiple pathologies. In: Kwang WJ, (ed) International Review of Cell and Molecular Biology, vol 297. Academic Press, 83–116
27. Niedermeyer J, Garin-Chesa P, Kriz M, Hilberg F, Mueller E, Bamberger U, Rettig WJ et al (2001) Expression of the fibroblast activation protein during mouse embryo development. Int J Dev Biol 45:445–447
28. Levy MT, McCaughan GW, Abbott CA, Park JE, Cunningham AM, Muller E, Rettig WJ et al (1999) Fibroblast activation protein: a cell surface dipeptidyl peptidase and gelatinase expressed by stellate cells at the tissue remodelling interface in human cirrhosis. Hepatology 29:1768–1778
29. Levy MT, McCaughan GW, Marinos G, Gorrell MD (2002) Intrahepatic expression of the hepatic stellate cell marker fibroblast activation protein correlates with the degree of fibrosis in hepatitis C virus infection. Liver Internat 22:93–101
30. Wang XM, Yu DMT, McCaughan GW, Gorrell MD (2005) Fibroblast activation protein increases apoptosis, cell adhesion and migration by the LX-2 human stellate cell line. Hepatology 42:935–945
31. Lee KN, Jackson KW, Christiansen VJ, Lee CS, Chun JG, McKee PA (2006) Antiplasmin-cleaving enzyme is a soluble form of fibroblast activation protein. Blood 107:1397–1404
32. Collins PJ, McMahon G, O'Brien P, O'Connor B (2004) Purification, identification and characterisation of seprase from bovine serum. Int J Biochem Cell Biol 36:2320–2333
33. Keane FM, Nadvi NA, Yao T-W, Gorrell MD (2011) Neuropeptide Y, B-type natriuretic peptide, substance P and peptide YY are novel substrates of fibroblast activation protein-α. FEBS J 278:1316–1332
34. Wong PF, Gall MG, Bachovchin WW, McCaughan GW, Keane FM, Gorrell MD (2016) Neuropeptide Y is a physiological substrate of fibroblast activation protein: enzyme kinetics in blood plasma and expression of Y2R and Y5R in human liver cirrhosis and hepatocellular carcinoma. Peptides 75:80–95
35. Park JE, Lenter MC, Zimmermann RN, Garin-Chesa P, Old LJ, Rettig WJ (1999) Fibroblast activation protein: a dual-specificity serine protease expressed in reactive human tumor stromal fibroblasts. J Biol Chem 274:36505–36512
36. Sulda ML, Abbott CA, Macardle PJ, Hall RK, Kuss BJ (2010) Expression and prognostic assessment of dipeptidyl peptidase IV and related enzymes in B-cell chronic lymphocytic leukemia. Cancer Biol Ther 10:180–189
37. Hamson EJ, Keane FM, Tholen S, Schilling O, Gorrell MD (2014) Understanding fibroblast activation protein (FAP): substrates, activities, expression and targeting for cancer therapy. Proteomics Clin Appl 8:454–463
38. Henderson JM, Zhang HE, Polak N, Gorrell MD (2016) Hepatocellular carcinoma: mouse models and the potential roles of proteases. Cancer Lett (In press)
39. Abbott CA, Gorrell MD (2002) The family of CD26/DPIV and related ectopeptidases. In: Langner J, Ansorge S (eds) Ectopeptidases: CD13/aminopeptidase N and CD26/dipeptidyl peptidase IV in medicine and biology. Kluwer/Plenum, NY, pp 171–195
40. Ren X, Hayashi Y, Yoshimura N, Takimoto K (2005) Transmembrane interaction mediates complex formation between peptidase homologues and Kv4 channels. Mol Cell Neurosci 29:320–332
41. Cronin S, Berger S, Ding J, Schymick JC, Washecka N, Hernandez DG, Greenway MJ et al (2008) A genome-wide association study of sporadic ALS in a homogenous Irish population. Hum Mol Genet 17:768–774

42. Abbott CA, Yu DMT, Woollatt E, Sutherland GR, McCaughan GW, Gorrell MD (2000) Cloning, expression and chromosomal localization of a novel human dipeptidyl peptidase (DPP) IV homolog, DPP8. Eur J Biochem 267:6140–6150
43. Olsen C, Wagtmann N (2002) Identification and characterization of human Dpp9, a novel homologue of dipeptidyl peptidase IV. Gene 299:185–193
44. Zhang H, Chen Y, Keane FM, Gorrell MD (2013) Advances in understanding the expression and function of dipeptidyl peptidase 8 and 9. Mol Cancer Res 11:1487–1496
45. Park J, Knott HM, Nadvi NA, Collyer CA, Wang XM, Church WB, Gorrell MD (2008) Reversible inactivation of human dipeptidyl peptidases 8 and 9 by oxidation. TOEIJ 1:52–61
46. Rummey C, Metz G (2007) Homology models of dipeptidyl peptidases 8 and 9 with a focus on loop predictions near the active site. Proteins 66:160–171
47. Pitman MR, Menz RI, Abbott CA (2010) Hydrophilic residues surrounding the S1 and S2 pockets contribute to dimerisation and catalysis in human dipeptidyl peptidase 8 (DPP8). Biol Chem 391:959–972
48. Tang H-K, Tang H-Y, Hsu SC, Chu JR, Chien CH, Shu CH, Chen X (2009) Biochemical properties and expression profile of human prolyl dipeptidase DPP9. Arch Biochem Biophys 485:120–127
49. Zhu H, Zhou ZM, Lu L, Xu M, Wang H, Li JM, Sha JH (2005) Expression of a novel dipeptidyl peptidase 8 (DPP8) transcript variant, DPP8-v3, in human testis. Asian J Androl 7:245–255
50. Abbott C, Gorrell M (2013) Dipeptidyl peptidase 8. In: Rawlings NL, Salvesen G (eds) Handbook of proteolytic enzymes, 3rd edn. Elsevier, San Diego, pp 3379–3384
51. Justa-Schuch D, Möller U, Geiss-Friedlander R (2014) The amino terminus extension in the long dipeptidyl peptidase 9 isoform contains a nuclear localization signal targeting the active peptidase to the nucleus. Cell Mol Life Sci 71:3611–3626
52. Gall MG, Chen Y, Ribeiro AJVd, Zhang H, Bailey CG, Spielman D, Yu DM et al (2013) Targeted inactivation of Dipeptidyl peptidase 9 enzyme activity causes mouse neonate lethality. PLoS ONE 8:e0078378
53. Wagner L, Hoffmann T, Rahfeld JU, Demuth HU (2006) Distribution of dipeptidyl peptidase IV-like activity enzymes in canine and porcine tissue sections by RT-PCR. Adv Exp Med Biol 575:109–116
54. Ansorge S, Bank U, Heimburg A, Helmuth M, Koch G, Tadje J, Lendeckel U et al (2009) Recent insights into the role of dipeptidyl aminopeptidase IV (DPIV) and aminopeptidase N (APN) families in immune functions. Clin Chem Lab Med 47:253–261
55. Dubois V, Ginneken CV, De Cock H, Lambeir A-M, Van der Veken P, Augustyns K, Chen X et al (2009) Enzyme activity and immunohistochemical localization of dipeptidyl peptidase 8 and 9 in male reproductive tissues. J Histochem Cytochem 57:531–541
56. Yu DMT, Ajami K, Gall MG, Park J, Lee CS, Evans KA, McLaughlin EA et al (2009) The in vivo expression of dipeptidyl peptidases 8 and 9. J Histochem Cytochem 57:1025–1040
57. Dubois V, Lambeir A-M, Vandamme S, Matheeussen V, Guisez Y, Scharpé S, De Meester I (2010) Dipeptidyl peptidase 9 (DPP9) from bovine testes: identification and characterization as the short form by mass spectrometry. BBA Proteins Proteomics 1804:781–788
58. Stremenova J, Krepela E, Mares V, Trim J, Dbaly V, Marek J, Vanickova Z et al (2007) Expression and enzymatic activity of dipeptidyl peptidase-IV in human astrocytic tumours are associated with tumour grade. Int J Oncol 31:785–792
59. Schade J, Stephan M, Schmiedl A, Wagner L, Niestroj AJ, Demuth HU, Frerker N et al (2008) Regulation of expression and function of dipeptidyl peptidase 4 (DP4), DP8/9, and DP10 in allergic responses of the lung in rats. J Histochem Cytochem 56:147–155
60. Matheeussen V, Baerts L, De Meyer G, De Keulenaer G, Van Der Veken P, Augustyns K, Dubois V et al (2011) Expression and spatial heterogeneity of dipeptidyl peptidases in endothelial cells of conduct vessels and capillaries. Biol Chem 392:189–198

61. Stremenova J, Mares V, Lisa V, Hilser M, Krepela E, Vanickova Z, Syrucek M et al (2010) Expression of dipeptidyl peptidase-IV activity and/or structure homologs in human meningiomas. Int J Oncol 36:351–358
62. Yao T-W, Kim W-S, Yu DM, Sharbeen G, McCaughan GW, Choi K-Y, Xia P et al (2011) A novel role of Dipeptidyl peptidase 9 in epidermal growth factor signaling. Mol Cancer Res 9:948–959
63. Chowdhury S, Chen Y, Yao T-W, Ajami K, Wang XM, Popov Y, Schuppan D et al (2013) Regulation of dipeptidyl peptidase 8 and 9 expression in activated lymphocytes and injured liver. World J Gastroenterol 19:2883–2893
64. Waumans Y, Vliegen G, Maes L, Rombouts M, Declerck K, Veken PVD, Berghe WV et al (2016) The dipeptidyl peptidases 4, 8, and 9 in mouse monocytes and macrophages: DPP8/9 inhibition attenuates M1 macrophage activation in mice. Inflammation 39:413–424
65. Maes M-B, Dubois V, Brandt I, Lambeir A-M, Van der Veken P, Augustyns K, Cheng JD et al (2007) Dipeptidyl peptidase 8/9-like activity in human leukocytes. J Leukoc Biol 81:1252–1257
66. Wilson C, Abbott C (2012) Expression profiling of dipeptidyl peptidase 8 and 9 in breast and ovarian carcinoma cell lines. Int J Oncol 41:919–932
67. Bjelke JR, Christensen J, Nielsen PF, Branner S, Kanstrup AB, Wagtmann N, Rasmussen HB (2006) Dipeptidyl peptidase 8 and 9 specificity and molecular characterization compared to dipeptidyl peptidase IV. Biochem J 396:391–399
68. Lee HJ, Chen YS, Chou CY, Chien CH, Lin CH, Chang GG, Chen X (2006) Investigation of the dimer interface and substrate specificity of prolyl dipeptidase DPP8. J Biol Chem 281:38653–38662
69. Geiss-Friedlander R, Parmentier N, Moeller U, Urlaub H, Van den Eynde BJ, Melchior F (2009) The cytoplasmic peptidase DPP9 Is rate-limiting for degradation of proline-containing peptides. J Biol Chem 284:27211–27219
70. Pilla E, Möller U, Sauer G, Mattiroli F, Melchior F, Geiss-Friedlander R (2012) A novel SUMO1-specific interacting motif in Dipeptidyl peptidase 9 (DPP9) that is important for enzymatic regulation. J Biol Chem 287:44320–44329
71. Pilla E, Kilisch M, Lenz C, Urlaub H, Geiss-Friedlander R (2013) The SUMO1-E67 interacting loop peptide is an allosteric inhibitor of the dipeptidyl peptidases 8 and 9. J Biol Chem 288:32787–32796
72. Wilson CH, Indarto D, Doucet A, Pogson LD, Pitman MR, Menz RI, McNicholas K et al (2013) Identifying natural substrates for dipeptidyl peptidase 8 (DP8) and DP9 using terminal amine isotopic labelling of substrates, TAILS, reveals in vivo roles in cellular homeostasis and energy metabolism. J Biol Chem 288:13936–13949
73. Zhang H, Maqsudi S, Rainczuk A, Duffield N, Lawrence J, Keane FM, Justa-Schuch D et al (2015) Identification of novel dipeptidyl peptidase 9 substrates by two-dimensional differential in-gel electrophoresis. FEBS J 282:3737–3757
74. Ajami K, Pitman MR, Wilson CH, Park J, Menz RI, Starr AE, Cox JH et al (2008) Stromal cell-derived factors 1 alpha and 1 beta, inflammatory protein-10 and interferon-inducible T cell chemo-attractant are novel substrates of dipeptidyl peptidase 8. FEBS Lett 582:819–825
75. Mentlein R (1999) Dipeptidyl-peptidase IV (CD26): role in the inactivation of regulatory peptides. Regul Pept 85:9–24
76. Frerker N, Wagner L, Wolf R, Heiser U, Hoffmann T, Rahfeld J-U, Schade J et al (2007) Neuropeptide Y (NPY) cleaving enzymes: structural and functional homologues of dipeptidyl peptidase 4. Peptides 28:257–268
77. Lu C, Tilan JU, Everhart L, Czarnecka M, Soldin SJ, Mendu DR, Jeha D et al (2011) Dipeptidyl peptidases as survival factors in ewing sarcoma family of tumors: implications for tumor biology and therapy. J Biol Chem 286:27494–27505
78. Yu DMT, Wang XM, McCaughan GW, Gorrell MD (2006) Extra-enzymatic functions of the dipeptidyl peptidase (DP) IV related proteins DP8 and DP9 in cell adhesion, migration and apoptosis. FEBS J 273:2447–2461

79. Zhang H, Chen Y, Wadham C, McCaughan GW, Keane FM, Gorrell MD (2015) Dipeptidyl peptidase 9 subcellular localization and a role in cell adhesion involving focal adhesion kinase and paxillin. BBA Mol Cell Res 1853:470–480
80. Lankas G, Leiting B, Roy R, Eiermann G, Beconi M, Biftu T, Chan C et al (2005) Dipeptidyl peptidase IV inhibition for the treatment of type 2 diabetes—Potential importance of selectivity over dipeptidyl peptidases 8 and 9. Diabetes 54:2988–2994
81. Wu J-J, Tang H-K, Yeh T-K, Chen C-M, Shy H-S, Chu Y-R, Chien C-H et al (2009) Biochemistry, pharmacokinetics, and toxicology of a potent and selective DPP8/9 inhibitor. Biochem Pharmacol 78:203–210
82. Jiaang WT, Chen YS, Hsu T, Wu SH, Chien CH, Chang CN, Chang SP et al (2005) Novel isoindoline compounds for potent and selective inhibition of prolyl dipeptidase DPP8. Bioorg Med Chem Lett 15:687–691
83. Van der Veken P, De Meester I, Dubois V, Soroka A, Van Goethem S, Maes MB, Brandt I et al (2008) Inhibitors of dipeptidyl peptidase 8 and dipeptidyl peptidase 9. part 1: identification of dipeptide derived leads. Bioorg Med Chem Lett 18:4154–4158
84. Van Goethem S, Matheeussen V, Joossens J, Lambeir AM, Chen X, De Meester I, Haemers A et al (2011) Structure-activity relationship studies on isoindoline inhibitors of dipeptidyl peptidases 8 and 9 (DPP8, DPP9): is DPP8-selectivity an attainable goal? J Med Chem 54:5737–5746
85. Wu W, Liu Y, Milo LJ Jr, Shu Y, Zhao P, Li Y, Woznica I et al (2012) 4-Substituted boro-proline dipeptides: Synthesis, characterization, and dipeptidyl peptidase IV, 8, and 9 activities. Bioorg Med Chem Lett 22:5536–5540
86. Van der Veken P, Soroka A, Brandt I, Chen YS, Maes MB, Lambeir AM, Chen X et al (2007) Irreversible inhibition of dipeptidyl peptidase 8 by dipeptide-derived diaryl phosphonates. J Med Chem 50:5568–5570
87. Burkey BF, Hoffmann PK, Hassiepen U, Trappe J, Juedes M, Foley JE (2008) Adverse effects of dipeptidyl peptidases 8 and 9 inhibition in rodents revisited. Diabetes Obes Metab 10:1057–1061
88. Ohnuma K, Ishii T, Iwata S, Hosono O, Kawasaki H, Uchiyama M, Tanaka H et al (2002) G1/S cell cycle arrest provoked in human T cells by antibody to CD26. Immunology 107:325–333
89. Ho L, Aytac U, Stephens LC, Ohnuma K, Mills GB, McKee KS, Neumann C et al (2001) In vitro and in vivo antitumor effect of the anti-CD26 monoclonal antibody 1F7 on human CD30 + anaplastic large cell T-cell lymphoma Karpas 299. Clin Cancer Res 7:2031–2040
90. Inamoto T, Yamochi T, Ohnuma K, Iwata S, Kina S, Inamoto S, Tachibana M et al (2006) Anti-CD26 monoclonal antibody-mediated G1-S arrest of human renal clear cell carcinoma Caki-2 is associated with retinoblastoma substrate dephosphorylation, cyclin-dependent kinase 2 reduction, p27kip1 enhancement, and disruption of binding to the extracellular matrix. Clin Cancer Res 12:3470–3477
91. Wrenger S, Hoffmann T, Faust J, Mrestaniklaus C, Brandt W, Neubert K, Kraft M et al (1997) The N-Terminal structure of HIV-1 tat is required for suppression of CD26-dependent T cell growth. J Biol Chem 272:30283–30288
92. Bank U, Heimburg A, Wohlfarth A, Koch G, Nordhoff K, Julius H, Helmuth M et al (2011) Outside or inside: role of the subcellular localization of DP4-like enzymes for substrate conversion and inhibitor effects. Biol Chem 392:169–187
93. Matheeussen V, Waumans Y, Martinet W, Van Goethem S, Van der Veken P, Scharpe S, Augustyns K et al (2013) Dipeptidyl peptidases in atherosclerosis: expression and role in macrophage differentiation, activation and apoptosis. Basic Res Cardiol 108:350
94. Bühling F, Kunz D, Reinhold D, Ulmer AJ, Ernst M, Flad HD, Ansorge S (1994) Expression and functional role of dipeptidyl peptidase IV (CD26) on human natural killer cells. Nat Immun 13:270–279

95. Madueno JA, Munoz E, Blazquez V, Gonzalez R, Aparicio P, Pena J (1993) The CD26 antigen is coupled to protein tyrosine phosphorylation and implicated in CD16-mediated lysis in natural killer cells. Scand J Immunol 37:425–429
96. Shingu K, Helfritz A, Zielinska-Skowronek M, Meyer-Olson D, Jacobs R, Schmidt RE, Mentlein R et al (2003) CD26 expression determines lung metastasis in mutant F344 rats: involvement of NK cell function and soluble CD26. Cancer Immunol Immunother 52:546–554
97. Reinhold D, Goihl A, Wrenger S, Reinhold A, Kühlmann UC, Faust J, Neubert K et al (2009) Role of dipeptidyl peptidase IV (DPIV)-like enzymes in T lymphocyte activation: investigations in DPIV/CD26 knockout mice. Clin Chem Lab Med 47:268–274
98. Milner JM, Kevorkian L, Young DA, Jones D, Wait R, Donell ST, Barksby E et al (2006) Fibroblast activation protein alpha is expressed by chondrocytes following a pro-inflammatory stimulus and is elevated in osteoarthritis. Arthritis Res Ther 8:R23
99. Thielitz A, Vetter RW, Schultze B, Wrenger S, Simeoni L, Ansorge S, Neubert K et al (2008) Inhibitors of dipeptidyl peptidase IV-like activity mediate antifibrotic effects in normal and keloid-derived skin fibroblasts. J Invest Dermatol 128:855–866
100. Spagnuolo PA, Hurren R, Gronda M, Maclean N, Datti A, Basheer A, Lin FH et al (2013) Inhibition of intracellular dipeptidyl peptidases 8 and 9 enhances parthenolide's anti-leukemic activity. Leukemia 27:1236–1244
101. Baird S, Rigopoulos A, Cao D, Allan L, Renner C, Scott F, Scott A (2015) Integral membrane protease fibroblast activation protein sensitizes fibrosarcoma to chemotherapy and alters cell death mechanisms. Apoptosis 20:1483–1498
102. Kurokawa Y, Matoba R, Takemasa I, Nakamori S, Tsujie M, Nagano H, Dono K et al (2003) Molecular features of non-B, non-C hepatocellular carcinoma: a PCR-array gene expression profiling study. J Hepatol 39:1004–1012
103. Yazbeck R, Sulda ML, Howarth GS, Bleich A, Raber K, von Hörsten S, Holst JJ et al (2010) Dipeptidyl peptidase expression during experimental colitis in mice. Inflamm Bowel Dis 16:1340–1351
104. Milner JM, Patel A, Rowan AD (2008) Emerging role of serine proteinases in tissue turnover in arthritis. Arthritis Rheum 58:3644–3656
105. Röhnert P, Schmidt W, Emmerlich P, Goihl A, Wrenger S, Bank U, Nordhoff K et al (2012) Dipeptidyl peptidase IV, aminopeptidase N and DPIV/APN-like proteases in cerebral ischemia. J Neuroinflammation 9:44
106. Matsubara J, Sugiyama S, Sugamura K, Nakamura T, Fujiwara Y, Akiyama E, Kurokawa H et al (2012) A dipeptidyl peptidase-4 inhibitor, des-fluoro-sitagliptin, improves endothelial function and reduces atherosclerotic lesion formation in apolipoprotein E-deficient mice. J Am Coll Cardiol 59:265–276
107. Fingerlin TE, Murphy E, Zhang W, Peljto AL, Brown KK, Steele MP, Loyd JE et al (2013) Genome-wide association study identifies multiple susceptibility loci for pulmonary fibrosis. Nat Genet 45:613–620
108. Marguet D, Baggio L, Kobayashi T, Bernard AM, Pierres M, Nielsen PF, Ribel U et al (2000) Enhanced insulin secretion and improved glucose tolerance in mice lacking CD26. Proc Natl Acad Sci USA 97:6874–6879
109. Niedermeyer J, Kriz M, Hilberg F, Garin-Chesa P, Bamberger U, Lenter MC, Park J et al (2000) Targeted disruption of mouse fibroblast activation protein. Mol Cell Biol 20:1089–1094
110. Chen Y, Gall MG, Zhang H, Keane FM, McCaughan GW, Yu DM, Gorrell MD (2016) Dipeptidyl peptidase 9 enzymatic activity influences the expression of neonatal metabolic genes. Exp Cell Res 342:72–82

The γ-Secretase Protease Complexes in Neurodegeneration, Cancer and Immunity

Caroline Coleman-Vaughan, Arijit Mal, Abhijit De and Justin V. McCarthy

Abstract

The intramembrane-cleaving proteases (I-CliPs) are necessary for the proteolytic cleavage of several transmembrane proteins and initiation of divergent signalling events. The tetrameric γ-secretase protease complexes, comprised of presenilin and three other subunits, represent a major subclass of the I-CliPs. The γ-secretase protease complexes are involved in regulated intramembrane proteolysis, an evolutionary conserved and important signal transduction process encompassing the sequential proteolysis of transmembrane substrates that are central to many physiological processes, including embryonic development, haematopoiesis, and normal functioning of the nervous and immune systems. Deregulated intramembrane proteolysis of certain substrates is proposed to be associated with neurodegeneration, cancer and impaired immune function. In this chapter, we summarise the major biochemical and functional properties (structure, catalytic mechanisms, substrate specificities, and regulation) of the γ-secretase protease complexes. We also present evidence for a role of γ-secretase protease complexes in neurodegeneration, cancer and inflammatory disease and consider the use of γ-secretase inhibitors as prospective therapeutics in several diseases.

C. Coleman-Vaughan · J.V. McCarthy (✉)
Signal Transduction Laboratory, School of Biochemistry & Cell Biology, University College Cork, ABCRF, Western Gateway Building, Cork, Ireland
e-mail: jv.mccarthy@ucc.ie

A. Mal · A. De
Molecular Functional Imaging Laboratory, Advanced Centre for Treatment, Research and Education in Cancer, Tata Memorial Centre, Kharghar, Navi Mumbai, India

J.V. McCarthy
School of Biochemistry & Cell Biology, University College Cork, ABCRF, 3.41 Western Gateway Building, Western Road, Cork, Ireland

S. Chakraborti and N.S. Dhalla (eds.), *Pathophysiological Aspects of Proteases*,
DOI 10.1007/978-981-10-6141-7_3

Keywords
γ-secretase protease · Regulated intramembrane proteolysis · Presenilin Neurodegeneration · Alzheimer's disease · Cancer · Innate immune signalling Immunity · Signalling · γ-secretase inhibitor

1 Introduction

With the identification and characterization of the γ-secretase protease complexes and parallel acceptance that specific enzymes can catalyse the proteolysis of proteins within the cell membrane, the concept of "regulated intramembrane proteolysis" was defined [1, 2]. Regulated intramembrane proteolysis (RIP) is now considered an evolutionary conserved and fundamental signal transduction process central to a diversity of cellular processes, such as cell differentiation, cell adhesion, cellular stress responses, protein degradation, growth factor secretion, axon guidance, neurite outgrowth, mitophagy, lipid metabolism and transcriptional regulation [3–8]. RIP is vital for several biological events, including embryonic development, haematopoiesis, and normal functioning of the nervous system and the immune system. Highlighting its biological importance, RIP is under austere regulation [6, 9], and deregulation is associated with the pathogenesis of diseases, such as Alzheimer's disease (AD), cancer and autoimmune disorders.

RIP is a two-step proteolytic process initiated following the proteolytic cleavage of type I or type II transmembrane proteins (Fig. 1). Referred to as ectodomain shedding, this cleavage occurs either constitutively or in response to ligand binding, close to the transmembrane domain (TMD) within the substrate ectodomain. Proteases collectively termed "sheddases" [10] including members of the a disintegrin and metalloproteases (ADAMs) [11], matrix metalloproteases (MMP) [12], neutrophil derived proteases (SDPS) and the aspartyl proteases β-site APP-cleaving enzymes 1 and 2 (BACE1 and BACE2) [13], perform the ectodomain shedding of most reported substrates. This initial cleavage results in extracellular release of a soluble ectodomain and generation of a membrane-anchored stub, which is subsequently cleaved within its TMD, called intramembrane proteolysis. Members of the family of intramembrane cleaving proteases (I-CliPs) are responsible for the hydrolysis of peptide bonds in the hydrophobic lipid bilayer of many transmembrane proteins and the release of extracellular peptides and an intracellular domain (ICD) into the cytosol (Fig. 1) [14].

An ever-increasing number of transmembrane protein substrates that undergo RIP have been reported, including growth factors, cytokines, cell adhesion proteins, growth factor and cytokine receptors, viral proteins and signal peptides [5, 8, 15]. Amongst the best studied and characterised RIP substrates are functionally important proteins such as amyloid-β precursor protein (APP) [16], Notch [17, 18], E-Cadherin [19], tumour necrosis factor-α (TNFα) [20], interleukin-1 (IL-1)

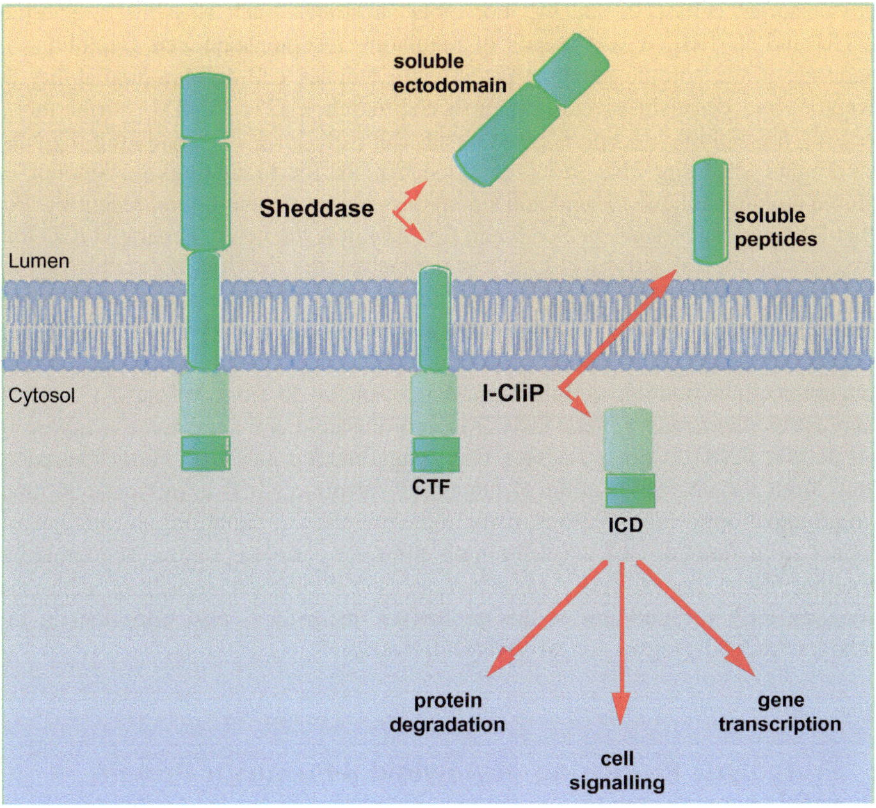

Fig. 1 Schematic depicting regulated intramembrane proteolysis. The sequential proteolysis of single-span transmembrane protein with type-I or type-II orientation, involving intramembrane-cleaving proteases (I-CliPs) is referred to as regulated intramembrane proteolysis. First, cleavage within the substrate extracellular domain, close to the transmembrane domain, by proteases collectively termed "sheddases" release the soluble ectodomain and generate a membrane-bound fragment, which subsequently undergoes a second cleavage within its transmembrane domain, called intramembrane proteolysis. Members of the family of I-CliPs are responsible for the hydrolysis of peptide bonds in the hydrophobic lipid bilayer of many transmembrane proteins and the release of extracellular soluble peptides and an intracellular domain (ICD) in the cytosol. The ICD can possess substrate-specific biological activities, including roles as a cytosolic effector in cell signalling and nuclear gene transcription

receptor, type I and II (IL-1RI and IL-1RII) [21–23], Insulin-like growth factor 1 (IGF-1) receptor [24], epidermal growth factor (EGF) [25], ErbB4 [26], p75 neurotrophin receptor ($p75^{NTR}$) [27–29], CD44 [30] TREM2 [31] and epithelial cell adhesion molecule (EpCAM) [32] amongst others, (reviewed in [5, 8, 15]). Arising from RIP, the soluble protein fragments generated; ectodomains and ICDs can possess biological activities with signalling characteristics distinct from the full-length substrate or are rapidly degraded (Fig. 1). For example, the shedding of growth factors provides for the generation of a soluble signalling molecule that can activate target cells [20, 33, 34]. For other substrates, such as p75^{NTR}, TNFR1, IL-1RI and IL-1RII, it is a means of regulating receptor-mediated signalling, as shedding of the soluble receptor ectodomain reduces cell surface availability of receptors and desensitises cells to further stimulation [21, 22, 35]. Furthermore, because the soluble receptor ectodomain can still bind to circulating ligands, ectodomain shedding also provides a mechanism for regulating the amount of soluble ligand available to bind and signal through membrane-bound receptors. For other RIP substrates, such as Notch and EpCAM, it is the newly generated ICD that possesses biological activity [17, 32]. In both cases, the EpCAM ICD (EpICD) and Notch ICD (NICD) translocate to the nucleus and facilitate the transcriptional regulation of target genes. For other RIP substrates, such as TNFα which is cleaved by SPP, both the soluble ectodomain and ICD are biologically active and initiate both extracellular and intracellular signalling events in different cells [20, 33, 34]. In many cases, the newly generated ICD is very unstable and short-lived, suggesting that certain ICDs do not possess a signalling function and are rapidly degraded. Thus, from the characterisation of numerous substrates, RIP is proposed to have two principal opposing functions, namely the initiation of signalling events that are distinct from functions of the full-length substrate and degradation of membrane proteins. Given the diversity of substrates and biological signalling pathways involving RIP, deregulation of this proteolytic pathway is also important in the pathogenesis and progression of several diseases.

2 Proteolytic Processing of Amyloid β-Precursor Protein (APP)

Much of our understanding of γ-secretase and RIP come from studies focused on the proteolytic processing of APP and generation of amyloid-β (Aβ) peptides, a defining characteristic in the molecular aetiology of AD. APP, a type I transmembrane glycoprotein, has three major isoforms, of 695, 751 or 770 amino acids, which are derived from alternative splicing of a single gene product (reviewed in [36]). In neurons APP695 is the predominantly expressed form, while APP751 and APP770 are mainly expressed in glia cells and other non-neuronal cells. APP undergoes consecutive cleavage events in its extracellular domain by α-secretase (ADAMs) or β-secretase (BACE1) and intramembrane proteolysis by γ-secretase to generate AICD. Aβ peptides are generated from APP during the late secretory pathway, in which active BACE1 is highly concentrated (reviewed in [37]). Therefore, like many other RIP substrates, the proteolysis of APP is closely associated to APP trafficking within the cell and the spatial distribution of specific proteases.

APP can undergo either proteolysis in its extracellular domain by α-secretase (ADAMs) or β-secretase (BACE1), which are referred to as the non-amyloidogenic and amyloidogenic pathways, respectively (Fig. 2). In neurons, the amyloidogenic pathway is important in the pathogenesis of AD, while the non-amyloidogenic pathway is more prevalent in most cell types. In this pathway, APP is first cleaved by α-secretase, resulting in the release of a soluble N-terminal fragment (sAPPα) and formation of a membrane-bound 83 amino acid C-terminal fragment (αCTF or C83). It has been shown that γ-secretase completes sequential endopeptidase-like and carboxypeptidase-like cleavage events (reviewed in [38]), which are exemplified in the proteolysis of APP. First the APP αCTF undergoes endoproteolysis by γ-secretase at the juxtamembrane region, which is referred to as the ε-cleavage site, generating the APP intracellular domain (AICD) and the p3 peptide. In the amyloidogenic pathway, following an initial cleavage by β-secretase (BACE1), which releases sAPPβ and generates a membrane-bound 89/99 amino acid βCTF, which is then proteolytically cleaved by the same γ-secretase at the ε-cleavage site to produce AICD and two major longer Aβ products (Aβ49 and Aβ48). Next γ-secretase effects carboxypeptidase-like processive cleavage of Aβ49 and Aβ48 resulting in the generation of Aβ peptides of different sizes arises due to a series of progressive γ-secretase cleavage steps, every 3–4 residues. Starting with ε-cleavage and generation of Aβ49, which is processed primarily at the ζ-cleavage site to generate Aβ46 and γ-cleavage sites to generate Aβ43 and Aβ40 peptides. In a similar manner Aβ48 is cleaved to produce Aβ45, Aβ42 and Aβ38 respectively, [39–41] (Fig. 2). This proposed model clearly explains that successive proteolysis of substrates by γ-secretase and highlights the importance of the ε-cleavage site in determining the size of Aβ peptides that will be produced following subsequent carboxypeptidase-like processive cleavage by γ-secretase, which ultimately contribute to the formation of amyloid plaques in AD. Like APP, many other substrates undergo γ-secretase cleavage at ε- and γ-cleavage sites. Therefore, like APP, it has been demonstrated for the Notch receptor that extracellular small peptides, referred to as Notch Aβ-like peptides (Nβ), are produced and secreted during RIP of Notch [42]. This also leads to the suggestion that the release of Aβ- or Aβ-like peptides is a common feature of γ-secretase activities during RIP.

3 Intramembrane Cleaving Proteases

Four major classes of I-CLiPs have been discovered each with distinct membrane topologies and embedded active sites within their TMDs (reviewed in [43, 44]). These include the rhomboid serine proteases, the metalloprotease site 2 protease (S2P) family and the intramembrane aspartyl proteases, including the signal peptide peptidase (SPP)-like (SPPL) family and the γ-secretase proteases. All the I-CLiPs are presumed to be enzymatically active as monomeric or dimeric proteins, with the exception of γ-secretases, which is a tetrameric protein complex, including the catalytic subunit presenilin [45]. Signature catalytic motifs are preserved across family members in the TMD of the I-CliP proteases. The S2P catalytic activity is dependent on its conserved HExxH motif, which contributes two histidines and one glutamate, and a distally located conserved aspartate residue [45]. Rhomboids contain a (Ser, His, Asn) catalytic triad, however others suggest that the Asn residue

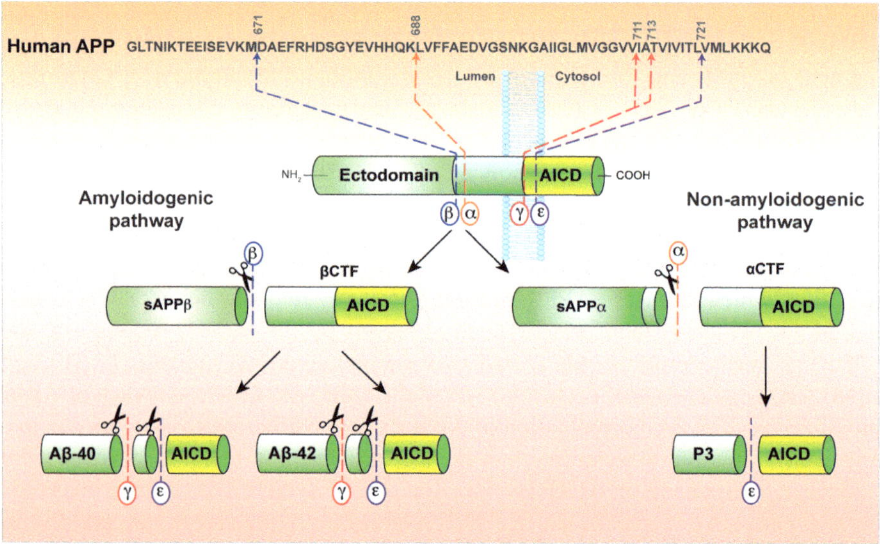

Fig. 2 Regulated Intramembrane Proteolysis of the Amyloid β-Precursor Protein. The amino-acid sequences adjacent to and within the transmembrane domain of APP are shown together with the principal sites of cleavage that occurs during regulated intramembrane proteolysis. In the non-amyloidogenic pathway, α-secretase cleavages APP 16 residues from the cell surface to release the soluble α-cleaved N-terminal fragment (sAPPα) and generate the 83 amino acid membrane-bound α-cleaved C-terminal fragment αCTF (C83). In the amyloidogenic pathway, β-secretase cleaves APP to release the shorter soluble β-cleaved N-terminal fragment (sAPPβ) and the longer 99 amino acid membrane anchored β-cleaved C-terminal fragment, βCTF (C99). The αCTF and βCTF remnant are further processed by the γ-secretase protease. APP αCTF is cleaved by γ-secretase at the ε-cleavage site, near residue 721 of APP, generating AICD and the p3 peptide. In the amyloidogenic pathway, βCTF is cleaved at the ε-cleavage site to produce AICD and two major longer Aβ products (Aβ49 and Aβ48) which undergo γ-secretase mediated processive cleavage to generate Aβ peptides of different sizes The γ-secretase cleavage of APP near residues 711 and 713 in APP, when combined with cleavage at the β-secretase site, generates amyloid-β peptides $A\beta_{40}$ and $A\beta_{42}$, respectively

is not necessary for enzymatic activity, and favour a catalytic dyad model of proteolysis [4]. SPP and the γ-secretase proteases are polytopic membrane proteins and members of the GxGD-type aspartyl protease family with two conserved aspartate residues that define the catalytic site [46]. With the exception of the rhomboid proteases, in general all I-CliPs require a preceding ectodomain cleavage step to generate a truncated version of the initial substrate [47]. However, recently substrates of γ-secretase that require direct cleavage without ectodomain shedding have also been reported [48]. Although structures of representative members of the I-CliP family have been solved, the full mechanism of substrate recognition and cleavage are still poorly understood [47].

Given that both γ-secretase and SPP are members of the same I-CliP subfamily, it's not surprising that they share many similarities, however many differences have also been identified (Fig. 3) (reviewed in [43, 44, 49]). Briefly, SPP functions

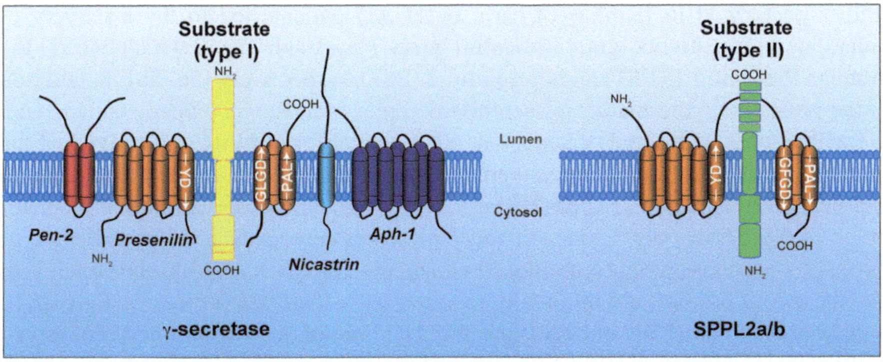

Fig. 3 Schematic comparing the membrane topologies of γ-secretase and the signal peptide peptidase (SPP) intramembrane cleaving aspartyl proteases. Topologies of γ-secretase tetrameric protease complex (*left*) and of SPPL2a/b, a member of the signal peptide peptidase (SPP/SPPL) family of aspartyl proteases. The γ-secretase protease is composed of four proteins including, presenilin (PS1 or PS2), nicastrin, Pen-2 and Aph-1 (Aph-1a/Aph-1b). Presenilin requires endoproteolysis to produce catalytically active presenilin NTF/CTF heterodimers. Note the opposite membrane orientation of type I (γ-secretase) and type II (SPPL2a/b) substrates. *Arrows* indicate the inverted orientations of their respective GxGD and YD active sites and PAL motif

without the requirement for additional cofactors for their activity, but may exist as an active homodimer [50–52]. However, the presenilins require endoproteolysis and at least three other additional proteins to form an active γ-secretase protease. The two proteases display distinctive substrate recognition profiles, where γ-secretase cleaves type I transmembrane substrates, SPP cleaves type II transmembrane proteins. Both share the highly conserved catalytic motifs (GxGD and YD), containing the two critical catalytic aspartate residues located on adjacent TMDs [20], and the C-terminus PAL motif, which is necessary for catalytic activity [53, 54]. SPP and γ-secretase do however display different membrane topologies, which account for the inverted orientation of the conserved catalytic motifs and PAL motif, and which may account for the difference in substrate specificity (Fig. 3). Certain specific γ-secretase inhibitors such as L-685,458 also significantly reduced SPP activity, and L-685,458 and helical peptides mimicking the substrate-docking site block SPP activity supporting the conclusions that these proteases have similar biochemical and structural similarities (reviewed in [49]).

4 γ-Secretase Protease Composition and Structure

In the early 1990s Haass and Selkoe first described "γ-secretase" a proteolytic enzyme, which in combination with BACE1 cleaved APP within the TMD resulting in the generation of Aβ peptides associated with AD pathology [55]. Subsequent studies performed in families of early onset AD patients led to the discovery of mutations in two homologous Presenilin genes *Psen1* and *Psen2* which encode the proteins Presenilin 1 (PS1) and Presenilin 2 (PS2), respectively [56–58]. In addition to the presenilins, the mature γ-secretase complexes have since been characterized to include three additional proteins that are essential for stability and activity of the γ-secretase protease: nicastrin, anterior pharynx-defective-1 or 2 (Aph-1/Aph-2) and presenilin enhancer-2 (Pen-2) (Fig. 2) [59–63]. The observation that reconstitution of these four polypeptides in *Saccharomyces cerevisiae*, a cell type devoid of intrinsic γ-secretase activity, produced enzymatic activity has led to the conclusion that these four proteins are minimally required for γ-secretase protease activity [64].

The presenilins (PS1 and PS2) are highly homologous evolutionary conserved multi-transmembrane aspartyl proteases with a proposed nine TMD topology with the amino-terminus in the cytosol and the carboxyl-terminus within the luminal/extracellular space [62, 63]. The presenilins are synthesised as catalytically inactive full-length proteins with a short half-life. Catalytic activity and incorporation into γ-secretase protease complexes requires endoproteolytic cleavage between TM6 and TM7 to produce a 27–28 kDa amino-terminal fragment (NTF) and 16–17 kDa carboxyl-terminal fragment (CTF) [65, 66], which together form stable and active presenilin NTF/CTF heterodimers. Human presenilins contain two highly conserved aspartate residues on adjacent TMDs, TM6 and TM7, which are located at D257 and D385 in human PS1 and D263 and D366 in human PS2, respectively [67], which are indispensable for γ-secretase protease activity. The C-terminus PAL motif in TM9 of presenilins also contributes to the catalytic activity and the formation of the γ-secretase complex [53, 54]. Despite sharing 67% amino acid identity, PS1 and PS2 are not redundant. Deletion of PS1 is lethal in mice while PS2 deficiency is associated with a very mild phenotype linked to pulmonary fibrosis [68, 69].

Nicastrin encodes a 78 kDa type I membrane protein with a large extracellular domain, transmembrane helix and smaller cytoplasmic domain. Nicastrin is highly glycosylated, resulting in an apparent molecular weight of ~130 kDa. It is proposed that glycosylation of nicastrin is required for proper trafficking of the mature complex [70, 71]. It is often considered the 'gatekeeper' of the γ-secretase protease complex [72, 73]. The ectodomain is proposed to bind to the free amino-terminal of ectodomain-shed proteins, essentially acting as a scaffold protein important in substrate recognition. The amino-terminus of nicastrin interacts with the carboxyl-terminus of presenilin, facilitating assembly of the complex [69, 74]. The final two subunits, Pen-2 and Aph-1, are less well studied. Pen-2 is a ~12 kDa protein that spans the membrane twice, with N- and C-terminus domains facing the lumen of the endoplasmic reticulum (ER). The incorporation of Pen-2 into the complex is critical for the final maturation of γ-secretase as it triggers endoproteolysis of the presenilin holoprotein into its stable heterodimer [69]. Binding of Pen-2 is mediated through the fourth TMD of presenilin proteins and this event subsequently prompts the final steps of the γ-secretase maturation process with the concomitant final glycosylation of nicastrin and the release of the γ-secretase enzyme complex from the ER to the *trans-Golgi* [75]. Aph-1 is a ~25 kDa protein with seven TMDs. Aph-1 is proposed to function in γ-secretase complex stability, activity and selectivity [66, 76]. Two human *Aph-1* homologous genes have been identified on human chromosomes 1 (*Aph-1a*) and 15 (*Aph-1b*) [60, 61], and *Aph-1a* is also expressed in two isoforms (*Aph-1aL* and *Aph-1aS*) via alternative splicing [77]. One additional homologue *Aph-1c* was identified in mice [78].

This quartet of proteins exists as a 1:1:1:1 heterodimer and several gain and loss of function studies, reconstitution experiments in yeast and mammalian cells have been performed, which collectively suggest that all these proteins stabilize each other in the assembly of catalytically active γ-secretase protease complexes [77]. Assembly of γ-secretase into functional complexes occurs in a stepwise manner in the ER, where nicastrin maintains stability of the assembling complex and regulates its intracellular trafficking. Aph-1, is required for initiating assembly of the complex and proteolytic activity [77]. Deficiency of either *Nicastrin* (*Nct*) or *Aph-1* decreases the levels of Pen-2 and PS1 NTF/CTF heterodimers, and disrupts their intracellular trafficking [77]. Pen-2 is destabilized in conditions of *Psen* deficiency, and in conditions of *Pen-2* deficiency full length PS1 accumulates. Subcellular localization motifs present in the TMDs of PS1, Pen-2 and nicastrin responsible for ER retrieval/retention prevent trafficking of unassembled subunits beyond the ER [79, 80]. These signals are concealed in properly assembled γ-secretase complexes thereby allowing trafficking to the *Golgi* apparatus and final subcellular localizations. These observations demonstrate that the four proteins cross-regulate each other and can influence the assembly, stabilization, and trafficking of the protease complex. Until recently details on the structure have been hampered due to difficulties in expressing and purifying intact γ-secretase [81]. Recently a 3D structure of an intact human γ-secretase complex has been shown at 4.5 angstrom resolution [82], and an atomic-resolution (3.4 angstrom) cryo-EM structure was also reported [83].

Like the presenilins, *Aph-1* homologues share overlapping but non-redundant roles. While deficiency of *Psen1* or *Aph-1A* is associated with embryonic lethality [68, 69, 84], *Psen2*- or *Aph-1b*-knockout mice mainly develop as normal. Thus, given the diversity of subunit homologues and distinct phenotypes from knockout mice, it is proposed that at least six distinct γ-secretase complexes can exist in humans [85, 86]. Consistent with this, the levels of expression of the presenilins and Aph-1a/Aph-1b can differ between cell types and tissues, and different γ-secretase complexes can also co-exist within the cell [78, 87]. Recent studies have demonstrated that PS1-containing γ-secretase complexes are mainly localized to the plasma membrane and cell surface [88, 89], while PS2-containing γ-secretase complexes are restricted and compartmentalised in the late endosome/lysosome [88, 89]. Additionally, the subcellular distribution of PS2-containing γ-secretase complexes differs according to the Aph-1 subunit composition, where PS2/Aph-1aS- and PS2-Aph-1b-containing γ-secretase complexes were observed in late endosomes and lysosomes, while PS2/Aph-1aL containing complexes were never detected in this subcellular space [88]. Biochemical evidence also shows that the different γ-secretase complexes differentially cleave substrates, including APP [89, 90] and Notch [89]. More specifically, PS1-containing γ-secretase complexes were shown to favour cell surface substrates, such as N-Cadherin, while PS2-containing complexes preferentially cleaved substrates that localize to the late endosome/lysosome, such as tyrosinase-related-protein (TRP1) and premelanosome protein (PEML) [89]. Similarly, while both Aph-1a and Aph-1b containing complexes are equally capable of cleaving synthetic APP and Notch, from in vivo studies it is clear that the Aph-1a-containing complexes are essential for Notch proteolysis during embryogenesis, while Aph-1b-containing complexes apparently do not have a substantial role in Notch signalling in peripheral tissue. Therefore, since the more restricted compartmentalization of PS2-containing complexes contribute to the generation of the AD-related intracellular pool of Aβ peptides [91–93], and given that *Aph-1b* is predominantly expressed in the brain with limited effects on Notch signalling, this potentially highlights PS2-containing complexes and/or Aph-1b-containing complexes as attractive drug target for AD.

5 γ-Secretase Enzyme Regulation

The cleavage of a substrate is a strictly controlled activity that involves either the direct regulation of the protease activity or substrate modifications (localization or post-translational modifications) that inhibit or enable proteolysis. The γ-secretase proteases cleave over 100 type-I transmembrane proteins (reviewed in [7, 15]). Despite its apparent promiscuous nature, γ-secretase enzymatic activity is strictly controlled by a number of mechanisms as intricate and diverse as its functions. As discussed previously, γ-secretase activity is dependent on the accurate assembly of its four indispensable subunits. The removal of any of the four subunits results in a reduction of active complex formation and loss of enzymatic activity. Furthermore, many proteins that interact with the complexes can influence enzymatic activity [94], including CD147, phospholipase D1, TMP21, GPR3, γ-secretase activating protein (GSAP), syntaxin-1, Arc, voltage-dependent anion channel 1 (VDAC1), contactin-associated protein 1 (CNTNAP1), TPPP, NDUFS7, Erlin-2, β-arrestin-1, β-arrestin-2, Hif-1α and Nexin 27 [94–96].

Post-translational modifications (PTM) of γ-secretase complex components or their substrates can enable or inhibit substrate recognition and cleavage by the γ-secretase proteases (reviewed in [7, 97]). Post-translational modifications can also regulate trafficking of substrates and their γ-secretase-generated cleavage products between distinct subcellular membrane compartments. The presenilin proteins undergo proteolysis (endoproteolysis and caspase cleavage), phosphorylation and ubiquitination [7, 98, 99]. These modifications have been shown to not only determine subcellular localization [100], and selectivity of binding-partners [101], but also regulate γ-secretase complex assembly and enzyme activity [67]. For example, PS1 is ubiquitinated by SEL-10 and tumour necrosis factor receptor associated factor 6 (TRAF6), which regulate γ-secretase activity and substrate cleavage [27, 98, 102]. Additionally, glycosylation of nicastrin is also essential for the trafficking and correct subcellular localization of γ-secretase complex components [68].

As summarized earlier, subsequent to APP ectodomain shedding, the remaining membrane-anchored APP-CTFs can be processed by intramembrane proteolysis or proteasomal degradation [103]. It has been reported that the heterogeneity of APP-CTFs is in part attributed to post-translational modification, specifically APP-CTF phosphorylation (reviewed in [104, 105]). APP has eight potential phosphorylation sites [106], seven of which are phosphorylated in AD brains, i.e., Y653, S655, T668, S675, Y682, T686, and Y687 (APP695 isoform numbering) [106, 107]. The phosphorylation of APP at T668 is mediated by neuronal cyclin-dependent protein kinase 5 (cdk5) [107], p34cdc2 protein kinase (cdc2) [108], glycogen synthase kinase 3β (GSK-3β) [109], or c-jun N-terminal kinases (JNK) [110–112], while phosphorylation of APP-CTFs by JNK has been shown to enable association with PS1, enhance their proteolysis by γ-secretase [112], and regulate the nuclear translocation of AICD [113]. Similar to APP, post-translational modification of other γ-secretase substrates has been shown to affect intramembrane proteolysis. For example, palmitoylation of $p75^{NTR}$ and CD44-CTFs, ubiquitination of IL-1R1 [23], $P75^{NTR}$ [27] and Notch, and phosphorylation of CD44-CTFs haves been shown to regulate their stability to enhance processing by γ-secretase [114].

The specificity of γ-secretase complexes is also determined by the compartmentalization and spatial co-expression of complexes and substrates. γ-secretase activity is observed at the *trans-Golgi* network, endocytic organelles and plasma membrane (reviewed in [115]). It was recently shown that the N-terminus of PS2 contains a highly conserved unique acidic-dileucine sorting sequence ($E_{16}RTSLM_{21}$) [89], which fits the acidic-dileucine motif [D/E]xxxL[L/I/M] present in other proteins that are transported to late endosomes/lysosomes [116]. This motif enables PS2 to bind to the γ1-σ1 (AP1G1-AP1S1) hemi-complex of the AP-1 adaptor complex, which enable sorting and localization of proteins at different stages on the endomembrane system. It was shown that phosphorylation of the $E_{16}RTSLM_{21}$ motif regulated PS2 interaction with AP-1 and restricted the spatial distribution of PS2 to late endosomal/lysosomal compartments [89]. In contrast, PS-1 containing γ-secretase complexes that lack the AP-1 interaction motif are trafficked along a different route, frequenting Rab-11-positive recycling endosomes, and are more prevalent at the plasma membrane. These observations also support the long-standing proposal that intracellular trafficking and localization of γ-secretase proteases also regulates the enzyme activities of γ-secretase proteases (reviewed in [9]). PS1-containing γ-secretase complexes are enriched in lipid rafts and their activity is modulated by lipid composition [117–121], further implying that the subcellular localization of γ-secretase is critical to its proteolytic activity.

A growing body of evidence also supports the proposal that the subcellular localization of γ-secretase substrates is important for intramembrane proteolysis and subsequent signalling functions. It has been shown that following cell surface shedding of Notch ectodomain, internalization of Notch is required for intramembrane proteolysis and generation of NICD [122, 123]. Similarly, subsequent to NGF ligand-induced ectodomain shedding, $p75^{NTR}$ C-terminal fragments are localized in early endosomes where they are proposed to undergo γ-secretase-dependent intramembrane proteolysis [124]. Other examples from our studies include the proinflammatory cytokine receptors TNFR1 and IL-1R1, which were recently identified as γ-secretase substrates [21, 125]. In our studies we have demonstrated that inhibition of dynamin-dependent endocytosis increases cell surface shedding of TNFR1 [125], and antagonises γ-secretase-dependent intramembrane proteolysis of TNFR1 [125], and IL-1R1 (unpublished). In the case of APP, inhibition of internalization was shown to antagonize the generation of sAPPβ (reduce BACE cleavage) and enhance cell surface generation of sAPPα. Of pathological relevance, endocytic disturbance has been described as an early characteristic of AD pathology (reviewed in [126]), and proposed to significantly induce endosomal accumulation of APP and BACE1, which may lead to the exacerbation of Aβ pathology [127]. Collectively, these examples highlight the importance of proper spatial distribution and the compartmentalization requirement of both γ-secretase complexes and substrates for appropriate intramembrane proteolysis, and suggest the existence of similar tissue, cell or subcellular spatial requirements for RIP of other substrates.

6 Alzheimer's Disease and Neurodegeneration

AD is a multifactorial disorder, associated with progressive and debilitating loss of cognition, memory and language, and is now considered the main cause of dementia and one of the greatest health-care challenges of the Twenty-first Century. Most AD cases are sporadic and late onset (LOAD), although genetic mutations found in patients with the dominantly inherited form of early onset familial AD (FAD), account for approximately >5% of AD cases, and are caused by autosomal dominant mutations in *App* or either of the presenilin genes, *Psen1* and *Psen2*. At least 266 AD-associated mutations have been mapped to the presenilin genes, with 277 identified in *Psen1* and 39 in *Psen2* (http://www.alzforum.org/mutations). Pathologically, AD is defined by the occurrence of profuse proteinaceous deposits; extraneuronal amyloid plaques and intraneuronal neurofibrillary tangles [128]. Amyloid plaques are a fibrous aggregate principally composed of Aβ peptides, whereas neurofibrillary tangles are composed of aggregated, hyperphosphorylated tau, a cytosolic protein normally associated with microtubules. Over the past 30 years of AD research extensive evidence has been presented to support the hypothesis that disproportionate accumulation of abnormally folded Aβ and tau are casually related to the neurodegenerative processes associated with AD. Genetic mutations found in patients with FAD also cause increased production and/or aggregation of Aβ, suggesting that Aβ is involved in the pathogenesis of AD. In particular, Aβ42 is predominantly deposited in the brains of AD patients. For this reason, the generation of Aβ peptides has been the primary target towards developing potential AD therapeutics, and inhibiting the enzymes that produce Aβ has been a priority and favoured strategy for AD drug discovery (reviewed in [52, 129–131]). However, the complexity of γ-secretase proteolysis and its functionally divergent substrates has recently discouraged some efforts to directly inhibit their proteolytic activity, which can have deleterious and toxic side effects.

Prior to the identification of β-secretase and γ-secretase [16, 132], the proteases responsible for the generation of Aβ peptides, several pharmaceutical companies identified and developed distinct classes of γ-secretase inhibitors (GSIs) with favourable pharmacokinetic profiles [133–135]. These included the active site-directed small molecule transition state analogue inhibitors, which do not discriminate between different γ-secretase complexes or substrates, and none have progressed to AD clinical trials [130, 136]. The non-transition state analogue inhibitors include the well-studied N-[N-(3,5-Difluorophenacetyl)-L-alanyl]-S-phenylglycine t-butyl ester (DAPT) and LY450,139 (semagacestat). However, in addition to inhibiting γ-secretase cleavage of APP and reducing Aβ peptide formation, these GSIs also inhibited the cleavage of Notch, in some cases to a similar or greater extent that APP (reviewed in [130, 131, 134, 137]). Recently, Eli Lilly were forced to prematurely terminate a phase III trial of LY450,139 (semagacestat) in a large cohort of patients with probable AD, due to severe side effects including, weight loss, worsening cognitive function scores, infections and increased incidences of skin cancer [138], some likely due to Notch-related toxicities [139]. Clinical trials with second-generation GSIs designed to have reduced antagonistic effects on Notch (referred to as Notch-sparing compounds or γ-secretase modulators (GSMs)), have been reported for avagacestat [140] and begacestat [141], however both were discontinued. Similarly, clinical trials with, ELND006 and ELND007 were prematurely stopped due to non-mechanism-based liver toxicity [142]. However, clinical trials with other GSM compounds, NIC51- and CHF-5075, which exhibited acceptable safety profiles in phase I trials, are now progressing through phase 2 clinical trials (http://clinicaltrials.gov). Because GSMs have less effect on Notch or other substrates, they should have fewer undesirable side effects and discovering more potent and safe GSMs is still a key strategy towards the development of a γ-secretase-based therapy for AD.

Another class of compounds, the nonsteroidal anti-inflammatory drugs (NSAIDs) were fortuitously identified to lower Aβ42 levels in vitro [143]. One such NSAID, ibuprofen was also shown to reduce plaque burden and brain inflammation in a mouse model of AD [144]. Subsequently a decade of research into the clinical suitability of NSAIDs and derivatives thereof in the treatment of AD has lead to several small clinical trials in AD with inconsistent or limited clinical benefits (reviewed in [129, 130]). Others have developed peptidomimetic compounds that interact with a proposed substrate-binding site on γ-secretase, however these have not progressed to clinical trials [145]. These failures have forced researchers and pharmaceutical companies to reanalyse the validity of γ-secretase as a viable therapeutic target and to search for more γ-secretase complex-specific inhibitors, substrate-specific inhibitors, or inhibitors that target the carboxypeptidase-like activity but spare the endopeptidase activity (reviewed in [130]). These efforts should deliver improved and safer therapeutics for AD and other disorders with which unfavourable γ-secretase activity have been recognised (reviewed in [5]).

7 Cancer

The γ-secretase protease complexes have an important role in onset and progression of many cancers, as many of its substrates are highly implicated in disease pathogenesis and malignancy. Nearly a century after its discovery [146], the evolutionarily conservation and functional significance of the Notch signalling cascade was well established in multiple cellular processes, such as embryonic development, neurogenesis, cell fate determination, proliferation, differentiation, and self-renewal in adult tissue [18, 147]. In mammals there are four Notch receptors (Notch 1–4), which are synthesised in the ER as inactive single peptide precursors. Three sequential proteolytic cleavage steps are required for canonical Notch receptor signalling [18]. First, the Notch peptide precursor is cleaved in the *trans-Golgi* by furin convertase (S1 cleavage), which produces non-covalently bound heterodimers composed of two fragments; an N-terminal fragment which is accessible to Notch ligands, Notch extracellular domain (NECD) and a C-terminal transmembrane fragment (NEXT), which is comprised of an extracellular stub, the transmembrane domain and NICD. There are five Notch ligands in mammals, Delta-like-ligand (DLL) 1, 3, 4 and Jagged (JAG) 1 and 2 that are transmembrane proteins expressed on the surface of neighbouring cells. Upon ligand binding to the NECD, a second proteolytic event (S2 cleavage) by ADAM10 or ADAM17 releases most of the NECD from the cell. Subsequently γ-secretase cleaves the NEXT fragment (S3 cleavage), generating the transcriptionally active NICD. As NICD enters nucleus, it displaces the co-repressor from the transcription factor CSL (CBF 1/Sn(H)/Lag 1). Upon binding with CSL, NICD forms a ternary complex with MML (Mastermind like protein 1) and further recruits transcriptional co-activators like CBP/P300 to increase the expression of target genes like *Hes1*, *C-myc* and *Igf1r* [148]. It is this S3 cleavage that can be targeted by γ-secretase inhibitor to block Notch signalling by inhibiting the release of NICD in the cytoplasm [149].

The first indication that Notch signalling had a role in human malignancy was unravelled in T-cell acute lymphoblastic leukaemia (T-ALL) cells. In the early 1990s a rare chromosomal translocation was identified involving the human Notch gene in T-ALL [150, 151]. Later, gain of function mutations in Notch1 receptors was identified in over 50% of T-ALL tumours [148, 152, 153]. Mutations were detected in the extracellular homodimerization (HD) domain and the C-terminal PEST domain in the Notch 1 receptor. Mutations in the HD domain allowed ligand-independent activation of Notch signalling, while mutations in the PEST domain stabilises the NICD and abruptly prolong Notch activation [154]. Notch signalling also plays a crucial role in tumorigenesis of several solid tumours like breast [155], colon [156, 157], pancreas [158], prostrate [159] and glioblastoma [160]. Aberrant activation of Notch signalling is mainly due to overexpression of ligand or loss of negative regulators of the pathway. Oncogenic *Ras* up-regulates Notch 1 ligand *Dll1,* and *Psen1* resulting in increased Notch 1 proteolysis and signalling [161]. Aberrant activation of Notch signalling was associated with loss of a negative regulator of Notch pathway, called Numb [162]. Notch also promotes angiogenesis and epithelial-mesenchymal transition in tumours. It also confers radio- and chemo-resistance to cancer cells [163]. Interestingly, Notch may act as a tumour suppressor as reported in hepatocellular carcinoma, small cell lung cancer, and cutaneous squamous cell carcinoma [164]. These findings validate the targeting of Notch signalling as a possible anti-cancer therapeutic [165].

EpCAM is a 37 kDa type I transmembrane glycoprotein [166, 167], localised to the basolateral membrane in the majority of normal epithelial tissues and is over-expressed in the majority of human epithelial cancers, making it an attractive diagnostic and therapeutic target in oncology [168, 169]. EpCAM has also been identified as a marker for cancer stem cells in various tumours including breast cancer, raising further interest in its use for tumour targeting and therapy [170]. EpCAM is overexpressed in 35–42% of patients with invasive breast cancer and is a predictor of poor prognosis and survival [171]. Given this, EpCAM is a widely used immunohistochemical marker for epithelial human malignancies, and has also been used to enrich circulating tumour cells before microscopic evaluation in the only FDA-approved assay for detection of circulating tumour cells in breast cancer. Similar to Notch, EpCAM oncogenic signalling requires intramembrane proteolysis of EpCAM by γ-secretase and generation of a biologically active EpICD that triggers more aggressive oncogenesis [32, 172]. More recently it has been shown that activation of EpCAM signalling relies on the homophilic aggregation of EpCAM via cell-cell contact, resulting in regulated intramembrane proteolysis of EpCAM [173], which has recently been shown to act as the mitogenic signal transducer of EpCAM in vitro and in vivo [32]. Sequential proteolysis by the TACE/ADAM17 and PS2-containing γ-secretase complexes facilitates the generation of soluble EpEx and the biologically active cytosolic EpICD [32, 172]. EpCAM interacts with Claudin 7, which is necessary for EpCAM recruitment to tetraspanin-enriched domain (TEM) [174], where EpCAM associates with CD9, CO-029 and CD44v6 [175]. EpCAM recruitment to TEM is necessary for its intramembrane proteolysis and signalling. In human carcinoma cells, upon cleavage of EpCAM cytosolic EpICD couples with FHL2 (four-and-a-half LIM domain protein) and Wnt pathway components; β-catenin and Lef-1 forming a nuclear complex that binds DNA at Lef-1 consensus sites and induces gene transcription of target genes, including *c-Myc* and *cyclin D1*, leading to increased cell proliferation and oncogenesis in immunodeficient mice [32].

Clinically profound evidence suggests that EpCAM expression is conserved with cancer progression and distant metastasis [176, 177]. Overexpression of EpCAM in breast cancer cells results in increased proliferation and altering EpCAM expression affects in vivo tumorigenesis [178]. Knockdown of *EpCAM* by siRNA or aptamer siRNA chimeras suppress tumour initiation and tumorigenesis in mice [178, 179], and the invasive potential of breast cancer cells [180]. EpCAM is also known to confer breast cancer cells chemo-resistance [181]. In thyroid carcinoma, nuclear and cytoplasmic accumulation of EpICD is used as an index for aggressiveness and poor prognosis [182, 183], and as a marker to distinguish between metastatic and non-metastatic papillary thyroid carcinoma [184]. With distinct differential expression of EpCAM with various cancer conditions, EpCAM is also an attractive tumour-associated surface antigen used for diagnostic imaging or for targeted drug delivery [185–191]. Thus, in view of the tremendous heterogeneity in solid tumours and the novel role of EpCAM as an oncogenic signal transducer [32], therapeutic targeting of γ-secretase mediated cleavage of EpCAM in human cancers is under consideration.

Structurally, $p75^{NTR}$ is a member of to the TNF receptor superfamily and binds all four mammalian neurotrophins. As a heterodimer with the Trk receptors, $p75^{NTR}$ activates MAPK (mitogen activated protein kinase) and PI-3K (phosphatidylinositol 3-kinase) signalling pathways to regulate cell cycle progression, survival, neurite outgrowth and differentiation reviewed in, [192, 193]. However, $p75^{NTR}$ homodimers can also activate the NF-κB (nuclear factor kappa-light chain-enhancer of activated B cells) and Akt pathways to promote cell survival and cell cycle progression [194, 195]. $p75^{NTR}$ is subjected to ectodomain shedding by ADAM17 [196, 197] and γ-secretase-mediated proteolysis [198], resulting in the generation of the $p75^{NTR}$ ICD and transcriptional control of gene expression [199], as well as induction of apoptosis [200, 201], and inhibition of neurite outgrowth [202].

Deregulation of $p75^{NTR}$ expression and signalling is associated with the pathogenesis and progression of several cancers. For example, in prostate cancer loss of $p75^{NTR}$ has been coupled to the progression of tumour growth [203, 204], $p75^{NTR}$ overexpression promotes tumour migration in melanoma and medulloblastoma [205, 206]. Indeed, $p75^{NTR}$ is used as a biomarker to recognize and isolate melanoma tumour stem cells [207]. Overexpression of $p75^{NTR}$ has also been implicated in the progression of melanoma [208], glioma [209] and medulloblastoma [210]. Expression of $p75^{NTR}$ in human gliomas can transform non-invasive cells into highly invasive gliomas [211]. Activation of $p75^{NTR}$ by NGF also induces breast cancer cell survival [212, 213]. $p75^{NTR}$ undergoes RIP in higher-grade glioma and in invasive glioblastoma cell lines [211, 214], and inhibition of γ-secretase antagonizes migration and metastasis in medulloblastoma and glioblastoma cell lines [28, 210, 214]. Similarly, inhibition of γ-secretase activity prevented spinal metastasis of medulloblastoma [210]. Collectively these studies support the development of γ-secretase inhibitors as anti-cancer therapeutics [211, 214].

As outlined above, many γ-secretase substrates have a critical function in the control of normal cell proliferation, differentiation and apoptosis; however, they are also overexpressed in many cancers. Therefore, in order to selectively target γ-secretase substrates, very specific and selective therapeutics need to be developed. Following positive proof-of-concept experiments using in vitro human cancer cell lines [215–219], and in vivo xenograft models [220], many GSIs initially developed as AD therapeutics have advanced as anti-cancer therapeutics. For example, γ-secretase inhibitors have been shown to increase survival in mouse models of leukaemia [221], to reduce osteosarcomas in vivo in mice [222], and to enhance the effect of chemotherapy in human colon adenocarcinoma cell lines [223]. Many GSIs have now progressed to phase I or II clinical trials (https://clinicaltrial.gov), alone or in combination with other standard chemotherapeutic strategies in adult patients with advanced haematological and solid tumours. A complete list of GSIs in clinical trials for cancer was recently reviewed in, [224, 225].

A list of some GSIs used in different in vitro, in vivo or clinical studies, are provided in Table 1, for a detailed listing please refer to [225]. Among these GSIs, extensive phase I/II clinical trials of RO4929097 (Roche) [226, 227], PF-03084014 (Pfizer) [228, 229], and MK0752 (Merck) [230–232] for advanced solid tumours (breast, pancreatic, nervous system and colorectal) and haematological malignancies are being conducted where maximum tolerated dose, efficacy, toxicity and effect on overall survival rates in different cancers are being determined [224, 225, 233–237]. Phase I trials of MK-0752, an oral GSI, show that it is also well-tolerated and exhibited Notch inhibition in children with refractory and recurrent central nervous system malignancies [230, 233]. A phase I study of the non-competitive, reversible GSI PF-03084014, in patients with T-cell acute lymphoblastic leukaemia (T-ALL) and T-cell lymphoblastic lymphoma (T-LBL) demonstrated a favourable pharmacokinetic profile and target inhibition, supporting further evaluation of PF-03084014 in patients with T-ALL or T-LBL.

GSIs can also be combined with other cancer therapeutic strategies such as radiotherapy and chemotherapy to achieve a better curative effect. For example, RO4929097 has also been administered in combination with other anti-cancer agents including capecitabine (ovarian cancer), gemcitabinein (pancreas, tracheal and breast cancers), cediranib (advanced solid tumours) and temsirolimus (advanced solid tumours) (reviewed in [224]). In combination with gemcitabine, PF-03084014 treatment caused tumour regression in pancreatic cancer xenograft models and target inhibition in patients with advanced solid tumours [229], encouraging further development of PF-03084014 for the treatment of patients with advanced solid tumours. GSIs are known to strongly synergize with the anti-cancer effects of several chemotherapeutic agents. In vivo studies have shown chemotherapeutic drugs in combination with GSIs have a better potency in killing cancer cells [238]. GSIs augment anti-cancer drug mediated apoptosis following mitotic arrest [239, 240], and increase the efficacy of chemotherapeutic drugs in eliminating the cancer stem cells [241, 242]. Radiotherapy is one of the important therapeutic strategies in treatment of cancer cells. Patients often receive it as a part of therapeutic or palliative treatment in combination with surgery and chemotherapy. While most of the patients respond well to radiation, sometimes reoccurrence of cancer arises. The relapse of the tumour may be result of tumour heterogeneity or acquired and intrinsic radiation resistance. Many signalling pathways requiring γ-secretase substrates are accountable for radiation treatment resistance [243–245]. Thus targeting intramembrane proteolysis of these substrates by inhibiting γ-secretase activity can increase the efficacy of radiation [246, 247]. In vitro studies have shown that cancer cells pre-treated with GSIs enhances susceptibility to apoptosis on exposure to ionising radiation [248]. GSIs increase radiation-induced cell death not only by affecting the DNA damage repair mechanism but also can act as radio-sensitizers for cancer stem cells (CSCs) [249–251]. Pre-treatments with the GSI RO4929097 also sensitized breast cancer stem cells to ionizing radiation.

Table 1 Common γ-secretase inhibitors and use in clinical trials for cancer therapy

Name	Chemical name Formula	Class	Type	Use in cancer clinical trial [165]
Z-Leu-Leu-Nle-CHO	Benzyloxycarbonyl-L-Leucyl-L-Leucine-L-Norleucinal **$C_{26}H_{41}N_3O_5$**	Peptide isostere	TSA	N.A.
L-685458	(5*S*)-(*tert*-Butoxycarbonylamino)-6-phenyl-(4*R*)-hydroxy-(2*R*)-benzylhexanoyl)-L-leucyl-L-phenylalaninamide **$C_{39}H_{52}O_6N_4$**	Azepine	TSA	N.A.
DAPT	N-[N-(3,5-Difluorophenacetyl)-L-alanyl]-S-phenylglycine t-butyl ester **$C_{23}H_{26}F_2N_2O_4$**	Azepine	Non-TSA	N.A.
Compound E	*N*-[(1*S*)-2-[[(3*S*)-2,3-Dihydro-1-methyl-2-oxo-5-phenyl-1*H*-1,4-benzodiazepin-3-yl]amino]-1-methyl-2-oxoethyl]-3,5-difluorobenzeneacetamide **$C_{27}H_{24}F_2N_4O_3$**	Azepine	Non-TSA	N.A.
LY-411575	N2-[(2S)-2-(3,5-Difluorophenyl)-2-hydroxyethanoyl]-N1-[(7S)-5-methyl-6-oxo-6,7-dihydro-5H-dibenzo[b,d]azepin-7-yl]-L-alaninamide **$C_{26}H_{23}F_2N_3O_4$**	Azepine	TSA	N.A.
Semagacestat (LY450139)	(2*S*)-2-hydroxy-3-methyl-*N*-((1*S*)-1-methyl-2-{[(1*S*)-3-methyl-2-oxo-2,3,4,5-tetrahydro-1*H*-3-benzazepin-1-yl]amino}-2-oxoethyl)butanamide **$C_{19}H_{27}N_3O_4$**	Azepine	TSA	N.A.
RO-4929097	2,2-dimethyl-N-[(7S)-6-oxo-5,7-dihydrobenzo[d] [1] benzazepin-7-yl]-N′-(2,2,3,3,3-pentafluoropropyl)propanediamide **$C_{22}H_{20}F_5N_3O_3$**	Azepine	TSA	– Advanced solid tumours – Refractory solid tumours – Epithelial ovarian cancer – Melanoma – Pancreatic adenocarcinoma – Colorectal cancer –Renal cell carcinoma – Progressive and malignant glioblastoma – Metastatic sarcoma

(continued)

Table 1 (continued)

Name	Chemical name Formula	Class	Type	Use in cancer clinical trial [165]
MRK-003	(3′R,6R,9R)-5′-(2,2,2-trifluoroethyl)-2-((E)-3-(4-(trifluoromethyl)piperidin-1-yl)prop-1-en-1-yl)-5,6,7,8,9,10-hexahydrospiro[6,9-methanobenzo [8] annulene-11,3′-[1, 2, 5] thiadiazolidine] 1′,1′-dioxide $\mathbf{C_{25}H_{31}F_6N_3O_2S}$	Sulfonamide	Non-TSA	N.A.
MK-0752	3-[4-(4-chlorophenyl)sulfonyl-4-(2,5-difluorophenyl)cyclohexyl]propanoic acid $\mathbf{C_{21}H_{21}ClF_2O_4S}$	Sulfonamide	Non-TSA	– Advanced solid tumour – Breast cancer – Pancreatic cancer – Paediatric brain tumour (CNS) – Ovarian cancer – Recurrent and metastatic colorectal cancer
PF-03084014	(2S)-2-[[(2S)-6,8-difluoro-1,2,3,4-tetrahydronaphthalen-2-yl]amino]-N-[1-[1-(2,2-dimethylpropylamino)-2-methylpropan-2-yl]imidazol-4-yl]pentanamide $\mathbf{C_{27}H_{41}F_2N_5O}$	Selective	Non-TSA	– T-ALL – T cell lymphoblastic lymphoma – Advanced solid tumour

Arising from the high recurrence rates and availability of limited therapeutic options, the prognosis for glioblastoma patients is very poor [252, 253]. Thus γ-secretase-dependent Notch and the $p75^{NTR}$ signalling have been implicated in the invasive phenotype of glioma, and in vivo studies demonstrate that γ-secretase inhibition may have direct clinical application for the treatment of malignant glioma. Indeed, in a glioblastoma multiform mouse model, combination of the clinically approved GSI RO4929097 with standard of care (radiotherapy and chemotherapy with temozolomide) reduced tumour growth and prolongs survival compared to dual combinations [244, 254]. Similarly, by combining treatment of the GSI MRK003, with the late stage autophagy inhibitor chloroquine, a significant induction in apoptosis and reduction in growth was noted in glioma neurospheres [255]. Most encouraging are results from a phase I trials of MK-0752, in children with refractory and recurrent central nervous system malignancies [230, 233], which show inhibition of Notch target and good drug tolerance. These findings indicate that GSIs alone or combined with standard of care treatment have an anti-glioma effect and encourages further translational and clinical studies.

Common side effects of treatment with these drugs include nausea, fatigue, diarrhoea, or anaemia. One major side effect of systemic administration of GSI in mice is diarrhoea [256]. Notch inhibition on GSI treatment in mice resulted in rapid increase of secretary goblet cell at the expense of adsorptive cells and consequently diarrhoea. However, diarrhoea can be prevented by co-administration of glucocorticoid [257]. GSI treatment also affects the immune function by altering B-cell versus T-cell decision within the earliest uncommitted lymphocyte stem cell population in the thymus. This defect in the immune system as a side effect of GSIs treatment is primarily due to inhibition of Notch signalling, however inhibition of E-cadherin and CD44 are also implicated. Inhibition of Notch signalling for cancer treatment can paradoxically increase the risk of skin cancer [258]. Given that only phase I or II trials are currently being conducted, progression of GSIs to phase III trials in large multicentre cohorts of cancer patients and the results of these trials are required before the full potential clinical benefits of GSIs can be determined.

8 γ-Secretase-Mediated Intramembrane Proteolysis in the Immune System

There is increasing evidence that suggests that the presenilins and γ-secretase-mediated proteolysis play an important role in regulating the immune system with the γ-secretase complex components preferentially expressed within the cells and tissues of the immune system including dendritic, myeloid cells and monocytes, bone marrow, tonsils, and spleen, respectively [5, 8, 259, 260]. A number of studies using in vivo murine models of presenilin-deletion or -mutations have shown a role for presenilin in neuroinflammation and the peripheral immune system, and clearly indicate the existence of a critical minimum presenilin threshold, below which a reduction in *Psen* gene dosage and function result in serious deficits in innate and adaptive immunity. The partial-deficient mice (*Psen1*$^{+/-}$*Psen2*$^{-/-}$), are normal until approximately 6-months of age, where after the majority of mice develop age-associated myeloproliferative and autoimmune diseases [68]. These mice develop pathologies such as dermatitis, splenomegaly, glomerulonephritis, keratitis and vasculitis, with increased $CD4^{+}/CD8^{+}$ T-cell ratio and B-cell infiltrates in several tissues similar to symptoms observed in human systemic lupus erythematous [261]. Similarly, reducing levels of γ-secretase by at least 30% in nicastrin-deficient mice (*Nct*$^{+/-}$ or *Nct*$^{+/-}$*Psen1*$^{+/-}$) is associated with age dependent (>15 months) hyper-proliferation of granulocytes and reduced T-cell populations, both indicators of splenomegaly [262]. In contrast, *Psen1*$^{+/-}$*Psen2*$^{+/-}$ mice remained relatively healthy.

Furthermore, conditional-deficiency in *Psen1 and Psen2* in T- and B-lymphocytes, cells of the adaptive immune system, antagonized the selection and development of T-cells by disrupting T-cell receptor (TCR) signalling [263], while B-cells presented with a deficit in both lipopolysaccharide (LPS) and B-cell antigen receptor-induced proliferation and signal transduction events [264]. To study the role of *Psen1 and Psen2* in adulthood, tissue-specific *Psen* knockout models were needed to circumvent the developmental abnormalities carried by *Psen1*-deletion in the whole animal. Several groups created postnatal *Psen* conditional double knockout mice (cDKO) where they had ubiquitous loss of *Psen2* in all the tissues but only lacked *Psen1* in the forebrain. These mice have impaired hippocampal memory formation and reduced synaptic plasticity with neurodegeneration apparent by 6-months of age. Microarray analysis of gene expression profile of cDKO mouse cerebral cortex revealed differential expression of several inflammatory genes including, complement components, MHC histocompatibility proteins, cathepsins, chemokines, CD antigens and GFAP, implying that neurodegeneration in cDKO mice may not be due to Aβ accumulation, but might be regulated by neuroinflammation [265, 266]. Notably, presenilin deficiency in the brain is associated with inflammation in the brain and periphery, which are associated with elevated leukocytes and pro-inflammatory mediators in the serum [267]. These cDKO mice studies have also identified a role for *Psen2* in negatively regulating the microglia-mediated inflammatory response, with interferon (IFN)-selectively up-regulating *Psen2* expression and *Psen2* deficiency resulting in enhanced cytokine expression [268]. Most of the hematopoietic defects in γ-secretase defective murine models have been attributed to the Notch signalling pathway, being consisted with the requirement for Notch in proper haematopoiesis.

Immune function depends on rapid signalling events that enable the activation and amplification of signalling responses to antigen. Proteolytic processing by ADAMs and MMPs are established regulatory processes associated with both innate and adaptive immunity. Both ADAMs and MMPs are involved in aspects of membrane protein ectodomain shedding, immune-cell migration, cytokine and chemokine maturation, lymphocyte maturation, clonal expansion, migration and effector function (reviewed in [269–271]). As outlined earlier, the proteolytic activity of sheddases is often followed by γ-secretase intramembrane proteolysis, and many innate immune receptors and ligands have now been identified as γ-secretase substrates (Table 2), although the direct function of many of their cleavage products remains to be determined. Nonetheless, this is proposed to further enable cells of the immune system to proteolytically regulate cell-type specific immune activities (reviewed in [272]), and emphasise an important role for γ-secretase proteolysis in the pathophysiology of infectious diseases (reviewed in [273]).

Work in our group has demonstrated that several pro-inflammatory cytokine receptors undergo γ-secretase proteolysis. We have shown that following TACE-mediated ectodomain shedding of TNFR1, the membrane-bound

Table 2 List of γ-secretase substrates relevant to innate and adaptive immune system

Substrate	Known immune functions	References
Signalling receptors		
IL-IRI	Pro-inflammatory cytokine, interleukin-1 (IL-1) receptor and signalling	[21]
IL-1RII	Interleukin-1 (IL-1) receptor (antagonist)	[22]
IL-6R	Interleukin-6 (IL-6) receptor. Regulation of the immune response, acute-phase reactions and haematopoiesis	[275]
CSF-1	Colony stimulating factor-1, also known as macrophage colony-stimulating factor (M-CSF). A secreted cytokine that influences hematopoietic stem cells to differentiate into macrophages or other related cell types	[276]
IFNαR2	Subunit of Type 1 interferon-α (IFN-α) receptor	[301]
TNFR1	Pro-inflammatory cytokine, tumour necrosis factor (TNFα) receptor. Pro-inflammatory and cell death signalling pathways	[125]
TREM2	Triggering receptor expressed on myeloid cells 2 (TREM-2). Role in chronic inflammations and may stimulate production of constitutive rather than inflammatory chemokines and cytokines. Homozygous mutations in TREM2 cause rare forms of dementia. Missense mutation (R47H substitution) confers a significant risk of Alzheimer's disease	[31]
Cytokines		
CXC16	Transmembrane chemokine Scavenger receptor on macrophages induces chemotactic response	[282]
CX3CL1 (fractalkine)	Transmembrane chemokine	[282]
Immune antigens		
CD43 (Leukosialin)	Cell adhesion molecule. Major glycoproteins of thymocytes and T lymphocytes	[283]
CD44	Cell adhesion molecule. Cell surface glycoprotein, hyaluronan receptor	[284] [286]
HLA	Human leucocyte antigen (HLA). MHC Class I protein	[302]

(continued)

Table 2 (continued)

Substrate	Known immune functions	References
HLA-A2	MHC Class I protein Immune response, T-cell development, synaptic plasticity and refinement	[303]
NOTCH 1-4	Regulation of cell differentiation and embryonic development	[304–306]
Pathogen receptors		
SDC1	Syndecan 1 (SDC1), Herpes simplex receptor. Receptor for HIV-1, Hepatitis E and human papilloma virus (HPV)	[302]
SDC2	Syndecan 2 (SDC2), Herpes simplex receptor	[302]
SDC3	Syndecan 3 (SDC3), Herpes simplex receptor. Receptor for HIV-1 and human papilloma virus (HPV)	[302, 289, 288]
NOTCH 1-4	Signalling receptors, Cell proliferation and differentiation Activated by Epstein Barr Virus	[304] [305, 306]
CD44	Cell adhesion molecule Entry receptor *for C. neoformans,* lymphocyte activation, recirculation and homing	[286]
CD46	Receptor for adenoviruses, measles virus, human Herpes Virus 6 (HHV-6), Streptococci, and Neisseria. Also cleaved by a protease secreted by *P. gingivalis*. Complement and T cell regulatory functions	[291, 290]
Desmoglein-2	Adenovirus receptor	[302, 307]
LRP1	Lipoprotein receptor 1 (LPR1). Receptor for rhinovirus, *Pseudomonas aeruginosa* and exotoxin A	[308]
LDLR	Low density lipoprotein receptor (LDLR). Receptor for rhinovirus, Hepatitis C and vesicular stomatitis virus	[302]
VLDLR	Very low density lipoprotein receptor (VLDLR). Receptor for rhinovirus	[309]
ERBB4	Epidermal growth factor receptor tyrosine kinase. Receptor for vaccinia and other poxviruses	[310]
NGFR ($P75^{NTR}$)	Neurotrophin receptor. Receptor for rabies virus	[29, 311]
Nectin-1α	Immunoglobulin superfamily member that mediates cell-cell adhesion. Receptor for herpes-simplex virus (HSV) and pseudorabies virus	[312]
ADAM10	A Disintegrin and metalloproteinase domain-containing protein 10 (ADAM10). Cytotoxin *Staphylococcus aureus* α-hemolysis-mediated cell injury	[313, 314]
Dystroglycan	Member of multiproteindystrophin-glycoprotein complex Receptor for the lymphocytic choriomeningitis, Lassa fever viruses and *Mycobacterium leprae* (the Leprosy pathogen)	[302]

TNFR1 CTF is subsequently cleaved by γ-secretase, and internalization of TNFR1 CTF is a requirement for proteolysis. Furthermore, using in vitro and in vivo model systems we have shown that loss of presenilin expression and γ-secretase activity, inhibits TNFα-induced c-Jun N-terminal kinase (JNK) activation, and reduces TNF-induced apoptosis [125]. We have also demonstrated that the

cytokine receptor IL-1R1, which is required for IL-1α/β pro-inflammatory signalling, undergoes constitutive or metalloprotease-mediated ectodomain shedding and is a substrate for γ-secretase cleavage [21]. We reported that inhibition of γ-secretase activity antagonises IL-1β-induced cytokine secretion, suggesting that γ-secretase regulates cytokine responsiveness. Previously IL-1R1 has been localised to the nucleus and within the IL-1R1 ICD nuclear localisation sequences have been reported. Based on these findings we hypothesise as with other substrates that the IL-IR1 ICD could act as a transcription factor that affects IL-1β-induced gene expression in the nucleus in conjunction with the usual signal transduction pathway [21]. IL-1RII, which is a decoy receptor that binds excess soluble IL-1β, and antagonises IL-1R1 signalling, is also cleaved by α- and β-secretase and is a substrate for γ-secretase proteolysis [22]. Elevated soluble IL-1RII levels are observed in the cerebrospinal fluid of patients with AD, which may be explained by the similar proteolytic processing as APP by γ-secretase [274]. Another member of the interleukin receptor family, the IL-6 receptor (IL-6R) is sequentially cleaved by ADAM17/TACE and γ-secretase at the plasma membrane. IL-6 mediated signals are important for the coordination of the immune system and IL-6 deregulation can lead to chronic inflammation [275].

The colony-stimulating factor 1 (CSF-1) receptor has an essential role in the haematopoiesis of cells of the immune system especially mononuclear phagocytes, such as macrophages and monocytes. CSF-1 is cleaved by γ-secretase and is regulated by Toll-like receptor (TLR) activation [276]. A similar system could be in operation in microglia and the activation of TLRs in response to Aβ peptides could lead to microglial activation via promotion of CSF-1 receptor proteolysis by TACE and γ-secretase. In our lab we have shown in bone marrow derived macrophages that down regulation of *Psen2* transcription is paralleled by reduced *tlr4* mRNA levels and loss of LPS-induced *tlr4* mRNA transcription regulation with reduced inflammatory cytokine expression [277]. Conversely, in studies with brain-specific knockouts of the presenilins, it was found that *Psen2* was critical in inhibiting microglia-mediated inflammation and with *Psen2* deficiency resulting in enhanced cytokine expression [268]. From the data, a model is emerging whereby the functions of TLRs and γ-secretase in inflammation and microglial activation combine to regulate the inflammatory response. Dysfunction of this system in the brain could be fundamental to the pathophysiology of AD.

Of particular relevance to the pathogenesis of AD, recently two independent groups identified variants of the *Trem2* gene, which encodes triggering receptor expressed on myeloid cells-2 (TREM2), which increase susceptibility to late onset AD [31]. TREM2 undergoes regulated intramembrane proteolysis [31]. A model has been proposed by which *Trem2* variants contribute to the pathogenesis of AD by deregulation of Aβ phagocytic ability and altering the pro-inflammatory response of microglia [278]. It has also been shown that haplodeficiency of *Trem2* or *Trem2* variants reduced the ability of microglia to surround and compact early Aβ deposits, thereby developing a neuro-protective barrier between early Aβ deposits and vulnerable neuronal cells [279, 280]. Importantly, a recent study reported that inhibition of γ-secretase activity interferes with TREM2-dependent signalling and phagocytic activity [281], further highlighting an important functional connection between γ-secretase, TREM2 and the pathogenesis of AD.

Recruitment of leukocytes to a site of infection is an early immune response following acute or chronic inflammation. The proteolytic activity of ADAMs and MMPs is a consistent occurrence in the generation of chemokine and cytokine gradients for immune cell recruitment (as reviewed in [270]). The ADAMs (ADAM10 and ADAM17/TACE), are involved in the shedding of biologically active chemokines, such as CX3CL1 (fractalkine) and CXCL16. Importantly both CX3CL1 (fractalkine) and CXCL16 have also been identified as γ-secretase substrates [282]. Ectodomain shedding of CX3CL1 and CXCL16 serves to enhance neutrophil recruitment to sites of infection. The generation of a chemokine gradient is often associated with the cell-adhesion properties of immune and endothelial cells. CD43 is a major glycoprotein of thymocytes and lymphocytes that is cleaved by γ-secretase during neutrophil activation. Unlike other substrates CD43 is first cleaved by cathepsin G, not a metalloprotease [283]. The extracellular domain of CD43 can regulate inflammation by preventing neutrophil adhesion on endothelial cells. CD43 cleavage by γ-secretase was found to be required for neutrophil adhesion, thereby regulating CD43-mediated signalling [283]. CD44 is a widely distributed cell surface adhesion antigen that is also a γ-secretase substrate [284]. It has been shown that the CD44 ICD translocates to the nucleus of macrophages and promotes the activation of NF-κB and macrophage fusion.

As the γ-secretase complex is localized to dendritic cells that are primarily responsible for the processing of antigens and presentation to B and T cells, it has been shown that many γ-secretase substrates are important in the lifecycle of many pathogens, with some γ-secretase substrates acting as an entry receptors or as scavenger receptors by binding to pathogen membrane components, such as binding of phosphatidylserine to CXCL16 [285]. The earlier mentioned CD44 glycoprotein is also an entry receptor for the opportunistic yeast *Cryptococcus neoformans* the causative agent of the infection cryptococcosis that can manifest as a secondary infection for AIDS patients [286]. The γ-secretase substrates nectin-1α, syndecan 1 and 2 are herpes simplex viral receptors [287], and the HIV-1 and human papilloma viruses also attach to dendritic cells via the syndecans, SDC1 and SDC3 [288], both also reported to be substrates for intramembrane proteolysis [273, 289]. CD46 another γ-secretase substrate of the immune system, which is responsible for protecting the cell against complement-mediated injury, also has a role in differentiation of $CD4^+$ into T-regulatory 1 cells that secrete IL-10, supressing immune response and therefore preventing autoimmunity. Similar to CD44, CD46 is also a target of many pathogens including adenoviruses, measles virus, human herpes virus 6 (HHV-6) and Streptococci [290]. Elevated γ-secretase cleavage of CD46 stimulates CD46-dependent T cell responses in response to *Neisseria gonorrhoeae* and *Neisseria meningitidis* infection [291].

There are several recent experimental studies that have demonstrated that GSIs may be potential therapies for inflammatory disease. For example, in vitro and in vivo studies have shown that GSIs can antagonise lipopolysaccharide-induced Notch signalling and attenuated the LPS-induced inflammatory response [292].

GSIs were shown to alleviate acute airway inflammation of allergic asthma in mice by down-regulating Notch-dependent Th17 cell differentiation [293]. Collectively these findings suggest a strong link between γ-secretase activity and the immune system and may in the future provide therapeutic potential in the treatment of immune-related diseases.

9 Conclusions

It was generally accepted that signalling events-mediated by cell surface receptors, including Notch, EpCAM, G-protein-coupled receptors, TLRs, $p75^{NTR}$ and death receptors (TNFR1, FasR and TRAIL-R1/2) are initiated and terminated at the plasma membrane. This dogma has now been challenged [294–300], and a new model proposes that following cell surface receptor activation and propagation of a signalling event, through the combined enzymatic activity of "sheddase" and γ-secretase protease complexes, receptors can also propagate distinct signalling events facilitated by the generation of soluble receptor ectodomains and intracellular ICD. The combined enzymatic activity of "sheddase" and γ-secretase protease complexes have defined regulated intramembrane proteolysis, a fundamental signal transduction process.

Since the discovery of the presenilin genes and their association with FAD, great progress has been made in understanding the structure, assembly, catalytic mechanisms, substrate specificities, and regulation of the γ-secretase protease complexes. A greater appreciation for the high degree of regulation surrounds assembly, localization, expression and activity of the γ-secretase protease complexes, and how γ-secretase protease complexes and their substrates affects cellular processes have emerged and reinvigorated efforts to develop therapeutic modulators and inhibitors of γ-secretase protease activity. Challenges for the future include the advancement of our understanding of the diversity of γ-secretase protease complexes in humans and translation of this knowledge into practical therapeutics. These efforts should not only provide better and safer therapeutics for AD, but also other diseases such as breast and haematological cancers, immune disorders, acne inversa, cardiomyopathy and kidney disorders, with which adverse γ-secretase activity and signalling have been attributed.

Acknowledgements We apologise to all colleagues whose work has not been discussed or cited owing to space limitations. This work was supported and funded by grants from Science Foundation Ireland (02/IN1/B218 and 09/IN.1/B2624) and a student bursary to CCV from the College of Science Engineering & Food Science, University College Cork.

References

1. Brown MS, Ye J, Rawson RB et al (2000) Regulated intramembrane proteolysis: a control mechanism conserved from bacteria to humans. Cell 100(4):391–398

2. Lichtenthaler SF, Haass C, Steiner H (2011) Regulated intramembrane proteolysis—lessons from amyloid precursor protein processing. J Neurochem 117(5):779–796
3. Schroder B, Saftig P (2016) Intramembrane proteolysis within lysosomes. Ageing Res Rev
4. Lemberg MK, Menendez J, Misik A et al (2005) Mechanism of intramembrane proteolysis investigated with purified rhomboid proteases. EMBO J 24(3):464–472
5. Jurisch-Yaksi N, Sannerud R, Annaert W (2013) A fast growing spectrum of biological functions of gamma-secretase in development and disease. Biochim Biophys Acta 1828 (12):2815–2827
6. Langosch D, Scharnagl C, Steiner H et al (2015) Understanding intramembrane proteolysis: from protein dynamics to reaction kinetics. Trends Biochem Sci 40(6):318–327
7. Duggan SP, McCarthy JV (2016) Beyond gamma-secretase activity: the multifunctional nature of presenilins in cell signalling pathways. Cell Signal 28(1):1–11
8. Hurst TP, Coleman-Vaughan C, Patwal I et al (2016) Regulated intramembrane proteolysis, innate immunity and therapeutic targets in Alzheimer's disease. AIMS Mol Sci 3(2):138–157
9. Sannerud R, Annaert W (2009) Trafficking, a key player in regulated intramembrane proteolysis. Semin Cell Dev Biol 20(2):183–190
10. Lichtenthaler SF, Steiner H (2007) Sheddases and intramembrane-cleaving proteases: RIPpers of the membrane. Symposium on Regulated Intramembrane Proteolysis. EMBO Rep 8(6):537–541
11. Edwards DR, Handsley MM, Pennington CJ (2008) The ADAM metalloproteinases. Mol Aspects Med 29(5):258–289
12. van Hinsbergh VWM, Koolwijk P (2008) Endothelial sprouting and angiogenesis: matrix metalloproteinases in the lead. Cardiovasc Res 78(2):203–212
13. Lin CY, Tseng IC, Chou FP et al (2008) Zymogen activation, inhibition, and ectodomain shedding of matriptase. Front Biosci 13:621–635
14. Morohashi Y, Tomita T (2013) Protein trafficking and maturation regulate intramembrane proteolysis. Biochimica et Biophysica Acta (BBA)—Biomembranes 1828(12):2855–2861
15. McCarthy JV, Twomey C, Wujek P (2009) Presenilin-dependent regulated intramembrane proteolysis and gamma-secretase activity. Cell Mol Life Sci 66(9):1534–1555
16. De Strooper B, Saftig P, Craessaerts K et al (1998) Deficiency of presenilin-1 inhibits the normal cleavage of amyloid precursor protein. Nature 391(6665):387–390
17. De Strooper B, Annaert W, Cupers P et al (1999) A presenilin-1-dependent [gamma]-secretase-like protease mediates release of Notch intracellular domain. Nature 398 (6727):518–522
18. Guruharsha KG, Kankel MW, Artavanis-Tsakonas S (2012) The Notch signalling system: recent insights into the complexity of a conserved pathway. Nat Rev Genet 13(9):654–666
19. Marambaud P, Shioi J, Serban G et al (2002) A presenilin-1/gamma-secretase cleavage releases the E-cadherin intracellular domain and regulates disassembly of adherens junctions. EMBO J 21(8):1948–1956
20. Fluhrer R, Grammer G, Israel L et al (2006) A [gamma]-secretase-like intramembrane cleavage of TNF[alpha] by the GxGD aspartyl protease SPPL2b. Nat Cell Biol 8(8):894–896
21. Elzinga BM, Twomey C, Powell JC et al (2009) Interleukin-1 receptor type 1 is a substrate for gamma-secretase-dependent regulated intramembrane proteolysis. J Biol Chem 284 (3):1394–1409
22. Kuhn PH, Marjaux E, Imhof A et al (2007) Regulated intramembrane proteolysis of the interleukin-1 receptor II by alpha-, beta-, and gamma-secretase. J Biol Chem 282 (16):11982–11995
23. Twomey C, Qian S, McCarthy JV (2009) TRAF6 promotes ubiquitination and regulated intramembrane proteolysis of IL-1R1. Biochem Biophys Res Commun 381(3):418–423
24. McElroy B, Powell JC, McCarthy JV (2007) The insulin-like growth factor 1 (IGF-1) receptor is a substrate for gamma-secretase-mediated intramembrane proteolysis. Biochem Biophys Res Commun 358(4):1136–1141

25. Sturtevant MA, Roark M, Bier E (1993) The Drosophila rhomboid gene mediates the localized formation of wing veins and interacts genetically with components of the EGF-R signaling pathway. Genes Dev 7(6):961–973
26. Ni CY, Murphy MP, Golde TE et al (2001) gamma-secretase cleavage and nuclear localization of ErbB-4 receptor tyrosine kinase. Science 294(5549):2179–2181
27. Powell JC, Twomey C, Jain R et al (2009) Association between Presenilin-1 and TRAF6 modulates regulated intramembrane proteolysis of the p75NTR neurotrophin receptor. J Neurochem 108(1):216–230
28. Berghoff J, Jaisimha AV, Duggan S et al (2015) Gamma-secretase-independent role for cadherin-11 in neurotrophin receptor p75 (p75(NTR)) mediated glioblastoma cell migration. Mol Cell Neurosci 69:41–53
29. Zampieri N, Xu CF, Neubert TA et al (2005) Cleavage of p75 neurotrophin receptor by alpha-secretase and gamma-secretase requires specific receptor domains. J Biol Chem 280 (15):14563–14571
30. Lammich S, Okochi M, Takeda M et al (2002) Presenilin-dependent Intramembrane Proteolysis of CD44 Leads to the Liberation of Its Intracellular Domain and the Secretion of an Aβ-like Peptide. J Biol Chem 277(47):44754–44759
31. Wunderlich P, Glebov K, Kemmerling N et al (2013) Sequential proteolytic processing of the triggering receptor expressed on myeloid cells-2 (TREM2) protein by ectodomain shedding and gamma-secretase-dependent intramembranous cleavage. J Biol Chem 288 (46):33027–33036
32. Maetzel D, Denzel S, Mack B et al (2009) Nuclear signalling by tumour-associated antigen EpCAM. Nat Cell Biol 11(2):162–171
33. Black RA, Rauch CT, Kozlosky CJ et al (1997) A metalloproteinase disintegrin that releases tumour-necrosis factor-alpha from cells. Nature 385(6618):729–733
34. Friedmann E, Hauben E, Maylandt K et al (2006) SPPL2a and SPPL2b promote intramembrane proteolysis of TNFalpha in activated dendritic cells to trigger IL-12 production. Nat Cell Biol 8(8):843–848
35. Chhibber-Goel J, Coleman-Vaughan C, Agrawal V et al (2016) γ-secretase activity is required for regulated intramembrane proteolysis of tumor necrosis factor (TNF) receptor 1 and TNF-mediated pro-apoptotic signaling. J Biol Chem 291(11):5971–5985
36. Selkoe DJ (2001) Alzheimer's disease: genes, proteins, and therapy. Physiol Rev 81(2): 741–766
37. Small SA, Gandy S (2006) Sorting through the cell biology of Alzheimer's disease: intracellular pathways to pathogenesis. Neuron 52(1):15–31
38. Tomita T, Iwatsubo T (2013) Structural biology of presenilins and signal peptide peptidases. J Biol Chem 288(21):14673–14680
39. Weidemann A, Eggert S, Reinhard FB et al (2002) A novel epsilon-cleavage within the transmembrane domain of the Alzheimer amyloid precursor protein demonstrates homology with Notch processing. Biochemistry 41(8):2825–2835
40. Zhao G, Cui MZ, Mao G et al (2005) γ-cleavage is dependent on zeta-cleavage during the proteolytic processing of amyloid precursor protein within its transmembrane domain. J Biol Chem 280(45):37689–37697
41. Zhao G, Mao G, Tan J et al (2004) Identification of a new presenilin-dependent zeta-cleavage site within the transmembrane domain of amyloid precursor protein. J Biol Chem 279(49):50647–50650
42. Okochi M, Fukumori A, Jiang J et al (2006) Secretion of the Notch-1 Aβ-like peptide during notch signaling. J Biol Chem 281(12):7890–7898
43. Beel AJ, Sanders CR (2008) Substrate specificity of gamma-secretase and other intramembrane proteases. Cell Mol Life Sci 65(9):1311–1334
44. Sun L, Li X, Shi Y (2016) Structural biology of intramembrane proteases: mechanistic insights from rhomboid and S2P to γ-secretase. Curr Opin Struct Biol 37:97–107
45. Wolfe MS (2009) Intramembrane-cleaving Proteases. J Biol Chem 284(21):13969–13973

46. Martoglio B (2003) Intramembrane proteolysis and post-targeting functions of signal peptides. Biochem Soc Trans 31(Pt 6):1243–1247
47. Strisovsky K (2013) Structural and mechanistic principles of intramembrane proteolysis—lessons from rhomboids. FEBS J 280(7):1579–1603
48. Laurent SA, Hoffmann FS, Kuhn PH et al (2015) γ-secretase directly sheds the survival receptor BCMA from plasma cells. Nat Commun 6:7333
49. Fluhrer R, Steiner H, Haass C (2009) Intramembrane proteolysis by signal peptide peptidases: a comparative discussion of GXGD-type aspartyl proteases. J Biol Chem 284 (21):13975–13979
50. Nyborg AC, Kornilova AY, Jansen K et al (2004) Signal peptide peptidase forms a homodimer that is labeled by an active site-directed γ-secretase inhibitor. J Biol Chem 279 (15):15153–15160
51. Nyborg AC, Herl L, Berezovska O et al (2006) Signal peptide peptidase (SPP) dimer formation as assessed by fluorescence lifetime imaging microscopy (FLIM) in intact cells. Mol Neurodegeneration 1(1):1–8
52. Gertsik N, Chau D-M, Li Y-M (2015) γ-secretase inhibitors and modulators induce distinct conformational changes in the active sites of γ-secretase and signal peptide peptidase. ACS Chem Biol 10(8):1925–1931
53. Tomita T, Watabiki T, Takikawa R et al (2001) The first proline of PALP motif at the C terminus of presenilins is obligatory for stabilization, complex formation, and gamma-secretase activities of presenilins. J Biol Chem 276(35):33273–33281
54. Sato C, Takagi S, Tomita T et al (2008) The C-terminal PAL motif and transmembrane domain 9 of presenilin 1 are involved in the formation of the catalytic pore of the gamma-secretase. J Neurosci 28(24):6264–6271
55. Haass C, Selkoe DJ (1993) Cellular processing of beta-amyloid precursor protein and the genesis of amyloid beta-peptide. Cell 75(6):1039–1042
56. Sherrington R, Rogaev EI, Liang Y et al (1995) Cloning of a gene bearing missense mutations in early-onset familial Alzheimer's disease. Nature 375(6534):754–760
57. Tanzi RE, Kovacs DM, Kim TW et al (1996) The gene defects responsible for familial Alzheimer's disease. Neurobiol Dis 3(3):159–168
58. Levy-Lahad E, Wasco W, Poorkaj P et al (1995) Candidate gene for the chromosome 1 familial Alzheimer's disease locus. Science 269(5226):973–977
59. Yu G, Nishimura M, Arawaka S et al (2000) Nicastrin modulates presenilin-mediated notch/glp-1 signal transduction and betaAPP processing. Nature 407(6800):48–54
60. Francis R, McGrath G, Zhang J et al (2002) aph-1 and pen-2 are required for Notch pathway signaling, gamma-secretase cleavage of betaAPP, and presenilin protein accumulation. Dev Cell 3(1):85–97
61. Goutte C, Tsunozaki M, Hale VA et al (2002) APH-1 is a multipass membrane protein essential for the Notch signaling pathway in Caenorhabditis elegans embryos. Proc Natl Acad Sci USA 99(2):775–779
62. Kimberly WT, LaVoie MJ, Ostaszewski BL et al (2003) Gamma-secretase is a membrane protein complex comprised of presenilin, nicastrin, Aph-1, and Pen-2. Proc Natl Acad Sci USA 100(11):6382–6387
63. De Strooper B (2003) Aph-1, Pen-2, and nicastrin with presenilin generate an active γ-secretase complex. Neuron 38(1):9–12
64. Kaether C, Haass C, Steiner H (2006) Assembly, trafficking and function of gamma-secretase. Neurodegener Dis 3(4–5):275–283
65. Ahn K, Shelton CC, Tian Y et al (2010) Activation and intrinsic γ-secretase activity of presenilin 1. Proc Natl Acad Sci 107(50):21435–21440
66. Takasugi N, Tomita T, Hayashi I et al (2003) The role of presenilin cofactors in the gamma-secretase complex. Nature 422(6930):438–441

67. Wolfe MS, Xia W, Ostaszewski BL et al (1999) Two transmembrane aspartates in presenilin-1 required for presenilin endoproteolysis and gamma-secretase activity. Nature 398(6727):513–517
68. Herreman A, Hartmann D, Annaert W et al (1999) Presenilin 2 deficiency causes a mild pulmonary phenotype and no changes in amyloid precursor protein processing but enhances the embryonic lethal phenotype of presenilin 1 deficiency. Proc Natl Acad Sci USA 96 (21):11872–11877
69. Dries DR, Yu G (2008) Assembly, maturation, and trafficking of the gamma-secretase complex in Alzheimer's disease. Curr Alzheimer Res 5(2):132–146
70. Yang DS, Tandon A, Chen F et al (2002) Mature glycosylation and trafficking of nicastrin modulate its binding to presenilins. J Biol Chem 277(31):28135–28142
71. Kimberly WT, LaVoie MJ, Ostaszewski BL et al (2002) Complex N-linked glycosylated nicastrin associates with active gamma-secretase and undergoes tight cellular regulation. J Biol Chem 277(38):35113–35117
72. De Strooper B (2005) Nicastrin: gatekeeper of the gamma-secretase complex. Cell 122 (3):318–320
73. Bolduc DM, Montagna DR, Gu Y et al (2016) Nicastrin functions to sterically hinder gamma-secretase-substrate interactions driven by substrate transmembrane domain. Proc Natl Acad Sci USA 113(5):E509–E518
74. Shah S, Lee SF, Tabuchi K et al (2005) Nicastrin functions as a gamma-secretase-substrate receptor. Cell 122(3):435–447
75. Capell A, Beher D, Prokop S et al (2005) Gamma-secretase complex assembly within the early secretory pathway. J Biol Chem 280(8):6471–6478
76. Lee SF, Shah S, Li H et al (2002) Mammalian APH-1 interacts with presenilin and nicastrin and is required for intramembrane proteolysis of amyloid-beta precursor protein and Notch. J Biol Chem 277(47):45013–45019
77. Zhang X, Li Y, Xu H et al (2014) The γ-secretase complex: from structure to function. Front Cell Neurosci 8
78. Hébert SS, Serneels L, Dejaegere T et al (2004) Coordinated and widespread expression of γ-secretase in vivo: evidence for size and molecular heterogeneity. Neurobiol Dis 17 (2):260–272
79. Kaether C, Scheuermann J, Fassler M et al (2007) Endoplasmic reticulum retention of the gamma-secretase complex component Pen2 by Rer1. EMBO Rep 8(8):743–748
80. Spasic D, Raemaekers T, Dillen K et al (2007) Rer1p competes with APH-1 for binding to nicastrin and regulates gamma-secretase complex assembly in the early secretory pathway. J Cell Biol 176(5):629–640
81. Dang S, Wu S, Wang J et al (2015) Cleavage of amyloid precursor protein by an archaeal presenilin homologue PSH. Proc Natl Acad Sci USA 112(11):3344–3349
82. Lu P, Bai XC, Ma D et al (2014) Three-dimensional structure of human gamma-secretase. Nature 512(7513):166–170
83. X-c Bai, Yan C, Yang G et al (2015) An atomic structure of human γ-secretase. Nature 525 (7568):212–217
84. Ma G, Li T, Price DL et al (2005) APH-1a is the principal mammalian APH-1 isoform present in gamma-secretase complexes during embryonic development. J Neurosci 25 (1):192–198
85. Mastrangelo P, Mathews PM, Chishti MA et al (2005) Dissociated phenotypes in presenilin transgenic mice define functionally distinct gamma-secretases. Proc Natl Acad Sci USA 102 (25):8972–8977
86. Shirotani K, Tomioka M, Kremmer E et al (2007) Pathological activity of familial Alzheimer's disease-associated mutant presenilin can be executed by six different gamma-secretase complexes. Neurobiol Dis 27(1):102–107
87. Shirotani K, Edbauer D, Prokop S et al (2004) Identification of distinct gamma-secretase complexes with different APH-1 variants. J Biol Chem 279(40):41340–41345

88. Meckler X, Checler F (2014) Visualization of specific gamma-secretase complexes using bimolecular fluorescence complementation. J Alzheimers Dis 40(1):161–176
89. Sannerud R, Esselens C, Ejsmont P et al (2016) Restricted location of PSEN2/gamma-Secretase determines substrate specificity and generates an intracellular aβ pool. Cell
90. Acx H, Chavez-Gutierrez L, Serneels L et al (2014) Signature amyloid beta profiles are produced by different gamma-secretase complexes. J Biol Chem 289(7):4346–4355
91. Bayer TA, Wirths O, Majtenyi K et al (2001) Key factors in Alzheimer's disease: beta-amyloid precursor protein processing, metabolism and intraneuronal transport. Brain Pathol 11(1):1–11
92. Gouras GK, Tampellini D, Takahashi RH et al (2010) Intraneuronal beta-amyloid accumulation and synapse pathology in Alzheimer's disease. Acta Neuropathol 119 (5):523–541
93. Pensalfini A, Albay R 3rd, Rasool S et al (2014) Intracellular amyloid and the neuronal origin of Alzheimer neuritic plaques. Neurobiol Dis 71:53–61
94. St George-Hyslop P, Fraser PE (2012) Assembly of the presenilin gamma-/epsilon-secretase complex. J Neurochem 120(1):84–88
95. Villa JC, Chiu D, Brandes AH et al (2014) Nontranscriptional role of Hif-1alpha in activation of gamma-secretase and notch signaling in breast cancer. Cell Rep 8(4):1077–1092
96. Wang X, Huang T, Zhao Y et al (2014) Sorting nexin 27 regulates Abeta production through modulating gamma-secretase activity. Cell Rep 9(3):1023–1033
97. Yan R, McCarthy JV (2010) Presenilin and γ-secretase activity: a viable therapeutic target for Alzheimer's disease? Curr Signal Transduct Ther 5(2):128–140
98. Yan R, Farrelly S, McCarthy JV (2013) Presenilins are novel substrates for TRAF6-mediated ubiquitination. Cell Signal 25(9):1769–1779
99. Duggan SP, Yan R, McCarthy JV (2015) A ubiquitin-binding CUE domain in presenilin-1 enables interaction with K63-linked polyubiquitin chains. FEBS Lett 589(9):1001–1008
100. Uemura K, Kuzuya A, Shimozono Y et al (2007) GSK3beta activity modifies the localization and function of presenilin 1. J Biol Chem 282(21):15823–15832
101. Massey LK, Mah AL, Monteiro MJ (2005) Ubiquilin regulates presenilin endoproteolysis and modulates gamma-secretase components, Pen-2 and nicastrin. Biochem J 391(3):513–525
102. Li J, Pauley AM, Myers RL et al (2002) SEL-10 interacts with presenilin 1, facilitates its ubiquitination, and alters A-beta peptide production. J Neurochem 82(6):1540–1548
103. Nunan J, Shearman MS, Checler F et al (2001) The C-terminal fragment of the Alzheimer's disease amyloid protein precursor is degraded by a proteasome-dependent mechanism distinct from gamma-secretase. Eur J Biochem 268(20):5329–5336
104. Suzuki T, Nakaya T (2008) Regulation of amyloid beta-protein precursor by phosphorylation and protein interactions. J Biol Chem 283(44):29633–29637
105. Rebelo S, Domingues SC, Santos M et al (2013) Identification of a novel complex AbetaPP: Fe65:PP1 that regulates AbetaPP Thr668 phosphorylation levels. J Alzheimers Dis 35 (4):761–775
106. Lee M-S, Kao S-C, Lemere CA et al (2003) APP processing is regulated by cytoplasmic phosphorylation. The Journal of Cell Biology 163(1):83–95
107. Iijima K, Ando K, Takeda S et al (2000) Neuron-specific phosphorylation of Alzheimer's beta-amyloid precursor protein by cyclin-dependent kinase 5. J Neurochem 75(3):1085–1091
108. Suzuki N, Cheung TT, Cai XD et al (1994) An increased percentage of long amyloid beta protein secreted by familial amyloid beta protein precursor (beta APP717) mutants. Science 264(5163):1336–1340

109. Aplin AE, Gibb GM, Jacobsen JS et al (1996) In vitro phosphorylation of the cytoplasmic domain of the amyloid precursor protein by glycogen synthase kinase-3beta. J Neurochem 67(2):699–707
110. Standen CL, Brownlees J, Grierson AJ et al (2001) Phosphorylation of thr668 in the cytoplasmic domain of the Alzheimer's disease amyloid precursor protein by stress-activated protein kinase 1b (Jun N-terminal kinase-3). J Neurochem 76(1):316–320
111. Taru H, Iijima K, Hase M et al (2002) Interaction of Alzheimer's beta-amyloid precursor family proteins with scaffold proteins of the JNK signaling cascade. J Biol Chem 277 (22):20070–20078
112. Vingtdeux V, Hamdane M, Gompel M et al (2005) Phosphorylation of amyloid precursor carboxy-terminal fragments enhances their processing by a gamma-secretase-dependent mechanism. Neurobiol Dis 20(2):625–637
113. Chang K-A, Kim H-S, Ha T-Y et al (2006) Phosphorylation of amyloid precursor protein (APP) at Thr668 regulates the nuclear translocation of the APP intracellular domain and induces neurodegeneration. Mol Cell Biol 26(11):4327–4338
114. Parra LM, Hartmann M, Schubach S et al (2015) Distinct intracellular domain substrate modifications selectively regulate ectodomain cleavage of NRG1 or CD44. Mol Cell Biol 35 (19):3381–3395
115. Annaert W, De Strooper B (2002) A cell biological perspective on Alzheimer's disease. Annu Rev Cell Dev Biol 18:25–51
116. Traub LM, Bonifacino JS (2013) Cargo recognition in clathrin-mediated endocytosis. Cold Spring Harb Perspect Biol 5(11):a016790
117. Wakabayashi T, Craessaerts K, Bammens L et al (2009) Analysis of the [gamma]-secretase interactome and validation of its association with tetraspanin-enriched microdomains. Nat Cell Biol 11(11):1340–1346
118. Osenkowski P, Ye W, Wang R et al (2008) Direct and potent regulation of gamma-secretase by its lipid microenvironment. J Biol Chem 283(33):22529–22540
119. Urano Y, Hayashi I, Isoo N et al (2005) Association of active gamma-secretase complex with lipid rafts. J Lipid Res 46(5):904–912
120. Vetrivel KS, Cheng H, Lin W et al (2004) Association of gamma-secretase with lipid rafts in post-Golgi and endosome membranes. J Biol Chem 279(43):44945–44954
121. Hur JY, Welander H, Behbahani H et al (2008) Active gamma-secretase is localized to detergent-resistant membranes in human brain. FEBS J 275(6):1174–1187
122. Vaccari T, Lu H, Kanwar R et al (2008) Endosomal entry regulates Notch receptor activation in Drosophila melanogaster. J Cell Biol 180(4):755–762
123. Gupta-Rossi N, Six E, LeBail O et al (2004) Monoubiquitination and endocytosis direct gamma-secretase cleavage of activated Notch receptor. J Cell Biol 166(1):73–83
124. Urra S, Escudero CA, Ramos P et al (2007) TrkA receptor activation by nerve growth factor induces shedding of the p75 neurotrophin receptor followed by endosomal gamma-secretase-mediated release of the p75 intracellular domain. J Biol Chem 282 (10):7606–7615
125. Chhibber-Goel J, Coleman-Vaughan C, Agrawal V et al (2016) γ-secretase activity is required for regulated intramembrane proteolysis of tumor necrosis factor (TNF) receptor 1 and TNF-mediated pro-apoptotic signaling. J Biol Chem
126. Peric A, Annaert W (2015) Early etiology of Alzheimer's disease: tipping the balance toward autophagy or endosomal dysfunction? Acta Neuropathol 129(3):363–381
127. Ueda N, Tomita T, Yanagisawa K et al (2016) Retromer and Rab2-dependent trafficking mediate PS1 degradation by proteasomes in endocytic disturbance. J Neurochem 137 (4):647–658
128. St George-Hyslop PH (2000) Genetic factors in the genesis of Alzheimer's disease. Ann NY Acad Sci 924:1–7
129. Godyń J, Jończyk J, Panek D et al (2016) Therapeutic strategies for Alzheimer's disease in clinical trials. Pharmacol Rep 68(1):127–138

130. De Strooper B, Chavez Gutierrez L (2015) Learning by failing: ideas and concepts to tackle gamma-secretases in Alzheimer's disease and beyond. Annu Rev Pharmacol Toxicol 55:419–437
131. De Strooper B (2014) Lessons from a failed γ-secretase Alzheimer trial. Cell 159(4):721–726
132. Vassar R, Bennett BD, Babu-Khan S et al (1999) β-secretase cleavage of Alzheimer's amyloid precursor protein by the transmembrane aspartic protease BACE. Science 286 (5440):735–741
133. Crump CJ, Johnson DS, Li Y-M (2013) Development and mechanism of γ-secretase modulators for Alzheimer's disease. Biochemistry 52(19):3197–3216
134. Golde TE, Koo EH, Felsenstein KM et al (1828) γ-secretase inhibitors and modulators. Biochimica et Biophysica Acta (BBA)—Biomembranes 12:2898–2907 (2013)
135. Josien H (2002) Recent advances in the development of gamma-secretase inhibitors. Curr Opin Drug Discov Devel 5(4):513–525
136. Kreft AF, Martone R, Porte A (2009) Recent advances in the identification of gamma-secretase inhibitors to clinically test the Abeta oligomer hypothesis of Alzheimer's disease. J Med Chem 52(20):6169–6188
137. Karran E, Hardy J (2014) Antiamyloid therapy for Alzheimer's disease—are we on the right road? N Engl J Med 370(4):377–378
138. Doody RS, Raman R, Farlow M et al (2013) A phase 3 trial of semagacestat for treatment of Alzheimer's disease. N Engl J Med 369(4):341–350
139. Henley DB, Sundell KL, Sethuraman G et al (2014) Safety profile of semagacestat, a gamma-secretase inhibitor: identity trial findings. Curr Med Res Opin 30(10):2021–2032
140. Coric V, van Dyck CH, Salloway S et al (2012) Safety and tolerability of the γ-secretase inhibitor avagacestat in a phase 2 study of mild to moderate alzheimer disease. Arch Neurol 69(11):1430–1440
141. Albright CF, Dockens RC, Meredith JE et al (2013) Pharmacodynamics of selective inhibition of γ-secretase by avagacestat. J Pharmacol Exp Ther 344(3):686–695
142. Probst G, Aubele DL, Bowers S et al (2013) Discovery of (R)-4-Cyclopropyl-7,8-difluoro-5-(4-(trifluoromethyl)phenylsulfonyl)-4,5-dihydro-1H-pyrazolo[4,3-c]quinoline (ELND006) and (R)-4-Cyclopropyl-8-fluoro-5-(6-(trifluoromethyl)pyridin-3-ylsulfonyl)-4,5-dihydro-2H-pyrazolo[4,3-c]quinoline (ELND007): metabolically stable γ-secretase inhibitors that selectively inhibit the production of amyloid-β over Notch. J Med Chem 56(13):5261–5274
143. Weggen S, Eriksen JL, Das P et al (2001) A subset of NSAIDs lower amyloidogenic A[beta] 42 independently of cyclooxygenase activity. Nature 414(6860):212–216
144. Lim GP, Yang F, Chu T et al (2000) Ibuprofen suppresses plaque pathology and inflammation in a mouse model for Alzheimer's disease. J Neurosci 20(15):5709–5714
145. Das C, Berezovska O, Diehl TS et al (2003) Designed helical peptides inhibit an intramembrane protease. J Am Chem Soc 125(39):11794–11795
146. Mohr OL (1919) character changes caused by mutation of an entire region of a chromosome in Drosophila. Genetics 4(3):275–282
147. Groeneweg JW, Foster R, Growdon WB et al (2014) Notch signaling in serous ovarian cancer. J Ovarian Res 7:95
148. Hales EC, Taub JW, Matherly LH (2014) New insights into Notch1 regulation of the PI3 K-AKT-mTOR1 signaling axis: targeted therapy of gamma-secretase inhibitor resistant T-cell acute lymphoblastic leukemia. Cell Signal 26(1):149–161
149. Takebe N, Nguyen D, Yang SX (2014) Targeting notch signaling pathway in cancer: clinical development advances and challenges. Pharmacol Ther 141(2):140–149
150. Reynolds TC, Smith SD, Sklar J (1987) Analysis of DNA surrounding the breakpoints of chromosomal translocations involving the beta T cell receptor gene in human lymphoblastic neoplasms. Cell 50(1):107–117

151. Ellisen LW, Bird J, West DC et al (1991) TAN-1, the human homolog of the Drosophila notch gene, is broken by chromosomal translocations in T lymphoblastic neoplasms. Cell 66 (4):649–661
152. Mao L (2015) NOTCH mutations: multiple faces in human malignancies. Cancer Prev Res (Phila) 8(4):259–261
153. Mutvei AP, Fredlund E, Lendahl U (2015) Frequency and distribution of Notch mutations in tumor cell lines. BMC Cancer 15:311
154. Weng AP, Ferrando AA, Lee W et al (2004) Activating mutations of Notch1 in human T cell acute lymphoblastic leukemia. Science 306(5694):269–271
155. Liu J, Shen JX, Wen XF et al (2016) Targeting Notch degradation system provides promise for breast cancer therapeutics. Crit Rev Oncol Hematol
156. Guilmeau S (2012) Notch signaling and intestinal cancer. Adv Exp Med Biol 727:272–288
157. Bertrand FE, Angus CW, Partis WJ et al (2012) Developmental pathways in colon cancer: crosstalk between WNT, BMP, Hedgehog and Notch. Cell Cycle 11(23):4344–4351
158. Avila JL, Kissil JL (2013) Notch signaling in pancreatic cancer: oncogene or tumor suppressor? Trends Mol Med 19(5):320–327
159. Su Q, Xin L (2016) Notch signaling in prostate cancer: refining a therapeutic opportunity. Histol Histopathol 31(2):149–157
160. Lino MM, Merlo A, Boulay JL (2010) Notch signaling in glioblastoma: a developmental drug target? BMC Med 8:72
161. Weijzen S, Rizzo P, Braid M et al (2002) Activation of Notch-1 signaling maintains the neoplastic phenotype in human Ras-transformed cells. Nat Med 8(9):979–986
162. Stylianou S, Clarke RB, Brennan K (2006) Aberrant activation of notch signaling in human breast cancer. Cancer Res 66(3):1517–1525
163. Ranganathan P, Weaver KL, Capobianco AJ (2011) Notch signalling in solid tumours: a little bit of everything but not all the time. Nat Rev Cancer 11(5):338–351
164. Koch U, Radtke F (2007) Notch and cancer: a double-edged sword. Cell Mol Life Sci 64 (21):2746–2762
165. Andersson ER, Lendahl U (2014) Therapeutic modulation of Notch signalling—are we there yet? Nat Rev Drug Discov 13(5):357–378
166. Gottlinger H, Johnson J, Riethmuller G (1986) Biochemical and epitope analysis of the 17-1A membrane antigen. Hybridoma 5(1):S29–S37
167. Herlyn M, Steplewski Z, Herlyn D et al (1979) Colorectal carcinoma-specific antigen: detection by means of monoclonal antibodies. Proc Natl Acad Sci USA 76(3):1438–1442
168. Schnell U, Cirulli V (1828) Giepmans BN (2013) EpCAM: structure and function in health and disease. Biochim Biophys Acta 8:1989–2001
169. Osta WA, Chen Y, Mikhitarian K et al (2004) EpCAM is overexpressed in breast cancer and is a potential target for breast cancer gene therapy. Cancer Res 64(16):5818–5824
170. Al-Hajj M, Wicha MS, Benito-Hernandez A et al (2003) Prospective identification of tumorigenic breast cancer cells. Proc Natl Acad Sci USA 100(7):3983–3988
171. Spizzo G, Went P, Dirnhofer S et al (2004) High Ep-CAM expression is associated with poor prognosis in node-positive breast cancer. Breast Cancer Res Treat 86(3):207–213
172. Hachmeister M, Bobowski KD, Hogl S et al (2013) Regulated intramembrane proteolysis and degradation of murine epithelial cell adhesion molecule mEpCAM. PLoS ONE 8(8): e71836
173. Denzel S, Maetzel D, Mack B et al (2009) Initial activation of EpCAM cleavage via cell-to-cell contact. BMC Cancer 9:402
174. Nubel T, Preobraschenski J, Tuncay H et al (2009) Claudin-7 regulates EpCAM-mediated functions in tumor progression. Mol Cancer Res 7(3):285–299
175. Kuhn S, Koch M, Nubel T et al (2007) A complex of EpCAM, claudin-7, CD44 variant isoforms, and tetraspanins promotes colorectal cancer progression. Mol Cancer Res 5 (6):553–567

176. van der Gun BT, Melchers LJ, Ruiters MH et al (2010) EpCAM in carcinogenesis: the good, the bad or the ugly. Carcinogenesis 31(11):1913–1921
177. Patriarca C, Macchi RM, Marschner AK et al (2012) Epithelial cell adhesion molecule expression (CD326) in cancer: a short review. Cancer Treat Rev 38(1):68–75
178. Gao J, Liu X, Yang F et al (2015) By inhibiting Ras/Raf/ERK and MMP-9, knockdown of EpCAM inhibits breast cancer cell growth and metastasis. Oncotarget 6(29):27187–27198
179. Gilboa-Geffen A, Hamar P, Le MT et al (2015) Gene knockdown by EpCAM Aptamer-siRNA chimeras suppresses epithelial breast cancers and their tumor-initiating cells. Mol Cancer Ther 14(10):2279–2291
180. Sankpal NV, Mayfield JD, Willman MW et al (2011) Activator protein 1 (AP-1) contributes to EpCAM-dependent breast cancer invasion. Breast Cancer Res 13(6):R124
181. Gao J, Yan Q, Liu S et al (2014) Knockdown of EpCAM enhances the chemosensitivity of breast cancer cells to 5-fluorouracil by downregulating the antiapoptotic factor Bcl-2. PLoS ONE 9(7):e102590
182. Ralhan R, Cao J, Lim T et al (2010) EpCAM nuclear localization identifies aggressive thyroid cancer and is a marker for poor prognosis. BMC Cancer 10:331
183. He HC, Kashat L, Kak I et al (2012) An Ep-ICD based index is a marker of aggressiveness and poor prognosis in thyroid carcinoma. PLoS ONE 7(9):e42893
184. Kunavisarut T, Kak I, Macmillan C et al (2012) Immunohistochemical analysis based Ep-ICD subcellular localization index (ESLI) is a novel marker for metastatic papillary thyroid microcarcinoma. BMC Cancer 12:523
185. Chopra A (2004) 64Cu-Labeled 1,4,7,10-tetraazacyclododecane-1,4,7,10-tetraacetic acid (DOTA)-conjugated IRDye 800CW (a near-infrared fluorescence dye) coupled to mAb7, an anti-epithelial cell adhesion molecule monoclonal antibody. In: Molecular imaging and contrast agent database (MICAD). Bethesda (MD)
186. Leung K (2004) 68 Ga-Labeled anti-EpCAM diabody against epithelial cell adhesion molecule. In: Molecular Imaging and Contrast Agent Database (MICAD). Bethesda (MD)
187. Leung K (2004) DiD-Labeled anti-EpCAM-directed NK-92-scFv(MOC31) zeta cells. In: Molecular imaging and contrast agent database (MICAD). Bethesda (MD)
188. Rybalov M, Ananias HJ, Hoving HD et al (2014) PSMA, EpCAM, VEGF and GRPR as imaging targets in locally recurrent prostate cancer after radiotherapy. Int J Mol Sci 15 (4):6046–6061
189. Kanwar JR, Roy K, Kanwar RK (2011) Chimeric aptamers in cancer cell-targeted drug delivery. Crit Rev Biochem Mol Biol 46(6):459–477
190. Flatmark K, Guldvik IJ, Svensson H et al (2013) Immunotoxin targeting EpCAM effectively inhibits peritoneal tumor growth in experimental models of mucinous peritoneal surface malignancies. Int J Cancer 133(6):1497–1506
191. Zhu B, Wu G, Robinson H et al (2013) Tumor margin detection using quantitative NIRF molecular imaging targeting EpCAM validated by far red gene reporter iRFP. Mol Imaging Biol 15(5):560–568
192. Reichardt L (2006) Neurotrophin-regulated signalling pathways. Philos Trans R Soc Lond B Biol Sci 361(1473):1545–1564
193. Roux P, Barker P (2002) Neurotrophin signaling through the p75 neurotrophin receptor. Prog Neurobiol 67(3):203–233
194. Hamanoue M, Middleton G, Wyatt S et al (1999) p75-mediated NF-kappaB activation enhances the survival response of developing sensory neurons to nerve growth factor. Mol Cell Neurosci 14(1):28–40
195. Roux P, Bhakar A, Kennedy T et al (2001) The p75 neurotrophin receptor activates Akt (protein kinase B) through a phosphatidylinositol 3-kinase-dependent pathway. J Biol Chem 276(25):23097–23104
196. Weskamp G, Schlöndorff J, Lum L et al (2004) Evidence for a critical role of the tumor necrosis factor alpha convertase (TACE) in ectodomain shedding of the p75 neurotrophin receptor (p75NTR). J Biol Chem 279(6):4241–4249

197. Kanning K, Hudson M, Amieux P et al (2003) Proteolytic processing of the p75 neurotrophin receptor and two homologs generates C-terminal fragments with signaling capability. J Neurosci: Official J Soc Neurosci 23(13):5425–5436
198. Jung K-M, Tan S, Landman N et al (2003) Regulated intramembrane proteolysis of the p75 neurotrophin receptor modulates its association with the TrkA receptor. J Biol Chem 278 (43):42161–42169
199. Parkhurst C, Zampieri N, Chao M (2010) Nuclear localization of the p75 neurotrophin receptor intracellular domain. J Biol Chem 285(8):5361–5368
200. Podlesniy P, Kichev A, Pedraza C et al (2006) Pro-NGF from Alzheimer's disease and normal human brain displays distinctive abilities to induce processing and nuclear translocation of intracellular domain of p75NTR and apoptosis. Am J Pathol 169(1):119–131
201. Kenchappa R, Zampieri N, Chao M et al (2006) Ligand-dependent cleavage of the P75 neurotrophin receptor is necessary for NRIF nuclear translocation and apoptosis in sympathetic neurons. Neuron 50(2):219–232
202. Domeniconi M, Zampieri N, Spencer T et al (2005) MAG induces regulated intramembrane proteolysis of the p75 neurotrophin receptor to inhibit neurite outgrowth. Neuron 46(6):849–855
203. Krygier S, Djakiew D (2002) Neurotrophin receptor p75(NTR) suppresses growth and nerve growth factor-mediated metastasis of human prostate cancer cells. Int J Cancer 98(1):1–7
204. Arrighi N, Bodei S, Zani D et al (2010) Nerve growth factor signaling in prostate health and disease. Growth Factors 28(3):191–201
205. Marchetti D, Aucoin R, Blust J et al (2004) p75 neurotrophin receptor functions as a survival receptor in brain-metastatic melanoma cells. J Cell Biochem 91(1):206–215
206. Marchetti D, Mrak R, Paulsen D et al (2007) Neurotrophin receptors and heparanase: a functional axis in human medulloblastoma invasion. J Exp Clin Cancer Res CR 26(1):5–23
207. Boiko A, Razorenova O, van de Rijn M et al (2010) Human melanoma-initiating cells express neural crest nerve growth factor receptor CD271. Nature 466(7302):133–137
208. Denkins Y, Reiland J, Roy M et al (2004) Brain metastases in melanoma: roles of neurotrophins. Neuro-oncology 6(2):154–165
209. Johnston A, Lun X, Rahn J et al (2007) The p 75 neurotrophin receptor is a central regulator of glioma invasion. PLoS Bio l5(8)
210. Wang X, Cui M, Wang L et al (2010) Inhibition of neurotrophin receptor p75 intramembran proteolysis by gamma-secretase inhibitor reduces medulloblastoma spinal metastasis. Biochem Biophys Res Commun 403(3–4):264–269
211. Forsyth PA, Krishna N, Lawn S et al (2014) p75 neurotrophin receptor cleavage by alpha- and gamma-secretases is required for neurotrophin-mediated proliferation of brain tumor-initiating cells. J Biol Chem 289(12):8067–8085
212. Descamps S, Toillon R, Adriaenssens E et al (2001) Nerve growth factor stimulates proliferation and survival of human breast cancer cells through two distinct signaling pathways. J Biol Chem 276(21):17864–17870
213. Verbeke S, Meignan S, Lagadec C et al (2010) Overexpression of p75(NTR) increases survival of breast cancer cells through p21(waf1). Cell Signal 22(12):1864–1873
214. Wang L, Rahn J, Lun X et al (2008) Gamma-secretase represents a therapeutic target for the treatment of invasive glioma mediated by the p 75 neurotrophin receptor. PLoS Biol 6(11)
215. Zhou J, Jain S, Azad AK et al (2016) Notch and TGFbeta form a positive regulatory loop and regulate EMT in epithelial ovarian cancer cells. Cell Signal 28(8):838–849
216. Dinicola S, Pasqualato A, Proietti S et al (2016) Paradoxical E-cadherin increase in 5FU-resistant colon cancer is unaffected during mesenchymal-epithelial reversion induced by gamma-secretase inhibition. Life Sci 145:174–183
217. Rosati E, Sabatini R, De Falco F et al (2013) gamma-Secretase inhibitor I induces apoptosis in chronic lymphocytic leukemia cells by proteasome inhibition, endoplasmic reticulum stress increase and notch down-regulation. Int J Cancer 132(8):1940–1953

218. Zou YH, Cao YQ, Wang LX et al (2011) γ-secretase inhibitor up-regulates vascular endothelial growth factor receptor-2 and endothelial nitric oxide synthase. Exp Ther Med 2 (4):725–729
219. Kalantari E, Saeidi H, Kia NS et al (2013) Effect of DAPT, a gamma secretase inhibitor, on tumor angiogenesis in control mice. Adv Biomed Res 2:83
220. Maraver A, Fernandez-Marcos PJ, Herranz D et al (2012) Therapeutic effect of gamma-secretase inhibition in KrasG12V-driven non-small cell lung carcinoma by derepression of DUSP1 and inhibition of ERK. Cancer Cell 22(2):222–234
221. Cullion K, Draheim KM, Hermance N et al (2009) Targeting the Notch1 and mTOR pathways in a mouse T-ALL model. Blood 113(24):6172–6181
222. Engin F, Bertin T, Ma O et al (2009) Notch signaling contributes to the pathogenesis of human osteosarcomas. Hum Mol Genet 18(8):1464–1470
223. Meng RD, Shelton CC, Li Y-M et al (2009) γ-secretase inhibitors abrogate oxaliplatin-induced activation of the Notch-1 signaling pathway in colon cancer cells resulting in enhanced chemosensitivity. Can Res 69(2):573–582
224. Yuan X, Wu H, Xu H et al (2015) Notch signaling: an emerging therapeutic target for cancer treatment. Cancer Lett 369(1):20–27
225. Kumar R, Juillerat-Jeanneret L, Golshayan D (2016) Notch antagonists: potential modulators of cancer and inflammatory diseases. J Med Chem
226. De Jesus-Acosta A, Laheru D, Maitra A et al (2014) A phase II study of the gamma secretase inhibitor RO4929097 in patients with previously treated metastatic pancreatic adenocarcinoma. Invest New Drugs 32(4):739–745
227. Lee SM, Moon J, Redman BG et al (2015) Phase 2 study of RO4929097, a gamma-secretase inhibitor, in metastatic melanoma: SWOG 0933. Cancer 121(3):432–440
228. Papayannidis C, DeAngelo DJ, Stock W et al (2015) A Phase 1 study of the novel gamma-secretase inhibitor PF-03084014 in patients with T-cell acute lymphoblastic leukemia and T-cell lymphoblastic lymphoma. Blood Cancer J 5:e350
229. Messersmith WA, Shapiro GI, Cleary JM et al (2015) A Phase I, dose-finding study in patients with advanced solid malignancies of the oral gamma-secretase inhibitor PF-03084014. Clin Cancer Res 21(1):60–67
230. Fouladi M, Stewart CF, Olson J et al (2011) Phase I trial of MK-0752 in children with refractory CNS malignancies: a pediatric brain tumor consortium study. J Clin Oncol 29 (26):3529–3534
231. Krop I, Demuth T, Guthrie T et al (2012) Phase I pharmacologic and pharmacodynamic study of the gamma secretase (Notch) inhibitor MK-0752 in adult patients with advanced solid tumors. J Clin Oncol 30(19):2307–2313
232. Piha-Paul SA, Munster PN, Hollebecque A et al (2015) Results of a phase 1 trial combining ridaforolimus and MK-0752 in patients with advanced solid tumours. Eur J Cancer 51 (14):1865–1873
233. Hoffman LM, Fouladi M, Olson J et al (2015) Phase I trial of weekly MK-0752 in children with refractory central nervous system malignancies: a pediatric brain tumor consortium study. Childs Nerv Syst 31(8):1283–1289
234. LoConte NK, Razak AR, Ivy P et al (2015) A multicenter phase 1 study of gamma -secretase inhibitor RO4929097 in combination with capecitabine in refractory solid tumors. Invest New Drugs 33(1):169–176
235. Diaz-Padilla I, Hirte H, Oza AM et al (2013) A phase Ib combination study of RO4929097, a gamma-secretase inhibitor, and temsirolimus in patients with advanced solid tumors. Invest New Drugs 31(5):1182–1191
236. Richter S, Bedard PL, Chen EX et al (2014) A phase I study of the oral gamma secretase inhibitor R04929097 in combination with gemcitabine in patients with advanced solid tumors (PHL-078/CTEP 8575). Invest New Drugs 32(2):243–249

237. Tolcher AW, Messersmith WA, Mikulski SM et al (2012) Phase I study of RO4929097, a gamma secretase inhibitor of Notch signaling, in patients with refractory metastatic or locally advanced solid tumors. J Clin Oncol 30(19):2348–2353
238. McAuliffe SM, Morgan SL, Wyant GA et al (2012) Targeting Notch, a key pathway for ovarian cancer stem cells, sensitizes tumors to platinum therapy. Proc Natl Acad Sci USA 109(43):E2939–E2948
239. Singh A, Zapata MC, Choi YS et al (2014) GSI promotes vincristine-induced apoptosis by enhancing multi-polar spindle formation. Cell Cycle 13(1):157–166
240. Woorons X, Mollard P, Pichon A et al (2008) Effects of a 4-week training with voluntary hypoventilation carried out at low pulmonary volumes. Respir Physiol Neurobiol 160 (2):123–130
241. Arasada RR, Amann JM, Rahman MA et al (2014) EGFR blockade enriches for lung cancer stem-like cells through Notch3-dependent signaling. Cancer Res 74(19):5572–5584
242. Li LC, Wang DL, Wu YZ et al (2015) Gastric tumor-initiating CD44+ cells and epithelial-mesenchymal transition are inhibited by gamma-secretase inhibitor DAPT. Oncol Lett 10(5):3293–3299
243. Ni J, Cozzi P, Hao J et al (2013) Epithelial cell adhesion molecule (EpCAM) is associated with prostate cancer metastasis and chemo/radioresistance via the PI3K/Akt/mTOR signaling pathway. Int J Biochem Cell Biol 45(12):2736–2748
244. Yahyanejad S, Theys J, Vooijs M (2016) Targeting Notch to overcome radiation resistance. Oncotarget 7(7):7610–7628
245. Huang X, Qian Y, Wu H et al (2015) Aberrant expression of osteopontin and E-cadherin indicates radiation resistance and poor prognosis for patients with cervical carcinoma. J Histochem Cytochem 63(2):88–98
246. Taylor IC, Hutt-Cabezas M, Brandt WD et al (2015) Disrupting Notch slows diffuse intrinsic pontine glioma growth, enhances radiation sensitivity, and shows combinatorial efficacy with bromodomain inhibition. J Neuropathol Exp Neurol 74(8):778–790
247. Mizugaki H, Sakakibara-Konishi J, Ikezawa Y et al (2012) γ-secretase inhibitor enhances antitumour effect of radiation in Notch-expressing lung cancer. Br J Cancer 106(12):1953–1959
248. Vermezovic J, Adamowicz M, Santarpia L et al (2015) Notch is a direct negative regulator of the DNA-damage response. Nat Struct Mol Biol 22(5):417–424
249. Panaccione A, Chang MT, Carbone BE et al (2016) NOTCH1 and SOX10 are essential for proliferation and radiation resistance of cancer stem-like cells in adenoid cystic carcinoma. Clin Cancer Res 22(8):2083–2095
250. Debeb BG, Cohen EN, Boley K et al (2012) Pre-clinical studies of Notch signaling inhibitor RO4929097 in inflammatory breast cancer cells. Breast Cancer Res Treat 134(2):495–510
251. Jowett PL, Nicohlson SS, Gamble GA (1986) Tissue levels of atrazine in a case of bovine poisoning. Vet Hum Toxicol 28(6):539–540
252. Kanu OO, Hughes B, Di C et al (2009) Glioblastoma multiforme oncogenomics and signaling pathways. Clin Med Oncol 3:39–52
253. Kanu OO, Mehta A, Di C et al (2009) Glioblastoma multiforme: a review of therapeutic targets. Expert Opin Ther Targets 13(6):701–718
254. Yahyanejad S, King H, Iglesias VS et al (2016) NOTCH blockade combined with radiation therapy and temozolomide prolongs survival of orthotopic glioblastoma. Oncotarget
255. Natsumeda M, Maitani K, Liu Y et al (2015) Targeting Notch signaling and autophagy increases cytotoxicity in glioblastoma neurospheres. Brain Pathol
256. Lundy EG, Sorokin CF, Meltz SK et al (1977) Reversible inhibition of human peripheral lymphocyte DNA synthesis by an extract of breast cancer cell line SKBR-3. J Surg Res 22 (6):654–659
257. Han J, Shen Q (2012) Targeting gamma-secretase in breast cancer. Breast Cancer 4:83–90
258. Demehri S, Turkoz A, Kopan R (2009) Epidermal Notch1 loss promotes skin tumorigenesis by impacting the stromal microenvironment. Cancer Cell 16(1):55–66

259. Cheng YL, Choi Y, Sobey CG et al (2015) Emerging roles of the gamma-secretase-notch axis in inflammation. Pharmacol Ther 147:80–90
260. Saura CA (2010) Presenilin/γ-secretase and inflammation. Front Aging Neurosci 2:16
261. Tournoy J, Bossuyt X, Snellinx A et al (2004) Partial loss of presenilins causes seborrheic keratosis and autoimmune disease in mice. Hum Mol Genet 13(13):1321–1331
262. Li T, Wen H, Brayton C et al (2007) Moderate reduction of gamma-secretase attenuates amyloid burden and limits mechanism-based liabilities. J Neurosci 27(40):10849–10859
263. Maraver A, Tadokoro CE, Badura ML et al (2007) Effect of presenilins in the apoptosis of thymocytes and homeostasis of CD8+T cells. Blood 110(9):3218–3225
264. Yagi T, Giallourakis C, Mohanty S et al (2008) Defective signal transduction in B lymphocytes lacking presenilin proteins. Proc Natl Acad Sci USA 105(3):979–984
265. Beglopoulos V, Sun X, Saura CA et al (2004) Reduced β-amyloid production and increased inflammatory responses in presenilin conditional knock-out mice. J Biol Chem 279 (45):46907–46914
266. Dong S, Li C, Wu P et al (2007) Environment enrichment rescues the neurodegenerative phenotypes in presenilins-deficient mice. Eur J Neurosci 26(1):101–112
267. Jiang X, Zhang D, Shi J et al (2009) Increased inflammatory response both in brain and in periphery in presenilin 1 and presenilin 2 conditional double knock-out mice. J Alzheimers Dis 18(3):515–523
268. Jayadev S, Case A, Eastman AJ et al (2010) Presenilin 2 is the predominant γ-secretase in microglia and modulates cytokine release. PLoS ONE 5(12):e15743
269. Parks WC, Wilson CL, Lopez-Boado YS (2004) Matrix metalloproteinases as modulators of inflammation and innate immunity. Nat Rev Immunol 4(8):617–629
270. Garton KJ, Gough PJ, Raines EW (2006) Emerging roles for ectodomain shedding in the regulation of inflammatory responses. J Leukoc Biol 79(6):1105–1116
271. Murphy G, Murthy A, Khokha R (2008) Clipping, shedding and RIPping keep immunity on cue. Trends Immunol 29(2):75–82
272. Lleo A, Saura CA (2011) gamma-secretase substrates and their implications for drug development in Alzheimer's disease. Curr Top Med Chem 11(12):1513–1527
273. Carter C (2011) Alzheimer's disease: APP, Gamma Secretase, APOE, CLU, CR1, PICALM, ABCA7, BIN1, CD2AP, CD33, EPHA1, and MS4A2, and their relationships with herpes simplex, C. Pneumoniae, Other suspect pathogens, and the immune system. Int J Alzheimers Dis 2011:501862
274. Garlind A, Brauner A, Hojeberg B et al (1999) Soluble interleukin-1 receptor type II levels are elevated in cerebrospinal fluid in Alzheimer's disease patients. Brain Res 826(1):112–116
275. Chalaris A, Gewiese J, Paliga K et al (1803) ADAM17-mediated shedding of the IL6R induces cleavage of the membrane stub by gamma-secretase. Biochim Biophys Acta 2:234–245 (2010)
276. Glenn G, van der Geer P (2008) Toll-like receptors stimulate regulated intramembrane proteolysis of the CSF-1 receptor through Erk activation. FEBS Lett 582(6):911–915
277. Agrawal V, Sawhney N, Hickey E et al (2015) Loss of presenilin 2 function is associated with defective LPS-mediated innate immune responsiveness. Mol Neurobiol 53(5):3428–3438
278. Hickman SE, El Khoury J (2014) TREM2 and the neuroimmunology of Alzheimer's disease. Biochem Pharmacol 88(4):495–498
279. Wang Y, Ulland TK, Ulrich JD et al (2016) TREM2-mediated early microglial response limits diffusion and toxicity of amyloid plaques. J Exp Med 213(5):667–675
280. Yuan P, Condello C, Keene CD et al (2016) TREM2 haplodeficiency in mice and humans impairs the microglia barrier function leading to decreased amyloid compaction and severe axonal dystrophy. Neuron 90(4):724–739

281. Glebov K, Wunderlich P, Karaca I et al (2016) Functional involvement of γ-secretase in signaling of the triggering receptor expressed on myeloid cells-2 (TREM2). J Neuroinflammation 13(1):1–7
282. Schulte A, Schulz B, Andrzejewski MG et al (2007) Sequential processing of the transmembrane chemokines CX3CL1 and CXCL16 by alpha- and gamma-secretases. Biochem Biophys Res Commun 358(1):233–240
283. Mambole A, Baruch D, Nusbaum P et al (2008) The cleavage of neutrophil leukosialin (CD43) by cathepsin G releases its extracellular domain and triggers its intramembrane proteolysis by presenilin/gamma-secretase. J Biol Chem 283(35):23627–23635
284. Pelletier L, Guillaumot P, Freche B et al (2006) Gamma-secretase-dependent proteolysis of CD44 promotes neoplastic transformation of rat fibroblastic cells. Cancer Res 66(7):3681–3687
285. Fukumoto N, Shimaoka T, Fujimura H et al (2004) Critical roles of CXC chemokine ligand 16/scavenger receptor that binds phosphatidylserine and oxidized lipoprotein in the pathogenesis of both acute and adoptive transfer experimental autoimmune encephalomyelitis. J Immunol 173(3):1620–1627
286. Jong A, Wu CH, Shackleford GM et al (2008) Involvement of human CD44 during Cryptococcus neoformans infection of brain microvascular endothelial cells. Cell Microbiol 10(6):1313–1326
287. Bacsa S, Karasneh G, Dosa S et al (2011) Syndecan-1 and syndecan-2 play key roles in herpes simplex virus type-1 infection. J Gen Virol 92(Pt 4):733–743
288. de Witte L, Bobardt M, Chatterji U et al (2007) Syndecan-3 is a dendritic cell-specific attachment receptor for HIV-1. Proc Natl Acad Sci 104(49):19464–19469
289. Schulz JG, Annaert W, Vandekerckhove J et al (2003) Syndecan 3 intramembrane proteolysis is presenilin/gamma-secretase-dependent and modulates cytosolic signaling. J Biol Chem 278(49):48651–48657
290. Persson BD, Schmitz NB, Santiago C et al (2010) Structure of the extracellular portion of CD46 provides insights into its interactions with complement proteins and pathogens. PLoS Pathog 6(9):e1001122
291. Weyand NJ, Calton CM, Higashi DL et al (2010) Presenilin/gamma-secretase cleaves CD46 in response to Neisseria infection. J Immunol 184(2):694–701
292. Tsao PN, Wei SC, Huang MT et al (2011) Lipopolysaccharide-induced Notch signaling activation through JNK-dependent pathway regulates inflammatory response. J Biomed Sci 18:56
293. Zhang W, Zhang X, Sheng A, Weng C, Zhu T, Zhao W, Li C (2015) γ-secretase inhibitor alleviates acute airway inflammation of allergic asthma in mice by downregulating Th17 cell differentiation. Mediators of Inflammation 2015 (2015):258168
294. Varfolomeev E, Goncharov T, Vucic D (2015) Roles of c-IAP proteins in TNF receptor family activation of NF-kappaB signaling. Methods Mol Biol 1280:269–282
295. Muppidi JR, Tschopp J, Siegel RM (2004) Life and death decisions: secondary complexes and lipid rafts in TNF receptor family signal transduction. Immunity 21(4):461–465
296. Tchikov V, Bertsch U, Fritsch J et al (2011) Subcellular compartmentalization of TNF receptor-1 and CD95 signaling pathways. Eur J Cell Biol 90(6–7):467–475
297. Schneider-Brachert W, Heigl U, Ehrenschwender M (2013) Membrane trafficking of death receptors: implications on signalling. Int J Mol Sci 14(7):14475–14503
298. Cabal-Hierro L, Lazo PS (2012) Signal transduction by tumor necrosis factor receptors. Cell Signal 24(6):1297–1305
299. Irannejad R, von Zastrow M (2014) GPCR signaling along the endocytic pathway. Curr Opin Cell Biol 27:109–116
300. Tsvetanova NG, Irannejad R, von Zastrow M (2015) G protein-coupled receptor (GPCR) signaling via heterotrimeric G proteins from endosomes. J Biol Chem 290(11):6689–6696
301. Saleh AZ, Fang AT, Arch AE et al (2004) Regulated proteolysis of the IFNaR2 subunit of the interferon-alpha receptor. Oncogene 23(42):7076–7086

302. Hemming ML, Elias JE, Gygi SP et al (2008) Proteomic profiling of ?-secretase substrates and mapping of substrate requirements. PLoS Biol 6(10):e257
303. Carey BW, Kim DY, Kovacs DM (2007) Presenilin/gamma-secretase and alpha-secretase-like peptidases cleave human MHC Class I proteins. Biochem J 401 (1):121–127
304. Saxena MT, Schroeter EH, Mumm JS et al (2001) Murine notch homologs (N1-4) undergo presenilin-dependent proteolysis. J Biol Chem 276(43):40268–40273
305. Okochi M, Steiner H, Fukumori A et al (2002) Presenilins mediate a dual intramembranous γ-secretase cleavage of Notch-1. EMBO J 21(20):5408–5416
306. Shimizu K, Chiba S, Hosoya N et al (2000) Binding of Delta1, Jagged1, and Jagged2 to Notch2 rapidly induces cleavage, nuclear translocation, and hyperphosphorylation of Notch2. Mol Cell Biol 20(18):6913–6922
307. Wang H, Li ZY, Liu Y et al (2011) Desmoglein 2 is a receptor for adenovirus serotypes 3, 7, 11 and 14. Nat Med 17(1):96–104
308. May P, Reddy YK, Herz J (2002) Proteolytic processing of low density lipoprotein receptor-related protein mediates regulated release of its intracellular domain. J Biol Chem 277(21):18736–18743
309. Hoe HS, Rebeck GW (2005) Regulation of ApoE receptor proteolysis by ligand binding. Brain Res Mol Brain Res 137(1–2):31–39
310. Lee HJ, Jung KM, Huang YZ et al (2002) Presenilin-dependent gamma-secretase-like intramembrane cleavage of ErbB4. J Biol Chem 277(8):6318–6323
311. Tuffereau C, Bénéjean J, Blondel D et al (1998) Low-affinity nerve-growth factor receptor (P75NTR) can serve as a receptor for rabies virus. The EMBO J 17(24):7250–7259
312. Kim DY, Ingano LA, Kovacs DM (2002) Nectin-1alpha, an immunoglobulin-like receptor involved in the formation of synapses, is a substrate for presenilin/gamma-secretase-like cleavage. J Biol Chem 277(51):49976–49981
313. Tousseyn T, Thathiah A, Jorissen E et al (2009) ADAM10, the rate-limiting protease of regulated intramembrane proteolysis of Notch and other proteins, is processed by ADAMS-9, ADAMS-15, and the γ-secretase. J Biol Chem 284(17):11738–11747
314. Wilke GA, Bubeck Wardenburg J (2010) Role of a disintegrin and metalloprotease 10 in Staphylococcus aureus alpha-hemolysin-mediated cellular injury. Proc Natl Acad Sci USA 107(30):13473–13478

The Role of Urinary Proteases in Bladder Cancer

Paulo Bastos, Sandra Magalhães, Lúcio Lara Santos, Rita Ferreira and Rui Vitorino

Abstract

Bladder cancer (BCa) is one of the most prevalent malignancies worldwide. Risk factors for BCa are well established and include smoking and infections, which can lead to immune system activation, altered gene expression patterns, proteolytic activity, tissue damage, and, ultimately, cancer development. Urine has become one of the most attractive diagnosis samples, and, notably, urine profiling by mass spectrometry allows the simultaneously analysis of multiple enzymes and their interactors, substrates, inhibitors, and regulators, providing an integrative view of enzymatic dynamics. Most BCa-associated enzymatic alterations take place at the level of proteases, being MMP-9, MMP-2, urokinase-type plasminogen activator, cathepsin D, and cathepsin G already related to BCa development and progression. Herein, we overview the role of

Paulo Bastos and Sandra Magalhães are equally contributed.

P. Bastos · S. Magalhães · R. Vitorino (✉)
iBiMED—Institute for Research in Biomedicine, Department of Medical Sciences, University of Aveiro, Aveiro, Portugal
e-mail: rvitorino@ua.pt

S. Magalhães · R. Ferreira
QOPNA, Mass Spectrometry Center, Department of Chemistry, University of Aveiro, Aveiro, Portugal

L.L. Santos
Experimental Pathology and Therapeutics Group—Research Center, Portuguese Oncology Institute-Porto (IPO-Porto), Porto, Portugal

R. Vitorino
Department of Physiology and Cardiothoracic Surgery, Faculty of Medicine, University of Porto, Porto, Portugal

S. Chakraborti and N.S. Dhalla (eds.), *Pathophysiological Aspects of Proteases*,
DOI 10.1007/978-981-10-6141-7_4

proteases and the classes more studied in BCa pathogenesis, as well as the methodologies used for assessing protease amount and activity in urine samples, highlighting its advantages and limitations, and the value of urinary proteases as disease biomarkers.

1 Introduction

Bladder cancer (BCa) is the ninth most common cancer worldwide with an estimated incidence above 400,000 new cases *per* year. BCa incidence rises with aging and is more frequent in men, which account to be the seventh most common cancer in this gender [1]. In women, BCa is the seventeenth most common type of malignant disease [1]. Whereas men are three to four times more likely to develop BCa, women have worse prognosis [2, 3]. This sexual dimorphism may be related to differences in the exposure to carcinogens such as tobacco or to genetic, hormonal, and anatomical factors [2]. There are several risk factors involved in the pathogenesis of this disease [4]. According to World Health Organization (WHO), cigarette smoking and exposure to aromatic amines are the most important risk factors [5, 6]. In fact, the risk of developing BCa in smokers is about 2–6 times higher compared to nonsmokers [5]. In turn, occupational exposure to carcinogens like aromatic amines is responsible for 25% of all BCa cases [5–7].

BCa is more frequent in developed countries where, in 2012, approximately 10 among every 100,000 individuals were diagnosed with BCa. Among developed regions, higher incidence rates have been reported in both Europe and North America [1]. As a consequence of its high incidence, prevalence, and aggressiveness, BCa is one of the most expensive diseases to treat and to manage [8–11]. Most of this economic burden results from the high number of invasive procedures performed (e.g., cystoscopies) and from the several hospitalizations required for its management [8–11]. Taken together, incidence, mortality, prevalence, and costs of BCa make this disease a public health problem, even though it is not officially regarded as such.

Currently, BCa is classified as a multifactorial disease that can be divided into three histological subtypes: urothelial cell carcinomas, adenocarcinomas, and squamous cell carcinomas [5, 12]. In developed countries, namely USA, France, and Italy, urothelial cell carcinomas constitute about 90% of BCa cases, while the incidence of the remaining types of BCa is much lower, ranging from 1.1 to 2.8% for squamous cell carcinomas and from 1.5 to 1.9% in the case of adenocarcinomas [5]. Urothelial cell carcinomas can be classified into low-grade BCa and high-grade BCa, a classification that is based on the histological analysis of the bladder. Low-grade BCa is less aggressive and rarely invades bladder muscle tissue or metastasizes to other parts of the body, and patients rarely die from it. On the other hand, high-grade BCa commonly invades muscle wall and metastasizes, therefore requiring more aggressive treatments. Consequently, high-grade BCa is responsible for almost all BCa-related deaths [7]. Another way to classify BCa is based on the invasion of the bladder muscular layer. Nonmuscle-invasive BCa is often treated by the surgical removing the tumors, sometimes combined with localized chemotherapy [7, 13]. Patients with nonmuscle-invasive BCa have a 5-year survival rate ranging from 82 to 100% for early-stage BCa. On the other hand, muscle-invasive BCa is much more aggressive, and treatment involves surgical removal of the bladder as well as aggressive radiotherapy and chemotherapy treatments [7, 13]. Even with the complications associated with surgery, surgical intervention for tumor removal remains the most effective treatment, highlighting the lack of specific therapeutic targets and effective pharmacological agents. However, even complete resections are followed by recurrence rates of 70% for superficial bladder tumors, and intravesical chemotherapy can attenuate this extremely high recurrence rate by only 15% at most. Nevertheless, this type of BCa has a poorer prognosis, so that only 5–10% of patients with metastasis live more than 2 years after diagnosis [14].

The proportion of squamous cell and transitional cell bladder carcinomas has been changing, at least in some populations, most likely due to increased exposure to distinct etiologic factors, especially smoking [15]. In areas where schistosomiasis is endemic (e.g., Sudan, Egypt), squamous cell carcinoma accounts for 75% of all malignant bladder tumors. However, the increased smoking prevalence has been responsible for a significant shift toward urothelial/transitional cell carcinoma (six-fold increase over squamous cell carcinoma) [15, 16]. This observation has important repercussions as urothelial/transitional cell carcinoma patients are poorer candidates for surgical intervention and present increased risk of recurrence [16]. In addition, other risk factors for developing BCa include family history of the disease, genetic mutations in *HRAS*, *RB1*, *PTEN/MMAC1*, *NAT2*, and *GSTM1* genes, exposure to arsenic, and certain drugs used in chemotherapy like cyclophosphamide [5, 7].

There are several pathways involved in bladder tumorigenesis. At a molecular level, it is known that mutations in *FGFR3*, *TP53*, *RB1*, *ERBB2, and PTEN* genes are involved in bladder carcinogenesis [17, 18]. Also, some oncogenic miRNAs are upregulated, and tumor suppressor miRNAs are downregulated in BCa patients [17]. At the metabolome level, like in other types of cancer, the Warburg effect contributes to BCa development, progression, and aggressiveness [19]. Neoplastic cells have a strong dependence on glycolysis in conditions of normoxia, and BCa cells display an upregulation of genes involved in glucose uptake, like GLUT1 and GLUT3 transporters and Akt, which promotes the PI3 K/Akt/mTOR pathway [19]. There is also an overexpression of pyruvate-related and pentose phosphate pathway-related genes, and glycogen metabolism is enhanced during BCa development. Simultaneously, there are alterations in lipid metabolism, such as increased fatty acid synthesis and oxidation, changes in the tricarboxylic acid cycle and ketogenesis activity [19], as well as the hyper activation of proteases, which may lead to tissue damage and invasion, immune system and apoptosis evasion, tumorigenesis and cancer metastasis [20, 21].

Early detection of BCa is not always easy, since BCa is commonly asymptomatic at the early stages. Also, the first sign of BCa is usually painless hematuria, which is shared by other diseases [5, 12, 18]. Moreover, other symptoms may include abdominal pain, fatigue, weight loss, and urinary frequency and urgency. Thus, diagnosis of BCa can be performed by ultrasounds, computed tomography scans, and noninvasive urine cytology [5, 12]. Urinary cytology has the advantage of being a noninvasive approach, but despite its high specificity, it lacks sensitivity to detect low-grade tumors, so that cystoscopy is always done to corroborate the diagnosis [5, 12].

One of the biggest and most promising challenges in BCa research is the development of a diagnostic tool based on biofluids' profiling. Not surprisingly, urine is the most promising biological fluid, once it can be collected in high amounts by a noninvasive and simple procedure [12]. Also, in the case of BCa, urine better reflects alterations taking place in the bladder because it is in direct contact with the tissue [12]. Several studies have explored the diagnosis value of biomarkers for urine assay in BCa [22–25]. Among the identified putative biomarkers are a few proteases, reflecting the role of proteolytic systems in carcinogenesis and metastization [22, 23]. However, the huge interplay between multiple proteases and its inhibitors makes the assay of proteases in urine complex and unpredictable. In this chapter, we will explore the role of proteases in BCa pathophysiology and how these enzymes might be assessed in biofluids for the diagnosis and management of this disease.

2 An Update of the Existing Biomarkers for the Clinical Management of BCa

Currently, there are three FDA-approved urinary tests for the screening and surveillance of BCa. NMP22 Bladder Check Test® is an enzyme immunoassay for the detection of NMP22, a nuclear matrix protein recognized as an urothelial-specific cancer biomarker [26, 27]. This test has the advantage of being inexpensive, rapid, and operator-independent. However, it leads to a high number of false positives [23, 27]. The FDP test is another FDA-approved one that detects increased urinary concentration of fibrin/fibrinogen degradation products (FDPs), which are associated with malignant tumors [23, 27]. However, the NMP22 test lacks sensitivity (68%) and displays a high rate of false positives, particularly in patients with hematuria. So, it was recently removed from the market [23]. Lastly, BTA stat and BTA-TRAK are two tests that detect the bladder tumor antigen (BTA) in urine. BTA is the human complement factor H-related protein, and it is produced by human BCa cells but not by normal epithelial cells [26]. These tests were approved for surveillance only, since both display a high rate of false positives. Despite its high sensitivity for low-grade lesions, the sensitivity of BTA tests for high-grade lesions is worse than cystoscopy [23]. Even so, it is possible to improve the sensitivity of these tests by combining them, but the rate of false positives remains high [26].

Therefore, efforts continue to be made for the identification of biomarkers that help to improve the clinical management of BCa. Still, there is a general awareness that it will be difficult, if not impossible, to identify a single biomarker with high specificity and sensitivity (a golden bullet). Consequently, emphasis has been given to the combination of multiple protein markers that when analyzed together in a multimarker panel may allow the diagnosis of BCa and improve the clinical management of this type of cancer. Considering the role of the proteolytic systems in bladder carcinogenesis and metastization, proteases might be seen as markers to be included in such diagnosis panels. Because proteolytic activity is regulated at multiple levels, including gene expression, enzyme compartmentalization, regulation by modulators and degradation [28], the biological significance of urinary proteases, and quantitative alterations are difficult to interpret and not readily predictable. The urinary levels of primarily intracellularly acting proteases may not accurately reflect local mRNA expression since these proteases may be enriched in the tissues and poorly secreted in urine [29–40]. In turn, urinary levels of highly secreted proteases are subjected to the influence of multiple factors including the rate of secretion, tumor burden, and transcription regulation [41–43]. As such, the exploitation of quantitative changes in the urinary protease levels has revealed a far more challenging avenue than initially envisioned. Still, proteomics appears to be the most promising approach to characterize urinary proteases and promises to allow BCa screening, diagnosis, monitoring, therapy management, and the identification of novel pharmacological targets.

3 Urine Proteases Profiling for Biomarker Screening

When studying proteins and peptides in urine samples, different enrichment procedures are usually applied, given different but complementary results. The identification and characterization of putative urinary biomarkers may be performed with gel-based or gel-free approaches (Table 1). Gel-based approaches usually involve protein separation by 1D/2D Sodium Dodecyl Sulfate-Polyacrylamide Gel Electrophoresis (SDS-PAGE), whereas gel-free approaches include capillary electrophoresis (CE) or liquid chromatography (LC). Nevertheless, gel-free and gel-based approaches might be combined with each other into a strategy known as GeLC. Starting from a complex sample such as urine, 1D/2D SDS-PAGE allows accurate separation of proteins according to their molecular weight and/or isoelectric point, resolving hundreds of proteins, then subjected to MS-based analysis, immunoblot detection, and/or zymography-based activity assays (Table 1) [29–31, 34, 36–38]. One of the main shortcomings of gel-based approaches results from the high-salt concentrations in urine (mainly urea), as well as the presence of lipids and glycosaminoglycans, which interfere with proteomic analysis (particularly during isoelectric focusing in 2D SDS-PAGE). Despite the versatility of gel-based approaches, these may become daunting when dealing with complex samples, particularly for high-throughput analyses. Consequently, LC has become the platform of choice for protein separation from urine samples of patients with BCa, providing better resolution and allowing easier sample manipulation (Table 1) [30, 32, 33, 35–37]. Furthermore, distinct chromatography-based techniques such as dual-lectin affinity, strong-cation exchange, and reverse-phase chromatography exploit unique physicochemical properties, which allow the enrichment of particular urinary subproteomes and the identification of novel enzymes as well as enzyme inhibitors [24, 32]. Similar to gel-based strategies, gel-free approaches can also be followed by MS analysis (with or without prior digestion), immunoblotting, and/or zymography (Table 1).

Table 1 Procedures employed in bladder cancer urine proteomics

Procedure	Distinguishing features	References
1D/2D sodium dodecyl sulfate-Polyacrylamide gel electrophoresis	High-throughput technique (hundreds to thousands of proteins) Useful for highly abundant and soluble proteins, allowing the separation of complex mixtures Very acid or basic proteins can be accurately resolved Allows bands/spots excised for MS analysis Lacks dynamic range, reproducibility, and sensitivity Bad for poorly abundant (enzyme inhibitors) or poorly soluble (membrane) proteins Laboriously coupled to MS analyzers	[29–31, 34, 36–38]
Capillary electrophoresis	Allows the study of naturally occurring peptides Small amounts of sample required (nL) Very high speed (high voltage allowed) and resolution without band broadening due to high surface area and heat dissipation Easily coupled to MS analyzers	[55–58]
Zymography	Detects enzymatic activity Allows for multiple enzymes/substrates monitoring simultaneously Allows prior molecular weight- and isoelectric point-based protein resolution, tissue sample in situ *zymography*, and analyses using protease-activated fluorogenic probes that allow the monitoring of multiple MPPs/fluorophores and the mapping of MMPs' activity	[48–53]

(continued)

Table 1 (continued)

Procedure	Distinguishing features	References
Chromatographic separations	High-resolution techniques Separation based on several characteristics (mass, volume, isoelectric point, hydrophobicity) Susceptible to ion suppression by the presence of salts High retention (e.g., column-to-column) time variability Easily coupled to MS analyzers	[30, 32, 33, 35–37]
Western blot and other immunological assays	Allows relative and absolute quantification of multiple proteins High resolution and sensitivity Targeted approach High costs Targeted (misses much information) Does not separate isoforms Requires previous knowledge of its targets	[30, 36–38]

3.1 Immunoaffinity Assays

Immunoaffinity or antibody-based approaches can be used either to isolate intended targets or to deplete unwanted ones, and these techniques have been exploited during all steps of urinary biomarker discovery, from targeted sample depletion and protein isolation (e.g., immunoprecipitation) to biomarker validation (e.g., Western blot, Dot blot) [30, 33, 44]. In addition to be highly sensitive and specific, immunoaffinity assays are, above all, the methods of choice required for biomarker validation. Particularly, as will be discussed in the subsequent sections, immunoaffinity assays such as Western blot have been successfully used for biomarker validation using urine samples from BCa patients, including the proteases cathepsin D, cathepsin G, cathepsin L, matrix metalloproteinase-2 (MMP-2), matrix metalloproteinase-9 (MMP-9), a disintegrin and metalloproteinase with thrombospondin motifs 7 (ADAMTS7), metalloproteinase inhibitors (e.g., TIMP1), among others.

In recent years, classical blotting techniques have started to be replaced by surface plasmon resonance imaging, which employ antibody arrays for the multiplexed detection of protein biomarkers [45, 46]. This technology has been applied to the detection of BCa in its early stages, through the assessment of cathepsin D/protein ratio in both serum and urine [47].

3.2 Zymography

Zymography has assumed a leading role when studying urine proteases because it allows protein resolution based on molecular weight and isoelectric point, enzymatic activity analysis, and the screening of substrates and enzyme inhibitors. Despite this versatility, zymography-based studies of proteolytic alterations in BCa patients have relied almost exclusively on PAGE-based resolution of urinary proteins, incubation with a substrate more or less specific for a group of proteases (e.g., gelatin zymography to access the activity of MMP-2 and MMP-9), and estimation of enzymatic activity by substrate hydrolysis quantification [48–50]. Also, zymography allows multiple analyses to be performed, either simultaneously or sequentially, thus largely expanding the repertoire of possibilities for analysis [51, 52].

By analyzing enzymatic activity levels, when applied to specimens collected from BCa patients, zymography has provided biomarker-based models with specificities of up to 100%, identifying molecular players not amenable to be monitored by other techniques [48–50, 53]. The concentration of urinary proteases may not accurately reflect proteolytic activity [48, 49]. Notably, zymography allows to study and to quantify enzymatic activities in addition to enzymatic concentrations, thus providing information regarding quantitative changes as well as qualitative alterations. As such, zymography becomes an invaluable tool for the analysis of functional aspects, as seen for BCa where enzymatic trafficking is impaired and protease/inhibitor complexes or protease dimers may be frequent, with repercussions to the proteolytic function [48–50, 53].

3.3 Peptidomics

The screen of native peptides (peptidomics) in urine, particularly polar, charged, or chargeable small ones, has mostly relied on CE analysis. CE-based urine peptidomics aims to screen the peptides differentially expressed across distinct populations, providing an indirect window to assess the proteases modulated by BCa. Such indirect urine profiling approach has been proven capable of accurately discriminating cancer patients from patients with other organ-related nonmalignant conditions [54]. Together with bioinformatics tools, CE may serve for the discovery and validation of multimarker panels [55–57] and for the comparison of excreted urinary peptides to endogenous "housekeeping" peptides [55, 56]. CE-based native urinary peptide profiling has been applied for the detection of primary (with up to 91% sensitivity and 68% specificity) and recurrent (with up to 87% sensitivity and 51% specificity) urothelial BCa [57].

Peptides shown to most accurately allow the detection of BCa were collagen fragments ($\sim$60%), distantly followed by hemoglobin, apolipoprotein A-I peptides, membrane-specific heparin, sulfate proteoglycan core protein, a disintegrin and metalloproteinase domain-containing protein 22 (ADAM22), and a disintegrin and metalloproteinase with thrombospondin motifs 1 (ADAMTS1). Together, these peptides reflect increased proteolytic activity, extracellular matrix remodeling, collagen degradation, and hematuria [57].

However, there are some sampling issues that should be considered in the analysis of proteome data retrieved from proteome profiling of urine samples collected from patients with BCa. Urine proteomics applied to the search of BCa biomarkers have focused most extensively on older patients (over the age of 60) [32, 34, 36], which reflect the clinical phenotype of the majority of patients but reduce the likelihood of finding markers for early diagnosis purposes. Gender also influences the urine proteome profile. Females tend to present higher albumin and transferrin urinary concentration and a more complex urinary subproteome (composed by the less abundant proteins) [58]. Therefore, the pathophysiological role and the prognosis and diagnosis value of a given BCa-related urinary biomarker may not be suitable across all age-groups or across genders.

4 Urinary Proteases Associated with Bladder Cancer

Urine has become one of the most attractive diagnosis samples, involving noninvasive collection of unrestricted quantities [59]. Generally, less than 150 mg of proteins is excreted in urine, mainly due to glomerular filtration, and partially by tubular reabsorption, secretion, and degradation [60, 61]. Around 70% of urinary proteins are of kidney origin [62], a percentage modulated by diseases. In addition to kidney diseases, recent findings show that each and several diseases can be accurately discriminated by a unique protein fingerprint. From more than 2,500 proteins identified by MS-based urinalysis as differentially expressed across human diseases, approximately 15% (approximately 360 proteins) result from malignant diseases, particular BCa (58% or 209 proteins of these approximately 360 malignancy-associated differentially expressed urinary proteins) [63, 64]. Indeed, by pooling data retrieved from ten MS-based studies concerning differentially expressed urinary proteins from patients with BCa [29–38], 209 proteins were highlighted. Among these, 44 proteins were classified as enzymes, 21 of which are hydrolases and from these 13 are proteases (Tables 2 and 3). These proteases/peptidases are listed in Table 3 together with its function and expression variation (up/down) in BCa. Likewise, the expression patterns of several genes have been determined by microarrays in close to 1,500 BCa patients and revealed several proteases as putative diagnostic and prognostic BCa markers (Table 3) [65–75]. However, as also depicted in Table 3, alterations in proteases mRNA do not always reflect those concerning the concentrations of proteases neither their activity. For instance, while both levels and the activity of MMP-2 and cathepsin G are upregulated in urine samples from BCa patients, their mRNA expression levels are downregulated (Table 3), which may reflect increased stimulation-dependent activation (by, for instance, inflammatory and stress–response signaling pathways), stimulation-independent activation or enhanced intrinsic hydrolytic activity or alterations in enzymatic/lysosomal trafficking (see the following section and Table 3).

The role of proteases goes behind protein digestion with recognized important functions in several signaling pathways (Tables 2 and 3; Figs. 1 and 2), namely those involved in DNA replication and translation processes, cellular proliferation,

Table 2 Classes of enzymes identified in the urine of bladder cancer human patients detected by mass spectrometry and corresponding enriched biological processes

Type of enzyme	Number	Enriched biological processes	Enzymes
Oxidoreductase	6	Prostaglandin metabolic process, hypochlorous acid biosynthetic process, daunorubicin metabolic process, doxorubicin metabolic process, polyketide metabolic process, peroxisome fission	ACOX1, AKR1C2, CP, DHRS2, NOS2, MPO
Transferase	11	Peptidyl-pyroglutamic acid biosynthetic process, using glutaminyl-peptide cyclotransferase, activation of phospholipase A2 activity by calcium-mediated signaling, UDP-glycosylation, glutamate catabolic process to aspartate or 2-oxoglutarate, L-kynurenine metabolic process, double-strand break repair via alternative nonhomologous end joining, ectopic germ cell programmed cell death, positive regulation of catenin import into nucleus	GOT2, CD38, EGFR, F13A1, GSTT1, ART3, PLK2, PRKDC, QPCT, TGM4, UGGT1
Isomerase	2	DNA topological change, cyclooxygenase pathway	PTGDS, TOP1
Lyase	4	Positive regulation of neurotrophin production, double-strand break repair via classical nonhomologous end joining, regulation of organelle transport along microtubule, positive regulation of inositol 1,4,5-trisphosphate-sensitive calcium-release channel activity, protein heterotetramerization, DNA demethylation, DNA ligation	XRCC6, CA1, APEX1, ADCY2
Hydrolase	21	Regulation of humoral immune response, positive regulation of epidermal growth factor receptor signaling pathway, positive regulation of ERBB signaling pathway, regulation of complement activation, regulation of protein activation cascade, antibiotic metabolic process, positive regulation of vascular smooth muscle cell proliferation, negative regulation of tumor necrosis factor (ligand) superfamily member 11 production, positive regulation of B cell proliferation, endodermal cell differentiation, positive regulation of neurotrophin production, biofilm formation, response to lactam/staurosporine/fluoxetine, regulation of smooth muscle cell-matrix adhesion, negative regulation of cell cycle checkpoint, negative regulation of DNA damage checkpoint, negative regulation of cysteine-type endopeptidase activity involved in apoptotic signaling pathway	ADAMTS7, CTSD, CPN1, CTSG, CD38, CFB, C2, DDX39B, MASP2, MMP-2, MMP-9, PGA5, PON1, ALPP, PTPRC, LTF, DPEP1, VNN2, APEX1, PLAU, XRCC6

Table 3 Proteases detected in distinct protein and mRNA levels in the urine of bladder cancer human patients

Type of protease	Protein name	Gene name	Uniprot accession number	Protein variation	mRNA variation	Bladder cancer-associated biological processes/pathways/responses
Membrane dipeptidase	Dipeptidase 1	DPEP1	P16444	↓	↓	Dipeptides hydrolysis, eicosanoid synthesis
Lysine carboxypeptidase	Carboxypeptidase N catalytic chain	CPN1	P15169	↑	↓	Steroid biosynthesis
Serine endopeptidase	Cathepsin G	CTSG	P08311	↑	↓	ACE inhibitor pathway/renin–angiotensin system, lysosomal trafficking
	Complement C2	C2	P06681	↑	↑	Complement and coagulation cascades, pertussis, *Staphylococcus aureus* infection
	Complement factor B	CFB	P00751	↑	–	Complement and coagulation cascades, *Staphylococcus aureus* infection
	Urokinase-type plasminogen activator	PLAU	P00749	↓	↑	NF-kappa B signaling pathway, complement and coagulation cascades, Wnt signaling pathway, osteopontin signaling, transcriptional misregulation in cancer, proteoglycans in cancer
	Mannan-binding lectin serine protease 2	MASP2	O00187	↓	↓/↑	Complement and coagulation cascades, *Staphylococcus aureus* infection
Aspartic endopeptidase	Cathepsin D	CTSD	P07339	↑	↓/↑	Lysosomal trafficking
	Pepsin A-5	PGA5	P0DJD9	↑	–	Protein digestion and absorption
Metalloendopeptidase	A disintegrin and metalloproteinase with thrombospondin motifs 7	ADAMTS7	Q9UKP4	↑	–	Cartilage oligomeric matrix protein degradation
	Matrix metalloproteinase-9	MMP-9	P14780	↑	↑	TWEAK signaling pathway, osteopontin signaling, angiogenesis, AGE/RAGE pathway, matrix metalloproteinases signaling, leukocyte transendothelial migration, estrogen signaling pathway, transcriptional misregulation in cancer
	Matrix metalloproteinase-2	MMP-2	P08253	↑	↓	AGE/RAGE pathway, matrix metalloproteinases signaling, leukocyte transendothelial migration, GnRH signaling pathway, estrogen signaling pathway

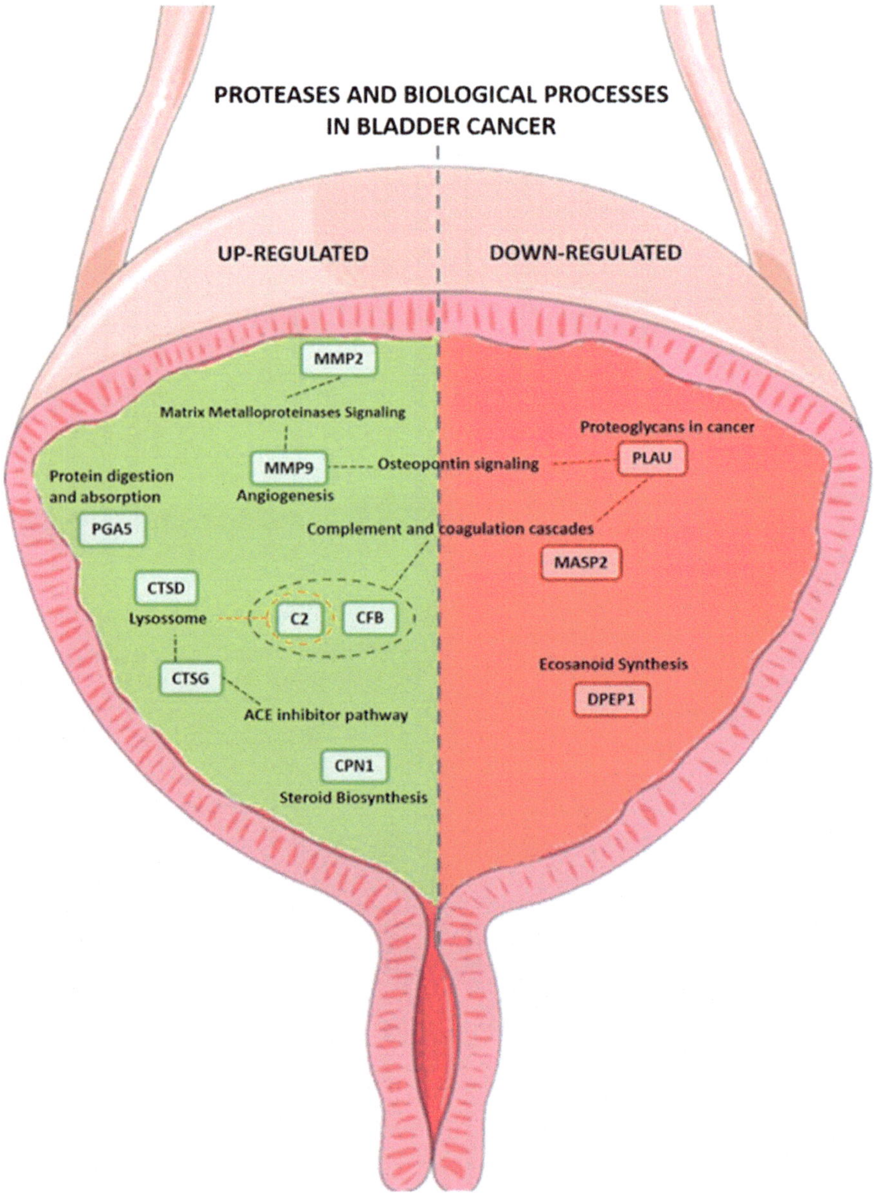

Fig. 1 Proteases modulated by bladder cancer and associated biological processes and putative biomarkers with diagnosis and prognosis value. Only a few representative examples are depicted. For a more comprehensive overview, see Table 3. Figure was designed using *Servier Medical Art*

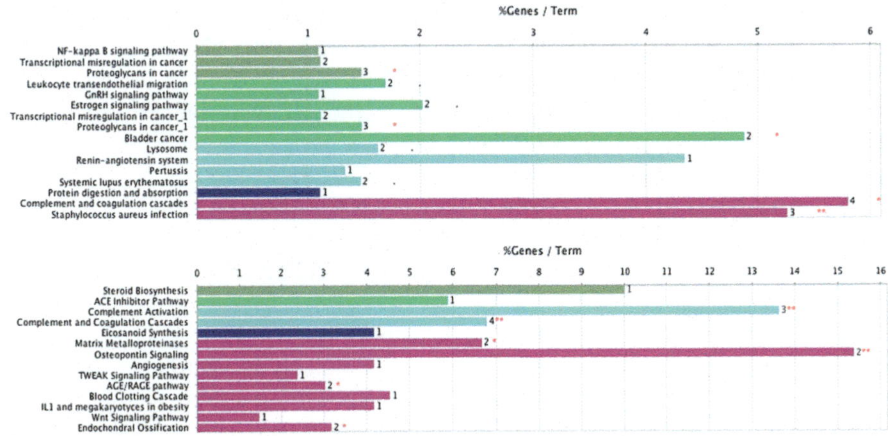

Fig. 2 Main pathways involving differentially expressed urinary proteases/peptidases modulated in bladder cancer human patients. Analysis performed using ClueGo plugin for Cytoscape. *Upper Graph* KEGG pathways. *Lower Graph* Wiki pathways. The number after bars corresponds to the number of genes contributing for those processes' enrichment. Bars marked with an asterisk (*) correspond to processes considered significantly changed (p-value < 0.05) by the built-in algorithm

tissue remodeling, angiogenesis, fertilization, hemostasis, blood coagulation, apoptosis, necrosis, immunity, organelle recycling, and cellular senescence [76–78]. Nevertheless, the human "degradome," which results from protein digestion, has been suggested to play a critical role in the pathogenesis of some diseases. Even thought, only 2% of protein coding genes contribute to such "degradome" [76–84]. These proteases are divided into cysteine proteases, serine proteases, aspartic proteases, threonine proteases, or metalloproteases [85]. Since the scope of this chapter is the biological role and clinical implications of urinary proteases in BCa, a brief description of proteases' families follows.

4.1 Cathepsins

Cathepsin proteases comprise a diverse family of hydrolytic enzymes, which are expressed at the cell surface and/or secreted into the extracellular milieu. These proteases are classically known for its role in lysosomal protein turnover and extracellular matrix degradation [86]. Despite this classical view, cathepsins encompass many and much more diverse functions. For instance, cathepsin S plays an important role in MHC-II-mediated antigen presentation, so that animal models lacking this enzyme cannot activate the MHC-II, displaying markedly impaired antigenic peptide binding, antibody class switching, and splenocytes/dendritic cells function [87]. Moreover, while trafficking through the endocytic pathway to allow invariant chain degradation and antigenic peptide loading onto MHC-II may be mediated by either one of several cathepsins in mice, this redundancy in humans is not observed, and therefore, alterations in the levels of one particular cathepsin may have more severe consequences than in animal models [88]. The involvement of cathepsins on progressive tissue remodeling [89] is also required for malignant disease development, progression, and metastasis [86, 90], which might explain the upregulation of these enzymes in BCa (Table 3).

Most members of this family of proteases (B, C, F, H, L, K, O, S, V, W, and X) display proteolytic activity at cysteine residues, while cathepsins A and G are serine carboxypeptidases and cathepsins D and E are aspartic proteases [91]. In turn, the activity of these proteases is regulated by its corresponding inhibitors cystatins and serpins [91].

Upregulated in the urine from BCa patients (Fig. 1), cathepsin G is a serine peptidase highly expressed in immune cells, most notably in neutrophils and mast cells, but also in monocytes and dendritic cells [92, 93]. Therefore, the high levels of cathepsin G, detected by MS profiling of urine from BCa patients (Table 3), are most likely of immune cell origin (rather than tumor cell origin) and may account for some of the biological alterations observed in BCa patients, such as NF-kappa B signaling pathway, eicosanoid synthesis, leukocyte transendothelial migration, and transcriptional deregulation (Fig. 2; Table 3).

Cathepsin G is known to cause extensive damage to host tissues due to its poor specificity [94]. Animal models of ischemia lacking cathepsin G can survive, while the condition is lethal to most wild-type animals. For instance, while neutrophil infiltration and the levels of CXCL1 and CXCL2 chemokines and myeloperoxidase are equal in both groups immediately after induction of ischemia, mice lacking cathepsin G do not suffer from the severe tubular necrosis, tubular cell apoptosis, and fibrosis characteristic of normal mice, showing that cathepsin G sustains tissue pathology and fibrosis after stress, inflammation, and injury induction [94].

Cathepsin G derived from neutrophils is coreleased into phagolysosomes, thereby helping to fight of infections [95]. In fact, studies with animal models suggest that cathepsin G activation is required for survival upon infection by *Aspergillus fumigatus* and *Staphylococcus aureus* [95, 96]. Likewise, both membrane-bound and secreted forms of cathepsin G colocalize and are coreleased with other peptidases [97, 98], possessing antimicrobial activity and being the main neutrophil-derived antibacterial mediators [95]. Accordingly, its antimicrobial activity seems to be partially enhanced by and dependent on other peptidases like elastase [95], and its surface expression may increase up to 30-fold upon stimulation [97]. Therefore, infections may trigger the upregulation of cathepsin G as a host protective mechanism, but may also cause tissue damage and promote tumorigenesis (Fig. 2) or metastization once cancer has been established.

Compared to other human proteases, cathepsin G is less selective by combining both chymotryptic and tryptic hydrolytic activities [99], and while mutations/evolutionary alterations responsible for this duality confer it a broader activity spectrum, these may also contribute for its lower potency [100]. Irrespectively, its broad spectrum of activity most likely contributes for extracellular matrix degradation, complement activation, proteoglycan degradation, and lysosomal signaling alterations in BCa (Fig. 2; Table 3). It should be noted that most of what is known about cathepsin G activity results from in vitro assays or animal studies, which may hamper the interpretation of its hypothetical implications in BCa. For instance, in contrast to human cathepsin G, murine cathepsin G displays a chymotryptic bias, lacks tryptic activity to a large extent, and is more specific, but significantly more potent [100].

The major activator of cathepsin G and related peptidases is the lysosomal cathepsin C (CTSC) or dipeptidyl peptidase I (DPEP1) [101], which was also found in increased levels in the urine from BCa patients (Fig. 1; Table 3). Increased levels of DPEP1 and the corresponding activated peptidases are associated with exaggerated immune responses and tissue damage. In contrast, animal models lacking DPEP1 are protected from these effects upon the induction of inflammatory response and display decreased expression of TNF-α and IL-1β [101], also involved in MMP-mediated BCa progression as discussed in the following section. Its tryptic activity, in turn, activates proteinase-activated receptors, calcium mobilization, and neutrophil–platelet interactions at sites of injury or inflammation [102], as well as complement proteins [103] and prourokinase plasminogen activator (PLAU) [104], which are in high levels in the urine of BCa patients (Table 3). Owing to its peptidase activity, human cathepsin G hydrolyzes angiotensin-(1-10) (angiotensin I) to yield angiotensin-(1-8) (angiotensin II) [105, 106] and activates metallopeptidases [107], thereby accounting for the alterations observed in the renin–angiotensin and lysosomal signaling (Fig. 2; Table 3). Such metallopeptidases include MMP-2, which is also in high content in the urine from BCa patients (as described in the next section), and the regulation of both substrates may therefore be intimately interconnected [107]. Taken together, cathepsin G, as well as its promoter and its substrates, is increased in urine samples from BCa patients and seems to be involved in the promotion of extracellular matrix component degradation (e.g., collagen, laminins) and local tissue damage, allowing malignant cells to infiltrate the bladder tissues and to disseminate into the blood circulation.

In contrast to cathepsin G (a serine protease), cathepsin D is an aspartyl protease involved in cellular components' recycling and apoptosis control. Procathepsin D is found in the Golgi complex and is proteolytically inactive but the intermediate and mature forms of cathepsin D are enzymatically active [108]. These forms are found in endosomes and lysosomes, respectively, and are responsible for the autophagy and apoptosis pathways [109], thus accounting for the enriched processes illustrated in Table 3 and Fig. 2. For instance, cathepsin D is known to function as an anti-apoptotic mediator in human malignant glioblastoma cells, inducing autophagy under stress conditions and conferring cancer cells resistance against genotoxicity mediated by oxidative agents such as hydrogen peroxide [110]. However, the proteolytic activity of cathepsin D also allows the release of growth factors that act by promoting tumor cell growth [109]. Cathepsin D mediates both metastasis and recurrence in breast cancer [111, 112] and metastasis/dissemination in laryngeal and pancreatic cancers [113, 114]. It was proposed as a marker of vascular and microvascular density in breast and ovarian tumors [115, 116]. However, in contrast to cathepsin G, which appears to be mostly of immune cells' origin, cathepsin D has been described as secreted primarily by cancer cells [117, 118]. In a small cohort of patients with BCa (15 patients), it was shown that the activity of cathepsin D was significantly increased in serum samples [119]. The urinary concentration of cathepsin D is approximately equal to that of serum in both BCa patients and healthy subjects [47]. In another small cohort of patients (21 subjects), tissue cathepsin D analyzed by immunohistochemistry displayed a significant but inverse correlation with tumor grade and stage. These data reinforce the involvement of this protease in the early stages of the disease and local tissue invasion, but its expression decreases once high grade and high invasiveness are attained [120]. A disengagement between the regulation of cathepsin D and its inhibitor cystatin C was also reported [120]. Subsequently, in a more comprehensive study comprising 68 BCa patients, both serum and urinary concentrations of cathepsin D were found significantly higher, using the surface plasmon resonance imaging (SPRI) biosensor [47]. The serum cathepsin D/creatinine ratio was reported to be significantly higher in superficial tumors (Ta+T1) compared to invasive tumors (T2+T3). This observation makes cathepsin D of particular value for detecting BCa in its early stages. Of uttermost importance, urinary cathepsin D/protein ratio was significantly higher in primary versus recurrent (2.24-fold) and low-grade versus high-grade (1.67-fold) tumors, reinforcing its diagnosis value for BCa in the initial stages [47]. Overall, BCa displayed significantly higher serum (eightfold) and urine (sevenfold) cathepsin D concentrations when compared with healthy controls, and differences remained significant even when normalized to total protein and creatinine levels [47].

Similar to cathepsins D and G, urinary cathepsin L is present in significantly higher urinary levels in patients with urothelial carcinoma compared to normal controls, even in patients with negative cytology [40]. Urinary cathepsin L levels were higher in patients with higher grade, being significantly associated with tumor invasiveness. Also, the associations with BCa presence and aggressiveness were maintained after adjusting cathepsin L levels for urinary creatinine content [40]. Perhaps of uttermost value, voided urinary cathepsin L was proven to be an independent predictor of BCa recurrence and invasiveness at advanced stages, outperforming both cytology and NMP22 (an FDA-approved marker for BCa at initial stages) [40]. These results were in accordance with increased cathepsin L mRNA expression in tissue samples from BCa patients [39]. Cathepsin L expression in tumor cells is under distinct control mechanisms compared to normal cells, as the transcription factors responsible for basal cathepsin L expression (NF-Y, Sp1, Sp2, and Sp3) are not responsible for its overexpression in tumor cells [121]. The 5′ region of the cathepsin L gene encompasses CpG islands, which regulate promoter activity, and demethylation of these CpG islands seems to be positively associated with cathepsin L-dependent types of malignant diseases, but not independent ones [121]. Therefore, one possible mechanism (albeit not tested to date) contributing for the upregulation of cathepsin L in BCa may encompass epigenetic alterations, particular DNA demethylation, explaining the transcriptional misregulation depicted in Table 3 and Fig. 2.

Overall, the upregulation of cathepsins and their inhibitors is known to mediate multiple stages of tumorigenesis, carcinogenesis, and angiogenesis [90, 122]. In addition to directly carrying out extracellular matrix and basement membrane degradation, cathepsins also disrupt intercellular adhesion proteins (e.g., E-cadherin, at adheren junctions) and activate proteolytic cascades in which the activity of other proteases such as MMPs and urokinase plasminogen activator is enhanced [86, 122]. Therefore, the levels of cathepsins and, more specifically, the cathepsin/CIP (cathepsin inhibitor) ratio may be used as a suitable marker of malignancy [39]. Nevertheless, among the cathepsin protein family, cathepsins G and D seem to be the best potential urinary biomarkers for BCa (Fig. 1; Table 3).

4.2 Matrix Metalloproteinases

MMPs are zinc-dependent endopeptidases crucial for many cellular processes, from physiological and developmental to pathological ones [123, 124]. Accordingly, these enzymes are particular important in regulating the local cellular and tissue microenvironment, not only by carrying out extracellular matrix remodeling but also by mediating several physiological processes through signaling regulation [125]. There are 23 human MMPs catalogued to date, most of which display the conserved propeptide, catalytic, and hemopexin-like C-terminal domains. Accordingly, MMPs are expressed as inactive propeptides with their zinc ions in the catalytic sites attached to the propeptide domain, being activated by the cleavage of this domain. Even though the cleavage of the prodomain requires the action of another proteolytic enzyme (intracellular furin or extracellular MMPs/serine Proteases), modifications taking place at the cysteine residue of the propeptide domain may also compromise such interactions and thus activate MMPs [126, 127].

In the setting of BCa, uncontrolled proliferation and compromised apoptosis are attributed in a large extent to the upregulation of zinc-dependent endopeptidases MMP-2 and MMP-9 (Fig. 1; Table 3) [128]. The proteolytic activity of MMPs is responsible for the cleavage of ligands, and corresponding receptors are responsible for conveying proapoptotic signals, thus allowing tumor cells to evade apoptosis [129]. These changes mediate both tumor cell migration and the development of new tumor-related blood vessels [53, 130]. So, MMP-2 as well as MMP-9 has been regarded as promising biomarkers for BCa [21, 49, 131, 132]. MMP-9 acts synergistically with ADAM metallopeptidases and epidermal growth factor receptor (EGFR) in order to degrade E-cadherin and liberate β-catenin, allowing its translocation into the nucleus and thus enhancing cellular proliferation [133, 134]. Also, tumor cells may use transforming growth factor-β (TGF-β) signaling via MMP-mediated proteolytic conversion, allowing tumors to evade immune surveillance and thus leading to local invasion and cancer metastasis [135, 136]. Similarly, ADAM-mediated proteolytic cleavage of tumor-associated major histocompatibility complex class I-related proteins MICA and MICB, which suppresses NK cell-mediated cytotoxicity, seems to require the activity of MMPs [137, 138]. Inflammatory cell-derived MMP-9 increases the bioavailability of vascular endothelial growth factor (VEGF), which is the most potent inducer of tumor angiogenesis [139].

High molecular weight forms of MMPs are also found in increased levels in the urine of BCa patients, most notably MMP-9 dimers and MMP-9 conjugated with its inhibitor TIMP1 [53], and MMP-2 conjugated with the MMP inhibitor TIMP2 [132]. The overexpression of MMP-9 promotes the upregulation of its inhibitor TIMP1 as a countermeasure. However, when MMP-9 is in excess over its inhibitor TIMP1, the formation of MMP-9 disulfide-bonded dimers takes place, forming a more stable and slowly activating MMP-9, which is less susceptible to inactivation [140]. MMP-9/NGAL conjugates seem to protect MMP-9 from autodegradation and are associated with the formation of cancer metastasis [53, 141, 142]. MMP-2, MMP-9, MMP-9/NGAL complexes, and MMP-9 dimer are overexpressed in high-grade BCa patients compared to low-grade ones, and MMP-9/TIMP1 complex is overexpressed in larger tumors compared to smaller ones [48]. Also, MMP-2 monomer, MMP-2 conjugated with its inhibitor, and MMP-2 fragments are significantly associated with high-grade/*lamina propria* invasion [132], while the presence of MMP-9/TIMP2 discriminates malignant tumors from benign ones [50]. The misbalance given by the ratio MMP/MMP inhibitor, rather than upregulation of any MMP independently, seems to be characteristic of BCa [30]. Moreover, in addition to enzyme monomers and enzyme complexes, immunoblotting approaches have revealed the presence of multiple MMP-2 fragments in the urine of BCa patients, which are also associated with the disease, particularly with transitional cell carcinoma cases [132].

More than measuring the levels of MMPs, MMPs' proteolytic activity assessed by zymography has assumed a paramount role in discerning the involvement of MMPs in BCa. In these studies of enzymatic activity, close to 65% of BCa patients present increased urinary proteolytic activity compared to healthy controls [48, 49]. The assessment of the proteolytic activity of multiple MMPs has provided specificities of up to 100% for BCa detection [48, 50]. Most of the proteolytic activity characteristic of urine samples from BCa patients seems to be carried out by MMP-9, followed by MMP-9/NGAL conjugates and MMP-2 and MMP-9 dimers. When MMP-9 dominates, the maximum sensitivity resulting from urine MMPs profiling for BCa detection seems to be attained with MMP-9 (62.1%), followed by MMP-9/NGAL conjugates (60.9%) and MMP-2 (54.5%) and MMP-9 dimers (53%) [79]. Irrespectively, MMPs are positively correlated with each other. So, the increased levels of a particular MMP upregulate the remaining one. In this regard, MMP-9 seems to suffer the highest positive influence, being more sensitivity and more susceptible to a larger number of factors [79, 141]. In fact, the expression of MMP-9 is enhanced by inflammatory cytokines such as TNF-α and interleukins, being its increase more susceptible to alterations triggered by inflammatory conditions (e.g., cystitis) and less specific to BCa itself [143], as illustrated by patients with cystitis who excrete significantly more MMP-9 but not more MMP-2 than controls [141]. In contrast, MMP-2 allows a better discrimination between BCa patients and healthy subjects [139, 143], being more specific and less influenced by cancer-unrelated factors [79]. However, MMP-2 and MMP-9 have failed to accurately identify BCa in the early stages/superficial bladder carcinoma (stages CIS, Ta, and T1). Urinary MMP-2 seems to be a suitable biomarker for high BCa tumor grade (G3), while both MMP-2 and MMP-9 are suitable for advanced BCa stage (T2 or greater) [21]. Moreover, data normalization to total protein instead to creatinine increases the sensitivity and specificity of these makers because total protein is less influenced by hematuria, which is commonly found in BCa patients [144]. Also, combining urine cytology with urine protease profiling increases sensitivity from approximately 75% (given by conventional screening approaches) to more than 95%, and if the MMP-9/TIMP-2 or MMP-2/TIMP-2 ratios are used, a sensitivity close 100% might be attained [50].

4.3 Serine Proteases

Serine proteases represent more than one-third of all identified proteases and include 40 families of well-known proteases, such as trypsin and thrombin [145, 146]. This family of enzymes, which have a serine in the catalytic site, is involved in several biological processes from food digestion to inflammation response [146]. Thrombin-like proteases are a class of serine proteases involved in blood coagulation and fibrinolysis and include thrombin, plasminogen activators, and plasmin [146]. Urokinase-type plasminogen activator (PLAU or uPA) is a serine protease that in addition to regulate fibrinolysis also modulates innate and adaptive immune responses (Table 3; Fig. 2), acting as an endogenous antibiotic (e.g., bacteriostatic against *Staphylococcus aureus*) [147] and leading to severe compromised T cell activation and proliferation if absent [148]. Its antimicrobial activity is mediated by the serine protease domain [149]. Moreover, PLAU is not only a marker of severe infection, but also a good predictor of mortality (76% sensitivity and 69% specificity for fatal sepsis) [150].

Despite the weak proteolytic activity, conversion of plasminogen to plasmin by uPA makes it a molecule of high interest for the study of tumors, since plasmin is related with various malignant properties of cancers [151]. In fact, the urokinase plasminogen activator system, which comprises PLAU, is involved in several steps of cancer development and metastization (Table 3; Figs. 1 and 2), including ECM degradation, cell proliferation, migration and adhesion, angiogenesis and intravasation [152–154]. PLAU can activate the precursor forms of MMP-9 and other matrix metalloproteases (such as MMP-3 and MMP-12), leading to ECM remodeling. ECM remodeling is essential to angiogenesis and consequent tumor growth [152]. However, a study in human prostate cancer cells shows that PLAU may have an antiangiogenic role, due to antiangiogenic activity of angiostatin, produced by PLAU/plasmin system [155]. While there are two types of plasminogen activator (PLAU and tissue-type plasminogen activator), only PLAU promotes cell migration and proliferation via its growth factor domain [156]. In contrast, large reductions in the levels of its receptor (approximately 70%) have been shown to induce a dormancy state in human epidermoid carcinoma (HEp3) cell lines [143]. PLAU also induces cell migration, via its kringle domain, by acting as a chemotactic agent [157]. Therefore, a continuum comprising infection-induced PLAU upregulation and PLAU-induced cellular proliferation/chemotaxis can be discerned. In BCa,

PLAU modulates many signaling pathways, including NK-kappa B, TWEAK signaling pathway, and Wnt signaling pathway, as highlighted in Table 3 and Fig. 2. Furthermore, PLAU levels were suggested as a parameter for BCa prognosis. In fact, evidences show that patients with BCa displaying higher PLAU levels have lower survival rates than patients with low levels of PLAU [158].

5 The Added Value of Urine Proteases for Bladder Cancer Diagnosis

In order to better integrate the role of proteases in the pathogenesis of BCa (Fig. 1) together with its clinical value, Table 3 overviews the differential involvement of urinary proteases in BCa development and progression. As can be depicted from this figure, MMP-2 and -9, PLAU, and cathepsins D and G seem to be involved in all stages of BCa, including tumorigenesis and angiogenesis, cancer progression, invasion, and metastasis. Therefore, the detection of these 4 proteases would most likely be indicative of BCa, independently of its stage. Moreover, these proteases might be seen as therapeutic targets in the set of BCa. However, the majority of clinical trials with MMP inhibitors has failed, most likely because these were introduced only in advanced stages of the disease [127]. In the search of more potent and specific agents, near-infrared polymer-based proteolytic beacons were tested in animal models and were able of detecting tumors as small as 0.01 cm^2. Techniques employing visible or near-infrared fluorescence resonance energy transfer fluorophores to detect and measure MMPs' proteolytic activity allowed the detection of tumors at early stages with increased sensitivity [159]. Packard and colleagues have demonstrated that using a high-resolution fluorescent biosensor based on MMP (2, 9, or 14)-mediated peptide cleavage and fluorescence emission, the protease activity can be localized at the polarized leading edge of migrating tumor cells (rather than further back on the cell body) [160]. Similarly, biocompatible fluorogenic MMP substrates allow assessing the efficacy of MMP inhibitors in tumors without the requirement for tissue biopsies [161]. Using this approach, MMP inhibition has been shown to take place within hours after treatment initiation using a potent MMP inhibitor as prinomastat [161]. Furthermore, radiolabeled imaging techniques such as positron emission tomography (PET) and single photon emission computed tomography (SPECT) may be employed for accessing MMPs' activity in order to screen patients and diagnose cancer [162, 163]. Nevertheless, as biomarkers assessed in biological fluids, MMPs and corresponding inhibitors are rather unspecific for Bca. So, envisioning to improve the clinical management of BCa, the most sensitive, specific, and robust panels of urinary biomarkers must combine: urinary MMP-2 and MMP-9, their inhibitors and ratios thereof, cathepsins G, D, L, and B, and PLAU (Table 3). However, one should keep in mind that these markers will definitively become more robust if combined with conventional markers of tumorigenesis and malignancy [164, 165].

6 Urinary Proteases as Therapeutic Targets in the Management of Bladder Cancer

Cathepsin, MMPs, or PLAU is not primary therapeutic targets in the clinical management of BCa, despite the available broad-spectrum inhibitors particularly for MMPs (e.g., batimastat, marismastat, salimatat, prinomastat, and tanomastat). These MMP inhibitors (MMPIs) have been tested for the treatment of cancers such as leukemias, lymphomas, testicular, lung, gastrointestinal, oropharangial, once inhibit malignant growth by enhancing fibrosis around malignant lesions. By doing so, MMPIs prevent tumor invasion, apoptosis, and angiogenesis. However, these therapies induce significant side defects, which limit its clinical applicability [166].

Despite not directly targeting protease activity, most of BCa therapies have impact on the regulation of proteases' expression and activity. For instance, tyrosine kinase inhibitors were reported to modulate proteases in the set of bladder cancer. For example, the treatment of bladder cancer cell lines with sunitinib impaired cathepsin B activation and stimulated a lysosomal-dependent necrosis, whereas pazopanib induced cathepsin B activation and autophagic cell death [167]. Glucocorticoids also impact the activity of proteases in BCa. Corticosterone, prednisone, and dexamethasone were shown to inhibit the expression of MMP-9 in bladder cancer cell lines with impact on cell invasion but marginal effect on cell growth [168].

Intravesical treatment with Bacillus Calmette-Guerin (BCG) is a clinically established and effective therapy for superficial bladder cancer and CIS though the mechanism of BCG immunotherapy is largely unknown. BCG was suggested to bind to fibronectin, near the carboxyl terminal region and adjacent to the heparin-binding domain, and protect this region of the molecule from tumor proteases [169]. More recently, the antitumor effect of BCG on bladder cancer was associated with BCG-induced cytokines TNF-α and INF-γ. Curiously, these cytokines were shown to induce MMP-9 expression and increase its enzyme activity in J-82 bladder cancer cells, potentially enhancing the invasiveness of bladder cancer in certain conditions [170]. Cathepsin B was also shown to mediate the BCG effect. The BCG-induced increase of this protease content and activity was related to the apoptosis of T24 and MB49 cell lines [171].

Immunotherapies with antibodies targeting PD-1 receptor expressed on T cells and its ligands, PD-L1 and PD-L2, have been recently proposed for the clinical management of BCa. Specifically, atezolizumab attaches to PD-L1 on the surface of tumor cells and prevents it from interacting with PD-1 receptors on immune cells and thus unleashes the immune system (ASCO 2016). These therapies also have impact on protease activity once targeting MMP-dependent cleavage of PD-1 ligands on fibroblasts, inhibiting inflammation in tissues [172]. Nevertheless, few experimental evidences exist on this topic once these immunotherapies are starting to be implemented for the treatment of BCa.

In vivo evidences of the impact of these therapies on the levels and activity of these proteases in urine will help to disclose the biomarker value of urinary proteases for monitoring patients' response to therapeutics.

7 Conclusions

Protease hyperactivation is a hallmark of BCa, leading to tissue damage and invasion, immune system and apoptosis evasion, tumorigenesis, and cancer metastasis. So, changes in the levels of proteases in biofluids as urine might help in the management of this malignancy. However, one should be aware that alterations in the gene expression patterns of proteases do not accurately reflect enzyme levels and activity. MS is particularly suitable for the detection of quantitative and qualitative alterations of proteases in urine samples during exploratory research, as it provides an integrative view of the multiple molecular mediators, promoters, and inhibitors that play a key role in BCa. In addition, zymography- and immunoaffinity-based approaches are required for activity assays and validation of proteases as biomarkers, respectively. The application of these methodological approaches has shown the involvement of several proteases in tumor development, progression, and dissemination. Nevertheless, these proteases have not been therapeutically targeted with success, but novel sensors are being developed and will hopefully allow the successful monitoring of urine proteases and therapy efficacy.

In conclusion, a fingerprint of proteolytic activity can be discerned in the urine of BCa patients, one comprising alterations in MMP-2 and MMP-9 activities, their inhibitors and ratios thereof, cathepsins G, D, L, and B, DPEP, and PLAU. Despite promising, the exploration of these proteases as biomarkers for the clinical management of BCa is still in its infancy. Future studies associating urine protease profile with BCa stage, patients' age and gender, and response to therapeutics are needed to support their diagnosis and prognosis value.

Acknowledgements This work was supported by the Portuguese Foundation for Science and Technology (FCT), European Union, QREN, FEDER, and COMPETE for funding the iBiMED, UnIC, QOPNA research units (project UID/BIM/04501/2013, UID/IC/00051/2013, UID/QUI/UI0062/2013), the Investigator Grant to RV (IF/00286/2015).

References

1. Ferlay J, Soerjomataram I, Ervik M et al (2014) GLOBOCAN 2012: estimated cancer incidence, mortality and prevalence worldwide in 2012: IARC CancerBase No. 11. Int. Agency Res. Cancer
2. Fajkovic H, Halpern JA, Cha EK et al (2011) Impact of gender on bladder cancer incidence, staging, and prognosis. World J Urol 29:457–463
3. Madeb R, Messing EM (2004) Gender, racial and age differences in bladder cancer incidence and mortality. Urol Oncol Semin Orig Investig 22:86–92

4. Bryan GT (1983) Pathogenesis of human urinary bladder cancer. Environ Health Perspect 49:201–207
5. Lopez-Beltran A, Sauter G, Gasser T et al (2004) World Health Organization classification of tumours: pathology and genetics of tumours of the urinary system and male genital organs. Pathol Genet tumors Urin Syst Male Genit Organs
6. WHO (2011) Environmental and occupational cancers. Fact sheet N°350
7. Bethesda (2016) PDQ adult treatment editorial board. bladder cancer treatment (PDQ®): health professional version. Natl Cancer Inst
8. Yeung C, Dinh T, Lee J (2014) The health economics of bladder cancer: an updated review of the published literature. Pharmacoeconomics 32:1093–1104
9. Johnson DC, Greene PS, Nielsen ME (2015) Surgical advances in bladder cancer: at what cost? Urol Clin North Am 42:235–252
10. Mossanen M, Gore JL (2014) The burden of bladder cancer care: direct and indirect costs. Curr Opin Urol 24:487–491
11. Svatek RS, Hollenbeck BK, Holmäng S et al (2014) The economics of bladder cancer: costs and considerations of caring for this disease. Eur Urol 66:253–262
12. Rodrigues D, Jerónimo C, Henrique R et al (2016) Biomarkers in bladder cancer: a metabolomic approach using in vitro and ex vivo model systems. Int J Cancer n/a–n/a
13. Stenzl A, Cowan NC, De Santis M et al (2011) Treatment of muscle-invasive and metastatic bladder cancer: update of the EAU guidelines. Eur Urol 59:1009–1018
14. Kamat AM, Lamm DL (2004) Antitumor activity of common antibiotics against superficial bladder cancer. Urology 63:457–460
15. Felix AS, Soliman AS, Khaled H et al (2008) The changing patterns of bladder cancer in Egypt over the past 26 years. Cancer Causes Control 19:421–429
16. Heyns CF, van der Merwe A (2008) Bladder cancer in Africa. Can J Urol 15:3899–3908
17. Nagata M, Muto S, Horie S (2016) Molecular biomarkers in bladder cancer: novel potential indicators of prognosis and treatment outcomes. Dis Mark 2016:8205836
18. Kaufman DS, Shipley WU, Feldman AS (18AD) Bladder cancer. Lancet 374:239–249
19. Massari F, Ciccarese C, Santoni M et al (2016) Metabolic phenotype of bladder cancer. Cancer Treat Rev 45:46–57
20. Lodillinsky C, Rodriguez V, Vauthay L et al (2009) Novel invasive orthotopic bladder cancer model with high cathepsin B activity resembling human bladder cancer. J Urol 182:749–755
21. Gerhards S, Jung K, Koenig F et al (2001) Excretion of matrix metalloproteinases 2 and 9 in urine is associated with a high stage and grade of bladder carcinoma. Urology 57:675–679
22. Rosser CJ, Chang M, Dai Y et al (2014) Urinary protein biomarker panel for the detection of recurrent bladder cancer. Cancer Epidemiol Biomark Prev 23:1340–1345
23. Lam T, Nabi G (2007) Potential of urinary biomarkers in early bladder cancer diagnosis. Expert Rev Anticancer Ther 7:1105–1115
24. Yang N, Feng S, Shedden K et al (2011) Urinary glycoprotein biomarker discovery for bladder cancer detection using LC/MS-MS and label-free quantification. Clin Cancer Res 17:3349–3359
25. Urquidi V, Goodison S, Cai Y et al (2012) A candidate molecular biomarker panel for the detection of bladder cancer. Cancer Epidemiol Biomark Prev 21:2149–2158
26. Ye F, Wang L, Castillo-Martin M et al (2014) Biomarkers for bladder cancer management: present and future. Am J Clin Exp Urol 2:1–14
27. Burchardt M, Burchardt T, Shabsigh A et al (2000) Current concepts in biomarker technology for bladder cancers. Clin Chem 46:595–605
28. Twining SS (1994) Regulation of proteolytic activity in tissues. Crit Rev Biochem Mol Biol 29:315–383
29. Li C, Li H, Zhang T et al (2014) Discovery of Apo-A1 as a potential bladder cancer biomarker by urine proteomics and analysis. Biochem Biophys Res Commun 446:1047–1052

30. Roy R, Louis G, Loughlin KR et al (2008) Tumor-specific urinary matrix metalloproteinase fingerprinting: identification of high molecular weight urinary matrix metalloproteinase species. Clin Cancer Res 14:6610–6617
31. Lei T, Zhao X, Jin S et al (2013) Discovery of potential bladder cancer biomarkers by comparative urine proteomics and analysis. Clin Genitourin Cancer 11:56–62
32. Chen CL, Lai YF, Tang P et al (2012) Comparative and targeted proteomic analyses of urinary microparticles from bladder cancer and hernia patients. J Proteome Res 11:5611–5629
33. Lindén M, Lind SB, Mayrhofer C et al (2012) Proteomic analysis of urinary biomarker candidates for nonmuscle invasive bladder cancer. Proteomics 12:135–144
34. Tsui K-H, Tang P, Lin C-Y et al (2010) Bikunin loss in urine as useful marker for bladder carcinoma. J Urol 183:339–344
35. Tan LB, Chen KT, Yuan YC et al (2010) Identification of urine PLK2 as a marker of bladder tumors by proteomic analysis. World J Urol 28:117–122
36. Feldman AS, Banyard J, Wu C-L et al (2009) Cystatin B as a tissue and urinary biomarker of bladder cancer recurrence and disease progression. Clin Cancer Res 15:1024–1031
37. Smalley DM, Sheman NE, Nelson K, Theodorescu D (2008) Isolation and identification of potential urinary microparticle biomarkers of bladder cancer. J Proteome Res 7:2088–2096
38. Lin C-Y, Tsui K-H, Yu C-C et al (2006) Searching cell-secreted proteomes for potential urinary bladder tumor markers. Proteomics 6:4381–4389
39. Staack A, Tolic D, Kristiansen G et al (2004) Expression of cathepsins B, H, and L and their inhibitors as markers of transitional cell carcinoma of the bladder. Urology 63:1089–1094
40. Svatek RS, Karam J, Karakiewicz PI et al (2008) Role of urinary cathepsin B and L in the detection of bladder urothelial cell carcinoma. J Urol 179:478–484. (discussion 484)
41. Ulrich F, Heisenberg CP (2009) Trafficking and cell migration. Traffic 10:811–818
42. Malemud CJ (2006) Matrix metalloproteinases (MMPs) in health and disease: an overview. Front Biosci 11:1696
43. Hadler-Olsen E, Fadnes B, Sylte I et al (2011) Regulation of matrix metalloproteinase activity in health and disease. FEBS J 278:28–45
44. Chen C-L, Lin T-S, Tsai C-H et al (2013) Identification of potential bladder cancer markers in urine by abundant-protein depletion coupled with quantitative proteomics. J Proteomics 85:28–43
45. Lee HJ, Nedelkov D, Corn RM (2006) Surface plasmon resonance imaging measurements of antibody arrays for the multiplexed detection of low molecular weight protein biomarkers. Anal Chem 78:6504–6510
46. Ladd J, Taylor AD, Piliarik M et al (2009) Label-free detection of cancer biomarker candidates using surface plasmon resonance imaging. Anal Bioanal Chem 393:1157–1163
47. Gorodkiewicz E, Guszcz T, Roszkowska-Jakimiec W, Kozłowski R (2014) Cathepsin D serum and urine concentration in superficial and invasive transitional bladder cancer as determined by surface plasmon resonance imaging. Oncol Lett 8:1323–1327
48. Mohammed MA, Seleim MF, Abdalla MS et al (2013) Urinary high molecular weight matrix metalloproteinases as non-invasive biomarker for detection of bladder cancer. BMC Urol 13:25
49. Õzdemir E, Kakehi Y, Okuno H, Yoshida O (1999) Role of matrix metalloproteinase-9 in the basement membrane destruction of superficial urothelial carcinomas. J Urol 161:1359–1363
50. Eissa S, Ali-Labib R, Swellam M et al (2007) Noninvasive diagnosis of bladder cancer by detection of matrix metalloproteinases (MMP-2 and MMP-9) and their inhibitor (TIMP-2) in urine. Eur Urol 52:1388–1397
51. Hattori S, Fujisaki H, Kiriyama T et al (2002) Real-time zymography and reverse zymography: a method for detecting activities of matrix metalloproteinases and their inhibitors using FITC-labeled collagen and casein as substrates. Anal Biochem 301:27–34

52. Vandooren J, Geurts N, Martens E et al (2013) Zymography methods for visualizing hydrolytic enzymes. Nat Methods 10:211–220
53. Moses MA, Wiederschain D, Loughlin KR et al (1998) Increased incidence of matrix metalloproteinases in urine of cancer patients. Cancer Res 58:1395–1399
54. Schönemeier B, Metzger J, Klein J, et al (2016) Urinary peptide analysis differentiates pancreatic cancer from chronic pancreatitis. Pancreas 45(7):1018–1026
55. Theodorescu D, Wittke S, Ross MM et al (2006) Discovery and validation of new protein biomarkers for urothelial cancer: a prospective analysis. Lancet Oncol 7:230–240
56. Jantos-Siwy J, Schiffer E, Brand K et al (2009) Quantitative urinary proteome analysis for biomarker evaluation in chronic kidney disease. J Proteome Res 8:268–281
57. Frantzi M, Van Kessel KE, Zwarthoff EC et al (2016) Development and validation of urine-based peptide biomarker panels for detecting bladder cancer in a multi-center study. Clin Cancer Res 22(16):4077–4086
58. Thongboonkerd V, Chutipongtanate S, Kanlaya R (2006) Systematic evaluation of sample preparation methods for gel-based human urinary proteomics: quantity, quality, and variability. J Proteome Res 5:183–191
59. Sedic M, Gethings LA, Vissers JPC et al (2014) Label-free mass spectrometric profiling of urinary proteins and metabolites from paediatric idiopathic nephrotic syndrome. Biochem Biophys Res Commun 452:21–26
60. Froehlich JW, Vaezzadeh AR, Kirchner M et al (2014) An in-depth comparison of the male pediatric and adult urinary proteomes. Biochim Biophys Acta 1844:1044–1050
61. Valente MAE, Damman K, Dunselman PHJM et al (2012) Urinary proteins in heart failure. Prog Cardiovasc Dis 55:44–55
62. Thongboonkerd V (2008) Urinary proteomics : towards biomarker discovery. Diagn Prognostics. 810–815
63. Lokeshwar VB, Habuchi T, Grossman HB et al (2005) Bladder tumor markers beyond cytology: international consensus panel on bladder tumor markers. Urology:35–63
64. Konety BR (2006) Molecular markers in bladder cancer: a critical appraisal. Urol Oncol 24:326–337
65. Dancik GM (2015) An online tool for evaluating diagnostic and prognostic gene expression biomarkers in bladder cancer. BMC Urol 15
66. Lindgren D, Sjödahl G, Lauss M et al (2012) integrated genomic and gene expression profiling identifies two major genomic circuits in urothelial carcinoma. PLoS ONE 7:e38863
67. Lindgren D, Frigyesi A, Gudjonsson S et al (2010) Combined gene expression and genomic profiling define two intrinsic molecular subtypes of urothelial carcinoma and gene signatures for molecular grading and outcome. Cancer Res 70:3463–3472
68. Kim W-J, Kim E-J, Kim S-K et al (2010) Predictive value of progression-related gene classifier in primary non-muscle invasive bladder cancer. Mol Cancer 9:3
69. Riester M, Taylor JM, Feifer A et al (2012) Combination of a novel gene expression signature with a clinical nomogram improves the prediction of survival in high-risk bladder cancer. Clin Cancer Res 18:1323–1333
70. Blaveri E, Simko JP, Korkola JE et al (2005) Bladder cancer outcome and subtype classification by gene expression. Clin Cancer Res 11:4044–4055
71. Dyrskjøt L, Zieger K, Real FX et al (2007) Gene expression signatures predict outcome in non–muscle-invasive bladder carcinoma: a multicenter validation study. Clin Cancer Res 13:3545–3551
72. Dyrskjøt L, Kruhøffer M, Thykjaer T et al (2004) Gene expression in the urinary bladder: a common carcinoma in situ gene expression signature exists disregarding histopathological classification. Cancer Res 64:4040–4048
73. Smith SC, Baras AS, Owens CR et al (2012) Transcriptional signatures of ral GTPase are associated with aggressive clinicopathologic characteristics in human cancer. Cancer Res 72:3480–3491

74. Sanchez-Carbayo M, Socci ND, Lozano J et al (2006) Defining molecular profiles of poor outcome in patients with invasive bladder cancer using oligonucleotide microarrays. J Clin Oncol 24:778–789
75. Choi W, Porten S, Kim S et al (2014) Identification of distinct basal and luminal subtypes of muscle-invasive bladder cancer with different sensitivities to frontline chemotherapy. Cancer Cell 25:152–165
76. Atapattu L, Lackmann M, Janes PW (2014) The role of proteases in regulating eph/ephrin signaling. Cell Adh Migr 8:294–307
77. Serim S, Haedke U, Verhelst SHL (2012) Activity-based probes for the study of proteases: recent advances and developments. ChemMedChem 7:1146–1159
78. López-Otín C, Bond JS (2008) Proteases: multifunctional enzymes in life and disease. J Biol Chem 283:30433–30437
79. Felix K, Gaida MM (2016) Neutrophil-derived proteases in the microenvironment of pancreatic cancer -active players in tumor progression. Int J Biol Sci 12:302–313
80. Pranjol MZI, Gutowski N, Hannemann M, Whatmore J (2015) The potential role of the proteases cathepsin D and cathepsin L in the progression and metastasis of epithelial ovarian cancer. Biomolecules 5:3260–3279
81. Olson OC, Joyce JA (2015) Cysteine cathepsin proteases: regulators of cancer progression and therapeutic response. Nat Rev Cancer 15:712–729
82. Drag M, Salvesen GS (2010) Emerging principles in protease-based drug discovery. Nat Rev Drug Discov 9:690–701
83. Turk B (2006) Targeting proteases : successes, failures and future prospects. Nat Rev Drug Discov 5(9):785–799
84. Vandooren J, Opdenakker G, Loadman PM, Edwards DR (2016) Proteases in cancer drug delivery. Adv Drug Deliv Rev 97:144–155
85. López-Otín C, Matrisian LM (2007) Emerging roles of proteases in tumour suppression. Nat Rev Cancer 7:800–808
86. Gocheva V, Joyce JA (2007) Cysteine cathepsins and the cutting edge of cancer invasion. Cell Cycle 6:60–64
87. Shi GP, Villadangos J a, Dranoff G et al (1999) Cathepsin S required for normal MHC class II peptide loading and germinal center development. Immunity 10:197–206
88. Bania J, Gatti E, Lelouard H et al (2003) Human cathepsin S, but not cathepsin L, degrades efficiently MHC class II-associated invariant chain in nonprofessional APCs. Proc Natl Acad Sci U S A 100:6664–6669
89. Stypmann J, Gläser K, Roth W et al (2002) Dilated cardiomyopathy in mice deficient for the lysosomal cysteine peptidase cathepsin L. Proc Natl Acad Sci U S A 99:6234–6239
90. Gocheva V, Zeng W, Ke D et al (2006) Distinct roles for cysteine cathepsin genes in multistage tumorigenesis. Genes Dev 20:543–556
91. Kuester D, Lippert H, Roessner A, Krueger S (2008) The cathepsin family and their role in colorectal cancer. Pathol Res Pract 204:491–500
92. Stoeckle C, Sommandas V, Adamopoulou E et al (2009) Cathepsin G is differentially expressed in primary human antigen-presenting cells. Cell Immunol 255:41–45
93. Kargi HA, Campbell EJ, Kuhn C 3rd (1990) Elastase and cathepsin G of human monocytes: heterogeneity and subcellular localization to peroxidase-positive granules. J Histochem Cytochem 38:1179–1186
94. Shimoda N, Fukazawa N, Nonomura K, Fairchild RL (2007) Cathepsin g is required for sustained inflammation and tissue injury after reperfusion of ischemic kidneys. Am J Pathol 170:930–940
95. Reeves EP, Lu H, Jacobs HL et al (2002) Killing activity of neutrophils is mediated through activation of proteases by K+flux. Nature 416:291–297
96. Tkalcevic J, Novelli M, Phylactides M et al (2000) Impaired immunity and enhanced resistance to endotoxin in the absence of neutrophil elastase and cathepsin G. Immunity 12:201–210

97. Owen CA, Campbell MA, Sannes PL et al (1995) Cell surface-bound elastase and cathepsin G on human neutrophils: a novel, non-oxidative mechanism by which neutrophils focus and preserve catalytic activity of serine proteinases. J Cell Biol 131:775–789
98. Brinkmann V, Reichard U, Goosmann C et al (2004) Neutrophil extracellular traps kill bacteria. Science (80) 303:1532–1535
99. Polanowska J, Krokoszynska I, Czapinska H et al (1998) Specificity of human cathepsin G. Biochim Biophys Acta 1386:189–198
100. Raymond WW, Trivedi NN, Makarova A et al (2010) How immune peptidases change specificity: cathepsin G gained tryptic function but lost efficiency during primate evolution. J Immunol 185:5360–5368
101. Adkison AM, Raptis SZ, Kelley DG, Pham CTN (2002) Dipeptidyl peptidase I activates neutrophil-derived serine proteases and regulates the development of acute experimental arthritis. J Clin Invest 109:363–371
102. Sambrano GR, Huang W, Faruqi T et al (2000) Cathepsin G activates protease-activated receptor-4 in human platelets. J Biol Chem 275:6819–6823
103. Maison CM, Villiers CL, Colomb MG (1991) Proteolysis of C3 on U937 cell plasma membranes. Purif Cathepsin G. J Immunol 147:921–926
104. Drag B, Petersen LC (1994) Activation of pro-urokinase by cathepsin G in the presence of glucosaminoglycans. Fibrinolysis 8:192–199
105. Reilly CF, Tewksbury DA, Schechter NM, Travis J (1982) Rapid conversion of angiotensin I to angiotensin II by neutrophil and mast cell proteinases. J Biol Chem 257:8619–8622
106. Klickstein LB, Kaempfer CE, Wintroub BU (1982) The granulocyte-angiotensin system. Angiotensin I-converting activity of cathepsin G. J Biol Chem 257:15042–15046
107. Shamamian P, Schwartz JD, Pocock BJZ et al (2001) Activation of progelatinase A (MMP-2) by neutrophil elastase, cathepsin G, and proteinase-3: a role for inflammatory cells in tumor invasion and angiogenesis. J Cell Physiol 189:197–206
108. Benes P, Vetvicka V, Fusek M (2008) Cathepsin D-Many functions of one aspartic protease. Crit Rev Oncol Hematol 68:12–28
109. Nicotra G, Castino R, Follo C et al (2010) The dilemma: Does tissue expression of cathepsin D reflect tumor malignancy? the question: does the assay truly mirror cathepsin D mis-function in the tumor? Cancer Biomarkers 7:47–64
110. Hah YS, Noh HS, Ha JH et al (2012) Cathepsin D inhibits oxidative stress-induced cell death via activation of autophagy in cancer cells. Cancer Lett 323:208–214
111. Dian D, Vrekoussis T, Shabani N et al (2012) Expression of cathepsin-D in primary breast cancer and corresponding local recurrence or metastasis: an immunohistochemical study. Anticancer Res 32:901–905
112. Lentari I, Segas I, Kandiloros D (2002) The importance of cathepsin's-D tissular detection in laryngeal squamous cell carcinoma. Acta Otorhinolaryngol Belg 56:383–389
113. Paksoy M, Hardal U, Caglar C (2011) Expression of Cathepsin D and E-Cadherin in primary laryngeal cancers correlation with neck lymph node involvement. J Cancer Res Clin Oncol 137:1371–1377
114. Dumartin L, Whiteman HJ, Weeks ME et al (2011) AGR2 is a novel surface antigen that promotes the dissemination of pancreatic cancer cells through regulation of cathepsins B and D. Cancer Res 71:7091–7102
115. Lösch A, Schindl M, Kohlberger P et al (2004) Cathepsin D in ovarian cancer: prognostic value and correlation with p 53 expression and microvessel density. Gynecol Oncol 92:545–552
116. González-Vela MC, Garijo MF, Fernández F et al (1999) Cathepsin D in host stromal cells is associated with more highly vascular and aggressive invasive breast carcinoma. Histopathology 34:35–42
117. Ohri SS, Vashishta A, Proctor M et al (2008) The propeptide of cathepsin D increases proliferation, invasion and metastasis of breast cancer cells. Int J Oncol 32:491–498

118. Vashishta A, Ohri SS, Proctor M et al (2006) Role of activation peptide of procathepsin D in proliferation and invasion of lung cancer cells. Anticancer Res 26:4163–4170
119. Szajda SD, Darewicz B, Kudelski J et al (2005) Cancer procoagulant and cathepsin D activity in blood serum in patients with bladder cancer. Pol Merkur Lek 18:651–653
120. Tokyol C, Köken T, Demirbas M et al (2006) Expression of cathepsin D in bladder carcinoma: correlation with pathological features and serum cystatin C levels. Tumori 92:230–235
121. Jean D, Rousselet N, Frade R (2006) Expression of cathepsin L in human tumor cells is under the control of distinct regulatory mechanisms. Oncogene 25:1474–1484
122. Tan G-J, Peng Z-K, Lu J-P, Tang F-Q (2013) Cathepsins mediate tumor metastasis. World J Biol Chem 4:91–101
123. Page-McCaw A, Ewald AJ, Werb Z (2007) Matrix metalloproteinases and the regulation of tissue remodelling. Nat Rev Mol Cell Biol 8:221–233
124. Egeblad M, Werb Z (2002) New functions for the matrix metalloproteinases in cancer progression. Nat Rev Cancer 2:161–174
125. Sun N, Zhao Q, Ye C et al (2014) Role of matrix metalloproteinase-1 (MMP-1)/ protease-activated receptor-1 (PAR-1) signaling pathway in the cervical cancer invasion. J Reprod Contracept 25:18–25
126. Sternlicht MD, Werb Z (2001) How matrix metalloproteinases regulate cell behavior. Annu Rev Cell Dev Biol 17:463–516
127. Kessenbrock K, Plaks V, Werb Z (2010) Matrix metalloproteinases: regulators of the tumor microenvironment. Cell 141:52–67
128. Vasala K, Turpeenniemi-Hujanen T (2007) Serum tissue inhibitor of metalloproteinase-2 (TIMP-2) and matrix metalloproteinase-2 in complex with the inhibitor (MMP-2:TIMP-2) as prognostic markers in bladder cancer. Clin Biochem 40:640–644
129. Mitsiades N, Yu WH, Poulaki V et al (2001) Matrix metalloproteinase-7-mediated cleavage of Fas ligand protects tumor cells from chemotherapeutic drug cytotoxicity. Cancer Res 61:577–581
130. Kleiner DE, Stetler-Stevenson WG (1999) Matrix metalloproteinases and metastasis. Cancer Chemother Pharmacol 43(Suppl):S42–S51
131. Hanemaaijer R, Sier CFM, Visser H et al (1999) MMP-9 activity in urine from patients with various tumors, as measured by a novel MMP activity assay using modified urokinase as a substrate. Ann N Y Acad Sci:141–149
132. Margulies IM, Hoyhtya M, Evans C et al (1992) Urinary type-IV collagenase—elevated levels are associated with bladder transitional cell-carcinoma. Cancer Epidemiol Biomark Prev 1:467–474
133. Cowden Dahl KD, Symowicz J, Ning Y et al (2008) Matrix metalloproteinase 9 is a mediator of epidermal growth factor-dependent E-cadherin loss in ovarian carcinoma cells. Cancer Res 68:4606–4613
134. Maretzky T, Reiss K, Ludwig A et al (2005) ADAM10 mediates E-cadherin shedding and regulates epithelial cell-cell adhesion, migration, and beta-catenin translocation. Proc Natl Acad Sci U S A 102:9182–9187
135. Massagué J (2008) TGF beta in Cancer. Cell 134:215–230
136. Yu Q, Stamenkovic I (2000) Cell surface-localized matrix metalloproteinase-9 proteolytically activates TGF-beta and promotes tumor invasion and angiogenesis. Genes Dev 14:163–176
137. Waldhauer I, Goehlsdorf D, Gieseke F et al (2008) Tumor-associated MICA is shed by ADAM proteases. Cancer Res 68:6368–6376
138. Le Maux Chansac B, Missé D, Richon C et al (2008) Potentiation of NK cell-mediated cytotoxicity in human lung adenocarcinoma: role of NKG2D-dependent pathway. Int Immunol 20:801–810
139. Bergers G, Brekken R, McMahon G et al (2000) Matrix metalloproteinase-9 triggers the angiogenic switch during carcinogenesis. Nat Cell Biol 2:737–744

140. Olson MW, Bernardo MM, Pietila M et al (2000) Characterization of the monomeric and dimeric forms of latent and active matrix metalloproteinase-9: differential rates for activation by stromelysin 1. J Biol Chem 275:2661–2668
141. Provatopoulou X, Gounaris A, Kalogera E et al (2009) Circulating levels of matrix metalloproteinase-9 (MMP-9), neutrophil gelatinase-associated lipocalin (NGAL) and their complex MMP-9/NGAL in breast cancer disease. BMC Cancer 9:390
142. Yan L, Borregaard N, Kjeldsen L, Moses MA (2001) The high molecular weight urinary matrix metalloproteinase (MMP) activity is a complex of gelatinase B/MMP-9 and neutrophil gelatinase-associated lipocalin (NGAL): Modulation of MMP-9 activity by NGAL. J Biol Chem 276:37258–37265
143. Aguirre Ghiso JA, Kovalski K, Ossowski L (1999) Tumor dormancy induced by downregulation of urokinase receptor in human carcinoma involves integrin and MAPK signaling. J Cell Biol 147:89–103
144. Eissa S, Ahmed MI, Said H et al (2004) Cell cycle regulators in bladder cancer: relationship to schistosomiasis. IUBMB Life 56:557–564
145. Di Cera E (2009) Serine Proteases. IUBMB Life 61:510–515. doi:10.1002/iub.186
146. Almonte AG, Sweatt JD (2011) Serine proteases, serine protease inhibitors, and protease-activated receptors: roles in synaptic function and behavior. Brain Res 1407:107–122
147. Jin T, Bokarewa M, Tarkowski A (2005) Urokinase-type plasminogen activator, an endogenous antibiotic. J Infect Dis 192:429–437. doi:10.1086/431600
148. Gyetko MR, Libre EA, Fuller JA et al (1999) Urokinase is required for T lymphocyte proliferation and activation in vitro. J Lab Clin Med 133:274–288
149. Vassalli JD (1985) A cellular binding site for the Mr 55,000 form of the human plasminogen activator, urokinase. J Cell Biol 100:86–92
150. Uusitalo-Seppälä R, Huttunen R, Tarkka M et al (2012) Soluble urokinase-type plasminogen activator receptor in patients with suspected infection in the emergency room: a prospective cohort study. J Intern Med 272:247–256
151. Jankun J, Skrzypczak-Jankun E (1999) Molecular basis of specific inhibition of urokinase plasminogen activator by amiloride. Cancer Biochem Biophys 17:109–123
152. Duffy MJ (2004) The urokinase plasminogen activator system: role in malignancy. Curr Pharm Des 10:39–49
153. Reuning U, Sperl S, Kopitz C et al Urokinase-type plasminogen activator (uPA) and its receptor (uPAR): development of antagonists of uPA/uPAR interaction and their effects in vitro and in vivo. Curr Pharm Des 9:1529–1543
154. Andreasen PA, Kjøller L, Christensen L, Duffy MJ (1997) The urokinase-type plasminogen activator system in cancer metastasis: a review. Int J Cancer 72:1–22
155. Gately S, Twardowski P, Stack MS et al (1996) Human prostate carcinoma cells express enzymatic activity that converts human plasminogen to the angiogenesis inhibitor, angiostatin. Cancer Res 56:4887–4890
156. Rabbani S, Mazar A, Bernier S et al (1992) Structural requirements for the growth factor activity of the amino-terminal domain of urokinase. J Biol Chem 267:14151–14156
157. Mukhina S, Stepanova V, Traktouev D et al (2000) The chemotactic action of urokinase on smooth muscle cells is dependent on its kringle domain. Characterization of interactions and contribution to chemotaxis. J Biol Chem 275:16450–16458
158. Hasui Y, Marutsuka K, Suzumiya J et al (1992) The content of urokinase-type plasminogen activator antigen as a prognostic factor in urinary bladder cancer. Int J Cancer 50:871–873
159. McIntyre JO, Matrisian LM (2009) Optical proteolytic beacons for in vivo detection of matrix metalloproteinase activity. Methods Mol Biol 539:155–174
160. Packard BZ, Artym VV, Komoriya A, Yamada KM (2009) Direct visualization of protease activity on cells migrating in three-dimensions. Matrix Biol 28:3–10
161. Bremer C, Tung CH, Weissleder R (2001) In vivo molecular target assessment of matrix metalloproteinase inhibition. Nat Med 7:743–748

162. Temma T, Sano K, Kuge Y et al (2009) Development of a radiolabeled probe for detecting membrane type-1 matrix metalloproteinase on malignant tumors. Biol Pharm Bull 32:1272–1277
163. Furumoto S, Takashima K, Kubota K et al (2003) Tumor detection using 18F-labeled matrix metalloproteinase-2 inhibitor. Nucl Med Biol 30:119–125
164. Smith SC, Theodorescu D (2009) The Ral GTPase pathway in metastatic bladder cancer: key mediator and therapeutic target. Urol Oncol Semin Orig Investig 27:42–47
165. Oxford G, Theodorescu D (2003) The role of Ras superfamily proteins in bladder cancer progression. J Urol 170:1987–1993
166. Chaudhary AK, Pandya S, Ghosh K, Nadkarni A (2013) Matrix metalloproteinase and its drug targets therapy in solid and hematological malignancies: an overview. Mutat Res—Rev Mutat Res 753:7–23
167. Santoni M, Amantini C, Morelli MB et al (2013) Pazopanib and sunitinib trigger autophagic and non-autophagic death of bladder tumour cells. Br J Cancer 109:1040–1050
168. Ishiguro H, Kawahara T, Zheng Y et al (2014) Differential regulation of bladder cancer growth by various glucocorticoids: corticosterone and prednisone inhibit cell invasion without promoting cell proliferation or reducing cisplatin cytotoxicity. Cancer Chemother Pharmacol 74:249–255
169. Cheng DLW, Shu WP, Choi JCS et al (1994) Bacillus Calmette-Guerin interacts with the carboxyl-terminal heparin binding domain of fibronectin: implications for BCG-mediated antitumor activity. J Urol 152:1275–1280
170. Belotti D, Paganoni P, Manenti L et al (2003) Matrix metalloproteinases (MMP9 and MMP2) induce the release of vascular endothelial growth factor (VEGF) by ovarian carcinoma cells: implications for ascites formation. Cancer Res 63:5224–5229
171. Sandes E, Lodillinsky C, Cwirenbaum R et al (2007) Cathepsin B is involved in the apoptosis intrinsic pathway induced by *Bacillus* Calmette-Guerin in transitional cancer cell lines. Int J Mol Med 20:823–828
172. Dezutter-Dambuyant C, Durand I, Alberti L et al (2016) A novel regulation of PD-1 ligands on mesenchymal stromal cells through MMP-mediated proteolytic cleavage. Oncoimmunology 5:e1091146

Endogenous Proteases in Tumoral Progression

Dalton Dittz Júnior, Fernanda de Oliveira Lemos and Ana Candida Araujo e Silva

1 Introduction

Tumor progression is a well-controlled process which comprises the steps from the origin of a tumor cell to its ability to invade and metastasize. During invasion, tumor cells need to detach from primary tumor and, as a result of a local proteolysis, to cross the basement membrane and migrate to vascular system [1]. The local proteolysis is facilitated by releasing proteases bound to cell surface or secreted from cells (tumoral and non-tumoral) present in tumor microenvironment. Thus, endogenous proteases have been associated with tumor progression because of their contribution during invasion and metastasis [2]. When tumor cells reach the circulatory system (intravasation), they disseminate through both vascular and lymphatic vessels, move out from the vasculature (extravasation), and colonize the surrounding tissue (metastasis) [3]. The balance between proteases and antiproteases, in vivo, seems to be the key for many diseases like cancer. During tumor initiation, growth, invasion, and metastasis, all five classes of endogenous proteases (serine, cysteine, aspartate, threonine, and metalloproteases) are involved [4]. In this review, we will discuss these protease classes separately, considering its participation in tumor progression.

D. Dittz Júnior (✉)
Pharmacology Department, Institute of Biological Sciences, Federal University of Minas Gerais, Belo Horizonte, Minas Gerais, Brazil
e-mail: dalton.dittz@gmail.com

F. de Oliveira Lemos
Pharmacology Department, School of Medicine, Yale University, New Haven, CT, USA

A.C. Araujo e Silva
Mucuri Medicine School, Federal University of Jequitinhonha and Mucuri Valleys, Teófilo Otoni, Minas Gerais, Brazil

S. Chakraborti and N.S. Dhalla (eds.), *Pathophysiological Aspects of Proteases*,
DOI 10.1007/978-981-10-6141-7_5

2 Serine Proteases

Over one-third of all known proteolytic enzymes are serine proteases. The name is due the presence of a nucleophilic serine (Ser) in the active site of these enzymes. According to MEROPS classification that divides proteases into clans based on catalytic mechanism and families on the basis of common ancestry, serine proteases are grouped into 13 clans and 40 families [5]. The family S1, clan PA, is the largest one and perhaps the best characterized group of proteases. These enzymes share structural similarities with trypsin and chymotrypsin, also members of this family. In the active site, there is a triad formed by Ser 195, His 57, and Asp 102 that is essential for catalytic activity [6, 7]. Trypsin and chymotrypsin cleave polypeptide chains at positively charged (Arg/Lys) or large hydrophobic (Phe/Trp/Tyr) residues, respectively. Most clan PA proteases have trypsin-like substrate specificity and prefer Arg or Lys side chains at the P1 position of substrate [8].

The family S1 is partitioned into two subfamilies, S1A and S1B, that share a common architecture but are philogenetically distinct and play different functions. The S1B proteases control the intracellular protein turnover and have a limited distribution in plants, prokaryotes, and the archaea. In contrast, the S1A proteases participate in a variety of extracellular processes and controlling many physiological processes as digestion, coagulation, immunity [9]. Examples of S1A members include trypsin, chymotrypsin, elastase, thrombin, urokinase-type plasminogen activator (uPA) system, tryptase, chymase, cathepsin G, matriptase, granzymes, and kallikrein [4, 7].

Activation of the most trypsin-like proteases requires proteolytic cleavage of an inactive zymogen precursor, involving the removal of one or more peptide ligations in N-terminal portion. In all known members of the family, the cleavage occurs in the same position (between residues 15- and 16-chymotrypsinogen numbering) [7]. The proteolysis is controlled by about 20 natural structurally diverse inhibitor families. Inhibitors are grouped according their mechanism of action in three different types: canonical (standard mechanism) and non-canonical inhibitors, and serpins. Serpins, similar to the canonical inhibitors, interact with their target proteases in a substrate-like manner; however, cleavage leads to an irreversible covalent acyl–enzyme complex, huge conformational changes in inhibitor, and disruption of protease active site [10]. A strict regulation of serine protease activities is essential to maintain physiological process. The abnormal activity of proteolytic enzymes promotes excessive tissue destruction and/or the aberrant processing of other proteins and peptides resulting in pathological disorders, as chronic inflammation, coagulation disturbs, and cancer [4, 11]. Following, we discuss about the role of the main serine proteases in cancer.

2.1 Trypsin

For a long time, trypsin was known only as a digestive enzyme. However, the detection of trypsin in patients that undergone pancreatectomy suggested that this serine protease is produced in other sites [12]. Later, trypsin expression was reported in epithelial cells in stomach, small intestine, colon, lung, liver, skin, kidney, neuronal cells, spleen, and inflammatory cells [13]. Trypsin is secreted as zymogen (trypsinogen) and in four isoforms with great structural homology: Trypsinogen-1, -2, and -3 are common in epithelial tissues and trypsinogen-4 in the brain [13, 14]. Trypsin activation is dependent on trypsinogen cleavage by enteropeptidases present on duodenal lumen and enterocytes and can be prevented by a antiprotease mediator pancreatic secretory trypsin inhibitor (PSTI), also named Kazal inhibitor [15]. Loss of balance in trypsin/PSTI activation leads to pathological conditions, as pancreatitis, and indicates increased risk for developing pancreatic adenocarcinoma [13]. The same peptide is secreted by tumor cells and is often referred to as "tumor-associated trypsin inhibitor" (TATI). TATI is a low molecular weight (6 kDa) coexpressed with trypsin in many cancer forms, and an elevated serum level is associated with poor prognosis. In colorectal cancer, TATI was indicated as a prognostic marker and its presence indicates poor tumor differentiation [16].

Trypsin is also overexpressed in many tumor cell type of ovary, prostate, lung, stomach, colon where it involved with proliferation and metastasis [4, 17]. These effects are mediated through activation of protease-activated receptor-2 (PAR-2). Proteinase-activated receptors (PARs) are G-protein-coupled receptors activated by many proteases. They comprise of four receptors that are activated by thrombin (PAR-1, -3, and -4) or by multiple trypsin-like enzymes, such as trypsin itself and mast cell tryptase (PAR-2). The activation is an irreversible phenomenon involving cleavage of the amino-terminal exodomain of the receptor by the protease, producing a new amino-terminal sequence that binds the core receptor and activates transmembrane signaling through G-proteins [17].

Cellular proliferation and differentiation are stimulated through PAR-2-driven signals and could contribute to the uncontrolled cellular growth in gastric and colorectal mucosa and invasion of pancreatic cancer by expressions of MMP-2, MMP-7, and MMP-9. These proteases promote degradation of extracellular matrix (ECM), facilitating the invasion, angiogenesis, and metastasis. The participation of PAR-2 in cancer invasion is suggested by the increased activity of MMP-2 and occurs via MMP/EGFR/MAPK/ERK1/2 pathway [17]. PAR-2 may also promote prostaglandin E_2 release through Ca^{2+} channel activation resulting in EGFR-stimulated cell proliferation [13, 18]. PAR-2 is also overexpressed by breast, gallbladder, lung, kidney, uterine, and cervical cancers and glioblastoma tumors, but the exact contribution of this receptor in the pathogenesis of these cancers remains unclear. In contrast, PAR-2 exerts a tumor-protective role in skin carcinogenesis [19]. An in vitro study with pancreatic carcinoma cells showed trypsin-induced proliferation and migration. These events were promoted via PAR-2 activation and the consequent increase of MMP-2 expression [18]. Further, recent findings showed that trypsin is overexpressed in head and neck cancer cells and can promote pain via PAR-2 activation [20].

2.2 Thrombin and Urokinase-Type Plasminogen Activator (UPA) System

Thrombin and uPA systems are known for their role in regulating hemostatic and fibrinolytic pathways. In addition, these enzymes also control many other processes including wound healing, ECM remodeling, angiogenesis, tumor invasion, and metastasis [21–23]. In addition to cleaving fibrinogen, thrombin activates cells through a proteolytic mechanism involving PAR-1, PAR-3, or PAR-4 activations [17]. PAR-1, -3, and -4 are expressed in epithelium, neurons, astrocytes, and immune cells [24, 25].

The expression of PARs (mainly PAR-1 and PAR-2) has been implicated in the occurrence of several types of human cancers and correlated with degree of invasiveness and prognostic [17, 19, 26]. PAR-1 expression has been detected in human pancreatic and breast cancers, colon adenocarcinoma, laryngeal carcinoma cell line, and melanoma [17, 26, 27]. In breast cancer, PAR-1 expression levels were directly correlated with the degree of invasiveness in primary breast tissue, especially in infiltrating ductal carcinoma. PAR-1 activates MMP-1 that stimulates cell migration and invasion of breast and gastric cancer cells, facilitating metastasis processes. Previous studies demonstrated that thrombin acts as a mitogen for vascular smooth muscle cells by activation of the nuclear factor kappa B (NF-κB) or driven by the release of bFGF by these cells. In osteosarcoma cell, thrombin induced in vitro expression and association of beta 1-integrin with MMP-9 in the cell surface localization of the protease by the integrin promotes tumor cell invasion [28].

The uPA system has been related to the tumor cell in tumor cell invasion and metastasis in cancer, specially breast, lung, and ovarian types [22, 29, 30]. This system consists of the serine protease urokinase (uPA), its cell surface receptor (uPAR), the serpin plasminogen activator inhibitors (PAI-1 and PAI-2), and the serine protease precursor plasminogen. Urokinase is produced as zymogen, denoted prourokinase, that is converted into its active form by plasmic cleavage. Active uPA cleaves plasminogen, thereby generating proteolytically active plasmin that in turn degrades components of the ECM as fibrin, fibronectin, laminin, and proteoglycans or activates certain matrix-degrading enzymes, as MMPs [30].

Based on this ability to inhibit uPA activity, it was expected that PAI-1 would suppress cancer progression. However, consistent data from clinical studies suggest that PAI-1 is also involved in mediating cancer progression. A recent investigation reported that a specific inhibitor of PAI-1 blocked angiogenesis and tumor progression in mice, evidencing a possible protumor effect of PAI-1 [31]. This is corroborated by a previous study that demonstrated defective angiogenesis occurred in PAI-1 deficient mice [32]. In addition to angiogenic promotion, PAI-1 can display antiapoptosis activity. In vitro, recombinant PAI-1 inhibited cytotoxic drug-induced apoptosis in prostate and leukemia cell line [33] and protects endothelial and other types of cells from apoptosis [34].

Finally, the overexpression of uPA, PAI-1, or uPAR is associated with advanced breast, lung, and ovarian cancer and can be used as an independent prognostic factor in several studies. They have been correlated with a short progression free and overall survival time for patients [35, 36]. Indeed, uPA and PAI-1 are among the best validated prognostic biomarkers available for breast cancer, and their dosage can drive the treatment. Low levels of both uPA and PAI-1 in lymph node-negative patients indicate a low risk of disease recurrence, and thus, patient may be spared from the toxic side effects and costs of adjuvant chemotherapy. In the opposite, lymph node-negative patients with high levels of these markers should be treated with adjuvant chemotherapy [30].

2.3 Tryptase, Chymase, and Cathepsin G

Tryptase, chymase, and cathepsin G (cath-G) are neutral serine protease trypsin-like present in secretory granules of mast cell (MC) and are involved in inflammatory process, angiogenesis, tissue repair, host defense, linking adaptive and innate immunity. Tryptase is stored in four different fully active forms in MC granules: α-(released from MCs in the bloodstream); β-(concentrated in the secretory granules of MCs and released only after degranulation—major content); γ- and δ-tryptase [37].

Human chymase is synthesized as an inactive pre-proenzyme, which is then converted to the proenzyme. The activation occurs into the secretory granules and a thiol proteinase; dipeptidyl peptidase I is necessary in this process. Chymase is enzymatically inactive at pH 5.5, and its optimum enzymatic activity is achieved at a pH of 7–9 following release into the interstitial tissues. Connective tissue and the submucosa are the major sites of MC-containing chymases [38].

Tryptase and chymase are involved in angiogenesis by activating MMPs as MMP-9, plasminogen activators or releasing growth factors by degrading ECM components or release matrix-associated growth factors. All these actions contribute to angiogenesis as well as invasion and metastasis of tumor cells. Tryptase also is a mitogen for human dermal microvascular endothelial cells with a significant increase of capillary growth, which is suppressed by specific tryptase inhibitors, and stimulates releasing of proangiogenic factors such as IL-1, IL-6, IL-8, stem cell factor, and TNF-α [37].

An evidence of the participation of tryptase in tumor progression is that tryptase-positive MCs increase in number and vascularization increases in a linear fashion in solid tumors, such as human malignant melanoma, endometrial carcinoma, breast cancer, uterine leiomyomas, gastric cancer, colorectal cancer, pancreatic ductal adenocarcinoma. In B cell non-Hodgkin's lynphomas and benign lymphadenopathies, tryptase-positive mast cell counts correlate with the development of angiogenesis and both with malignancy grades [37, 39–43].

Cath-G is a 26-kDa neutral serine protease found in the azurophil granules of neutrophils and a subset of monocytes. It is the major serine protease released by activated neutrophils, acting through hydrolysis of a list of proteins (such as chemoattractants), ECM, and hormonal factors, playing an important role in inflammation [44, 45]. Cath-G is found in lung and breast tumors and has been implicated in tumor progression and metastasis by facilitating angiogenesis and tumor cell dissemination. A mechanism purposed is the stimulus of VEGF and MCP-1 secretion via TNF-beta signaling activation [46]. Another mechanism is the facilitation in forming cell aggregates as verified in vitro with MCF-7 mammary adenocarcinoma cells is exposed to a supernatant from culture of neutrophil activated, or in the presence of these cells, there is spontaneous formation of aggregates. It is postulated that these aggregates disseminate in the bloodstream and form tumor emboli at distant sites, potentially leading to the establishment of metastatic foci. In the presence of activated neutrophil supernatant, with inhibition of cath-G, the aggregation is abolished, suggesting the participation of cath-G on this process and the dependence on proteolytical activity [47, 48].

2.4 Kallikreins

Kallikreins form a group of 15 members (KLK 1-15) of serine protease S1 family, sharing similarities with trypsin and chymotrypsin. They are present in diverse tissues and biological fluids, playing important roles in inflammation, reproductive function, blood clotting, fibrinolysis, and cancer. These enzymes are categorized in tissue kallikrein and plasma kallikrein, differing significantly in molecular weight, substrate specificity, immunological characteristics, gene structure, and type of kinin released from kininogen cleavage [49, 50].

Currently, the potential roles of kallikreins in cancer progression have been investigated. These enzymes are thought to promote invasion of tumor cells by degrading ECM components or cleaving cell–cell adhesion proteins (e.g., E-cadherin). Indeed, KLK1, 2, 4, 5, 6, and 14 can activate protease-activated receptors (PARs) and stimulate cellular migration, contributing to tumor spreading. In vitro studies showed that KLK1, 3, 10, and 13 increase the invasiveness of tumor cells [51] and KLK6 also is involved in human melanoma progression. Although KLK6 was not detectable in tumor cells, KLK6 protein expression was detected in adjacent cells to primary melanomas, cutaneous metastatic lesions, and benign nevi lesions, indicating a possible paracrine function of extracellular KLK6 during cancer development. This was evidenced by the observation of a recombinant KLK6 protein-promoted melanoma cell migration and invasion, with an intense intracellular Ca^{2+}-flux, through PAR1 activation [52]. Despite activities in angiogenesis and proliferation, some kallikreins can play dual role: as promoter or suppressor tumor. It is possible that KLK3 protein regulates oxygen balance in tumors, inhibiting cell migration, and tube formation of endothelial cells in vitro and presents effects on inflammation. T cell proliferation and dendritic cell maturation may be inhibited by KLK3, indicating immunosuppressive effects of this protein. By the way KLK3 seems to stimulate IFN-γ secretion by natural killer cells, displaying a proinflammatory action. Thus, classifying kallikreins as tumor promoters or suppressors is still difficult, needing other functional studies [51].

Because of the number and roles of kallikreins, there is an increasing interest for use as diagnostic markers and for therapeutic targets. Currently, for prostate cancer diagnoses, prostate-specific antigen (PSA) is a useful biomarker and KLK2 and KLK11 have been emerging for this purpose. In ovarian cancer, KLK5, 6, 7, 10, 11, and 14 have been considered as biomarkers [53]. Despite their known importance, many of the functions of kallikreins need to be enlightened [52].

2.5 Matriptase

Matriptase is a type II transmembrane serine protease expressed in endothelial, neural, and white blood cells and in a wide range of epithelial tissues, such as gastrointestinal and respiratory tracts, and in the epidermis. Structurally, matriptase contains a serine protease domain, four low-density lipoprotein–receptor domains, a transmembrane domain, and two CUB domains [54–56]. It plays a critical role in skin formation, epidermal differentiation, and skin function. Matriptase-ablated mice die shortly after birth, due to a severe dehydration that results from an impaired epidermal barrier function [57, 58]. Matriptase is nearly ubiquitously coexpressed with the Kunitz-type transmembrane serine protease inhibitors, HAI-1 and HAI-2, in adult and embryonic tissues [9], and both HAI-1 and HAI-2 display potent matriptase inhibitory activity in purified systems [59, 60].

Overexpression of matriptase mRNA or protein was detected in a wide variety of benign and malignant tumors of epithelial origin, as gastric, pancreatic, breast, cervical, ovarian, endometrial, and prostate carcinomas [61–67]. Overexpression of matriptase is a significant predictor for poor prognosis in endometrial breast cancer [66] and correlated with tumor aggressiveness in ovarian cancer [68]. An increased matriptase/HAI-1 ratio was indicative of the poor clinical outcome of advanced-stage tumors, suggesting that loss of protease inhibition may play a role in the late stages of the disease [61]. It has been suggested that the ratio of these two gene products may serve as a promising biomarker for prostate cancer progression and a potential marker for establishing the efficacy of therapeutic and chemopreventive interventions [69]. The molecular mechanisms of matriptase-induced epithelial carcinogenesis remain to be determined. In cell-free or cell-based assays, matriptase can activate several proteins that have been previously associated with malignant progression, including prohepatocyte growth factor (pro-HGF), pro-uPA, PAR-2, suggesting that matriptase can be involved in angiogenesis and degradation of ECM [7, 56, 70].

3 Cysteine Protease

3.1 Cathepsin

Cathepsins (cath) B, C, F, H, K, L, O, S, V, W, and X form a class of the cysteine proteases which contain a highly conserved cysteine residue in their active sites. Cathepsins are optimally active and stable at slightly acidic pH and inactive at neutral pH, with the exception of cath-S. Cathepsins are synthesized as preproenzymes, and the signal peptide is removed during the translocation to the endoplasmic reticulum, where cathepsins are also *N*-glycosylated. Procathepsins are transported to the endosomal/lysosomal compartments and proteolytically processed to the active forms, either autocatalytically or with the help of other proteases [71–73]. It has already been described that cysteine cathepsins have an important role in the cancer progression of solid tumors and metastasis development. In agreement with this, the levels of cystatins, natural cysteine protease inhibitors, can also vary quite widely in different cancers [74].

Cathepsin expression is frequently increased in patient sera and in tumors in comparison with normal tissue and is significantly associated with poor prognosis in breast, lung, head and neck, and colorectal cancers, melanoma, and many other cancers [75]. Cathepsins most commonly associated with cancer are cath-B, C, H, L, S, and X [72].

Numerous studies have been reported that cath-B and -L are capable of degrading the proteins of the basement membrane and extracellular matrix (ECM), such as laminin, fibronectin, and collagen IV, thus facilitating tumor cell invasion and metastasis. In addition, these cathepsins can activate other enzymes, such as metalloproteases and urokinase-type plasminogen activator, which act downstream in the proteolytic cascade and cause even more extensive degradation of the ECM [71, 75–81]. Cysteine cathepsins also degrade cell adhesion molecules, such as E-cadherin. E-cadherin can be specifically cleaved and removed from the cell surface by the cath-B, -L and -S, and the decrease of this cell adhesion protein correlates with a more invasive tumor phenotype [71].

Cathepsins can also regulate angiogenesis. As their participation in ECM degradation facilitates capillary formation, cysteine cathepsins are involved in the generation of pro- and antiangiogenic factors and exposure of cryptic protein sequences that play a functional role in angiogenesis. Tumor neovascular formation was directly related to cath-B, S, L and X overexpression [71, 75, 77, 81].

Besides participation in invasion, metastasis, and angiogenesis, cysteine proteases have a pivotal role in the regulation of innate and adaptive antitumor immune response. Their high concentration in tumor-infiltrating innate immune cells, such as macrophages, mast cells, and granulocytes, aids tumor cells in tissue remodeling, angiogenesis, and consequently, metastasis [76].

In contrast to tumorigenic effect of cathepsins, these enzymes have a proapoptotic activity in several tumor cell lines. Cathepsins have been shown to participate in apoptosis event promoted by tumor necrosis factor-a (TNF-a) or TNF-related apoptosis-inducing ligand (TRAIL) in human cervical carcinoma, murine fibrosarcoma, ovarian cancer, prostatic cancer, and hepatoma cells. It is important to highlight that apoptosis induced by TRAIL shows to specifically target cancer cells without damaging normal cells and tissues, interesting observation for cancer therapy [81].

3.2 Calpain

Calpain is a well-conserved family of Ca^{2+}-dependent cysteine proteases that the enzymatic activity is regulated by cytosolic calcium [82–84]. Differently from other proteases that totally degrade the substrate, calpains conduct a limited process, after which the modified target protein may acquire an additional basic function or a novel function. Fifteen different calpain isoform genes have already been identified in human [82, 85]. The representative members of the calpain family, μ–calpain and m–calpain, are heterodimers consisting of a catalytic (80 kDa) subunit and a regulatory subunit (28 kDa). Some isoforms of calpain are ubiquitously expressed, such as μ–calpain and m–calpain, and others are expressed in a tissue-specific manner, such as calpain 9, which is found in the digestive tract [82–84].

Aberrant expression of calpain has been implicated in tumorigenesis. Overexpression of calpain was observed in different cancer type, such as schwannomas and meningiomas, renal cell carcinoma and colorectal adenocarcinomas, uterine cervical neoplasia, sarcomas, and carcinosarcomas. However, the downregulation of calpain is also related to cancer, e.g., in melanoma and gastric cancer [86]. In addition, aberrant level of calpastatin, and ubiquitously expressed endogenous inhibitor of μ–calpain and m–calpain, has been also observed in tumor tissue, e.g., in endometrial cancer [83]. The relative level of expression of various calpain system members within a number of different tumors is implicated in disease prognosis and progression in a number of cancers, including breast, ovarian, pancreatic, and gastro-esophageal tumors [84].

The calpain family has an established role in mediating cellular migration through their ability to regulate focal adhesion dynamics in an enzymatic-dependent manner. Calpain associates with focal adhesion structures and directly cleaves focal adhesion proteins, e.g., FAK (Focal Adhesion Kinase), talin, paxillin, fodrin, ezrin, vinculin, the actin-regulating protein cortactin, the intermediate filament protein vimentin, the protein tyrosine phosphatase PTPN23, and the cell–cell adhesion molecule PTPμ [82–84]. Calpain-2, up-regulated in localized and in metastatic prostate cancer, can also disrupt the E-cadherin adhesion complex and facilitate tumor cells invasion [82]. Inhibition of calpain has been shown to stabilize peripheral focal adhesions and subsequently decrease the rate of retraction at the rear of the cell, inhibiting migration. Calpain can also regulate the membrane protrusion at the leading edge of migrating cells by modifying actin cytoskeleton, such as cortactin [83]. Several studies also suggest that calpains are involved in matrix metalloprotease (MMP) expression and secretion, in particular MMP-2 and

MMP-9, which are important for the degradation of the extracellular matrix [82, 84].

Calpain has also been shown to play a role in the regulation of the cell cycle at a number of important steps. In breast cancer, calpain has been shown to cleave cyclin E, a protein required for the transition from G1 to S phase, to a hyperactive low molecular weight form. In addition, cyclin E overexpression has been shown to activate m-calpain [83–85]. Calpain activity also promotes hyperphosphorylation of retinoblastoma (RB) and alters the levels of cyclins A and D, as well as cyclin-dependent kinase 2 (CDK2). Calpain can cleave the CDK inhibitor p27 in a MAPK-dependent process; the cleavage of p27 is involved in G1/S transition. Furthermore, calpain can interfere with the interaction between protein phosphatase 2A (PP2A) and AKT to prevent forkhead box O (FOXO)-mediated cell death. Interestingly, PP2A can also negatively regulate calpain during cell migration [83].

Calpain has been implicated in both cell death and survival. As a prosurvival effect, calpain is able to cleave wild-type p53, regulating protein stability to prevent p53-dependent apoptosis. Growth arrest-specific protein 2 (GAS2), a protein which is cleaved during apoptosis to allow rearrangement of the actin cytoskeleton, can physically associate with calpain to prevent p53 cleavage resulting in enhanced p53 stability. In addition, calpain can promote survival through activation of NF–κB by cleavage of its inhibitor IκBα [82–85]. Calpain is responsible for the proteolysis of various substrates that can sensitize cells to apoptosis, such as the transcription factor Myc. Cleavage of Myc by calpain converts the active protein to Myc-nick, which is a transcriptionally inactive [83, 84].

In addition to survival, calpain has a clear role in some cellular apoptotic pathways. Calpain-mediated proteolysis of a number of caspases (caspase-3, -7, -9, -10, and -12) has been demonstrated; however, this can result in activation (caspase-3, -7, -10, and -12) or inactivation (caspase-9) of these cysteine proteases. Furthermore, calpain is implicated in proapoptotic pathways by the cleavage of a number of other apoptosis-associated proteins such as BAX, BCL-2, JNK, JUN, CDK5, APAF1, and FOS [82–85]. Besides, mitochondrial-located calpain is implicated in caspase-independent apoptosis—calpain cleavage of apoptosis-inducing factor (AIF) allows the mitochondrial release of AIF. Autophagy-related gene 5 (Atg5) protein is an ubiquitin ligase required for the formation of autophagosomes. Calpain can cleave Atg5 to switch autophagy to apoptosis via Atg5 translocation to the mitochondria, which results in cytochrome c release [83, 84].

3.3 Caspases

Caspases are a family of cysteine-dependent aspartate-specific proteases, which contain cysteine residue at their active site and cleave their substrate at position next to aspartate residue. Protein cleavage by nucleophilic cysteine residues is a well-conserved mechanism among proteases, but the specific affinity for aspartate is unique to caspases [87, 88]. Caspases are largely known for their role in controlling cell death (caspase-2, -3, -6, -7, -8, -9, and -10) and inflammation (caspase-1, -4, -5,

-11, and -12) [89, 90]. The apoptotic caspases are subdivided into the initiators (caspase-2, -8, -9, and -10) and the effectors (caspase-3, -6, and -7) in which the initiators are activated by dimerization, while the executioner caspases require proteolytic cleavage of prodomains by activated upstream caspases. Activated caspases result in intracellular proteolysis of cytoskeletal components and proteins vital for organelle integrity toward cell execution (apoptosis). The two main pathways signaling the execution of apoptosis are the extrinsic (death receptor-mediated) and intrinsic (mitochondrial) [87, 90, 91]. Considering the importance of caspase to the apoptosis, it is intuitive that loss of these proteases promotes tumor development. However, caspases can also interfere in other tumorigenic events, such as proliferation, invasion, and migration.

Caspase-1 and -2 are tumor suppressor enzymes frequently downregulated in human cancers. Caspase-1 is especially downregulated in prostate cancer. It was reported that androgen-independent prostate cancer cell lines overexpressing caspase-1 are more prone to death induced by irradiation. About caspase-2, the low expression of this caspase is correlated with poor prognosis in some cancers, such as gastric and colorectal cancers, whereas the high caspase-2 levels are associated with remission and survival in adults with acute lymphoblastic leukemia and acute myeloid leukemia [89].

About caspase-3, it was observed that these gene variant alleles are associated with squamous cell carcinomas of the head and neck, endometrial cancer, non-Hodgkin's lymphoma, and multiple myeloma. Upregulation of caspase-3 was observed in malignant breast cancer compared to non-malignant breast cancer tissue. Despite the increase of tumor cell apoptosis, the increase of caspase-3 exacerbates the secretion of PGE2, causing the repopulation of tissue with tumor cells and resistance to radiotherapy [89, 90, 92].

Despite mutations in caspase-6 and caspase-7 are uncommon in most human cancers, they are especially important to colon and gastric cancers. Somatic mutations in caspase-6 and -7 genes associated with reduced apoptosis can promote susceptibility to lung cancer and endometrial cancer [89, 92]. Differently from caspase-6 to -7, mutations in caspase-8 gene are not rare, being found in 5% of invasive colorectal carcinomas, but not adenomas, and in hepatocellular carcinomas, advanced gastric cancers, malignant neuroendocrine tumors, and small cell lung carcinoma. Intriguingly, a specific polymorphism in the caspase-8 gene, D302H, is associated with significantly reduced overall risk breast cancer [89, 90].

About caspase-9, it was found a positive correlation between polymorphisms in caspase-9 genes and the risk of lung cancer in smokers, evidencing a case of gene–environment interaction [92]. Regarding caspase-10, it is known that inactivating caspase-10 genes has been detected in non-Hodgkin's lymphoma, acute lymphoblastic leukemia, and multiple myeloma, as well as colon, breast, lung, hepatocellular carcinoma, and gastric cancers [89].

4 Aspartate Protease

4.1 Cathepsin D

Cathepsin D (cath-D) is a soluble lysosomal aspartic endoproteinase that requires an acidic pH to be proteolytically active (pH 4.5–5.0) [93–95]. Human cath-D is synthesized as a 52-kDa pre-procathepsin-D that is converted to a procathepsin-D (pcath-D) 48-kDa single-chain intermediate within endosomes and, then, to the fully active mature protease in lysosomes. In some physiological and pathological conditions, pcath-D/cath-D escapes normal targeting mechanism and is secreted from the cells [94]. In contrast to other tissue proteases (e.g., serine proteases and metalloproteinases), no endogenous cath-D tissue inhibitor is known in mammal [95].

High levels of cath-D and pcath-D were reported in several human tumors such as prostate, breast, endometrial, and ovarian tumors; colorectal, lung, bladder, and gastric carcinomas and melanoma. Clinical reports have even made positive associations between level of cath-D or pcath-D and tumor size, tumor grade, tumor aggressiveness, incidence of metastasis, prognosis, and a degree of chemoresistance [94].

Several reports have indicated that cath-D and pcath-D stimulate cancer cell proliferation. The mitogenic effect of cath-D can be explained by its enzymatic activity. Several growth factors, growth factor receptors, and extracellular matrix (ECM) components have been found among cath-D substrates. Moreover, cath-D digests various chemokines and may therefore attenuate antitumoral immune response. Regarding pcath-D mitogenic effect, the secreted pcath-D could be converted to the active enzyme in an extracellular milieu because of the fact that the extracellular pH of tumors is moderately acidic. The cell proliferation effect of cath-D can be also explained independently of enzymatic activity. Some studies demonstrated that enzymatically inactive pcath-D mutants stimulate growth of cancer cells in vitro and in vivo in the same manner as wild-type pcath-D, and the growth-promoting effect of pcath-D on cancer cells is not inhibited by pepstatin (an aspartic protease inhibitor). Then, it is hypothesized that pcath-D secreted from cancer cells serves as an autocrine growth factor for cancer cells since pcath-D can bind to cell surface [93–96]. The mitogenic activity of cath-D and pcath-D can also explain the vessel formation observed in different solid tumors. Cath-D can release proangiogenic growth factors from ECM, e.g., bFGF. In addition, a direct stimulation of endothelial cell growth is speculated, via a yet unidentified cell surface receptor, that could be present on both cancer cells or endothelial cells [94, 95]. In contrast to be observed in several tumor tissues, pcath-D secreted by prostate cancer cells was shown to have a possible role in generating angiostatin via proteolysis—a specific inhibitor of angiogenesis in vitro as well as in vivo [96].

In addition to the mitogenic effects of cath-D on cancer cells and endothelial cells, numerous studies demonstrated the involvement of cath-D in cancer invasion and metastasis. Different in vitro and in vivo tumor models showed that metastasis stimulation by cath-D seems to be more dependent on cell proliferation, favoring the growth of micrometastases, rather than on the invasive potential [95].

About the involvement of cath-D in apoptosis, it has already described that cath-D can either promote or prevent apoptosis. The enzymatic activity is implicated in the mechanism of proapoptosis effect of cath-D. Cath-D mediates apoptosis induced by many apoptotic agents since it is released from the lysosome to cytosol, leading in turn to the mitochondrial release of cytochrome c into the cytosol, activation of procaspase-9 and -3, cleavage of Bid at pH 6, or Bax activation independently of Bid cleavage. Numerous studies have shown that pepstatin A, an aspartic protease inhibitor, could partially delay apoptosis induced by different proapoptotic agents [93, 95]. Regarding the antiapoptotic effect of cath-D, the mechanism has not been elucidated yet. Some studies have shown that xenografts of cancer cells overexpressing cath-D displayed less tumor apoptosis than mock-transfected cancer cells and that cath-D protected human neuroblastoma cells from doxorubicin-induced cell death [93].

4.2 Cathepsin E

Cathepsin E (cath-E) is an intracellular aspartic protease that is highly homologous to the cath-D. However, differently from cath-D, cath-E is a non-lysosomal protease predominantly expressed in cells of the immune system such as macrophages, lymphocytes, microglia, and dendritic cells. Cath-E is localized in plasma membranes, endosomal structures, endoplasmic reticulum, and Golgi apparatus [97–99].

Several adenocarcinomas and squamous cell carcinoma show high levels of cath-E expression and secretion, including lung, gastric, pancreatic ductal and bladder adenocarcinomas, and lung squamous cell carcinoma [97, 99–101]. However, the clinical significance of the increased cath-E expression in tumorous tissue has been controversial. The high expression of cath-E is related to a poor prognosis in some premalignant tumors, such as lung carcinoma, pancreatic ductal adenocarcinoma, and colorectal adenomas, whereas cath-E is associated with higher survival rates in hepatocellular carcinoma, breast and bladder cancers [99, 102].

The positive outcome in cancers that overexpress cath-E can be explained by the fact that cath-E prevents tumor growth and metastasis by inhibiting angiogenesis and augmenting the immune response [98, 100]. In carcinoma cell lines, cath-E induces tumor growth arrest and apoptosis by catalyzing the release of tumor necrosis factor-related apoptosis-inducing ligand (TRAIL) from the cell surface and by increasing the expression of interleukin-12 and endostatin, antiangiogenic proteins [97].

5 Threonine Proteases (Proteasome)

The ubiquitin proteasome system (UPS) is responsible for 80–90% of the degradation of all eukaryotic cellular proteins [103, 104]. This system is composed by ubiquitin, ubiquitination enzymes, deubiquitination enzymes (DUBs), and proteasome. First, ubiquitin monomers are conjugated to other proteins by a ubiquitin-activating enzyme (E1), resulting in their activation. The complex formed is transferred to a ubiquitin-conjugating enzyme (E2 or UBC) and then to the target protein. An ubiquitin ligase (E3) is required in the last step, which is responsible by selecting a protein for ubiquitination [105].

Human 26S proteasome consists of a proteolytic core particle (20S proteasome) and 19S regulatory particles (19S regulatory complex) that cap 20S at both ends. The proteasome is a threonine protease, where the N-terminal threonine of the β-subunit bonds in target proteins [105]. A polyubiquitinated protein is recognized by the 19S proteasome that unfolds and liberates the protein from the polyubiquitin chain. Following, 19S promotes the translocation of the protein into a chamber inside 20S proteasome for degradation. Through DUB action, ubiquitin molecules are recycled and the peptides generated are used for antigen presentation or degradated into amino acids that are used for new protein synthesis [103, 105].

The UPS is involved in every cellular function, including cell proliferation, apoptosis, migration, and invasion, and its deregulation may lead to cancer development. Overexpression of UPS has been identified in many types of cancer, including colorectal and gastric cancers [104]; novel antitumor agents based on inhibition of proteasome inhibitors have been studied, and some were approved to treat multiple myelomas (e.g., carfilzomib, marizomib, ixazomib, oprozomib) [103]. Programmed cell death is promoted when proteasome is inhibited, due to accumulation of misfolded and ubiquitinated intracellular debris, and the minor degradation of proapoptotic factors. NF-kB is a cytosolic protein that is inactivated by IkB family inhibitors. When IkB is phosphorylated, it becomes a target for degradation by the 26S proteasome. Thus, NF-kB is able to translocate to the nucleus which displays an antiapoptotic effect. If proteasome is inhibited, IkB availability is bigger, resulting in consistent inhibition of NF-kB leading to an increase in apoptosis [103, 105].

6 Matrix Metalloproteinases

Matrix metalloproteinase (MMP) is a family of enzymes with conserved methionine residue in the active site and depends on the zinc ion for enzymatic reactions. Members of MMP family include the "classical" matrix metalloproteinases (MMPs), a disintegrin and metalloproteinases (ADAMs), and ADAMs with thrombospondin motifs (ADAMTS) [106]. According to their substrate specificity, MMPs are classified as collagenases (MMP-1, -8 and -13), gelatinases (MMP-2 and -9), stromelysins (MMP-3, -10, and -11) and a heterogeneous group containing matrilysin (MMP-7), metalloelastase (MMP-12), enamelysin (MMP-20), endometase (MMP-26), and epilysin (MMP-28). In addition, a set of MMPs are anchored to the cell surface (referred as MT-MMP-14, -15, -16, -17, -24, and -25) with their catalytic site exposed to the extracellular space [107]. The MMP activity is also controlled by inhibitors in the ECM and in the circulation such as the four tissue inhibitors of MMPs (TIMP-1–4) and α2-macroglobulin [108].

6.1 Metalloproteases in Cell Proliferation

One of the main tumor cell characteristics is the constant progression through cell cycle, which is sustained by the release of growth-stimulating signals from itself and neighboring cells. Although the primary function of MMPs is the remodeling of ECM, they also act on non-matrix substrates as cell surface. As a result of its proteolytic activity, MMPs and also ADAMs participate in the releasing of cell surface growth factors as TGF-β, TGF-α, TNF-α, EGF, and FGF which will stimulate cell proliferation. The membrane-bound enzymes capable of cleaving extracellular portions of transmembrane proteins, releasing the soluble ectodomains (as growth factor) from cell surface, are known as "sheddase." The balance between MMPs and TIMPs governs the cleavage and the release of many important growth factors and cell surface receptors [109, 110].

The MMP-7, matrilysin, has been associated with the release of soluble HB-EGF (heparin-binding epidermal growth factor) and the cleavage of all six members of the insulin-like growth factor-binding protein (IGF-BP) family releasing these free grow factors which promote cell growth and survival [111, 112]. MMP-8 is associated with human ovarian tumor proliferation as a result of degradation of type I collagen, and its overexpression seems to be an indicator for poor prognosis [113, 114]. On the other hand, MMP-8 prevents metastasis in melanoma and lung carcinoma in mice through the modulation of cell adhesion and invasion [115, 116]. Among MMPs that act as "sheddases," ADAM is the main family with this characteristic. In epidermal keratinocytes, ADAM-10 is involved in the releasing of the chemoattractant CXC-chemokine ligand 16 (CXCL16) from cell membrane. In addition to the sheddase activity of ADAM-10, epithelial cells with this metalloproteinase overexpressed have the proliferation enhanced as a result of an increase in the activity of β-catenin [117]. ADAM-12 regulates cell proliferation and migration by shedding heparin-binding epidermal growth factor-like growth factor (HB-EGFR) allowing it to bind to EGFR [118, 119]. Although ADAM-13 is known by its sheddase property, it also degrades the ECM, and its protease activity is determinant for neural crest cell migration [120].

The primary activity of TIMPs is the regulation of MMP activity, but its participation in cell signaling events also proliferation and apoptosis (as inhibitor) has been described [121, 122]. TIMP1 and TIMP2 inhibit FGF-2- and VEGF-induced proliferation in endothelial cells by increasing the activity of phosphatases on FGFR1 and VEGFR [123, 124]. The mitogenic activity of TIMP3 and TIMP4 is not as extensive as for other TIMPs, although these inhibitors are expressed in tumor cells [125, 126].

6.2 Metalloproteases in Invasion

During tumor invasion cells, even those from stroma exchange enzymes and cytokines that modulate the local ECM and stimulate cell migration. One of the ECM modulations is the breakdown of connective tissue barriers made by collagens, laminins, fibronectin, vitronectin, and heparan sulfate proteoglycans, which requires the MMP activity [109]. Boire et al. [127] reported that MMP-1 cleaves PAR-1 receptor in breast cancer cells promoting its invasion and tumorigenesis and contributing to their metastatic potential. MMP-7 and MMP-3 may further promote tumor invasion by cleavage of E-cadherin, a cell adhesion molecule, leading to decreased cellular adhesion and increased cell mobility [107]. On the contrary, in experimental squamous cell carcinoma mouse models, MMP-8 was found to have antitumor and antimetastatic through the modulation of tumor cell adhesion and invasion [128, 129], also observed in melanoma cells by enhancing their adhesion to the ECM [115]. In breast cancer and melanoma, MMP-9 leads to the activation of TGF-β and breakdown of type IV collagen contributing to cell invasion and proliferation [130, 131]. Although MMP-11 is not capable of degrading the major ECM, it promotes tumor cell invasion, in a proteolytic-dependent manner [132]. After a vascular injury, MMP-14 is capable of degrading fibrin interrupting the mechanism of repair and allowing endothelial cell invasion [133]. Importantly, MMP-14 can serve as prognostic factor, and its expression is strongly associated with cancer progression and metastasis and poor prognosis of patients with melanoma [134]. Interestingly, expression of MMP-16 is associated with rapid fibrin and poor collagen invasion suggesting that MMP-16 might be important for infiltration of melanoma cells in perivascular space which is frequently abundant with fibrin [135].

ADAMs and ADAMTS, detected in numerous tumors, also contribute to their invasion and metastasis by degrading ECM [136]. Through an G-protein-coupled receptor–EGFR transactivation, ADAM10, ADAM15, and ADAM17 have been shown to promote tumorigenesis and cell invasion [137]. In contrast, Ungerer et al. [138] showed that overexpressed ADAM-15 in melanoma cells reduced invasion and growth in vitro, suggesting a tumor suppressor role for ADAM-15 in melanoma. In pancreatic cancer, the overexpression of ADAMTS-1 seems to be involved in tumor progression characterized by a local invasion and lymph node metastasis [139]. Moreover, an upregulation of ADAMTS-1 levels is achieved by breast tissue-associated fibroblasts contributing to cancer cell invasion [140]. Protumor and metastatic activity of ADAMTS-1 also may be associated with proteolytic action of MMP-1 and the release of the membrane-bound epithelial growth factor (EGF-like) factors and heparin-binding epidermal growth factor (HB-EGF) [141, 142]. Protumor activities have also been reported for ADAMTS-12 which potentiate trophoblast invasion by disrupting $\alpha_v\beta_3$ integrin-mediated cell adhesion to ECM [143].

The balance between MMP and TIMP plays a crucial role in tumor progression and is associated with an inhibitory effect of TIMPs on tumor cell invasion and metastasis [144]. In astrocytoma cells, TIMP-1 reduced the growth rate and inhibited cell invasion [145], as well as prevented metastasis of gastric cancer cells [146] and invasion and growth of mammary carcinoma cells [147]. Recombinant TIMP-2 also inhibited the invasion but had no effect on growth of fibrosarcoma cells [148, 149]. The overexpression of TIMP-4 in breast carcinoma has been associated with a reduction in lymph node and lung metastasis by suppressing tumor growth and invasion [150]. Thus, the expression pattern of individual TIMPs as well as the cell type may define the role of these inhibitors in tumor progression in vivo.

6.3 Metalloproteases in Angiogenesis

The process of angiogenesis involves a complex and dynamic interaction between endothelial cells and extracellular environment and is modulated by soluble angiogenic factors, cytokines, and insoluble ECM components [151].

MMPs play a wide and complex role in angiogenesis since endothelial cells produce many MMPs during the formation of new blood vessels in physiological and pathological conditions. In this context, the interaction between MMP-2 and $\alpha_v\beta_3$ integrin is essential to restrict the enzyme activity to the surface of newly forming vessels [152]. Through cleavage of laminin-5, MMP-2 increases endothelial cell migration [153] and the release of VEGF that stimulates angiogenesis, in physiological conditions and also during tumor development. The release of VEGF which switches cells from vascular quiescence to angiogenesis has also been shown for MMP-9 and MMP-14 in tumors [154, 155]. Complementarily, MMP-9 promotes proteolytic degradation of vascular basement membrane proteins during the formation of new blood vessels enhancing cell survival and angiogenesis [156]. The expression in the peritumoral area of MMP-3, MMP-7, MMP-9, and MMP-12 has been shown to generate angiostatin from plasminogen, acting as a limit for tumor-induced angiogenesis [157–159].

TIMPs have been proposed to play a regulatory role during angiogenesis as inhibitors of MMPs. In Burkitt's lymphoma, Ehrlich cells and Kaposi's sarcoma-derived KS-IMM cells which overexpress TIMP1 in tumor cells show decreased tumor angiogenesis and growth [160–162]. In cutaneous and squamous cell cancers, the stromal and adjacent to the tumor expression of TIMP-1, TIMP-2, and TIMP-3, represents a limit to tumor invasion and tumor-induced angiogenesis [163–165]. Reed et al. [166] showed that TIMP1 can also inhibit angiogenesis without changes in collagenase or gelatinase activity. TIMP-2 (in melanoma cells) and TIMP-3 (in human breast carcinoma cells) have been associated with the reduction of FGF-2- and VEGF-induced proliferation through blockage of its receptors contributing to decrease of invasion and angiogenesis in these cell lines [167, 168].

References

1. Liotta L, Rao CN, Rao C, Barsky SH, (1983) Tumor invasion and the extracellular matrix. Lab Invest 49(6):636-649
2. Lopez-Otin C, Matrisian LM (2007) Emerging roles of proteases in tumour suppression. Nat Rev Cancer 7(10):800–808
3. Koblinski JE, Ahram M, Sloane BF (2000) Unraveling the role of proteases in cancer. Clinica Chimica Acta 291(2):113–135. doi:http://dx.doi.org/10.1016/S0009-8981(99)00224-7
4. Rakashanda S, Rana F, Rafiq S, Masood A, Amin S (2012) Role of proteases in cancer: a review. Biotechnol Mol Biol Rev 7(4):90–101. doi:10.5897/BMBR11.027
5. Rawlings ND, Morton FR, Kok CY, Kong J, Barrett AJ (2008) MEROPS: the peptidase database. Nucleic Acids Res 36 (Database issue):D320–D325. doi:10.1093/nar/gkm954
6. Yousef GM, Elliott MB, Kopolovic AD, Serry E, Diamandis EP (2004) Sequence and evolutionary analysis of the human trypsin subfamily of serine peptidases. Biochim et Biophys Acta (BBA)—Proteins and Proteomics 1698(1):77–86. doi:http://dx.doi.org/10.1016/j.bbapap.2003.10.008
7. Di Cera E (2009) Serine proteases. IUBMB life 61(5):510–515. doi:10.1002/iub.186
8. Schechter I (2012) Reprint of "on the size of the active site in proteases. I. Papain". Biochem Biophys Re Commun 425(3):497–502. doi:http://dx.doi.org/10.1016/j.bbrc.2012.08.015
9. Antalis TM, Bugge T, Wu Q (2011) Membrane-anchored serine proteases in health and disease. Prog Mol Biol Transl Sci 99:1–50. doi:10.1016/B978-0-12-385504-6.00001-4
10. Krowarsch D, Cierpicki T, Jelen F, Otlewski J (2003) Canonical protein inhibitors of serine proteases. Cell Mol Life Sci CMLS 60(11):2427–2444. doi:10.1007/s00018-003-3120-x
11. Walker B, Lynas JF (2001) Strategies for the inhibition of serine proteases. Cell Mol Life Sci CMLS 58(4):596–624. doi:10.1007/PL00000884
12. Itkonen O, Stenman U-H, Osman S, Koivunen E, Halila H, Schröder T (1996) Serum samples from pancreatectomized patients contain trypsinogen immunoreactivity. J Lab Clin Med 128(1):98–102. doi:10.1016/S0022-2143(96)90118-3
13. Soreide K, Janssen EA, Körner H, Baak JPA (2006) Trypsin in colorectal cancer: molecular biological mechanisms of proliferation, invasion, and metastasis. J Pathol 209(2):147–156. doi:10.1002/path.1999
14. Wiegand U, Corbach S, Minn A, Jie K, Müller-Hill B (1993) Cloning of the cDNA encoding human brain trypsinogen and characterization of its product. Gene 136(1–2):167–175. doi:http://dx.doi.org/10.1016/0378-1119(93)90460-K
15. Kemik O, Kemik A, Sümer A, Almali N, Gurluler E, Gures N, Purisa S, Adas G, Dogan Y, Tuzun S (2013) The relationship between serum tumor-associated trypsin inhibitor levels and clinicopathological parameters in patients with gastric cancer. Eur Rev Med Pharmacol Sci 17:6
16. Koskensalo S, Hagström J, Louhimo J, Stenman UH, Haglund C (2012) Tumour-associated trypsin inhibitor TATI Is a prognostic marker in colorectal cancer. Oncology 82(4):234–241
17. Wojtukiewicz MZ, Hempel D, Sierko E, Tucker SC, Honn KV (2015) Protease-activated receptors (PARs)—biology and role in cancer invasion and metastasis. Cancer Metastasis Rev 34:775–796. doi:10.1007/s10555-015-9599-4
18. Xie L, Duan Z, Liu C, Zheng Y, Zhou J (2015) Protease-activated receptor 2 agonist increases cell proliferation and invasion of human pancreatic cancer cells. Exp Ther Med 9:6
19. Sedda S, Marafini I, Caruso R, Pallone F, Monteleone G (2014) Proteinase activated-receptors-associated signaling in the control of gastric cancer. World J Gastroenterol: WJG 20(34):11977–11984. doi:10.3748/wjg.v20.i34.11977
20. Lam DK, Dang D, Zhang J, Dolan JC, Schmidt BL (2012) Novel animal models of acute and chronic cancer pain: a pivotal role for PAR2. J Neurosci: Official J Soc Neurosci 32 (41):14178–14183. doi:10.1523/JNEUROSCI.2399-12.2012

21. Kahn ML, Nakanishi-Matsui M, Shapiro MJ, Ishihara H, Coughlin SR (1999) Protease-activated receptors 1 and 4 mediate activation of human platelets by thrombin. J Clin Invest 103(6):879–887
22. Andreasen P, Egelund R, Petersen H (2000) The plasminogen activation system in tumor growth, invasion, and metastasis. Cell Mol Life Sci CMLS 57(1):25–40. doi:10.1007/s000180050497
23. Esther Z, Juan G-E, Francisco E, Luis AR, Raul C, Amparo E (2008) Fibrinolysis: the key to new pathogenetic mechanisms. Curr Med Chem 15(9):923–929. doi:http://dx.doi.org/10.2174/092986708783955455
24. Ossovskaya VS, Bunnett NW (2004) Protease-activated receptors: contribution to physiology and disease. Physiol Rev 84(2):579–621
25. Austin KM, Covic L, Kuliopulos A (2013) Matrix metalloproteases and PAR1 activation. Blood 121(3):431–439. doi:10.1182/blood-2012-09-355958
26. Fujimoto D, Hirono Y, Goi T, Katayama K, Yamaguchi A (2008) Prognostic value of protease-activated receptor-1 (PAR-1) and matrix metalloproteinase-1 (MMP-1) in gastric cancer. Anticancer Res 28(2A):847–854
27. Tas F, Bilgin E, Karabulut S, Erturk K, Duranyildiz D (2016) Clinical significance of serum protease-activated receptor-1 (PAR-1) levels in patients with cutaneous melanoma. BBA Clinical 5:166–169. doi:10.1016/j.bbacli.2016.04.001
28. Radjabi AR, Sawada K, Jagadeeswaran S, Eichbichler A, Kenny HA, Montag A, Bruno K, Lengyel E (2008) Thrombin Induces tumor Invasion through the induction and association of matrix metalloproteinase-9 and β1-integrin on the cell surface. J Biol Chem 283(5):2822–2834. doi:10.1074/jbc.M704855200
29. Henneke I, Greschus S, Savai R, Korfei M, Markart P, Mahavadi P, Schermuly RT, Wygrecka M, Stürzebecher J, Seeger W, Günther A, Ruppert C (2010) Inhibition of urokinase activity reduces primary tumor growth and metastasis formation in a murine lung carcinoma model. Am J Respir Crit Care Med 181(6):611–619. doi:10.1164/rccm.200903-0342OC
30. Duffy MJ, McGowan PM, Harbeck N, Thomssen C, Schmitt M (2014) uPA and PAI-1 as biomarkers in breast cancer: validated for clinical use in level-of-evidence-1 studies. Breast Cancer Res 16(4):1–10. doi:10.1186/s13058-014-0428-4
31. Masuda T, Hattori N, Senoo T, Akita S, Ishikawa N, Fujitaka K, Haruta Y, Murai H, Kohno N (2013) SK-216, an inhibitor of plasminogen activator inhibitor-1, limits tumor progression and angiogenesis. Mol Cancer Ther 12. doi:10.1158/1535-7163.mct-13-0041
32. Bajou K, Noel A, Gerard RD, Masson V, Brunner N, Holst-Hansen C, Skobe M, Fusenig NE, Carmeliet P, Collen D, Foidart JM (1998) Absence of host plasminogen activator inhibitor 1 prevents cancer invasion and vascularization. Nature Med 4. doi:10.1038/nm0898-923
33. Kwaan HC, Wang J, Svoboda K, Declerck PJ (2000) Plasminogen activator inhibitor 1 may promote tumour growth through inhibition of apoptosis. Br J Cancer 82. doi:10.1054/bjoc.2000.1207
34. Fang H, Placencio VR, DeClerck YA (2012) Protumorigenic activity of plasminogen activator inhibitor-1 through an antiapoptotic function. J Natl Cancer Inst 104. doi:10.1093/jnci/djs377
35. Chen H, Hao J, Wang L, Li Y (2009) Coexpression of invasive markers (uPA, CD44) and multiple drug-resistance proteins (MDR1, MRP2) is correlated with epithelial ovarian cancer progression. Br J Cancer 101(3):432–440
36. Zhang W, Ling D, Tan J, Zhang J, Li L (2013) Expression of urokinase plasminogen activator and plasminogen activator inhibitor type-1 in ovarian cancer and its clinical significance. Oncol Rep 29(2):9. doi:10.3892/or.2012.2148
37. Ribatti D, Ranieri G (2015) Tryptase, a novel angiogenic factor stored in mast cell granules. Exp Cell Res 332(2):157-162. doi:http://dx.doi.org/10.1016/j.yexcr.2014.11.014

38. Heuston S, Hyland NP (2012) Chymase inhibition as a pharmacological target: a role in inflammatory and functional gastrointestinal disorders? Br J Pharmacol 167(4):732–740. doi:10.1111/j.1476-5381.2012.02055.x
39. Benítez-Bribiesca L, Wong A, Utrera D, Castellanos E (2001) The Role of mast cell tryptase in neoangiogenesis of premalignant and malignant lesions of the uterine cervix. J Histochem Cytochem 49(8):1061–1062. doi:10.1177/002215540104900816
40. Ribatti D, Ennas MG, Vacca A, Ferreli F, Nico B, Orru S, Sirigu P (2003) Tumor vascularity and tryptase-positive mast cells correlate with a poor prognosis in melanoma. Eur J Clin Invest 33(5):420–425. doi:10.1046/j.1365-2362.2003.01152.x
41. Esposito I, Menicagli M, Funel N, Bergmann F, Boggi U, Mosca F, Bevilacqua G, Campani D (2004) Inflammatory cells contribute to the generation of an angiogenic phenotype in pancreatic ductal adenocarcinoma. J Clin Pathol 57(6):630–636. doi:10.1136/jcp.2003.014498
42. Ribatti D, Guidolin D, Marzullo A, Nico B, Annese T, Benagiano V, Crivellato E (2010) Mast cells and angiogenesis in gastric carcinoma. Int J Exp Pathol 91(4):350–356. doi:10.1111/j.1365-2613.2010.00714.x
43. Marech I, Ammendola M, Sacco R, Capriuolo GS, Patruno R, Rubini R, Luposella M, Zuccalà V, Savino E, Gadaleta CD, Ribatti D, Ranieri G (2014) Serum tryptase, mast cells positive to tryptase and microvascular density evaluation in early breast cancer patients: possible translational significance. BMC Cancer 14(1):1–7. doi:10.1186/1471-2407-14-534
44. Senior RM, Campbell EJ (1984) Cathepsin G in human mononuclear phagocytes: comparisons between monocytes and U937 monocyte-like cells. J Immunol 132(5):2547–2551
45. El Rayes T, Catena R, Lee S, Stawowczyk M, Joshi N, Fischbach C, Powell CA, Dannenberg AJ, Altorki NK, Gao D, Mittal V (2015) Lung inflammation promotes metastasis through neutrophil protease-mediated degradation of Tsp-1. Proc Natl Acad Sci U S A 112(52):16000–16005. doi:10.1073/pnas.1507294112
46. Wilson TJ, Nannuru KC, Futakuchi M, Singh RK (2010) Cathepsin G-mediated enhanced TGF-β signaling promotes angiogenesis via upregulation of VEGF and MCP-1. Cancer Lett 288(2):162. doi:10.1016/j.canlet.2009.06.035
47. Morimoto-Kamata R, S-i Mizoguchi, Ichisugi T, Yui S (2012) Cathepsin G induces cell aggregation of human breast cancer MCF-7 cells via a 2-step mechanism: catalytic site-independent binding to the cell surface and enzymatic activity-dependent induction of the cell aggregation. Mediat Inflamm 2012:456462. doi:10.1155/2012/456462
48. Cools-Lartigue J, Spicer J, Najmeh S, Ferri L (2014) Neutrophil extracellular traps in cancer progression. Cell Mol Life Sci 71(21):4179–4194. doi:10.1007/s00018-014-1683-3
49. Yousef GM, Diamandis EP (2001) the new human tissue kallikrein gene family: structure, function, and association to disease. Endocr Rev 22(2):184–204. doi:10.1210/edrv.22.2.0424
50. Thorek DLJ, Evans MJ, Carlsson SV, Ulmert D, Lilja H (2013) Prostate-specific kallikrein-related peptidases and their relation to prostate cancer biology and detection. Established relevance and emerging roles. Thromb Haemost 110(3):484–492. doi:10.1160/TH13-04-0275
51. Lawrence MG, Lai J, Clements JA (2010) Kallikreins on steroids: structure, function, and hormonal regulation of prostate-specific antigen and the extended kallikrein locus. Endocr Rev 31(4):407–446. doi:10.1210/er.2009-0034
52. Krenzer S, Peterziel H, Mauch C, Blaber SI, Blaber M, Angel P, Hess J (2011) Expression and function of the kallikrein-related peptidase 6 in the human melanoma microenvironment. J Invest Dermatol 131(11):2281–2288. doi:10.1038/jid.2011.190
53. Paliouras M, Borgono C, Diamandis EP (2007) Human tissue kallikreins: the cancer biomarker family. Cancer Lett 249(1):61–79. doi:10.1016/j.canlet.2006.12.018

54. Lin C-Y, Anders J, Johnson M, Sang QA, Dickson RB (1999) Molecular cloning of cDNA for matriptase, a matrix-degrading serine protease with trypsin-like activity. J Biol Chem 274 (26):18231–18236. doi:10.1074/jbc.274.26.18231
55. Takeuchi T, Shuman MA, Craik CS (1999) Reverse biochemistry: use of macromolecular protease inhibitors to dissect complex biological processes and identify a membrane-type serine protease in epithelial cancer and normal tissue. Proc Natl Acad Sci 96(20):11054–11061. doi:10.1073/pnas.96.20.11054
56. Takeuchi T, Harris JL, Huang W, Yan KW, Coughlin SR, Craik CS (2000) Cellular localization of membrane-type serine protease 1 and identification of protease-activated receptor-2 and single-CHAIN urokinase-type plasminogen activator as substrates. J Biol Chem 275(34):26333–26342. doi:10.1074/jbc.M002941200
57. List K, Szabo R, Wertz PW, Segre J, Haudenschild CC, Kim S-Y, Bugge TH (2003) Loss of proteolytically processed filaggrin caused by epidermal deletion of matriptase/MT-SP1. J Cell Biol 163(4):901–910. doi:10.1083/jcb.200304161
58. Kamata Y, Taniguchi A, Yamamoto M, Nomura J, Ishihara K, Takahara H, Hibino T, Takeda A (2009) Neutral cysteine protease bleomycin hydrolase is essential for the breakdown of deiminated filaggrin into amino acids. J Biol Chem 284(19):12829–12836. doi:10.1074/jbc.M807908200
59. Oberst MD, Williams CA, Dickson RB, Johnson MD, Lin C-Y (2003) The activation of matriptase requires its noncatalytic domains, serine protease domain, and its cognate inhibitor. J Biol Chem 278(29):26773–26779. doi:10.1074/jbc.M304282200
60. Szabo R, Hobson JP, List K, Molinolo A, Lin C-Y, Bugge TH (2008) Potent inhibition and global co-localization implicate the transmembrane kunitz-type serine protease inhibitor hepatocyte growth factor activator inhibitor-2 in the regulation of epithelial matriptase activity. J Biol Chem 283(43):29495–29504. doi:10.1074/jbc.M801970200
61. Oberst MD, Johnson MD, Dickson RB, Lin C-Y, Singh B, Stewart M, Williams A, al-Nafussi A, Smyth JF, Gabra H, Sellar GC (2002) Expression of the serine protease matriptase and Its inhibitor HAI-1 in epithelial ovarian cancer. Am Assoc Cancer Res 8 (4):1101–1107
62. Kang JY, Dolled-Filhart M, Ocal IT, Singh B, Lin C-Y, Dickson RB, Rimm DL, Camp RL (2003) Tissue microarray analysis of hepatocyte growth factor/met pathway components reveals a role for met, matriptase, and hepatocyte growth factor activator inhibitor 1 in the progression of node-negative breast cancer. Cancer Res 63(5):1101–1105
63. Lee J-W, Yong Song S, Choi J-J, Lee S-J, Kim B-G, Park C-S, Lee J-H, Lin C-Y, Dickson RB, Bae D-S (2005) Increased expression of matriptase is associated with histopathologic grades of cervical neoplasia. Hum Pathol 36(6):626–633. doi:10.1016/j.humpath.2005.03.003
64. Tanimoto H, Shigemasa K, Tian X, Gu L, Beard JB, Sawasaki T, O'Brien TJ (2005) Transmembrane serine protease TADG-15 (ST14/matriptase/MT-SP1): expression and prognostic value in ovarian cancer. Br J Cancer 92(2):278–283. doi:10.1038/sj.bjc.6602320
65. Riddick ACP, Shukla CJ, Pennington CJ, Bass R, Nuttall RK, Hogan A, Sethia KK, Ellis V, Collins AT, Maitland NJ, Ball RY, Edwards DR (2005) Identification of degradome components associated with prostate cancer progression by expression analysis of human prostatic tissues. Br J Cancer 92(12):2171–2180
66. Nakamura K, Hongo A, Kodama J, Abarzua F, Yasutomo N, Kumon H, Hiramatsu Y (2009) Expression of matriptase and clinical outcome of human endometrial cancer. Anticancer Res 29(5):1685–1690
67. List K (2009) Matriptase: a culprit in cancer? Future Oncol 5(1):97–104. doi:10.2217/14796694.5.1.97
68. Jin J-S, Hsieh D-S, Loh S-H, Chen A, Yao C-W, Yen C-Y (2006) Increasing expression of serine protease matriptase in ovarian tumors: tissue microarray analysis of immunostaining score with clinicopathological parameters. Mod Pathol 19(3):447–452

69. Saleem M, Adhami VM, Zhong W, Longley BJ, Lin C-Y, Dickson RB, Reagan-Shaw S, Jarrard DF, Mukhtar H (2006) A Novel biomarker for staging human prostate adenocarcinoma: overexpression of matriptase with concomitant loss of its inhibitor, hepatocyte growth factor activator inhibitor-1. Cancer Epidemiol Biomark Prev 15(2):217–227. doi:10.1158/1055-9965.epi-05-0737
70. Kawaguchi M, Kataoka H (2014) Mechanisms of hepatocyte growth factor activation in cancer tissues. Cancers 6(4):1890–1904. doi:10.3390/cancers6041890
71. Loser R, Pietzsch J (2015) Cysteine cathepsins: their role in tumor progression and recent trends in the development of imaging probes. Front Chem 3:37. doi:10.3389/fchem.2015.00037
72. Fonovic M, Turk B (2014) Cysteine cathepsins and their potential in clinical therapy and biomarker discovery. Proteomics Clin Appl 8(5–6):416–426. doi:10.1002/prca.201300085
73. Mohamed MM, Sloane BF (2006) Cysteine cathepsins: multifunctional enzymes in cancer. Nat Rev Cancer 6(10):764–775. doi:10.1038/nrc1949
74. Cox JL (2009) Cystatins and cancer. Front Biosci 14:463–474 (Landmark Ed)
75. Olson OC, Joyce JA (2015) Cysteine cathepsin proteases: regulators of cancer progression and therapeutic response. Nat Rev Cancer 15(12):712–729. doi:10.1038/nrc4027
76. Obermajer N, Doljak B, Kos J (2006) Cysteine cathepsins: regulators of antitumour immune response. Expert Opin Biol Ther 6(12):1295–1309. doi:10.1517/14712598.6.12.1295
77. Tan GJ, Peng ZK, Lu JP, Tang FQ (2013) Cathepsins mediate tumor metastasis. World J Biol Chem 4(4):91–101. doi:10.4331/wjbc.v4.i4.91
78. Fonovic M, Turk B (2014) Cysteine cathepsins and extracellular matrix degradation. Biochim Biophys Acta 1840(8):2560–2570. doi:10.1016/j.bbagen.2014.03.017
79. Palermo C, Joyce JA (2008) Cysteine cathepsin proteases as pharmacological targets in cancer. Trends Pharmacol Sci 29(1):22–28. doi:10.1016/j.tips.2007.10.011
80. Gocheva V, Joyce JA (2007) Cysteine cathepsins and the cutting edge of cancer invasion. Cell Cycle 6(1):60–64. doi:10.4161/cc.6.1.3669
81. Vasiljeva O, Turk B (2008) Dual contrasting roles of cysteine cathepsins in cancer progression: apoptosis versus tumour invasion. Biochimie 90(2):380–386. doi:10.1016/j.biochi.2007.10.004
82. Moretti D, Del Bello B, Allavena G, Maellaro E (2014) Calpains and cancer: friends or enemies? Arch Biochem Biophys 564:26–36. doi:10.1016/j.abb.2014.09.018
83. Storr SJ, Carragher NO, Frame MC, Parr T, Martin SG (2011) The calpain system and cancer. Nat Rev Cancer 11(5):364–374. doi:10.1038/nrc3050
84. Storr SJ, Thompson N, Pu X, Zhang Y, Martin SG (2015) Calpain in breast cancer: role in disease progression and treatment response. Pathobiology 82(3–4):133–141. doi:10.1159/000430464
85. Leloup L, Wells A (2011) Calpains as potential anti-cancer targets. Expert Opin Ther Targets 15(3):309–323. doi:10.1517/14728222.2011.553611
86. Peng P, Wu W, Zhao J, Song S, Wang X, Jia D, Shao M, Zhang M, Li L, Wang L, Duan F, Zhao R, Yang C, Wu H, Zhang J, Shen Z, Ruan Y, Gu J (2016) Decreased expression of calpain-9 predicts unfavorable prognosis in patients with gastric cancer. Sci Rep 6:29604. doi:10.1038/srep29604
87. Hensley P, Mishra M, Kyprianou N (2013) Targeting caspases in cancer therapeutics. Biol Chem 394(7):831–843. doi:10.1515/hsz-2013-0128
88. Fiandalo MV, Kyprianou N (2012) Caspase control: protagonists of cancer cell apoptosis. Exp Oncol 34(3):165–175
89. Shalini S, Dorstyn L, Dawar S, Kumar S (2015) Old, new and emerging functions of caspases. Cell Death Differ 22(4):526–539. doi:10.1038/cdd.2014.216
90. MacKenzie SH, Clark AC (2008) Targeting cell death in tumors by activating caspases. Curr Cancer Drug Targets 8(2):98–109
91. Fernald K, Kurokawa M (2013) Evading apoptosis in cancer. Trends Cell Biol 23(12):620–633. doi:10.1016/j.tcb.2013.07.006

92. Ghavami S, Hashemi M, Ande SR, Yeganeh B, Xiao W, Eshraghi M, Bus CJ, Kadkhoda K, Wiechec E, Halayko AJ, Los M (2009) Apoptosis and cancer: mutations within caspase genes. J Med Genet 46(8):497–510. doi:10.1136/jmg.2009.066944
93. Masson O, Bach AS, Derocq D, Prebois C, Laurent-Matha V, Pattingre S, Liaudet-Coopman E (2010) Pathophysiological functions of cathepsin D: targeting its catalytic activity versus its protein binding activity? Biochimie 92(11):1635–1643. doi:10.1016/j.biochi.2010.05.009
94. Benes P, Vetvicka V, Fusek M (2008) Cathepsin D–many functions of one aspartic protease. Crit Rev Oncol Hematol 68(1):12–28. doi:10.1016/j.critrevonc.2008.02.008
95. Liaudet-Coopman E, Beaujouin M, Derocq D, Garcia M, Glondu-Lassis M, Laurent-Matha V, Prebois C, Rochefort H, Vignon F (2006) Cathepsin D: newly discovered functions of a long-standing aspartic protease in cancer and apoptosis. Cancer Lett 237(2):167–179. doi:10.1016/j.canlet.2005.06.007
96. Pranjol MZ, Gutowski N, Hannemann M, Whatmore J (2015) The potential role of the proteases cathepsin D and cathepsin L in the progression and metastasis of epithelial ovarian cancer. Biomolecules 5(4):3260–3279. doi:10.3390/biom5043260
97. Zaidi N, Hermann C, Herrmann T, Kalbacher H (2008) Emerging functional roles of cathepsin E. Biochem Biophys Res Commun 377(2):327–330. doi:10.1016/j.bbrc.2008.10.034
98. Shin M, Kadowaki T, Iwata J, Kawakubo T, Yamaguchi N, Takii R, Tsukuba T, Yamamoto K (2007) Association of cathepsin E with tumor growth arrest through angiogenesis inhibition and enhanced immune responses. Biol Chem 388(11):1173–1181. doi:10.1515/BC.2007.154
99. Yamamoto K, Kawakubo T, Yasukochi A, Tsukuba T (2012) Emerging roles of cathepsin E in host defense mechanisms. Biochim Biophys Acta 1(1824):105–112. doi:10.1016/j.bbapap.2011.05.022
100. Fristrup N, Ulhoi BP, Birkenkamp-Demtroder K, Mansilla F, Sanchez-Carbayo M, Segersten U, Malmstrom PU, Hartmann A, Palou J, Alvarez-Mugica M, Zieger K, Borre M, Orntoft TF, Dyrskjot L (2012) Cathepsin E, maspin, Plk1, and survivin are promising prognostic protein markers for progression in non-muscle invasive bladder cancer. Am J Pathol 180(5):1824–1834. doi:10.1016/j.ajpath.2012.01.023
101. Konno-Shimizu M, Yamamichi N, Inada K, Kageyama-Yahara N, Shiogama K, Takahashi Y, Asada-Hirayama I, Yamamichi-Nishina M, Nakayama C, Ono S, Kodashima S, Fujishiro M, Tsutsumi Y, Ichinose M, Koike K (2013) Cathepsin E is a marker of gastric differentiation and signet-ring cell carcinoma of stomach: a novel suggestion on gastric tumorigenesis. PLoS ONE 8(2):e56766. doi:10.1371/journal.pone.0056766
102. Kawakubo T, Yasukochi A, Toyama T, Takahashi S, Okamoto K, Tsukuba T, Nakamura S, Ozaki Y, Nishigaki K, Yamashita H, Yamamoto K (2014) Repression of cathepsin E expression increases the risk of mammary carcinogenesis and links to poor prognosis in breast cancer. Carcinogenesis 35(3):714–726. doi:10.1093/carcin/bgt373
103. Naymagon L, Abdul-Hay M (2016) Novel agents in the treatment of multiple myeloma: a review about the future. J Hematol Oncol 9:52. doi:10.1186/s13045-016-0282-1
104. Zhong J-L, Huang C-Z (2016) Ubiquitin proteasome system research in gastrointestinal cancer. World J Gastrointest Oncol 8(2):198–206. doi:10.4251/wjgo.v8.i2.198
105. Almond J, Cohen G (2002) The proteasome: a novel target for cancer chemotherapy. Leukemia 16:11
106. Moro N, Mauch C, Zigrino P (2014) Metalloproteinases in melanoma. Eur J Cell Biol 93(1–2):23–29. doi:http://dx.doi.org/10.1016/j.ejcb.2014.01.002
107. Klein T, Bischoff R (2011) Physiology and pathophysiology of matrix metalloproteases. Amino Acids 41(2):271–290. doi:10.1007/s00726-010-0689-x
108. Hadler-Olsen E, Winberg J-O, Uhlin-Hansen L (2013) Matrix metalloproteinases in cancer: their value as diagnostic and prognostic markers and therapeutic targets. Tumor Biol 34 (4):2041–2051. doi:10.1007/s13277-013-0842-8

109. Chang C, Werb Z (2001) The many faces of metalloproteases: cell growth, invasion, angiogenesis and metastasis. Trends Cell Biol 11(11):S37–S43. doi:10.1016/S0962-8924(01)02122-5
110. Primakoff P, Myles DG (2000) The ADAM gene family: surface proteins with adhesion and protease activity. Trends Genet 16(2):83–87. doi:10.1016/S0168-9525(99)01926-5
111. Yu W-H, Woessner JF, McNeish JD, Stamenkovic I (2002) CD44 anchors the assembly of matrilysin/MMP-7 with heparin-binding epidermal growth factor precursor and ErbB4 and regulates female reproductive organ remodeling. Genes Dev 16(3):307–323. doi:10.1101/gad.925702
112. Nakamura M, Miyamoto S, Maeda H, Ishii G, Hasebe T, Chiba T, Asaka M, Ochiai A (2005) Matrix metalloproteinase-7 degrades all insulin-like growth factor binding proteins and facilitates insulin-like growth factor bioavailability. Biochem Biophys Res Commun 333(3):1011–1016. doi:http://dx.doi.org/10.1016/j.bbrc.2005.06.010
113. Stadlmann S, Pollheimer J, Moser PL, Raggi A, Amberger A, Margreiter R, Offner FA, Mikuz G, Dirnhofer S, Moch H (2003) Cytokine-regulated expression of collagenase-2 (MMP-8) is involved in the progression of ovarian cancer. Eur J Cancer 39(17):2499–2505. doi:10.1016/j.ejca.2003.08.011
114. Stenman M, Paju A, Hanemaaijer R, Tervahartiala T, Leminen A, Stenman UH, Konttinen YT, Sorsa T (2003) Collagenases (MMP-1, -8 and -13) and trypsinogen-2 in fluid from benign and malignant ovarian cysts. Tumor Biol 24 (1010–4283 (Print)):9–12
115. Gutiérrez-Fernández A, Fueyo A, Folgueras AR, Garabaya C, Pennington CJ, Pilgrim S, Edwards DR, Holliday DL, Jones JL, Span PN, Sweep FCGJ, Puente XS, López-Otín C (2008) Matrix metalloproteinase-8 functions as a metastasis suppressor through modulation of tumor cell adhesion and invasion. Cancer Res 68(8):2755–2763
116. Korpi JT, Kervinen V, Maklin H, Vaananen A, Lahtinen M, Laara E, Ristimaki A, Thomas G, Ylipalosaari M, Astrom P, Lopez-Otin C, Sorsa T, Kantola S, Pirila E, Salo T (2008) Collagenase-2 (matrix metalloproteinase-8) plays a protective role in tongue cancer. Br J Cancer 98(4):766–775
117. Scholz F, Schulte A, Adamski F, Hundhausen C, Mittag J, Schwarz A, Kruse M-L, Proksch E, Ludwig A (2007) Constitutive expression and regulated release of the transmembrane chemokine CXCL16 in human and murine skin. J Invest Dermatol 127(6):1444–1455. doi:10.1038/sj.jid.5700751
118. Mori S, Tanaka M, Nanba D, Nishiwaki E, Ishiguro H, Higashiyama S, Matsuura N (2003) PACSIN3 Binds ADAM12/meltrin α and up-regulates ectodomain shedding of heparin-binding epidermal growth factor-like growth factor. J Biol Chem 278(46):46029–46034. doi:10.1074/jbc.M306393200
119. Tanaka M, Nanba D, Mori S, Shiba F, Ishiguro H, Yoshino K, Matsuura N, Higashiyama S (2004) ADAM binding protein Eve-1 Is required for ectodomain shedding of epidermal growth factor receptor ligands. J Biol Chem 279(40):41950–41959. doi:10.1074/jbc.M400086200
120. Alfandari D, Cousin H, Gaultier A, Smith K, White JM, Darribère T, DeSimone DW (2001) *Xenopus* ADAM 13 is a metalloprotease required for cranial neural crest-cell migration. Curr Biol 11(12):918–930. doi:10.1016/S0960-9822(01)00263-9
121. Murphy G (2011) Tissue inhibitors of metalloproteinases. Genome Biol 12(11):1–7. doi:10.1186/gb-2011-12-11-233
122. Stetler-Stevenson WG (2008) The tumor microenvironment: regulation by MMP-independent effects of tissue inhibitor of metalloproteinases-2. Cancer Metastasis Rev 27. doi:10.1007/s10555-007-9105-8
123. Porter JF, Shen S, Denhardt DT (2004) Tissue inhibitor of metalloproteinase-1 stimulates proliferation of human cancer cells by inhibiting a metalloproteinase. Br J Cancer 90(2):463–470. doi:10.1038/sj.bjc.6601533

124. Hoegy SE, Oh H-R, Corcoran ML, Stetler-Stevenson WG (2001) Tissue inhibitor of metalloproteinases-2 (TIMP-2) suppresses TKR-growth factor signaling independent of metalloproteinase inhibition. J Biol Chem 276(5):3203–3214. doi:10.1074/jbc.M008157200
125. Lizárraga F, Maldonado V, Meléndez-Zajgla J (2004) Tissue inhibitor of metalloproteinases-2 growth-stimulatory activity is mediated by nuclear factor-kappa B in A549 lung epithelial cells. Int J Biochem Cell Biol 36(8):1655–1663. doi:http://dx.doi.org/10.1016/j.biocel.2004.02.004
126. Brew K, Nagase H (2010) The tissue inhibitors of metalloproteinases (TIMPs): an ancient family with structural and functional diversity. Biochim Biophys Acta 1803. doi:10.1016/j.bbamcr.2010.01.003
127. Boire A, Covic L, Agarwal A, Jacques S, Sherifi S, Kuliopulos A (2005) PAR1 is a matrix metalloprotease-1 receptor that promotes invasion and tumorigenesis of breast cancer cells. Cell 120(3):303–313. doi:10.1016/j.cell.2004.12.018
128. Balbin M, Fueyo A, Tester AM, Pendas AM, Pitiot AS, Astudillo A, Overall CM, Shapiro SD, Lopez-Otin C (2003) Loss of collagenase-2 confers increased skin tumor susceptibility to male mice. Nat Genet 35(3):252–257. doi:http://www.nature.com/ng/journal/v35/n3/suppinfo/ng1249_S1.html
129. López-Otín C, Palavalli LH, Samuels Y (2009) Protective roles of matrix metalloproteinases: from mouse models to human cancer. Cell Cycle (Georgetown, Tex) 8(22):3657–3662
130. Yu Q, Stamenkovic I (2000) Cell surface-localized matrix metalloproteinase-9 proteolytically activates TGF-β and promotes tumor invasion and angiogenesis. Genes Dev 14 (2):163–176. doi:10.1101/gad.14.2.163
131. Yu Q, Stamenkovic I (1999) Localization of matrix metalloproteinase 9 to the cell surface provides a mechanism for CD44-mediated tumor invasion. Genes Dev 13(1):35–48
132. Noel A, Boulay A, Kebers F, Kannan R, Hajitou A, Calberg-Bacq CM, Basset P, Rio MC, Foidart JM (2000) Demonstration in vivo that stromelysin-3 functions through its proteolytic activity. Oncogene 19(12):1605–1612
133. Hiraoka N, Allen E, Apel IJ, Gyetko MR, Weiss SJ (1998) Matrix metalloproteinases regulate neovascularization by acting as pericellular fibrinolysins. Cell 95(3):365–377. doi:10.1016/S0092-8674(00)81768-7
134. Iida J, Wilhelmson KL, Price MA, Wilson CM, Pei D, Furcht LT, McCarthy JB (2004) Membrane type-1 matrix metalloproteinase promotes human melanoma invasion and growth. J Invest Dermatol 122(1):167–176. doi:10.1046/j.0022-202X.2003.22114.x
135. Tatti O, Arjama M, Ranki A, Weiss SJ, Keski-Oja J, Lehti K (2011) Membrane-Type-3 matrix metalloproteinase (MT3-MMP) functions as a matrix composition-dependent effector of melanoma cell invasion. PLoS ONE 6(12):e28325. doi:10.1371/journal.pone.0028325
136. Shao S, Li Z, Gao W, Yu G, Liu D, Pan F (2014) ADAM-12 as a diagnostic marker for the proliferation, migration and invasion in patients with small cell lung cancer. PLoS ONE 9(1):e85936. doi:10.1371/journal.pone.0085936
137. Huovila A-PJ, Turner AJ, Pelto-Huikko M, Kärkkäinen I, Ortiz RM (2005) Shedding light on ADAM metalloproteinases. Trends Biochem Sci 30(7):413–422. doi:10.1016/j.tibs.2005.05.006
138. Ungerer C, Doberstein K, Bürger C, Hardt K, Boehncke W-H, Böhm B, Pfeilschifter J, Dummer R, Mihic-Probst D, Gutwein P (2010) ADAM15 expression is downregulated in melanoma metastasis compared to primary melanoma. Biochem Biophys Res Commun 401 (3):363–369. doi:http://dx.doi.org/10.1016/j.bbrc.2010.09.055
139. Masui T, Hosotani R, Tsuji S, Miyamoto Y, Yasuda S, Ida J, Nakajima S, Kawaguchi M, Kobayashi H, Koizumi M, Toyoda E, Tulachan S, Arii S, Doi R, Imamura M (2001) Expression of METH-1 and METH-2 in pancreatic cancer. Am Assoc Cancer Res 7 (11):3437–3443
140. Tyan S-W, Hsu C-H, Peng K-L, Chen C-C, Kuo W-H, Lee EYHP, Shew J-Y, Chang K-J, Juan L-J, Lee W-H (2012) Breast cancer cells induce stromal fibroblasts to secrete

ADAMTS1 for cancer Invasion through an epigenetic change. PLoS ONE 7(4):e35128. doi:10.1371/journal.pone.0035128
141. Lu X, Wang Q, Hu G, Van Poznak C, Fleisher M, Reiss M, Massagué J, Kang Y (2009) ADAMTS1 and MMP1 proteolytically engage EGF-like ligands in an osteolytic signaling cascade for bone metastasis. Genes Dev 23(16):1882–1894. doi:10.1101/gad.1824809
142. Liu YJ, Xu Y, Yu Q (2006) Full-length ADAMTS-1 and the ADAMTS-1 fragments display pro- and antimetastatic activity, respectively. Oncogene 25(17):2452–2467. doi:10.1038/sj.onc.1209287
143. Beristain AG, Zhu H, Leung PCK (2011) Regulated expression of ADAMTS-12 in human trophoblastic cells: a role for ADAMTS-12 in epithelial cell invasion? PLoS ONE 6(4): e18473. doi:10.1371/journal.pone.0018473
144. Cruz-Munoz W, Khokha R (2008) The role of tissue inhibitors of metalloproteinases in tumorigenesis and metastasis. Crit Rev Clin Lab Sci 45. doi:10.1080/10408360801973244
145. Matsuzawa K, Fukuyama K, Hubbard SL, Dirks PB, Rutka JT (1996) Transfection of an invasive human astrocytoma cell line with a TIMP-1 cDNA: modulation of astrocytoma invasive potential. J Neuropathol Exp Neurol 55(1):88–96
146. Watanabe M, Takahashi Y, Ohta T, Mai M, Sasaki T, Seiki M, (1996) Inhibition of metastasis in human gastric cancer cells transfected with tissue inhibitor of metalloproteinase 1 gene in nude mice. Cancer 77(8):1676–1680
147. Alonso DF, Skilton G, De Lorenzo MS, Scursoni AM, Yoshiji H, Gomez DE (1998) Histopathological findings in a highly invasive mouse mammary carcinoma transfected with human tissue inhibitor of metalloproteinases-1. Oncol Rep 5(5):1083–1087
148. Albini A, Melchiori A, Santi L, Liotta LA, Brown PD, Stetler-Stevenson WG (1991) Tumor cell invasion inhibited by TIMP-2. J Natl Cancer Inst 83(11):775–779. doi:10.1093/jnci/83.11.775
149. DeClerck YA, Yean T-D, Chan D, Shimada H, Langley KE (1991) Inhibition of tumor invasion of smooth muscle cell layers by recombinant human metalloproteinase inhibitor. Cancer Res 51(8):2151–2157
150. Wang M, Liu Y, Greene J, Sheng S, Fuchs A, Rosen EM, Shi YE (1997) Inhibition of tumor growth and metastasis of human breast cancer cells transfected with tissue inhibitor of metalloproteinase 4. Oncogene 14(23):2767–2774
151. Carmeliet P, Jain RK (2000) Angiogenesis in cancer and other diseases. Nature 407 (6801):249–257
152. Brooks PC, Clark RA, Cheresh DA (1994) Requirement of vascular integrin alpha v beta 3 for angiogenesis. Science 264(5158):569–571
153. Risau W (1997) Mechanisms of angiogenesis. Nature 386(6626):671–674
154. Bergers G, Brekken R, McMahon G, Vu TH, Itoh T, Tamaki K, Tanzawa K, Thorpe P, Itohara S, Werb Z, Hanahan D (2000) Matrix metalloproteinase-9 triggers the angiogenic switch during carcinogenesis. Nat Cell Biol 2(10):737–744. doi:10.1038/35036374
155. Sounni NE, Roghi C, Chabottaux V, Janssen M, Munaut C, Maquoi E, Galvez BG, Gilles C, Frankenne F, Murphy G, Foidart J-M, Noel A (2004) Up-regulation of vascular endothelial growth factor-A by active membrane-type 1 matrix metalloproteinase through activation of src-tyrosine kinases. J Biol Chem 279(14):13564–13574. doi:10.1074/jbc.M307688200
156. Shapiro S, Khodalev O, Bitterman H, Auslender R, Lahat N (2010) Different activation forms of MMP-2 oppositely affect the fate of endothelial cells. Am J Physiol—Cell Physiol 298(4):C942–C951
157. Dong Z, Kumar R, Yang X, Fidler IJ (1997) Macrophage-derived metalloelastase is responsible for the generation of angiostatin in lewis lung carcinoma. Cell 88(6):801–810. doi:10.1016/S0092-8674(00)81926-1
158. Patterson BC, Sang QA (1997) Angiostatin-converting enzyme activities of human matrilysin (MMP-7) and gelatinase B/Type IV collagenase (MMP-9). J Biol Chem 272 (46):28823–28825

159. Lijnen HR, Ugwu F, Bini A, Collen D (1998) Generation of an angiostatin-like fragment from plasminogen by stromelysin-1 (MMP-3). Biochemistry 37(14):4699–4702. doi:10.1021/bi9731798
160. Guedez L, McMarlin AJ, Kingma DW, Bennett TA, Stetler-Stevenson M, Stetler-Stevenson WG (2001) Tissue inhibitor of metalloproteinase-1 alters the tumorigenicity of burkitt's lymphoma via divergent effects on tumor growth and angiogenesis. Am J Pathol 158(4):1207–1215
161. Ikenaka Y, Yoshiji H, Kuriyama S, Yoshii J, Noguchi R, Tsujinoue H, Yanase K, Namisaki T, Imazu H, Masaki T, Fukui H (2003) Tissue inhibitor of metalloproteinases-1 (TIMP-1) inhibits tumor growth and angiogenesis in the TIMP-1 transgenic mouse model. Int J Cancer 105(3):340–346. doi:10.1002/ijc.11094
162. Zacchigna S, Zentilin L, Morini M, Dell'Eva R, Noonan DM, Albini A, Giacca M (2004) AAV-mediated gene transfer of tissue inhibitor of metalloproteinases-1 inhibits vascular tumor growth and angiogenesis in vivo. Cancer Gene Ther 11(1):73–80
163. Wagner SN, Ockenfels HM, Wagner C, Peter Soyer H, Goos M (1996) Differential expression of tissue inhibitor of metalloproteinases-2 by cutaneous squamous and basal cell carcinomas. J Invest Dermatol 106(2):321–326. doi:10.1111/1523-1747.ep12342979
164. Airola K, Ahonen M, Johansson N, Heikkilä P, Kere J, Kähäri V-M, Saarialho-Kere UK (1998) Human TIMP-3 Is expressed during fetal development, hair growth cycle, and cancer progression. J Histochem Cytochem 46(4):437–447
165. Sutinen M, Kainulainen T, Hurskainen T, Vesterlund E, Alexander JP, Overall CM, Sorsa T, Salo T (1998) Expression of matrix metalloproteinases (MMP-1 and -2) and their inhibitors (TIMP-1, -2 and -3) in oral lichen planus, dysplasia, squamous cell carcinoma and lymph node metastasis. Br J Cancer 77(12):2239–2245
166. Reed MJ, Koike T, Sadoun E, Sage EH, Puolakkainen P (2003) Inhibition of TIMP1 enhances angiogenesis in vivo and cell migration in vitro. Microvasc Res 65(1):9–17. doi: http://dx.doi.org/10.1016/S0026-2862(02)00026-2
167. Valente P, Fassina G, Melchiori A, Masiello L, Cilli M, Vacca A, Onisto M, Santi L, Stetler-Stevenson WG, Albini A (1998) TIMP-2 over-expression reduces invasion and angiogenesis and protects B16F10 melanoma cells from apoptosis. Int J Cancer 75(2):246–253. doi:10.1002/(SICI)1097-0215(19980119)75:2<246:AID-IJC13>3.0.CO;2-B
168. Qi JH, Ebrahem Q, Moore N, Murphy G, Claesson-Welsh L, Bond M, Baker A, Anand-Apte B (2003) A novel function for tissue inhibitor of metalloproteinases-3 (TIMP3): inhibition of angiogenesis by blockage of VEGF binding to VEGF receptor-2. Nat Med 9(4):407–415

Tripeptidyl Peptidase I and Its Role in Neurodegenerative and Tumor Diseases

Mashenka B. Dimitrova, Dimitrinka Y. Atanasova and Nikolai E. Lazarov

Abstract

Tripeptidyl peptidase I (TPPI) is a lysosomal enzyme widely distributed in mammals and humans. Its genetically determined deficiency causes the classical late-infantile form of neuronal ceroid lipofuscinosis, a fatal hereditary neurodegenerative disease associated with severe symptoms and early death, usually in the second decade of life. Many studies also show that TPPI is differentially regulated under various pathological conditions such as malignancy, neurodegeneration, ischemia, and inflammation, pointing at possible enzyme involvement in the pathogeneses of these entities. This chapter focuses on the TPPI participation in neurodegenerative and neoplastic diseases.

Keywords

Tripeptidyl peptidase I · Genetically determined diseases · Neurodegeneration Tumor diseases

M.B. Dimitrova
Institute of Experimental Morphology, Pathology and Anthropology with Museum, Bulgarian Academy of Sciences, 1113 Sofia, Bulgaria

D.Y. Atanasova · N.E. Lazarov
Institute of Neurobiology, Bulgarian Academy of Sciences, 1113 Sofia, Bulgaria

N.E. Lazarov (✉)
Department of Anatomy and Histology, Faculty of Medicine, Medical University of Sofia, 2, Zdrave Street, 1431 Sofia, Bulgaria
e-mail: nlazarov@medfac.acad.bg

D.Y. Atanasova
Department of Anatomy, Faculty of Medicine, Trakia University, 6003 Stara Zagora, Bulgaria

S. Chakraborti and N.S. Dhalla (eds.), *Pathophysiological Aspects of Proteases*,
DOI 10.1007/978-981-10-6141-7_6

1 Introduction

Tripeptidyl peptidase I (TPPI; EC 3.4.14.9) is a serine-type lysosomal protease sequentially removing tripeptides from the unmodified *N*-terminal of polypeptides and small proteins at pH around 4.5 [1, 2]. The enzyme can also act as an endopeptidase depending on the acidity of the medium [3, 4]. It is the first mammalian member of the S53 family of pepstatin-insensitive sedolisin serine-carboxyl peptidases (reviewed in [5, 6]). In humans, TPPI is encoded by the cln2 gene mapped to 11p15 [7], which mutations are associated with the fatal autosomal recessive disease classical late-infantile neuronal ceroid lipofuscinosis [8]. Therefore, TPPI is also called CLN2 protein. Different approaches have been applied to study the enzyme structure, catalytic mechanism and maturation, including studies on recombinant human TPPI [9–11], site-directed mutagenesis [12–14], and crystallographic structures of TPPI pro-enzyme [15, 16]. According to those studies, the TPPI molecule possesses a subtilisin-like fold and an octahedrally coordinated Ca^{2+}-binding site [14]. Its catalytic domain contains the active nucleophile Ser475 in a catalytic triad Ser475-Glu272-Asp360, distinct from the one of the classical serine proteases (Ser-His-Asp) but typical of sedolisin proteases. Furthermore, Asp360 and Asp276 participate in the formation of the oxyanion hole and are responsible for the acidic pH optimum of TPPI [15, 16], whereas Trp542 is critical for maintaining the proper tertiary structure of both the precursor and mature TPPI [14]. Studies on the *N*-glycosylation of the enzyme show that TPPI is a highly glycosylated protein with the oligosaccharide at Asn286 being crucial for folding and lysosomal targeting [12, 17]. Like most lysosomal proteases, TPPI is synthesized as an inactive precursor and is targeted to lysosomes through a mannose-6-phosphate-dependent pathway, where its maturation takes part involving autoprocessing or proteolytic activation [10, 11]. The prosegment of TPPI that is removed in the lysosomal compartment acts as a slow binding potent inhibitor of the parent enzyme [18]. Among the synthetic inhibitors, the most powerful and enzyme-specific inhibition is found for Ala-Ala-Phe-chloromethyl ketone [2, 19]. The substrate specificity of TPPI is thoroughly studied to show that the enzyme prefers bulky hydrophobic amino acid residues at P1 position and positively charged amino acids at P3 [20] and that its activity is much higher to substrates containing amino acid at P1' compared with the commonly used synthetic substrates like methylcoumaryl amides [4]. Comprehensive studies on the enzyme specificity, including ones from our laboratory, provide guidance for the synthesis of specific enzyme substrates [21–23]. Natural substrates of TPPI are not fully identified, but it is known that the enzyme participates in the hydrolysis of collagen type I [1, 2], cholecystokinin-5 and cholecystokinin-8 [24, 25], angiotensin II and glucagon [2], angiotensin III and neuromedin B [26, 27], substance P [19], and the subunit c of the mitochondrial ATP synthase [19, 28]. Analyses on the presence of orthologs of human TPPI in the genomic sequences of a number of eukaryotic species have been performed to show that the enzyme is highly conservative and widely distributed in higher organisms [29].

Localization and activity of TPPI in different organs and tissues of laboratory rats and mice have been studied using biochemical assays, immunohistochemistry, and enzyme histochemistry. Our results, and data from other laboratories, show high enzyme activity in the neurons and glial cells in the central nervous system (CNS) [30–32] as well as in a number of peripheral organs such as the spleen, kidney, liver, pancreas, lungs [30, 33], male and female reproductive organs [34], and the carotid body [35]. Similar studies in human autopsy materials [36, 37] reveal an equally wide tissue distribution of the enzyme in humans. Furthermore, it has been shown that the enzyme is developmentally regulated in rats [32, 38] and mice [39] as well as in humans [36, 37]. According to Kopan et al. [27], TPPI participates in the degradation of neuropeptides entering cells through receptor-mediated endocytosis. Other authors assume a possible involvement of the enzyme in the breakdown of unnecessary mitochondria by intracellular autophagy in the stomach and in intracytoplasmic digestion of the protein content in secretory granules in excess through crinophagy in the pancreas [30]. Recently, it has also been shown that TPPI participates in receptor-mediated apoptosis [40, 41]. In addition, other studies have demonstrated that TPPI is differentially regulated in pathological processes (reviewed in [42]). The aim of the present chapter, therefore, is to summarize the current state of understanding on the role of TPPI in the pathogenesis of neurodegenerative and oncological diseases.

2 Genetically Determined Deficiency of TPPI

Genetically determined deficiency of TPPI causes the hereditary neurodegenerative disorder classical late-infantile neuronal ceroid lipofuscinosis (LINCL). It is inherited in a recessive autosomal manner and represents one of the forms of neuronal ceroid lipofuscinoses (NCLs). NCLs are a group of at least 14 genetic diseases caused by mutations in the genes encoding the so-called CLN proteins (for a recent review see [43]). NCLs are classified according to the age of onset, clinical course, and the mutated gene involved [44, 45]. They share some common features like accumulation of ceroid lipopigment in membrane-bound lysosomal residual bodies, a loss of neurons in the CNS, mental retardation, motor deterioration, seizures, and shortened life expectancy [46]. Despite the similarities, there are also substantial clinical variations in NCLs depending on the affected genes and specific types of mutations. Comprehensive reviews on NCLs have been published [47–54]. Latest achievements in the studies of NCLs are recently collected in a special issue of Biochimica et Biophysica ACTA (BBA)—Molecular Basis of Disease (BBA Volume 1852, Issue 10, Part B, Pages 2235–2338, 2015)—entitled "Current Research on the Neuronal Ceroid Lipofuscinoses (Batten Disease)" [55].

The symptoms of LINCL appear at the age of 2–4 years (late infancy) when the enzyme reaches its adult activity levels in the CNS and most probably becomes crucial for the neuronal functions [36, 37]. The symptoms develop gradually and include epileptic seizures, mental decline, impaired locomotor functions, visual loss, and usually death in the second decade of life [48, 56]. Some mutations have been documented to cause a later onset and considerably milder phenotype [57, 58] due to the presence of residual TPPI activity [58, 59]. Earlier diagnoses of LINCL as well as other NCLs have been largely based on histopathological technics including ultrastructural analyses of the lipopigment deposits. Most of the storage material represents the subunit c of the mitochondrial ATP synthase [19, 28] although other proteins, lipids, and various components are also included [60]. In LINCL, storage bodies form curvilinear profiles sometimes accompanied by fingerprint-like deposits [49]. Those curvilinear bodies can be observed both in the CNS and extraneural tissues. Recent methods of diagnosis are much more complex and include electroencephalography, electrophysiological studies, electroretinography, enzyme assays in cultured skin fibroblasts as well as genetic analysis (reviewed in [50]). Although LINCL is still incurable, therapies are available, mostly directed at reducing the symptoms. Animal models of the disease could serve an important and useful tool for developing novel therapies. The available animal models for all types of NCLs including LINCL have been recently reviewed by Bond et al. [61]. Mice bred through biotechnological modifications targeting cln2 gene [62] have been used to test an enzyme replacement therapy [63, 64] and adeno-associated virus gene therapy [65, 66]. The results show attenuated neuropathology and a longer lifespan. Experimental therapies are promising and need careful clinical trials to achieve progress in the treatment of LINCL patents [67].

3 TPPI in Other Neurodegenerative Diseases

The similarities in NCLs have led to the search of some functional relationships or direct interactions between CLN proteins, which are deficient in individual diseases (reviewed in [53]). The first attempt to identify possible protein–protein interactions was made using a yeast two-hybrid system, and they revealed no evidence for molecular connections between CLN proteins [68]. Later on, co-immunoprecipitation and in vitro binding assays of CLN proteins from transfected COS-1 cells over-expressing them, provided evidence for a direct interaction of the TPPI and CLN5 protein. The latter is responsible for the Finnish variant of LINCL [69, 70]. Recently, however, based on uncertainties in the experimental procedures, these results were questioned [53]. As a matter of fact, it is still unclear whether possible interactions between CLN proteins might be of any biological significance [53].

Immunohistochemical studies on autopsy materials have shown that TPPI is over-expressed in both glial cells and neurons of individuals suffering from the infantile (CLN1) and juvenile (CLN3) forms of NCLs when compared to age-matched controls [37]. According to the authors, this increase may be due to either a brain reaction to the lack of important lysosomal proteins or phagocytosis of cellular components of damaged tissue or both. Using a biochemical method, an increased TPPI activity has also been detected in brains of patients suffering from infantile (CLN1) and adult (CLN4) NCLs [71]. Similarly, TPPI activity in CLN5-deficient fibroblasts (Finnish variant of LINCL) has been found to increase by about 35% as compared to normal fibroblasts [69]. All these results indicate a possible involvement of TPPI in compensatory mechanisms of other forms of NCLs.

Mutations in the cln2 gene encoding TPPI have recently been identified to cause autosomal recessive spinocerebellar ataxia 7 (SCAR7), a rare, slow progressive hereditary disease [72, 73]. Only six cases have been described so far, five of which are in members of the same family. The clinical symptoms of SCAR7 resemble to some extent those of LINCL but are much milder and slowly progressing, resulting in an almost normal life expectancy. The symptoms include dysarthria, limb ataxia, and cerebellar atrophy but no eye anomalies or epilepsy. Ultrastructurally, curvilinear profiles have not been detected either. It has been supposed that mutations in cln2 causing a reduction of the TPPI activity, and not a dramatic loss of the enzyme functions, are responsible for SCAR7 [73].

TPPI has been reported to degrade the synthetic amyloid-β (Aβ) peptides 1–42 and 1–28 in vitro [19]. Since senile plaques in Alzheimer disease (AD) contain Aβ42, and studies have been performed on the TPPI localization and activity in AD. Earlier studies have shown a presence of amyloid precursor protein (APP) in the storage material in different forms of NCLs [74]. The authors suppose that CLN proteins are involved in the degradation pathway of APP and their deficiency leads to an abnormal APP processing. However, CLN-enzymes are lysosomal and therefore are inactive in neutral pH media outside the lysosomes. Using biochemical studies with a TPPI selective substrate, the enzyme activity has been found to increase about two-fold in AD brains when compared with normal or age-matched controls [71]. On the other hand, using immunohistochemistry it has been shown that TPPI possesses decreased expression levels in brains of patients with AD as well as in aged controls, particularly in areas of senile plaques and in neurons with neurofibrillary tangles [37]. In this study, high TPPI expression has been detected only in glial cells associated with certain senile plaques. These results are not necessarily controversial since the biochemical increase in the enzyme activity might for instance be due to gliosis. According to Kida et al. [37], TPPI may be involved in the intracellular degradation of Aβ. Recent research has shown that TPPI-deficient fibroblasts from LINCL patients have a decreased degradation by macroautophagy and impaired formation of autophagosomes resulting from an up-regulation of Akt-mTOR signaling pathway [75]. Thus, TPPI appears to be an important participant in this lysosomal pathway for the turnover of organelles and long-lived proteins. Furthermore, it is known that macroautophagy is abnormally activated in AD neurons and is one of the earliest events observed even before the formation of senile plaques. According to similar studies, it might also represent an additional pathway for generation of Aβ [76]. Whether TPPI is implicated in such pathogenic processes in AD is yet to be explored in further studies.

Hypoxic stress has been recognized to be one of the main causes of neurodegeneration to various extents. It may result from different disorders like cardiac arrest, brain injury, stroke, or even ingestion of harmful chemicals. Recent studies

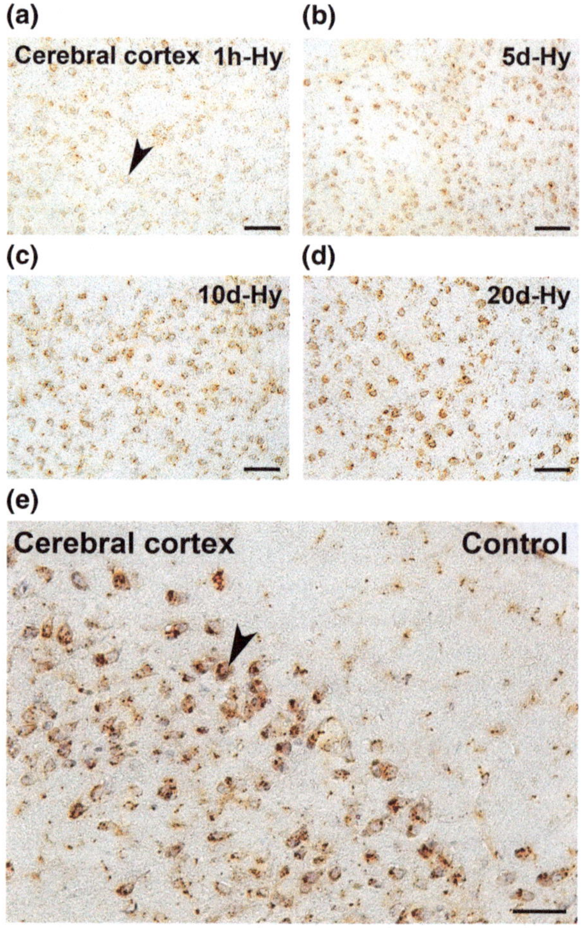

Fig. 1 Histochemical imaging of TPPI activity in cryostat sections of the rat cerebral cortex at different time periods after sodium nitrite-induced hypoxic shock using the substrate glycyl-L-prolyl-L-metionyl-5-chloro-1-anthraquinonylhydrazide. The enzyme activity in cortical neurons (*arrowheads*) is very low or missing one hour post-treatment (**a**). Later on, it increases slowly at the fifth (**b**), tenth (**c**), and twentieth (**d**) day after animals treatment to almost reach its control levels (**e**). Scale bars = 50 μm

have shown that TPPI-deficiency is accompanied by a decrease in the catalase activity and intensive accumulation of reactive oxygen species [75] as well as a substantial mitochondrial fragmentation in response to oxidative stress [41]. Furthermore, it has been revealed that TPPI activity increases in the embryonic cells of goldfish after 24 h of austere hypoxia, and the enzyme has been recognized as one of the possible target genes of hypoxia-inducible factor-1 in fish [77]. Based on these studies, recently we followed up the consecutive alterations of TPPI activity in the brain of rats, which have undergone an acute hypoxic stress throughout a single injection of sodium nitrite, a simple and efficient model to study the impact of hypoxic insult on the brain [78, 79]. We found reduced activity levels of the enzyme in the CNS neurons almost immediately after the hypoxic event which slowly returned to their normal values 2–3 weeks after the treatment (Fig. 1). These results may be representative for a possible participation of TPPI in the brain response to hypoxia [80]. In this respect, studies on TPPI expression pattern in autopsied human brains after stroke have revealed a very weak enzyme expression in the neurons adjacent to the necrotic area but substantial enzyme levels in the microglia and astroglia [37]. Stroke results from a severe hypoxic shock leading to necrotic foci in the affected areas, and in line with this, neurons in less affected adjacent areas show a similar low TPPI expression which coincides with our results. The biological significance of TPPI activity reduction in brain hypoxia should be a subject of more detailed studies.

4 TPPI in Tumor Diseases

Most of the studies on TPPI deal with LINCL or other neurodegenerative diseases, whereas the enzyme involvement in tumors is relatively less explored. In fact, one of the earliest studies on TPPI was performed on the enzyme isolated and purified from human osteoclastomas [1]. It has been shown that TPPI participates in the degradation of collagen and probably other matrix proteins. Moreover, the TPPI specific inhibitor Ala-Ala-Phe-chloromethyl ketone reduces the ability of isolated osteoclasts to form resorption lacunae in bone slices in vitro. This might give an opportunity for the use of specific enzyme inhibitors as therapeutics for inhibiting excessive bone resorption in osteoclastoma.

Further, the enzyme activity levels have been measured in a large number of primary breast carcinomas isolated at surgery and compared to those in normal breast tissue specimen extracted at reductive mammoplasty [81]. The results reveal up to 17-fold higher enzyme activity in the tumor tissues which correlates positively with the established biomarkers in breast cancer such as cathepsin D, estrogen, and progesterone receptors. Specific methods for measuring the enzyme activity in tissue homogenates and in blood samples have also been proposed [82] in order to use them for diagnostic or prognostic purposes in breast cancer.

Additionally, in an experiment designed to establish the individual enzyme profiles and possible biochemical differences between adenocarcinomas of the gastroesophageal junction and squamous cell carcinomas of the lower third of the esophagus, the activities of a number of lysosomal enzymes including TPPI have been measured [83]. The findings have demonstrated higher TPPI activity levels in tumor tissues in the lower esophagus as compared to those in the bordering intact mucosa. According to the authors, lysosomal enzymes act synergistically to facilitate the tumor invasion into adjacent tissues.

Moreover, TPPI expression has been evaluated in colorectal carcinomas at various clinical stages using immunoblot analyses, immunohistochemistry, and quantitative RT-PCR [84]. Immunocytochemical studies have shown that the enzyme is localized mainly in the invasive front of the tumor and has a significantly elevated expression in both liver and lymph node metastatic foci. The authors suggest that TPPI is expressed in cancer metastases as a matrix protease. Since the expression levels of TPPI are associated with advanced clinical stages and occurrence of distant metastases, a diagnostic and prognostic value of TPPI has been proposed for colorectal carcinomas.

Increased TPPI expression has also been reported in other tumors, such as thyroid adenocarcinoma, liver cancer, meningioma, mesothelioma [37]. Using enzyme histochemical method in a rare malignant epithelial neoplasm, i.e., pancreatic acinar cell carcinoma, we revealed a higher TPPI activity in tumor acinar cells with varying grades of differentiation and intensity of the stromal cell reaction (Fig. 2).

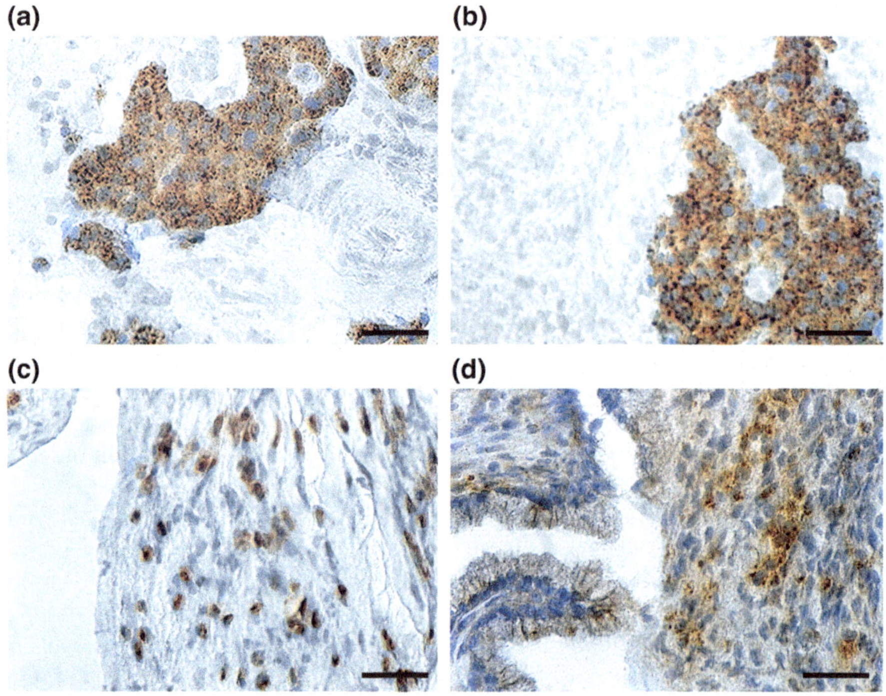

Fig. 2 TPPI activity in acinar cell carcinoma of the pancreas. The tumor acinar cells in well-differentiated (**a**), moderately differentiated (**b**) and undifferentiated (anaplastic) cancer (**c**) exhibit enzyme activity. Note the diffuse mononuclear cell infiltration around the neoplastic focus (**b**) and the marked periductal stromal cell reaction (**d**). Scale bars = 50 μm

All the studies on solid tumors described above show substantially elevated TPPI expression in the areas of tumor invasion into the adjacent tissues and/or in metastases. In this respect, it seems tempting to conclude that the enzyme is involved in the degradation of the extracellular matrix components, thus opening free spaces for the tumor growth and metastasizing, and probably for the formation of tumor vessels. However, the lysosomal localization and acidic pH optimum of TPPI are not in favor of such an assumption. On the other hand, secretion of TPPI from cells over-expressing it has been documented although the precise mechanisms involved are not fully understood [10]. Similarly, secretion of other acid hydrolases has also been reported and different mechanisms have been proposed [85, 86]. Therefore, TPPI in the extracellular matrix obviously appears in an inactive form as a pro-enzyme unable of autoactivation in neutral media [10]. More recent studies of Golabek et al. [87] have shown that the TPPI pro-enzyme binds to polyanionic glycosaminoglycans such as heparin, dextran sulfate, heparan sulfate, and chondroitin sulfate B, and these substances exert a protective effect on the enzyme molecule. They not only protect TPPI from heat- or alkaline pH-induced degradation, but also facilitate the enzyme activation at pH up to 6.0. Since sulfated polyanionic glycosaminoglycans may be present in the extracellular matrix, these findings may explain the possible TPPI role as an extracellular (matrix) protease in tumor diseases.

5 Conclusion

In summary, TPPI activity is critical for neuronal functions. Its genetically determined deficiency leads to classical late-infantile neuronal ceroid lipofuscinosis, a lysosomal storage disorder which is associated with a progressive loss of neurons and photoreceptor cells, severe symptoms, and short lifespan. Some mutations in the TPPI encoding gene lead to autosomal recessive spinocerebellar ataxia 7, a milder hereditary neurodegenerative disease. A number of studies also show that the enzyme is involved in the pathogenic mechanisms of other neurodegenerative diseases, e.g., Alzheimer disease, as well as in neurodegeneration induced by brain hypoxia. TPPI involvement in tumors is less explored, but available studies reveal its possible diagnostic or prognostic value in different types of solid tumors. The role of the enzyme in pathological processes such as tumor invasion and metastasizing are to be elucidated in further studies.

Acknowledgements The authors would like to thank Dr. Enrico Marani and Dr. Angel Dandov for their helpful comments and critical reading of the manuscript.

References

1. Page AE, Fuller K, Chambers TJ, Warburton MJ (1993) Purification and characterization of a tripeptidyl peptidase I from human osteoclastomas: evidence for its role in bone resorption. Arch Biochem Biophys 306:354–359
2. Vines D, Varburton MJ (1998) Purification and characterisation of a tripeptidyl aminopeptidase I from rat spleen. Biochim Biophys Acta 1384:233–242
3. Ezaki J, Takeda-Ezaki M, Oda K, Kominami E (2000) Characterization of endopeptidase activity of tripeptidyl peptidase-I/CLN2 protein which is deficient in classical late infantile neuronal ceroid lipofuscinosis. J Biochem Biophys Res Commun 268:904–908
4. Kondo MY, Gouvea IE, Okamoto DN, Santos JAN, Souccar C, Oda K, Juliano L, Juliano MA (2016) Analysis of catalytic properties of tripeptidyl peptidase I (TTP-I), a serine carboxyl lysosomal protease, and its detection in tissue extracts using selective FRET peptide substrate. Peptides 76:80–86
5. Wlodawer A, Li M, Gustchina A, Oyama H, Dunn BM, Oda K (2003) Structural and enzymatic properties of the sedolisin family of serine-carboxyl peptidases. Acta Biochim Polon 50:81–102
6. Oda K (2012) New families of carboxyl peptidases: serine-carboxyl peptidases and glutamic peptidases. J Biochem 151:13–25
7. Sharp JD, Wheeler RB, Lake BD, Savukoski M, Jarvela IE, Peltonen L, Gardiner RM, Williams RE (1997) Hum Mol Genet 6:591–595
8. Sleat DE, Donnelly JR, Lackland H, Lin GC, Sohar I, Pullarkat RK, Lobel P (1997) Association of mutations in a lysosomal protein with classical late-infantile neuronal ceroid lipofuscinosis. Science 277:1802–1805
9. Lin L, Sohar I, Lackland H, Lobel P (2001) The human CLN2 protein/tripeptidyl peptidase I is a serine protease that autoactivates at acidic pH. J Biol Chem 276:2249–2255
10. Golabek A, Kida E, Walus M, Wujek P, Mentha P, Wisniewski K (2003) Human tripeptidyl-peptidase I: biosynthesis, glycosylation and enzymatic processing in vivo. J Biol Chem 278:7135–7145
11. Golabek A, Wujek P, Walus M, Bieler S, Soto C, Wisniewski K, Kida E (2004) Maturation of human tripeptidyl-peptidase I in vitro. J Biol Chem 279:31058–31067
12. Wujek P, Kida E, Walus M, Wisniewski K, Golabek A (2004) N-glycosylation is crucial for folding, trafficking, and stability of human tripeptidyl-peptidase I. J Biol Chem 279:12827–12839
13. Walus M, Kida E, Wisniewski K, Golabek A (2005) Ser475, Glu272, Asp276, Asp327 and Asp360 are involved in catalytic activity of human tripeptidyl-peptydase I. FEBS Lett 579:1383–1388
14. Kuizon S, DiMaiuta K, Walus M, Jenkins EC Jr, Kuizon M, Kida E, Golabek AA, Espinoza DO, Pullarkat RK, Junaid MA (2010) A critical tryptophan and Ca^{2+} in activation and catalysis of TPPI, the enzyme deficient in classic late-infantile neuronal ceroid lipofuscinosis. PLoS ONE 5(8):e11929
15. Pal A, Kraetzner R, Gruene T, Grapp M, Schreiber K, Groenborg M, Urlaub H, Becker S, Asif AR, Gaertner J, Sheldrick GM, Steinfeld R (2009) Structure of tripeptidyl-peptidase I provides insight into the molecular basis of late infantile neuronal ceroid lipofuscinosis. J Biol Chem 284:3976–3984
16. Guhaniyogi J, Sohar I, Das K, Stock A, Lobel P (2009) Crystal structure and autoactivation pathway of the precursor form of human tripeptidyl-peptidase I, the enzyme deficient in late infantile ceroid lipofuscinosis. J Biol Chem 284:3985–3997
17. Tsiakas K, Steinfeld R, Storch S, Ezaki J, Lukacs Z, Kominami E, Kohlschutter A, Ullrich K, Braulke T (2004) Mutation of the glycosylated asparagines residue 286 in human CLN2 protein results in loss of enzymatic activity. Glycobiology 14:1C–5C

18. Golabek AA, Dolzhanskaya N, Walus M, Wisniewski KE, Kida E (2008) Prosegment of tripeptidyl peptidase I is a potent, slow-binding inhibitor of its cognate enzyme. J Biol Chem 283:16497–16504
19. Junaid M, Wu G, Pullarkat R (2000) Purification and characterization of bovine brain lysosomal pepstatin-insensitive proteinase, the gene product deficient in the human late-infantile neuronal ceroid lipofuscinosis. J Neurochem 74:287–294
20. Tian Y, Sohar I, Taylor JW, Lobel P (2006) Determination of the substrate specificity of tripeptidyl-peptidase I using combinatorial peptide libraries and development of improved fluorogenic substrates. J Biol Chem 281:6559–6572
21. Dikov A, Dimitrova M, Ivanov I, Krieg R, Halbhuber K-J (2000) Original method for the histochemical demonstration of tripeptidyl aminopeptidase I. Cell Mol Biol 46:1219–1225
22. Steinfeld R, Fuhrmann J, Gartner J (2006) Detection of tripeptidyl peptidase I activity in living cells by fluorogenic substrates. J Histochem Cytochem 54:991–996
23. Ivanov I, Tascheva D, Todorova R, Dimitrova M (2009) Synthesis and use of 4-peptidylhydrazido-N-hexyl-1,8-naphthalimides as fluorogenic histochemical substrates for dipeptidyl peptidase IV and tripeptidyl peptidase I. Eur J Med Chem 44:384–392
24. Bernardini F, Warburton M (2002) Lysosomal degradation of cholecystokinin-(29–33)-amide in mouse brain is dependent of tripeptidyl peptidase—I: implications for the degradation and storage of peptides in classical late-infantile neuronal ceroid lipofuscinosis. Biochem J 366:521–529
25. Warburton MJ, Bernardini F (2002) Tripeptidyl peptidase-I is essential for the degradation of sulphated cholecystokinin-8 (CCK-8S) by mouse brain lysosomes. Neurosci Lett 331:99–102
26. Du P, Kato S, Li Y, Maeda T, Yamane T, Yamamoto S, Fujiwara, Yamamoto Y, Nishi K, Ohkubo I (2001) Rat tripeptidyl peptidase I: molecular cloning, functional expression, tissue localization and enzymatic characterization. Biol Chem 382:1715–1725
27. Kopan S, Sivasubramaniam U, Warburton M (2004) The lysosomal degradation of neuromedin B is dependent on tripeptydil peptidase—I: evidence for the impairment of neuropeptide degradation in late-infantile neuronal ceroid lipofuscinosis. Biochem Biophys Res Commun 319:58–65
28. Ezaki J, Tanida I, Kanehagi N, Kominami E (1999) A lysosomal proteinase, the late infantile neuronal ceroid lipofuscinosis gene (CLN2) product, is essential for degradation of a hydrophobic protein, the subunit c of ATP synthase. J Neurochem 72:2573–2582
29. Wlodawer A, Durell SR, Li M, Oyama H, Oda K, Dunn B (2003) A model of tripeptidyl peptidase I (CLN2), a ubiquitous and highly conserved member of the sedolisin family of serine-carboxyl peptidases. BMC Struct Biol 3:8–18
30. Koike M, Shibata M, Ohsawa Y, Kametaka S, Waguri S, Kominami E, Uchiyama Y (2002) The expression of trypeptidyl peptidase I in various tissues of rats and mice. Arch Histol Cytol 65:219–232
31. Dimitrova M, Ivanov I, Deleva D (2009) Distribution of tripeptidyl peptidase I activity in the rat brain and spinal cord. Comp Rend Acad Bulg Sci 62:729–734
32. Dimitrova M, Deleva D (2009) Histochemical study of the changes in trypeptidyl peptidase I activity in developing rat brain and spinal cord. Compt Rend Acad Bulg Sci 62:1407–1412
33. Yayoi Y, Ohsawa Y, Koike M, Zhang GQ, Kominami E, Uchiyama Y (2001) Specific localization of lysosomal aminopeptidases in type II alveolar epithelial cells of the rat lung. Arch Histol Cytol 64:89–97
34. Dimitrova M, Ivanov I, Todorova R, Tsenova V (2008) Fluorescent localization of tripeptidyl peptidase I activity in tissue sections of Balb/c mice reproductive organs. Compt Rend Acad Bulg Sci 61:349–356
35. Atanasova D, Lazarov N (2015) Histochemical demonstration of tripeptidyl aminopeptidase I in the rat carotid body. Acta Histochem 117:219–222
36. Kurachi Y, Oka A, Itoh M, Mizuguchi M, Hayashi M, Takashima S (2001) Distribution and development of CLN-2 protein, the late-infantile neuronal ceroid lipofuscinosis gene product. Acta Neuropathol 102:20–26

37. Kida E, Golabek A, Walus M, Wujek P, Kaczmarski W, Wisniewski EK (2001) Distribution of tripeptidyl peptidase I in human tissues under normal and pathological conditions. J Neuropathol Exper Neurol 60:280–292
38. Suopanki J, Partanen S, Ezaki J, Baumann M, Kominami E, Tyynela J (2000) Developmental changes in the expression of neuronal ceroid lipofuscinoses-linked proteins. Mol Genet Metab 71:190–194
39. Dimitrova M, Deleva D, Pavlova V, Ivanov I (2011) Developmental study of tripeptidyl peptidase I activity in the mouse central nervous system and peripheral organs. Cell Tissue Res 346:141–149
40. Autefage H, Albinet V, Garcia V, Berges H, Nicolau ML, Therville N, Altie MF, Caillaud C, Levade T, Andrieu-Abadie N (2009) Lysosomal serine protease CLN2 regulates tumor necrosis factor-alfa-mediated apoptosis in a bid-dependent manner. J Biol Chem 284:11507–11516
41. Van Beersel G, Tihon E, Demine S, Hamer I, Jadot M, Arnould T (2013) Different molecular mechanisms involved in spontaneous and oxidative stress-induced mitochondrial fragmentation in tripeptidyl peptidase-1 (TPP-1)-deficient fibroblasts. Biosci Rep 33:243–258
42. Golabek AA, Kida E (2006) Tripeptidyl-peptidase I in health and disease. Biol Chem 387:1091–1099
43. Marani E, Lazarov N (2016) Lipofuscin and lipofuscinosis. Neuroscience and Biobehavioral Psychology. Elsevier, Nr. 02594 (in press)
44. Williams RE, Mole SE (2012) New nomenclature and classification scheme for the neuronal ceroid lipofuscinoses. Neurology 79:183–191
45. Kousi M, Lehesjoki AE, Mole SE (2012) Update of the mutation spectrum and clinical correlations of over 360 mutations in eight genes that underlie the neuronal ceroid lipofuscinoses. Hum Mutat 33:42–63
46. Mole SE, Williams RE, Goebel HH (2011) The neuronal ceroid lipofuscinoses (Batten disease), contemporary neurology series. Oxford University Press, Oxford, p 480
47. Wisniewski KE, Kida E, Connell F, Zhong N (2000) Neuronal ceroid lipofuscinoses: research update. Neurol Sci 21:S49–S56
48. Wisniewski KE, Kida E, Golabek AA, Kaszmarski W, Connell F, Zhong N (2001) Neuronal ceroid lipofuscinoses: classification and diagnosis. Adv Genet 45:1–34
49. Mole SE, Williams RE, Goebel HH (2005) Correlations between genotype, ultrastructural morphology and clinical phenotype in the neuronal ceroid lipofuscinoses. Neurogenetics 6:107–126
50. Williams RE, Aberg L, Autti T, Goebel HH, Kohlschütter A, Lönnqvist T (2006) Diagnosis of the neuronal ceroid lipofuscinoses: an update. Biochim Biophys Acta 1762:865–872
51. Persaud-Sawin D-A, Mousallem T, Wang C, Zucker A, Kominami E Boustany R-MN (2007) Neuronal ceroid lipofuscinosis: a common pathway? Pediatric Res 61:146–152
52. Pierret C, Morrison JA, Kirk MD (2008) Treatment of lysosomal storage disorders: focus on the neuronal ceroid-lipofuscinoses. Acta Neurobiol Exp 68(429):442
53. Getty AL, Pearce DA (2011) Interactions of the proteins of neuronal ceroid lipofuscinosis: clues to function. Cell Mol Life Sci 68:453–474
54. Warrier V, Vieira M, Mole SE (2013) Genetic basis and phenotypic correlations of the neuronal ceroid lipofuscinoses. Biochim Biophys Acta 1832:1827–1830
55. Kohan R, Mole SE, Cotman SL (eds) (2015) Current research on the neuronal ceroid lipofuscinoses (Batten disease). Biochim Biophys Acta (BBA)—Mol Basis Dis 1852(10),Part B:2235–2338
56. Steinfeld R, Heim P, von Gregory H, Meyer K, Ullrich K, Goebel HH, Kohlschutter A (2002) Late infantile neuronal ceroid lipofuscinosis: quantitative description of the clinical course in patients with CLN2 mutations. Am J Med Genet 112:347–354
57. Steinfeld R, Steinke H-B, Isbrandt D, Kohlschuaetter A, Gaertner J (2004) Mutations in classical late infantile neuronal ceroid lipofuscinosis disrupt transport of tripeptidyl-peptidase I to lysosomes. Hum Mol Genet 13:2483–2491

58. Kohan R, Carabelos MN, Xin W, Sims K, Guelbert N, Cismondi IA, Pons P, Alonso GI, Troncoso M, Witting S, Pearce DA, Dodelson de Kremer R, Oller-Ramírez AM, Noher de Halac I (2013) Neuronal ceroid lipofuscinosis type CLN2: a new rationale for the construction of phenotypic subgroups based on a survey of 25 cases in South America. Gene 516:114–121
59. Sleat DE, El-Banna M, Sohar I, Kim K-H, Dobrenis K, Walkley SU, Lobel P (2008) Residual levels of tripeptidyl-peptidase I activity dramatically ameliorate disease in late-infantile neuronal ceroid lipofuscinosis. Mol Genet Metab 94:222–233
60. Palmer DN, Oswald MJ, Westlake VJ, Kay GW (2002) The origin of fluorescence in the neuronal ceroid lipofuscinoses (Batten disease) and neuron cultures from affected sheep for studies of neurodegeneration. Arch Gerontol Geriatr 34:343–357
61. Bond M, Holthaus S-M, Tammen I, Tear G, Russell C (2013) Use of model organisms for the study of neuronal ceroid lipofuscinosis. Biochim Biophys Acta 1832:1842–1865
62. Sleat DE, Wiseman JA, El-Banna M, Kim KH, Mao Q, Price S, Macauley SL, Sidman RL, Shen M, Zhao Q, Passini MA, Davidson BL, Stewart GR, Lobel P (2004) A mouse model of classical late-infantile neuronal ceroid lipofuscinosis based on targeted disruption of the CLN2 gene results in a loss of tripeptidylpeptidase I activity and progressive neurodegeneration. J Neurosci 24:9117–9126
63. Chang M, Cooper JD, Sleat DE, Cheng SH, Dodge JC, Passini MA, Lobel P, Davidson BL (2008) Intraventricular enzyme replacement improves disease phenotypes in a mouse model of late infantile neuronal ceroid lipofuscinosis. Mol Ther 16:649–656
64. Xu S, Wang L, El-Banna M, Sohar I, Sleat DE, Lobel P (2011) Large-volume intrathecal enzyme delivery increases survival of a mouse model of late infantile neuronal ceroid lipofuscinosis. Mol Ther 19:1842–1848
65. Passini M, Dodge J, Bu J, Yang W, Zhao Q, Sondhi D, Hackett N, Kaminsky S, Mao Q, Shihabuddin L, Cheng S, Sleat D, Stewart G, Davidson B, Lobel P, Crystal R (2006) Intracranial delivery of CLN2 reduces brain pathology in a mouse model of classical late infantile neuronal ceroid lipofuscinosis. J Neurosci 26:1334–1342
66. Sondhi D, Hackett NR, Peterson DA, Stratton J, Baad M, Travis KM, Wilson JM, Crystal RG (2007) Enhanced survival of the LINCL mouse following CLN2 gene transfer using the rh. 10 rhesus macaque-derived adeno-associated virus vector. Mol Ther 15:481–491
67. Neverman NJ, Best HL, Hofmann SL, Hughes SM (2015) Experimental therapies in the neuronal ceroid lipofuscinoses. Biochim Biophys Acta 1852:2292–2300
68. Zhong NA, Moroziewicz DN, Ju W, Wisniewski KE, Jurkiewicz A, Brown WT (2000) CLN-encoded proteins do not interact with each other. Neurogenetics 3:41–44
69. Vesa J, Chin MH, Oelgeschlaeger K, Isosomppi J, DellAngelica EC, Jalanko A, Peltonen L (2002) Neuronal ceroid lipofuscinoses are connected at molecular level: interaction of CLN5 protein with CLN2 and CLN3. Mol Biol Cell 13:2410–2420
70. Lyly A, von Schantz C, Heine C, Schmiedt ML, Sipilae T, Jalanko A, Kyttaelae A (2009) Novel interactions of CLN5 support molecular networking between neuronal ceroid lipofuscinosis proteins. BMC Cell Biol 10:83
71. Junaid MA, Pullarkat RK (1999) Increased brain lysosomal pepstatin-insensitive proteinase activity in patients with neurodegenerative diseases. Neurosci Lett 264:157–160
72. Breedveld GJ, van Wetten B, Raa GD, Brusse E, van Swieten JC, Oostra BA, Maat-Kievit JA (2004) A new locus for a childhood onset, slowly progressive autosomal recessive spinocerebellar ataxia maps to chromosome 11p15. (Lett) J Med Genet 41:858–866
73. Sun Y, Almomani R, Breedveld GJ, Santen GWE, Aten E, Lefeber DJ, Hoff JI, Brusse E, Verheijen FW, Verdijk RM, Kriek M, Oostra B, Breuning MH, Losekoot M, den Dunnen JT, van de Warrenburg BP, Maat-Kievit AJA (2013) Autosomal recessive spinocerebellar ataxia 7 (SCAR7) is caused by variants in TPP1, the gene involved in classic late-infantile neuronal ceroid lipofuscinosis 2 disease (CLN2 disease). Hum Mutat 34:706–713
74. Wisniewski KE, Maslinska D, Kitaguchi T, Kim KS, Goebel HH, Haltia M (1990) Topographic heterogeneity of amyloid B-protein epitopes in brains with various forms of

neuronal ceroid lipofuscinoses suggesting defective processing of amyloid precursor protein. Acta Neuropathol 1990:26–34
75. Vidal-Donet JM, Carcel-Trullols J, Casanova B, Aguado C, Knecht E (2013) Alterations in ROS activity and lysosomal pH account for distinct patterns of macroautophagy in LINCL and JNCL fibroblasts. PLoS ONE 8(2):e55526
76. Yu WH, Cuervo AM, Kumar A et al (2005) Macroautophagy—a novel beta-amyloid peptide-generating pathway activated in Alzheimer's disease. J Cell Biol 171:87–98
77. Zhong X-P, Wang D, Zhang Y-B, Gui J-F (2009) Identification and characterization of hypoxia-induced genes in *Carassius auratus* blastulae embryonic cells using suppression subtractive hybridization. Comp Biochem Phys B 152:161–170
78. Zaidi ZF (2010) Sodium nitrite-induced hypoxic injury in rat hippocampus. Pak J Med Sci 26:532–537
79. Zaidi ZF (2010) Effects of sodium nitrite-induced hypoxia on cerebellar Purkinje cells in adult rats. Pak J Med Sci 26:261–266
80. Petrova EB, Dimitrova MB, Ivanov IP, Pavlova VG, Dimitrova SG, Kadiysky DS (2016) Effect of acute hypoxic shock on the rat brain morphology and tripeptidyl peptidase I activity. Acta Histochem 118:496–504
81. Junaid MA, Clark GM, Pullarkat RK (2000) A lysosomal pepstatin-insensitive proteinase as a novel biomarker for breast carcinoma. Int J Biol Markers 15:129–134
82. Pullarkat RK, Junaid MA (2001) A lysosomal pepstatin-insensitive proteinase as a novel biomarker for detecting and diagnosing breast cancer W02001069260:A2
83. Altorjay A, Paal B, Sohar N, Kiss J, Szanto I, Sohar I (2005) Significance and prognostic value of lysosomal enzyme activities measured in surgically operated adenocarcinomas of the gastroesophageal junction and squamous cell carcinomas of the lower third of esophagus. World J Gastroenterol 11:5751–5756
84. Tsukamato T, Lida J, Dobashi Y, Furukawa T, Konishi F (2006) Overexpression in colorectal carcinoma of two lysosomal enzymes, CLN2 and CLN1, involved in neuronal ceroid lipofuscinosis. Cancer 106:1489–1497
85. Ioannou YA, Bishop DF, Desnick RJ (1992) Overexpression of human alpha-galactosidase A results in its intracellular aggregation, crystallization in lysosomes, and selective secretion. J Cell Biol 119:1137–1150
86. Andrews NW (2002) Lysosomes and the plasma membrane: trypanosomes reveal a secret relationship. J Cell Biol 158:389–394
87. Golabek AA, Walus M, Wisniewski KE, Kida E (2005) Glycosaminoglycans modulate activation, activity, and stability of tripeptidyl-peptidase I in vitro and in vivo. J Biol Chem 280:7550–7560

Calpain Activity and Expression in Human Colonic Tumors

Rajendra K. Sharma, Umashankar Das, Sreejit Parameswaran, Sujeet Kumar and Jonathan R. Dimmock

Abstract

Families of cysteine proteases which are activated by Ca^{2+} are known as calpains. These enzymes can contribute to the metastatic behavior of certain tumors. In particular, we have found from Western Blotting and immunohistochemistry that the activity and expression of *m*-calpain is higher in colorectal tumors than non-malignant colonic mucosa. Furthermore, the expression of calpastin and high molecular weight calmodulin-binding protein, which are inhibitors of calpains, is reduced in colon cancers. These findings are of great importance in the design and development of anticancer drugs as well as providing diagnostic value as an indicator of the development of colorectal cancer. This chapter is devoted principally to the discoveries in our laboratories of calpain and its inhibitors in relation to human colonic neoplasms.

Keywords

Calpains · Calpastatin · High molecular weight calmodulin-binding protein Colonic tumors

R.K. Sharma (✉) · S. Parameswaran · S. Kumar
Department of Pathology and Laboratory Medicine, College of Medicine,
University of Saskatchewan, Saskatoon, SK, S7N 5E5, Canada
e-mail: rajendra.sharma@usask.ca

U. Das · J.R. Dimmock
College of Pharmacy and Nutrition, University of Saskatchewan,
Saskatoon, SK, S7N 5C9, Canada

S. Chakraborti and N.S. Dhalla (eds.), *Pathophysiological Aspects of Proteases*,
DOI 10.1007/978-981-10-6141-7_7

1 Introduction

Cancer is one of the insidious diseases and leading causes of death worldwide. Cancer is characterized by uncontrolled growth and spread of abnormal cells that can result in death if the spread is not controlled. In the Western world, major cancer deaths occur due to colon cancer [1]. The sequencing of the human genome had made it possible to understand the genetic basis of cancer. The underlying mutations and the lack of certain gene functions are some of the leading causes of cancer. The introduction or manipulation of genes into cancer cells where the gene function has been impaired can restore gene function and inhibit tumor progression.

2 Calpains

Recently, the calpains are recognized for their important role in various diseases including cancer [2, 3]. Calpains are intracellular Ca^{2+}-activated proteases which act as important mediators of the action Ca^{2+} [4, 5]. The calpain family (EC 3.4.22.17; Clan CA, family CO_2) consists of at least 16 proteins, all of which are coded for by an independent gene [6, 7]. These genes lead to the formation of various calpain homologs which have such functional domains as Zn-finger and Ca^{2+}-binding domains [4].

Cysteine proteases, μ-calpain and *m*-calpain (also known as calpains I and II), require micromolar and millimolar concentrations of Ca^{2+}, respectively [4]. Both of these enzymes have a subunit of 28 kDa which has a regulatory function as well as a catalytic subunit of 80 kDa. There are different ways of regulating *m*-calpain, and its activity is inhibited by calpastatin [8–10]. The high molecular weight calmodulin-binding protein (HMWCaMBP) was discovered in our laboratory [11] and shown to be homologous to calpastatin based on calpain inhibition, amino acid analysis, sequence homology, and antibody reactivity [12].

Calpain has four catalytic domains and two regulatory domains [4–10]. Activation by Ca^{2+} leads to autolysis of the N-terminus of domain 1, and thus, the requirement of Ca^{2+} is less. This process leads to the subunits dissociating. Hence, the regulation of calpain specificity and activity includes autolysis [4]. When Ca^{2+} is absent, there are two subdomains IIa and IIb in the protease domain which are folded into one Ca^{2+} binding site [4–10]. In the calpain family of enzymes, this binding site is the most conserved which suggests that it has an essential role in bioactivity. Both μ- and *m*-calpains have protease domains which display Ca^{2+}-dependent protease activity and do not require other domains. Structural studies revealed Ca^{2+} binding to domains IIa and IIb. In fact, all domains, i.e., IIa, IIb, III, IV, and VI, bind at least one Ca^{2+} although with different affinities [4–10].

The calpains have been implicated in the phenomenon of metastasis [2, 3], as well as cell cycle control, mitosis, ischemia reperfusion, muscle contraction, myoblast fusion in addition to cell migration, spreading, and adhesion [13–17]. An important way that post-translational modification occurs is by proteolysis mediated by calpain which affects cell proliferation and migration, apoptosis, and other physiological processes [10]. Substrate activity may be changed by the proteolysis of substrates by calpains. The intramolecular signals which lead to fast proteolytic degradation by *m*-calpain are the PEST sequences which are segments along the polypeptide chain that are rich in proline (P), glutamate (E), serine (S) and threonine (T). Proteolysis of calcineurin (CaN), which is mediated by *m*-calpain, resulted in increased activity and strong immunostaining of CaN in ischemic-perfused rat hearts [18, 19]. Both μ- and *m*-calpain react strongly with CaN in epileptic chickens but not normal birds [20, 21], while *m*-calpain interacts with N-myristoyltransferase (NMT) in epileptic chickens [21, 22]. It is of interest to note that of the two isoforms of NMT, it is NMT2 which reacts more strongly with *m*-calpain than NMT1 [21, 22].

3 Calpains in Pathological Conditions

The calpains cause a number of proteins in cells to undergo proteolysis [4] such as tyrosine phosphatase 1B, pp60c-Src, protein kinase, and various signaling enzymes [23–25]. Most of these proteins are believed to be involved in the development of human tumors. Thus, the calpains appear to have an important regulatory role to exercise in cancer formation. For example, greater amounts of μ-calpain were found in human renal cell carcinomas which had metastasized to peripheral lymph nodes than in tumors which had not undergone metastasis [26]. The invasiveness of some human prostate cancers has been activated epigenetically by *m*-calpain, and targeting this protein reduces tumor progression [27]. Caspase-9 which is specific for gastric cells is down-regulated in some cancers although its effect on differential status or the development of cancers has yet to be clarified [28, 29]. In chronic lymphocytic leukemia cells, the activity and expression of μ-calpain but not *m*-calpain nor calpastatin were elevated compared to non-malignant cells [30].

Various proteins are substrates of calpain including the non-receptor kinase focal adhesion kinase (FAK) and protein kinase C both of which are signaling molecules [31–33]. In addition, calpain is a substrate for several cytoskeletal proteins, the protooncogenes c-fos and c-jun as well as the tumor suppressor protein p53 [31–33]. V-Src-induced morphological changes are caused by a focal adhesion disassembly and cleavage of FAK which is mediated by calpain [33]. This v-Src activation leads to alterations in cellular proliferation, survival, motility, adhesion, and morphology [34, 35]. In chicken embryo fibroblasts, this activation leads to increases in the concentrations of *m*-calpain and decreased amounts of calpastatin [36]. The same study revealed that the activation of the v-Src oncogene led to a feedback loop mechanism of calpain activation [36]. In various human colorectal and breast solid tumors, activation of Src has taken place [37, 38]. Various protein–protein interactions have been observed in our laboratories. Thus, *m*-calpain interacts with NMT1 and caspase-3 couples with NMT2 in human colorectal tumors [39, 40]. Furthermore, NMT1 is inactivated by *m*-calpain by proteolysis [41].

While apoptosis occurs in healthy individuals, its reduced occurrence is found in a number of pathological conditions including cancer [4–10]. To some extent, apoptosis and cell proliferation may be controlled by calpains which do so by determining the level and duration of the transduction signals [8]. Nevertheless, the role of calpains in apoptosis is debatable. A number of factors such as stress conditions, expression of various oncogenes, and cell adhesion regulate the level of p53 which is a tumor suppressor gene and influences the susceptibility of a cell to apoptosis [42, 43]. In this regard, calpains have cleaved p53 in vitro by proteolysis [32, 44]. Furthermore in the case of colorectal cancer, apoptosis was reduced which may have been due to the increased level of *m*-calpain acting on p53. Similarly, the cleavage of Bax by calpain enhances the pro-apoptotic effect of Bax [45]. In addition, the transformation of procaspase-3 and procaspase-7 into caspase-3 and caspase-7 is mediated by calpain [46, 47], and the reduced levels of caspase-3, caspase-7 and caspase-9 in human colon cancers have been reported [48]. On the other hand, when the calpain inhibitor II was added to neoplastic lymphoid cells, the caspase system was activated, thereby enhancing apoptosis [49]. In colon cancers, apoptosis appears to be controlled by the extent of cross-talking between calpain and caspases.

This chapter summarizes the overexpression and elevated activity of calpain in human colonic tumors that is carried in our laboratory.

4 Calpain and Its Inhibitor Calpastatin in Colon Cancer

We have found that in human colorectal adenocarcinoma, there is an increased activity and expression of *m*-calpain as well as a reduction in the expression of calpastatin [50]. The activity of calpain and protein expression in human colorectal adenocarcinoma has been examined with a view to finding the role of *m*-calpain in this tumor. Fifty patients were evaluated although only six cases were reported. Calpain activity was greater in this tumor than in non-malignant mucosa in 83% of the cases ($p < 0.05$, Fig. 1a). In particular, the activity of *m*-calpain was greater in polyps than non-malignant tissues but less so than in cancers (Fig. 1a). Western Blot analysis revealed that *m*-calpain was weakly expressed in non-malignant mucosa, but a higher expression was found in the tumors (Fig. 1b). The expression of *m*-calpain was 2- to 3-fold greater in colorectal neoplasms than in the respective non-malignant mucosa ($p < 0.05$) as determined by a quantitative analysis of the 80 kDa band (Fig. 1b). On the other hand, there was no change in the expression of *m*-calpain at the 28 kDa small subunit [50]. The expression of *m*-calpain was higher in polyps than in non-malignant tissues, but no significant changes were noted in the other normal tissues (Fig. 1b, c).

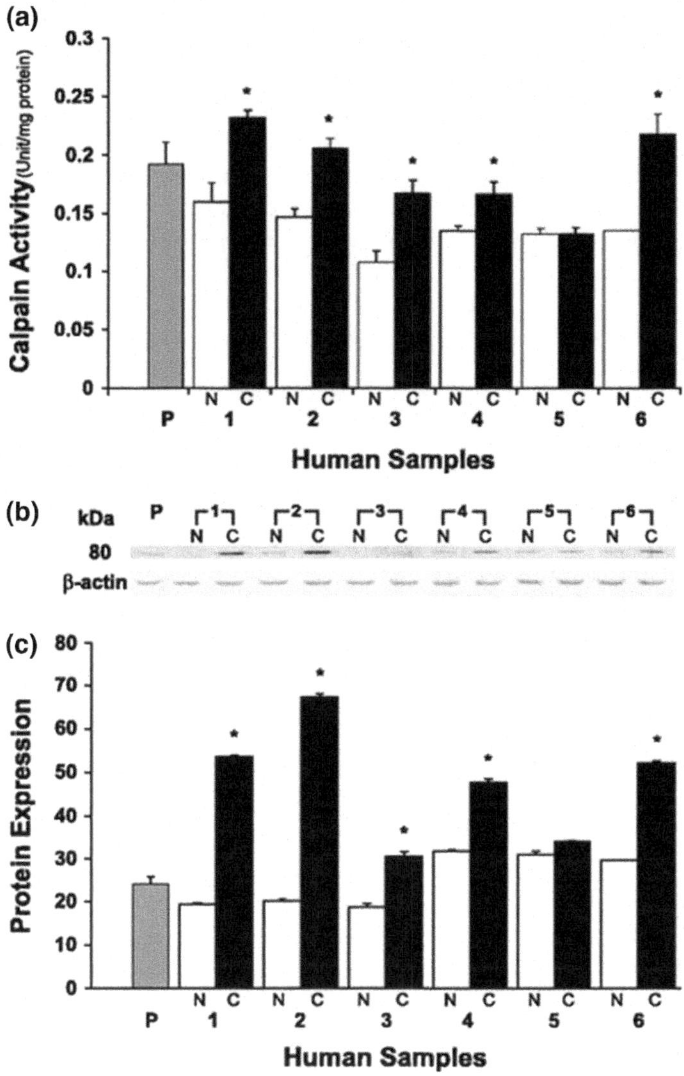

Fig. 1 Calpain activity and protein expression in human colorectal patients. **a** Calpain activity in normal (N), polyp (P), and cancer (C) tissue. **b** Western Blot analysis of *m*-calpain in human colorectal patients in normal (N), polyp (P) and cancer (C) tissue, **c** quantitative analysis of **b**. For details see Lakshmikuttyamma et al. [50]

The strong staining for *m*-calpain in colorectal adenocarcinoma was demonstrated by immunohistochemistry (Fig. 2c). When sections of the mucosa which were remote from the tumors were examined, small reactivity (<10% of the expression of the proteins) was observed (Fig. 2a). In contrast to tumors, both immunoreactivity and staining were lower in polyps (Fig. 2b). These investigations clearly reveal that *m*-calpain activity and protein expression are higher in human colonic tumors.

Calpastatin is an endogenous polypeptide which controls the activity of calpain [9, 10, 12] by inhibiting both of the major forms of calpain [51]. Therefore, we examined the expression of calpastatin to determine whether it regulates calpain activity in colorectal adenocarcinomas [50]. In non-malignant mucosa, Western Blot analysis indicated strong expression of calpastatin whereas in colorectal neoplasms there was a weak expression of this endogenous polypeptide (Fig. 3a).

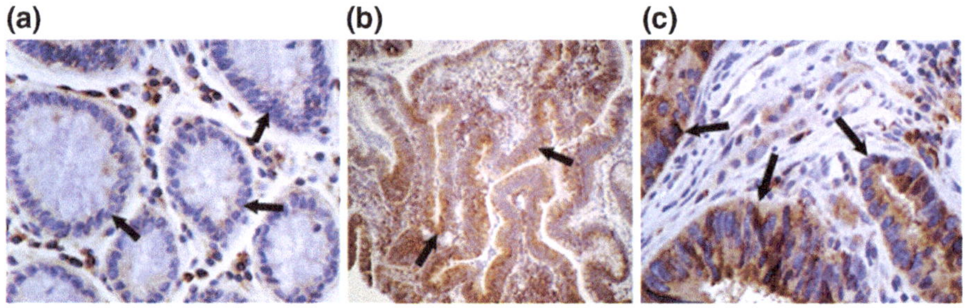

Fig. 2 Immunohistochemical analysis of *m*-calpain using anti-*m*-calpain. **a** Normal mucosa showing a mild degree of focal staining. **b** Polyps showing moderate staining. **c** Colorectal adenocarcinoma showing a positive staining. Original magnification: ×120. For details see Lakshmikuttyamma et al. [50]

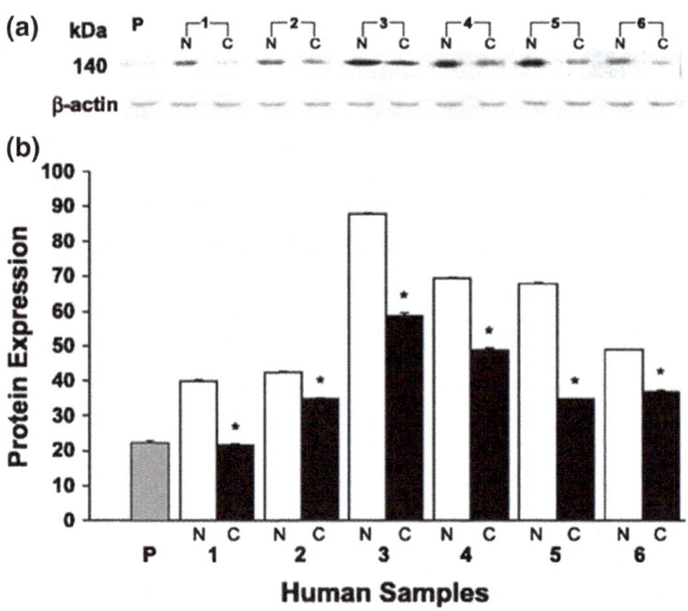

Fig. 3 Expression of calpastatin in human colorectal patients. **a** Western Blot analysis of calpastatin in human colorectal patients in normal (N), polyp (P), and cancer (C) tissue. **b** Quantitative analysis of **a**. For details see Lakshmikuttyamma et al. [50]

In fact, quantitative analysis demonstrated that there is a twofold increase in the expression of calpastatin in non-malignant cells compared to neoplastic tissues (Fig. 3b). Immunochemistry revealed that calpastatin was expressed weakly in colon tumors (Fig. 4b) while non-malignant mucosa displayed moderate to strong staining (Fig. 4a). A weak cytoplasmic positivity was noted for calpastatin in invasive carcinoma with lowered intensity in the invasive component (Fig. 4b).

As discussed earlier, our laboratory was the first to discover HMWCaMBP [11]. We have established that HMWCaMBP is homologous with calpastatin based on its ability to inhibit calpain as well as antibody reactivity, amino acid analysis and sequence homology [12]. However, it is interesting to observe that HMWCaMBP not calpasatin contains the calmodulin-binding domain [11]. Therefore, we compared the immunohistochemical studies of HMWCaMBP with calpastatin in human colonic tumors. In addition, there is a mild reactivity of HMWCaMBP in colorectal adenocarcinoma as demonstrated by histochemistry (Fig. 5b). On the other hand, there was strong staining of the mucosa which was remote from the tumor (Fig. 5a).

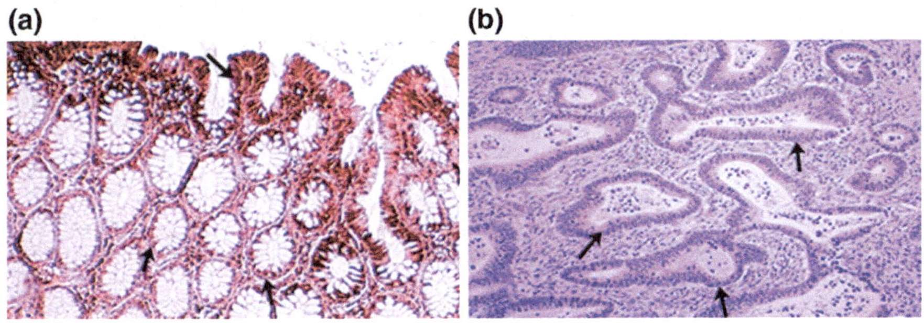

Fig. 4 Immunohistochemical staining for calpastatin using anti-calpastatin. **a** Normal mucosa showing a marked degree of focal staining. **b** Colorectal adenocarcinoma showing a mild degree of reactivity. Original magnification: ×120. For details see Lakshmikuttyamma et al. [50]

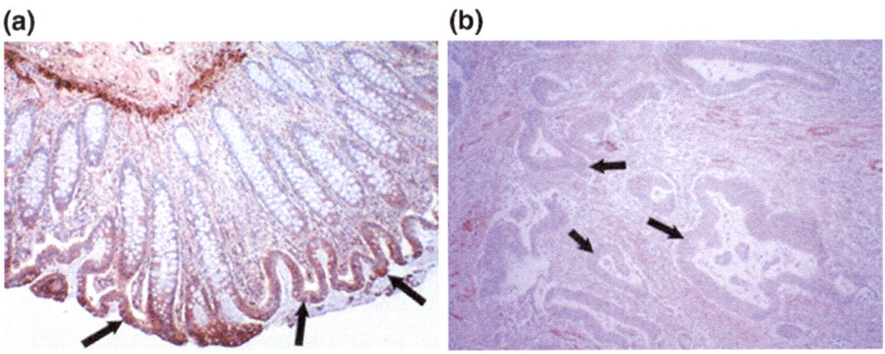

Fig. 5 Immunohistochemical staining for HMWCaMBP using anti-HMWCaMBP. **a** Normal mucosa showing a marked degree of focal staining. **b** Colorectal adenocarcinoma showing a mild degree of reactivity. Original magnification: ×120. For details see Lakshmikuttyamma et al. [50]

A correlation was noted between the lowered expression of HMWCaMBP and calpastatin in human colorectal adenocarcinomas on the one hand and the increased activity and expression of *m*-calpain on the other hand.

5 Conclusion

Our findings suggest that the development of colorectal adenocarcinoma may be influenced by the increased expression of *m*-calpain. There are many issues to be addressed regarding the proliferation of cells in colonic neoplasms including the ways in which calpain expression is increased and the effect of calpain on cell signaling. The increased activity and moderate staining of *m*-calpain in polyps reveal its use as an immunological marker for the early detection of colorectal adenocarcinoma. It is conceivable that based on their ability to inhibit calpain, the overexpression of HMWCaMBP and/or calpastatin may be a useful procedure in treating colon cancer.

Acknowledgements This study was supported by the Canadian Institutes of Health Research, Canada.

References

1. Stewart B, Wild C (2014) World cancer report 2014. International agency for research on cancer (WHO Press)
2. Storr SJ, Carragher NO, Frame MC, Parr T, Martin SG (2011) The calpain system and cancer. Nat Rev Cancer 11:364–374
3. Lelou L, Wells A (2011) Calpains as potential anti-cancer targets. Expert Opin Ther Targets 15:309–323
4. Goll DE, Thompson VF, Li H, Wei W, Cong J (2002) The calpain system. Physiol Rev 83:731–801
5. Parameswaran S, Kumar S, Sharma RK (2013) Role of calpains in calmodulin regulated system. In: Chakraborty S, Dhalla NS (eds) Proteases in health and disease. Springer Publisher, pp 33–48
6. Cong J, Thompson V, Goll D (2002) Immunoaffinity purification of the calpains. Protein Expr Purify 25:283–290
7. Sorimachi H, Hata S, Ono Y (2011) Calpain chronicle-an enzyme family under multidisciplinary characterization. Proc Jpn Acad Ser B Phys Biol Sci 87:287–327
8. Ono Y, Sorimachi H, Suzuki K (1998) Structure and physiology of calpain, an enigmatic protease. Biochem Biophys Res Commun 245:289–294
9. Maki M, Ma H, Takano E, Adachi Y, Lee WJ, Hatanaka M, Murachi T (1991) Calpastatins: biochemical and molecular biological studies. Biomed Biochim Acta 50:509–516
10. Carragher NO (2006) Calpain inhibition: a therapeutic strategy targeting multiple disease states. Curr Pharm Des 12:615–638
11. Sharma RK (1990) Purification and characterization of novel calmodulin-binding protein from cardiac muscle. J Biol Chem 265:1152–1157
12. Kakkar R, Raju RV, Mellgren RL, Radhi J, Sharma RK (1997) Cardiac high molecular weight calmodulin binding protein contains calpastatin activity. Biochemistry 36:11550–11555

13. Carragher NO, Frame MC (2002) Calpain: a role in cell transformation and migration. Int J Biochem Cell Biol 34:1539–1543
14. Huttenlocher A, Palecek SP, Lu Q, Zhang W, Mellgren RL, Lauffenburger DA, Ginsberg MH, Horwitz AF (1997) Regulation of cell migration by the calcium-dependent protease calpain. J Biol Chem 272:32719–32722
15. Patel YM, Lane MD (2000) Mitotic clonal expansion during pre-adipocyte differentiation: calpain-mediated turnover of p27. J Biol Chem 275:17653–17660
16. Parameswaran S, Sharma RK (2014) Altered expression of calcineurin, calpain, calpastatin and HMWCaMBP in cardiac cells following ischemia and reperfusion. Biochem Biophys Res Commun 443:604–609
17. Parameswaran S, Sharma RK (2014) Ischemia and reperfusion induce differential expression of calpastatin and its homologue high molecular weight calmodulin-binding protein in murine cardiomyocytes. PLoS ONE 9:e114653
18. Lakshmikuttyamma A, Selvakumar P, Kakkar R, Kanthan R, Wang R, Sharma RK (2003) Activation of calcineurin expression in ischemia-reperfused rat heart and in human ischemic myocardium. J Cell Biochem 90:987–997
19. Lakshmikuttyamma A, Selvakumar P, Ratan Sharma A, Anderson DH, Sharma RK (2004) In vitro proteolytic degradation of bovine brain calcineurin by m-calpain. Neurochem Res 29:1913–1921
20. Lakshmikuttyamma A, Selvakumar P, Charavaryamath C, Singh B, Tuchek J, Sharma RK (2006) Expression of calcineurin and its interacting proteins in epileptic fowl. J Neurochem 96:366–373
21. Lakshmikuttyamma A, Selvakumar P, Tuchek J, Sharma RK (2008) Myristoyltransferase and calcineurin: novel molecular therapeutic target for epilepsy. Prog Neurobiol 84:77–84
22. Selvakumar P, Lakshmikuttyamma A, Charavaryamath C, Singh B, Tuchek J, Sharma RK (2005) Expression of myristoyltransferase and its interacting proteins in epilepsy. Biochem Biophys Res Commun 335:1132–1139
23. Kishimoto A, Kajikawa N, Shiota M, Nishizuka Y (1983) Preoteolytic activation of calcium-activated, phospholipid-dependent protein kinase by calcium-dependent neutral protease. J Biol Chem 258:1156–1164
24. Oda A, Druker BJ, Ariyoshi H, Smith M, Salzman EW (1993) pp60src is an endogenous substrate for calpain in human blood platelets. J Biol Chem 268:12603–12608
25. Frangioni JV, Oda A, Smith M, Salzman EW, Neel BG (1993) Calpain-catalyzed cleavage and subcellular relocation of protein phosphotyrosine phosphatase 1B PTB-1B in human platelets. EMBO J 12:4843–4856
26. Braun C, Engel M, Seifert M, Theisinger B, Seitz G, Zang KD, Welter C (1999) Expression of calpain I messenger RNA in human renal cell carcinoma: correlation with lymph node metastasis and histological type. Int J Cancer 84:6–9
27. Mamoune A, Luo JH, Lauffenburger DA, Wells A (2003) Calpain-2 is a target for limiting prostate cancer invasion. Cancer Res 63:4632–4640
28. Yoshikawa Y, Mukai H, Hino F, Asada K, Kato I (2000) Isolation of two novel genes, down-regulated in gastric cancer. Cancer Sci 91:459–463
29. Liu K, Li L, Cohen SN (2000) Antisense RNA-mediated deficiency of the calpain protease, nCL-4, in NH3T3cells is associated with neoplastic transformation and tumorigenesis. J Biol Chem 275:31093–31098
30. Witkowski JM, Zmuda-Trzebiatowska E, Swiercz JM, Cichorek M, Ciepluch H, Lewandowski K, Bryl E, Hellmann A (2002) Modulation of the activity of calcium-activated neutral proteases calpains in chronic lymphocytic leukemia B-CLL cells. Blood 100:1802–1809
31. Carillo S, Pariat M, Steff AM, Roux P, Etienne-Julan M, Lorca T, Piechaczyk M (1994) Differential sensitivity of FOS and JUN family members to calpains. Oncogene 9:1679–1689
32. Kubbutat M, Vousden KH (1997) Proteolytic cleavage of human p53 by calpain: a potential regulator of protein stability. Mol Cell Biol 17:460–468

33. Carragher NO, Fincham VJ, Riley D, Frame MC (2001) Cleavage of focal adhesion kinase by different proteases during SRC-regulated transformation and apoptosis. Distinct roles for calpain and caspases. J Biol Chem 276:4270–4275
34. Johnson D, Frame MC, Wyke JA (1998) Expression of the v-Src oncoprotein in fibroblasts disrupts normal regulation of the CDK inhibitor p 27 and inhibits quiescence. Oncogene 16:2017–2028
35. Kellie S (1988) Cellular transformation, tyrosine kinase oncogenes and the cellular adhesion plaque. BioEssays 8:25–30
36. Carragher NO, Westhoff MA, Riley D, Potter DA, Dutt P, Elce JS, Greer PA, Frame MC (2002) v-Src-induced modulation of the calpain-calpastatin proteolytic system regulates transformation. Mol Cell Biol 22:257–269
37. Bolen JB, Veillette A, Schwartz AM, DeSeau V, Rosen N (1987) Activation of pp60c-src protein kinase activity in human colon carcinoma. Proc Natl Acad Sci USA 84:2251–2255
38. Ottenhoff-Kalff AE, Rijksen G, Van Beurden EA, Hennipman A, Michels AA, Staal GE (1992) Characterization of protein tyrosine kinases from human breast cancer: involvement of the c-src oncogene product. Cancer Res 52:4773–4778
39. Selvakumar P, Smith-Windsor E, Bonham K, Sharma RK (2006) N-myristoyl- transferase 2 expression in human colon cancer: cross-talk between the calpain and caspase system. FEBS Lett 580:2021–2026
40. Sharma RK, Kumar S, Parameswaran S, Dimmock JR (2014) Regulation of N myristoyl-transferase by the calpain and caspase systems. Indian J Biochem Biophys 51:506–511
41. Raju RV, Kakkar R, Datla RS, Radhi J, Sharma RK (1998) Myristoyl-coA: protein N-myristoyltransferase from bovine cardiac muscle: molecular cloning, kinetic analysis and in vitro proteolytic cleavage by *m*-calpain. Exp Cell Res 241:23–35
42. Levine AJ (1997) p53, the cellular gatekeeper for growth and division. Cell 88:323–331
43. Giaccia AJ, Kastan MB (1998) The complexity of p53 modulation: emerging patterns from divergent signals. Genes Dev 12:2973–2983
44. Gonen H, Shkedy D, Barnoy S, Kosower NS, Ciechanover A (1997) On the involvement of calpains in the degradation of the tumor suppressor protein p53. FEBS Lett 406:17–22
45. Toyota H, Yanase N, Yoshimoto T, Moriyama M, Sudo T, Mizuguchi J (2003) Calpain-induced Bax-cleavage product is a more potent inducer of apoptotic cell death than wild-type Bax. Cancer Lett 189:221–230
46. Blomgren K, Zhu C, Wang X, Karlsson JO, Leverin AL, Bahr BA, Mallard C, Hagberg H (2001) Synergistic activation of caspase-3 by *m*-calpain after neonatal hypoxia-ischemia: a mechanism of ‘pathological apoptosis’? J Biol Chem 276:10191–10198
47. Ruiz-Vela A, de Buitrago GG, Martínez-A C (1999) Implication of calpain in caspase activation during B cell clonal deletion. EMBO J 18:4988–4998
48. Chen T, Yang I, Irby R, Shain KH, Wang HG, Quackenbush J, Coppola D, Cheng JQ, Yeatman TJ (2003) Regulation of caspase expression and apoptosis by adenomatous polyposis coli. Cancer Res 63:4368–4374
49. Zhu DM, Uckun FM (2000) Calpain inhibitor II induces caspase-dependent apoptosis in human acute lymphoblastic leukemia and non-Hodgkin’s lymphoma cells as well as some solid tumor cells. Clin Cancer Res 6:2456–2463
50. Lakshmikuttyamma A, Selvakumar P, Kanthan R, Kanthan SC, Sharma RK (2004) Overexpression of m-calpain in human colorectal adenocarcinomas. Cancer Epidemiol Biomarkers Prev 13:1604–1609
51. Murachi T (1989) Intracellular regulatory system involving calpain and calpastatin. Biochem Int 18:263–294

Role of Cathepsins, in Particular Cathepsins B and D in Breast Cancer: Mechanisms and Clinical Implications

Anuradha Ratna and Salil K. Das

Abstract

Consistent with the biological relevance of cathepsins B and D, their overexpression, translational or post-translational modifications, and mitogenic effects are hallmarks of breast cancer. Several studies have established a link between increased expression of cathepsins B and D and tumor growth, invasion, and metastasis. This review outlines the potential role of these two cathepsins, specifically in breast cancer, and different mechanisms involved during tumor progression and metastasis. The development of various therapeutic strategies, including specific inhibitors, gene targeted therapy, and antibodies against cathepsins in an attempt to curb breast tumor progression holds a promising future. However, from a translational perspective, further extensive studies are needed in order to unravel the therapeutic abilities of cathepsins B and D in controlling breast cancer.

Keywords

Cathepsin B · Cathepsin D · Breast cancer · Overexpression · Metastasis Biomarker · Cancer therapy

---

A. Ratna · S.K. Das (✉)
Department of Biochemistry and Cancer Biology, Meharry Medical College, Nashville, TN, USA
e-mail: sdas@mmc.edu

S. Chakraborti and N.S. Dhalla (eds.), *Pathophysiological Aspects of Proteases*,
DOI 10.1007/978-981-10-6141-7_8

1 Introduction

The term "cathepsin" originated from the Greek word for "digesting" [1]. Originally it was isolated from the stomach mucous membrane as an intracellular acidic protein. Cathepsin(s) were purified from spleen in 1940, and their lysosomal association was established in 1955 [2, 3]. These intracellular peptide hydrolases are a class of globular proteases, although several cathepsins also have extracellular functions [4]. Active forms of cathepsins are localized in endosomal or lysosomal vesicles, cell membranes, and/or secreted and localized in pericellular environments as soluble enzymes involved in cleaving the extracellular matrix (ECM) proteins laminin and type IV collagen, and cell-adhesion proteins such as E-cadherin and matricellular proteins [5–8]. According to their active site amino acid, cathepsins can be divided into three sub-groups: cysteine (B, C, H, F, K, L, O, S, V, W and X/Z), aspartate (D and E) and serine (G) cathepsins [9]. The lysosome relies on these protein hydrolases and other enzymes to carry out intracellular degradation before recycling cellular constituents. The functional implication of cathepsins was made possible by various genomic, proteomic, and imaging tools as well as by the generation and in-depth analysis of knockout and transgenic mice [10]. According to these studies, cathepsins act as not only redundant, homeostatic enzymes involved in the turnover of proteins delivered to the lysosome by endocytosis or autophagocytosis, but also play a critical role in the proteolytic processing of specific substrates. Thus, cathepsins contribute to distinct physiologic processes, such as antigen presentation in the immune system [11], collagen turnover in bone and cartilage [12, 13], and neuropeptide and hormone processing [14]. In addition, ectopic or excessive expression and activity of cathepsins promotes the development of several common diseases in humans, including cancer. Human cysteine cathepsins are highly up-regulated in a wide variety of cancers. Several studies have shown a correlation between cathepsin proteolytic activity and neoplastic transformation, tumor invasion, and metastasis through the destruction of ECM components and basement membranes [15]. Here we emphasize specifically, cathepsins B and D as these are the most thoroughly studied enzymes. However, neither the functions of these two cathepsins in cancer, nor their roles in tumor cells and the tumor-associated cells that contribute to neoplastic progression have been clearly defined. This chapter focusses on the role played by cathepsins B and D and their therapeutic potential, particularly in breast cancer.

2 Cellular Location of Cathepsin B and Its Function

Cathepsin B (E.C. 3.4.22.1) is a lysosomal cysteine protease belonging to the papain family of enzymes which is constitutively expressed in normal cells [16]. Cathepsin B is localized on cell membranes, and secreted and localized in endosomal or lysosomal vesicles, suggesting that its enzymatic substrates and functions might change according to its localization. Cathepsin B possesses both endopeptidase, and exopeptidase (carboxydipeptidase) activity [17, 18]. The presence of an occluding loop which interferes with the binding of substrates to the active site contributes to the dual activity of cathepsin B [16]. At acidic pH, the occluding loop, partially blocks the active site of the molecule and prevents the entry of large substrates, yet allows the access of carboxy terminus of proteins. Thus, cathepsin B displays its carboxydipeptidase activity at acidic pH. Among the extracellular roles, cathepsin B is involved in the cleavage of ECM proteins such as laminin and type IV collagen [19], tenascin C [20], and cell-adhesion proteins such as E-cadherin [21]. At neutral pH, cathepsin B functions as an endopeptidase [22]. The loop is displaced leaving the active site to be available for the entry of large substrates. Cathepsin B is overexpressed in many human malignancies by tumor cells and tumor-associated cells at the mRNA and protein levels [23–25]. Using transgenic murine models of pancreatic and mammary carcinomas, a causal relationship has been established between cathepsin B and tumor initiation, proliferation, angiogenesis, invasion, and metastasis [26, 27]. Cathepsin B was found to promote growth of both tumor cells and tumor-associated macrophages in transgenic models. Unravelling the detailed mechanisms underlying increased expression of cathepsin B in tumors is a requisite for therapeutic targeting of cathepsin B.

3 Cellular Location of Cathepsin D and Its Function

Cathepsin D (E.C. 3.4.23.5) is a member of the aspartic proteases family that function in intracellular catabolism at lysosomal compartments. Initially, the inactive pre-pro-cathepsin D (43 kDa) is synthesized in rough endoplasmic reticulum, which is in turn cleaved and glycosylated to form pro-cathepsin D (52 kDa) containing N-linked oligosaccharides with mannose-6-phosphate (M6P) residues. Pro-cathepsin D is then directed to intracellular vesicular structures such as lysosomes, endosomes and phagosomes where it undergoes a sequence of proteolytic cleavage and autocatalysis to form mature cathepsin D containing a heavy chain (34 kDa) and a light chain (14 kDa) [28]. Further, lysosomal trafficking of cathepsin D may be either dependent on the M6P receptors or independent of M6P receptors where it undergoes endocytosis (Fig. 1).

In the past two decades, beyond its original "housekeeping" role, the participation of cathepsin D in a multitude of biological processes has been documented. It aids in the proteolytic events that are critical in controlling diverse biological processes, such as apoptosis, cell cycle, cellular differentiation, morphogenesis and tissue remodeling, immunological processes, fertilization, neuronal outgrowth, and angiogenesis [29]. Various pathological conditions, including atherosclerosis, cancer, neurological and skin disorders may occur due to the deficiency, altered regulation, or post-translational modification of cathepsin D [29]. In particular, deregulated synthesis, post-translational modifications, and over-secretion of cathepsin D, along with its mitogenic effects, are well established characteristics of cancer [29]. Overexpression and hyper-secretion of Cathepsin D is evident in

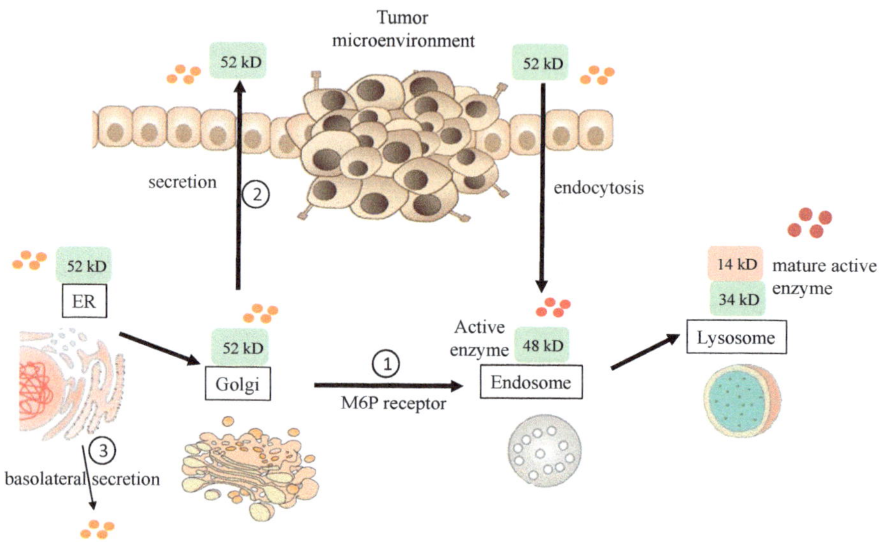

Fig. 1 Model depicting localization of cathepsin D in cancer cells. The majority of these pathways is found to be altered in breast cancer. (*1*) In cancer cells, the reduced acidification of endosomal/lysosomal compartment results in the improper processing of cathepsin D, leading to increased secretion of pro-cathepsin D. Pathways (*2*) and (*3*) are highly elevated causing excessive degradation of ECM [29, 30]. *ECM* extracellular matrix; *ER* endoplasmic reticulum; *M6P* mannose 6-phosphate; *kD* kiloDalton

different types of cancer (breast cancer, ovarian cancer, endometrial cancer, head and neck cancer, bladder cancer, malignant glioma, melanoma). Overexpression of Cathepsin D by cancer cells leads to its accumulation in the cells where it may modify the tumor micro-environment by affecting stromal cell behavior and/or degrading ECM components, and the pro-cathepsin D is hyper-secreted in the tumor micro-environment [30]. Therefore, cathepsin D activity can be blocked by targeting extracellular cathepsin D located in the cytosol, and/or intracellular cathepsin D located in intracellular vesicles or nucleus (Fig. 1). Cathepsin D is being explored extensively due to its overexpression in advanced stages of solid tumors causing aggressive metastasis and deteriorating the pathological condition.

4 Role of Cathepsins B and D in Breast Cancer

In cancer cells, lysosomes are redistributed from the perinuclear area to the cellular periphery, where they can release cathepsins or be secreted into the extracellular space [24, 31]. The secreted cathepsins facilitate tumor cells to disintegrate the stroma and intravasate into lymphatic or blood vessels, thus promoting metastasis. Cathepsins promote invasion and migration of cancer cell through ECM

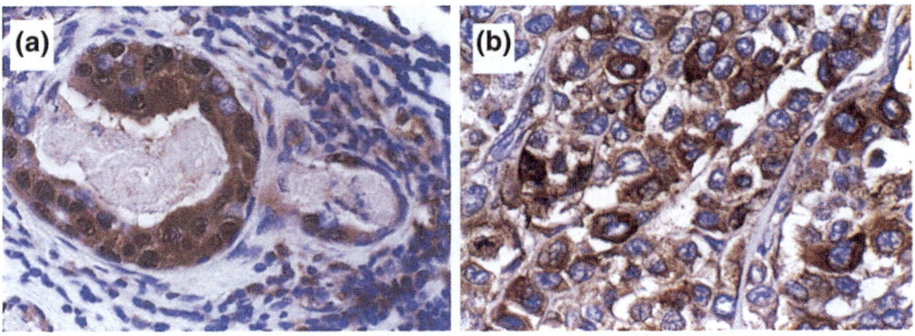

Fig. 2 Immunohistochemical staining of infiltrating ductal breast carcinoma. **a** Section stained with anti-cathepsin B antibody (original modification ×400). **b** Section stained with anti-cathepsin D antibody (original modification ×400) [38]

components in different types of cancer, including breast cancer [32]. It has also been reported that cathepsins initiate proteolytic cascades of serine-proteases and matrix metalloproteinase (MMP) [33, 34]. Besides, cathepsins may enable tumoral invasion, through proteolysis of E-cadherin, and resulting in a loss of its function in different tumors [35, 36]. Consequently, it leads to ECM remodeling, angiogenesis, invasion cancer metastasis. Expression, redistribution and/or secretion of cathepsins B and D have been reported to parallel the malignant progression [37]. The immunohistochemical study done by Castiglioni et al. reported an over expression of cathepsins B and D in most of the tumor samples as compared with normal breast epithelium from the same patients (Fig. 2) [38]. The mechanisms underlying cathepsin upregulation ranges from gene amplification to post-transcriptional modification [24]. However, the exact molecular mechanisms by which cathepsins affect cancer progression still remains elusive. The potential roles of cathepsins B and D are depicted in Fig. 3.

Aberrant overexpression and redistribution of cathepsin B has been reported in invasive and metastatic cancers, including breast cancer, melanoma and colorectal cancer [23–25]. There are reports that link cathepsin B to apoptosis, tumor-associated inflammation, angiogenesis and metastasis by contributing to the altered intracellular protein metabolism of cancer cells and to proteolytic cascades in the tumor microenvironment [39]. Cathepsin B is a prognostic marker in several types of cancer and its increased expression by tumor cells is correlated with poor outcomes in breast cancer [23, 40–42]. Cathepsin B interacts with cystatins [43, 44] and annexin II tetramer (p11) [45] and thus, responsible for the proteolytic activation and subsequent degradation of ECM.

Different mechanisms have been attributed to the upregulation of cathepsin B in tumors. Chromosomal loci for various cysteine cathepsins are prone to mutations; however, a novel amplicon has been reported only for cathepsin B. The overexpression of cathepsin B in adenocarcinomas of the esophagus has been demonstrated where tumor cells express transcript variants of cathepsins B due to the use of alternative promoters and alternative splicing [46–48]. A comprehensive study on

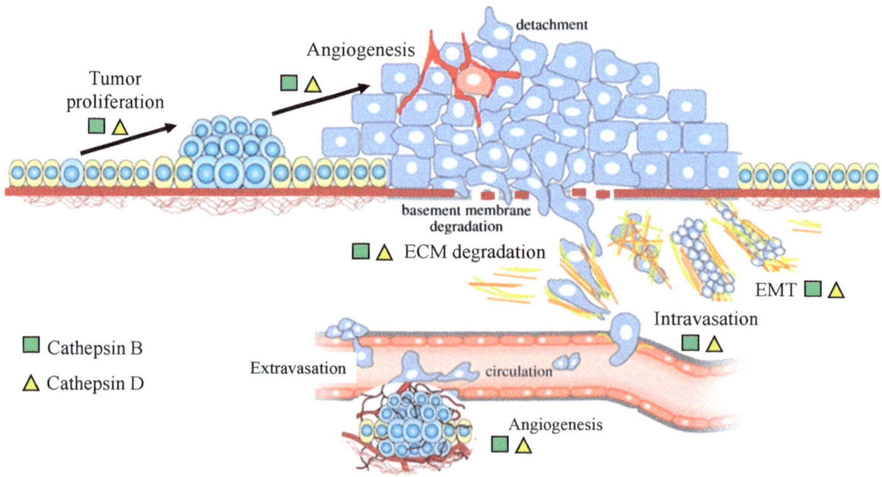

Fig. 3 Potential role of cathepsins B and D in breast cancer progression. *ECM* extracellular matrix; *EMT* epithelial-mesenchymal transition

the transcriptional regulation and the presence of transcript variants of cathepsin B in human cancers, is necessitated. One of the study showed that cathepsin B promoter contains many Sp1 sites [49], and these Sp1 sites increase in tumor cells [50]. Some other independent studies concluded that the varying expression of cathepsin B depends mainly on the different levels of Sp1 [51]. Cathepsin B also plays a role in autophagy and tumor cannibalism that could be possibly involved in specific immune resistance [52], enabling tumor cell proliferation and infiltration. This could be a possible reason for the necrotic cores of highly infiltrative tumors. Cathepsin B is responsible for initiating the proteolytic cascade involving uPA (urokinase plasminogen activator), plasminogen and plasmin [53]. Cathepsin B suppresses the inhibitors of active proteases, thus it appears to support a promalignant phenotype. For example, in human articular chondrocytes, cathepsin B destroys the inhibitors of MMPs [e.g., TIMP1 (tissue inhibitor of metalloproteinases 1) and TIMP2] and maintains its high level, leading to ECM degradation and angiogenesis [54]. It was also reported that inducers of monocytic and granulocytic differentiation [e.g., calcitriol (D3), sodium butyrate (NaB) and all-trans retinoic acid (RA)] increased cathepsin B mRNA levels in a dose-dependent manner [55]. From all these findings, we can conclude cathepsin B as a multifaceted molecule with diverse function and is regulated at various cellular levels which are still not clearly understood.

Numerous studies have documented that cathepsin D promotes breast cancer cell proliferation [56, 57], and metastasis [58, 59]. Cathepsin D expression is increased by 2- to 50-fold in breast cancer cells, compared to other cells, such as fibroblasts or normal mammary glands [60]. To investigate the causes of elevated secretion of cathepsin D in tumor cells, multiple studies have focused on and compared the protein structure, glycosylation, and proteolytic activity of cathepsin D in normal and cancer cells. It has been anticipated that cathepsin D can activate pro-cathepsin B and/or degrade cystatins which triggers the proteolytic cascade leading to ECM degradation [61]. Nevertheless, many studies have highlighted the role of cathepsin D in promoting distant metastasis via an indirect mitogenic activity rather than stimulation of local invasion and extravasation [32]. The in vitro mitogenic potential of cathepsin D in MCF7 cells when cultured using purified pro-cathepsin D [56] is debated [62] but confirmed in vivo following cDNA transfection [32, 63]. Cathepsin D has been suggested to be an independent prognostic factor in women with stage I breast cancer [64].

Different molecular mechanisms have been put forward to explain the mitogenic activity of cathepsin D. Studies performed in human breast cancer cells have explained the mitogenicity of pro-cathepsin D. Cathepsin D acts as a competitor and interferes with the interaction between IGF2 (insulin-like growth factor) and M6P moieties of the M6P/IGF-2 receptor. As a result, IGF2 is displaced from the IGF1 receptor, leading to the activation of mitogenic IGF1 receptor pathway [57, 65]. Alternatively, it has also been proposed that the catalytic activity of secreted cathepsin D may be implicated in releasing growth factors, such as FGF2 (fibroblast growth factor), from the ECM [66]. In the case of ER (estrogen receptor) positive breast cancer, an increased number of large acid vesicles containing cathepsin D cause tumor invasion. Elevated level of cathepsin D in primary tumors is indicative of local recurrence or distant metastasis [67]. In ER-positive breast cancer, both estrogen and growth factors stimulate cathepsin D protein and mRNA accumulation [68–70]. Like other steroid-responsive genes, estrogen induced accumulation of cathepsin D mRNA is mainly due to increased initiation of transcription [70]. Since estrogen and growth factors stimulate the growth of ER-positive tumors, the induction of cathepsin D appears to be associated as well. Contrary to other estrogen-induced genes such as pS2, the progesterone receptor and wild type BrCa1, cathepsin D is also overexpressed in ER-negative breast cancer [68]. In triple-negative breast cancer cathepsin D is reported to be a key biological marker along with a high Ki-67 index [71]. It has been shown that cathepsin D is constitutively overexpressed in ER-negative breast cancer cell lines. The molecular mechanism behind overexpression of cathepsin D in ER-negative breast cancer cells may engage local reorganization of the chromatin structure of the cathepsin D promoter [72].

The prognostic significance of cathepsin D and optimal methodologies to measure cathepsin D in breast cancers is controversial. There exists ample evidence associating an elevated level of cathepsin D in tumor extracts with poor overall and relapse-free survival [73, 74]. Henry et al. reported that immunohistochemically assessed cathepsin D in tumor cells was associated with a favorable prognosis in node-positive patients [75]. In a separate study based on immunoassays and enzymatic assays, cathepsin D was also reported as a prognostic indicator for node-negative breast cancer patients [76]. In contrast, Isola et al. in a study involving 262 node-negative patients showed that tumor cell-associated cathepsin D expression was associated with a poor prognosis [77]. Another group suggested that cathepsin D determinations were not of prognostic value using immunoblotting and immunohistochemical methods [78]. After using the double antibody immunoradiometric (IRMA) assay for measuring cathepsin levels, the same group concluded that cathepsin D is of doubtful value in predicting risk of early relapse or death for patients with newly diagnosed invasive breast cancer. Thus, the literature suggests that either cathepsin D is not a potent prognostic readout or the techniques for its optimal measurement has not been unambiguously defined.

5 Therapeutic Targeting of Cathepsins B and D in Breast Cancer: Panoply of Possibilities

The role of cathepsins in the progression and metastasis of cancer has been well documented. There is growing evidence that cathepsins may have a therapeutic potential for reducing the malignant progression of tumor cells and for treating some kinds of metastatic cancer because ablation or inhibition of cathepsins in tumor models decreased or delayed metastasis [27, 79, 80]. A range of therapeutic strategies has been employed to inhibit the proteolytic activity of cathepsins with the aim to restrain the metastatic infiltration mediated by them (Fig. 4). Among the strategies developed, chemical inhibitors, gene therapy, and use of cathepsin antibodies against cancer have shown promising developments.

5.1 Cathepsin Inhibitors

Each cathepsin has a different cleavage-bond specificity for substrate proteins, thus allowing for the development of cathepsin—specific inhibitors for targeting different family members [81]. These inhibitors are capable of selectively suppressing the cathepsin activity and inhibiting tumor growth. For example, some studies have reported the therapeutic efficacy of anti-cathepsin B and D in cancer [81, 82]. Zhang and colleagues showed for the first time that cystatin M significantly delayed breast tumor growth in mammary fat pads and lowered the metastatic burden at secondary

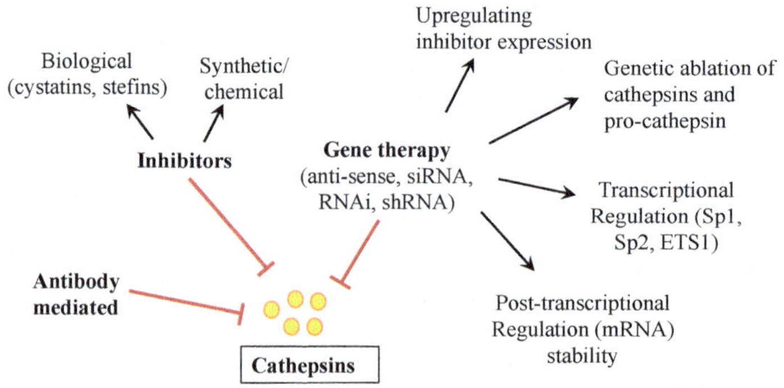

Fig. 4 Schematic representation of strategies employed to target cathepsins in breast cancer. *siRNA* small interfering RNA; *RNAi* RNA interference; *shRNA* short hairpin RNA; *Sp1* specificity protein 1; *ETS1* E26 transformation-specific 1

sites in SCID mice orthotopically implanted with breast cancer cells expressing cystatin M.

Small molecule inhibitors that target cathepsins B and D have been tested in vitro. Multiple in vitro experiments has reported that the inhibition of specific cysteine cathepsins results in a reduced invasion of tumor cells through matrigel or ECM in different types of cancers, including melanoma, glioblastoma, colon, prostate, and lung cancers [83, 84]. Studies have shown that both intracellular and extracellular inhibitions of cathepsins can inhibit invasion, suggesting that both the secretion of active enzyme and intracellular degradation pathways are important. Due to the extensive study of cathepsin B and D functions, and progress in the atomic understanding of cathepsin structure, a variety of inhibitors have been developed to combat these enzymes in breast cancer. However, even with different inhibitors being identified and developed, none of them has been proven as a successful therapeutic candidate.

One study conducted in nude mice reported that intraperitoneal administration of a highly selective cathepsin B inhibitor, such as CA-074, reduced the metastatic potential of breast cancer cells [80]. In another study, a novel approach termed activity-based protein profiling (ABPP) was developed to screen compound libraries for activity, inhibiting multiple cathepsins in defined tissues simultaneously [85]. This approach has highlighted a particular compound, ASM7, which has good activity against cathepsin B (IC50: 40–70 nM), and could be useful to reevaluate the available inhibitors for required specificity. Parker, et al. showed that expression of cystatin A may lead to the inhibition of cathepsin B, subsequently causing a reduction in distant metastasis in breast tumor cells. [86]. Another research group showed that depletion of cathepsin B was able to completely reverse the invasive phenotype of MCF7 cells and HER2-expressing SKBR-3 and MDA-MB-453 cells [40]. In addition, combining cathepsin inhibitors with other approaches may lead to better clinical strategies to treat breast cancers and metastasis. Shree, et al. demonstrated a tumor protective role of cathepsin B in breast tumors, and they showed that the use of compound JPM (research grade cathepsin B inhibitor) and the chemotherapeutic drug Taxol, significantly impaired tumor growth [87].

Nowadays, interest in the development of potent inhibitors of aspartic peptidases such as cathepsin D is increasing. Unlike other cathepsins (e.g. cysteine or serine), no mammalian endogenous lysosomal or cytoplasmic cathepsin D inhibitor has been identified, whereas the majority of exogenous cathepsin D inhibitors is synthetic compounds produced by microorganisms, plants and lower animals [88, 89]. Numerous studies are being conducted to discover novel cathepsin D inhibitors for the treatment of breast cancer. Earlier, in an attempt to develop cathepsin D inhibitors with desirable biological activity, McConnell et al. constructed a library of cathepsin D inhibitors with varying physical properties that improved their potency and half-life in vivo [90]. Anantaraju et al. performed an energy-based pharmacophore virtual screening and molecular docking studies to discover novel cathepsin D inhibitors [91]. These molecules were efficient to inhibit cell growth in the triple-positive and TNBC (triple-negative breast cancer) cell lines. Recently, one group used the structure-activity relationship of a vectorized cathepsin D inhibitor, JMV4463, a pepstatin bioconjugate containing a vector composed of four AMPA (AMPA, ortho-aminomethylphenylacetyl) units, to inhibit the proteolysis of critical cathepsin D substrates involved in tumor cell proliferation [92]. Although many of the cathepsin D inhibitors will be limited in their therapeutic usefulness, with a large enough pool of active compounds it is possible that a few may someday prove to be drug candidates for the treatment of cancer. To date, cathepsin D inhibitors have not reached clinical trials, and there remains more to explore about their role and involvement in breast cancer chemotherapy.

5.2 Gene Therapy

Breast cancer is a very heterogeneous and multigenic disease clearly requiring new therapeutic strategies, targeting specific genes and proteins actively engaged in pathophysiology of breast cancer rather than conventional cytotoxic chemotherapy, which is often hampered by cytotoxicity or resistance. In order to combat this treatment barrier, attempts at specialized, targeted therapies are emerging. There are several prospective strategies using gene therapy, including: (a) expression of a gene to induce apoptosis or enhance tumor sensitivity to conventional drug/radiation therapy; (b) introduction of a tumor suppressor gene to compensate for its loss/deregulation; (c) blockade of an oncogene expression by using siRNA/shRNA approach; and (d) enhancement of tumor immunogenicity to stimulate immune cell recognition [93].

Besides, the use of chemical and biological inhibitors of cathepsin, a more advanced tool such as gene therapy is being used. The reduction in tumor growth, invasion, metastasis, and angiogenes due to either increased expression of the endogenous inhibitors of cysteine cathepsins or downregulated expression of cathepsin B lays the foundation for targeting cysteine cathepsins genes in cancer. Here we mention some of these genetic approaches carried out to downregulate or silence the expression of cathepsin B in various carcinomas, including breast cancer. Gene targeting strategies is being used towards cathepsin B expression either directly or indirectly through endogenous inhibitors (the cystatins and stefins) or associated transcription factors in human colon and prostate carcinomas, gliomas, and melanomas [94, 95]. In the literature, there are reports of genetic downmodulation of cathepsin B carried out in transgenic mice, however, studies employing gene-targeted approaches in human breast cancer are scarce. The antisense, siRNA, and shRNA technologies have been carried out in different transgenic murine models of carcinomas in order to downregulate the expression of cathepsin B. Studies have reported a reduction in motility and invasion of osteosarcoma cells due to antisense downregulation of cathepsin B, [96] and reduction in bone metastasis in a murine mammary carcinoma model due to shRNA downregulation of cathepsin B [80].

Disruption of cathepsin B gene production in the mouse model of mammary carcinoma (MMTV-PyMT) was found to delay the onset, reduce the growth of primary mammary tumors and decrease the volume of the lung metastases [26]. Studies have aimed at the downregulation of cathepsin B function by increasing the expression of the endogenous inhibitor [24]. Alternatively, studies in which cystatin C and stefin/cystatin A were overexpressed in tumor cell lines demonstrated increased invasion and metastasis of tumor cells of epithelial and mesenchymal origins [97, 98]. In line with these reports, silencing cystatin M expression in a metastatic oral cancer cell line resulted in the inhibition of cysteine cathepsin activity, further causing an increase in the migration and invasion of cancerous cells [99]. Genetic ablation of cystatin C in the RIP1–Tag2 mouse model of pancreatic cancer has shown to increase the number and size of pre-malignant lesions, and tumor vascularity [100].

Different classes of proteases (cathepsins, MMPs, uPAR) interact with each other and take part in proteolytic pathway. Strategies targeting more than one class of proteases during tumor progression might prove to be more efficient than targeting a single cathepsin. Rao and colleagues carried out simultaneous downregulation of cathepsin B and MMP9 (34) and of cathepsin B and uPAR140 in SNB19 glioblastoma cells [101]. They reported that intratumoral injections of plasmid DNA expressing hairpin RNAs for these enzymes resulted in regression of pre-established tumors. Using a similar approach, Gondi and Rao showed that RNAi-mediated downregulation of cathepsin B and uPAR inhibited cell proliferation and initiated a partial extrinsic apoptotic cascade [102]. These studies demonstrate the effectiveness of RNAi-mediated downregulation of cathepsin B and other proteases in retarding tumor burden, invasion, and angiogenesis. It is evident that the downregulation of cathepsin B and other tumor-associated molecule simultaneously exhibit an additive effect when compared to the downregulation of cathepsin B alone.

Targeting the transcriptional regulation of cathepsins could be another promising approach. Transcription factors Sp1, Sp3, and those of the ETS family have been associated with the transcriptional regulation of cysteine cathepsins [24]. ETS1, known to be expressed in invasive tumors, regulates cathepsin B transcription. Therefore, exploiting transcriptional regulation of cathepsins might considerably decrease the expression of cathepsins in tumors and tumor-associated cells. However, clinically relevant and more robust studies need to be done to develop cathepsin B in the human breast cancer treatment.

Similarly, different approaches have been made to inhibit cathepsin D production or its action in breast cancer. Cathepsin D is associated with breast cancer progression, as well as it acts as a rate-limiting factor in stimulating in vitro and in vivo tumor growth. One study reported that reduced expression of cathepsin D in MDA-MB-231 breast cancer cells, by transfection of antisense cathepsin D cDNA, significantly decreased anchorage-independent growth, tumorigenicity, and lung colonization in nude mice as compared to mock-transfected cells [103]. An additional targeting strategy is based on inhibition of proteolytically inactive pro-forms of cathepsins. Overexpression of pro-cathepsin D, has been associated with highly invasive malignancies, including breast cancer. Various in vitro and animal studies have shown that the transfection of pro-cathepsin D affects the tumor cell development, and that the rate of tumor growth was directly linked to the level of pro-cathepsin D expression [32, 104]. Vashishta et al. [58] tested the efficacy of the anti-pro-cathepsin D hammerhead ribozyme as a potential gene therapy agent in human breast cancer. Anti-pro-cathepsin D ribozyme used specifically to inhibit pro-cathepsin D in breast cancer cells not only inhibited invasion and growth but also induced apoptosis of MDA-MB-231 cells [58]. These findings reveal that cathepsin D is a potential candidate for new breast cancer therapy. The development of such novel therapies in addition to other existing therapies targeting genetic alteration in cancers might prove to be more efficacious in clinical setting and patient selection for this therapy would rely on the current assay for cathepsin D in primary breast tumors. Despite preclinical advancement with respect to both enhanced targeting and expression in a tumor-selective manner, multiple hurdles, including non-specific expression, low-efficiency delivery, and biosafety prevent the success of cathepsin and their inactive pro-forms in the clinic.

5.3 Antibody-Targeted Therapy

In the last decade, the use of antibody therapy in the field of oncology has shown very promising results [105]. One of the most promising and exciting fields in modern anti-cancer therapy involves the use of monoclonal antibodies which, once administered to the patient, will selectively and efficiently target a particular protein involved with the proliferation of tumor cells. Due to their high specificity, antibodies represent a promising method for interfering with a single target molecule, such as cathepsins, with high selectivity. Although research into neutralizing antibodies against protease targets is still budding, some examples of this drug class in preclinical and early clinical development are available. These include neutralizing antibodies against cathepsin B (Mab 2A2) [106] and uPA [107] for cancer treatment. Mab 2A2 was capable of impairing the intracellular and extracellular activities of cathepsin B and thus reducing the invasion of MCF-10A neo T cells (ras-transformed human breast epithelial cells). Significant inhibition of cell invasion by cathepsin B neutralizing Mab 2A2 suggests its application in antitumor therapy.

The involvement of pro-cathepsin D and cathepsin D in cancer progression is well documented in the literature, and their effects can be inhibited both in vitro and in vivo by antibodies targeting proenzyme or active enzymes. Breast cancer cells secrete enzymatically inactive pro-cathepsin D and the region of pro-cathepsin D responsible for its mitogenic activity was localized to its activation peptide corresponding to amino acids 27-44 [108]. The proteolytic cleavage of the activation peptide generates enzymatically active cathepsin D but it retains a portion of activation peptide [109]. There is data showing regression in tumor growth when the activation peptide is blocked by specific antibodies. One study demonstrated that using both anti-activation peptide and anti-27-44 peptide monoclonal antibodies inhibited its mitogenic potential and also inhibited the growth of human breast tumors in athymic nude mice [108]. Additionally, specific inhibition of pro-cathepsin D by addition of anti-pro-cathepsin D antibodies specifically inhibited its mitogenic activities [110]. Blocking the activities of either the activation peptide of pro-enzymes or the active form of cathepsins by specific antibodies or analogs functioning as antagonists might be a valuable tool in breast cancer inhibition. With the emergence of new technologies, and as the detailed physiological function of cathepsins in cancer are revealed and updated, it should become easier to develop specific antibodies against them.

6 Conclusions

The broad influence of cathepsins B and D on various aspects of breast tumor development and progression proves them as an attractive target for breast cancer therapy. It is evident that targeting these cathepsins has significant therapeutic potential. From a translational perspective, the integration of cathepsin-targeting strategies into established chemotherapeutic regimens seems to hold substantial promise. However, more detailed studies are necessitated to determine if focussing these two cathepsins, may be useful in attenuating the progression of breast cancer. For cathepsins to emerge as a successful druggable target, a detailed knowledge is required about the tumor type, individual patient, and the course of tumor progression when anti-cathepsin therapies might be most effective. It still remains elusive whether the functions of cathepsins B and D change during the course of tumor progression; whether these enzymes come from tumor cells, or tumor-associated cells, or both; the mechanisms by which cathepsins interact with the tumor, and tumor microenvironment. In the next decade, we hope to achieve an integrated understanding of cathepsin activity in the tumor microenvironment and determine specific instances for which the reliance of cancers upon cathepsins represents a fatal, exploitable weakness. If cathepsins are to be targeted therapeutically, extensive studies are still needed to define the multiple and temporal roles of cathepsins B and D in breast cancer, including its cellular localization and interactions with other pathways contributing to malignancy.

Acknowledgements The authors acknowledge the support from Fuji Oil Company, Osaka, Japan, and the Meharry Translational Research Center (MeTRC: NIH grant 5U54MD007593). The authors like to thank the Meharry Office for Scientific Editing and Publications (NIH endowment S21MD000104) for proofreading.

References

1. Willstatter R, Bamann E (1929) Über die Proteasen der Magenschleimhaut. Erste Abhandlung über die Enzyme der Leukozyten. Hoppe-Seylers Z Physiol Chem 180: 127–143
2. Anson ML (1940) The Purification of Cathepsin. J Gen Physiol 23:695–704

3. De Duve C, Pressman BC, Gianetto R, Wattiaux R, Appelmans F (1955) Tissue fractionation studies. 6. Intracellular distribution patterns of enzymes in rat-liver tissue. Biochem J 60:604–617
4. Kuester D, Lippert H, Roessner A, Krueger S (2008) The cathepsin family and their role in colorectal cancer. Pathol Res Pract 204:491–500
5. Brix K, Dunkhorst A, Mayer K, Jordans S (2008) Cysteine cathepsins: cellular roadmap to different functions. Biochimie 90:194–207
6. Brix K, Jordans S (2005) Watching proteases in action. Nat Chem Biol 1:186–187
7. Lecaille F, Bromme D, Lalmanach G (2008) Biochemical properties and regulation of cathepsin K activity. Biochimie 90:208–226
8. Fonovic M, Turk B (2014) Cysteine cathepsins and their potential in clinical therapy and biomarker discovery. Proteomics Clin Appl 8:416–426
9. Rawlings ND, Barrett AJ, Bateman A (2012) MEROPS: the database of proteolytic enzymes, their substrates and inhibitors. Nucleic Acids Res 40:D343–D350
10. Vasiljeva O, Reinheckel T, Peters C, Turk D, Turk V, Turk B (2007) Emerging roles of cysteine cathepsins in disease and their potential as drug targets. Curr Pharm Des 13: 387–403
11. Honey K, Rudensky AY (2003) Lysosomal cysteine proteases regulate antigen presentation. Nat Rev Immunol 3:472–482
12. Stoch SA, Wagner JA (2008) Cathepsin K inhibitors: a novel target for osteoporosis therapy. Clin Pharmacol Ther 83:172–176
13. Deal C (2009) Potential new drug targets for osteoporosis. Nat Clin Pract Rheumatol 5: 20–27
14. Funkelstein L, Toneff T, Mosier C, Hwang SR, Beuschlein F, Lichtenauer UD, Reinheckel T, Peters C, Hook V (2008) Major role of cathepsin L for producing the peptide hormones ACTH, beta-endorphin, and alpha-MSH, illustrated by protease gene knockout and expression. J Biol Chem 283:35652–35659
15. Kayser K, Richter N, Hufnagl P, Kayser G, Kos J, Werle B (2003) Expression, proliferation activity and clinical significance of cathepsin B and cathepsin L in operated lung cancer. Anticancer Res 23:2767–2772
16. Illy C, Quraishi O, Wang J, Purisima E, Vernet T, Mort JS (1997) Role of the occluding loop in cathepsin B activity. J Biol Chem 272:1197–1202
17. Musil D, Zucic D, Turk D, Engh RA, Mayr I, Huber R, Popovic T, Turk V, Towatari T, Katunuma N et al (1991) The refined 2.15 A X-ray crystal structure of human liver cathepsin B: the structural basis for its specificity. EMBO J 10:2321–2330
18. Keppler D, Sloane BF (1996) Cathepsin B: multiple enzyme forms from a single gene and their relation to cancer. Enzym Protein 49:94–105
19. Buck MR, Karustis DG, Day NA, Honn KV, Sloane BF (1992) Degradation of extracellular-matrix proteins by human cathepsin B from normal and tumour tissues. Biochem J 282:273–278
20. Mai J, Sameni M, Mikkelsen T, Sloane BF (2002) Degradation of extracellular matrix protein tenascin-C by cathepsin B: an interaction involved in the progression of gliomas. Biol Chem 383:1407–1413
21. Gocheva V, Zeng W, Ke D, Klimstra D, Reinheckel T, Peters C, Hanahan D, Joyce JA (2006) Distinct roles for cysteine cathepsin genes in multistage tumorigenesis. Genes Dev 20:543–556
22. Quraishi O, Nagler DK, Fox T, Sivaraman J, Cygler M, Mort JS, Storer AC (1999) The occluding loop in cathepsin B defines the pH dependence of inhibition by its propeptide. Biochemistry 38:5017–5023
23. Podgorski I, Sloane BF (2003) Cathepsin B and its role(s) in cancer progression. Biochem Soc Symp 263–276
24. Mohamed MM, Sloane BF (2006) Cysteine cathepsins: multifunctional enzymes in cancer. Nat Rev Cancer 6:764–775

25. Andl CD, McCowan KM, Allison GL, Rustgi AK (2010) Cathepsin B is the driving force of esophageal cell invasion in a fibroblast-dependent manner. Neoplasia 12:485–498
26. Vasiljeva O, Papazoglou A, Kruger A, Brodoefel H, Korovin M, Deussing J, Augustin N, Nielsen BS, Almholt K, Bogyo M, Peters C, Reinheckel T (2006) Tumor cell-derived and macrophage-derived cathepsin B promotes progression and lung metastasis of mammary cancer. Cancer Res 66:5242–5250
27. Vasiljeva O, Korovin M, Gajda M, Brodoefel H, Bojic L, Kruger A, Schurigt U, Sevenich L, Turk B, Peters C, Reinheckel T (2008) Reduced tumour cell proliferation and delayed development of high-grade mammary carcinomas in cathepsin B-deficient mice. Oncogene 27:4191–4199
28. Benes P, Vetvicka V, Fusek M (2008) Cathepsin D-many functions of one aspartic protease. Crit Rev Oncol Hematol 68:12–28
29. Khalkhali-Ellis Z, Hendrix MJ (2014) Two faces of cathepsin D: physiological guardian angel and pathological demon. Biol Med 6:206
30. Masson O, Bach AS, Derocq D, Prebois C, Laurent-Matha V, Pattingre S, Liaudet-Coopman E (2010) Pathophysiological functions of cathepsin D: targeting its catalytic activity versus its protein binding activity? Biochimie 92:1635–1643
31. Bussemakers MJ, Schalken JA (1996) The role of cell adhesion molecules and proteases in tumor invasion and metastasis. World J Urol 14:151–156
32. Garcia M, Derocq D, Pujol P, Rochefort H (1990) Overexpression of transfected cathepsin D in transformed cells increases their malignant phenotype and metastatic potency. Oncogene 5:1809–1814
33. Koblinski JE, Ahram M, Sloane BF (2000) Unraveling the role of proteases in cancer. Clin Chim Acta 291:113–135
34. Rao JS (2003) Molecular mechanisms of glioma invasiveness: the role of proteases. Nat Rev Cancer 3:489–501
35. Strathdee G (2002) Epigenetic versus genetic alterations in the inactivation of E-cadherin. Semin Cancer Biol 12:373–379
36. Joyce JA, Baruch A, Chehade K, Meyer-Morse N, Giraudo E, Tsai FY, Greenbaum DC, Hager JH, Bogyo M, Hanahan D (2004) Cathepsin cysteine proteases are effectors of invasive growth and angiogenesis during multistage tumorigenesis. Cancer Cell 5:443–453
37. Sameni M, Elliott E, Ziegler G, Fortgens PH, Dennison C, Sloane BF (1995) Cathepsin B and D are localized at the surface of human breast cancer cells. Pathol Oncol Res 1:43–53
38. Castiglioni T, Merino MJ, Elsner B, Lah TT, Sloane BF, Emmert-Buck MR (1994) Immunohistochemical analysis of cathepsins D, B, and L in human breast cancer. Hum Pathol 25:857–862
39. Weber E, Barbulescu E, Medek R, Reinheckel T, Sameni M, Anbalagan A, Moin K, Sloane BF (2015) Cathepsin B-deficient mice as source of monoclonal anti-cathepsin B antibodies. Biol Chem 396:277–281
40. Rafn B, Nielsen CF, Andersen SH, Szyniarowski P, Corcelle-Termeau E, Valo E, Fehrenbacher N, Olsen CJ, Daugaard M, Egebjerg C, Bottzauw T, Kohonen P, Nylandsted J, Hautaniemi S, Moreira J, Jaattela M, Kallunki T (2012) ErbB2-driven breast cancer cell invasion depends on a complex signaling network activating myeloid zinc finger-1-dependent cathepsin B expression. Mol Cell 45:764–776
41. Gopinathan A, Denicola GM, Frese KK, Cook N, Karreth FA, Mayerle J, Lerch MM, Reinheckel T, Tuveson DA (2012) Cathepsin B promotes the progression of pancreatic ductal adenocarcinoma in mice. Gut 61:877–884
42. Sevenich L, Werner F, Gajda M, Schurigt U, Sieber C, Muller S, Follo M, Peters C, Reinheckel T (2011) Transgenic expression of human cathepsin B promotes progression and metastasis of polyoma-middle-T-induced breast cancer in mice. Oncogene 30:54–64
43. Pavlova A, Bjork I (2003) Grafting of features of cystatins C or B into the N-terminal region or second binding loop of cystatin A (stefin A) substantially enhances inhibition of cysteine proteinases. Biochemistry 42:11326–11333

44. Pol E, Bjork I (2001) Role of the single cysteine residue, Cys 3, of human and bovine cystatin B (stefin B) in the inhibition of cysteine proteinases. Protein Sci 10:1729–1738
45. Mai J, Finley RL Jr, Waisman DM, Sloane BF (2000) Human procathepsin B interacts with the annexin II tetramer on the surface of tumor cells. J Biol Chem 275:12806–12812
46. Hughes SJ, Glover TW, Zhu XX, Kuick R, Thoraval D, Orringer MB, Beer DG, Hanash S (1998) A novel amplicon at 8p22-23 results in overexpression of cathepsin B in esophageal adenocarcinoma. Proc Natl Acad Sci U S A 95:12410–12415
47. Seth P, Mahajan VS, Chauhan SS (2003) Transcription of human cathepsin L mRNA species hCATL B from a novel alternative promoter in the first intron of its gene. Gene 321:83–91
48. Yan S, Sloane BF (2003) Molecular regulation of human cathepsin B: implication in pathologies. Biol Chem 384:845–854
49. Qian F, Frankfater A, Chan SJ, Steiner DF (1991) The structure of the mouse cathepsin B gene and its putative promoter. DNA Cell Biol 10:159–168
50. Konduri S, Lakka SS, Tasiou A, Yanamandra N, Gondi CS, Dinh DH, Olivero WC, Gujrati M, Rao JS (2001) Elevated levels of cathepsin B in human glioblastoma cell lines. Int J Oncol 19:519–524
51. Sitabkhan Y, Frankfater A (2007) Differences in the expression of cathepsin B in B16 melanoma metastatic variants depend on transcription factor Sp1. DNA Cell Biol 26: 673–682
52. Fais S (2007) Cannibalism: a way to feed on metastatic tumors. Cancer Lett 258:155–164
53. Somanna A, Mundodi V, Gedamu L (2002) Functional analysis of cathepsin B-like cysteine proteases from Leishmania donovani complex. Evidence for the activation of latent transforming growth factor beta. J Biol Chem 277:25305–25312
54. Kostoulas G, Lang A, Nagase H, Baici A (1999) Stimulation of angiogenesis through cathepsin B inactivation of the tissue inhibitors of matrix metalloproteinases. FEBS Lett 455:286–290
55. Berquin IM, Yan S, Katiyar K, Huang L, Sloane BF, Troen BR (1999) Differentiating agents regulate cathepsin B gene expression in HL-60 cells. J Leukoc Biol 66:609–616
56. Vignon F, Capony F, Chambon M, Freiss G, Garcia M, Rochefort H (1986) Autocrine growth stimulation of the MCF 7 breast cancer cells by the estrogen-regulated 52 K protein. Endocrinology 118:1537–1545
57. Faridi JS, Mohan S, De Leon DD (2004) Modulation of cathepsin D routing by IGF-II involves IGF-II binding to IGF-II/M6P receptor in MCF-7 breast cancer cells. Growth Factors 22:169–177
58. Vashishta A, Ohri SS, Proctor M, Fusek M, Vetvicka V (2007) Ribozyme-targeting procathepsin D and its effect on invasion and growth of breast cancer cells: an implication in breast cancer therapy. Int J Oncol 30:1223–1230
59. Ohri SS, Vashishta A, Proctor M, Fusek M, Vetvicka V (2007) Depletion of procathepsin D gene expression by RNA interference: a potential therapeutic target for breast cancer. Cancer Biol Ther 6:1081–1087
60. Capony F, Rougeot C, Montcourrier P, Cavailles V, Salazar G, Rochefort H (1989) Increased secretion, altered processing, and glycosylation of pro-cathepsin D in human mammary cancer cells. Cancer Res 49:3904–3909
61. Lenarcic B, Kos J, Dolenc I, Lucovnik P, Krizaj I, Turk V (1988) Cathepsin D inactivates cysteine proteinase inhibitors, cystatins. Biochem Biophys Res Commun 154:765–772
62. Stewart AJ, Piggott NH, May FE, Westley BR (1994) Mitogenic activity of procathepsin D purified from conditioned medium of breast-cancer cells by affinity chromatography on pepstatinyl agarose. Int J Cancer 57:715–718
63. Liaudet E, Derocq D, Rochefort H, Garcia M (1995) Transfected cathepsin D stimulates high density cancer cell growth by inactivating secreted growth inhibitors. Cell Growth Differ 6:1045–1052

64. Spyratos F, Maudelonde T, Brouillet JP, Brunet M, Defrenne A, Andrieu C, Hacene K, Desplaces A, Rouesse J, Rochefort H (1989) Cathepsin D: an independent prognostic factor for metastasis of breast cancer. Lancet 2:1115–1118
65. Mathieu M, Rochefort H, Barenton B, Prebois C, Vignon F (1990) Interactions of cathepsin-D and insulin-like growth factor-II (IGF-II) on the IGF-II/mannose-6-phosphate receptor in human breast cancer cells and possible consequences on mitogenic activity of IGF-II. Mol Endocrinol 4:1327–1335
66. Briozzo P, Badet J, Capony F, Pieri I, Montcourrier P, Barritault D, Rochefort H (1991) MCF7 mammary cancer cells respond to bFGF and internalize it following its release from extracellular matrix: a permissive role of cathepsin D. Exp Cell Res 194:252–259
67. Fehrenbacher N, Jaattela M (2005) Lysosomes as targets for cancer therapy. Cancer Res 65:2993–2995
68. Rochefort H, Cavailles V, Augereau P, Capony F, Maudelonde T, Touitou I, Garcia M (1989) Overexpression and hormonal regulation of pro-cathepsin D in mammary and endometrial cancer. J Steroid Biochem 34:177–182
69. Cavailles V, Garcia M, Rochefort H (1989) Regulation of cathepsin-D and pS2 gene expression by growth factors in MCF7 human breast cancer cells. Mol Endocrinol 3: 552–558
70. Cavailles V, Augereau P, Rochefort H (1993) Cathepsin D gene is controlled by a mixed promoter, and estrogens stimulate only TATA-dependent transcription in breast cancer cells. Proc Natl Acad Sci USA 90:203–207
71. Huang L, Liu Z, Chen S, Liu Y, Shao Z (2013) A prognostic model for triple-negative breast cancer patients based on node status, cathepsin-D and Ki-67 index. PLoS ONE 8:e83081
72. Giamarchi C, Solanas M, Chailleux C, Augereau P, Vignon F, Rochefort H, Richard-Foy H (1999) Chromatin structure of the regulatory regions of pS2 and cathepsin D genes in hormone-dependent and -independent breast cancer cell lines. Oncogene 18:533–541
73. Foekens JA, Look MP, Bolt-de Vries J, Meijer-van Gelder ME, van Putten WL, Klijn JG (1999) Cathepsin-D in primary breast cancer: prognostic evaluation involving 2810 patients. Br J Cancer 79:300–307
74. Westley BR, May FE (1996) Cathepsin D and breast cancer. Eur J Cancer 32A:15–24
75. Henry JA, McCarthy AL, Angus B, Westley BR, May FE, Nicholson S, Cairns J, Harris AL, Horne CH (1990) Prognostic significance of the estrogen-regulated protein, cathepsin D, in breast cancer. An immunohistochemical study. Cancer 65:265–271
76. Kute TE, Shao ZM, Sugg NK, Long RT, Russell GB, Case LD (1992) Cathepsin D as a prognostic indicator for node-negative breast cancer patients using both immunoassays and enzymatic assays. Cancer Res 52:5198–5203
77. Isola J, Weitz S, Visakorpi T, Holli K, Shea R, Khabbaz N, Kallioniemi OP (1993) Cathepsin D expression detected by immunohistochemistry has independent prognostic value in axillary node-negative breast cancer. J Clin Oncol 11:36–43
78. Ravdin PM, de Moor CA, Hilsenbeck SG, Samoszuk MK, Vendely PM, Clark GM (1997) Lack of prognostic value of cathepsin D levels for predicting short term outcomes of breast cancer patients. Cancer Lett 116:177–183
79. Rothberg JM, Bailey KM, Wojtkowiak JW, Ben-Nun Y, Bogyo M, Weber E, Moin K, Blum G, Mattingly RR, Gillies RJ, Sloane BF (2013) Acid-mediated tumor proteolysis: contribution of cysteine cathepsins. Neoplasia 15:1125–1137
80. Withana NP, Blum G, Sameni M, Slaney C, Anbalagan A, Olive MB, Bidwell BN, Edgington L, Wang L, Moin K, Sloane BF, Anderson RL, Bogyo MS, Parker BS (2012) Cathepsin B inhibition limits bone metastasis in breast cancer. Cancer Res 72:1199–1209
81. Turk V, Stoka V, Vasiljeva O, Renko M, Sun T, Turk B, Turk D (2012) Cysteine cathepsins: from structure, function and regulation to new frontiers. Biochim Biophys Acta 1824:68–88
82. Tsukuba T, Okamoto K, Yasuda Y, Morikawa W, Nakanishi H, Yamamoto K (2000) New functional aspects of cathepsin D and cathepsin E. Mol Cells 10:601–611

83. Coulibaly S, Schwihla H, Abrahamson M, Albini A, Cerni C, Clark JL, Ng KM, Katunuma N, Schlappack O, Glossl J, Mach L (1999) Modulation of invasive properties of murine squamous carcinoma cells by heterologous expression of cathepsin B and cystatin C. Int J Cancer 83:526–531
84. Levicar N, Strojnik T, Kos J, Dewey RA, Pilkington GJ, Lah TT (2002) Lysosomal enzymes, cathepsins in brain tumour invasion. J Neurooncol 58:21–32
85. Sadaghiani AM, Verhelst SH, Gocheva V, Hill K, Majerova E, Stinson S, Joyce JA, Bogyo M (2007) Design, synthesis, and evaluation of in vivo potency and selectivity of epoxysuccinyl-based inhibitors of papain-family cysteine proteases. Chem Biol 14:499–511
86. Parker BS, Ciocca DR, Bidwell BN, Gago FE, Fanelli MA, George J, Slavin JL, Moller A, Steel R, Pouliot N, Eckhardt B, Henderson MA, Anderson RL (2008) Primary tumour expression of the cysteine cathepsin inhibitor Stefin A inhibits distant metastasis in breast cancer. J Pathol 214:337–346
87. Shree T, Olson OC, Elie BT, Kester JC, Garfall AL, Simpson K, Bell-McGuinn KM, Zabor EC, Brogi E, Joyce JA (2011) Macrophages and cathepsin proteases blunt chemotherapeutic response in breast cancer. Genes Dev 25:2465–2479
88. Minarowska A, Karwowska A, Gacko M (2009) Quantitative determination and localization of cathepsin D and its inhibitors. Folia Histochem Cytobiol 47:153–177
89. Gacko M, Minarowska A, Karwowska A, Minarowski L (2007) Cathepsin D inhibitors. Folia Histochem Cytobiol 45:291–313
90. McConnell RM, Godwin WE, Sayyar K, Trana C, Green A, McConnell M, Young A, Young L, Hatfield SE (2005) Synthesis and evaluation of new cathepsin D Inhibitors. J Ark Acad Sci 59:122–131
91. Anantaraju HS, Battu MB, Viswanadha S, Sriram D, Yogeeswari P (2016) Cathepsin D inhibitors as potential therapeutics for breast cancer treatment: molecular docking and bioevaluation against triple-negative and triple-positive breast cancers. Mol Divers 20: 521–535
92. Vezenkov LL, Sanchez CA, Bellet V, Martin V, Maynadier M, Bettache N, Lisowski V, Martinez J, Garcia M, Amblard M, Hernandez JF (2016) Structure-activity relationships of JMV4463, a vectorized cathepsin D inhibitor with antiproliferative properties: the unique role of the AMPA-based vector. Chem Med Chem 11:302–308
93. Das SK, Menezes ME, Bhatia S, Wang XY, Emdad L, Sarkar D, Fisher PB (2015) Gene therapies for cancer: strategies, challenges and successes. J Cell Physiol 230:259–271
94. Hizel C, Ferrara M, Cure H, Pezet D, Dechelotte P, Chipponi J, Rio P, Bignon YJ, Bernard-Gallon D (1998) Evaluation of the 5′ spliced form of human cathepsin B mRNA in colorectal mucosa and tumors. Oncol Rep 5:31–34
95. Gong Q, Chan SJ, Bajkowski AS, Steiner DF, Frankfater A (1993) Characterization of the cathepsin B gene and multiple mRNAs in human tissues: evidence for alternative splicing of cathepsin B pre-mRNA. DNA Cell Biol 12:299–309
96. Krueger S, Haeckel C, Buehling F, Roessner A (1999) Inhibitory effects of antisense cathepsin B cDNA transfection on invasion and motility in a human osteosarcoma cell line. Cancer Res 59:6010–6014
97. Sokol JP, Schiemann WP (2004) Cystatin C antagonizes transforming growth factor beta signaling in normal and cancer cells. Mol Cancer Res 2:183–195
98. Li W, Ding F, Zhang L, Liu Z, Wu Y, Luo A, Wu M, Wang M, Zhan Q, Liu Z (2005) Overexpression of stefin A in human esophageal squamous cell carcinoma cells inhibits tumor cell growth, angiogenesis, invasion, and metastasis. Clin Cancer Res 11:8753–8762
99. Vigneswaran N, Wu J, Nagaraj N, James R, Zeeuwen P, Zacharias W (2006) Silencing of cystatin M in metastatic oral cancer cell line MDA-686Ln by siRNA increases cysteine proteinases and legumain activities, cell proliferation and in vitro invasion. Life Sci 78: 898–907

100. Wang B, Sun J, Kitamoto S, Yang M, Grubb A, Chapman HA, Kalluri R, Shi GP (2006) Cathepsin S controls angiogenesis and tumor growth via matrix-derived angiogenic factors. J Biol Chem 281:6020–6029
101. Gondi CS, Lakka SS, Dinh DH, Olivero WC, Gujrati M, Rao JS (2004) RNAi-mediated inhibition of cathepsin B and uPAR leads to decreased cell invasion, angiogenesis and tumor growth in gliomas. Oncogene 23:8486–8496
102. Gondi CS, Rao JS (2013) Cathepsin B as a cancer target. Expert Opin Ther Targets 17: 281–291
103. Glondu M, Liaudet-Coopman E, Derocq D, Platet N, Rochefort H, Garcia M (2002) Down-regulation of cathepsin-D expression by antisense gene transfer inhibits tumor growth and experimental lung metastasis of human breast cancer cells. Oncogene 21:5127–5134
104. Vetvicka V, Benes P, Fusek M (2002) Procathepsin D in breast cancer: what do we know? Effects of ribozymes and other inhibitors. Cancer Gene Ther 9:854–863
105. Fauvel B, Yasri A (2014) Antibodies directed against receptor tyrosine kinases: current and future strategies to fight cancer. MAbs 6:838–851
106. Premzl A, Zavasnik-Bergant V, Turk V, Kos J (2003) Intracellular and extracellular cathepsin B facilitate invasion of MCF-10A neoT cells through reconstituted extracellular matrix in vitro. Exp Cell Res 283:206–214
107. Reuning U, Sperl S, Kopitz C, Kessler H, Kruger A, Schmitt M, Magdolen V (2003) Urokinase-type plasminogen activator (uPA) and its receptor (uPAR): development of antagonists of uPA/uPAR interaction and their effects in vitro and in vivo. Curr Pharm Des 9:1529–1543
108. Vetvicka V, Vetvickova J, Fusek M (1999) Anti-human procathepsin D activation peptide antibodies inhibit breast cancer development. Breast Cancer Res Treat 57:261–269
109. Glondu M, Coopman P, Laurent-Matha V, Garcia M, Rochefort H, Liaudet-Coopman E (2001) A mutated cathepsin-D devoid of its catalytic activity stimulates the growth of cancer cells. Oncogene 20:6920–6929
110. Vetvicka V (2012) Procathepsin D in cancer development. J Cancer Ther Res 1:22

PAR1-Mediated Apoptosis and Tumor Regression of Breast Cancer Cells by *Vibrio cholerae* Hemagglutinin Protease

Tanusree Ray and Amit Pal

Abstract

Recent therapies of cancer have different side effects, so present scenario demands new therapeutic strategies. Bacterial toxins have been reported to show promising results in cancer treatment. *Vibrio cholerae* hemagglutinin protease (HAP) regresses tumor growth in Swiss albino mice by programmed cell death of mouse breast cancer cells. Treatment with HAP (one μg/week for three successive weeks) reduced solid tumor and enhanced survival of Ehrlich ascites carcinoma (EAC)-induced Swiss albino mice. A good therapeutic agent should specifically kill malignant cells without any effect on the survival of healthy normal cells. HAP induced PAR1 activation in mouse breast cancer cells (EAC). Overexpression of PAR1 has been reported in different malignant cells when compared to normal cells. HAP-induced PAR1 activated the downstream signaling pathways by nuclear translocation of p50-p65 and the phosphorylation of p38 which caused the activation of NFκB and MAP kinase pathways. The NFκB and MAP kinase activation enhanced the cellular ROS levels. The basal ROS level is reported to be higher in malignant cells as compared to normal healthy cells. Malignant cells cross the threshold level of ROS faster than the normal cells and switch on the cascades of cellular apoptosis. HAP-induced PAR1-mediated apoptosis of malignant cells without altering normal healthy cells makes it a good therapeutic agent for cancer treatment.

Keywords

HAP · Apoptosis · Tumor regression · PAR1 · NFκB · MAP kinase

T. Ray · A. Pal (✉)
Division of Pathophysiology, National Institute of Cholera and Enteric Diseases,
P-33, CIT Road, Scheme-XM, Beliaghata, Kolkata 700010, India
e-mail: apal7364@rediffmail.com

S. Chakraborti and N.S. Dhalla (eds.), *Pathophysiological Aspects of Proteases*,
DOI 10.1007/978-981-10-6141-7_9

1 Introduction

Present-day cancer therapies affect both malignant and healthy cells. These therapies are also responsible for different side effects and may also develop drug resistance. Therapies that can kill specifically malignant cells, without damaging the healthy cells, can overcome these problems. Different targeted strategies are used in cancer therapy such as gene therapy, hormone therapy, inhibitors of signal transduction, and apoptosis inducers [1]. Malignant cells have strategies to bypass apoptosis [1, 2]. Apoptosis inducers destroy these strategies and kill malignant cells. Pediatric leukemia and certain solid tumors are diminished by induction of apoptosis by different drugs' treatments [3]. In a recent study, we have shown that *Vibrio cholerae* hemagglutinin protease (HAP) induced apoptosis in mouse breast cancer cells Ehrlich ascites carcinoma (EAC) by targeting specific PAR1 receptor.

Bacterial toxins have been used for cancer therapies. Bacterial toxins can control cell proliferation, differentiation, and apoptosis. These properties are associated with malignancies and inhibit normal cell controls. Bacterial toxins per se are reported to show its efficacy for cancer therapy [1]. *Clostridium perfringens* type A strain secretes Clostridium perfringens enterotoxin (CPE) [2, 3], which can kill pancreatic cancer cells and inhibit tumor growth in vivo [4–6]. Adenylate cyclase (AC) toxin from *Bordetella pertussis* induced apoptosis [7] in different malignant cells. Proteases play important roles in cellular processes and are widely related to different pathological conditions including cancer. Proteases are associated with cancer prognosis, cellular invasion, and metastasis [8]. A protease from *Serratia marcescens* kums 3958 showed antitumor property [9]. In our recent studies, we have shown that *V. cholerae* hemagglutinin protease (HAP) enhanced cellular ROS level and induced the intrinsic pathway of apoptosis and also regresses tumor growth in mice model [4]. HAP is a major protease in *V. cholerae*, and purified HAP showed cytotoxic effects on HeLa cells and also showed hemorrhagic fluid accumulation in rabbit ileal loop [10].

Protease signaling is mostly regulated by protease activated receptors (PARs). Protease signaling is initiated by the activation of PARs through the protease-mediated cleavage of PARs at its N-terminal domain [11]. Proteases have been reported to enhance tumor progression by activating the invasive property of malignant cell leading to metastasis [12]. However, recent studies also suggested its role in tumor regression [9]. PARs are the class of G-protein-coupled receptors (GPCRs) [13, 14]. There are four different subtypes of PARs, PAR1 – PAR4. The expression of PARs is generally higher in cancerous cells than in normal cells [15]. PARs are involved in both progression of cancer and apoptosis of cancerous cells depending on different stimuli [13, 14, 16].

PAR1-induced apoptosis has been reported in different neuronal, fibroblasts, endothelial and epithelial tumor cells. PAR1 expression is higher in breast cancer cells when compared to normal cells [17]. The role of PAR1-mediated cancer progression and its therapeutic property in cancer prevention are still not clear. The functions of microbial proteases are diverse, and they are involved as antimicrobial peptides [18], destruction of the protective mucosal barrier [19], and in apoptosis of target cells [20].

Vibrio cholerae hemagglutinin protease (HAP) caused PAR1 activation and triggered its downstream NFκB and MAP kinase signaling pathways. These pathways enhanced cellular ROS level and induced apoptosis. These strategies of HAP make it a good therapeutic precursor for cancer therapy [21].

2 HAP Induces Intrinsic Apoptotic Pathway in Malignant Cells and Causes Tumor Regression in Murine Model

2.1 Purified HAP Induces Intrinsic Pathway of Apoptosis in Malignant Cells Without Effecting the Normal Healthy Cells

In presence of HAP, different malignant cells like EAC and RAW 264.7 cells show apoptotic response (Fig. 1a–d). 1 μg/ml of purified HAP induced apoptotic response in EAC and RAW 264.7 cells, but interestingly, this dose of HAP is unable to induce apoptotic response in normal mice peritoneum macrophages (Fig. 1e, f). HAP induced overexpression of p53. Moreover, it increased the ratio of proapoptotic Bax to antiapoptotic Bcl-2 in EAC cells. HAP treatment results in activation of caspase 3 and 9 with significant increase in cytochrome c level in mitochondria-free cytosolic fraction of HAP-treated cells. HAP induces intrinsic apoptotic pathway in mouse breast cancer cells (Fig. 1g, h).

2.2 HAP Treatment Regresses Solid Tumor in Mice

Solid tumor was developed subcutaneously at the right thigh muscle by injecting 2×10^6 number of EAC cells. Once the solid tumor reaches 15 mm in diameter (1.76 cc), 1 μg of HAP was inoculated into the tumor site. After three successive weeks of treatment with 1 μg of HAP, the tumor size diminished significantly, whereas in positive control group, the tumor size increased from 15 to 35 mm in 60 days (Fig. 2a). HAP treatment significantly decreased the tumor volume. In positive control group, the tumor volume increased significantly from 1.76 cc (at 0 day) to 14.75 cc (at 60 days). In the HAP-treated group, the tumor volume decreased from 1.76 to 0 cc at 30 days (Fig. 2b). When HAP was inhibited with EDTA, it failed to diminish tumor volume. Similar effect was observed when only buffer was used instead of HAP (Fig. 2a, b).

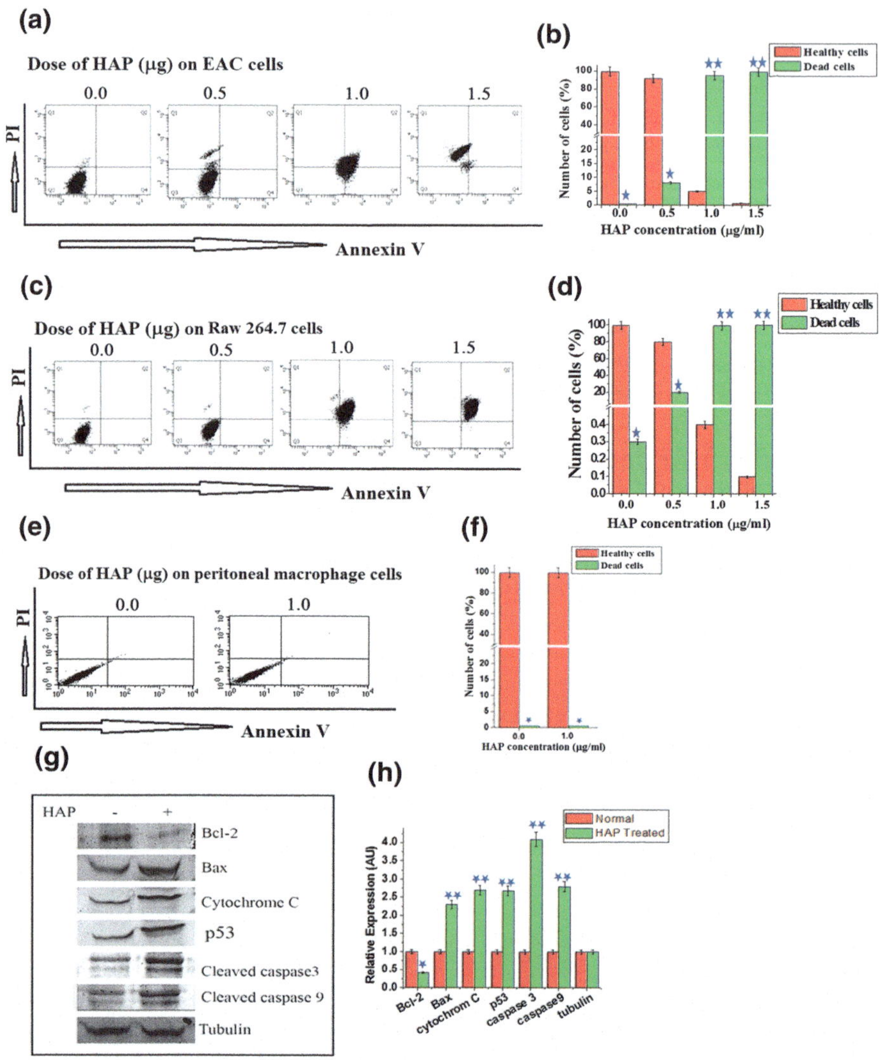

Fig. 1 HAP induced intrinsic pathway of apoptosis in mice macrophages and breast cancer cells. **a**, **c**, **e** Dose-dependent response of HAP on EAC, RAW 264.7, and mouse peritoneal macrophage cells analyzed by FACS to study cellular apoptosis. **b**, **d**, **f** The above results are graphically represented in the bar diagram. In all panels, $*p < 0.05$, $**p < 0.005$, and $***p < 0.0005$. **g** Western blot and **h** densitometric analysis on expression levels of different proteins of apoptotic pathway in HAP-induced and HAP-uninduced cells

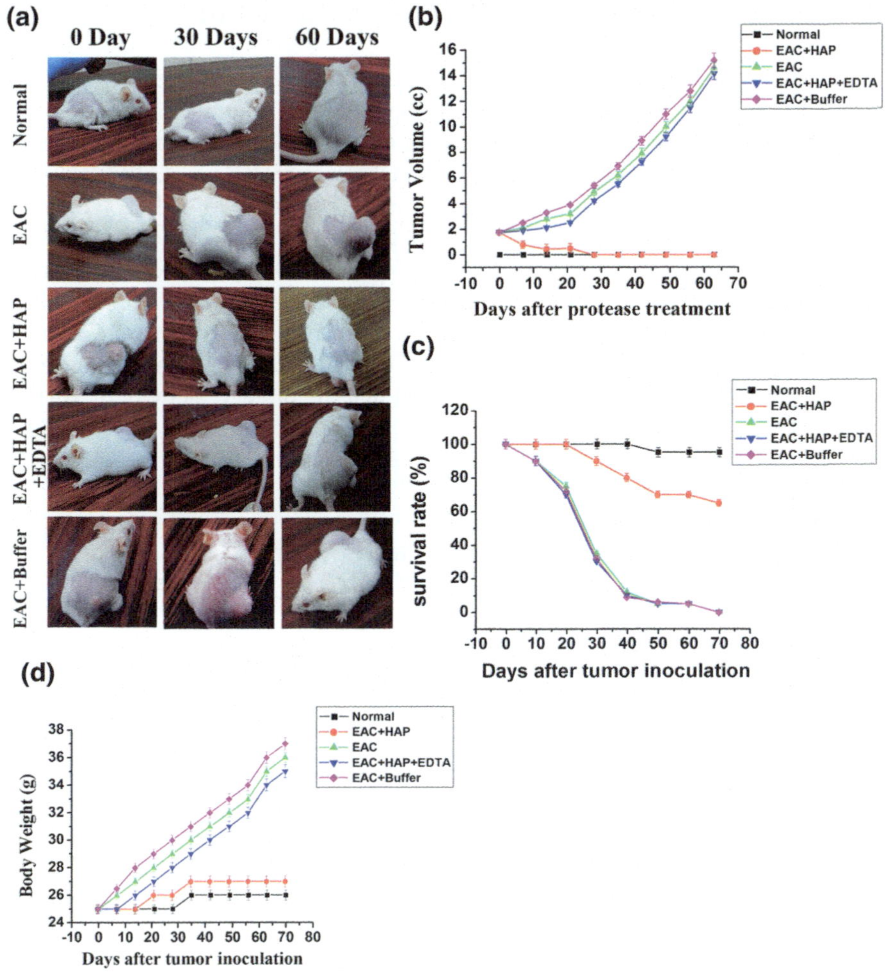

Fig. 2 Effect of HAP on EAC-induced solid tumor and intraperitoneum mice model. **a** Antitumor effect of HAP on EAC-induced solid tumor. The 0 day indicates tumor size of 15 mm in diameter which is diminished after HAP treatment. Tumor control group, buffer control group, and EDTA mediated HAP inhibited group showed significant increase in tumor size. **b** The above results are graphically represented showing the reduction in tumor volume with respect to time. HAP in presence of EDTA and only buffer showed no effect on tumor regression. **c** Effect of HAP in EAC-induced intraperitoneum mice model showed significant increase in survival rate. HAP in presence of EDTA and only buffer showed similar survival rate as tumor control group. **d** Effect of HAP in EAC-induced intraperitoneum mice model showed no significant increase in body weight. Tumor control group and EDTA treated in presence of HAP showed increase in body weight. In each panel, error bars were calculated based on results obtained from minimum of three independent experiments

2.3 HAP Increased the Survival of EAC-Induced Mice

The survival kinetics and the changes in body weight were observed in intraperitoneal EAC-induced mice. Tumor was developed with 10^6 numbers of EAC cells. In the tumor control group, the survival rate of mice was decreased to 30% and then to 5% after 30 and 60 days, respectively. HAP treatment (1 μg HAP/weekly for successive 3 weeks) maintained the survival rate of mice to 90 and 70% after 30 and 60 days, respectively (Fig. 2c). The body weight of tumor control group increased with time (from 25 g at 0 day to 45 g at 60 days), and HAP-treated EAC-induced mice did not show significant changes in body weight (Fig. 2d). Inactivated HAP (by EDTA) treatment and the buffer (25 mM Tris, pH 7.4) control group failed to increase the survival rate of mice (Fig. 2c) and also showed increase in body weight (Fig. 2d). These results proved that the protease activity of HAP is responsible for its antitumor effect.

3 Effect of HAP on Liver, Kidney, and Thigh Muscle of Intraperitoneally EAC-Induced Mice

Histological examination (using hematoxylin-eosin staining) revealed healthy cellular structure of liver, kidney, and thigh muscle fibers. However, in tumor control group, damaged and destroyed cell morphology was found in liver, kidney, and thigh muscles. HAP treatment recovered this damage and helped to regain healthy tissue structure in liver, kidney, and thigh muscle (Fig. 3a).

Presence of apoptosis or necrotic cells was identified by ethidium bromide (EB) and acridine orange (AO) staining of thigh, liver, and kidney tissues of normal, tumor-bearing, and HAP-treated animals. The presence of EB-positive apoptotic or necrotic cells was highly present in tumor-bearing animals as compared to normal and HAP-treated animals (Fig. 3b).

Ki-67 is a nuclear protein that is present in the nucleus of all stages of proliferating cells except G_0 stage and dead cells. The absence of Ki-67 in EAC control group indicates the measure of thigh, liver, and kidney damage. The presence of Ki-67 in the nucleus of normal and HAP-treated cells proved the recovery of damaged cells after HAP treatment (Fig. 3c).

The liver function was tested by the measuring of ALT and AST. Kidney functionality was measured by urea and creatinine levels in the serum sample of animals. The level of ALT and AST was increased significantly, and urea and creatinine levels were significantly decreased in the tumor control group compared to the normal group. There was no significant change in urea, creatinine, ALT and AST levels in HAP-treated group as compared to the normal group (Fig. 3d).

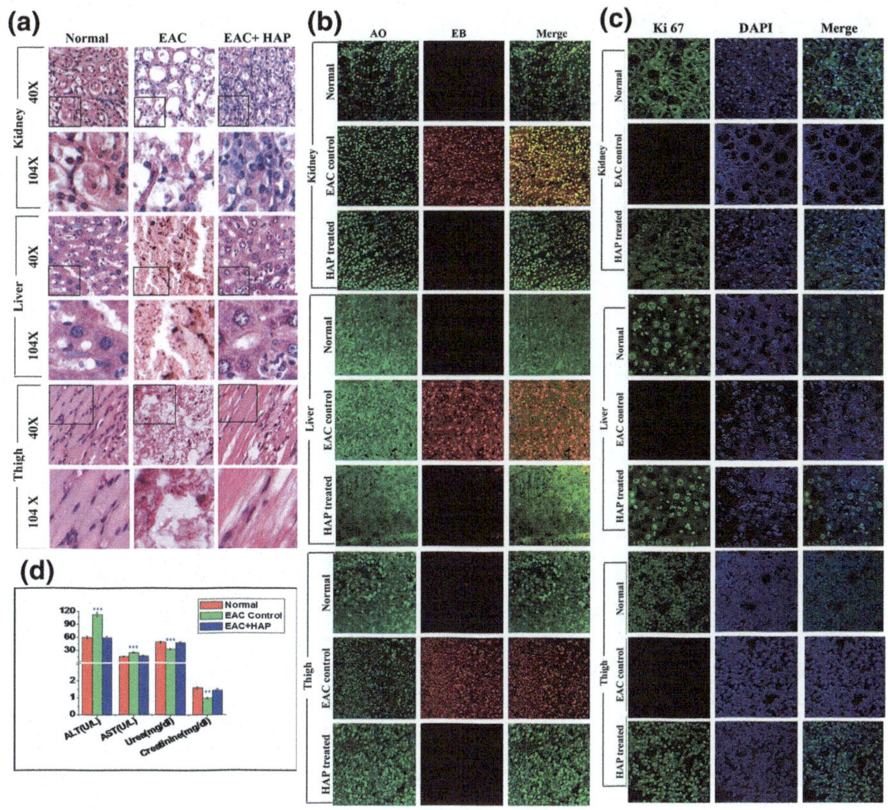

Fig. 3 Effect of HAP on histological and biochemical parameters of tumor induced mice. **a** Histological studies by H&E staining showed that treatment with purified HAP improved the damaged architecture of liver, kidney, and thigh muscle caused by tumor induction. **b** AO and EB staining showed that treatment with purified HAP improved the cell viability in liver, kidney, and thigh muscle caused by tumor induction. **c** Immunohistochemistry of Ki-67 indicated that HAP treatment caused cellular proliferation of damaged liver, kidney, and thigh muscle. **d** Biochemical tests of liver and kidney functionality (ALT, AST, urea and creatinine) indicated the recovery of liver and kidney cells from EAC-mediated damage after HAP treatment

4 Molecular Mechanism of HAP Induced Apoptosis of Malignant Cells

4.1 HAP Causes PAR1 Activation and Induces Downstream Signaling of NFκB and MAP Kinase Pathway

Real-time PCR revealed 30-fold overexpression of PAR1 RNA level as compared to the other PARs (Fig. 4a, c) in RAW 264.7 (Fig. 4a) and in EAC cells (Fig. 4c) after 1 h treatment with 1 μg/ml of HAP. HAP treatment did not cause any

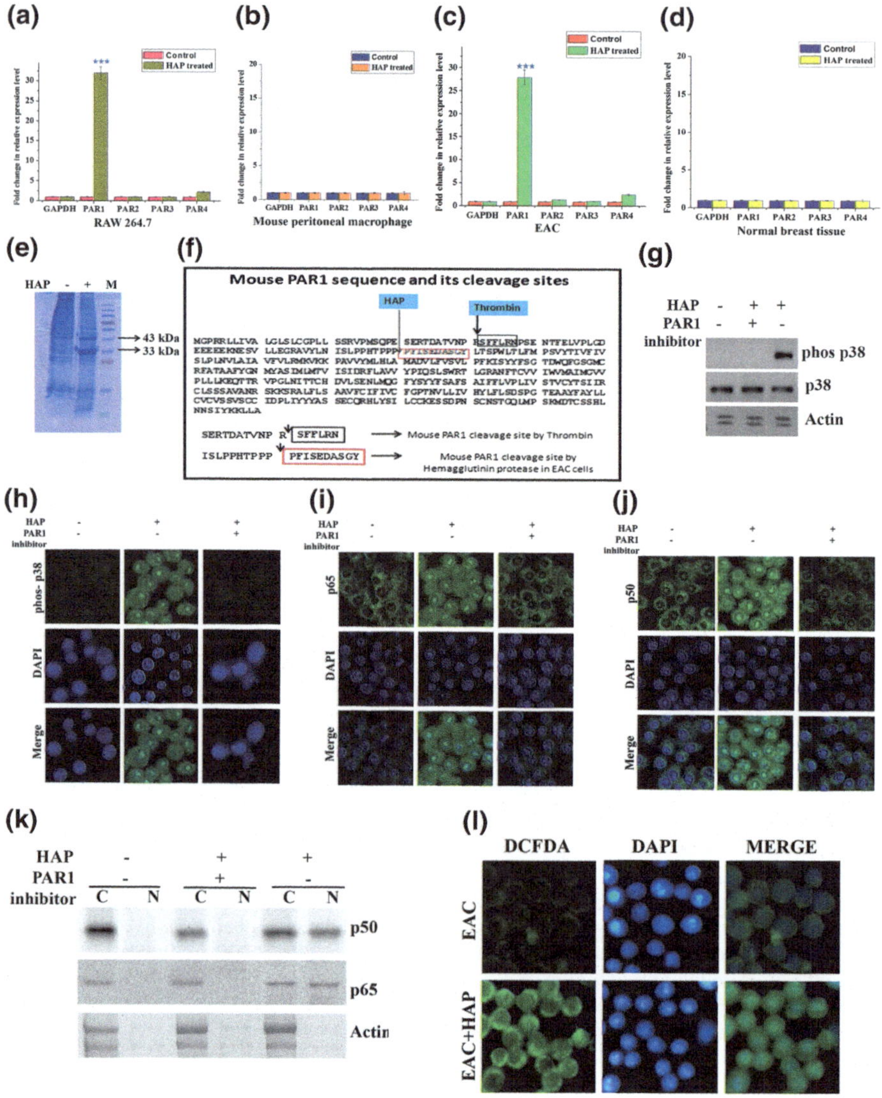

Fig. 4 HAP-induced PAR1 activation causes downstream signaling of NFκB and MAP kinase pathway. **a**, **c** Real-time PCR showed almost 30-fold overexpression of PAR1 RNA compared to the other PARs in RAW 264.7 and EAC cells upon HAP treatment. **b**, **d** There was no significant change in PARs expression level in mouse peritoneal macrophage cells and in normal breast tissues on HAP treatment. **e** HAP treatment causes overexpression of 43 and 33 kDa protein bands in EAC cells compared to untreated cells. **f** N-terminal sequencing of both the bands revealed homology with PAR1. The N-terminal sequence of the cleaved 33 kDa band is "PFISEDASGY". **h–j** Immunofluorescence assay revealed that HAP-mediated p38 phosphorylation and the nuclear translocation of p50 and p65 were inhibited by PAR1 inhibitor. **g**, **k** Western blot analysis confirmed the inhibition of HAP-mediated phosphorylation of p38 and the nuclear translocation of p50 and p65 by PAR1 inhibitor. **l** DCFDA staining observed under fluorescence showed that HAP increased the cellular ROS level in EAC cells

alteration of PARs' expression level in mouse peritoneal macrophage cells (Fig. 4b) and in normal mouse breast tissue (Fig. 4d). PAR1 activation has been well studied with thrombin, and it has been shown to induce tumorigenesis and metastasis. In our study, we have found HAP causes PAR1 activation. HAP treatment caused overexpression of 43 and 33 kDa protein bands when compared to normal EAC cells (Fig. 4e). N-terminal sequencing of 43 and 33 kDa bands showed homology with PAR1, whereas the 33 kDa band showed a new N-terminal sequence "PFI-SEDASGY" (Fig. 4f).

HAP causes NFκB activation by the nuclear translocation of p65 and p50. HAP also induces MAP kinase pathways by phosphorylation of p38. Use of PAR 1 inhibitor blocked both the pathways, which indicates HAP-mediated PAR1 activation induces NFkB (Fig. 4i–k) and MAP kinase pathway (Fig. 4g, h). Several evidences showed that p38 and NFκB enhance cellular ROS level and induced intrinsic apoptotic pathway [22].

4.2 HAP Treatment Increased Cellular ROS Level that Reduced the Viability of Malignant Cells

Cellular ROS level is higher in malignant cells when compared to normal cells [23, 24]. HAP treatment enhanced cellular ROS level, and as a result after HAP treatment, malignant cells crossed their threshold ROS level faster than the normal cells. Increased ROS level induced apoptosis [25, 26]. DCFDA is a nonpolar compound that readily diffuses into cells, where it is hydrolyzed to the non-fluorescent polar derivative DCFH and is thereby trapped within the cells. In the presence of ROS, DCFH is oxidized to highly fluorescent DCF. DCFDA staining showed that HAP treatment enhanced ROS level in EAC cells compared to untreated cells and caused apoptosis (Fig. 4l).

5 Discussion

Breast cancer is one of the most common malignancies in females [27]. Present-day therapeutic strategies have several flaws for complete remission of the disease [28, 29]. The use of targeted inhibitor of oncogenic proteins and cytotoxic agents has increased the rate of success of cancer treatment [30, 31].

Malignant cells divide faster than normal cells, and chemotherapeutic drugs target these cells. In this process, they also kill proliferating healthy cells. Natural products isolated from plants, marine organism and microorganism, are reported to show anticancer activity [32, 33]. *V. cholerae* hemagglutinin protease showed anticancer property by inducing intrinsic apoptotic pathway in malignant cells. It diminished the solid tumor in EAC-induced mice model and enhanced the life span of the cancerous mice. Protease from *S. marcescens* kums 3958 also showed antitumor activity [9], and its degenerating effect of tumor tissue was due to its proteolytic activity [34]. HAP is required in a very low dose to show its antitumor activity. 1 μg/ml of HAP induced apoptosis in malignant cells. HAP treatment caused overexpression of proapoptotic Bax and downregulation of antiapoptotic Bcl-2. The release of cytochrome c into the cytosol from the mitochondria and the activation of caspase 9 along with caspase 3 suggest that HAP triggers activation of intrinsic pathway of apoptosis [35, 36].

EAC cells are widely used for tumor development and to study the anticancer activity of compounds in vivo [37, 38]. HAP treatment significantly diminished the tumor size in mice and increased the survival rate (70%) of the mice when compared to the tumor control group. HAP treatment affected the viability of malignant cells. The histological studies revealed that HAP treatment improved the cellular morphology and architecture of tissues in HAP-treated group compared to tumor-bearing group, suggesting regression of tumor in mice models.

Cellular ROS level of cancerous cells is higher than normal cells [23, 24]. HAP treatment increased cellular ROS level in EAC-induced mice; as a result, malignant EAC cells crossed the threshold level of cellular ROS earlier than normal cells. Increased ROS initiates the process of apoptosis [25, 26]. Due to generation of high ROS, there is a change in mitochondrial membrane potential [39] resulting in the release of cytochrome c into the cytosol [40, 41]. The ratio of proapoptotic Bax to antiapoptotic Bcl-2 is increased leading to the activation of caspase 9 and caspase 3. HAP treatment activates the intrinsic pathway of apoptosis of EAC cells.

Protease signaling is regulated via the activation of protease activated receptors (PARs). PARs function in cell proliferation, metastasis, and apoptosis depending on different stimuli. PARs are G-coupled receptor. PAR1 is reported to play a role in metastatic process of different cancers [42–45]. Thrombin is a well-known activator of PAR1 [46, 47]. PAR1 becomes activated when thrombin cleaves a specific site on the N-terminal domain (R41-S42). Synthetic peptides that are derived from the first few amino acids of thrombin-mediated cleaved N-terminus (SFLLRN) can act as intramolecular ligand of PAR1.

HAP induced PAR1-mediated apoptosis of EAC cells. HAP cleaved the N-terminal sequence of PAR1 at a novel site. The sequence of this novel N-terminal cleaved site is "PFISEDASGY". The role of this peptide to induce the downstream signaling of PAR1 is still unknown.

Thrombin-mediated PAR1 activation has been well studied [48]. PAR1 activation-induced different downstream signaling pathways such as PI3K [49], protein kinase C [50], c-Jun N-terminal kinase (JNK) [51], p38 MAPK [52], RhoGTPase [53], and nuclear factor NFκB [54] have been reported. HAP enhanced ROS level that activates the intrinsic apoptotic pathway [55] in EAC cells.

This is for the first time we have shown that *V. cholerae* hemagglutinin protease (HAP) induced apoptosis by PAR1 activation in cancer cells. HAP causes PAR1 activation that induces NFκB and MAP kinase signaling. These pathways enhance cellular ROS level that activates the intrinsic apoptotic pathways (Fig. 5). PAR1 expression in normal cell is lower than that in the malignant cells [48]. Thus, HAP is considered as a good therapeutic agent to specifically kill malignant cells, whereas the normal healthy cells remain unaltered in the same environment.

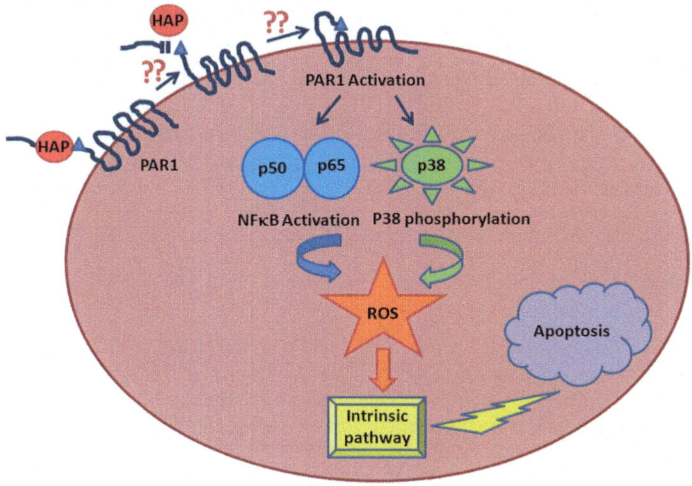

Fig. 5 HAP induced PAR1-mediated apoptosis of breast cancer cells by activation of NFκB and MAP kinase pathways. HAP induced PAR1 cleavage and activation, though the exact "tethered ligand" that binds intramolecularly to trigger transmembrane signaling is unknown but this receptor activation induced phosphorylation of p38 and nuclear translocation of p50 and p65 in breast cancer cells. The activation of MAP kinase and NFκB signaling enhance the cellular ROS level and triggered the intrinsic pathways of apoptosis

6 Conclusion

Conventional tumor therapies are relatively non-specific and show dose-limiting side effects and also develop drug resistance. Several proteases and peptides have been reported to have antitumor activity [56–58]. For being a good therapeutic agent, it should kill malignant cells specifically without altering the survival of normal healthy cells. This is possible by targeted strategies that specifically disrupt oncogenically active cell surface receptors and endogenous signaling molecules [42]. Protease signaling is mostly regulated via the activation of protease-activated receptors (PARs). PARs showed the property for cell proliferation and apoptosis of malignant cells depending on stimuli [15, 45]. In our earlier study, *V. cholerae* hemagglutinin protease (HAP) showed PAR1 induced apoptosis via MAP kinase and NFκB pathways [21]. We have found HAP induced a new PAR1 cleavage site "PFISEDASGY" but the exact "tethered ligand" that interacts with the receptor itself and induces downstream signaling is still unknown. Since HAP-mediated PAR1 activation caused apoptosis in different mouse malignant cells, so we hypothesized that the "tethered ligand" sequence will provide a novel proapoptotic peptide that may be able to induce PAR1-mediated apoptosis. Due to the advancement in the large-scale synthesis of peptides, it will be possible to make peptide-based anticancer drugs more affordable to patients, though this technique needs more practice and research. We are hopeful that this peptide will kill malignant cells without altering the survival of normal healthy cells via targeting PAR1 that is known to be overexpressed in malignant cells. This proapoptotic peptide can be used for an alternate strategy for cancer therapy.

Acknowledgements All animal experiments were conducted after getting permission from Institutional Animal Ethical Committee (PRO/125/April 2016–March 2019) and the Institutional bio-safety committee NICED/BS/AP-003/2016.

References

1. Carswell EA, Old LJ, Kassel RL, Green S, Fiore N, Williamson B (1975) An endotoxin-induced serum factor that causes necrosis of tumors. Proc Natl Acad Sci 72:3666–3670
2. Kokai KJF, Mcclane BA (1997) Determination of functional regions of *Clostridium perfringens* enterotoxin through deletion analysis. Clin Infect Dis 25:S165–S167. doi:10.1086/516246
3. Kokai KJF, Benton K, Wieckowski EU, Mcclane BA (1999) Identification of a *Clostridium perfringens* enterotoxin region required for large complex formation and cytotoxicity by random mutagenesis. Infect Immun 67:5634–5641
4. Michl P, Buchholz M, Rolke M, Kunsch S, Löhr M, McClane B et al (2001) Claudin-4: a new target for pancreatic cancer treatment using *Clostridium perfringens* enterotoxin. Gasrtoenterology 121:678–684. doi:10.1053/gast.2001.27124
5. Hough CD, Sherman- Baust CA, Pizer ES, Montz FJ, Im DD, Rosenshein NB (2000) Large scale serial analysis of gene expression reveals genes differentially expressed in ovarian cancer. Cancer Res 60:6281–6287
6. Kominsky SL, Vali M, Korz D, Gabig TG, Weitzman SA, Argani P et al (2004) *Clostridium perfringens* enterotoxin elicits rapid and specific cytolysis of breast carcinoma cells mediated through tight junction proteins claudin 3 and 4. Am J Pathol 164:1627–1633. doi:10.1016/S0002-9440(10)63721-2
7. Nougayrede JP, Taieb F, De Rycke J, Oswald E (2005) Cyclomodulins: bacterial effectors that modulate the eukaryotic cell cycle. Trends Microbiol 13:103–110. doi:10.1016/j.tim.2005.01.002
8. Mohamed MM, Sloane BF (2006) Cysteine cathepsins: multifunctional enzymes in cancer. Nat Rev Cancer 6:764–775. doi:10.1038/nrc1949
9. Maeda H, Matsumura Y, Molla A (1987) Antitumor activity of some bacterial proteases: eradication of solid tumors in mice by intratumor injection. Cancer Res 47:563–566
10. Ghosh A, Saha DR, Hoque KM, Asakuna M, Yamasaki S, Koley H et al (2006) Enterotoxigenicity of mature 45-kilodalton and processed 35-kilodalton forms of hemagglutinin protease purified from a cholera toxin gene-negative *Vibrio cholerae* non-O1, non-O139 strain. Infect Immun 74(5):2937–2946. doi:10.1128/IAI.74.5.2937-2946.2006
11. Turk B (2006) Targeting proteases: successes, failures and future prospects. Nat Rev Drug Discov 5:785–799
12. DeClerck YA, Mercurio AM, Stack MS, Chapman HA, Zutter MM, Muschel RJ et al (2004) Proteases, extracellular matrix, and cancer. Am J Pathol 164(4):1131–1139. doi:10.1016/S0002-9440(10)63200-2
13. Han N, Jin K, He K, Cao J, Teng L (2011) Protease-activated receptors in cancer: a systematic review. Oncol Lett 2:599–608. doi:10.3892/ol.2011.291
14. Soh UJ, Dores MR, Chen B, Trejo J (2010) Signal transduction by protease-activated receptors. Br J Pharmacol 160(2):191–203. doi:10.1111/j.1476-5381.2010.00705.x
15. Flynn AN, Buret AG (2004) Proteinase-activated receptor 1 (PAR1) and cell apoptosis. Apoptosis 6:729–737

16. Turk B, Turk D, Turk V (2012) Protease signalling: the cutting edge. EMBO J 31(7):1630–1643. doi:10.1038/emboj.2012.42
17. Even-Ram S, Uziely B, Cohen P et al (1998) Thrombin receptor overexpression in malignant and physiological invasion processes. Nat Med 4:909–914
18. Taggart CC, Greene CM, Smith SG, Levine RL, McCray PB Jr, O'Neill S et al (2003) Inactivation of human beta-defensins 2 and 3 by elastolytic cathepsins. J Immunol 171:931–937
19. Moncada D, Keller K, Chadee K (2003) Entamoeba histolytica cysteine proteinases disrupt the polymeric structure of colonic mucin and alter its protective function. Infect Immun 71(2):838–844
20. Denecker G, Declercq W, Geuijen CA, Boland A, Benabdillah R, Gurp MV et al (2001) *Yersinia enterocolitica* YopP-induced apoptosis of macrophages involves the apoptotic signaling cascade upstream of bid. J Biol Chem 276:19706–19714
21. Ray T, Pal A (2016) PAR1 mediated apoptosis of breast cancer cells by *V. cholerae* hemagglutinin protease. Apoptosis 21(5):609–620. doi:10.1007/s10495
22. Duan WJ, Li QS, Xia MY, Tashiro S, Onodera S, Ikejima T (2011) Silibinin activated ROS-p 38-NF-κB positive feedback and induced autophagic death in human fibrosarcoma HT1080 cells. J Asian Nat Prod Res 13(1):27–35. doi:10.1080/10286020.2010.540757
23. Klaunig JE, Kamendulis LM, Hocevar BA (2010) Oxidative stress and oxidative damage in carcinogenesis. Toxicol Pathol 38:96–109. doi:10.1177/0192623309356453
24. Trachootham D, Alexandre J, Huang P (2009) Targeting cancer cells by ROS-mediated mechanisms: a radical therapeutic approach? Nat Rev Drug Discov 8:579–591. doi:10.1038/nrd2803
25. Higuchi M, Honda T, Proske RJ, Yeh ET (1998) Regulation of reactive oxygen species-induced apoptosis and necrosis by caspase 3-like proteases. Oncogene 17:2753–2760
26. Harsdorf RV, Li PF, Dietz R (1999) signaling pathways in reactive oxygen species-induced cardiomyocyte apoptosis. Circulation 99:2934–2941. doi:10.1161/01.CIR.99.22.2934
27. Matsen CB, Neumayer LA (2013) Breast cancer: a review for the general surgeon. JAMA Surg 148(10):971–979. doi:10.1001/jamasurg.2013.3393
28. Greenberg PA, Hortobagyi GN, Smith TL, Ziegler LD, Frye DK, Buzdar AU (1996) Long-term follow-up of patients with complete remission following combination chemotherapy for metastatic breast cancer. J Clin Oncol 14:2197–2205
29. Orlando L, Colleoni M, Fedele P, Cusmai A, Rizzo P, D'Amico M et al (2007) Management of advanced breast cancer. Ann Oncol 18(Suppl 6):vi74–vi76. doi:10.1093/annonc/mdm230
30. Ihemelandu CU, Leffall LD Jr, Dewitty RL, Naab TJ, Mezghebe HM, Makambi KH et al (2007) Molecular breast cancer subtypes in premenopausal and postmenopausal African-American women: age-specific prevalence and survival. J Surg Res 143:109–118. doi:10.1016/j.jss.2007.03.085
31. Dalerba P, Cho RW, Clarke MF (2007) Cancer stem cells: models and concepts. Annu Rev Med 58:267–284. doi:10.1146/annurev.med.58.062105.204854
32. Cragg GM, Kingston D, Newman DJ (2005) Anticancer agents from natural products. Brunner-Routledge Psychology Press, London, pp 186–205
33. Newman DJ, Cragg GM, Snader KM (2003) Natural products as a source of new drugs over the period 1981–2002. J Nat Prod 66:1022–1037. doi:10.1021/np030096l
34. Bhattacharyya A, Choudhuri T, Pal S, Chattopadhyay S, Datta GK, Sa G et al (2003) Apoptogenic effects of black tea on Ehrlich's ascites carcinoma cell. Carcinogenesis 24(1):75–80. doi:10.1093/carcin/24.1.75
35. Koopman G, Reutelingsperger CP, Kuijten GA, Keehnen RM, Pals ST, Oers MHV (1994) Annexin V for flow cytometric detection of phosphatidylserine expression on B cells undergoing apoptosis. Blood 84:1415–1420
36. Yang J, Liu X, Bhalla K, Kim CN, Ibrado AM, Cai J et al (1997) Prevention of apoptosis by Bcl-2: release of cytochrome C from mitochondria blocked. Science 275:1129–1132. doi:10.1126/science.275.5303.1129

37. Somasagara RR, Hegde M, Chiruvella KK, Musini A, Choudhary B, Raghavan SC (2012) Extracts of strawberry fruits induce intrinsic pathway of apoptosis in breast cancer cells and inhibits tumor progression in mice. PLoS ONE 7(10):1–11. doi:10.1371/journal.pone.0047021
38. Srivastava M, Nambiar M, Sharma S, Karki SS, Goldsmith G, Hegde M et al (2012) An inhibitor of nonhomologous end-joining abrogates double-strand break repair and impedes cancer progression. Cell 151:1474–1487. doi:10.1016/j.cell.2012.11.054
39 Zamzami N, Hirsch T, Dallaporta B, Petit PX, Kroemer G (1997) Mitochondrial implication in accidental and programmed cell death: apoptosis and necrosis. J Bioenerg Biomembr 29:185–193. doi:10.1023/A:1022694131572
40 Marzo I, Brenner C, Zamzami N, Jurgensmeier JM, Susin SA, Vieira HL et al (1998) Bax and adenine nucleotide translocator cooperate in the mitochondrial control of apoptosis. Science 281:2027–2031. doi:10.1126/science.281.5385.2027
41 Shimizu S, Narita M, Tsujimoto Y (1999) Bcl-2 family proteins regulate the release of apoptogenic cytochrome c by the mitochondrial channel VDAC. Nature 399:483–487. doi:10.1038/20959
42 Granovsky-Grisaru S, Zaidoun S, Grisaru D, Yekel Y, Prus D, Beller U et al (2006) The pattern of protease activated receptor 1 (PAR1) expression in endometrial carcinoma. Gynecol Oncol 103:802–806. doi:10.1016/j.ygyno.2006.05.048
43 Granovsky-Grisaru S, Salah Z, Maoz M, Pruss D, Beller U, Bar-Shavit R (2005) Differential expression of protease activated receptor 1 (Par1) and pY397FAK in benign and malignant human ovarian tissue samples. Int J Cancer 113:372–378. doi:10.1002/ijc.20607
44 Agarwal A, Covic L, Sevigny LM, Kaneider NC, Lazarides K, Azabdaftari G et al (2008) Targeting a metalloprotease-PAR1 signaling system with cell-penetrating pepducins inhibits angiogenesis, ascites, and progression of ovarian cancer. Mol Cancer Ther 7:2746–2757. doi:10.1158/1535-7163.MCT-08-0177
45 Boire A, Covic L, Agarwal A, Jacques S, Sherifi S, Kuliopulos A (2005) PAR1 is a matrix metalloprotease-1 receptor that promotes invasion and tumorigenesis of breast cancer cells. Cell 120:303–313. doi:10.1016/j.cell.2004.12.018
46 Kuliopulos A, Covic L, Seeley SK, Sheridan PJ, Helin J, Costello CE (1999) Plasmin desensitization of the PAR1 thrombin receptor: kinetics, sites of truncation, and implications for thrombolytic therapy. Biochemistry 38:4572–4585. doi:10.1021/bi9824792
47 Ossovskaya VS, Bunnett NW (2004) Protease-activated receptors: contribution to physiology and disease. Physiol Rev 84:579–621. doi:10.1152/physrev.00028.2003
48 Coughlin SR (1999) How the protease thrombin talks to cells. Proc Natl Acad Sci USA 96:11023–11027. doi:10.1073/pnas.96.20.11023
49 Chalmers CJ, Balmanno K, Hadfield K, Ley R, Cook SJ (2003) Thrombin inhibits Bim (Bcl-2-interacting mediator of cell death) expression and prevents serum-withdrawal-induced apoptosis via protease-activated receptor 1. Biochem J 375:99–109. doi:10.1042/bj20030346
50 Lidington EA, Haskard DO, Mason JC (2000) Induction of decayaccelerating factor by thrombin through a protease-activated receptor 1 and protein kinase C-dependent pathway protects vascular endothelial cells from complement-mediated injury. Blood 96:2784–2792
51 Mitsui H, Maruyama T, Kimura S, Takuwa Y (1998) Thrombin activates two stress-activated protein kinases, c-Jun N-terminal kinase and p38, in HepG2 cells. Hepatology 27:1362–1367. doi:10.1002/hep.510270524
52 Marin V, Farnarier C, Gres S, Kaplanski S, Su MS, Dinarello CA, Kaplanski G (2001) The p38 mitogenactivated protein kinase pathway plays a critical role in thrombin-induced endothelial chemokine production and leukocyte recruitment. Blood 98:667–673
53 Ming XF, Barandier C, Viswambharan H, Kwak BR, Mach F, Mazzolai L, Hayoz D, Ruffieux J, Rusconi S, Montani JP, Yang Z (2004) Thrombin stimulates human endothelial arginase enzymatic activity via RhoA/ROCK pathway: implications for atherosclerotic endothelial dysfunction. Circulation 110:3708–3714. doi:10.1161/01.CIR.0000142867.26182.32

54 Rahman A, True AL, Anwar KN, Ye RD, Voyno-Yasenetskaya TA, Malik AB (2002) Galpha (q) and Gbetagamma regulate PAR1 signaling of thrombin-induced NF-kappaB activation and ICAM-1 transcription in endothelial cells. Circ Res 91:398–405. doi:10.1161/01.RES. 0000033520.95242.A2
55 Ray T, Chakrabarti MK, Pal A (2015) Hemagglutinin protease secreted by *V. cholerae* induced apoptosis in breast cancer cells by ROS mediated intrinsic pathway and regresses tumor growth in mice model. Apoptosis 1–12. doi:10.1007/s10495-015-1194-11-12
56 Sharma S, Panjamurthy K, Choudhary B, Srivastava M, Shahabuddin MS, Giri R (2011) A novel DNA intercalator, 8-methoxy pyrimido[4′,5′:4,5]thieno (2,3-b)quinoline-4(3H)-one induces apoptosis in cancer cells, inhibits the tumor progression and enhances lifespan in mice with tumor. Mol Carcinog 52(6):413–425. doi:10.1002/mc.21867
57 Kumagai H, Mukaisho K, Sugihara H, Miwa K, Yamamoto G, Hattori T (2004) Thioproline inhibits development of esophageal adenocarcinoma induced by gastroduodenal reflux in rats. Carcinogenesis 25:723–727. doi:10.1093/carcin/bgh067
58 Ribble D, Goldstein NB, Norris DA, Shellman YG (2005) A simple technique for quantifying apoptosis in 96-well plates. BMC Biotechnol 10:5–12. doi:10.1186/1472-6750-5-12

Matrix Metalloproteinases (MMPs) in Cancer Initiation and Progression

Sanjeev Kumar Maurya, Nitesh Poddar, Pallavi Tandon and Ajit Kumar Yadav

Abstract

Matrix metalloproteinases (MMPs) are calcium-dependent zinc-containing peptidehydrolase, which are actively involved in degradation of the extracellular matrix (ECM), organ development, and tissue remodeling to maintain homeostasis of tissue. Through degradation of ECM in tumor, MMPs provide fundamental base for further tumor cell metastasis. The complex constitution of tumor microenvironment permits various types of regulatory mechanism and expression of cascades of MMPs. Through which various functions of MMPs can be determined. The physiological role of MMP enzymes can be determined by their location and time frame of its activity during tumor progression. According to the recent studies which have revealed the diverse functions of MMPs other than ECM degradation, MMPs are known to play a major role in regulation of many signaling pathways. Their participation in such pathways helps in altering cell physiology as well as in combating disease. MMPs regulate initiation of apoptosis in tumor cells through cleavage of ligands or receptors. There are evidences which support MMPs role in angiogenic and lymph-angiogenic processes. Most of the studies suggest the major involvement of MMP-2, MMP-9, and MMP-14 in tumor angiogenesis, and to a smaller extent, MMP-1 and MMP-7 are also known to be involved. MMPs also play a prominent role in generation of growth signals, apoptosis regulation, tumor vasculature, initiation of neoplastic progression, invasion and metastasis, metastatic niche formation, and MMPs orchestrate inflammation in cancer. Some other non-proteolytic

S.K. Maurya (✉) · N. Poddar · P. Tandon
Department of Biotechnology, Invertis University, Bareilly, Uttar Pradesh, India
e-mail: sanjeev.m@invertis.org

A.K. Yadav
Department of Pharmacy, Invertis University, Bareilly, Uttar Pradesh, India

S. Chakraborti and N.S. Dhalla (eds.), *Pathophysiological Aspects of Proteases*,
DOI 10.1007/978-981-10-6141-7_10

functions of MMPs are also important to be considered in cancer. Wide range of MMPs role in cancer initiation and progression also provides wide range of therapeutic opportunity for cancer treatment.

Keywords
Matrix metalloproteinase · Oncogenesis · Vasculature · Angiogenesis
Tumor growth · Metastasis · Invasion

1 Introduction

Wide range of physiological, biochemical pathways and genes is involved in maintenance of normal physiology of the cell. Undesirable alterations in vital genes, which control important metabolic/physiologic pathways, in cell lead to understand growth of tissue cells [1]. The cell microenvironment is analogous to the inflammatory response, which reassures angiogenesis, alterations of the extracellular matrix (ECM), and tumor cell mobilizations [2]. Knowing the complex interplay among malignant cancer cells and the adjoining stroma, the molecular mechanisms represent some of the major challenges in cancer research. The recent studies highlight the role of extracellular proteinases, such as matrix metalloproteinases (MMPs) that facilitate changes in the microenvironment of tumor cells during cancer progression. These enzymes control arrange of physiological processes and signaling pathways, and thus they act as great mediators in molecular communication between tumor and stroma. Concerning the failure of MMP inhibitors as targets for anticancer therapy in different clinical trials, development of the new understandings into the roles of these extracellular proteinases in cancer, which depends on the circumstances, may either conqueror endorse tumorigenesis or even act freely of their proteolytic action. In Fig. 1, the interest in MMPs research continually increases with better understanding and knowledge about its potential for diagnosis and therapy in different type of tumors.

In this book chapter, we collected literatures from current developments and studies regarding MMPs and its role in tumor initiation and progression. This chapter will help to answer various questions that are very relevant to understand the role of MMPs in special reference to cancer. What are characteristics of MMPs family? What are the mechanisms of MMPs activity? How they play a role in cancer initiation, progression, growth signaling, apoptosis regulation, tumor vasculature, neoplastic progression, tissue invasion and metastasis, metastatic niche formation, inflammatory response, and other non-proteolytic function? These questions when answered with all available experimental and literature data will increase our understanding about the role of MMPs and its utilization for treatment of cancer.

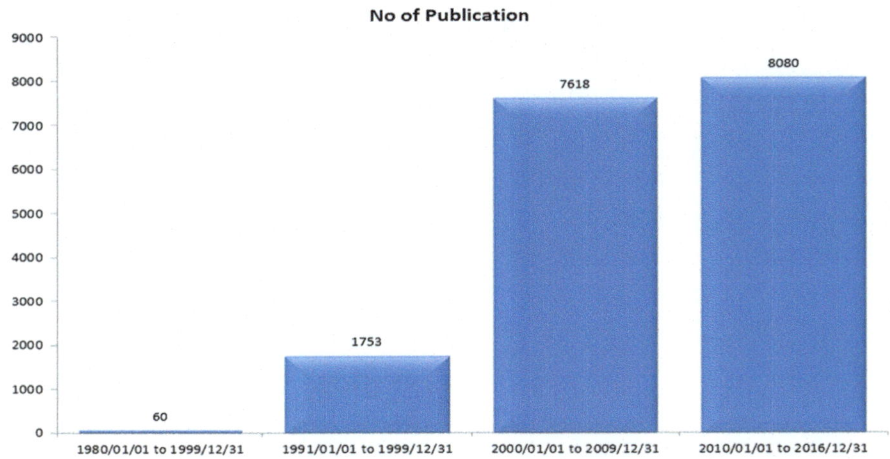

Fig. 1 No. of publication: initially, the scientific interest in MMPs was low but as potential of MMPs discovered MMPs research shoot-up

2 Introduction About MMPs and Its Family

MMPs are a member of zinc-dependent endopeptidases family, firstly defined nearly half a century ago [3] in amphibian during metamorphosis. In the face of intensive study in vitro, in cell culture and in animal models, the standard physiological roles of these extracellular proteases have been identified. They play a crucial role in numerous physiological courses including tissue remodeling and organ development [4], in the regulation of inflammatory processes [5], act on pro-inflammatory cytokines and chemokines, in degradation of the extracellular matrix and in diseases such as cancer [6] (detailed classification described in Fig. 2).

Till now 26 human MMPs are known. On the basis of their structure and specificity, these MMPs are classified into five classes: collagenases, gelatinases, stromelysins, matrilysins, and membrane-type (MT-MMPs) (described in Table 1).

The broad structural outline of MMPs illustrates three domains which are common to nearly all MMPs, the N-terminal pro-domain, the catalytic domain, and the hemopexin-like domain at C-terminal which is connected via a flexible hinge region to the catalytic domain. The membrane-type MMPs (MT-MMPs) combine with a supplementary transmembrane domain that anchors the main cell surface plasma membrane. Pro-MMPs are primarily inactive state because of structural protection by cysteine amino acid present on pro-domain with the zinc ion of the catalytic site. After digestion of this interface through cysteines which are typically arbitrated by proteolytic elimination of the pro-domain cysteine residue. After this modification, the enzyme gets converted into proteolytically active enzyme. Sequence-specific cleavage by convertases occurs in the pro-domain that contains a

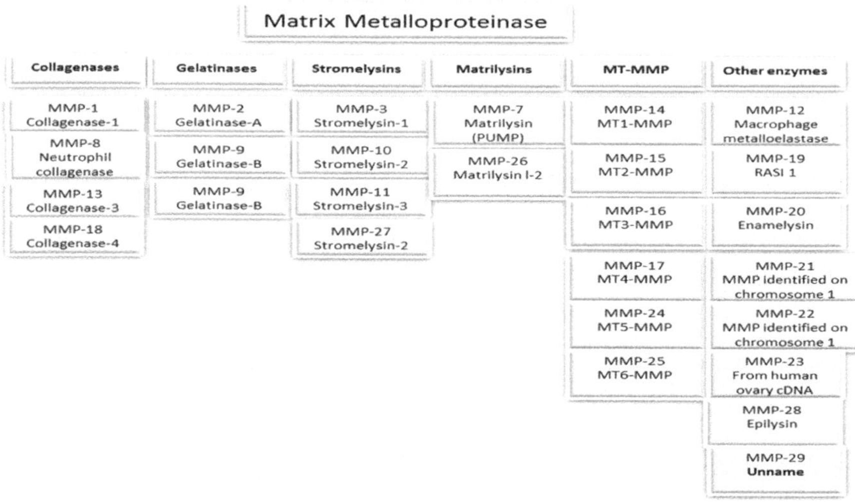

Fig. 2 Classification of MMPs: MMPs are classified in 6 groups depending on their mechanism of action, preferred substrate, and structure

Table 1 The matrix metalloproteinase family: different class of MMPs and their natural substrate in or out of the cells in a tissue

Enzymes	MMPs	Position	Primary substrate
Interstitial collagenase	MMP1	Secreted	Substrates include Col I, II, III, VII, VIII, X, gelatin
Gelatinase-A, 72 kDa gelatinase	MMP2	Secreted	Substrates include gelatin, Col I, II, III, IV, VII, X
Stromelysin 1	MMP3	Secreted	Substrates include Col II, IV, IX, X, XI, gelatin
Matrilysin, PUMP 1	MMP7	Secreted	Membrane associated through binding to cholesterol sulfate in cell membranes, substrates include: fibronectin, laminin, Col IV, gelatin
Neutrophil collagenase	MMP8	Secreted	Substrates include Col I, II, III, VII, VIII, X, aggrecan, gelatin
Gelatinase-B, 92 kDa gelatinase	MMP9	Secreted	Substrates include Gelatin, Col IV, V
Stromelysin 2	MMP10	Secreted	Substrates include Col IV, laminin, fibronectin, elastin
Stromelysin 3	MMP11	Secreted	MMP-11 shows more similarity to the MT-MMPs, is convertase-activatable and is secreted therefore usually associated to convertase-activatable MMPs. Substrates include Col IV, fibronectin, laminin, aggrecan

(continued)

Table 1 (continued)

Enzymes	MMPs	Position	Primary substrate
Macrophage metalloelastase	MMP12	Secreted	Substrates include elastin, fibronectin, Col IV
Collagenase 3	MMP13	Secreted	Substrates include Col I, II, III, IV, IX, X, XIV, gelatin
MT1-MMP	MMP14	Membrane-bound	Type-I transmembrane MMP; substrates include gelatin, fibronectin, laminin
MT2-MMP	MMP15	Membrane-bound	Type-I transmembrane MMP; substrates include gelatin, fibronectin, laminin
MT3-MMP	MMP16	Membrane-bound	Type-I transmembrane MMP; substrates include gelatin, fibronectin, laminin
MT4-MMP	MMP17	Membrane-bound	Glycosyl phosphatidylinositol-attached; substrates include fibrinogen, fibrin
Collagenase 4, xcol4, xenopuscollagenase	MMP18	Not known	Not known
RASI-1, occasionally referred to as stromelysin-4	MMP19	Not known	Not known
Enamelysin	MMP20	Secreted	Not known
X-MMP	MMP21	Secreted	Not known
CA-MMP	MMP23A and B	Membrane-bound	Type-II transmembrane cysteine array
MT5-MMP	MMP24	Membrane-bound	Type-I transmembrane MMP
MT6-MMP	MMP25	Membrane-bound	Glycosyl phosphatidylinositol-attached
Matrilysin-2, endometase	MMP26	Secreted	
MMP-22, C-MMP	MMP27	Secreted	
Epilysin	MMP28	secreted	Discovered in 2001 and given its name due to have been discovered in human keratinocytes. Unlike other MMPs this enzyme is constitutively expressed in testis and at lower levels in lung, heart, brain, colon, intestine, placenta, salivary glands, uterus, skin.

consensus sequence, in which is degraded intracellularly by furin or extracellularly by different MMPs or serine proteinases, for example, plasmin [7, 8]. Some proteins functionally correlated with MMPs are termed as a disintegrin and metalloproteinase (ADAM) and a disintegrin and metalloproteinase with thrombospondin motifs (ADAMTS) families of metzincin proteinases. ADAMs play an important role in fertilization, development, and tumor [9, 10]. Most ADAMs are membrane-bound and perform their functions in the pericellular space. Even though all of the ADAMs has a metalloproteinase domain, but only half of them display proteolytic activity, signifying that ADAMs perform their role by shedding connection accomplices or by interceding the organic capacity in a non-proteolytic manner. The ADAMTS enzymes consist of a protease domain, disintegrin domain, and one or more than one thrombospondin domains [11].

ADAMTS enzymes are generally secreted and are soluble in nature. These enzymes perform central roles in ECM arrangement with other members, ovulation, and tumor [12]. The physiological function of MMPs is based on the nearby adjust among different MMPs and their physiological inhibitors. Considerable amount of energy reserves of the human body are assigned for the inhibition of unregulated extracellular proteolysis by MMPs and different proteinases. For instance, high centralizations of the proteinase inhibitors α2-macroglobulin (α2-MG), α1-proteinase-inhibitor (α1-PI), and α1-chymotrypsin (α1-CT) are combined in the liver and sent into the blood, where these predicaments to the dynamic site of various proteinases [13–15]. The resultant proteinase-inhibitor complexes are targeted to a scavenger receptor and are finally engulfed by macrophages. Most significant physiological inhibitors of MMP are tissue inhibitors of metalloproteinases (TIMPs), which are moreover differentially expressed in tumor cells, and stromal cells [16–18]. TIMP-1, TIMP-2, TIMP-3, and TIMP-4 form 1:1 stochiometric buildings with dynamic MMPs prompting hindrance of proteolytic action. Like MMPs, the proteolytic ADAM and ADAMTS relatives are additionally restrained by particular TIMPs [19].

The outflow of metalloproteinases and their inhibitors in the tumor smaller scale condition are stopped adversely (Fig. 3). Although different cancer cells/tissues express various types of MMPs, ADAMs, and TIMPs family, the major sources of these proteinases are the stromal cells infiltrating the cancer [20]. The various types

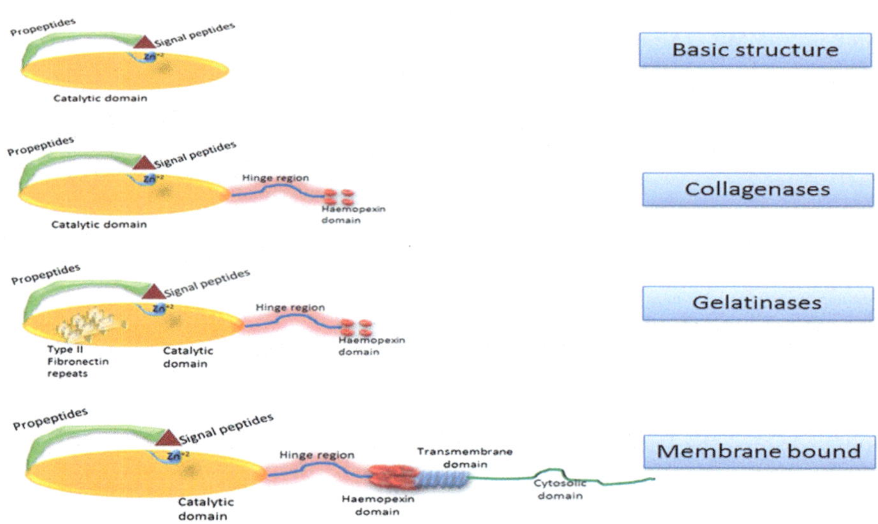

Fig. 3 General structure of MMPs

of stromal cells secrete a particular pair of proteinases and proteinases inhibitors, which are finally discharged into the extracellular space, and help to change the microenvironment around the tumor cells. The MMPs function and activity potential may be governed by their cellular sources, for example, infiltrating neutrophil derived MMP-9 is more readily active in contrast to other cell types [21].

3 Leading MMP Activity

The complex constitution of cancer microenvironment permits various types of regulatory mechanism as well as expression of cascades of MMPs which help us to decide their numerous roles. MMPs activity, proteolysis can be regulated at various levels like compartmentalization, transformation from proactive zymogen to fully active functional catalyst, and at last the existence of precise inhibitors. The most crucial step of MMPs activity is the conversion of inactive enzymes, known as pro-enzymes such as zymogen into a regulatory proteolytic enzyme. Though the understanding of MMPs physiological activity is increasing at a good pace, but still we require a lot of experimental data supported by factual data to get clear-cut knowledge of MMPs activation as well as its regulatory mechanism. There are numerous proteinases that facilitate MMP activation such as plasmin, furin, and other active MMPs [22–24]. Inflammatory cells releasing hyperactive proteinases causing significant injuries in surrounding tissues [5]. For instance, MMPs also degrade plasmin-suppressing serpin proteinases inhibitors. This initiates the changeover of pro-MMPs into active MMPs. Such as, α2-AP is inactivated by MMP-3, whereas serpins, for example, α1-PI and α1-CT are inactivated by several MMPs, this delays the proteolytic action of extracellular proteinases that are restrained by these particles [25]. In another illustration, solidified working of MMP-9, α1-PI, and neutrophil elastase happens in skin rankle arrangement, in which MMP-9 capably destroys α1-PI, an utilitarian serpin inhibitor of neutrophil elastase and further serine proteinases [26, 27]. This supports elastase intervened grid debasement that begins in dermal–epidermal division and rankling [27]. On the other hand, their active oxygen species (ROS) influence working of MMPs. By which huge amounts of ROS occur at the tumor site due to the inflammatory response. These ROS are secreted by regulatory neutrophils and macrophage [28]. This leads to the active function of MMPs via route of oxidation of the pro-domain cysteine [29] while the inactivation occurs by hypochlorous acid (HOCl) through modification of amino acids of catalytic domain. A hemeprotein myeloperoxidase, secreted by neutrophils, monocytes, and macrophages, uses hydrogen peroxide to generate hypochlorous acid [30]. The compartmentalization of MMPs at normal physiological circumstances frequently commands their natural capacities. A few MMPs connect with outside receptors, for example, integrin or think to specific zones of ECM which upgrades MMP's action by expanding their nearby focus and furthermore may limit their accessibility to endogenous inhibitors [31]. The trial confirmations proposed that connection of MMP-2 to integrin αvβ3 by means of its hemopexin space is basic structure's cell intrusive activity [32]. For instance, high convergences of active MMP-14 on the cell surface of obtrusive malignancy cells demonstrate critical characters in cell relocation [33–35]. Likewise, there may be different instruments to aggregate extracellular proteinases at clear places around tumor small-scale condition. Neutrophilic granulocytes, amid cell initiation, are framed and known as neutrophil extracellular traps (NETs). These are web-like organizations with high proteinases convergence of MMP-9 and leukocyte elastase limited on the extracellular chromatin framework [36, 37]. These NETs demonstrate chiefly to battle bacterial contaminations and pathogenesis of immune system infections [22, 38]. Whether the wellsprings of NETs are neutrophils, in case of tumor smaller scale condition, is not clear yet. Likewise, their contribution in malignancy is yet to be determined by compartmentalization and confining proteinase action to certain tumor locales [39]. Mechanical powers (hydrostatic pressure, shear stress, compression, and tension) add to tumor movement [40–42] perhaps by hardening proteolysis of ECM components. Above powers may loosen up and open the basic compliance of substrate proteins, in this way allowing acknowledgment and cleavage by MMPs.

The well-known major controller of primary hemostasis and blood clotting is the Von Willebrand factor (VWF) which is produced in the form of ultra-long chain consisting of hundreds of VWF monomers. The bulky size of this primarily secreted multimeric complex makes it more prone to attacking by high shear forces in the blood stream at spots of rapture. These shear forces cause conformational variation in multimeric complex which leads to exposure of broke site, recognized by ADAMTS-13. The resultant smaller fragments of VWF begin blood clotting process [43]. Tumor advancement is frequently described by expanded tissues stiffness, raised interstitial liquid weight, and changing blood stream conditions [40]. Thus, it is likely that comparable to instruments subject to mechanical strengths are turned out to be the overseeing components for MMP work in the tumor-miniaturized scale condition.

4 MMP Function in Cancer

The physiological role of MMP enzymes is determined by their location and time frame of their activity during cancer progression. In many tumors, the facts still remain hidden that from where and when these enzymes display their proteolytic activity in the cancer microenvironment. In this way, there is a developing need to build up some non-intrusive innovative techniques to manage these inquiries. These techniques would utilize imaging tests in light of MMP-particular exercises. It will be simpler to beat specialized obstacles of clinical criticalness, for example, the finding of early-stage tumors with enhanced affectability by controlling the proteolytic movement of MMPs, the recognizable proof of tumors that open new plausibility to utilize metalloproteinase inhibitors (MPIs) as anticancer medications.

Due to technical advancements in recent time, imaging MMP movement is generally in light of fluorescent optical imaging modalities including fluorescent resonance energy transfer (FRET), radio marked imaging, for example, positron discharge tomography (PET), single photon emanation figured tomography (SPECT), and attractive reverberation imaging (MRI) [44]. Fluorogenic MMP molecules in vivo display that cancers have elevated MMP activity compared to non-tumor-bearing animals [45]. Thus, the evaluation of the effectiveness of MPIs on MMPs activity can take place in whole tumors in vivo [46]. Quantifiable imaging techniques like PET and SPECT are routinely used for detection of cancer, classification, and staging in the treatment as well as in the detection of on spot MMP activity in vivo. Such as, an 18 F-labeled PETMMP-2 inhibitor that specifically binds with cells in tumor models of breast cancer in mice using a specific MMP activity in cancer detection has been used [47]. Radiolabeled ^{123}I-MMP inhibitors specific for the MMPs can recognize MMP-2 and MMP-9 motion using SPECT [48]. Tc-anti-MMP-14 a radiolabeled 99 m monoclonal antibody used as a probe is detected by radiolabeled SPECT in cells bearing malignant mammary tumors in rodents [49]. Gadolinium-based paramagnetic complexity specialists that convey affectability to the MMP proteolytic cleavage turn out to be less hydrophilic on proteolytic cleavage bringing about a discernible change interestingly with MMP-2 [50]. MRI facilities are normally accessible in clinical focuses and are used to detect tumors. This is the most commonly used method as a diagnostic tool for cancer patients. Now, due to these technologies it is easy to constrict the site of MMPs activity in tumor microenvironment by the help of higher resolution probes. Scientists use cell-penetrating peptides linked with fluorescent cargo that are activated by proteolysis and accumulated in area of high MMPs activity [51]. This experiment has been successfully used to visualize MMP-2 and MMP-9 activity in mouse xenograft model and cell culture systems [51, 52]. Known as proteolytic beacons, fluorogenic substrates built on self-quenched and near-infrared FRET pairs, were developed to reduce absorption and scattering and increase tissue penetration. It showed critical increasing MMPs activity in experimental models of colon and pancreatic cancers [53, 54]. For instance, MMP-7 activity imaging in vivo by a specific wavelength of infrared polymer-based proteolytic beacon can sense tumors of 0.01 cm^2 in rodent model [44, 53]. It is hypothesized that such probes can be developed to image MMPs activity at subcellular level in extracellular matrices in vitro. This new type of high-resolution investigation mechanism provides site-specific reporting about the insights of protease activity and cells migration through the extracellular matrices. Imaging of MMPs activity is a promising strategy particularly in the development of MPIs as anticancer drugs. Although many technical challenges with further in vivo imaging of MMP activity are in the way, yet according to the survey, most of the clinical trials with MPIs became unsuccessful, probably due to handling of advanced-stage tumors. Imaging with profound MMP probes will greatly help to govern the time span of MMP activity in which MPI administration is required to be operative. As the understanding and technological advances will grow, this technique may gain a new level.

5 Role of MMPs in Cancer

From last 40 years, MMPs are known to act as a major factor in cancer progression as MMP-mediated ECM degradation leads to cancer cell invasion and metastasis [55]. The idea of MMPs inhibition leading to the suppression of tumor invasive potential was a major turn in MMPs research and its implementation into clinical trials. After so many years of research, the actual tumor suppression potential is yet to be unraveled, but prominence of MMPs in tumor progression cannot be ignored [2]. The recent studies suggest wide range of roles that MMPs perform other than ECM proteolysis. It regulates and participates in many signaling pathways which alter cell physiology and disease. MMPs arbitrate an extensive variety of biological properties on their neighboring tissue (Fig. 4). In the subsequent segments, we will be dealing with the consolidated knowledge about the physiological or pathological processes moderated by MMPs, which are known to have a great influence on the tumor microenvironment.

5.1 MMPs Affect Growth Signals

General feature of tumor cells is uncontrolled proliferation. Principally, there are two basic ways to accomplish this condition: by obtaining independence of external growth promoting signals (growth activator) or by becoming insensitive to anti-growth signals (growth suppressor). The exact balance of two processes decides fate of cells. Any imbalance leads to the occurrence of diseases. In the ECM microenvironment, MMPs may play important role in disturbing the

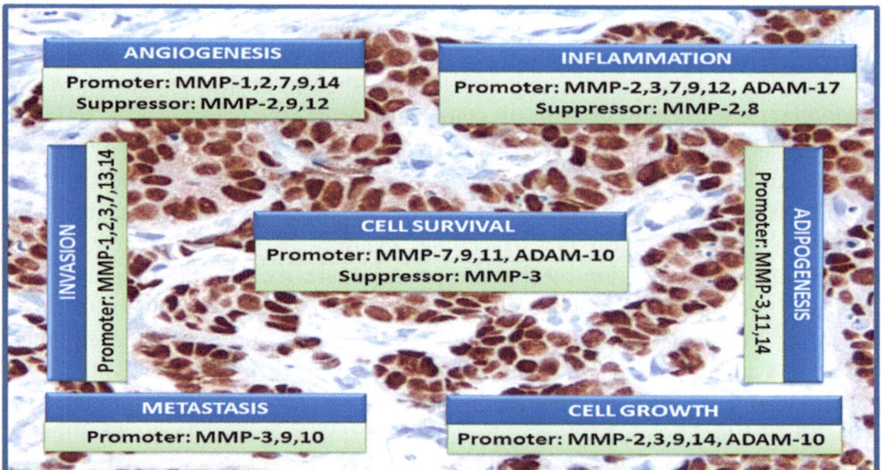

Fig. 4 Summary of different MMPs functions in cancer initiation and progression

balance between growth and anti-growth signals. For instance, TGF-β usually exerts tumor-suppressive properties by applying cytostasis and differentiation in one central signaling pathway that plays a very crucial role in tissue homeostasis.

During the course of malignant tumor progression, mutations accumulated in genome lead to unresponsiveness of TGF-β receptor system to TGF-β factor which results in increased invasion and metastasis [56, 57]. Proactive TGF-β converts into active TGF-β form due to proteolytic conversion by furin or other proteinases, such as MMP-9, which generally comes into play during cellular inflammation [58]. Similarly, TGF-β1 is proteolytically activated by MMP-14 and MMP-2 [59]. Then again, MMP-2, MMP-9, and MMP-14 indirectly modify TGF-β bioactivity by chopping the ECM component latent TGF-β-binding protein1 (LTBP-1), thus making ECM-bound TGF-β free [60–62]. Provided the tumor cells frequently obtain non-responsiveness to TGF-β, this proposes that proteolytic activation of TGF-β by MMPs has tumor-promoting effect leading to increased development of stroma-mediated invasion and metastasis of the tumor. In another example, epidermal growth factor receptor (EGFR) ligands are effective driver so f cell proliferation and important watch dog of tissue homeostasis. Genetic mutations leading to malfunctioning of the molecules involved in this system are commonly witnessed in breast cancer and other malignant diseases [63, 64]. ADAM proteinases are known for their ability to regulate EGFR pathway. For instance, ADAM-10 starts discharge of soluble EGF while ADAM-17 acts as a converter of preforms of TGF-α and epiregulin into their respective active forms. Activation of EGFR upregulates MMP-9 which further degrades E-cadherin. It is a strong regulator of numerous cellular functions like cell-to-cell adhesion and differentiation. Relationship inbetween EGFR, MMP-9, and E-cadherin plays a significant role in ovarian tumor and metastasis due to the co-localization specificity of EGFR and MMP-9 in an area of reduced E-cadherin [65].

The proteolytic cleavage of E-cadherin by MMPs or ADAM has a strong influence on cancer cell proliferation. As a matter of fact, ADAM-10 helps in the detachment of E-cadherin which consequently leads to changeover of β-catenin to the nucleus, causing increased cell proliferation [66]. In addition to the role of ADAM-10, overexpression of MMP-3 in mammary epithelium initiates a cascade of subsequent events including the cleavage of E-cadherin which leads to epithelial–mesenchymal transition [67, 68]. Here, the MMPs inhibitors may slowdown hyperactive MMPs and may prevent tumor growth and metastasis. Consolidated action mechanism of inhibitors of these metalloproteinases with inhibitors of EGFR and HER-2/neu kinases synergistically prevents the growth of human breast cancer xenografts [69]. Above studies help to unravel the hidden facts that correspond with proteolytic speeding up of cellular growth and put forth that exact inhibition of these metalloproteinases can prove to be beneficial in inhibiting unregulated cell growth and uncontrolled cell proliferation.

5.2 MMPs and Apoptosis

Non-occurrence of programmed cell death or apoptosis leads to increased cell number and tumor size. Fas receptors, one of the extracellular receptors, initiate proteolytic cascades of intracellular caspases after binding with Fas ligand which therefore prompts the selective corruption of subcellular part and nuclear DNA, i.e., the conduction of normal apoptosis. MMPs obstruct initiation of apoptosis in tumor cells through cleavage of ligands or receptors. In doxorubicin-treated cancer cells, MMP-7 degrades Fas ligand from the surface [70, 71]. This leads to the lowered effect of chemotherapy on tumor due to hindered apoptotic process. Non-small cell lung cancer patients show resistance to chemotherapy that can be predicted with the help of predictive markers such as MMP-7 expression [72]. Likewise, ADAM-10 hinders/halts apoptosis initiation by cytotoxic lymphocytes via the deprivation of Fas ligand, thus Fas receptor-triggered cell death of target cells is interfered [73]. In pancreatic ductal adenocarcinoma, MMP-7 and Fas ligand interaction may play important role. According to the experimental studies, mice with deficiency of MMP-7 or having non-functional Fas ligand showed reduced metaplasia [74, 75].

MICA and MICB, tumor-associated major histocompatibility complex class-I protein, can be proteolytically degraded by ADAM-17 resulting into the suppression of NK-cell mediated cytotoxicity toward the cancer cells, and hence, ADAM-17 can potentially interfere with anti-tumor directed immune response [76]. It remains unknown whether MMPs show same interfering mechanism like ADAM-17, but the studies have suggested that use of MMPs inhibitors in combination with interleukin-15 in small cell lung cancer cells can act in overcoming tumor resistance issue [77]. Above experimental demonstration shows tumor-promoting role of MMPs by blocking receptor transmitted or lymphocyte-mediated apoptosis.

5.3 MMPs and Angiogenesis in Tumor

Solid tumor vasculature is fundamentally different from that of regular vasculature, and special features of tumor vessels serve as decent targets for cancer therapies. Circulating vasculogenic progenitor cells emerging from local blood vessels form tumor vasculature.

Tumor cells can penetrate these newly formed blood vessels because they are often irregular and leaky in nature due to lack of pericyte cover. In addition to blood capillaries, lymphatic endothelial cells have less-developed junctions, large inter-endothelial gaps, and impaired basement membranes. Angiogenic factor and invasive margin are critical for stimulation of angiogenesis and lymph-angiogenesis in tumor [78]. In following paragraphs, we have cited the evidences that support of MMPs role in angiogenic and lymph-angiogenic processes [79]. According to most of the studies, MMPs involved in tumor angiogenesis are MMP-2, MMP-9, and MMP-14, and to a smaller extent MMP-1 and MMP-7. There are a number of MMPs known for their expression in all tumors. Each MMPs contribute to distinct vascular events in same tumor [45]. The bioavailability of vascular endothelial growth factor (VEGF), one of the potent factors of tumor angiogenesis and therapeutic target, is regulated by MMP-9 [80, 81] for its receptor VEGFR2. This process is known as angiogenic switching. Angiogenic switching by MMP-9 comprises a complex interaction related factors. In a mouse experimental model of glioblastoma, CD45-positive bone marrow-derived cells are recruited by the hypoxia inducible factor-1α (HIF-1α) to initiate angiogenesis. MMP-9 action provided by CD45-positive myeloid cells is vital and adequate for the angiogenic switching by enhancing VEGF bioavailability [82].

It is found that VEGF obstructs tumor cell movement along blood vessels, but it encourages perivascular tumor cell infiltration into the brain parenchyma [83]. MMP-3, MMP-7, MMP-9, or MMP-16 cleave matrix bound VEGF which leads to the formation of modified VEGF molecules with modulated bioavailability, this changes vascular patterning of tumors in vivo. In MMP-9 deficient mice, tumor transplanted to irradiated tissue did not show angiogenesis in vivo, while the tumor growth was restored by transplanting CD11b-positive myeloid cells from the bone marrow of MMP-9 sufficient mice. This experiment proved that MMP-9 is undoubtedly very much required for tumor vasculogenesis [84]. Therefore, during chemotherapy and radiotherapy of tumor, MMP-9 could become an important target for adjunct therapy to increase the response of tumors to radiotherapy.

Neutrophilic granulocytes deliver MMP-9 for performing a specific role. Neutrophil derived pro-MMP-9 does not form complex with another protein TIMP-1 and thus, it is easily activated to initiate tumor angiogenesis [85]. The angiogenic function of MMP-9, derived from neutrophil, needs its active and hemopexin domain, and finally activates basic fibroblast growth factor (FGF)-2 [86]. It is interesting to note that MMPs produced by different cells types can function differently. The release of neutrophil derived TIMP-1 free MMP-9 may be an important factor in the pro-angiogenic effect. So it can be put forward like this, the number of neutrophils is directly proportional to the increase in the intensity of occurrence of angiogenesis. Same as, a high number of neutrophils in fibro sarcoma correlates with high microvessel density in tumor, and [87] while reducing neutrophils in pancreatic cancer significantly decreases angiogenic switching in dysplasia [88]. Above results provide an insight of vital role of infiltrating neutrophils in the initiation of tumor angiogenesis. The degradation of ECM components and other extracellular molecules may generate different fragments with different bioactivities, which may inhibit angiogenesis [89].

For instance, naturally active endostatinis produced via proteolysis of type XVIII collagen by MMP-3, MMP-7, MMP-9, MMP-13, and MMP-20 [90, 91]. Further, the proteolysis of collagen Ivα3 through MMP-9 generates the monomeric NC1 domain called tumstatin, a powerful suppressor of neovascularization. This is established in tumor angiogenesis and amplified tumor growth in MMP-9-lacking mice [92]. Significant amounts of angiostatin can be produced by the degradation of plasminogen by MMP-2, MMP-9, and MMP-12, a product with anti-angiogenic function [93]. Angiostatin production by MMP-12 shows suppressive effects on outgrowth of lung metastases [94].

Collectively, MMPs can produce both anti-angiogenic as well as promoting signals. The effects of MMPs on angiogenesis might be diverse depending on the time frame of MMP expression and the availability of substrates. MMPs regulate vascular stability and permeability. Predominantly, MMP-14 seems to facilitate the vascular reaction to tissue injury and tumor development by activation of TGF-β [95]. Lymph-angiogenesis is known to play a very important role in tumor biology, provided it is directly related to the development of lymphatic metastases. MMPs definitely have overall influence on lymph-angiogenesis as supported by the application of broad-spectrum MMP inhibitors [96]. For lymph metastases, active VEGF bioavailability is an important factor which is provided by MMPs, especially by MMP-9. The three-dimensional culture system using mouse thoracic duct fragments embedded in a collagen gel provided experimental proof of the involvement of MMPs in lymph-angiogenesis.

Amplified MMP-1, MMP-2 [98], and MMP-3 [99] expression are correlated with lymphatic invasion and lymph-node metastases. Attenuation of angiogenesis and lymph-angiogenesis is achieved by inhibition of MMP-2, MMP-9, and MMP-14 [96]. MMP-2 deletion may affect lymphatic vasculature in a negative way, but may not affect aortic vasculature [97]. The literature and experimental evidences are insufficient to prove the link between lymph-angiogenesis and MMPs. Further detailed studies are required to reveal the facts.

5.4 MMPs and Adipocyte in Cancer Progression

Adipocytes are well-known constituents of tumor stroma and play a vital role in cancer progression. Adipokines, the adipose secretory products, are involved in negative consequences of adipose tissue expansion in diabetes, cardiovascular disease, and cancer [100, 101]. Further, adipokines such as leptin control the secretion and activation of MMPs, for instance, leptin encourages MMP-13 expression in glioma cells causing enhanced migration and tumor invasion [102]. Cancer cell and human adipose tissue stem cell are co-cultured to produce CCL5, which consequently increases breast cancer cell invasion associated with MMP-9 activity [103]. MMP-3 is known to be greatly involved in adipocyte differentiation in mammary gland during post-lactational involution. MMPs or TIMPs are expressed by fibroblastic adipogenic progenitor cells. Specific pre-adipocytes induce transcription of a number of MMP's and TIMP's mRNA (MMP-2, MMP-3, MMP-13, and MMP-14 and TIMP-1, TIMP-2, and TIMP-3) while fully developed adipocytes are involved in the expression of MMP-2 [104].

It has been observed that mice carrying a mutated MMP-3 or mammary glands of transgenic mice that overexpress MMPs inhibitors showed increased differentiation and hypertrophy of adipocytes [105]. Development of lipodystrophic null mice occurs due to aborted white adipose tissue development because of the absence of MMP-14.

MMP-14 also contributes to the coordination of adipocyte differentiation [106]. Recently secretome, a member of adipocyte, is found to improve MMP-2 activity [107].

Remodeling of extracellular milieu of adipogenesis and tumor microenvironments, where adipocytes are the key stromal components, is governed by Wdnm-1-like protein (a distant member of the whey acidic protein/four disulfide core [WAP/4-DSC] [108]. So far, only MMP-11 induced by adipose tissue is known to directly affect cancer progression. MMP-11 controls adipogenesis by decreasing pre-adipocyte differentiation and reverting adipocyte differentiation on tumor infiltrates the adjacent environment. Adipocyte dedifferentiation favors cancer cell survival and tumor progression due to collection of no-malignant peritumoral fibroblast-like cells at specific sites. It was speculated that this MMP-11-mediated crosstalk between infiltrating cancer cells and nearby adipocytes/pre-adipocytes emphasizes on its special role during tumor desmoplasia and signifies a molecular correlation between obesity and cancer [109].

5.5 MMPS and Neoplastic Progression

Loss of cell–cell adhesion and enhanced cellular mobility leads to cancer progression due to the combined effects of tumor invasion and epithelial–mesenchymal transition (EMT) programmers. Over expression of several MMPs, including MMP-3, MMP-7 and MMP-14 leads to carcinoma formation [6]. MMP-3 overexpression enhances EMT and induces genomic instability in cultured mammary epithelial cells leading to all stages of malignant transformation, neoplastic progression, and mammary carcinomas in transgenic mice [67, 110]. These things are related with the occurrence of an alternatives splice product of Rac1 that later on encourages the secretion of reactive oxygen species by mitochondria [68].

5.6 Tissue Invasion and Metastasis

The occurrence of metastasis is the resultant outcome of majority of cancers due to the infection of cancer cells from the primary tumor outgrowth to distant sites, i.e., secondary tumors (Fig. 5). The beginning of metastasis includes the invasion of the tumor into the outer tissue leading to intravasation of cancer cells into blood or lymphatic vessels from where they distribute into secondary organs. For a tumor cell to metastasize, four events are known to occur. They are:

(i) Dispassion from the primary tumor and later invasion into tissue stroma.
(ii) Entry to the blood vessel, such as the endothelial basement membrane or lymphatic system (intravasation) to cross other sites in the body.
(iii) Exudation from the circulation (extravasation).
(iv) Development to metastatic colonies. For distant setting it needs receptive sites called pre-metastatic niche.

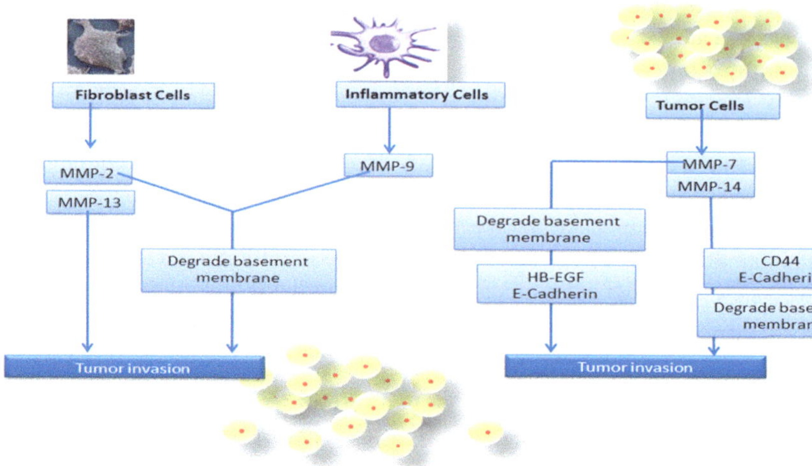

Fig. 5 Representative path and involvement of MMPs in cancer invasion and metastasis

This process has a complex relationship with inflammatory cells and hematopoietic progenitor cells [111]. In the presence of MMP-1, a special mechanism of MMP-mediated signal transduction is initiated which facilitates the detachment of primary tumor cells. In breast cancer model, MMP-1, extracted from tumor infiltrating fibroblasts, cleave PAR-1 which initiates cancer cell migration and invasion of tumor cells [112].

This demonstrates a prominent role of stroma-derived proteinases in progression of tumorigenesis, supported by particular signal transduction on cancer cells. One of the most common locations for metastasis is bone, often leading to mortality. Osteolysis and dissemination into bone tissue depend on MMPs expressed at the interface between tumor and stromal cells. In a rodent model of prostate cancer, osteoclasts expressed by MMP-7 at the tumor–bone interface twitches osteolysis and subsequently bone metastasis [113]. Close interaction between activated osteoclasts and osteoblasts permits binding of RANKL to its receptor RANK on osteoclast progenitors leading to osteoclast differentiation. Highly metastatic cell line, MDA-MB-231, of human breast cancer, was prepared by knocking down MMP-1 and ADAMTS-1 which showed similar effects like MMP-7 action [114]. Osteolysis and metastasis in bone are the results of activation of RANKL pathway, and this pathway is activated by ADAMTS-1 and MMP-1 which proteolytically engage EGF-like ligands. These studies support the idea of MMP-1 gene as being a piece of the multigenic program that mediates occurrence of bone metastasis in breast cancer cells [115].

Altogether, these proteinases can become potential therapeutic targets to prevent metastasis to the bone in breast or prostate cancer. Though MMPs play diverse roles in biological aspects yet these are known to play central role in degradation and remodeling of ECM creating a way through the marginal tissue for invasion and metastasis. Current studies by high-resolution multimodal microscopy have validated the significance of ECM remodeling by MMP-14-driven pericellular proteolysis, which competently prepares the tissue to facilitate single-cell and subsequently collective-cell migration and invasion [34].

A number of proteolytic ECM enzymes, like MMP-1, MMP-2, MMP-13, and MMP-14 and cathepsins B, K, and L have been involved in this process; but MMP-14 may be solely very important and rate-limiting in collagen turn over [34, 116]. It is known that metastatic cancer cells become protease-independent from protease-dependent invasion by using an ameboid migration mode [34]. However, it is not clear that whether the ameboid migration is only applicable under the in vitro conditions [116]. This advocates that MMPs produced by tumor-associated macrophages facilitate intravasation of cancer cells into the blood stream. Also MMPs derived from myeloid cells help in intra and extravasation of metastatic cancer cells.

5.7 MMPS and Metastasis Preparation

Metastatic cancer cells most likely have their preference for their distant locations, i.e., secondary sites. L most-preferred secondary sites for metastasis are lungs, liver, or bone. These migrating cells require some open condition, known as a metastatic specialty that is particularly suited for tumor cells at the far-off organ. During the development of this microenvironment, MMPs and other proteinases become active in the formation of a metastatic niche. Fibroblasts produce soluble factors which trigger the formation of a metastatic niche for secondary tumors [111]. These secretions, before disseminated tumor cells, become noticeable at distant organs; therefore the name of this process is given "pre-metastatic niche." Infiltration of VEGFR1-positive cells that confirm metastasis-supporting microenvironment occurs at these secondary metastatic sites due to the secretion of increased levels of fibronectin. According to recent studies, the factors, for example, fibronectin and so forth are discharged by the essential tumors that start the metastatic specialty development. Primary tumors secrete VEGFA, TGF-β, and tumor necrosis factor-α (TNF-α) that trigger the expression of S100 chemokines by lung endothelium that helps in myeloid cells movement to various sites [117]. The mechanisms and proper reasons for why above factor triggers chemokines expressions in specific tissues are not known. Another factor playing a significant role is serum amyloid A3(SAA3) which works as an upstream regulator of chemokine S100 throughout this process.

Infiltrating myeloid cell's mediate inflammatory response that initiates formation of pre-metastatic niche.

It is speculated that SAA3 prompts toll-like receptor4 (TLR4) signaling in penetrating myeloid cells leading to beginning of the nuclear factor-κ-B (NF-κB) pathway [118]. NF-κB pathway activates production of MMPs by specific cells like, stromal cells [119], which helps various cellular microenvironment alterations.

Undoubtedly, MMP-9 plays an unprecedented role in development of the metastatic niche [111], which is closely and most probably related with its stability to release VEGF and thus promotes angiogenesis [80]. MMP-9 discharges solvent Kit-ligand to enlist stem and progenitor cells from the bone marrow [120] which may have a specific significance in this unique situation; provided the niche-forming progenitor cells express c-Kit [111]. Hypoxic breast cancer cells secrete Lysyloxidase (LOX) which can play role in metastatic site creation due to crosslinking of collagen IV proteins in basement membrane structures. This leads to the employment of tumor cells to these locations [121], perhaps owing to changed tissues stiffness [122]. This infiltrating tumor cell exudes MMPs that destroy collagen fiber and discharge peptides that can pave the bone marrow driven cells and metastasizing tumor cells to these destinations. Simultaneous production of MMP-3 and MMP-10 is upregulated altogether with angiogenic modulator angiopoietin-2 in prematastatic tissue even before myeloid cells are utilized to these destinations [123]. Silencing by in vivo, RNA impedance of MMP-3, MMP-10 and angiopoietin-2 diminish vascular penetrability and penetration of myeloid cells in lungs. It creates the impression that these three variables synergize in the destabilization of the pneumonic vasculature and in this way advancing metastasis. These outcomes demonstrate that essential part of extracellular proteolysis in pre-metastatic specialty arrangement. Though thorough studies are required to explain these extremely important pathways modulated by these proteinases yet it appears that the action of MMPs in this example critically involves the inflection of inflammatory responses.

5.8 MMPs and Inflammation

Recent findings suggest that MMPs are the chief controllers of innate and acquired immunity. As speculated, inflammatory responses are one of the contributing factors to cancer. Numerous chronic inflammatory diseases (like pancreatitis and Crohn's disease) are known to increase chances of occurrence of cancer [124]. In the consequent portion, we will explain the part of MMPs and their capacities and how MMPs adjust the capacity of cytokines and chemokines and what impacts this insusceptible administrative capacity may have on tumor microenvironment. Knockout mice experimental model gives strong evidence that MMPs play a crucial role in acute and chronic inflammation [5]. TNF-α is the most studied genius provocative cytokine communicated as layer bound precursor(pro-TNF-α) on macrophages and T cells. Translation of pro-TNF-α to TNF-α requires proteolytic cleavage by ADAM-17/TACE (TNF converting enzyme), or by MMPs including MMP-1, MMP-2, MMP-3, MMP-9, MMP-12, MMP-14, MMP-15, and MMP-17 [125]. It is found that in some special circumstances, MMP-7 can act as TNF-α-generating convertase instead of ADAM-17 which is known to play a major role in generation of above convertase [126]. In many tumors TNF-α is upregulated many folds and it advances growth cell survival in NF-κB-subordinate way [127].

These suggest that generation of functional TNF-α by MMPs and ADAM-17 is an essential venture to begin tumor-advancing course. A significant number of individuals from the CCL/monocyte chemoattractant protein (MCP) group of chemokines are severed by MMPs, which especially abandon them into non-activating receptor antagonists with inflammation-dampening effects [128]. For example, CCL8/MCP-2 is processed by MMP-1 and MMP-3. Undoubtedly, the proteolytic cleavage of CCL8 can hinder the anti-tumor ability of this chemokine in a melanoma model [129]. MMP-8, MMP-9, and MMP-12 alter the bioactivity of CXCL11/I-TAC, a strong Th1 lymphocyte-pulling in chemokine [130]. This literature reveals that proteolytic cleavage of a chemokine can probably have a strong impact in a clinically significant location of tumor development. It is found out after vigorous studies that MMP-mediated C-terminal cleavage abolishes the antagonist function and removes heparin-binding capacity of CXCL11 while N-terminal truncation of CXCL11 leads to inactivation of the chemokine. This creates a potent receptor antagonist, thereby solubilizing the chemokine from the ECM. These studies reveal the importance of myeloid cell resulting MMPs in the direction of T cell reactions, due to which they may have significant consequences on the adaptive anti-tumor immune response. Expression of CXCR7, biological receptor of CXCL11, on several tumor cells can generate signals for growth as well as survival [131]. Proteolytic fragments of MMPs cleaved ECM components exert chemotactic properties; similarly, macrophage elastase MMP-12 produces neutrophil-attracting peptides by degrading elastin [132]. A fragment N-acetyl Pro-Gly-Pro (PGP) is generated from ECM which stops working during lungs inflammation. This tripeptide shows demonstrate chemotactic action through enactment of CXC chemokine receptors on neutrophilic granulocytes [133]. Another factor known as chemotactic PGP, generated by MMP-8, manages neutrophil enrollment at inflammation site [134, 135]. Also MMP-8 secreted by neutrophils plays a vital and phenomenal tumor stifling part in a mouse model of cancer-causing agent initiated skin disease [136]. MMP-8 deficient mice have a tendency to develop chronic inflammation and therefore, the loss of MMP-8 contributes mechanistically to expanded weakness of skin adenocarcinoma and melanoma arrangement in people [137]. It is found out that MMP-8 expression in tumor cells tightens their linkage to the ECM, and therefore it may lead to the repression of metastatic activities of tumor cells [138]. Neutrophils penetration in tumor-miniaturized scale condition as a rule connects with poor anticipation [139]. Like other inflammatory cells, neutrophils have one of a kind capacity of detecting the focus slope of chemokines of CXCL1/KC, a homolog of CXCL8 that is known to shape buildings with the heparan sulfate proteoglycan syndecan-1 on interstitial cell surfaces. MMP-7 continuously changes the CXCL1 bioactivity by cutting syndecan-1 from the cell exterior, consequently discharging chemotactic complexes of syndecan-1 and CXCL1 [140]. CXXL1 syndecan-1 forms focus inclination of dissolvable chemotactic complex in ECM. Consequently, MMPs coordinate the employment of leukocytes as a vital segment of tumor-related irritations.

6 Non-proteolytic Function of MMPs

Nowadays, the most evolving zone of attention is the non-synergist capacity of MMPs. It is not surprising to realize that about portion of all ADAMs indicates proteolytic capacity, whereas other half capacities in non-proteolytic way [141]. The structural domain of MMPs, hemopexin, is very imperative for their non-proteolytic usefulness. TIMP-1 and TIMP-2 bind on hemopexin domain of MMPs and perform hemopexin-mediated function. MMP-2 activation needs [142] TIMP-2 which is attached with single molecule of MMP-14 through its synergist area and furthermore is associated with pro-MMP-2 through its hemopexin sphere. At last, a moment atom of MMP-14 then chemically actuates MMP-2 [143]. During immune or cancer cell migration, numerous member of MMPs family take part to mediate this migration. Recent literatures suggest that they intervene chemotaxis even without utilizing their proteolytic area. Inactive precursors of MMP-2 and MMP-9 increase cell movement in a trans-well chamber assay. MMP-9 hemopexin domain is required for MMP-9 interceded epithelial cell movement in this measure [144] MMP-9. Upregulation of MAP kinase and PI3 kinase pathways is speculated to be included in this non-proteolytic capacity of MMP-9. MMP-14's cytoplasmic tail performs a relocation advancing capacity on macrophages. When this cytoplasmic tail is removed, it impairs the movement of macrophages amid in vitro relocation through Matrigel [145]. Recent analysis done by using the technique of hereditary adjustment of one of the two MMP qualities of Drosophila melanogaster revealed that in spite of the fact that the synergist area is required for all MMP capacities, the hemopexin space is extraordinarily required in tissue intrusion amid transformation; however, not for tracheal redesigning [146]. Disfunctioning of MMP-12 in mice prompt flawed finish bacterial clearance and upgraded mortality when contaminated with gram-negative and gram-positive microorganisms. One of a kind four amino corrosive grouping inside the hemopexin-like C-terminal space of MMP-12 shows antimicrobial properties. A few MMPs may interface with other extracellular parts without instigating proteolytic cleavage. MMP-14 associates with C1q segment of supplement framework in a ligand-receptor way without making C1qr and C1qs proteinase action. It is proposed that this connection counteract initiation of supplement proteinase course [147]. The exact purpose of binding of MMP-14 and C1q is still not clear and also how it helps in tumorigenesis is yet to be elucidated. In ECM, MMPs bind with integrin family proteins on cell surface receptors. Pro-MMP-1 interacts with $\alpha 2\beta 1$ integrin in a non-proteolytic manner on neurons [147]. Reduced AKT dephosphorylation after MMP-1 incubation, recommends that integrin binding, rather than proteinase activity, is significant for MMP-1-mediated cytotoxicity. MMP-9 is known to promote B cell survival, in chronic lymphocytic leukemia, in a non-proteolytic fashion via its hemopexin domain by getting attached to the surface receptors $\alpha 4\beta 1$ and CD44v which actuate intracellular flagging including Lyn enactment and STAT3 phosphorylation that forestalls B cell apoptosis [148]. Consolidating the facts, these experimental data are just providing a suitable start to understand the non-proteolytic functions of MMPs. More elucidations and revelations of facts are required to clarify this principle and to assess its role under in vivo conditions. However, it is tempting to find out that non-proteolytic functions of MMPs, may be, were the main reasons behind the failure of early clinical trials using inhibitors of the MMP catalytic domains as anticancer therapeutics.

7 Conclusion and Future Prospective

Organization of MMPs with numerous cancers makes them as potent therapeutic target during the treatment of cancer. The idea of obstructing of MMP-interceded angiogenesis and metastasis laid the reason for advancement of first medication, and additionally, various little atom metalloproteinase inhibitor (MPI) sedates in stage III clinical trials. MPIs in these trials did not prove to be potential drugs as they failed to treat cancer patients. Many reasons were found out, and it was speculated that MPIs are, may be, just restricted to be beneficial in early stages of cancer. MPIs were used as anticancer drugs because of their ability to truncate interstitial relocation of metastatic cancer cells. Later on, it was found that these cancers can change to an ameboid-like protease-independent movement mode by shaping actin-rich bulges [34]. Thus, MPIs lost their ability to inhibit the transitory conduct of metastatic tumor cells. But the fact remains hidden regarding the migration mode whether that is actually relevant for tumor cell relocation under in vivo situation in the nearness of a normally crosslinked collagen lattice.

According to a study supported by strong evidences, MMP-14 plays a major role in the migration and invasion of metastatic tumor cells; hence, MMP-14 is a promise target for tumor cells [116]. These all findings proved that researchers can use MPIs as effective anti-invasive drug that can inhibit MMP-14. Additionally, MMPs are likewise known to meddle with apoptosis acceptance, particularly after chemotherapy, as they break Fas ligand from the surface of disease cells as appeared for MMP-7 [70]. In experimental trials, MPIs were given to patients who were at the last stage of cancer. This was done to check whether even at the last stage these MPIs can actually help in survival of patient. Due to such studies drugs like Bevacizumab (Avastin, anti-VEGF monoclonal antibody) were developed. These were known to interfere with the tumor vasculature, thus helped to inhibit tumor growth. These drugs were approved by FDA to use in combination with chemotherapy for the treatment of various metastatic cancers. The recent studies on MMP-9 suggest that it plays a dominant role in angiogenesis switch. MMP-9 also helps in the regulation of bioavailability of VEGF tumors, thus unravels the beneficial effect of MPIs on tumor. A few reviews likewise propose that MMP-9 additionally fills in as a powerful ECM fragment generator like tumstatin which goes about as a suppressor of tumor vasculature arrangement. This prompts expanded tumor development in MMP-9-insufficient mice [92]. This leads to the discovery of the fact that one MMP can have contradicting effects in different tumor types. This highlights the necessity for the use of specific type of MPIs with proper consideration and detailed examined assessment of various types of cancers. The preliminary study suggested that the major sources of MMPs are non-malignant stromal cells. These studies were based on the usage of cancer cell lines that could overexpress certain members of the MMPs family. But these studies were not applicable under in vivo conditions. In fact, the cellular source of each MMP is given due importance as the activity of the released enzyme varies substantially between cell types. When assessing the expression patterns of MMPs in cancer types, these cellular sources of each MMP are taken into consideration to select which type of MPI should be used for the treatment. In this manner, the part of MMPs has stretched out from being powerful proteinases that debase ECM to particular modulators of angiogenesis as well as fine-tuners of cell flagging pathways and the incendiary reaction. Discovery of several regulatory effects of MMPs in the microenvironment of stromal cells has been proved to be the major breakthroughs in MMPs research. MMPs influence adipocyte work, which is likely to be involved in fat-rich tumor destinations, for example, breast. The Inflammatory response is directed in different ways subsequently the enrollment of incendiary cells is encouraged prompting the modification of the capacity of chemokines and the bioavailability of imperative pro-inflammatory cytokines. Talking about a settled connection among irritation and malignancy [124], it was hypothesized that intercession with MMP-interceded immunoregulatory capacities could wind up plainly worthwhile for disease patients. For example, the movement of a few sorts of tumor is contributed by TNF-alpha factor [148]. Inhibition of TNF-α activation using MPIs may help in reducing the intensity of inflammatory response in the tumor microenvironment. The generation of the pre-metastatic niche helps us to know the effects of MMPs. MMP-2, MMP-3, and MMP-9 are well-known because of their capacity to add to the foundation of metastasis-inclined destinations at tumor far-off organs [121]. These analyses help us to understand that MPIs can be used in many cancers at preliminary stages, i.e., prior to the full initiation of tumor-associated inflammation and before the infection reaches to the distant organs. The failure of various broad-spectrum MPIs as anticancer drugs is attributed to the various tumor-suppressing functions of these MMPs [149] and for the link between MMP-8, loss-of-function mutations and melanoma in humans. Some MMPs are known to carry out biological functions other than proteolytic cleavage as they specifically bind to certain target molecules, for instance via their hemopexin domain. Small molecule MMP inhibitors are not as effective as the bigger molecules of MMP as they are certainly ineffective to interfere with a non-proteolytic role of MMPs.

The ultimate goal of many future researchers is that they can really prove to be a breakthrough in MMPs studies, which is the advance in active site-directed inhibitors or antibodies those were precise for single MMPs without any interaction with other MMPs. The relocation and incursion of endothelial cells in fibrin and collagen gels can be inhibited by monoclonal antibodies which are raised against the catalytic domain of MMP-14. These antibodies are speculated to specifically site those functional non-catalytic domains of MMPs. Cytotoxic agents such as anthrax toxin can be activated with the help of MMP to target the tumor vasculature. The validation of these agents is done by testing them in MMP-deficient animals. With the help of specific probe designed for MMPs and their activity can view by these imaging probes. Due to the usage of probes along with minimal invasive imaging techniques will help us to visualize MMP function during progression of tumors.

It should be noted that these imaging techniques are very useful as they help us to understand the tumor progression at all instants of time. The cell signaling pathways which are responsible for the control of cell growth and cell fat gets deregulated due to the accumulation of multiple independent mutations which lead to the development of oncogenic state. Defining the cancer subtypes, recurrence of disease and response to specific therapies has become possible now due to DNA microarray-based gene expression techniques. Various gene expression profiles for the analysis of oncogenic pathways are also used nowadays. Here, expression level of genes reflects the activation status of several oncogenic pathways. When a large number of human cancers are studied, these gene expression signatures can help in identifying patterns of pathway deregulation occurring in tumors and also relevant associations with disease outcomes can also be found. When these individual gene expression signatures are combined across several pathways then the coordinated patterns of pathway deregulation that distinguish between specific cancers and tumor subtypes can be obtained. Combining tumors based on pathway signatures further define prognosis in respective patient subsets that indicate the patterns of oncogenic pathway deregulation underlying the development of the oncogenic phenotype which further reflect the biology and outcome of specific cancers. This associated knowledge of pathway deregulation and therapeutic action of therapeutic agents helps in providing an opportunity to make use of these oncogenic pathway signatures to guide the use of targeted therapeutics.

References

1. Hanahan D, Weinberg RA (2000) The hallmarks of cancer. Cell 100:57–70
2. Coussens LM, Fingleton B, Matrisian LM (2002) Matrix metalloproteinase inhibitors and cancer: trials and tribulations. Science 295:2387–2392
3. Gross J, Lapiere CM (1962) Collagenolytic activity in amphibian tissues: a tissue culture assay. Proc Natl Acad Sci USA 48:1014–1022
4. Page-McCaw A, Ewald AJ, Werb Z (2007) Matrix metalloproteinases and the regulation of tissue remodelling. Nat Rev Mol Cell Biol 8:221–233
5. Parks WC, Wilson CL, Lopez-Boado YS (2004) Matrix metalloproteinases as modulators of inflammation and innate immunity. Nat Rev Immunol 4:617–629
6. Egeblad M, Werb Z (2002) New functions for the matrix metalloproteinases in cancer progression. Nat Rev Cancer 2:161–174
7. Hideaki N (1998) Cell surface activation of progelatinaseA (proMMP-2) and cell migration. Cell Res 8:179–186
8. Sternlicht MD, Werb Z (2001) How matrix metalloproteinases regulate cell behavior. Annu Rev Cell Dev Biol 17:463–516
9. Edwards DR, Handsley MM, Pennington CJ (2008) The ADAM metalloproteinases. Mol Aspects Med 29:258–289

10. Tong W, Ye F, He L et al (2016) Serum biomarker panels for diagnosis of gastric cancer. Onco Targets Ther 26(9):2455–2463
11. Kelwick R, Desanlis I, Wheeler GN, Edwards DR (2015) The ADAMTS (a disintegrin and metalloproteinase with thrombospondin motifs) family. Genome Biol 16:113
12. Kelwick R, Wagstaff L, Decock J et al (2015) Metalloproteinase-dependent and -independent processes contribute to inhibition of breast cancer cell migration, angiogenesis and liver metastasis by a disintegrin and metalloproteinase with thrombospondin motifs-15. Int J Cancer 136(4):E14–E26
13. Arpino V, Brock M, Gill SE (2015) The role of TIMPs in regulation of extracellular matrix proteolysis. Matrix Biol 44–46:247–254
14. Wang S, Wei X, Zhou J et al (2014) Identification of α2-macroglobulin as a master inhibitor of cartilage-degrading factors that attenuates the progression of posttraumatic osteoarthritis. Arthritis Rheumatol 66(7):1843–1853
15. Misra UK, Pizzo SV (2015) Activated α2-macroglobulin binding to human prostate cancer cells triggers insulin-like responses. J Biol Chem 290(15):9571–9587
16. Deryugina EI, Quigley JP (2006) Matrix metalloproteinases and tumor metastasis. Cancer Metastasis Rev 25:9–34
17. Duan JX, Rapti M, Tsigkou A, Lee MH (2015) Expanding the Activity of tissue inhibitors of metalloproteinase (TIMP)-1 against surface-anchored metalloproteinases by the replacement of its C-terminal domain: implications for anti-cancer effects. PLoS ONE 10(8):e0136384
18. Asadzadeh R, Khosravi S, Zavareh S, Ghorbanian MT, Paylakhi SH, Mohebbi SR (2008) Vitrification affects the expression of matrix metalloproteinases and their tissue inhibitors of mouse ovarian tissue. Int J Reprod Biomed (Yazd) 14(3):173–180
19. Murphy G (2008) The ADAMs: signalling scissors in the tumour microenvironment. Nat Rev Cancer 8:932–941
20. Egeblad M, Ewald AJ, Askautrud HA et al (2008) Visualizing stromal cell dynamics in different tumor microenvironments by spinning disk confocal microscopy. Dis Model Mech 1:155–167
21. Kessenbrock K, Plaks V, Werb Z (2010) Matrix metalloproteinases: regulators of the tumor microenvironment. Cell 141(1):52–67. doi:10.1016/j.cell.2010.03.015
22. Kessenbrock K, Krumbholz M, Schonermarck U et al (2009) Netting neutrophils in autoimmune small-vessel vasculitis. Nat Med 15:623–625
23. Peres R, Furuya H, Pagano I, Shimizu Y, Hokutan K, Rosser CJ (2016) Angiogenin contributes to bladder cancer tumorigenesis by DNMT3b-mediated MMP2 activation. Oncotarget. doi:10.18632/oncotarget.10097
24. Basagiannis D, Christoforidis S (2016) Constitutive endocytosis of VEGFR2 protects the receptor against shedding. J Biol Chem. pii:jbc.M116.730309
25. Weiss A, Joerss H, Brockmeyer J (2014) Structural and functional characterization of cleavage and inactivation of human serine protease inhibitors by the bacterial SPATE protease EspPα from enterohemorrhagic *E. coli*. PLoS ONE 9(10):e111363
26. Grinnell F, Zhu M (1996) Fibronectin degradation in chronic wounds depends on the relative levels of elastase, alpha1-proteinase inhibitor, and alpha2-macroglobulin. J Invest Dermatol 106(2):335–341
27. Liu Z, Zhou X, Shapiro SD et al (2000) The serpin alpha1-proteinase inhibitor is a critical substrate for gelatinase B/MMP-9 in vivo. Cell 102(5):647–655
28. Lian S, Xia Y, Khoi PN et al (2015) Cadmium induces matrix metalloproteinase-9 expression via ROS-dependent EGFR, NF-κB, and AP-1 pathways in human endothelial cells. Toxicology 338:104–116. doi:10.1016/j.tox.2015.10.008
29. Weiss SJ, Peppin G, Ortiz X, Ragsdale C, Test ST (1985) Oxidative autoactivation of latent collagenase by human neutrophils. Science 227:747–749
30. Fu X, Kassim SY, Parks WC, Heinecke JW (2003) Hypochlorous acid generated by myeloperoxidase modifies adjacent tryptophan and glycine residues in the catalytic domain

of matrix metalloproteinase-7 (matrilysin): an oxidative mechanism for restraining proteolytic activity during inflammation. J Biol Chem 278:28403–28409
31. Nagase H, Visse R, Murphy G (2006) Structure and function of matrix metalloproteinases and TIMPs. Cardiovasc Res 69:562–573
32. Rupp PA, Visconti RP, Czirok A, Cheresh DA, Little CD (2008) Matrix metalloproteinase 2-integrin alpha(v)beta3 binding is required for mesenchymal cell invasive activity but not epithelial locomotion: a computational time-lapse study. Mol Biol Cell 19:5529–5540
33. Sabeh F, Ota I, Holmbeck K et al (2004) Tumor cell traffic through the extracellular matrix is controlled by the membrane-anchored collagenase MT1-MMP. J Cell Biol 167:769–781
34. Wolf K, Wu YI, Liu Y, Geiger J, Tam E, Overall C, Stack MS, Friedl P (2007) Multi-step pericellular proteolysis controls the transition from individual to collective cancer cell invasion. Nat Cell Biol 9:893–904
35. Friedl P, Wolf K (2008) Tube travel: the role of proteases in individual and collective cancer cell invasion. Cancer Res 68:7247–7249
36. Brinkmann V, Reichard U, Goosmann C et al (2004) Neutrophil extracellular traps kill bacteria. Science 303:1532–1535
37. Carmona-Rivera C, Zhao W, Yalavarthi S, Kaplan MJ (2015) Neutrophil extracellular traps induce endothelial dysfunction in systemic lupus erythematosus through the activation of matrix metalloproteinase-2. Ann Rheum Dis 74(7):1417–1424
38. Boone BA, Orlichenko L, Schapiro NE et al (2015) The receptor for advanced glycation end products (RAGE) enhances autophagy and neutrophil extracellular traps in pancreatic cancer. Cancer Gene Ther 22(6):326–334
39. Cedervall J, Zhang Y, Huang H, Zhang L, Femel J, Dimberg A et al (2015) Neutrophil extracellular traps accumulate in peripheral blood vessels and compromise organ function in tumor-bearing animals. Cancer Res 75(13):2653–2662
40. Butcher DT, Alliston T, Weaver VM (2009) A tense situation: forcing tumour progression. Nat Rev Cancer 9:108–122
41. Kopanska KS, Alcheikh Y, Staneva R, Vignjevic D, Betz T (2016) Tensile forces originating from cancer spheroids facilitate tumor invasion. PLoS ONE 11(6):e0156442
42. Jerrell RJ, Parekh A (2016) Matrix rigidity differentially regulates invadopodia activity through ROCK1 and ROCK2. Biomaterials 84:119–129
43. Zhang X, Halvorsen K, Zhang CZ, Wong WP, Springer TA (2009) Mechanoenzymatic cleavage of the ultralarge vascular protein von Willebrand factor. Science 324:1330–1334
44. Scherer RL, McIntyre JO, Matrisian LM (2008) Imaging matrix metalloproteinases in cancer. Cancer Metastasis Rev 27:679–690
45. Littlepage LE, Sternlicht MD, Rougier N et al (2010) Matrix metalloproteinases contribute distinct roles in neuroendocrine prostate carcinogenesis, metastasis, and angiogenesis progression. Cancer Res 70:2224–2234
46. Bremer C, Tung CH, Weissleder R (2001) In vivo molecular target assessment of matrix metalloproteinase inhibition. Nat Med 7:743–748
47. Furumoto S, Takashima K, Kubota K, Ido T, Iwata R, Fukuda H (2003) Tumor detection using 18F-labeled matrix metalloproteinase-2 inhibitor. Nucl Med Biol 30:119–125
48. Schafers M, Riemann B, Kopka K, Breyholz HJ, Wagner S, Schafers KP, Law MP, Schober O, Levkau B (2004) Scintigraphic imaging of matrix metalloproteinase activity in the arterial wall in vivo. Circulation 109:2554–2559
49. Temma T, Sano K, Kuge Y, Kamihashi J, Takai N, Ogawa Y et al (2009) Development of a radiolabeled probe for detecting membrane type-1 matrix metalloproteinase on malignant tumors. Biol Pharm Bull 32:1272–1277
50. Jastrzebska B, Lebel R, Therriault H et al (2009) New enzyme-activated solubility-switchable contrast agent for magnetic resonance imaging: from synthesis to in vivo imaging. J Med Chem 52:1576–1581
51. Olson ES, Aguilera AT, Jiang T et al (2009) In vivo characterization of activatable cell penetrating pepides for targeting protease activity in cancer. Integr Biol 1:382–393

52. Cho H, Bhatti FU, Yoon TW, Hasty KA, Stuart JM, Yi AK (2016) Non-invasive dual fluorescence in vivo imaging for detection of macrophage infiltration and matrix metalloproteinase (MMP) activity in inflammatory arthritic joints. Biomed Opt Express 7 (5):1842–1852
53. Scherer RL, VanSaun MN, McIntyre JO, Matrisian LM (2008) Optical imaging of matrix metalloproteinase-7 activity in vivo using a proteolytic nanobeacon. Mol Imaging 7(3):118–131
54. Hugenberg V, Hermann S, Galla F et al (2016) Radiolabeled hydroxamate-based matrix metalloproteinase inhibitors: how chemical modifications affect pharmacokinetics and metabolic stability. Nucl Med Biol 43(7):424–437
55. Liotta LA, Tryggvason K, Garbisa S, Hart I, Foltz CM, Shafie S (1980) Metastatic potential correlates with enzymatic degradation of basement membrane collagen. Nature 284:67–68
56. Massague J (2008) TGFbeta in Cancer. Cell 134:215–230
57. Lei R, Zhang K, Liu K et al (2016) Transferrin receptor facilitates TGF-β and BMP signaling activation to control craniofacial morphogenesis. Cell Death Dis 7(6):e2282
58. Yu Q, Stamenkovic I (2000) Cell surface-localized matrix metalloproteinase-9 proteolytically activates TGF-beta and promotes tumor invasion and angiogenesis. Genes Dev 14:163–176
59. Mu D, Cambier S, Fjellbirkeland L et al (2002) The integrin alpha(v)beta8 mediates epithelial homeostasis through MT1-MMP-dependent activation of TGF-beta1. J Cell Biol 157:493–507
60. Dallas SL, Rosser JL, Mundy GR, Bonewald LF (2002) Proteolysis of latent transforming growth factor-beta (TGF-beta)-binding protein-1 by osteoclasts. A cellular mechanism for release of TGF-beta from bone matrix. J Biol Chem 277:21352–21360
61. Tatti O, Vehvilainen P, Lehti K, Keski-Oja J (2008) MT1-MMP releases latent TGF-beta1 from endothelial cell extracellular matrix via proteolytic processing of LTBP-1. Exp Cell Res 314:2501–2514
62. De Vlaeminck Y, González-Rascón A, Goyvaerts C, Breckpot K (2016) Cancer-associated myeloid regulatory cells. Front Immunol 7:113
63. Hynes NE, Lane HA (2005) ERBB receptors and cancer: the complexity of targeted inhibitors. Nat Rev Cancer 5:341–354
64. Wang ZQ, Faddaoui A, Bachvarova M et al (2015) BCAT1 expression associates with ovarian cancer progression: possible implications in altered disease metabolism. Oncotarget 6(31):31522–31543
65. Cowden Dahl KD, Symowicz J, Ning Y et al (2008) Matrix metalloproteinase 9 is a mediator of epidermal growth factor-dependent e-cadherin loss in ovarian carcinoma cells. Cancer Res 68:4606–4613
66. Maretzky T, Reiss K, Ludwig A et al (2005) ADAM10 mediates Ecadherin shedding and regulates epithelial cell-cell adhesion, migration, and beta-catenin translocation. Proc Natl Acad Sci USA 102:9182–9187
67. Lochter A, Galosy S, Muschler J, Freedman N, Werb Z, Bissell MJ (1997) Matrix metalloproteinase stromelysin-1 triggers a cascade of molecular alterations that leads to stable epithelial-to-mesenchymal conversion and a premalignant phenotype in mammary epithelial cells. J Cell Biol 139:1861–1872
68. Radisky DC, Levy DD, Littlepage LE et al (2005) Rac1b and reactive oxygen species mediate MMP-3-induced EMT and genomic instability. Nature 436:123–127
69. Witters L, Scherle P, Friedman S et al (2008) Synergistic inhibition with a dual epidermal growth factor receptor/HER-2/neu tyrosine kinase inhibitor and a disintegrin and metalloprotease inhibitor. Cancer Res 68:7083–7089
70. Mitsiades N, Yu WH, Poulaki V, Tsokos M, Stamenkovic I (2001) Matrix metalloproteinase-7-mediated cleavage of Fas ligand protects tumor cells from chemotherapeutic drug cytotoxicity. Cancer Res 61:577–581

71. Khamis Zahraa I, Man YG et al (2016) Evidence for a proapoptotic role of matrix metalloproteinase-26 in human prostate cancer cells and tissues. J Cancer 7(1):80
72. Liu H, Zhang T, Li X et al (2008) Predictive value of MMP-7 expression for response to chemotherapy and survival in patients with non-small cell lung cancer. Cancer Sci 99:2185–2192
73. Schulte M, Reiss K, Lettau M, Maretzky T, Ludwig A, Hartmann D, De Strooper B, Janssen O, Saftig P (2007) ADAM10 regulates FasL cell surface expression and modulates FasL-induced cytotoxicity and activation induced cell death. Cell Death Differ 14:1040–1049
74. Crawford HC, Scoggins CR, Washington MK, Matrisian LM, Leach SD (2002) Matrix metalloproteinase-7 is expressed by pancreatic cancer precursors and regulates acinar-to-ductal metaplasia in exocrine pancreas. J Clin Invest 109:1437–1444
75. Liao Jung Chun, You BJ et al (2015) Raf/ERK/Nrf2 signaling pathway and MMP-7 expression involvement in the trigonelline-mediated inhibition of hepatocarcinoma cell migration. Food Nutr Res 59:29884. doi:10.3402/fnr.v59.29884
76. Waldhauer I, Goehlsdorf D, Gieseke F et al (2008) Tumorassociated MICA is shed by ADAM proteases. Cancer Res 68:6368–6376
77. Le MauxChansac B, Misse D, Richon C et al (2008) Potentiation of NK cell-mediated cytotoxicity in human lung adenocarcinoma: role of NKG2D-dependent pathway. Int Immunol 20:801–810
78. Padera TP, Kadambi A, Di Tomaso E et al (2002) Lymphatic metastasis in the absence of functional intratumorlymphatics. Science 296:1883–1886
79. Abdelfattah NS, Amgad M, Zayed AA, Hussein H, Abd El-Baky N (2016) Molecular underpinnings of corneal angiogenesis: advances over the past decade. Int J Ophthalmol 9 (5):768–779
80. Bergers G, Brekken R, McMahon G et al (2000) Matrix metalloproteinase-9 triggers the angiogenic switch during carcinogenesis. Nat Cell Biol 2(10):737–744
81. Asanuma K, Yoshikawa T, Yoshida K et al (2016) Argatroban more effectively inhibits the thrombin activity in synovial fluid than naturally occurring thrombin inhibitors. Cell Mol Biol 62(6):27–32
82. Han YH, Gao B, Huang JH et al (2015) Expression of CD147, PCNA, VEGF, MMPs and their clinical significance in the giant cell tumor of bones. Int J Clin Exp Pathol 8(7):8446–8452
83. Du R, Lu KV, Petritsch C et al (2008) HIF1alpha induces the recruitment of bone marrow-derived vascular modulatory cells to regulate tumor angiogenesis and invasion. Cancer Cell 13:206–220
84. Ahn GO, Brown JM (2008) Matrix metalloproteinase-9 is required for tumor vasculogenesis but not for angiogenesis: role of bone marrow-derived myelomonocytic cells. Cancer Cell 13:193–205
85. Ardi VC, Kupriyanova TA, Deryugina EI, Quigley JP (2007) Human neutrophils uniquely release TIMP-free MMP-9 to provide a potent catalytic stimulator of angiogenesis. Proc Natl Acad Sci USA 104:20262–20267
86. Ardi VC, Van den Steen PE, Opdenakker G, Schweighofer B, Deryugina EI, Quigley JP (2009) Neutrophil MMP-9 Proenzyme, unencumbered by TIMP-1, undergoes efficient activation in vivo and catalytically induces an giogenesis via a basic fibroblast growth factor (FGF-2)/FGFR-2 pathway. J Biol Chem 284:25854–25866
87. Mentzel T, Brown LF, Dvorak HF et al (2001) The association between tumour progression and vascularity in myxofibrosarcoma and myxoid/round cell liposarcoma. Virchows Arch 438:13–22
88. Nozawa H, Chiu C, Hanahan D (2006) Infiltrating neutrophils mediate the initial angiogenic switch in a mouse model of multistage carcinogenesis. Proc Natl Acad Sci USA 103:12493–12498

89. Ribatti D (2009) Endogenous inhibitors of angiogenesis: a historical review. Leuk Res 33:638–644
90. Heljasvaara R, Nyberg P, Luostarinen J et al (2005) Generation of biologically active endostatin fragments from human collagen XVIII by distinct matrix metalloproteases. Exp Cell Res 307:292–304
91. Behl Tapan, Kotwani A (2015) Possible role of endostatin in the antiangiogenic therapy of diabetic retinopathy. Life Sci 135:131–137
92. Hamano Y, Zeisberg M, Sugimoto H et al (2003) Physiological levels of tumstatin, a fragment of collagen IV alpha3 chain, are generated by MMP-9 proteolysis and suppress angiogenesis via alphaV beta3 integrin. Cancer Cell 3:589–601
93. Cornelius LA, Nehring LC, Harding E et al (1998) Matrix metalloproteinases generate angiostatin: effects on neovascularization. J Immunol 161:6845–6852
94. Houghton AM, Grisolano JL, Baumann ML et al (2006) Macrophage elastase (matrix metalloproteinase-12) suppresses growth of lung metastases. Cancer Res 66:6149–6155
95. Sounni NE, Dehne K, van Kempen L, Egeblad M, Affara NI, Cuevas I, Wiesen J, Junankar S, Korets L, Lee J et al (2010) Stromal regulation of vessel stability by MMP14 and TGFbeta. Dis Model Mech 6:317–332
96. Nakamura ES, Koizumi K, Kobayashi M, Saiki I (2004) Inhibition of lymphangiogenesis-related properties of murine lymphatic endothelial cells and lymph node metastasis of lung cancer by the matrix metalloproteinase inhibitor MMI270. Cancer Sci 95:25–31
97. Bruyere F, Melen-Lamalle L, Blacher S et al (2008) Modeling lymphangiogenesis in a three-dimensional culture system. Nat Methods 5:431–437
98. Langenskiold M, Holmdahl L, Falk P, Ivarsson ML (2005) Increased plasma MMP-2 protein expression in lymph node-positive patients with colorectal cancer. Int J Colorectal Dis 20:245–252
99. Islekel H, Oktay G, Terzi C, Canda AE, Fuzun M, Kupelioglu A (2007) Matrix metalloproteinase-9,-3 and tissue inhibitor of matrix metalloproteinase-1 in colorectal cancer: relationship to clinicopathological variables. Cell Biochem Funct 25:433–441
100. Rutkowski JM, Davis KE, Scherer PE (2009) Mechanisms of obesity and related pathologies: the macro- and microcirculation of adipose tissue. FEBS J 276:5738–5746
101. Tsiklauri L, Werner J, Frommer KW, Müller-Ladner U, Wenisch S, Neumann E (2016) A4. 09 Adipokines affect differentiation of osteoarthritis and osteoporosis spongiosa-derivedmesenchymal stromal cells. Rheum Dis (Suppl 1): A40–A40
102. Yeh WL, Lu DY, Lee MJ, Fu WM (2009) Leptin induces migration and invasion of glioma cells through MMP-13 production. Glia 57:454–464
103. Pinilla S, Alt E, Abdul Khalek FJ et al (2009) Tissue resident stem cells produce CCL5 under the influence of cancer cells and thereby promote breast cancer cell invasion. Cancer Lett 284:80–85
104. Nakayama A, Aoki S, Uchihashi K et al (2016) Interaction between esophageal squamous cell carcinoma and adipose tissue in vitro. Am J Pathol 186(5):1180–1194
105. Alexander CM, Selvarajan S, Mudgett J, Werb Z (2001) Stromelysin-1 regulates adipogenesis during mammary gland involution. J Cell Biol 152:693–703
106. Chun TH, Hotary KB, Sabeh F, Saltiel AR, Allen ED, Weiss SJ (2006) A pericellular collagenase directs the 3-dimensional development of white adipose tissue. Cell 125: 577–591
107. Peters-Hall Jennifer R, Brown KJ et al (2015) Quantitative proteomics reveals an altered cystic fibrosis in vitro bronchial epithelial secretome. Am J of Res Cell and Mol Biol 53:22–32
108. Wu Y, Smas CM (2008) WSdnm1-like, a new adipokine with a role in MMP-2 activation. Am J Phys Endocrinol Metab 295:E205–E215
109. Motrescu ER, Rio MC (2008) Cancer cells, adipocytes and matrix metalloproteinase 11: a vicious tumor progression cycle. Biol Chem 389:1037–1041

110. Sternlicht MD, Lochter A, Sympson CJ, Huey B, Rougier JP, Gray JW, Pinkel D, Bissell MJ, Werb Z (1999) The stromal proteinase MMP3/stromelysin-1 promotes mammary carcinogenesis. Cell 98:137–146
111. Kaplan RN, Riba RD, Zacharoulis S et al (2005) VEGFR1-positive haematopoietic bone marrow progenitors initiate the pre-metastatic niche. Nature 438:820–827
112. Boire A, Covic L, Agarwal A, Jacques S, Sherifi S, Kuliopulos A (2005) PAR1 is a matrix metalloprotease-1 receptor that promotes invasion and tumorigenesis of breast cancer cells. Cell 120:303–313
113. Lynch CC, Hikosaka A, Acuff HB et al (2005) MMP-7 promotes prostate cancer-induced osteolysis via the solubilization of RANKL. Cancer Cell 7:485–496
114. Lu X, Wang Q, Hu G, Van Poznak C, Fleisher M, Reiss M, Massague J, Kang Y (2009) ADAMTS1 and MMP1 proteolytically engage EGF-like ligands in an osteolytic signaling cascade for bone metastasis. Genes Dev 23:1882–1894
115. Kang Y, Siegel PM, Shu W et al (2003) Amultigenic program mediating breast cancer metastasis to bone. Cancer Cell 3:537–549
116. Zeitoun AH, Ibrahim SS, Bagowski CP (2012) Identifying the common interaction networks of amoeboid motility and cancer cell metastasis. Netw Biol 2(2):45
117. Hiratsuka S, Watanabe A, Aburatani H, Maru Y (2006) Tumourmediatedupregulation of chemoattractants and recruitment of myeloid cells predetermines lung metastasis. Nat Cell Biol 8:1369–1375
118. Hiratsuka S, Watanabe A, Sakurai Y et al (2008) The S100A8-serum amyloid A3-TLR4 paracrine cascade establishes a pre-metastatic phase. Nat Cell Biol 10:1349–1355
119. Bond M, Fabunmi RP, Baker AH, Newby AC (1998) Synergistic upregulation of metalloproteinase-9 by growth factors and inflammatory cytokines: an absolute requirement for transcription factor NF-kappa B. FEBS Lett 435:29–34
120. Heissig B, Hattori K, Dias S et al (2002) Recruitment of stem and progenitor cells from the bone marrow niche requires MMP-9 mediated release of kit-ligand. Cell 109:625–637
121. Erler JT, Bennewith KL, Cox TR et al (2009) Hypoxia-induced lysyl oxidase is a critical mediator of bone marrow cell recruitment to form the premetastatic niche. Cancer Cell 15:35–44
122. Levental KR, Yu H, Kass L et al (2009) Matrix crosslinking forces tumor progression by enhancing integrin signaling. Cell 139:891–906
123. Huang Y, Song N, Ding Y et al (2009) Pulmonary vascular destabilization in the premetastatic phase facilitates lung metastasis. Cancer Res 69:7529–7537
124. Lin WW, Karin M (2007) A cytokine-mediated link between innate immunity, inflammation, and cancer. J Clin Invest 117:1175–1183
125. Manicone AM, Guire JK (2008) Matrix metalloproteinases as modulators of inflammation. Semin Cell Dev Biol 19:34–41
126. Haro H, Crawford HC, Fingleton B, Shinomiya K, Spengler DM, Matrisian LM (2000) Matrix metalloproteinase-7-dependent release of tumor necrosis factor-alpha in a model of herniated disc resorption. J Clin Invest 105:143–150
127. Luo JL, Maeda S, Hsu LC, Yagita H, Karin M (2004) Inhibition of NF-kappaB in cancer cells converts inflammation-induced tumor growth mediated by TNFalpha to TRAIL-mediated tumor regression. Cancer Cell 6:297–305
128. McQuibban GA, Gong JH, Wong JP, Wallace JL, Clark-Lewis I, Overall CM (2002) Matrix metalloproteinase processing of monocyte chemoattractant proteins generates CC chemokine receptor antagonists with anti-inflammatory properties in vivo. Blood 100:1160–1167
129. Struyf S, Proost P, Vandercappellen J et al (2009) Synergistic up-regulation of MCP-2/CCL8 activity is counteracted by chemokine cleavage, limiting its inflammatory and anti-tumoral effects. Eur J Immunol 39:843–857
130. Cox JH, Dean RA, Roberts CR, Overall CM (2008) Matrix metalloproteinase processing of CXCL11/I-TAC results in loss of chemoattractant activity and altered glycosaminoglycan binding. J BiolChem 283:19389–19399

131. Burns JM, Summers BC, Wang Y et al (2006) A novel chemokine receptor for SDF-1 and I-TAC involved in cell survival, cell adhesion, and tumor development. J Exp Med 203:2201–2213
132. Houghton AM, Quintero PA, Perkins DL et al (2006) Elastin fragments drive disease progression in a murine model of emphysema. J Clin Invest 116:753–759
133. Weathington NM, van Houwelingen AH, Noerager BD et al (2006) A novel peptide CXCR ligand derived from extracellular matrix degradation during airway inflammation. Nat Med 12:317–323
134. Rocks N, Paulissen G, Quesada-Calvo F et al (2008) ADAMTS-1 metalloproteinase promotes tumor development through the induction of a stromal reaction in vivo. Cancer Res 68:9541–9550
135. Gutierrez-Fernandez A, Inada M, Balbin M et al (2007) Increased inflammation delays wound healing in mice deficient in collagenase-2 (MMP-8). FASEB J 21:2580–2591
136. Balbin M, Fueyo A, Tester AM et al (2003) Loss of collagenase-2 confers increased skin tumor susceptibility to male mice. Nat Genet 35:252–257
137. Palavalli LH, Prickett TD, Wunderlich JR et al (2009) Analysis of the matrix metalloproteinase family reveals that MMP8 is often mutated in melanoma. Nat Genet 41:518–520
138. Gutierrez-Fernandez A, Fueyo A, Folgueras AR et al (2008) Matrix metalloproteinase-8 functions as a metastasis suppressor through modulation of tumor cell adhesion and invasion. Cancer Res 68:2755–2763
139. De Visser KE, Eichten A, Coussens LM (2006) Paradoxical roles of the immune system during cancer development. Nat Rev Cancer 6:24–37
140. Li Q, Park PW, Wilson CL, Parks WC (2002) Matrilysin shedding of syndecan-1 regulates chemokine mobilization and transepithelial efflux of neutrophils in acute lung injury. Cell 111:635–646
141. Polverino F, Zhang L, Laucho-Contreras M, Owen C (2015) Surface-bound TIMP-1 on PMNs promotes pericellular proteolysis: a new culprit in COPD? Eur Respir J 46:PA905
142. Leśniak W, Agnieszka Hrabia A (2016) Involvement of matrix metalloproteinases (MMP-2,-7,-9) and their tissue inhibitors (TIMP-2,-3) in the chicken oviduct regression and recrudescence. Cell Tissue Res 5:1–12
143. Strongin AY, Collier I, Bannikov G, Marmer BL, Grant GA, Goldberg GI (1995) Mechanism of cell surface activation of 72-kDa type IV collagenase. Isolation of the activated form of the membrane metalloprotease. J Biol Chem 270:5331–5338
144. Dufour A, Sampson NS, Zucker S, Cao J (2008) Role of the hemopexin domain of matrix metalloproteinases in cell migration. J Cell Physiol 217:643–651
145. Sakamoto T, Seiki M (2009) Cytoplasmic tail of MT1-MMP regulates macrophage motility independently from its protease activity. Genes Cells 14:617–626
146. Glasheen BM, Kabra AT, Page-McCaw A (2009) Distinct functions for the catalytic and hemopexin domains of a Drosophila matrix metalloproteinase. Proc Natl Acad Sci USA 106:2659–2664
147. Conant K, St Hillaire C, Nagase H et al (2004) Matrix metalloproteinase 1 interacts with neuronal integrins and stimulates dephosphorylation of Akt. J Biol Chem 279:8056–8062
148. Redondo-Munoz J, Ugarte-Berzal E, Terol MJ et al (2010) Matrix metalloproteinase-9 promotes chronic lymphocytic leukemia b cell survival through its hemopexin domain. Cancer Cell 17:160–172
149. Lopez-Otin C, Matrisian LM (2007) Emerging roles of proteases in tumour suppression. Nat Rev Cancer 7:800–808

Proteolytic Networks at the Crossroads of Cancer Cell Life and Death: Cancer Stem Cell Deciding Cell Fate

Poulami Khan, Shruti Banerjee, Apoorva Bhattacharya, Dipanwita Dutta Chowdhury, Apratim Dutta and Tanya Das

Abstract

Over the years, proteases have been implicated in the development of tumors. The proteolytic network, which critically modulates the functioning of a normal cell, is often dysregulated in cancers. In the recent past, the identification of a subpopulation of cancer cells, termed as cancer stem cells (CSCs), has helped gain a better understanding of the complex mechanisms involved in cancer development, progression, as well as recurrence. In this context, it is of considerable importance to comprehend the pivotal role of proteases in regulating the fate of cancer cells via the CSCs. In fact, the proteolytic network influences cancer cell's fate via CSC and its associated niche, which coordinates the functions of CSCs. In this chapter, we have emphasized on the dynamic role displayed by the proteases in regulating numerous steps of tumorigenesis commencing from tumor initiation, angiogenesis, invasion and metastasis. Apart from this, CSCs also execute a survival mechanism with the help of proteases, upon induction of apoptosis. We have also revisited the mechanisms underlying the contribution of proteases in tumor drug resistance, which ultimately leads to cancer relapse, and the role of CSCs in the same. Similarly, proteases are also intricately involved in inflammation and immune surveillance of CSCs. Given the important role of proteases in carcinogenesis, further development of antiprotease therapeutics may enable better treatment procedures and minimize the risk of recurrence. This chapter has, therefore, epitomized the complex crosstalk involving proteases, CSCs and its niche.

The authors Shruti Banerjee and Apoorva Bhattacharya have contributed equally towards this chapter.

P. Khan · S. Banerjee · A. Bhattacharya · D.D. Chowdhury · A. Dutta · T. Das (✉)
Division of Molecular Medicine, Bose Institute, P-1/12 CIT Scheme VII M, Kolkata 700054, India
e-mail: tanya@jcbose.ac.in

S. Chakraborti and N.S. Dhalla (eds.), *Pathophysiological Aspects of Proteases*,
DOI 10.1007/978-981-10-6141-7_11

Keywords
Angiogenesis · Cancer stem cells · Drug resistance · Immune surveillance Metastasis · Microenvironment · Proteases

1 Introduction

In the recent past, a stark rise has been observed globally in cancer incidence and mortality [1]. Despite considerable progress in cancer detection and treatment procedures, elimination of this deadly disease is becoming increasingly difficult. Cancer recurrence is an emerging problem which further complicates disease-free survival of patients. The recent concepts of cancer stem cells (CSCs) provide a better comprehension of various mechanisms exhibited by the tumor populations. CSCs are a subpopulation of cancer cells which exhibit stem-like characteristics and are solely responsible for imparting various traits which result in tumor aggression, resistance, and relapse [2–4]. It is interesting to note that the features exhibited by the CSCs are mainly driven by the tumor microenvironment [5] and the interaction between CSCs and tumor microenvironment facilitates tumor growth, development, and resistance. In fact, different cell types present in the CSC niche execute critical functions which enhance the survival and proliferation of tumor. The CSC niche accommodates fibroblasts, pericytes, endothelial cells, immune cells, etc., which interact in a convoluted manner and succor the CSCs for inducing malignancy [6–8]. Therefore, the microenvironment of the tumor plays an essential role in tumor initiation, angiogenesis, invasion and metastasis. Keeping in mind the critical contribution of CSC niche in maintaining tumor growth, it is important to discuss the means by which the supportive tumor environment executes its functions. It is well accepted that the stromal cells which are part of the tumor microenvironment regulate the tumor growth by secreting various cytokines, chemokines, growth factors, ECM proteins, proteases, etc. [9, 10]. Among which, the proteases exhibit a paramount role in malignant progression [11–13]. At present, we know that tumor progression is associated with enhanced expression, enhanced activity, and altered localization of proteases in cells. Consequently, here we would like to discuss the role played by the proteolytic networks in regulating the cancer cell fate via CSC niche.

2 Proteases

Proteolysis, which is one of the utmost required biological reactions, is accredited to a category of enzymes called proteases. In normal cells, proteases regulate different biological processes. In fact, the equilibrium between proteases and antiproteases plays important role in maintaining normal cellular function; disruption of which results in diseases like cancer both at primary and metastatic sites [14]. Deregulation of proteolytic activity has been elucidated in numerous cancer cases and often corresponds to poor patient prognosis [15]. Recent findings claim a positive

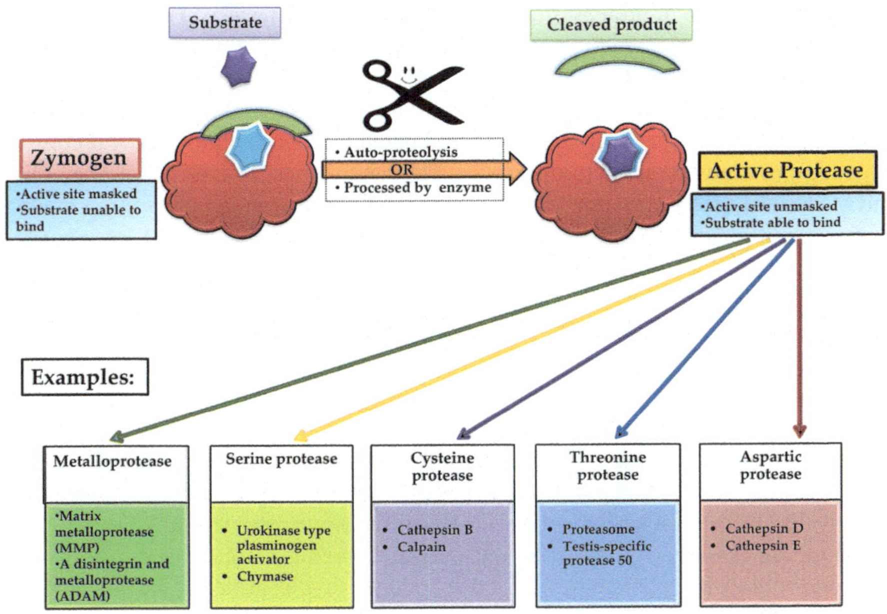

Fig. 1 Activation of a protease from its zymogen state by autoproteolysis or processing by other enzymes. The five major groups of proteases involved in cancer progression are depicted with examples

interconnection between the amount of secreted proteases and aggressiveness of tumor. It is often observed that the cancer cells have the ability to foster the expression of proteolytic enzymes in adjacent normal cells, thereby directing them in favor of tumor growth and maintenance [16]. The function of proteases in cancer progression has been defined as "cancer degradome" after recent identification of the entire gamut of proteases due to the accessibility of the genome sequence of different organisms [17]. There are at least 569 proteases in human degradome which have been categorized into five catalytic classes (Fig. 1): 194 metalloproteases—that use a metal, usually zinc (e.g., meprin, gelatinases A and B), 176 serine proteases—that use serine alcohol (e.g., urokinase-type plasminogen activator, chymase, plasmin), 150 cysteine proteases—that use a cysteine thiol (e.g., cathepsins B, L, S, K, Q; bleomycin hydrolase; caspases), 28 threonine proteases—that use a threonine secondary alcohol (e.g., proteasome), and 21 aspartic proteases—that use an aspartate carboxylic acid (e.g., cathepsins D and E) [18]. Proteases are normally synthesized as inactive precursor called a zymogen or proenzyme, which gets converted to mature, active enzymes by limited but regulated proteolysis of the proenzyme either by a different enzyme or by autoproteolysis (Fig. 1).

These enzymes target diverse substrates, thereby regulating numerous cellular functions that are indispensable for cell survival and death in organisms [13]. However, not all of them are linked with cancer [19, 20]. Actually, the complex process of tumor formation involves diverse changes in the normal cell, beginning from tumor initiation, angiogenesis, growth, metastasis, and invasion into secondary site as well as induction of multidrug resistance. Available reports mainly demonstrate the role of proteases in tumor invasion and metastasis. However, studies are now emanating which imply that proteases are involved in all these above-mentioned steps, thereby aiding tumorigenesis and its progression [18]. This discussion, therefore, raises the possibility of using protease inhibitors as anticancer drugs. Choi et al. [21] and Turk [22] have separately reviewed the pros and cons of targeting proteases in order to develop novel therapeutics. In contrast, López-Otín and Matrisian [23] have reviewed the potential role of proteases as tumor suppressors.

Further, it is interesting to note that the proteasome complexes influence induction, maintenance, and exit of self-renewal as well as pluripotency in both mouse and human [24]. The proteasome complex as well as the proteasomal degradation pathway represents a major system controlling many cellular processes. Even though most proteasomal substrates are required to be ubiquitinated prior to being degraded, there exist some deviations in this normal phenomenon, mainly when the proteasome engages in a normal role in the post-translational processing of the protein [25]. In addition, Yasutaka et al. [26] demonstrated ubiquitin proteasomal system as a key regulator of stem cell pluripotency. In connection with this, lung tumor spheres are reported to be enriched in cells with low proteasome activity [27].

Above discussion signifies that the proteases mediate a complex role in carcinogenesis and its management. Studies related to the proteasomal network in influencing cancer cell fate by regulating CSCs within tumor niche, therefore, highlight the emergence of a new and important field of research. Therefore, in this chapter, we would like to portray the role of the proteolytic networks which regulate tumor initiation, angiogenesis, invasion and metastasis, etc., via the CSC niche.

3 Cancer Stem Cells

Recently, there has been a paradigm shift in the concept of cancer biology, where accumulating evidences show that a minority subpopulation of the cancer cells residing inside a tumor are the root cause of several facets of tumorigenesis including tumor initiation, angiogenesis, metastasis drug resistance, and tumor relapse (Fig. 2) [28–31]. These cells have the capacity to self-renew and to produce the heterogeneous lineages of cancer cells forming the bulk of tumor. Owing to their properties that they share with normal stem cells, e.g., self-renewal and differentiation, or their ability to initiate and promulgate tumors in vivo, this subpopulation of cells has been given the nomenclature as cancer stem cells (CSCs). The other synonymous terms that are used to describe CSCs are "tumor-initiating cell" and "tumorigenic cell" [32, 33]. CSCs having tumorigenic potential and drug-resistant phenotypes have been identified in many human malignancies. Extensive investigation conducted over the last few years on the existence of CSC has gained immense attention due to its profound impact on cancer progression.

In fact, in a tumor mass, CSCs retain the potential to create all the differentiated cell populations that form the original tumor. These tumor-initiating cells that represents less than 5% of the total tumor mass, have been observed to play vital

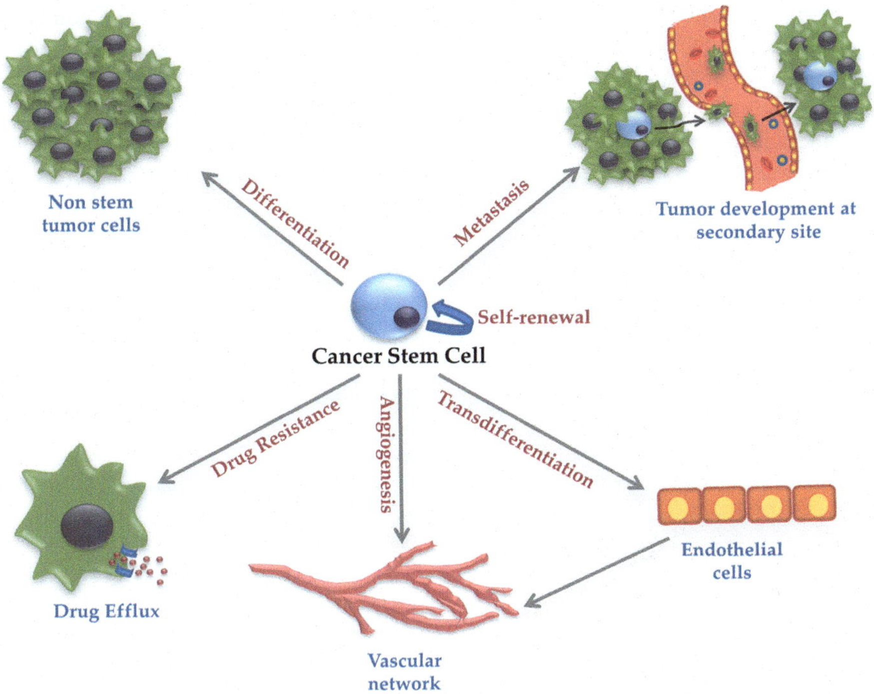

Fig. 2 Functional network of cancer stem cells: CSCs have the ability to self-renew as well as differentiate into the tumor cells to drive tumor initiation. CSCs are also the major culprit behind metastasis, drug resistance, and tumor angiogenesis

role in tumor-induced angiogenesis by shedding the major pro-angiogenic factor VEGF that invites endothelial cell proliferation, migration, and network formation [34]. Very recent reports signify that CSCs even contribute in VEGF-independent neo-angiogenesis by transdifferentiating into endothelial-like cells having potency for network formation [35, 36]. Melanoma-specific $CD133^+$ CSCs have been observed to transdifferentiate into an endothelial-like state when cultured under endothelial differentiation-promoting conditions [37]. Accumulating evidence suggests a considerable role of these self-renewing CSCs in inherent resistance to traditionally used chemotherapy [38]. In addition, clinically, the residual risk of recurrence has been found to be higher in cancer patients with higher number of CSCs that acquire chemoresistance [39]. Moreover, intrinsic CSCs are endowed with higher migration potential than non-stem cancer cells due to the suppression of the tumor suppressor, E-cadherin [40]. According to Mukherjee et al. [41], intrinsic non-migratory CSCs convert non-stem cancer cells to migratory $CXCR4^+$ CSCs with aggravated migration property, thereby shaping them for metastasis and recurrence in cancer patients. Besides, two major aspects of CSC physiopathology are in vivo quiescence and plasticity [42]. There exist different theories regarding the generation of cancer stem cells in a tissue by (i) the mutation in stem cells,

(ii) the mutation in non-stem cancer cells, and (iii) the dedifferentiation of non-stem cancer cells to CSCs. Recent findings also demonstrated that these undifferentiated, self-renewing drug-resistant cells themselves induce dedifferentiation of non-stem cancer cells to CSCs via paracrine signaling by shedding different growth factors, cytokines depending on the microenvironmental conditions [39, 41] and thereby increasing CSC pool in a tumor mass. Recent report demonstrates microRNA (miRNA) regulation of self-renewal and differentiation of CSCs of various types of cancers, e.g., lung cancer, head and neck cancer, breast cancer [43]. Development of miRNA-based therapeutics may, therefore, help control CSC population in diverse forms of cancers in the future. Furthermore, the CSCs are strongly regulated by the circumambient microenvironment, which also influences tumor responses. Above discussion signifies that CSCs play critical role in tumor development and progression. Therefore, it is necessary to understand how these clinically important CSCs are controlling the cell fate decision.

4 Cancer Stem Cell Niche

The highly specialized microenvironment where CSCs reside within the tumor and have their fates decided is referred to as the cancer stem cell niche. Within the tumor microenvironment, they have been identified as anatomically distinct population of cells. The CSC niche includes the non-stem cancer cells (NSCC) of the tumor, the adjacent stroma, blood vessels, fibroblasts, immune cells, extracellular matrix proteins, and secretory factors that decide cell proliferation, self-renewal, and fate of CSCs, either directly or indirectly (Fig. 3) [44]. Reportedly, any stem cell has limited function outside its predefined niche because of its dependence on the local microenvironment for signals determining cell growth, proliferation, and differentiation [45]. The niche of CSCs in the primary tumor potentially decides the progression of tumors to a more malignant state [46]. The CSC niche may be acquired by CSCs by either utilizing the tissue-specific somatic stem cell niche that is already present, or expressing the soluble factors which signal between the specific microenvironment and CSCs. The cells of the surrounding stroma provide the basic support framework essential for maintaining the tumor microenvironment. The paracrine and autocrine functions of the cytokines released by mesenchymal stem cells (MSCs), which are multipotent stromal cells, enhance cancer stemness. Impeded access to the blood vascular system results in lack of oxygen (hypoxia) for the tumor which enhances stemness through dedifferentiation and activation of stemness-promoting genes [47]. Moreover, the reactive oxygen species (ROS) produced due to hypoxia enhance epithelial to mesenchymal transition (EMT) and CSC survival [48]. Cancer-associated fibroblasts (CAFs) present in the CSC niche can cause increased proliferation, ECM production, and stemness via various signaling pathways like WNT and Notch [49]. Although Notch signaling impedes cells from responding to differentiation factors, it aids in stem cell maintenance and cell fate determination [50]. On the other hand, tumor-associated

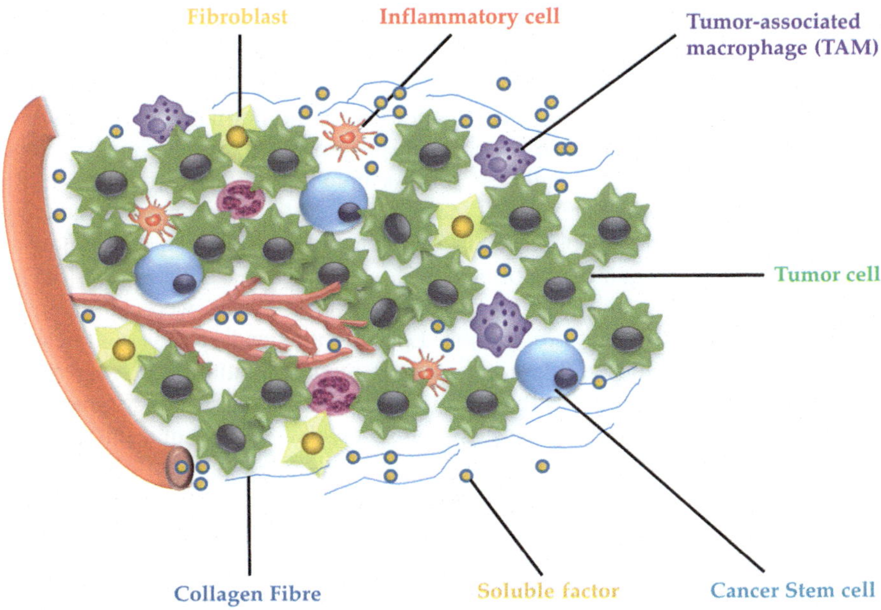

Fig. 3 Tumor microenvironment and the CSC niche depicting various cellular subtypes and extracellular matrix involved in CSC functional network

macrophages (TAMs) present in the tumor milieu secrete transforming growth factor beta (TGF-β) that can cause immunosuppression by recruiting regulatory T cells (Tregs) [51] TAMs show plasticity in their phenotypes and may act as alternatively activated M2 macrophages by supporting tumor growth. They also induce ECM breakdown, tumor invasion and metastasis [52, 53]. Matrix metalloproteinases (MMPs) degrade the ECM components in tumors and up-regulate invasion, angiogenesis, and metastasis in tumor cells by releasing cytokines from the ECM [54, 55].

Hence, all these numerous factors contribute to the final structure and function of the microenvironment. The CSCs are, therefore, armed with an arsenal of techniques through which it can manipulate its niche and the tumor microenvironment to assist with the initiation, invasion and metastasis of cancer. While on the other hand, the components of the CSC niche can regulate the growth, proliferation, self-renewal, and differentiation of CSCs via its vast array of signaling networks. An elevated interdependency is thus observed between these two essential facets of cancer: the CSC and its niche.

5 Tumor Initiation

Initiation of cancer is due to abnormal proliferation of neoplastic cells upon genomic alterations. Cancer-causing mutations alter the normal cell functioning, ultimately resulting in tumor development. The process of tumor initiation and subsequent progression is a multistep process involving different hallmarks [2]. Previous concepts mainly highlighted the role of genetic aberrations in inducing tumors. However, with the progress in research, the significant contribution of epigenetic changes has gained importance in context to cancer initiation [56]. In an interesting finding, the molecules involved in axon guidance were observed to regulate cancer initiation and progression. Eph receptors and ephrins significantly contribute in modulating the important steps of tumor initiation and progression. Upon subsequent knockdown of these molecules, tumor initiation was found to be perturbed [57]. Also, the immune system plays a crucial role in tumor initiation. In fact, immune system has been found to display a dual role in cancer initiation and progression, i.e., in certain cases, it exhibits tumoricidal response while in others furnishing tumor-enhancing effect [58]. The inflammatory response by the immune cells also significantly contributes to the process of tumor initiation and hence has been discussed in later sections. Further research in the domain of cancer biology has successfully identified a heterogeneous population of cells inside the tumor mass. Even though the cell of origin is unknown to a large extent, it is speculated that different subtypes represent discrete cells of origin during the time of tumor initiation [4]. The heterogeneity existing in the disease is largely attributed to the CSCs. These CSCs are responsible for the generation of diverse cell populations within the tumor [3]. Experimentally, CSCs have been identified as the cell population that has the ability to generate a tumor in xenograft models of human cancer [32]. Even though the detailed molecular mechanisms of cancer initiation are poorly understood, Boumahdi et al. [59] unveiled the essential role of SOX2 in controlling tumor initiation and CSC properties. Therefore, the CSCs contribute significantly in initiating the tumor. However, it is extremely important to discuss the role of the microenvironment which supports the initiation process. As already discussed, the CSC niche is a complex environment which houses various cell types that aid the process of carcinogenesis. It is acknowledged that the factors secreted by the CSC niche help in the self-renewal and maintenance of CSCs [60]. Moreover, the proteases secreted from the tumor microenvironment play a pivotal role in the initiation process. In this regard, reports have illustrated the role of Taspase1 in initiating cancer (Fig. 4). This protease has been reported to regulate the proliferation of cancer cells in which it is often up-regulated [61]. In contrast, down-regulation of Taspase1 disrupts cell cycle and enhances the expression of CDK inhibitors. This protease is mainly involved in the cleavage of MLL (mixed lineage leukemia) [62]. Therefore, Taspase1 has been identified as a novel therapeutic target. Further

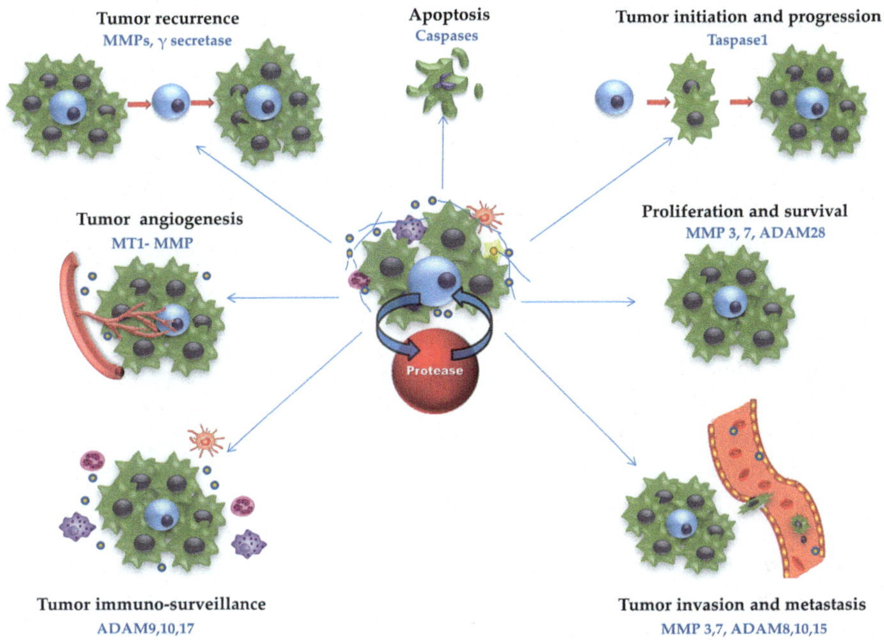

Fig. 4 Involvement of different proteases in CSCs that drives the hallmarks of cancer: tumor initiation and progression, proliferation and survival, invasion and metastasis, tumor angiogenesis, tumor recurrence, tumor immune surveillance, and regulation of apoptosis

research will help determine whether inhibitors of Taspase1 can be effective against CSCs [63]. Interestingly, Kumar et al. [64] have demonstrated using a biophysical model how extracellular matrix (ECM) proteolysis is responsible for changing the spatial distribution and number of CSCs. In this manner, ECM proteolysis plays a crucial role in regulating the growth of tumor as well as its relapse. In fact, ECM proteolysis has been observed to increase tumor growth. It is interesting to note that the CSCs in association with the tumor microenvironment play a critical role in ECM proteolysis. Pericellular proteolysis is known to be a leading process in generating a cancer-promoting extracellular milieu. The degradation of ECM components is achieved by the pericellular proteases [65]. Therefore, the regulation of these pericellular proteases is largely mediated by the tumor microenvironment. As mentioned above that ECM proteolysis regulates the distribution of CSCs, it will not be out of context to associate CSC and its niche with regulation of ECM proteolysis by pericellular proteases. Thus, the above discussion mainly highlights the critical role of CSCs and its associated niche in initiating carcinogenesis via proteases.

6 Tumor Angiogenesis

The development of tumor essentially depends on angiogenesis. Survival, proliferation, and metastatic spread of cancer require ample supply of nutrients, oxygen, and removal of waste material, thus making angiogenesis one of the most crucial steps in the development of cancer. In fact, angiogenesis is a prerequisite for metastasis since tumor cells must penetrate blood vessels and travel along the blood stream in order to reach secondary sites and colonize [66]. The means by which new blood vessels form is known as neo-angiogenesis, and it is mandatory for a continuous supply of oxygen, nutrients, and immune cells and also for the removal of waste products [66]. Since tumor growth and metastasis depend on angiogenesis, angiogenic factors are receiving a lot of attention. During the period of rapid growth, tumor cells release chemical signals that trigger angiogenesis [66]. In a study, done by Muthukkaruppan et al. [67], it was shown that when cancer cells were positioned in an area where angiogenesis was feasible, they grew in size (more than 2mm3), while they did not grow in an area devoid of angiogenesis. In the absence of adequate blood supply, tumor cells may become necrotic or even apoptotic [68, 69]. Hence, angiogenesis is essential for the progression of cancer. Activators and inhibitors are both involved in the regulation of angiogenesis. For example, tumor neo-angiogenesis is triggered not only by up-regulation of angiogenic factors but also by down-regulation of antiangiogenic molecules [70]. The CSC niche is also vital in inducing angiogenesis since endothelial cells, that control vascular function, are an essential part of the CSC niche [71, 72]. Many reports have depicted that CSCs via the CSC niche regulate the angiogenesis. Two most potent angiogenic factors are VEGF and bFGF [73]. Among the VEGF family members, VEGF-A has major role in angiogenesis. VEGF-A binds to VEGFR1 and VEGFR2 to trigger angiogenesis [74]. The role of VEGFR2 in angiogenesis is well defined while that of VEGFR1 is not that much understood. Recently, the role of CSCs in tumor angiogenesis has caught a lot of attention. This is due to the fact that CSCs survive various antiangiogenic therapies like Avastin (monoclonal antibody targeting VEGF-A) and Sunitinib (VEGFR inhibitor) due to the overexpression of drug efflux pumps, thereby causing tumor relapse. Bao et al. [75] demonstrated that glioma CSCs stimulate angiogenesis via VEGF. Recent studies by Tang et al. [35] have clearly depicted the role of CSCs in promoting angiogenesis. Ovarian CSCs have also been found to even transdifferentiate into endothelial cells and promote angiogenesis via CCL5 signaling. Given the important role of CSCs and its niche in angiogenesis, we would like to highlight the role of proteases in modulating angiogenesis.

Hypoxia has a major role in CSC-induced angiogenesis since experiments have shown that hypoxia enhances the CSC population in a tumor. Recently, the role of hypoxia in regulating the self-renewal property of CSCs by enhancing the activity of stemness proteins like Oct-4, Sox-2, and Nanog has been reported [76]. It is believed that hypoxia induces VEGF secretion from CSCs triggering angiogenesis though the exact mechanism is not clear. Hence now, it is widely believed that by targeting hypoxia, we can actually inhibit CSC-induced angiogenesis. In addition, the equilibrium between the proteases and their inhibitors plays the role of a critical determinant in angiogenesis [77]. Proteases are involved in the degradation of the ECM, thus allowing endothelial cells to invade the tissues and penetrate the tumor stroma. In this regard, pericellular proteases have been reported to play an essential role in angiogenesis (Fig. 4). They include serine proteases, membrane-type matrix metalloproteinases (MT-MMPs), membrane-bound aminopeptidases, and cysteine cathepsins [78]. Reports demonstrate major roles of MT1-MMP, MMP-2, and MMP-9 in initiating angiogenesis [79]. Among these metalloproteases, MT-MMPs are predominantly membrane-bound while MMP-2 has the ability to localize to the membrane by binding to $\alpha v\beta 3$-integrin and MMP-9 normally localizes to the membrane utilizing CD44 [80]. The overexpression of MT1-MMP has been found to be associated with increased tumor growth and vascularization in vivo [81]. The role of MT1-MMP in tumor angiogenesis is caused by various mechanisms like the activation of $\alpha v\beta 3$ integrin, which leads to protection of the endothelial cells from apoptosis by fibrinolytic activity [82]. It is well known that VEGF is a crucial regulator of angiogenesis [83]. A subset of MMPs like MMP9 is proposed to cleave the ECM components and the c terminal region of VEGF, thereby releasing the biologically active VEGF from the ECM. MMP9-mediated VEGF release leads to its higher interaction with VEGFR at the quiescent blood vessels, thereby turning on the angiogenic switch [65]. Proteases are also shown to play differential role in regulation of angiogenesis by cleaving plasminogen and collagen XVIII to release angiostatin and endostatin, respectively, which have inhibitory effect on angiogenesis [84, 85].

Proteases trigger angiogenesis by activation of angiogenic growth factors, followed by degradation of the endothelial cell matrix and ECM [86]. They are also involved in the modification of the properties of cytokines and angiogenic growth factors. Angiogenesis requires migration of endothelial cells, which in turn is dependent on the protease–protein complexes which congregate at the migrating cell front [65]. Multiprotein complexes are formed in the lamellipodia and focal adhesions of migrating cells by utilizing MT1-MMP and urokinase (u-PA), thus favoring proteolysis, which provides adequate support for the survival and migration of endothelial cell [87]. However, uncontrolled proteolysis results in the loss of endothelial cell–matrix interaction, thereby hampering neovascularization [88]. MMP-9 and cathepsin L recruit accessory cells, which are derived from blood or bone marrow, for augmenting neovascularization [89]. During proteolysis, fragments of ECM and homeostatic factors are generated and these possess antiangiogenic properties [90]. Understanding the role of proteases in neovascularization is essential for recognizing new targets for inhibiting angiogenesis. The proteases are also thought to play a critical role in the vasculogenic mimicry, which is recently reported as a tumor vascular paradigm, where the CSCs form vessel-like networks that provide adequate blood supply to the growing tumor, independent of angiogenesis [91]. Since increasing evidences suggest that the neovascularization and vasculogenic mimicry are associated with the CSCs, it can be well predicted that regulation of their niche by different proteases may be responsible for their angiogenic fate.

7 Tumor Invasion and Metastasis

Tumor invasion and metastasis are two major hurdles in the battle against cancer, which ultimately drive cancer relapse, poor prognosis, and high mortality. Metastasis is one of the most common types of complexities associated with cancer, treatment failure, and high mortality. During tumor progression, tumor cells undergo epithelial to mesenchymal transition (EMT) that involves reduced expression of epithelial markers and enhanced expression of the mesenchymal ones. The cascade of metastasis is a multistep process which includes local invasion, EMT—which induces stemness in cancer cells, intravasation, extravasation and colonization and rehabilitation at a distant location [92]. The ECM, being an important part of tumor niche and microenvironment, has captured the interest of researchers from a long time, since proteolysis and remodeling of ECM seem to work as key factors in cancer metastasis and invasion.

Interestingly, recent reports signify that although the tumor tissue comprises of heterogeneous population of cells, only CSCs exhibit selective growth advantage to form tumors at distant metastatic sites [93]. Thus, the crosstalk between CSC niche, epithelial mesenchymal transition (EMT) process, and CSC is coordinated for initiating the metastasis at secondary tissues is of great therapeutic interest. In this regard, the role of matrix metalloproteinases (MMPs) is being extensively studied because of their potential to degrade and remodel extracellular matrix (ECM) and hence leading to an increase in tumor invasion and metastasis (Fig. 4). Moreover, enhanced expression of pericellular proteases is seen in many cancers and has long been associated with poor prognosis and patient survival. A flock of pericellular proteases, which can degrade all ECM constituents including proteoglycans, collagens, and non-collagenous glycoproteins, have been well identified and characterized [55]. In addition to primary tumor progression and invasion, these pericellular proteases also play significant roles in tumor cell incorporation into the circulation. It also modulates invasion and tumor formation at the distant secondary sites. Disruption of vascular barriers by various pericellular proteases has been reported. For example, MMP17 facilitates the detachment of pericytes from blood vessels, thereby aiding intravasation of breast cancer cell leading to an increase in vascular leakage [94]. Besides degradation of ECM, disruption of intercellular junction proteins and cell adhesion molecules like E-cadherin is also brought about by proteases, thereby diminishing physical barriers. E-cadherin is shown to be cleaved by MMP 3, MMP 7, ADAM 10, ADAM 15, and also by Cysteine cathepsins B, L, S [95–98, 65]. Cleavage of the cell attachment molecules abolishes cell–cell adhesion, thus promoting tumor cell invasion and metastasis. MMP1 is known to cleave and activate protease-activated receptor-1 (PAR-1) present on endothelial cells, leading to an increase in endothelial permeability and ultimately transendothelial migration of cancer cells [99]. Degradation of Claudin and Occludin (important cell junction proteins) by MMP 2 and 9 breaks the blood–brain barrier and thereby promotes CNS invasion of cancer [100]. All these accumulated evidences are indicating toward a huge role of proteases, through a plethora of molecular mechanisms, in tumor invasion and metastasis.

Perceptually, $CD44^{+}CD24^{low/-}$ breast cancer stem cells showed increased expression of metastasis-related genes and induced lung metastasis in vivo [101]. Induction of EMT program in immortalized human mammary epithelial cells (HMLEs) results in an increased potential to form tumor spheres, and in the expression of stem cell-like markers [102]. The microenvironment of CSCs dictates the behavior and fate of the tumor [55]. CSCs interact with adjacent cells and ECM that regulate the rate of self-renewal, proliferation, and differentiation of CSCs and thus the destiny of tumor mass. MMPs have the ability to modulate adult stem cell niches by cleaving, degrading, and rearranging ECM molecules. For example, MMP14 (MT1-MMP) modulates HIF-mediated gene transcription of cytokines and chemokines inside the hematopoietic stem cell niche, thereby regulating its behavior [103]. In addition, MMP3 is shown to be involved in the sustenance of adult epithelial stem cells in the mammary gland by binding and inactivating non-canonical Wnt ligands [55]. The remarkable loss of tumor progression by knocking down of CSC marker CD44v6 (a variant of CD44 known for metastasis promotion) has been documented to depend on its capacity to collaborate with associating integrins and proteases [104, 105]. This demonstrated a novel role for the serine protease Granzyme B in up-modulating the epithelial-to-mesenchymal transition, which is an indicator of stemness gain, in colorectal cancer cells. In another study, the $CD133^{+}$ lung CSCs, that are responsible for the tumor metastasis and patients' survival, have been shown to promote the tumor invasion and metastasis via the up-regulation of MMP-9 expression [106].

Thus, the protease cascades, having intensive role in modulation of stem cell niches and remodeling of ECM, and "pre-metastatic niche formation," points toward its huge possibility in tumor invasion and metastasis by manipulating the cancer stem cells [23]. However, this area demands further attention of the researchers.

8 Proliferation and Survival

Cellular survival and proliferation are the major aspects of any developmental process including tumors. The course of tumor development involves maintenance of tumor cells and sustaining their proliferation. In this regard, CSCs are considered major contributors to cancer growth and progression although the precise roles of CSCs in governing tumor development, progression, and survival are under intense scrutiny. CSCs undergo symmetric and asymmetric divisions depending on the requirement of the daughter cells to maintain their self-renewal capacities. In asymmetric division, old mitochondria upon segregation get distributed to the daughter cells which undergo differentiation. Contrary to this, symmetric division involves equal distribution of both young and old mitochondria between the two daughter cells. Thus, metabolism and mitochondrial biology play significant roles in deciding the fate of stem cells. Since tissue homeostasis is directly affected by these decisions, it is essential to understand the regulatory mechanisms which govern cellular metabolism. A correlation between the survival and CSC population has been reported in highly aggressive cancers like glioblastoma [107]. In fact, self-renewal property of CSCs (i) ensures maintenance of CSC pool and the metabolic defects can shift cell symmetric division, thereby leading to stem cell exhaustion [108], and (ii) generates the heterogeneous population of non-stem cancer cells that comprise the tumor mass. This information signifies that CSCs play a pivotal role in survival and growth of the tumor mass. Recent reports demonstrate that the matrix degradation mediates the cell growth and survival by different protease-mediated mechanisms regulating bioavailability of growth factors. A variety of critical growth factors are released as part of a protein complex which prolongs their half-life. However, upon association with these binding proteins, their availability to the cells is limited [109]. Proteolytic modification of these binding proteins is therefore necessary to release the growth factors and permit their interactions with the corresponding receptors to induce cellular proliferation (Fig. 4). As an example, insulin-like growth factors (IGFs) are released upon cleavage of IGF binding proteins (IGFBPs) by MMP7 and ADAM 28 [110–112]. MMP7 cleaves pro-HB EGF to produce the mature HB EGF. ADAM17 mediates the shedding of ectodomain of TGF-α [113], while MMP9, MMP2, and cathepsin B are involved in the release of TGF-β [114]. Thus, proteolytic processing is required by various growth factors in order to gain receptor specificity and activity. Apart from that, pericellular proteases are involved in the regulation of tumor cell survival by processing ECM, and induction of apoptosis. An interesting example in this regard can be the proteolytic processing of the Fas ligand [115]. Contrariwise, proteases have the ability to minimize tumor cell proliferation and survival by cleaving the ecto-domain of the cytokine receptors, ultimately leading to its inactivation.

Moreover, MMP-3 is reported to act as a regulator of Wnt-signaling, whereas MMP-7 regulates Notch signaling—two important signaling pathways leading to maintenance of pluripotency of embryonic as well as cancer stem cells [116]. Furthermore, MMP9 localizes to the cell surface utilizing the surface receptor CD44 to proteolytically activate TGFβ. On the other hand, protease-activated receptor PAR2/GSK3β pathway has been demonstrated to play a crucial role in regulating proliferation and survival of colon cancer stem/progenitor cells [117]. According to Amoury et al. [118], Granzyme B-based cytolytic fusion protein targeting EpCAM, overexpression of which is associated with CSC phenotype, has been found to eliminate triple negative breast cancer cells in vitro and also to inhibit tumor growth in vivo.

These reports on the whole indicate the possibility of these proteases to regulate the fate of survival of tumor by orchestrating regulation of CSCs.

9 Tumor Resistance and Recurrence

The phenomenon of drug resistance exhibited by tumors over the years is an emerging problem. The efficacy of the treatment is hampered, and the resistance to the chemotherapeutic agents ultimately results in tumor recurrence [119]. Despite the tremendous effort for targeted therapies against cancer, development of resistance to genotoxic drugs is a predominant hurdle in successfully eliminating the aggressive tumors [120]. The resistance phenomenon that is exhibited in response to chemotherapeutics is normally subdivided into two broad classes: (1) intrinsic or innate resistance, which mainly refers to the resistance potential that exists in the tumor cells before any exposure to chemotherapy, and (2) acquired or adaptive resistance, which normally develops during the treatment procedure of tumors. Acquired resistance may occur due to mutations emerging during the treatment or alteration of signaling pathways [119]. Intensive research in the recent past has identified that the CSCs are predominantly responsible for tumor resistance and relapse. The conventional chemotherapeutics fail to eliminate the CSCs residing in the tumor bulk which ultimately results in cancer relapse [121]. In this context, Liu et al. [122] observed that the CSCs associated with glioblastoma exhibit immense resistance to chemotherapy. The drug resistance gene BCRP1 along with MGMT (DNA mismatch repair gene) was found to be significantly up-regulated. Apart from this, aberrant expression of ATP-binding cassette (ABC) drug transporters in CSCs is essentially responsible for drug resistance [121]. ABCG2 and multidrug resistance-associated protein-1 (MRP1) have been implicated in protecting the CSCs from chemotherapy [123, 124]. Recently, research work from our laboratory has reported that the CSCs, upon treatment with chemotherapeutic agents, foster an inflammatory environment dependent on NFκB-IL6-pathway that enhances multidrug resistance and initiates stemness in the non-stem cancer cells [39]. Therefore, it is clearly evident that the CSCs contribute significantly in drug resistance and relapse. In this context, we would like to highlight the role of proteases in modulating the phenomenon of drug resistance in cancer via CSCs. The aberrant activation of the Notch pathway is found to be crucial in maintaining CSC self-renewal [125–127]. Further, investigations have revealed that Notch activity is essentially involved in imparting radiation resistance in cancer [128], and hence, this pathway is gaining attention by representing a critical target for overcoming radiation resistance [129, 130]. Interestingly, γ-secretase is involved in the processing of Notch-1. The intracellular domain of Notch-1 is activated by the proteolytic activity of this protease which assists its translocation to the nucleus in order to exhibit its function. Further research has highlighted the role of γ-secretase inhibitors in negatively regulating the Notch activity. The processing of Notch was impeded by these protease inhibitors, thereby affecting the Notch pathway activation. Lagadec et al. [131] demonstrated that the expression of Notch ligands and receptors was significantly down-regulated, upon induction of radiation, when the mammospheres were treated with a γ-secretase inhibitor. Also, Mamaeva et al. [132] have depicted a novel delivery vehicle using nanoparticles that accommodates γ-secretase inhibitors for targeting CSCs. Thus, γ-secretase represents a potential target for essentially enhancing the effectiveness of radiation therapy. Apart from this, MMPs are significantly involved in activating protease-activated receptor 1 (PAR1) [133]. Further, Fujimoto et al. [134] highlighted the role of PAR1 in inducing multidrug resistance in gastric CSCs. Therefore, the above discussion showcases that proteases notably help the CSCs to become drug resistant which further assists in tumor recurrence (Fig. 4).

10 Apoptosis

Apoptosis or programmed cell death is a process that leads to changes in cell characteristics, thereby inducing cell death in multicellular organisms. However, a certain degree of resistance is observed in this intrinsic process during oncogenesis. Apoptosis is triggered via two pathways: (i) extrinsic pathway that involves the death receptor cascade and (ii) intrinsic pathway involving the mitochondria [135]. The two apoptotic pathways ultimately trigger the activation of cysteine proteases called caspases. Since the caspases are usually found in an inactive form, mild proteolysis of these proenzymes, either autoproteolysis or by other caspases, is essential for their enzymatic activation [136]. The crucial role of caspases in regulating apoptosis is evident from the investigations which concluded that the deficiency of FADD and Apaf-1, which assist the activation of caspases, results in impaired apoptosis in mice. Similarly, knockout of Apaf-1, which activates caspase-9, resulted in malformations in multiple tissues further characterized by developmental cell death in mice [137, 138]. Reports indicate that elevated levels of survivin mediate multidrug resistance and reduced apoptosis in breast CSCs [139]. The outcome of the WW domain containing oxidoreductase (WWOX) gene on apoptosis as well as cell cycle regulations in human ovarian CSCs was explored [140]. Results conferred that the WWOX protein was expressed stably in ovarian CSCs which inhibited its proliferation. The WWOX gene imparted its effects by down-regulating the expression levels of essential cell cycle regulators like cyclin D1-CDK4 and cyclin E-CDK2. In addition to this, WWOX gene can lead to the overexpression of JNK, Wnt-5α, and caspase-3 which contributes to the ovarian CSC apoptosis. The WWOX gene may thus prove to be an essential target for ovarian cancer therapeutic treatment imminently. An interesting finding by Jinesh et al. [141] portrayed a survival mechanism by CSCs, induced upon apoptosis. The cells undergoing apoptosis fuse to form "blebbishields" which form tumorigenic spheres. Activated caspases are involved in the formation of these spheres with studies highlighting the role of caspases in protecting CSCs from apoptosis by use of inhibitors. Similarly, reports have also indicated that Salinomycin causes apoptosis of breast CSCs by down-regulating the Hedgehog signaling modulators, Smo and Gli [142].

These are just a very few examples out of hundreds that demonstrate the fact that apoptosis of CSCs is essential in order to destroy the entire tumorigenic mass completely and have a better chance of survival. Apoptosis is an essential intrinsic mechanism of cell death in multicellular organisms that balances cell proliferation and cell death. As has been discussed above, cells which escape the cues that control apoptosis can lead to many debilitating and life-threatening diseases like cancer. However, this is just one side of the story. CSCs are master manipulators of their niche and local microenvironment. Here, they reign supreme. Thus, we need to also explore the fact that CSCs might themselves regulate the apoptosis of and sacrifice non-stem cancer cells (NSCCs) and other cells of the surrounding tissue when it proves advantageous to the CSCs. The cellular debris thus produced may help to occupy the attention of the immune cells and help the CSCs to escape immunosurveillance.

11 Inflammation and Immune Surveillance

Substantial clinical evidences have identified the role of inflammation in the development of cancer. Chronic inflammation has been found to foster an environment which is conducive to tumor development. Researchers have identified that in certain types of cancers the inflammatory environment is present before the occurrence or development of tumor. On the other hand, in certain tumor types, oncogenic insults result in an inflammatory tumor microenvironment [143]. This tumor microenvironment that houses an inflammatory state enhances the infiltration of tumor-associated macrophages (TAMs), nuclear factor kappa B (NFκB), inflammatory cytokines such as interleukin (IL)-6, IL-1, tumor necrosis factor (TNF), thereby consequently facilitating carcinogenesis [144]. The inflammation associated with cancer has been termed as the 7th hallmark of cancer, owing to its significant contribution in tumor development and maintenance [145]. Therefore, the "smouldering" inflammation associated with the tumor microenvironment contributes significantly in tumor development and proliferation [143]. In this context, the links between inflammation and CSCs are also being investigated. The immune cells present in the tumor microenvironment are responsible for the secretion of various cytokines which modulate cellular networks leading to CSC growth and maintenance [146]. Recent studies have highlighted the role of chemo-resistant CSCs in modulating the inflammatory states of tumor microenvironment [147]. Therefore, we can clearly understand the critical link between inflammation and CSC development. Interestingly, the CSC niche housing inflammatory immune cells orchestrates multiple functions via proteases. Hagemann et al. [148] have experimentally confirmed that the invasiveness of tumors is enhanced due to the secretion of matrix metalloproteinases (MMPs) by the TAMs. The observed overexpression of MMPs was dependent on tumor necrosis factor-α (TNF-α). Apart from this, it was also validated that breast cancer cells, overexpressing protease cathepsin B [149], enhanced the expression of protease cathepsin

B in tumor-associated monocytes in an IL-6-dependent manner [150]. Interestingly, cathepsin B was found to be involved in regulating the self-renewal of glioblastoma CSCs by inducing the expression of Sox2 and Bmi1 [151].

During tumor development, the cells of the immune system try to avert the occurrence of malignant cells, thereby inhibiting the progression of cancer. This is achieved by a mechanism termed as immune surveillance [152–154]. The tumor microenvironment plays a pivotal role in immune surveillance. Natural killer (NK) cells, which are part of the CSC niche, depict an essential role in regulating immune surveillance in cancer by secreting serine proteases called granzymes [155, 156]. Apart from NK cells, mast cells, present in the CSC niche, are also implicated in immune surveillance and granzyme secretion [157, 158]. Recent studies have also highlighted the ability of NK cells to preferentially kill CSCs via NKGD2-dependent pathway [159]. Interestingly, ADAM 10 (a disintegrin and metalloproteinase) and ADAM 17 proteases have been found to be involved in the shedding of NKGD2 ligands (NKGD2L) in tumor cells [160]. Recent investigations have revealed the role of tenascin-C, an extracellular matrix protein, in protecting CSCs from immune surveillance. The activation of T cells was prevented by this protein [161]. Consequently, it was also observed that ADAM 9 acted as a critical modulator in tenascin-C induced invasiveness in brain CSCs. The proteolytic activity of ADAM 9 facilitated the invasiveness of brain CSCs [162]. It is evident that the CSC niche modulates the functions of the immune cells via proteases. Thus, the above discussion provides an overview of the role played by proteases in modulating inflammation and immune surveillance in CSCs (Fig. 4).

12 Therapeutic Strategies

Proteases are involved in wide-ranging activities of tumorigenesis. As discussed above, they modulate numerous steps starting from tumor initiation, angiogenesis, survival proliferation, invasion and metastasis, which facilitate carcinogenesis. Therefore, involvement of different proteases in cancer paves the way to promising therapeutic approaches using protease inhibitors for patients suffering from metastatic cancer of poor prognosis. In a report, Remacle et al. showed that MMP14 influences tumor growth by its HPX domain in a breast cancer mouse model [163]. For this purpose, Developmental Therapeutics Program (NCI/NIH) virtual ligand screening compound library was utilized in order to determine specific inhibitors of the HPX domain of MMP14. But clinical trials organized with broad-spectrum MMP inhibitors (MMPIs) resulted in outcomes with unsatisfactory results. Alternative approaches include changing the proportion of proteases and protease inhibitors or restricting the inhibitory effects of individual proteases by using engineered protease inhibitors called TIMPs. Further, modification of TIMPs with the help of strategies involving mutagenesis enhanced the binding capacity for the target protease [116, 164]. An alternative strategy for directing TIMP activity to the plasma membrane involves using constructs of TIMPs fused to GPI anchors. This resulted in a shift in the TIMP-1 activity from ECM to the cell surface, ultimately resulting in apoptosis and decreased proliferation of renal carcinoma cells [165].

In a recent report by Darini et al. [166], it has been observed that HIV-protease inhibitors (HIV-PIs) selectively target CSCs from tumors with different origins. HIV-PIs were observed to reduce proliferation and induce cell death efficiently in a dose-dependent manner with enhanced specificity for CSCs in comparison with the entire cell population of cancer cells or embryonic stem cells. Lopinavir (LPV), the most effective HIV-PI among the studied, has been shown to impair self-renewal and induce apoptosis of CSCs, consequently inhibiting CSC-induced allograft formation in vivo. Killing of CSCs by LPV involved activation of caspase 3 (CASP3) expression and subsequent cleavage of poly (ADP-ribose) polymerase (PARP-1). Moreover, two key pharmacophores within the LPV structure were recognized, which enable LPV to selectively target CSCs. In addition to this, several protease inhibitors that are used in combination with other drugs to treat HIV infection may also be effective in targeting certain types of cancer [167]. Altogether, these reports identify protease inhibitors as compounds having selective toxicity for CSCs with a stemness signature.

13 Conclusion

With the advent of the CSC theory, the intriguing questions, relevant in the field of cancer biology, which remained unanswered for many years, have been addressed to certain extent. The CSCs have been predominantly identified as the master regulators in imparting distinct properties and characteristics to the tumors. It is implicated that the proteolytic networks are critically involved in the functioning of the CSCs. The niche of the CSC, which harbors different cell types, assists the CSCs in their activities. The proteases are involved in the regulation of distinctive cellular functions which ultimately modulates the development of cancer. The above discussions have highlighted the critical crosstalk between CSCs and proteases and how they are intricately involved in regulating the multiple steps of carcinogenesis starting from tumor initiation, angiogenesis, invasion and metastasis, proliferation, and survival. Apart from this, the CSCs also significantly modulate the process of apoptosis via proteases. The resistance exhibited by the tumors to various drugs ultimately leads to recurrence of the disease. In this context, the proteases play an essential role in drug resistance. On the other hand, the proteases have been found to display a dual role in regulating immune surveillance of CSCs. Therefore, our discussion emphasizes the widespread activities of proteases, which are majorly regulated by the CSCs and its associated niche, in determining the progress of tumor development and sustainability. Hence, antiprotease therapeutics may be regarded as a single bullet through which multifaceted proteolytic networks can be targeted.

References

1. Ferlay J, Soerjomataram I, Dikshit R et al (2015) Cancer incidence and mortality worldwide: sources, methods and major patterns in GLOBOCAN 2012. Int J Cancer 136:E359–E386
2. Hanahan D, Weinberg RA (2011) Hallmarks of cancer: the next generation. Cell 144 (5):646–674
3. Kreso A, Dick JE (2014) Evolution of the cancer stem cell model. Cell Stem Cell 14(3): 275–291
4. Visvader JE, Lindeman GJ (2008) Cancer stem cells in solid tumors: accumulating evidence and unresolved questions. Nat Rev Cancer 8:755–768
5. Kise K, Kinugasa-Katayama Y, Takakura N (2016) Tumor microenvironment for cancer stem cells. Adv Drug Deliv Rev 99:197–205
6. Pietras K, Ostman A (2010) Hallmarks of cancer: interactions with the tumor stroma. Exp Cell Res 316:1324–1331
7. Bhowmick NA, Neilson EG, Moses HL (2004) Stromal fibroblasts in cancer initiation and progression. Nature 432:332–337
8. Raza A, Franklin MJ, Dudek AZ (2010) Pericytes and vessel maturation during tumor angiogenesis and metastasis. Am J Hematol 85:593–598
9. Mao Y, Keller ET, Garfield DH et al (2013) Stromal cells in tumor microenvironment and breast cancer. Cancer Metastasis Rev 32:303–315
10. Hewitt R, Danø K (1996) Stromal cell expression of components of matrix-degrading protease systems in human cancer. Enzyme Protein 49:163–173
11. Mason SD, Joyce JA (2011) Proteolytic networks in cancer. Trends Cell Biol 21:228–237
12. Lopez-Otin C, Bond JS (2008) Proteases: multifunctional enzymes in life and disease. J BiolChem 283:30433–30437
13. López-Otín C, Hunter T (2010) The regulatory crosstalk between kinases and proteases in cancer. Nat Rev Cancer 10:278–292
14. Yang Y, Hao Hong H, Yin Zhang Y et al (2009) Molecular imaging of proteases in cancer. Cancer Growth Metastasis 2:13–27
15. Duffy MJ (1996) Proteases as prognostic markers in cancer. Clin Cancer Res 2:613–618
16. Zucker S, Cao J, Chen WT (2000) Critical appraisal of the use of matrix metalloproteinase inhibitors in cancer treatment. Oncogene 19:6642–6650
17. Lopez-Otin C, Overall CM (2002) Protease degradomics: a new challenge for proteomics. Nat Rev Mol Cell Biol 3:509–519
18. Rakashanda S, Rana F, Rafiq S et al (2012) Role of proteases in cancer: a review. Biotechnol Mol Biol Rev 7:90–101
19. Quesada V, Ordóñez GR, Sánchez LM et al (2009) The Degradome database: mammalian proteases and diseases of proteolysis. Nucleic Acids Res 37:D239–D243
20. Puente XS, Sánchez LM, Overall CM et al (2003) Human and mouse proteases: a comparative genomic approach. Nat Rev Genet 4:544–558
21. Choi KY, Swierczewska M, Lee S et al (2012) Protease-activated drug development. Theranostics 2:156–178
22. Turk B (2006) Targeting proteases: successes, failures and future prospects. Nat Rev Drug Discov 5:785–799
23. López-Otín C, Matrisian LM (2007) Emerging roles of proteases in tumor suppression. Nat Rev Cancer 7:800–808
24. Schroter F, Adjaye J (2014) The proteasome complex and the maintenance of pluripotency: sustain the fate by mopping up? Stem Cell Res Ther 5:24
25. Rape M, Jentsch S (2002) Taking a bite: proteasomal protein processing. Nature Cell Bio 4: E113–E116
26. Yasutaka O, Keiichi IN (2012) UPS delivers pluripotency. Cell Stem Cell 11:728–730
27. Pan J, Zhang Q, Wang Y et al (2010) 26S Proteasome activity is down-regulated in lung cancer stem-like cells propagated in vitro. PLoS ONE 5:e13298

28. Chinchar E, Makey KL, Gu, JW (2014) Sunitinib significantly suppresses the proliferation, migration, apoptosis resistance, tumor angiogenesis and growth of triple-negative breast cancers but increases breast cancer stem cells. Vasc Cell. 6
29. Hill RP, Marie-Egyptienne DT, Hedley DW (2009) Cancer stem cells, hypoxia and metastasis. Semin Radiat Oncol 19:106–111
30. Catalano V, Turdo A, Di Franco S et al (2013) Tumor and its microenvironment: a synergistic interplay. Semin Cancer Biol 23:522–532
31. Han L, Shi S, Gong T et al (2013) Cancer stem cells: therapeutic implications and perspectives in cancer therapy. Acta Pharm Sin B 3:65–75
32. Clarke MF, Dick JE, Dirks PB (2006) Cancer stem cells—perspectives on current status and future directions: ACCR workshop on cancer stem cells. Cancer Res 66:9339–9344
33. Baker M (2008) Cancer stem cells, becoming common. Nat Rep Stem Cells. doi:10.1038/stemcells.2008.153
34. Seton-Rogers S (2011) Cancer stem cells. VEGF promotes stemness. Nat Rev Cancer 11:831
35. Tang S, Xiang T, Huang S et al (2016) Ovarian cancer stem-like cells differentiate into endothelial cells and participate in tumor angiogenesis through autocrine CCL5 signaling. Cancer Lett 376:137–147
36. Bussolati B, Grange C, Sapino A et al (2009) Endothelial cell differentiation of human breast tumour stem/progenitor cells. J Cell Mol Med 13:309–319
37. Kumar D, Kumar S, Gorain M et al (2016) Notch1-MAPK signaling axis regulates CD133+ cancer stem cell-mediated melanoma growth and angiogenesis. J Invest Dermatol pii: S0022-202X(16)32232-1
38. Saha S, Mukherjee S, Mazumdar M et al (2014) Mithramycin A sensitizes therapy-resistant breast cancer stem cells toward genotoxic drug doxorubicin. Transl Res 165:558–577
39. Saha S, Mukherjee S, Khan P et al (2016) Aspirin suppress the acquisition of chemoresistance in breast cancer by disrupting an NFκB-IL6 signaing axis responsible for the generation of Cancer Stem Cells. Cancer Res 76:2000–2012
40. Mukherjee S, Mazumdar M, Chakraborty S et al (2014) Curcumin inhibits breast cancer stem cell migration by amplifying the E-cadherin/β-catenin negative feedback loop. Stem Cell Res Ther 5:116
41. Mukherjee S, Manna A, Bhattacharjee P et al (2016) Non-migratory tumorigenic intrinsic cancer stem cells ensure breast cancer metastasis by generation of CXCR4+ migrating cancer stem cells. Oncogene. doi:10.1038/onc.2016.26
42. Tang DG (2012) Understanding cancer stem cell heterogeneity and plasticity. Cell Res 22:457–472
43. Chakraborty C, Chin KY, Das S (2016) miRNA-regulated cancer stem cells: understanding the property and the role of miRNA in carcinogenesis. Tumor Biol. doi:10.1007/s13277-016-5156-1
44. Korkaya H, Liu S, Wicha MS (2006) Breast cancer stem cells, cytokine networks, and the tumor microenvironment. J Clin Invest 121:3804–3809
45. Scadden DT (2006) The stem-cell niche as an entity of action. Nature 441:1075–1079
46. Fessler E, Dijkgraaf FE, De Sousa E, Melo F et al (2013) Cancer stem cell dynamics in tumor progression and metastasis: is the microenvironment to blame? Cancer Lett 34:97–104
47. Bennewith KL, Durand RE (2004) Quantifying transient hypoxia in human tumorxenografts by flow cytometry. Cancer Res 64:6183–6189
48. Wong DJ, Liu H, Ridky TW et al (2008) Module map of stem cell genes guides creation of epithelial cancer stem cells. Cell Stem Cell 2:333–344
49. Vermeulen L, De Sousa E, Melo F, van der Heijden M (2010) Wnt activity defines colon cancer stem cells and is regulated by the microenvironment. Nat Cell Biol 12:468–476
50. Ohishi K, Varnum-Finney B, Bernstein ID (2002) The Notch pathway: modulation of cell fate decisions in hematopoiesis. Int J Hematol 75:449–459

51. Chanmee T, Ontong P, Mochizuki N et al (2014) Excessive hyaluronan production promotes acquisition of cancer stem cell signatures through the coordinated regulation of Twist and the transforming growth factor β (TGF-β)-Snail signaling axis. J BiolChem 289:26038–26056
52. Kitamura T, Qian BZ, Pollard JW (2015) Immune cell promotion of metastasis. Nat Rev Immunol 15:73–86
53. Noy R, Pollard JW (2014) Tumor-associated macrophages: from mechanisms to therapy. Immunity 41:49–61
54. Siefert SA, Sarkar R (2012) Matrix metalloproteinases in vascular physiology and disease. Vascular 20:210–216
55. Kessenbrock K, Dijkgraaf GJ, Lawson DA et al (2013) A role for matrix metalloproteinases in regulating mammary stem cell function via the Wnt signaling pathway. Cell Stem Cell 13:300–313
56. Jones PA, Baylin SB (2002) The fundamental role of epigenetic events in cancer. Nat Rev Genet 3:415–428
57. Chen J (2012) Regulation of tumor initiation and metastatic progression by Eph receptor tyrosine kinases. Adv Cancer Res 114:1–20
58. Hagerling C, Casbon AJ, Werb Z (2015) Balancing the innate immune system in tumor development. Trends Cell Biol 25:214–220
59. Boumahdi S, Driessens G, Lapouge G (2014) SOX2 controls tumor initiation and cancer stem-cell functions in squamous-cell carcinoma. Nature 511:246–250
60. Plaks V, Kong N, Werb Z (2015) The cancer stem cell niche: how essential is the niche in regulating stemness of tumor cells. Cell Stem Cell 16:225–238
61. Chen DY, Liu H, Takeda S et al (2010) Taspase1 functions as a non-oncogene addiction protease that coordinates cancer cell proliferation and apoptosis. Cancer Res 70:5358–5367
62. Hsieh JJ, Cheng EH, Korsmeyer SJ (2003) Taspase1: a threonine aspartase required for cleavage of MLL and proper HOX gene expression. Cell 115:293–303
63. Wünsch D, Hahlbrock A, Jung S et al (2016) Taspase1: a 'misunderstood' protease with translational cancer relevance. Oncogene 35:3351–3364
64. Kumar S, Kulkarni R, Sen S (2016) Cell motility and ECM proteolysis regulate tumor growth and tumor relapse by altering the fraction of cancer stem cells and their spatial scattering. PhysBiol 13:036001
65. Sevenich L, Joyce JA (2014) Pericellular proteolysis in cancer. Genes Dev 28:2331–2347
66. Folkman J (1971) Tumor angiogenesis: therapeutic implications. N Engl J Med 285:1182–1186
67. Muthukkaruppan VR, Kubai L, Auerbach R (1982) Tumor-induced neovascularization in the mouse eye. J Natl Cancer Inst 69:699–708
68. Holmgren L, O'Reilly MS, Folkman J (1995) Dormancy of micrometastases: balance proliferation and apoptosis in the presence of angiogenesis suppression. Nat Med 1:149–153
69. Parangi S, O'Reilly M, Christofori G et al (1996) Angiogenesis therapy of transgenic mice impairs de novo tumor growth. Proc Natl Acad Sci USA 93:2002–2007
70. Dameron KM, Volpert OV, Tainsky MA et al (1994) Control of angiogenesis in fibroblasts by p53 regulation of thorombospondin-1. Science 265:1582–1584
71. Gu JW, Rizzo P, Pannuti A et al (2012) Notch signals in the endothelium and cancer "stem-like" cells: opportunities for cancer therapy. Vascular Cell 4:7
72. Matsuda S, Yan T, Mizutani A et al (2014) Cancer stem cells maintain a hierarchy of differentiation by creating their niche. Int J Cancer 135:27–36
73. Folkman J, Klagsbrun M (1987) Angiogenic factors. Science 235:442–447
74. Kiba A, Yabana N, Shibuya M (2003) A set of loop-1 and -3 structures in the novel VEGF family member, VEGF-ENZ-7, is essential for the activation of VEGFR-2 signaling. J Biol Chem 278:13453–13461
75. Bao S, Wu Q, Sathornsumetee S et al (2006) Stem cell-like glioma cells promote tumor angiogenesis through vascular endothelial growth factor. Cancer Res 66:7843–7848

76. Hadjimichael C, Chanoumidou K, Papadopoulou N (2015) Common stemness regulators of embryonic and cancer stem cells. World J Stem Cells 7:1150–1184
77. Egeblad M, Werb Z (2002) New functions for the matrix metalloproteinases in cancer progression. Nat Rev Cancer 2:161–174
78. Yana I, Weiss SJ (2000) Regulation of membrane type-1 matrix metalloproteinase activation by proproteinconvertases. Mol Biol Cell 11:2387–2401
79. Handsley MM, Edwards DR (2005) Metalloproteinases and their inhibitors in tumor angiogenesis. Int J Cancer 115:849–860
80. Kajita M, Itoh Y, Chiba T et al (2001) Membrane-type 1 matrix metalloproteinase cleaves CD44 and promotes cell migration. J Cell Biol 153:893–904
81. Seiki M (2003) Membrane-type 1 matrix metalloproteinase: a key enzyme for tumor invasion. Cancer Lett 194:1–11
82. Brooks PC, Stromblad S, Sanders LC et al (1996) Localization of matrix metalloproteinase MMP-2 to the surface of invasive cells by interaction with integrin alpha v beta 3. Cell 85:683–693
83. Shibuya M (2011) Vascular endothelial growth factor (VEGF) and its receptor (VEGFR) signaling in angiogenesis: a crucial target for anti- and pro-angiogenic therapies. Genes Cancer 2:1097–1105
84. Brauer R, Beck IM, Roderfeld M et al (2011) Matrix metalloproteinase-19 inhibits growth of endothelial cells by generating angiostatin-like fragments from plasminogen. BMC Biochem 12:38
85. Felbor U, Dreier L, Bryant RA et al (2000) Secreted cathepsin L generates endostatin from collagen XVIII. EMBO J 19:1187–1194
86. Kessenbrock K, Plaks V, Werb Z (2010) Matrix metalloproteinases: regulators of the tumor microenvironment. Cell 141:52–67
87. Li S, Huang NF, Hsu S (2005) Mechanotransduction in endothelial cell migration. J CellBiochem 96:1110–1126
88. Iivanainen E, Kähäri VM, Heino J et al (2003) Endothelial cell-matrix interactions. Microsc Res Tech 60:13–22
89. Du R, Lu KV, Petritsch C (2008) HIF1alpha induces the recruitment of bone marrow-derived vascular modulatory cells to regulate tumor angiogenesis and invasion. Cancer Cell 13:206–220
90. Wojtukiewicz MZ, Sierko E, Klement P et al (2001) The hemostatic system and angiogenesis in malignancy. Neoplasia 3:371–384
91. Qiao L, Liang N, Zhang J et al (2015) Advanced research on vasculogenic mimicry in cancer. J Cell Mol Med 19:315–326
92. Zijl FV, Krupitza G, Mikulits W (2011) Initial steps of metastasis: cell invasion and endothelial transmigration. Mutat Res 728:23–34
93. Iqbal W, Alkarim S, AlHejin A et al (2016) Targeting signal transduction pathways of cancer stem cells for therapeutic opportunities of metastasis. Oncotarget. doi:10.18632/oncotarget.10942
94. Chabottaux V, Ricaud S, Host L et al (2009) Membrane-type 4 matrix metalloproteinase (MT4-MMP) induces lung metastasis by alteration of primary breast tumor vascular architecture. J Cell Mol Med 13:4002–4013
95. Klimstra D, Reinheckel T, Peters C et al (2006) Distinct roles for cysteine cathepsin genes in multistage tumorigenesis. Genes Dev 20:543–556
96. Noë V, Fingleton B, Jacobs K et al (2001) Release of an invasion promoter E-cadherin fragment by matrilysin and stromelysin-1. J Cell Sci 114:111–118
97. Najy AJ, Day KC, Day ML (2008) Theectodomain shedding of E-cadherin by ADAM15 supports ErbB receptor activation. J Biol Chem 283:18393–18401
98. Grabowska MM, Sandhu B, Day ML (2012) EGF promotes the shedding of soluble E-cadherin in an ADAM10-dependent manner in prostate epithelial cells. Cell Signal 24:532–538

99. Juncker-Jensen A, Deryugina EI, Rimann I (2013) Tumor MMP-1 activates endothelial PAR1 to facilitate vascular intravasation and metastatic dissemination. Cancer Res 73:4196–4211
100. Feng S, Cen J, Huang Y (2011) Matrix metalloproteinase-2 and -9 secreted by leukemic cells increase the permeability of blood-brain barrier by disrupting tight junction proteins. PLoS ONE 6:e20599
101. Tsai JH, Yang J (2013) Epithelial-mesenchymal plasticity in carcinoma metastasis. Genes Dev 27:2192–2206
102. Mani SA, Guo W, Liao MJ (2008) The epithelial-mesenchymal transition generates cells with properties of stem cells. Cell 133:704–715
103. Nishida C, Kusubata K, Tashiro Y et al (2011) MT1-MMP plays a critical role in hematopoiesis by regulating HIF-mediated chemokine/cytokine gene transcription within niche cells. Blood 119:5405–5416
104. Wang Z, von Au A, Schnölzer M et al (2016) CD44v6-competent tumorexosomes promote motility, invasion and cancer-initiating cell marker expression. Oncotarget. doi:10.18632/oncotarget.10580
105. D'Eliseo D, Di Rocco G, Loria R et al (2016) Epitelial-to-mesenchimal transition and invasion are upmodulated by tumor-expressed granzyme B and inhibited by docosahexaenoic acid in human colorectal cancer cells. J ExpClin Cancer Res. doi:10.1186/s13046-016-0302-6
106. Gao Y, Feng J, Wu L et al (2015) Expression and pathological mechanism of MMP-9 and HIF-2α in CD133(+) lung cancer stem cells. Zhonghua Yi XueZaZhi 95:2607–2611
107. Talukdar S, Das SK, Pradhan AK, Emdad et al (2016) Novel function of MDA-9/Syntenin (SDCBP) as a regulator of survival and stemness in glioma stem cells. Oncotarget. doi:10.18632/oncotarget.10851
108. Ito K, Ito K (2016) Metabolism and the control of cell fate decisions and stem cell renewal. Annu Rev Cell Dev Biol. doi:10.1146/annurev-cellbio-111315-125134
109. Baxter RC (2014) IGF binding proteins in cancer: mechanistic and clinical insights. Nat Rev Cancer 14:329–341
110. Miyamoto S, Yano K, Sugimoto S et al (2004) Matrix metalloproteinase-7 facilitates insulin-like growth factor bioavailability through its proteinase activity on insulin-like growth factor binding protein 3. Cancer Res 64:665–671
111. Hemers E, Duval C, McCaig C et al (2005) Insulin-like growth factor binding protein-5 is a target of matrix metalloproteinase-7: implications for epithelial-mesenchymalsignaling. Cancer Res 65:7363–7369
112. Mochizuki S, Shimoda M, Shiomi T et al (2004) ADAM28 is activated by MMP-7 (matrilysin-1) and cleaves insulin-like growth factor binding protein-3. Bio Chem Biophys Res Commun 315:79–84
113. Sylvain MG, Thorsten M, Priya D et al (2010) ADAM17 is regulated by a rapid and reversible mechanism that controls access to its catalytic site. J Cell Sci 123:3913–3922
114. Yu Q, Stamenkovic I (2000) Cell surface-localized matrix metalloproteinase-9 proteolytically activates TGF-beta and promotes tumor invasion and angiogenesis. Genes Dev 14:163–176
115. Lahiry L, Saha B, Chakraborty J et al (2010) Theaflavins target Fas/caspase-8 and Akt/pBad pathways to induce apoptosis in p53-mutated human breast cancer cells. Carcinogenesis 31:259–268
116. Kessenbrock K, Wang WY, Werb Z (2015) Matrix metalloproteinases in stem cell regulation and cancer. Matrix Biol 44:184–190
117. Nasri I, Bonnet D, Zwarycz B et al (2016) PAR2-dependent activation of GSK3β regulates the survival of colon stem/progenitor cells. Am J Physiol Gastrointest Liver Physiol 311: G221–G236

118. Amoury M, Kolberg K, Pham AT et al (2016) Granzyme B-based cytolytic fusion protein targeting EpCAM specifically kills triple negative breast cancer cells in vitro and inhibits tumor growth in a subcutaneous mouse tumor model. Cancer Lett 372:201–209
119. Longley DB, Johnston PG (2005) Molecular mechanisms of drug resistance. J Pathol 205:275–292
120. Mohanty S, Saha S, Md D, Hossain S (2014) ROS-PIASγ cross talk channelizes ATM signaling from resistance to apoptosis during chemosensitization of resistant tumors. Cell Death Dis 5:e1021
121. Dean M, Fojo T, Bates S (2005) Tumor stem cells and drug resistance. Nat Rev Cancer 5:275–284
122. Liu G, Yuan X, Zeng Z et al (2006) Analysis of gene expression and chemoresistance of CD133+ cancer stem cells in glioblastoma. Mol Cancer 5:67
123. Eckford PD, Sharom FJ (2009) ABC efflux pump-based resistance to chemotherapy drugs. Chem Rev 109:2989–3011
124. Scharenberg CW, Harkey MA, Torok-Storb B (2002) The ABCG2 transporter is an efficient Hoechst 33342 efflux pump and is preferentially expressed by immature human hematopoietic progenitors. Blood 99:507–512
125. Wang J, Sullenger BA, Rich JN (2012) Notch signaling in cancer stem cells. Adv Exp Med Biol 727:174–185
126. Wang Z, Da Silva TG, Jin K et al (2014) Notch signaling drives stemness and tumorigenicity of esophageal adenocarcinoma. Cancer Res 74:6364–6374
127. D'Angelo RC, Ouzounova M, Davis A et al (2015) Notch reporter activity in breast cancer cell lines identifies a subset of cells with stem cell activity. Mol Cancer Ther 14:779–787
128. Theys J, Yahyanejad S, Habets R et al (2013) High Notch activity induces radiation resistance in non small cell lung cancer. Radiother Oncol1 08:440–445
129. Yahyanejad S, Theys J, Vooijs M (2016) Targeting Notch to overcome radiation resistance. Oncotarget 7:7610–7628
130. Phillips TM, McBride WH, Pajonk F (2006) The response of CD24(-/low)/CD44+ breast cancer-initiating cells to radiation. J Natl Cancer Inst 98:1777–1785
131. Lagadec C, Vlashi E, Alhiyari Y et al (2013) Radiation-induced Notch signaling in breast cancer stem cells. Int J Radiat Oncol Biol Phys 87:609–618
132. Mamaeva V, Niemi R, Beck M et al (2016) Inhibiting Notch activity in breast cancer stem cells by glucose functionalized nanoparticles carrying γ-secretase Inhibitors. Mol Ther 24:926–936
133. Austin KM, Covic L, Kuliopulos A (2013) Matrix metalloproteases and PAR1 activation. Blood 121:431–439
134. Fujimoto D, Ueda Y, Hirono Y et al (2015) PAR1 participates in the ability of multidrug resistance and tumorigenesis by controlling Hippo-YAP pathway. Oncotarget 6:34788–34799
135. He YC, Zhou FL, Shen Y et al (2014) Apoptotic death of cancer stem cells for cancer therapy. Int J MolSci 15:8335–8351
136. Inohara N, Nuñez G (2001) The NOD: a signaling module that regulates apoptosis and host defense against pathogens. Oncogene 20:6473–6481
137. Cecconi F, Alvarez-Bolado G, Meyer BI et al (1998) Apaf1 (CED-4 homolog) regulates programmed cell death in mammalian development. Cell 94:727–737
138. Yoshida H, Kong YY, Yoshida R et al (1998) Apaf1 is required for mitochondrial pathways of apoptosis and brain development. Cell 94:739–750
139. Yu CJ, Ou JH, Wang ML et al (2015) Elevated survivin mediated multidrug resistance and reduced apoptosis in breast cancer stem cells. J BUON 20:1287–1294
140. Yan H, Tong J, Lin X et al (2015) Effect of the WWOX gene on the regulation of the cell cycle and apoptosis in human ovarian cancer stem cells. Mol Med Rep 12:1783–1788

141. Jinesh GG, Choi W, Shah JB et al (2013) Blebbishields, the emergency program for cancer stem cells: sphere formation and tumorigenesis after apoptosis. Cell Death Differ 20:382–395
142. Lu Y, Zhang C, Li Q et al (2015) Inhibitory effect of salinomycin on human breast cancer cells MDA-MB-231 proliferation through Hedgehog signaling pathway. Zhonghua Bing Li XueZaZhi 44:395–398
143. Mantovani A, Allavena P, Sica A et al (2008) Cancer-related inflammation. Nature 454 (436):444
144. Lowe DB, Storkus WJ (2011) Chronic inflammation and immunologic-based constraints in malignant disease. Immunotherapy 3:1265–1274
145. Colotta F, Allavena P, Sica A et al (2009) Cancer-related inflammation, the seventh hallmark of cancer: links to genetic instability. Carcinogenesis 30:1073–1081
146. Shigdar S, Li Y, Bhattacharya S et al (2014) Inflammation and cancer stem cells. Cancer Lett 345:271–278
147. Jinushi M (2014) Role of cancer stem cell-associated inflammation in creating pro-inflammatory tumorigenic microenvironments. Oncoimmunology 15:e28862
148. Hagemann T, Robinson SC, Schulz M et al (2004) Enhanced invasiveness of breast cancer cell lines upon co-cultivation with macrophages is due to TNF-alpha dependent up-regulation of matrix metalloproteases. Carcinogenesis 25:1543–1549
149. Bengsch F, Buck A, Günther SC et al (2014) Cell type-dependent pathogenic functions of overexpressed human cathepsin B in murine breast cancer progression. Oncogene 33:4474–4484
150. Mohamed MM, Cavallo-Medved D, Rudy D et al (2010) Interleukin-6 increases expression and secretion of cathepsin B by breast tumor-associated monocytes. Cell Physiol Biochem 25:315–324
151. Gopinath S, Malla R, Alapati K et al (2013) Cathepsin B and uPAR regulate self-renewal of glioma-initiating cells through GLI-regulated Sox2 and Bmi1 expression. Carcinogenesis 34:550–559
152. Dunn GP, Old LJ, Schreiber RD (2004) The immunobiology of cancer immunosurveillance and immunoediting. Immunity 21:137–148
153. Ryungsa K, Manabu E, Kazuaki T et al (2007) Cancer immunoediting from immune surveillance to immune escape. Immunology 121:1–14
154. Swann Jeremy B, Smyth Mark J (2007) Immune surveillance of tumors. J Clin Invest 117:1137–1146
155. Cullen SP, Brunet M, Martin SJ (2010) Granzymes in cancer and immunity. Cell Death Differ 17:616–623
156. Waldhauer I, Steinle A (2008) NK cells and cancer immuno surveillance. Oncogene 27:5932–5943
157. Strik MC, de Koning PJ, Kleijmeer MJ et al (2007) Human mast cells produce and release the cytotoxic lymphocyte associated protease granzyme B upon activation. MolImmunol 44:3462–3472
158. Dimitriadou V, Koutsilieris M (1997) Mast cell–tumor cell interactions: for or against tumor growth and metastasis. Anticancer Res 17:1541–1549
159. Ames E, Canter RJ, Grossenbacher SK et al (2015) NK cells preferentially target tumor cells with a cancer stem cell phenotype. J Immunol 195:4010–4019
160. Waldhauer I, Goehlsdorf D, Gieseke F et al (2008) Tumor-associated MICA is shed by ADAM proteases. Cancer Res 68:6368–6376
161. Jachetti E, Caputo S, Mazzoleni S et al (2015) Tenascin-C protects cancer stem-like cells from immune surveillance by arresting T-cell activation. Cancer Res 75:2095–2108
162. Sarkar S, Zemp FJ, Senger D et al (2015) ADAM-9 is a novel mediator of tenascin-C-stimulated invasiveness of brain tumor-initiating cells. Neuro Oncol 17:1095–1105

163. Remacle AG, Golubkov VS, Shiryaev SA, Dahl R, Stebbins JL, Chernov AV, Cheltsov AV, Pellecchia M, Strongin AY (2012) Novel MT1-MMP small-molecule inhibitors based on insights into hemopexin domain function in tumor growth. Cancer Res 72(9):2339–2349
164. Butler GS, Hutton M, Wattam BA et al (1999) The specificity of TIMP-2 for matrix metalloproteinases can be modified by single amino acid mutations. J BiolChem 274:20391–20396
165. Djafarzadeh R, Noessner E, Engelmann H et al (2006) GPI-anchored TIMP-1 treatment renders renal cell carcinoma sensitive to FAS-meditated killing. Oncogene 25:1496–1508
166. Darini CY, Martin P, Azoulay S et al (2013) Targeting cancer stem cells expressing an embryonic signature with anti-proteases to decrease their tumor potential. Cell Death Dis 4: e706
167. Bernstein WB, Dennis PA (2008) Repositioning HIV protease inhibitors as cancer therapeutics. CurrOpin HIV AIDS 3:666–675

Role of Proteases in Tumor Immune Evasion

Abir K. Panda, Sreeparna Chakraborty, Kirti Kajal, Dia Roy, Tania Sarkar and Gaurisankar Sa

Abstract

Proteolysis or protein degradation is an important biological function which has been attributed to a major class of enzymes called proteases. In the living system, a fine-tuning between the activity of proteases and their counterparts can be seen, and disruption of this balance leads to the occurrence of various diseases including cancer. Several studies indicate that protease activity highly correlated with cancer advancement. The hallmarks of cancer like tissue evasion and metastasis, apoptosis, angiogenesis largely depend on proteolytic degradation and protease activation. Protease also contributes to avoidance of immune system, one of the emerging hallmarks in cancer progression. The immune system can recognize intracellular pathogens including cancer and elicits an effective immune response to restrict their activity and subsequently protect the host. In cancer microenvironment, immune cells are unable to recognize tumor neoantigens as well as incapable of inducing proper immune response for the clearance of tumor cells. Recent studies found that proteases not only promote tumor cell migration and metastasis it also encourages to maintain the tolerogenic tumor microenvironment by modulation of immune cell functions. Proteases are involved in several immune responses such as antigen processing and presentation, lymphocytes and neutrophil infiltration, activation of dendritic cells. The anomalies of protease actions dampen immune cell activities and establish immune tolerance. Inappropriate protease activities are responsible for inhibiting immunosurveillance and maintaining immune tolerance that ultimately leads to tumor immune evasion. Involvement of proteases in cancer suggests the use of protease inhibitors as anticancer drugs to targets not only tumor cells as well as tolerogenic immune cells to reeducate immune responses

A.K. Panda · S. Chakraborty · K. Kajal · D. Roy · T. Sarkar · G. Sa (✉)
Division of Molecular Medicine, Bose Institute, P-1/12 CIT Scheme VII M, Kolkata 700054, India
e-mail: gauri@jcbose.ac.in

S. Chakraborti and N.S. Dhalla (eds.), *Pathophysiological Aspects of Proteases*,
DOI 10.1007/978-981-10-6141-7_12

and reestablish immune surveillance. This chapter summarizes the current understanding of the interplay between proteases and the immune cells of the body and their involvement in cancer progression.

1 Introduction

The immune system comprises of distinct specialized cells that act in a collaborative way to recognize the presence of foreign invaders and cancer outgrowth and induces specific effective immune responses to clear them from vertebrate's body. Although, the immune system is organized and orchestrated at various cellular, molecular, and systemic levels to discriminate and stimulate effective tumor immunity, but cancer also exploits several strategies to avoid immune surveillance and established immunosuppression that subsequently leads to cancer progression [1]. Tumor microenvironment mainly consists of tumor cells, stromal cells, extracellular matrix and immune cell that allow the cancer cells to foster proliferation, survival, metastasis, and evasion. Therefore, successful, effective immune responses against cancer thus hindered by several immune escape strategies such as tumor inducing tolerogenic environment, avoidance of immune systems, activation of negative regulatory pathways, and secretion of immunosuppressive factors [1, 2]. The immunosuppression can arise either from an outgrowth of escape tumor mutants or during tumor sculpting immune responses. The involvement of immune system in cancer development has long been witnessed. In 1867, Rudolf Virchow first observed that leukocyte infiltration highly correlated with fostering tumor inflammation. In 1957, Sir Macfarlane Thomas and Lewis Burnet postulated the evidence of tumor immune surveillance mechanisms: the immunological resistance mechanisms to eradicate cancer progression relentlessly. This hypothesis has been challenged for years due to lack of experimental validation. In 2003, Schreiber and colleagues put forward cancer immune editing hypothesis which consists of three different and integrated steps (a) elimination, (b) equilibrium, and (c) escape. This three "E"s of immune editing stages defined either cancer eradication (immunosurveillance) or cancer development (immune escape). Immunosurveillance takes place during the elimination process whereas Darwinian selection ensued in equilibrium phases and the clinically apparent tumor has been found in escape stages [3].

1.1 Elimination

In the elimination or cancer immunosurveillance stage, successful eradication of tumor development takes place without progression of subsequent stages. In the first phase of elimination, once solid tumors begin to grow invasively it requires an adequate amount of nutrients through the production of stromagenic and angiogenic factors by the formation of new blood vessels or angiogenesis [3, 4]. Invasive growth damages surrounding tissues and induces tumor-specific inflammatory signals that ultimately triggers the infiltration of innate immune cells (dendritic cells, macrophages, NK/NKT, γδ-T cells, etc.). The tumor-associated antigens displayed by transformed cells recognized by innate immune cells such as γδ-T cells, NK/NKT cells which are then stimulated to produce IFN-γ. In the second phase, IFN-γ that initially made by innate immune cells induces the limited amount of antiproliferative and apoptosis mediated tumor cell apoptosis. IFN-γ also activates to produce numerous chemokines and monokines which include CXCL-10, CXCL-11, and CXCL-9 from cancer cells and nearby surrounding normal tissues [3, 5]. Several monokines and chemokines have strong angiostatic activities, thus perturbing the formation of new blood vessels or angiogenesis process within the tumor which also induces tumor cell apoptosis. Tumor cell debris generated either direct or indirect consequences of IFN-γ. In the initial stages, tumor cells are ingested by local dendritic cells, which home to draining lymph nodes. Chemokines and monokines produced at tumor inflammation also recruit several immune cells such as macrophages, dendritic cells, NK cells at the tumor site. In the third phases, tumor-infiltrating NK cells and macrophages transactivate each other to produce IFN-γ and IL2 in a reciprocal way. Activated NK cells and macrophages also induce tumor cell killing through several mechanisms such as perforins, granzymes, reactive oxygen, nitrogen intermediates, and necrosis-related apoptosis-inducing ligand-mediated cell death. In the draining lymph nodes, newly immigrated activated dendritic cells also induce tumor-specific $CD4^+$ IFN-γ^+ T cells (Th1 cells) that in turn stimulated the development of $CD8^+$ IFN-γ^+ T cells or cytotoxic $CD8^+$ T cells (Tc cells). In the last phase of elimination, both Th1 and Tc cells infiltrate into tumor sites and destroy the remaining tumor antigen-bearing cancer cells that show enhanced immunogenicity after exposure of locally produced IFN-γ and other chemokines [3, 5, 6].

1.2 Equilibrium

In this stage, some transformed tumor cells have become resistant and survived in an elimination process that eventually enters into a dynamic equilibrium. In equilibrium-phase immune cells, chemokines and cytokines especially lymphocytes and IFN-γ exert strong selection pressure that is insufficient to complete clearance of mutant-transformed tumor cells. The tumor bed contains several genetically unstable and rapidly mutant cancer cells. The Darwinian selection process takes place during this period and many of the original preexistent tumor cells destroyed and cleared although newly transformed cells arisen that carry different mutations, eventually resistant to immune attacks and withstand to survive [3, 5, 6]. Tumor cells exploit several strategies to demolish the constant threat by immune cells and thereby maintained a quiescent, dormant state. In equilibrium phase, either immune surveillance or immune escape decision may take place [5, 7]. Depending on the situations, this equilibrium may swing toward either the elimination of tumor cells or their escape from immune surveillance. Equilibrium phase is the longest phase among three stages and may occur over a period of years.

1.3 Escape

In the escape phase, newly transformed mutant cells that acquired insensitivity and resistant to apoptosis escape from immunological detection constant immune attacks in immune surveillance stage. Mutant cells from escape stage acquired consecutive genetic and epigenetic alterations that constantly divide in an uncontrolled way that results clinically detected malignant cancer if left unchecked may cause the death of the host [5, 7, 8]. The highly malignant cancer cells modify themselves in such a way that are not detected by the immune system as a result effective immune response to complete clearance of them not successfully generated [9]. Moreover, avoidance of immune system ensues when cancer cells exploit several immunoevasion strategies like secretion of immunosuppressive factors such as TGFβ, IL-10, VEGF-A, PGE2, IDO, galectins that converts potent immunogenic immune cells become highly tolerogenic so that they cannot discriminate and destroy cancer cells. In addition to the death of T-effector and NK/NKT cells, shifting of M1 macrophages to M2 macrophages/tumor-associated macrophages (TAMs); conversion of immunogenic dendritic cells (iDCs) to plasmacytoid DCs; development of myeloid-derived stromal cells (MDSCs), and the generation of T-regulatory cells create a grave environment. Cancer cells completely paralyze the immune system and grow continuously in an uncontrolled manner [1, 7, 10].

Proteases play indispensable roles in several biological processes and are associated with several physiological conditions including cancer. Proteases are proteolytic enzymes that perform specific catalytic cleavage of peptide bonds and largely influence the cell behavior, cell death, and survival [11]. In 1946, Fisher first proposed degradation of extracellular matrix and successive tumor invasion into the surrounding normal tissue caused due to the tumor-associated proteolytic activities. The human degradome consists of at least 569 types of proteases that are divided into five subclasses. In human; 194 Metalloproteinases, 176 serine proteases, 150 cysteine proteases, 28 threonine proteases, and 21 aspartic proteases have been characterized to date [12]. The proteases can contribute all stages of tumor progression. Both intracellular and extracellular proteases can activate signaling cascades that are essential in cancer progression [13, 14]. Proteases regulate several biological activities which include proliferation, adhesion, apoptosis, autophagy, senescence, angiogenesis as well as immune evasion. The intracellular proteases are associated with aspartyl cathepsins and lysosomal cysteine-mediated degradation and removal of undesired or damaged endocytosed proteins. Moreover, cysteine proteases are responsible for tightly controlled apoptosis [13, 15]. The cysteine autophagins carry out self-eating that is mainly observed during starvation. Another cysteine protease deubiquitinases (DUBs) involved in the removal of ubiquitin modification or ubiquitin-related modifier (SUMO) from proteins. All of these intracellular proteases are involved in protective mechanisms. The loss of function mutation of these proteases is responsible for cancer development which clearly indicates that they serve as a tumor suppressors [15]. On the other hand, extracellular protease frequently gets overexpressed in several types of cancers by activating oncogenic transcriptional pathways. Matrix Metalloproteinases (MMPs) are the major class of extracellular proteases which are highly involved in several types of cancer progressions. Cysteine protease and serine protease also show antitumor activities. Extracellular proteases and pericellular which include MMPs, cathepsins, kallikreins, prostasin (PRSS8), neprilysin (MME), testisin (PRSS21), dipeptidyl peptidase 4 (DPP4), ADAMTSs (disintegrin-Metalloproteinases with thrombospondin domains) exert tumor-defying activities [15–18].

Both extracellular and intracellular proteases directly or indirectly involved in cancer progression as well as responsible for tumor immune suppression. Immune cells such as macrophages, neutrophils, polymorphonuclear leukocytes secret intense-level matrix Metalloproteinases (MMP-8, MMP-26) that induce cancer inflammation and metastasis. Protease executes several immunosuppressive strategies such as apoptosis, autophagy, immune cell infiltration into tumor site, immune cell dysfunction (dendritic cells, macrophages, NK, and NKT cells) antigen presentation which converts immunosurveillance to immune escape and established tumor immune evasion [19, 20].

2 Protease and Immune Systems in Cancer Immune Evasion

2.1 Apoptosis and Involvement of Protease in Immune Evasion

Apoptosis refers to an energy dependent form of programmed cell death that appears to be a crucial phenomenon of several distinct processes including the cycle of normal cell death and rebirth, hormonal regulation as well as immune system functioning. This coordinated catabolic process requires the coordinated involvement of an array of cysteine proteases namely endoproteases called "caspases." Apart from the two distinct apoptotic pathways, namely death receptor-mediated pathway called extrinsic pathway and mitochondrial pathway called intrinsic pathway, there is a third pathway involving T cell-mediated cytotoxicity and granzyme/perforin-mediated apoptosis [21]. All three modes of apoptosis converge to a common pathway involving the cleavage of caspase 3 which results in breakage of DNA and fragmentation of nuclear as well as cytoskeletal proteins. Caspase 3 is the most important effector caspase which is involved in the activation of the endonuclease CAD. CAD normally stays in its inactive form via forming a complex with ICAD. In apoptotic cells, caspase 3 releases CAD via cleavage of ICAD. CAD then causes DNA degradation and chromosomal condensation [22].

Caspases, expressed by most of the cells in their proactive forms, when get activated allows the activation of other pro-caspases, thereby initiating a caspase cascade. This chain reaction amplifies the apoptotic signal that in turn results in rapid cell death. Activation of caspases marks the irreversible commitment of the cell toward apoptosis. Till date, caspases can be categorized as initiators which include caspase 2, caspase 8, caspase 9, and caspase 10, effectors or executioner caspases, namely caspase 3, caspase 6 and caspase 7 as well as inflammatory caspases, namely caspase 1, caspase 4, and caspase 5. Apart from these three subtypes, few caspases like caspase 11 and caspase 12 play significant roles in the regulation of apoptosis during septic shock and endoplasmic–specific apoptosis, respectively.

2.1.1 Perforin-Granzyme Pathway–Potent Immunosurveillance Mechanism Against Tumor

T_c cells can exert their cytotoxic effects on the target cells via both extrinsic and intrinsic pathways. Apart from these, T_c cells secrete perforin that create pores on the target cells specifically the tumorigenic cells with the subsequent burst of cytoplasmic granules through the pores inside the target cells, thereby causing effective apoptosis of the target cells [23]. Two major players of this pathway are granzyme A and granzyme B. Apart from direct cytotoxicity, granzyme B can activate the execution-specific apoptotic caspase 3, thereby bypassing the initial step of apoptosis. Cytotoxicity mediated by granzyme B helps in the proliferation of Th2 cell types. Since Th2-type biasness is a marked feature of tumor progression, it can be fairly concluded that granzyme B-mediated apoptotic pathway can prevent Th2-biased and hence tumor progression.

Unlike granzyme B, granzyme A facilitates T cell-mediated apoptosis without the aid of caspase. Inside the cell, granzyme A cleaves the nucleosome assembly protein SET which in turn form complex with the tumor suppressor protein NM23-H1. Cleaving of SET restricts the inhibition of the tumor suppressor protein NM23-H1 which plays a major role in immunosurveillance to prevent cancer via the activation of tumor cell apoptosis mainly by causing DNA nicking [24].

2.1.2 Apoptosis in Maintaining T Cell Homeostasis

Caspase-dependent apoptotic pathway plays a significant role in numerous vital immunological processes that include lymphocyte development, positive and negative selection and homeostasis of T cells [25]. In the periphery, silencing of autoreactive lymphocytes occurs via inactivation also referred to as anergy or the autoreactive cells are destroyed by receptor-mediated apoptosis [26]. Furthermore, central tolerance relies mostly on the intrinsic mitochondria-based pathway of apoptosis stimulated by the T cell receptor-mediated pathway. Clonal deletion of the autoreactive T cells may be hampered if there occur some defects in the intrinsic mitochondria-based pathway of apoptosis. Thus, both the extrinsic and intrinsic pathways of apoptosis are vital for maintaining T cell homeostasis and both these pathways are executed via the activation of a cascade of proteolytic enzymes, as mentioned previously, called caspases [27].

In a nutshell, caspases, crucial for apoptosis, provide one of the robust mechanisms to safeguard body against cancer development. Reduced expression of caspases is validated in a variety of cancers as for example inactivating CASP8 mutations have been reported in different types of cancers. WT caspase 8 emerges out to be a potent tumor suppressor gene as it leads to death receptor-induced apoptosis. Thus, activation of caspases can act as a "safety catch" to be used in cancer therapy [28].

3 Autophagy and Role of Proteases in Tumor Immune Evasion

Autophagy emerges from Greek word "autophagy," meaning "eating self-components" refers to an array of highly regulated complex catabolic processes that lead to the degradation of surplus macromolecules or whole organelles, mostly mitochondria, thereby maintaining cellular homeostasis. The process involves the sequestration of the organelle within the double-membrane-bound vesicles called phagosomes which in turn fuse with a lysosome to form the autolysosome resulting in lysosomal degradation of the contents within the vesicles. This process encompasses the involvement of proteolytic enzymes that breaks the peptide bonds and thereby playing a major role in the initiation and progression of autophagy. Concomitantly proteases can also dampen autophagy under few circumstances [29]. In the current section, we will talk about an important protease family involved in autophagy-autophagins-4 mammalian orthologs of yeast Atg4 and its role in shaping the proper immune response of the body.

3.1 Role of Proteases in Autophagy

The initial step of macroautophagy involves the proteolytic cleavage of the autophagy-related proteins Atg8 by Atg4, a cysteine protease which in turn results in the fusion of Atg8 with phosphatidyl-ethanolamine and subsequently forms the autophagic vessel [30]. Based on the catalytic residues present in active site, Cathepsin A type of lysosomal hydrolase can be categorized into several groups, namely aspartic, cysteine, and serine cathepsins, which along with other factors cleaves the peptide bond of autophagy substrates facilitating the outflux of the products of autophagy. Apart from being a positive regulator of autophagy, some cathepsins, such as cathepsin L, help in degradation of specific protein constituents of the lysosomal membrane, namely GABARAP-II and LC3-II. Proteolytic processing of LAMP-2A by serine protease cathepsin A inhibits chaperon-mediated autophagy. Calpain, a non-lysosomal calcium-dependent protease, cleaves and inactivates major Atg proteins, thereby negatively influencing the execution of autophagy [31].

3.1.1 Mitophagy and Immune Cell Homeostasis in Context of Tumor Microenvironment

Mitochondria also referred to as the powerhouse of the cells plays an important function in various aspects of cell-mediated functions including production of energy, controller of apoptosis as well as a regulator of immune signaling. Therefore, maintenance of the functional attributes of mitochondria is of prime concern for the cell [32]. Mitophagy can be viewed as a specialized form of autophagy by which cell maintains an overall homeostasis of the mitochondrial population by identifying and repairing or eliminating the dysfunctional mitochondria from the system [33]. Mitophagy is also involved in supporting the immune system. Quiescent naive T cells after activation enter the log phase of growth. During log phase, the primed cells increase in size, undergo proliferation and finally differentiation to other subsets. Immune-suppressive regulatory T cell depends on fatty acid oxidation occurring in mitochondria for energy. Similarly, M2-like macrophages promoting the repair of tissue rely on beta-oxidation and oxidative phosphorylation for ATP production. Both regulatory T cells and M2 type of macrophages are predominant in the tumor microenvironment. Thus, dysfunction of mitophagy results in uncontrolled proliferation of regulatory T cells and tumor-associated macrophages (M2) which in turn results in tumor progression [34].

3.1.2 Role of Protease in Coordinating Mitophagy

Proteases regulate a number of proteolytic reactions in a signaling cascade, for a number of protein components. In this context, protein cleaving enzymes (AAA protease, Lon protease) are known to regulate the process of mitophagy. Comprising the first line of cellular defense, two membrane-bound AAA proteases maintain the integrity of the inner mitochondrial membrane by clearing the mitochondrial matrix from the damaged mitochondrial proteins. The Lon proteases remove the damaged proteins which constitute the matrix of the mitochondria. Apart from this, m-AAA and Lon proteases control mitochondrial biogenesis at the transcriptional and translational level, respectively [35]. Thus, overexpressing the proteases may be a way for prompt tumor progression as they positively regulate mitochondrial biogenesis and the large energy burst that enables a rapid proliferation of tumor-promoting immune cells (T-regulatory cells and M2 macrophages).

3.2 Autophagy and Innate Immunity in the Context of Protease System

Autophagy appears to be the one of the many arms of the body's first line of defense system and macrophages play an important role as a soldier of the innate immune system. In macrophages, secretion of active IL1b upon stimulation with LPS is controlled by autophagy [36]. ProIL1b is cleaved to its active IL1b form by the pro-inflammatory protease caspase 1. Strikingly, this activation step is controlled by an important autophagic protein Atg16 L1. It has been found that loss of expression of this autophagic protein leads to uncontrolled secretion of this

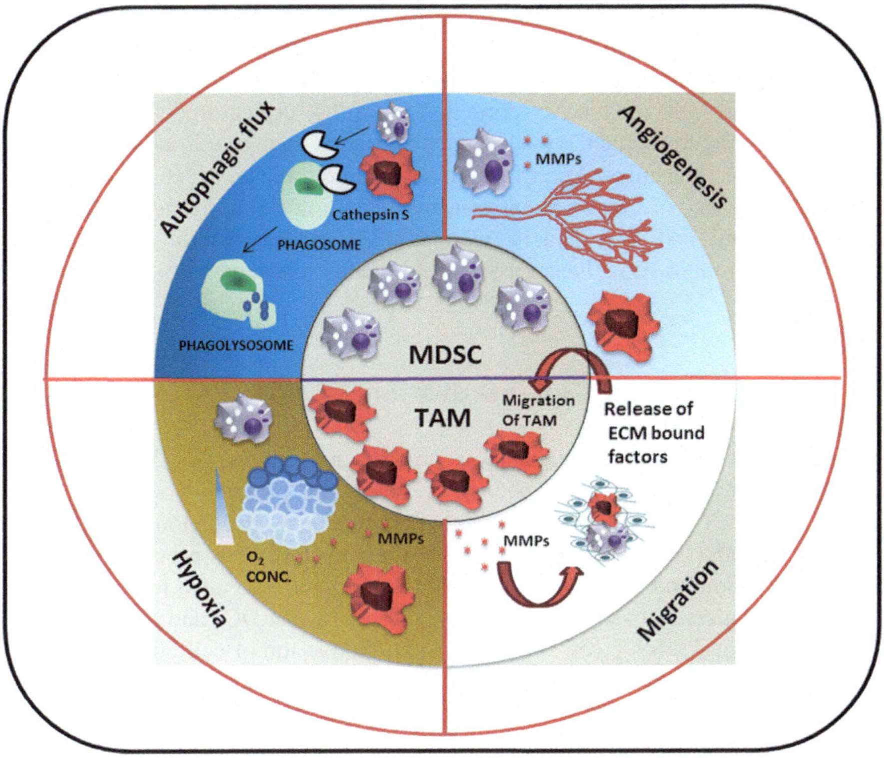

Fig. 1 **a** Autophagy and adaptive immunity: Lysosomal degradation of the antigen by cysteine protease-like cathepsin S produces small antigen peptides which are loaded onto the MHC-II molecules and are subsequently transported on to the cell surface. **b** Autophagy and innate immunity: Upon bacterial LPS stimulation, caspase 1 cleaves the pro-IL1ß to its active form. The functioning of caspase 1, in turn, is tightly controlled by an autophagy-related protein Atg16L1. Thus, the loss in activity of the autophagy-related protein leads to the uncontrolled secretion of IL1ß. **c** Tumor immunity and mitophagy: Membrane-bound AAA protease and Lon protease successfully remove the damaged misfolded proteins from the inner mitochondrial membrane and the mitochondrial matrix, respectively. This in turn positively regulates mitochondrial biogenesis and consecutive energy burst. This, in turn, leads to a rapid proliferation of regulatory T cells and tumor-associated macrophages and helps in tumor progression

inflammatory cytokine. Thus, tight regulation of autophagic process is of prime importance to prevent excessive inflammatory response which can be observed in autoimmune diseases as well as in many types of cancer (melanoma, prostate cancer) [37] (Fig. 1).

3.2.1 Autophagy: Role in MHC-Mediated Antigen Presentation in Context of Protease System

Autophagy plays a major role in modulation MHC class I and II antigen presentation which in turn regulates autoimmunity and tolerance as well as tumor immunity. Macroautophagy and cell-mediated autophagy turn out to be the major player in mediating antigen presentation by MHC molecules. MHC class I molecules present the antigens from endogenous viruses, tumors, or self-proteins to the $CD8^+$ T cells [38]. Cytosolic proteases such as proteasome degrade these endogenous antigens which are subsequently transported to ER and bind to the nascent MHC 1 molecules. Apart from this, cysteine proteases such as cathepsin S help in the process of cross-presentation by degrading the exogenous antigens in phagosome via TAP-independent manner [39]. This process is particularly important as immunosurveillance mechanism against the tumor [40].

4 Proteases and DC Cell Inactivation

During antigen presentation, endogenous proteins are presented on MHC-I of DCs to CD8+ T-lymphocytes and exogenous proteins are presented on MHC-II and activate CD4+ T-lymphocytes. In another mechanism, exogenous proteins are presented on MHC-I and stimulate CD8+ T-lymphocytes. This procedure is called cross-presentation and it is the crucial defense mechanism against tumor cells [41, 42]. The formation of an immunological synapse between DCs and T-lymphocytes is critical for successful T cell activation and effective tumor cell elimination [43]. Matrix metalloproteinase 13 (MMP-13) regulates dendritic cells (DCs) immune biology. DCs express MMP-13, and it has an important role in the context of the tumor. MMP-13 enhances the antitumorigenic function of DCs by up-regulating MHC-I surface expression, cytokine/chemokine secretion, and T-lymphocyte activation. MMP-9 is crucial for migration of DCs in response to tumor antigens, but MMP-13 has no role in the migratory capacity of DCs. Inhibition of MMP-13 lowers the capability of DCs to activate CD8+ T cells, through reducing MHC-I surface presentation [44]. MMP-13 increases the surface expression of CD11c and maintains the DC cytokine/chemokine profile such as IL-6, IL-12, and IL-23. Pro-inflammatory cytokine IL-12 increases the expansion and survival of $CD8^+$ T cells by driving the help of CD4+ T cells toward a Th1 phenotype [45]. IL-23, another powerful pro-inflammatory cytokine, encourages the generation of IL-17-producing Th17 cells which can effectively kill tumor cells. Inhibition of MMP-13 down-regulates all the above-mentioned DC-derived pro-inflammatory cytokines with detrimental consequences on NK cells, Th17 cells, memory T cells, and $CD8^+$ cytotoxic T cells subsequently leading to decreased tumor cell apoptosis. Hence, MMP-13 may be a promising target for a therapeutic approach in malignancy.

5 Proteases in Macrophages and Myeloid-Derived Stromal Cell (MDSC) Inactivation

Tumor-associated macrophages (TAM) and myeloid-derived suppressor cells (MDSCs) deserve special attention. Monocytes circulating in the blood mature to form macrophages. These macrophages can be classified into M1 or classically activated macrophages and M2 or alternatively activated macrophages [46]. Tumor-associated macrophages possess the phenotype of alternatively activated macrophages (M2 types) and these TAMs help in tumor cell proliferation and neovascularization. Apart from these, TAMs dampen the adaptive immune response and incessant matrix turnover which leads to tumor progression via neoplastic transformation and tumor immune evasion. Similar to macrophages, MDSCs represents a class of myeloid cells which facilitates the progression of the tumor [47]. MDSC can be categorized into two major subgroups: monocytic and granulocytic subtypes. Although TAMs and MDSCs represents two different immune cell types, but the demarcation lines between them are fuzzy and they share many similar attributes. Herein, we will discuss the role of these two subsets in tumor progression via protease secretion.

5.1 Signaling Pathway Mediating Protease Secretion in TAMs and MDSCs

Macrophage secreting matrix metalloproteinase 9 (MMP9) acts as a major armor in the progression of tumor and thus detailed overview of their regulation in macrophage deserves special mention. It has been reported that IL6 possesses a regulatory role on MMP9 expression. IL6-mediated induction of MMP9 is modulated by two pathways (a) Cox-2 $\rightarrow$ PGE2 $\rightarrow$ MMP-9 pathway as reported in macrophages of murine origin and (b) IL6 positively regulates MMP9 expression via activation of $MAPK^{erk1/2}$ whereas JAK-dependent activation of IL10 negatively regulates MMP9 expression [48]. IL6 positively regulates MMP9 expression by facilitating co-induction of microsomal PGE synthase and Cox2 and concomitantly inhibiting the expression of 15-hydroxyprostaglandin dehydrogenase. Inhibition of the above-mentioned dehydrogenase leads to increase in the expression of prostaglandin E2 which in turn increases the expression of matrix metalloproteinase 9 [49] (Fig. 2).

5.2 Role of Proteases in TAM- and MDSC-Mediated Angiogenesis

TAMs and MDSCs exert a crucial role that enhances the dynamic extracellular matrix remodeling and degradation of the basement membrane which are important prerequisites for angiogenesis. Through the production of proteolytic enzymes and

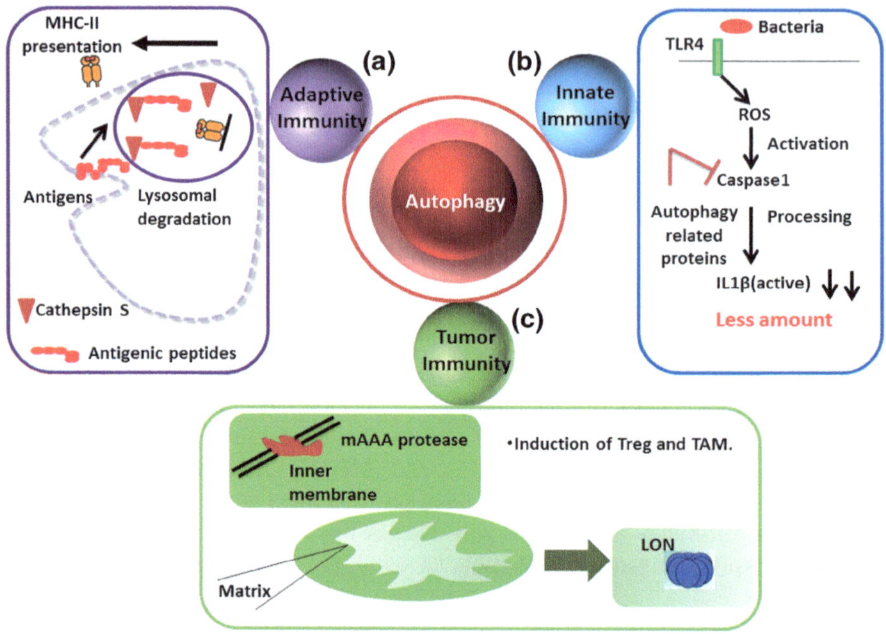

Fig. 2 Role of proteases in the context of MDSC(s) and TAM(s): Autophagic. Flux: Cathepsin S, a serine protease secreted by MDSC and TAM, facilitates the fusion of the phagosome with lysosome resulting in the formation of phagolysosome. Angiogenesis: Proteases secreted by MDSC(s) and TAM(s) help in the formation of new blood vessels, thereby playing a major role in tumor progression. Hypoxia: MMP1 and MMP7—two proteases secreted by MDSC(s) and TAM (s) create a hypoxic environment which in turn facilities tumor progression. Migration: Extracellular proteases secreted by TAM(s) and MDSC(s) play a major role in untangling the extracellular matrix, thereby facilitating the migration of these immune cells

MMPs, namely MMP-2, MMP-7, MMP-9 and MMP-12 [50], TAMs and MDSCs reorganize the extracellular matrix and degrade the basement membrane which helps in modulating angiogenesis and in turn provides a conduit for tumor cell extravasation. It has been reported that abrogation of MMP9 activity in MDSCs completely demolishes their role as tumor-promoting agent [51]. Heparin-bound growth factors like VEGF-A are released by the action of proteases like MMP-9 and cathepsin secreted by MDSCs and TAMs.

5.3 Role of Proteases in TAM and MDSC Migration

Apart from acting as angiogenesis modulating enzymes, these extracellular proteases (MMPs, cathepsins, and Urokinase-type plasminogen activator) facilitate macrophage and MDSC migration [52]. Two different molecular mechanisms involving migration exist. (a) Proteases untangle matrix proteins and/or release cryptic chemotactic factors that help in migration. (b) Proteases target macrophage cell surface proteins to regulate adhesion sites. Few extracellular matrix molecules and their respective proteolytic fragments play a distinct role in the recruitment of innate immune components and the expression of inflammation-inducing genes [53]. Fragmented collagen I and elastin generated by MMPs play a crucial role as a chemoattractant for monocytes and macrophages [54].

5.4 Protease-Mediated Autophagic Flux in Macrophages

Proteases expressed by tumor-associated macrophages (TAM) mediate autophagic flux in TAMs, thereby accelerating tumor development. In this context, cathepsins —cysteine lysosomal proteases, play a pivotal role [55]. Cathepsin S found predominantly in lymphatic tissue, macrophages, and other APC helps in the fusion processes of autophagosomes and lysosomes followed by the degradation of the contents present within the vacuoles. Thus in the context of the tumor microenvironment, the activity of CatS plays an important role in macrophage autophagy and also mediates M2 type transition of TAMs [56]. As mentioned before, macrophages activated by IL4, IL13, and IL10 are referred as M2 macrophages which apart from having a profound anti-inflammatory property also induce angiogenesis and tissue repair [57].

5.5 Protease-Mediated Hypoxia

Proteolytic enzymes mainly MMP 1 and MMP7 released by TAMs and MDSCs play a major role in creating hypoxic tumor microenvironment [58]. Apart from facilitating angiogenesis, hypoxia facilitates the polarization of the macrophages toward the pro-angiogenic (M2) phenotype [59]. HIF-1α and HIF-2α-2 hypoxia-inducible factors play a role in regulating M1/M2 polarization with HIF-1α-regulating NOS2 expression and the M1 state and HIF-2α arginase 1 expression and the M2 state [60].

In a nutshell, few key points can be highlighted:

- TAMs and MDSCs localizing in areas of tumor invasion appears to be a potent source of proteases which in turn play a critical role at multiple stages in the metastasis cascade, including the invasion and intravasation step.
- Proteases facilitate tumor cell migration by facilitating the breakdown of components of the extracellular matrices (ECM) and basement membrane.
- Among different types of proteases, MMPs and lysosomal proteinases mostly released by macrophages received most of the attention.

These proteases help in tumor progression and invasiveness as well as in the activation of growth factors and cytokines by prodomain cleavage. These products range from anergic products like CXCL1, and CXCL4, and antagonists like CCL7, or more potent chemoattractants like CXCL8, thereby modulating the composition of leukocyte population within a tumor.

6 Proteases in Antigen Presentation and Tumor Immune Evasion

6.1 Proteases and MHC-I-Mediated Antigen Presentation

T cell is one of the strongest arms of the adaptive immune response. Evolution of Adaptive immune system was done to recognize the products of intracellular partial proteolysis. $CD8^+$ T cells recognize peptides bonded with major histocompatibility complex (MHC) class I molecules (MHC-I); on the other hand, $CD4^+$ T cells recognize peptides bound to MHC class II molecules (MHC-II). MHC gene is composed of a large variety of polygenic stretch of which many conserved genes are there in addition to MHC-I and MHC-II molecules. Some of them express products important to MHC-I and MHC-II function. In different species, MHC encodes diverse MHC-I and MHC-II molecules. It is expected to have ascended by gene duplication. Structures of MHC-I and MHC-II have been identified which shows great variety in structures across species. The peptide-binding structure contains a membrane-distal groove. It is designed by two antiparallel α-helices overlapping an eight-strand β-sheet [61]. In the case of MHC-I, the groove resembles adjacent amino acid order designed by the N-terminal region of the single MHC-encoded subunit, α-chain, while for MHC-II it is formed by the apposition of the N-terminal regions of MHC-encoded α- and β-chains. The peptide-binding groove of MHC-II composed of conserved domain, one of the α-subunit and another of the β-subunit. For MHC-I, another protein, β2-microglobulin, a soluble product is non-covalently associated with the heavy α-chain. In the ER, assembly of the MHC class I occurs. It is assisted by chaperones like calnexin, calreticulin, the protein disulfide isomerase (PDI), and ERp57 [62].

Peptides are the products of proteolysis. The majority of proteolysis is occurring inside the proteasome of the cytosol. Proteasome is a cylindrical complex comprised of four stacked rings which form a central pore. The outer rings of the proteasome are composed of α-subunits. The middle two rings are composed of β-subunits. β1, β2, and β5 establish the active proteolytic machinery and function of α-subunits is to maintain a gateway through which proteins enter into the proteasomal barrel [63]. LMP2 and LMP7, subunits of a subset of proteasomes, are proteins encoded by MHC genes and believed to play a role in antigenic peptide formation. Interestingly, expression of LMP2 and LMP7 is enhanced by cytokine IFN-γ [64]. Precursor peptides produced from self- and nonself-proteins are then transported by heterodimeric transporter, viz TAP1 and TAP2, which are associated with antigen presentation, from the cytoplasm into the ER. Chaperone tapasin is a constituent of the multimeric peptide-loading complex (PLC), which is required for stabilization of TAP and eases peptide packing onto MHC class I molecules [65].

Another ER luminal component that is important for the correct expression of MHC-I-peptide complex is an aminopeptidase [66, 67]. These proteases are involved in the last and vital step of the production of MHC class I-binding peptides because only optimally clipped peptides from the antigen with a definite binding motif are precisely loaded onto MHC class I. Then, they are transported to the cell surface for recognition by $CD8^+$ T cells or CTLs. Hence, ER aminopeptidases are a crucial determinant of antigen processing and presentation. This aminopeptidase is named in the mouse as endoplasmic reticulum aminopeptidase associated with Ag processing (ERAAP) in the rat [66] and endoplasmic reticulum aminopeptidase 1 (ERAP1) in the human [67]. A structural change is required for the cleavage of antigenic protein by aminopeptidase which is specially induced by a longer peptide and thus prevents over the trimming of TAP-translocated peptides, to a size that would remove their skill to bind MHC-I [68]. The expression of aminopeptidase is also cordially regulated by different cytokines, viz IFN-γ, TNFα.

Distinct histology of human tumors shows that they express the very low level of MHC class I surface antigens, which may be due to variation or inhibition of the generation of numerous MHC class I-mediated antigen presentation machinery (APM) [65, 69]. Mutations and/or deletions in β2-microglobulin are noticed in colon carcinoma (21%), melanoma (15%), and other tumors (<5%) [69]. But in some neuroblastoma and melanoma mutations in TPN and LMP subunits are also reported. These mutations are either point mutations or base-pair deletion mutation, and this modifies their epigenetic, transcriptional, and posttranscriptional regulations. For example, in some esophageal squamous cell carcinoma, RCC and colon carcinoma-altered epigenetic modifications, viz methylation and histone deacetylation of APM components, are observed and could be renovated by handling with histone deacetylase inhibitors or DAC [70]. Apart from this, in some cases IFN-mediated induction of MHC class I APM is also repressed, which is caused by different anomalies in the IFN-γ signal transduction cascade [71]. Moreover, distinct expression profiles of both cytosolic and ER-resident aminopeptidases have been defined in some tumors [72], and the aberrant production of these peptidases affects MHC class I antigen presentation. In many human tumors, viz gastric cancer, thyroid cancer, renal cancer, melanoma, the distinct expression pattern of aminopeptidase N (APN) or CD13 is observed [73, 74], although the underlying molecular mechanisms associated with this differential expression pattern remains to be explored.

6.2 MHC-II-Mediated Antigen Presentation

MHC class II molecules present exogenous antigens to $CD4^+$T-lymphocytes. Three types of MHC-II molecules, viz HLA-DR, HLA-DQ, and HLA–DP, are present in human. The antigen-binding groove in MHC-II is consists of a position of the N-terminal regions of MHC-encoded two α- and β-chains with each side of the peptide can extend the binding groove. Assembly of MHC-II molecules occurs within the ER, smoothed by a specific chaperone, the invariant chain (I chain) whereas functional activation occurs in endosomal compartments where antigenic peptides are formed by proteolysis. I chain is a type-II transmembrane glycoprotein and non-polymorphic, which accumulates in the ER as homo- or heterotrimers, involving p33, p35, p41, and p43 in humans [75]. The function of this I chain is to prevent antigenic peptide binding with MHC-II αβ-dimers in their early biosynthesis stage within ER, and this prevents cross-presentation of the antigenic peptide specific for MHC-I [76]. Next, MHC-II-I chain complexes are routed from ER to the endocytic pathway by trans-Golgi network (TGN) and endocytosis from the cell membrane (1) which is directed by I chain di-leucine motifs [77]. In acidic endosomes, progressive proteolysis releases I chain, and an unevenly extended peptide of 20 residues remains bound to MHC-II antigenic binding groove which is known as CLIP [78]. MHC-II alleles have a low affinity for CLIP and are predisposed to the advancement of autoimmunity because premature release of CLIP favors the choice of epitopes from autoantigens or self-peptides within different endosomal compartments [79, 80]. The release of CLIP from MHC-II is favored by a different MHC-encoded glycoprotein, HLA-DM [81]. In the late endosomes, HLA-DM promotes a conformational change in MHC-II which induces dissociation of CLIP, similar to tapasin-mediated MHC-1 peptide editing. HLA-DM removes low-affinity peptides from MHC-II binding groove, and this ensures the buildup of MHC-II complexes with high-affinity peptides [82]. On the other hand, the function of HLA-DM is further inhibited by HLA-DO which is also MHC-encoded MHC-II-like αβ-heterodimer [83] (Fig. 3).

Many pathogens, viz viruses, bacteria, and fungi use endocytic pathways as a gateway into cells, but this favors their antigenic presentation as well as immune recognition. In many cases, disruptions of these pathways allow immune evasion [84]. Clathrin-mediated endocytosis or phagocytosis and macropinocytosis remain common routes that are involved in antigen internalization. They are then sorted to vesicular organelles for processing in endosome and presentation by MHC-II molecules (1). All of these pathways exist in all the professional antigen presenting cells (APCs), viz macrophages, DCs, and B lymphocytes, although some differences in competence and regulation persist among them. In phagocytosis, internalized antigens enter into the phagosome and early endosome, which then converted to late endosome containing reactive oxygen species (ROS), proteases, and antimicrobial agents by a series of event, which favors protein denaturation and proteolysis. Early endosomes are converted first into late endosomes then into lysosomes by a sequential increase in luminal acidification as well as by accumulation of different proteases delivered by TGN-derived vesicles. The functional activity of lysosomal is influenced by endosomal pH [61, 85].

Many endosomal proteases, viz cathepsins B, D, E, F, K, L, and S participate in I chain degradation as well as in antigen processing [86, 87]. Cathepsins B and D are two major aspartyl and cysteine proteases, respectively, that are involved in antigen degradation, but splenocytes which do not have abundant cysteine protease unexpectedly showed normal Ii degradation. These indicate that both cathepsins B and D are dispensable for MHC-II antigen presentation [88]. Alternatively, other two

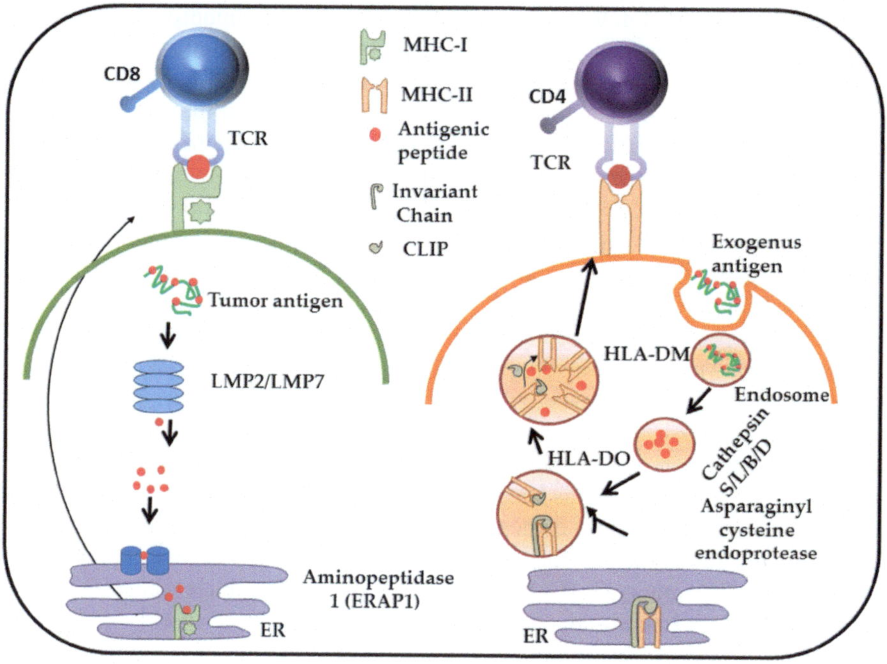

Fig. 3 Endogenous and exogenous antigen presentation pathways showing the involvement of CD8+ and CD4+ T cells, respectively. Aminopeptidase 1 (ERAP1) is involved in the processing of endogenous peptides; on the other hand, cathepsins S, L, B, and D and asparaginyl cysteine endoprotease are involved in the processing of exogenous peptides

lysosomal cysteine proteases, viz cathepsin S and cathepsin L, have distinct and indispensable roles in antigen presentation.

Cathepsin L is a papain-like cysteine protease, similar like cathepsins B, K, F, and W. These cysteine proteases contain a pro-region at their N-terminus within the active site which prevents its untimely activation during their journey through the ER and Golgi [89, 90]. The pro-region also played a role in the proper folding [90]. Mannose-6-phosphate receptor-dependent pathway is responsible for the transportation of the protease endosomes [69] and where the pro-region dissociates by the acidic environment within the endosome which finally results in the enzyme activation [91]. Cathepsin L or the human homolog, cathepsin L2 and cathepsin V, is abundantly expressed in the thymus [92]. Another member of the papain family of cysteine proteases is Cathepsin S which is mainly expressed in some professional APCs, viz DCs, macrophages, and B cells [93]. This protease is responsible for the generation of the different epitopes from the antigenic peptide. In the absence of cathepsin S germinal center, the formation is impaired [93].

Endosomal antigens are processed by many other exopeptidases and endopeptidases which directly affect the construction or destruction of epitopes that are recognized by $CD4^+$ T cells. Asparaginyl cysteine endoprotease (AEP) is first discovered in human B cells [94, 95]. This Asparaginyl cysteine protease induces the first cuts in the protein, which further makes the additional processing sites accessible to other cysteine proteases [95]. But this protease plays negative effect in the processing of some antigenic peptides, i.e., destructs some of the immunodominant epitopes of myelin basic protein (MBP) [96]. Apart from these, closely related to caspases and separases is AEP. It is capable of degrading I chain [97]. AEP is insensitive to leupeptin [95, 96] and plays a role in the first couple of steps of I chain cleavage and AEP-cleavage sites mutation in I chain inhibit the generation of intermediate fragments and radically curtailed the rate of MHC class II expression [95].

Lysosomal cysteine proteases which are the critical dictator of the multiple steps in the endosomal pathway are very precisely regulated. These cysteine proteases are appointed to the maturing phagosome in a very ordered fashion. For example, cathepsin S, cathepsin L, and cathepsin B are stable and active within a span of wide pH and they are incorporated specifically into late endosomes. Conversely, ap41 fragment of I chains binds keenly to the active site of cathepsin L [98, 99] and regulates the activity of this cathepsin in a negative feedback loop. Along with these, also the inflammatory mediators, viz IFN-γ, up-regulate the expression of cathepsins in different cells [100]. In macrophages, IFN-γ increased cathepsin S activity to mediate I chain degradation.

Different endogenous TAAs, viz transmembrane proteins or cytoplasmic and nuclear antigens, are endocytosed and sent to lysosomes for degradation and finally presented by MHC class II molecules [101]. But, this MHC class II antigen processing and presentation pathway is portentously altered as a part of tumorigenesis process. Impairments in the presentation of tumor-associated antigens (TAAs) to MHC class II molecules certainly favor the evasion of tumor immune "escape." Although MHC class II does not express by several solid tumors and infiltrating APCs engulf tumor cells and present the antigenic peptide to $CD4^+$ T cells. Some of the tumor cells can present antigens, but the absence of co-stimulatory molecules promotes tolerance. MHC class II molecules are often expressed in tumor cells in breast and colorectal cancer [102]. The MHC class II molecules do not typically express in normal breast epithelium, but MHC expression phenotype is induced in the presence of different hormones or cytokines [103]. However, diverse important constituents of the MHC class II pathway is often lost in MHC-II-expressing tumor cells. Despite high expression of surface MHC class II, B cells from B-cell chronic lymphocytic leukemia (B-CLL) patients have limited capacity to present a soluble antigen [104]. In different adenomas and carcinomas, I chain expression is very high which certainly indicates the malfunctioning in the degradation of I chain by different proteases, and this leads to the inability of MHC-II to present the associated TAAs [105]. While considering tumor vaccines, genetically engineered tumor cells express MHC class II, but they do not express Ii [106], and thus, the binding of a palette of antigens (including TAAs) to MHC class II molecules increases considerably over an extensive variety of compartments. Moreover, nowadays, milatuzumab, the MHC-I chain-specific monoclonal antibody is used as immunotherapeutic agent [107].

In various tumor cells, different cysteine proteases, viz cathepsins S, B, L, H and D, are highly expressed and predominantly involved metastasis by the degradation of ECM. In lung and prostate tumors, the cathepsin S level is elevated [108, 109]. But interestingly, apart from tumor cells, increased cathepsin S expression has been observed in alveolar macrophages, tumor-infiltrated macrophages and B lymphocytes, and a higher risk of death is associated with patients with low levels of cathepsin S [110]. These indicate that in lung tumors, cathepsin S fuels the defense mechanisms probably by enhancing the MHC-II-mediated antigen presentation.

7 Proteases in T Cell and Neutrophil Infiltration

The extracellular matrix (ECM) represents a structural support as well as an obstacle for moving cells, such as T-lymphocytes. T cell migration through ECM involves adhesive and proteolytic interactions with components of the ECM, including degradation of ECM by proteolytic enzymes such as matrix Metalloproteinases, serine proteases, and cathepsins. But rapidly moving T cells use non-proteolytic migration which depends on gliding and squeezing through the ECM which is independent of collagenase activity. ECM-degrading enzymes are expressed at mRNA level in rapidly moving T cells, but they do not express at the cell surface. Consequently, non-proteolytic T cell migration is insensitive to treatment with therapeutic protease inhibitors. All the information provides an interesting approach to MMP inhibitor therapy in cancer patients [111].

Polymorphonuclear neutrophils (PMNs) play an important role in the progression of pancreatic ductal adenocarcinoma (PDAC). PMNs have attracted to the tumor site by tumor-secreted chemokines which correlate with poor prognosis. PMN can kill tumor cells directly or indirectly by antibody-dependent cell-mediated cytotoxicity (ADCC), but on the other side, PMN has various detrimental effects, such as secretion of matrix-degrading proteases [111]. Once activated, these proteases promote turnover of the ECM components, thus facilitating angiogenesis and tumor cell migration. Neutrophil elastase, cathepsin G, proteinase 3, and ADAMs are the most powerful serine proteases that are secreted by neutrophils [112]. Neutrophil elastase degrades a variety of substrates such as elastin, collagen, cadherins, proteoglycan, fibronectin, platelet receptors, complement receptors, thrombomodulin, growth factors. Imbalance of elastase and its inhibitors plays a crucial role in liver, lung, or colorectal cancer advancement [113]. Infiltrated neutrophils are the major source of elastase in the pancreatic tumor, and an increased concentration of this protease is associated with poor prognosis [113]. Neutrophil elastase cleaves the tumor cell adhesion molecule E-cadherin, allowing tumor cell to separate from the primary tumor, followed by epithelial to mesenchymal transition, migration, and invasion [114]. On the other hand, tumorigenic activities of cathepsin G-included ECM degradation [115], receptor processing, or shedding [116–118], and increased tumor cell invasion through ECM. ADAM8 and ADAM10 expression by PMN is associated with increased invasiveness and

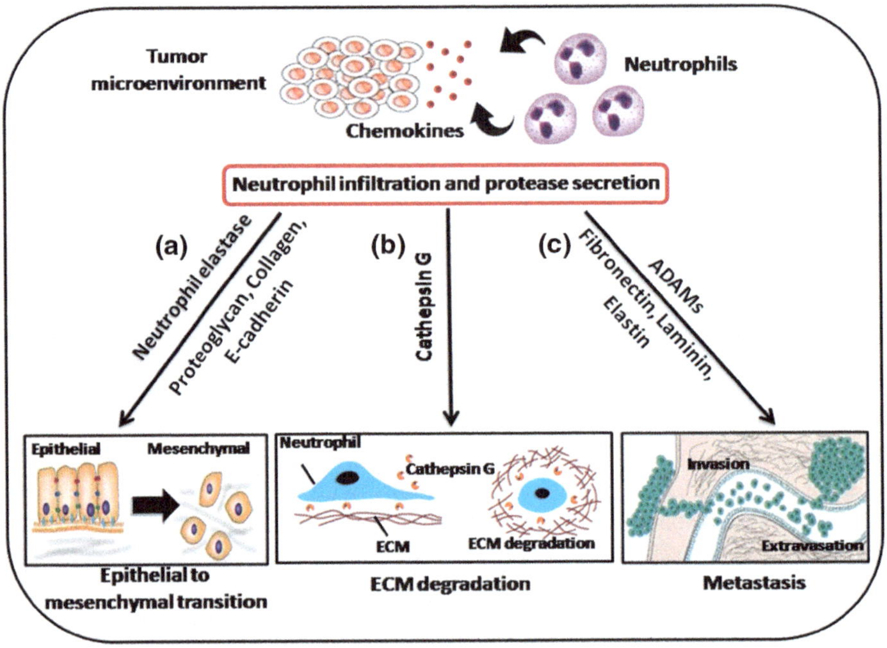

Fig. 4 Polymorphonuclear neutrophils are attracted to the tumor microenvironment by tumor cell-secreted chemokines, and these infiltrated neutrophils secrete varieties of proteases. Neutrophil elastase, cathepsin G, and ADAMs are the most important among these cytokines. **a** Neutrophil elastase cleaves proteoglycan, collagen, and E-cadherin subsequently leading to epithelial to mesenchymal transition or EMT. **b** Cathepsin G promotes ECM degradation. **c** ADAM family proteases cleave fibronectin, laminin, and elastin leading to metastasis

reduced patient survival [119]. ADAM9 degrades several extracellular matrix proteins, including fibronectin, entactin, laminin, and elastin. Zinc-endopeptidases ADAMTS (a disintegrin and metalloproteinase with thrombospondin motifs) are also released by neutrophils that cleave heparan sulfate proteoglycans syndecan-1, and glypican-1, ultimately leading to cancer progression [120, 121] (Fig. 4).

Cysteine cathepsins mostly act as lysosomal endopeptidases, but cathepsins B, C, H, and X are lysosomal exopeptidases. They are involved in tumor progression and consequent innate and adaptive immune responses. Cathepsins B, H, and L are expressed constitutively at the surface of most innate immune cells and play a significant role in cancer cell phagocytosis [122]. Other cysteine cathepsins (such as cathepsin X) are expressed more precisely on limited cells.

Cathepsin X has carboxypeptidase activity, and its expression is limited to monocytes, macrophages, T cells, and dendritic cells. It regulates dendritic cell maturation and T cell activation by its synergy with β2 integrins and heparan sulfate proteoglycans [123]. Integrins connect the cells with ECM proteins. Following adhesion, integrins recruit various cytoskeletal proteins, and anchor this protein complex to the actin filament. These sequentially guide the renovation of actin filament and the formation of typical adhesive complex, labeled as focal adhesions. Besides forming a structural connection between the actin cytoskeleton and ECM proteins, focal adhesions are extensive junctions of signal transduction that control cell migration, proliferation, and differentiation [124]. Active cathepsin X regulates β2-integrin-dependent adhesion and activation of T cells by interacting with lymphocyte function-associated antigen 1 (ITAL or LFA-1) and macrophage antigen 1 (Mac-1) which induces homotypic aggregation and cytoskeletal changes in T cells [125]. Likewise, in 2D and 3D in vitro Matrigel models that simulate the ECM, cathepsin X intensifies T-lymphocyte migration, invasiveness. On the other hand, Cathepsin S is involved in MHC class II-mediated antigen presentation. Cathepsin C is expressed by CD8+ cytotoxic T cells and NK cells where it is involved in converting pro-granzymes into proteolytically active granzymes, which elicit apoptosis in malignant cells. All these activities of cysteine cathepsins are restrained by endogenous cysteine protease inhibitors, cystatins.

8 Proteases and NK/NKT Cell Dysfunction

Cytolytic and cytokine-producing natural killer (NK) cells are one of the key components of the innate immune system, constituting the first line of defense against tumor cells. NK cells have a cytotoxic function which gets activated without any prior activation, something which T cells cannot do; so NK cells were initially named so. NK cells release varieties of cytokines and chemokines, viz IFN-γ, TNFα, IL-3, IL-10, GM-CSF, G-CSF, CXCL8, CCL2, CCL3, CCL4, CCL5. IL-2, IL-12, IL-15, IL-18, and IFNs are all powerful catalysts of NK cells activity. On the other hand, natural killer T (NKT) cells which originate in thymus show a combination of NK cell and T cell aspects. They express various NK cell markers and T cell receptor complex on their cell surfaces.

NK and NKT cells negotiate a strong pro-inflammatory response and are linked with progression of various cancers. NK cells terminate cancer cells without prior activation by binding to MHC class I molecules expressed on the surface of cancer cells. The efficacy of NK and NKT cells against cancer cells depends on the number of activated NK and NKT cells. It also depends on the expression of cytotoxic granules, inhibitory receptors, activating receptors and on the fine harmony between activating and inhibitory signals sent by surface receptors. For example, if the activating signal is robust, then NK cell will be activated; similarly, if the inhibitory signal is stronger, then NK cell activity will be inhibited. Activating receptors of NK and NKT cells are C-type lectin family receptors (Ly49, NKG2), NCR (natural cytotoxicity receptor) and CD16 (FcγIIIA) that can recognize self-molecules encoded by host's genome whose expression is up-regulated during cancer. On the other hand, inhibitory receptors (Killer-cell immunoglobulin-like receptor (KIR), Leukocyte inhibitory receptor (LIR)) are broadly recognized by MHC class I molecules.

Flawed NK and NKT cell activity can be seen in gastric cancer, colorectal cancer, and pancreatic cancer. Cancer cells by immunoediting cause NK cell modulation by up-regulation of inhibitory receptors (KIR3DL1, KIR2DL1/DS1) and down-regulation of cytotoxic granules (Perforin-Granzyme B), activating receptors (NKG2D, NCR), and decreased secretion of cytokines (TNFα, IFN-γ). The underlying mechanisms behind this are mostly unknown.

NK cell activating receptor NKG2D and its ligand MHC-I chain-related molecule (MIC) are one of the key players of tumor immunoediting. NKG2D is expressed by NK cells, NKT cells, $CD8^+$ T cells, and $\gamma\delta$ T cells. Ligands for NKG2D receptor are poorly expressed on the majority of normal cell surfaces but are up-regulated on the tumor cell surface [126]. These ligands can be released from the cell surface by proteolytic cleavage. Ligation of NKG2D receptor with its tumor-associated NKG2D ligands (NKG2DL) sends an activating signal to NK cells and co-stimulatory signal to CD8+ T cells and causes recognition of tumor cells by cytotoxic lymphocytes, resulting in cellular or genotoxic stress. But cancer cells evade immune attack by shedding membrane ligands for the NKG2D receptor; leading to desensitization of NK cells and it is a major countermechanism of cancer cells to overturn NKG2D-mediated immunosurveillance. NKG2DL MICA (MHC-I chain-related molecule A) is released by proteolytic cleavage in the stalk of the MICA ectodomain, and it involves the participation of "a disintegrin and metalloproteinase" (ADAM) family. ADAM10 and ADAM17 are crucially involved in the proteolytic release of MICA, thus helping tumor immune escape [127]. Thus, MICA shedding by tumor cells can be restricted by blocking ADAM10 and ADAM17 proteases. But MICA shedding by membrane-type matrix metalloproteinase (MT-MMP) MMP14 is independent of ADAMs. Silencing of MMP14 expression prevents MICA shedding and overexpression of MMP14 augments MICA shedding. Hence, therapeutic blockade of ADAM10, ADAM17, and MMP-14 is a promising treatment for cancer.

Matrix metalloproteinase 9 (MMP-9) is a 92-kDa type-IV collagenase and is released by mesenchymal stem cells (MSC). MMP-9 significantly down-regulates the cytotoxicity of NK and NKT cells [128, 129]. Indoleamine 2,3-dioxygenase (IDO) plays a major role in MMP-9-mediated immunosuppression by impeding NK cell accumulation in tumor milieu [130]. Hence, tumor-induced NK cell dysfunction is mainly carried out by IDO and MMP-9. Decreased infiltration of NK cells in tumor milieu is associated with increased COX-2 expression, which promotes tumor growth by producing prostaglandin E2 (PGE2) in NK cell-dependent manner [131–133]. Together, IDO, MMP-9, and PGE2 are potent factors in the synergy between NK cells and cancer cells in the tumor microenvironment [134].

NK and NKT cell cytolytic function against tumor cells can be induced directly by various receptor expressions on the cell surface. It can also be triggered through Fc receptors resolving antibody-dependent cellular cytotoxicity (ADCC) by antibodies. The efficacy of ADCC is directly parallel to CD16 receptor (FcγRIIIa) expression on NK cells. The CD16 expression is negatively altered by upsurged expression of MMPs. Inhibition of MMPs causes elevated CD16 expression by NK cells and strengthened ADCC activity against antibody-coated tumor cells [135].

Hence, MMP inhibitors enhance NK cell-mediated ADCC against tumor cells and can be exploited therapeutically.

9 Proteases and Ubiquitin System in Tumor Immune Evasion

Ubiquitin system is an integral part of immune tolerance, immune cell development, T cell differentiation, antigen or cytokine-induced signaling pathways, and hematopoiesis [136]. Ubiquitination is a three-step enzymatic reaction. E1, E2, and E3 ubiquitin ligases are enzymes which perform catalysis of different proteins. E1 enzyme forms thiol ester bond, and this leads to activation of ubiquitin. E1-activated ubiquitin gets transferred to an E2 enzyme. E2 complex interacts with the E3 enzyme. This interaction leads to isopeptide bond formation between glycine (C-terminal) of ubiquitin and an amino group of a lysine on substrate protein or ubiquitin [136, 137]. E1, E2, E3 ubiquitin ligase and deubiquitinases (DUBs) build, edit and remove ubiquitin chains [138]. E1, E2, E3, and DUBs play a central role in posttranslational modification of proteins. E3 ligases which play an important role in immune regulation are GRAIL (gene related to anergy in lymphocytes), CBL-B (Casitas B-cell lymphoma-B), and ITCH (AIP4). GRAIL and CBL-B contribute to T cell tolerance and peripheral T cell tolerance, respectively. ITCH plays an important role in T cell tolerance and TNF signaling. A20 (TNFAIP3) has a pivotal role in B-cell tolerance, germinal center selection and CD40 signaling, TNF, TLR, and NLR signaling [136, 138]. N-terminal domain of A20 acts as DUB while its C-terminal end acts as E3 ligase [138]. Ubiquitin system plays a pivotal role in antigen presentation, and this establishes proper coordination between innate and adaptive immune system. Innate immune system (dendritic cell, NK cells, and macrophages) interacts with the adaptive immune system so as to have prolonged protection against distinct microbes. The adaptive immune response evoked by pathogen leads to the formation of TCR (T cell receptor) and BCR (B-cell receptor) which are programmed to eradicate infectious pathogens [138].

Dys-regulation in proteasomal machinery will lead to malfunctioning of both adaptive and innate immune cells. These will create "danger-free zone" for cancer cells as there will be no checkpoint, and hence, tumor cells will grow profusely in the host system. Therefore, a glitch in Ubiquitin system leads to tumor immune evasion.

9.1 Ubiquitin system and Treg-Mediated Immunosuppression

Suppressive action of Treg cell is a characteristic marker of various types of cancers such as myeloma, sarcoma, melanoma. Transcription factor Foxp3 plays a pivotal role in imparting suppressive property to T-regulatory cells. Deubiquitinase USP7 regulates FOXP3 protein expression in Treg cells by deubiquitination of Foxp3 protein. These stabilize FOXP3 protein amounts. USP7 knockdown leads to impaired FOXP3 protein expression which affects Treg cell-suppressive property [139]. P5091 is a chemical inhibitor of USP7 and has been found to be effective against multiple myelomas [140].

9.2 Ubiquitination, Impairment of NFκβ Activation, and Antitumor Immune Responses

NFκβ/Rel is transcription factor that regulates both innate and adaptive immune system in eukaryotes. DUBs play an important role in NFκβ regulation. TNFAIP3 or A20 is one such DUB which regulates NFκβ. In both T and B lymphocytes, A20 acts as negative regulator of NFκβ signaling [139]. NFκβ activation in T and B lymphocytes is regulated by paracaspase activity of MALT1 and deubiquitinase action of A20. MALT1 cleaves A20 to weaken its NFκβ inhibitory function, and this activates NFκβ signaling in T cells [140]. Interactions between ubiquitinated MALT1 and IKK complex are impaired by deubiquitinase A20 [141]. These show remarkable equilibrium between MALT1 and DUB A20 whose coordination regulates NFκβ signaling. This balance gets dys-regulated during B-cell lymphomas. In B-cell lymphoma, paracaspase activity of MALT1 is constitutive in B-cell lymphomas and catalytic activity of MALT1 gets inhibited; this keeps A20 intact leading to inhibition of NFκβ [142]. TNFAIP3 might act as a tumor suppressor in various forms of B-cell lymphomas. Point and deletion mutation in A20 leads to DLBCL (diffuse large B-cell lymphoma), marginal zone lymphoma, MALT lymphoma, and Hodgkin lymphoma [142–144].

CYLD belongs to DUB family and acts as tumor suppressor and is an NFκβ regulator. An inactivating mutation in CYLD leads to multiple myelomas and T-ALL (T cell acute lymphoblastic leukemia). In T-ALL, Notch/Hes 1 pathway represses CYLD expression and these results constitutive IKK activation and prolonged cell survival [145]. A20 also plays a crucial role in MDSCs and negatively regulates apoptosis in many cell types. Silencing of A20 not only inhibited tumor growth but also decreased infiltration of MDSCs at the tumor site. Inhibition of A20 activates JNK pathway which subsequently induced caspase 3- and caspase 8-mediated apoptosis [146–148]. Moreover, another ubiquitination enzyme CBL-B negatively regulates IL2 production and T cell proliferation. CBL-B also controls TCR and CD28 signaling. Hence, knockdown of CBL-B in effector T cells may lead to CD28 expression and enhanced autocrine production of IL2. These will increase survival and proliferation in CD8+ T cells which will make them responsive to tumor antigen. Adoptive transfer of CBL-$B^{-/-}$ CD8 T cells proved to be effective against leukemia. Therefore, abrogation of CBL-B in CD8 T cells can become an immunotherapy modality to treat human malignancy [147–149].

10 Conclusion

Summing up the evidence discussed so forth, it can be concluded that proteases emerge out to be a crucial target for cancer therapeutics. Therefore, tight regulation of proteases emerges out to be a critical step in developing a successful antitumor response. Thinking about the potent role of proteases in tumor prognosis, pharmaceutical companies started developing MMP inhibitors (MMPI). Till date, two protease inhibitors, namely DX2400 and REGA 3G-12, have been tested. Both the inhibitors successfully reduce the tumor burden in a murine model. Another intriguing fact about these two inhibitor(s) is their substrate specificity. DX-2400 inhibits MMP14 whereas the latter inhibits MMP9. Apart from this, several clinical trials have been done to use doxycycline as MMP inhibitor in combinatorial cancer therapy. But the majority of current therapeutic approach targets nonspecifically wide array of any dividing cells. These lead to adverse immune response associated with chemotherapy. Thus, targeted therapy encompassing MMP inhibition is the new challenge to be faced. Apart from liposome-mediated drug delivery, targeted genome editing of the MMP gene can be explored as a successful anticancer therapeutic approach.

References

1. Rabinovich GA, Gabrilovich D, Sotomayor EM (2007) Immunosuppressive strategies that are mediated by tumor cells. Annu Rev Immunol 25:267–296
2. Dunn GP, Bruce AT, Ikeda H, Old LJ, Schreiber RD (2002) Cancer immunoediting: from immunosurveillance to tumor escape. Nat Immunol 3(11):991–998
3. Dunn GP, Old LJ, Schreiber RD (2004) The three Es of cancer immunoediting. Annu Rev Immunol 22:329–360
4. Hanahan D, Folkman J (1996) Patterns and emerging mechanisms of the angiogenic switch during tumorigenesis. Cell 86(3):353–364
5. Schreiber RD, Old LJ, Smyth MJ (2011) Cancer immunoediting: integrating immunity's roles in cancer suppression and promotion. Science 331(6024):1565–1570
6. Kim R, Emi M, Tanabe K (2007) Cancer immunoediting from immune surveillance to immune escape. Immunology 121(1):1–14
7. Zou W (2005) Immunosuppressive networks in the tumor environment and their therapeutic relevance. Nat Rev Cancer 5(4):263–274
8. Vinay DS, Ryan EP, Pawelec G et al (2015) Immune evasion in cancer: mechanistic basis and therapeutic strategies. Semin Cancer Biol 35:S185–S198
9. Hanahan D, Weinberg RA (2011) Hallmarks of cancer: the next generation. Cell 144 (5):646–674
10. Kim R, Emi M, Tanabe K (2006) Cancer immunosuppression and autoimmune disease: beyond immunosuppressive networks for tumor immunity. Immunology 119(2):254–264
11. López-Otín C, Overall CM (2002) Protease degradomics: a new challenge for proteomics. Nat Rev Mol Cell Biol 3(7):509–519
12. Fischer A (1946) Mechanism of the proteolytic activity of malignant tissue cells. Nature 6 (157):442
13. Egeblad M, Werb Z (2002) New functions for the matrix metalloproteinases in cancer progression. Nat Rev Cancer 2(3):161–174

14. Mohamed MM, Sloane BF (2006) Cysteine cathepsins: multifunctional enzymes in cancer. Nat Rev Cancer 6(10):764–775
15. López-Otín C, Matrisian LM (2007) Emerging roles of proteases in tumor suppression. Nat Rev Cancer 7(10):800–808
16. Borgoño CA, Diamandis EP (2004) The emerging roles of human tissue kallikreins in cancer. Nat Rev Cancer 4(11):876–890
17. Coussens LM, Fingleton B, Matrisian LM (2002) Matrix metalloproteinase inhibitors and cancer: trials and tribulations. Science 295(5564):2387–2392
18. Overall CM, Kleifeld O (2006) Tumour microenvironment—opinion: validating matrix metal lo-proteinases as drug targets and anti-targets for cancer therapy. Nat Rev Cancer 6(3):227–239
19. Turk B (2006) Targeting proteases: successes, failures and future prospects. Nat Rev Drug Discov 5(9):785–799
20. Goetzl EJ, Banda MJ, Leppert D (1996) Matrix metalloproteinases in immunity. J Immunol 156(1):1–4
21. Elmore S (2007) Apoptosis: a review of programmed cell death. Toxicol Pathol 35(4): 495–516
22. Hengartner MO (2000) The biochemistry of apoptosis. Nature 407:770–776
23. MacKenzie SH, Clark AC (2008) Targeting cell death in tumors by activating caspases. Curr Cancer Drug Targets 8(2):98–109
24. Cullen SP, Brunet M, Martin SJ (2010) Granzymes in cancer and immunity. Cell Death Differ 17:616–623
25. Walsh CM, Edinger AL (2010) The complex interplay between autophagy, apoptosis, and necrotic signals promotes T-cell homeostasis. Immunol Rev 236:95–109
26. Xing Y, Hogquist KA (2016) T-cell tolerance: central and peripheral. Cold Spring Harb Perspect Biol 4:a006957
27. Jameson SC et al (2005) Central tolerance: learning self-control in the thymus. Nat Rev Immunol 5:772–782
28. Ghavami S, Hashemi M, Ande SR, Yeganeh B, Xiao W, Eshraghi M, Bus CJ, Kadkhoda K, Wiechec E, Halayko AJ, Los M (2009) Apoptosis and cancer: mutations within caspase genes. J Med Genet 46:497–510
29. Mintern JD, Harris J et al (2015) Autophagy and immunity. Immunol Cell Biol 93:1–2
30. Nair U et al (2012) A role for Atg8–PE deconjugation in autophagosome biogenesis. Autophagy 8(5):780–793
31. Kaminskyy V, Zhivotovsky B et al (2012) Proteases in autophagy. Biochimi Biophys Acta (BBA)-Proteins Proteomics 1824(1):44–50
32. Yin F, Cadenas E et al (2015) Mitochondria: the cellular hub of the dynamic coordinated network. Antioxid Redox Signal 22(12):961–964
33. Lazarou M et al (2015) Keeping the immune system in check: a role for mitophagy. Immunol Cell Biol 93:3–10
34. Ma Y, Galluzzi L, Zitvogel L, Kroemer G et al (2013) Autophagy and cellular immune responses. Immunity 39
35. Bohovych I, Chan SS, Khalimonchuk O et al (2015) Mitochondrial protein quality control: the mechanisms guarding mitochondrial health. Antioxid Redox Signal 22(12):977–994
36. Brough D et al (2011) Understanding the mechanism of IL-1β secretion. Cytokine Growth Factor Rev 22(4):189–195
37. Wei L-X et al (2015) The role of autophagy induced by tumor microenvironment in different cells and stages of cancer. Cell Biosci 13578-015-0005-2
38. Crotzer VL, Blum JS et al (2009) Autophagy and its role in MHC-mediated antigen presentation. J Immunol 182(6):3335–3341
39. Rock KL, Farfán-Arribas DJ, Shen L et al (2010) Proteases in MHC class I presentation and cross-presentation. J Immunol 184(1):9–15

40. Mah LY, Ryan KM et al (2012) Autophagy and cancer. Cold Spring Harb Perspect Biol 4: a008821
41. Mellman I, Steinman RM (2001) Dendritic cells: specialized and regulated antigen processing machines. Cell 106:255–258
42. Norbury CC, Basta S, Donohue KB (2004) CD8+ T cell cross-priming via transfer of proteasome substrates. Science 304:1318–1321
43. Bromley SK, Burack WR, Johnson KG, Somersalo K, Sims TN, Sumen C, Davis MM, Shaw AS, Allen PM, Dustin ML (2001) The immunological synapse. Annu Rev Immunol 19:375
44. Bartmann J, Frankenberger M, Neurohr C, Eickelberg O, Noessner E, von Wulffen W (2016) A novel role of MMP-13 for murine DC function: its inhibition dampens T-cell activation. Int Immunol. doi:10.1093/intimm/dxw008
45. Trinchieri G (2003) Interleukin-12 and the regulation of innate resistance and adaptive immunity. Nat Rev Immunol 3:133–146
46. Quatromoni JG, Eruslanov E et al (2012) Tumor-associated macrophages: function, phenotype, and link to prognosis in human lung cancer. Am J Transl Res 4(4):376–389
47. Solinas G et al (2009) Tumor-associated macrophages (TAM) as major players of the cancer-related inflammation. J Leukoc Biol 86(5):1065–1073
48. Kothari P et al (2014) IL-6-mediated induction of MMP-9 is modulated by JAK-dependent IL-10 expression in macrophages. J Immunol 192(1): 10.4049
49. Trikha P, Carson WE et al (2014) Signaling pathways involved in MDSC regulation. Biochim Biophys Acta 1846(1):55–65
50. Kessenbrock K, Plaks V, Werb Z et al (2010) Matrix metalloproteinases: regulators of the tumor microenvironment. Cell 141(1):52–67
51. Albeituni SH, Yan J et al (2013) Hampering the immune suppressors: therapeutic targeting of myeloid-derived suppressor cells (MDSC) in cancer. Cancer J 19(6):490–501
52. Parker KH et al (2015) Myeloid-derived suppressor cells: critical cells driving immune suppression in the tumor microenvironment. Adv Cancer Res 128:95–139
53. Vérollet C, Charrière GM et al (2011) Extracellular proteolysis in macrophage migration: losing grip for a breakthrough. Eur J Immunol (10):2805–13
54. Chanmee T et al (2014) Tumor-associated macrophages as major players in the tumor microenvironment. Cancers (Basel) 6(3):1670–1690
55. Yang M et al (2014) Cathepsin S-mediated autophagic flux in tumor-associated macrophages accelerates tumor development by promoting M2 polarization. Mol Cancer 1476-4598-13-43
56. Shree T et al (2011) Macrophages and cathepsin proteases blunt chemotherapeutic response in breast cancer. Genes Dev 25(23):2465–2479
57. Rőszer T (2015) Understanding the mysterious M2 macrophage through activation markers and effector mechanisms. Mediat Inflamm 2015:816460
58. Nizet V et al (2009) Interdependence of hypoxic and innate immune responses. Nat Rev Immunol 9:609–617
59. Weidemann A, Johnson RS et al (2008) Biology of HIF-1α. Cell Death Differ 15:621–627
60. Lewis CE, Pollard JW et al (2006) Distinct role of macrophages in different tumor microenvironments. Cancer Res 66(2)
61. Blum JS, Wearsch PA, Cresswell P (2013) Pathways of antigen processing. Annu Rev Immunol 31:443–473
62. Seliger B (2008) Different regulation of MHC class I antigen processing components in human tumors. J Immunotoxicol 5:361–367
63. Maupin-Furlow J (2012) Proteasomes and protein conjugation across domains of life. Nat Rev Microbiol 10:100–111
64. Yewdell J, Lapham C, Bacik I, Spies T, Bennink J (1994) MHC-encoded proteasome subunits LMP2 and LMP7 are not required for efficient antigen presentation. J Immunol 152:1163–1170

65. Cabrera T, Maleno I, Collado A, Lopez Nevot MA, Tait BD, Garrido F (2007) Analysis of HLA class I alterations in tumors: choosing a strategy based on known patterns of underlying molecular mechanisms. Tissue Antigens 69(Suppl 1):264–268
66. Serwold T, Gonzalez F, Kim J, Jacob R, Shastri N (2002) ERAAP customizes peptides for MHC class I molecules in the endoplasmic reticulum. Nature 419:480–483
67. Saric T, Chang SC, Hattori A, York IA, Markant S, Rock KL, Tsujimoto M, Goldberg AL (2002) An IFN-gamma-induced aminopeptidase in the ER, ERAP1, trims precursors to MHC class I-presented peptides. Nat Immunol 3:1169–1176
68. Koopmann JO, Post M, Neefjes JJ, Hammerling GJ, Momburg F (1996) Translocation of long peptides by transporters associated with antigen processing (TAP). Eur J Immunol 26:1720–1728
69. Seliger B, Ritz U, Ferrone S (2006) Molecular mechanisms of HLA class I antigen abnormalities following viral infection and transformation. Int J Cancer 118:129–138
70. Nie Y, Yang G, Song Y, Zhao X, So C, Liao J, Wang LD, Yang CS (2001) DNA hypermethylation is a mechanism for loss of expression of the HLA class I genes in human esophageal squamous cell carcinomas. Carcinogenesis 22:1615–1623
71. Rodriguez T, Mendez R, Del Campo A, Jimenez P, Aptsiauri N, Garrido F, Ruiz-Cabello F (2007) Distinct mechanisms of loss of IFN-gamma mediated HLA class I inducibility in two melanoma cell lines. BMC Cancer 7:34
72. Fruci D, Ferracuti S, Limongi MZ, Cunsolo V, Giorda E, Fraioli R, Sibilio L, Carroll O, Hattori A, van Endert PM, Giacomini P (2006) Expression of endoplasmic reticulum aminopeptidases in EBV-B cell lines from healthy donors and in leukemia/lymphoma, carcinoma, and melanoma cell lines. J Immunol 176:4869–4879
73. Kehlen A, Lendeckel U, Dralle H, Langner J, Hoang-Vu C (2003) Biological significance of aminopeptidase N/CD13 in thyroid carcinomas. Cancer Res 63:8500–8506
74. Menrad A, Speicher D, Wacker J, Herlyn M (1993) Biochemical and functional characterization of aminopeptidase N expressed by human melanoma cells. Cancer Res 53:1450–1455
75. Marks MS, Blum JS, Cresswell P (1990) Invariant chain trimers are sequestered in the rough endoplasmic reticulum in the absence of association with HLA class II antigens. J Cell Biol 111:839–855
76. Roche PA, Cresswell P (1990) Invariant chain association with HLA-DR molecules inhibits immunogenic peptide binding. Nature 345:615–618
77. Landsverk OJ, Bakke O, Gregers TF (2009) MHC II and the endocytic pathway: regulation by invariant chain. Scand J Immunol 70:184–193
78. Riberdy JM, Newcomb JR, Surman MJ, Barbosa JA, Cresswell P (1992) HLA-DR molecules from an antigen-processing mutant cell line are associated with invariant chain peptides. Nature 360:474–477
79. Mohan JF, Petzold SJ, Unanue ER (2011) Register shifting of an insulin peptide-MHC complex allows diabetogenic T cells to escape thymic deletion. J Exp Med 208:2375–2383
80. Pu Z, Lovitch SB, Bikoff EK, Unanue ER (2004) T cells distinguish MHC-peptide complexes formed in separate vesicles and edited by H2-DM. Immunity 20:467–476
81. Denzin LK, Cresswell P (1995) HLA-DM induces CLIP dissociation from MHC class II alpha beta dimers and facilitates peptide loading. Cell 82:155–165
82. Kropshofer H, Vogt AB, Moldenhauer G, Hammer J, Blum JS, Hammerling GJ (1996) Editing of the HLA-DR-peptide repertoire by HLA-DM. EMBO J 15:6144–6154
83. Denzin LK, Sant'Angelo DB, Hammond C, Surman MJ, Cresswell P (1997) Negative regulation by HLA-DO of MHC class II-restricted antigen processing. Science 278:106–109
84. Ham H, Sreelatha A, Orth K (2011) Manipulation of host membranes by bacterial effectors. Nat Rev Microbiol 9:635–646
85. Muller S, Dennemarker J, Reinheckel T (2012) Specific functions of lysosomal proteases in endocytic and autophagic pathways. Biochim Biophys Acta 1824:34–43

86. Hsing LC, Rudensky AY (2005) The lysosomal cysteine proteases in MHC class II antigen presentation. Immunol Rev 207:229–241
87. Maric MA, Taylor MD, Blum JS (1994) Endosomal aspartic proteinases are required for invariant-chain processing. Proc Natl Acad Sci USA 91:2171–2175
88. Deussing J, Roth W, Saftig P, Peters C, Ploegh HL, Villadangos JA (1998) Cathepsins B and D are dispensable for major histocompatibility complex class II-mediated antigen presentation. Proc Natl Acad Sci USA 95:4516–4521
89. Tao K, Stearns NA, Dong J, Wu QL, Sahagian GG (1994) The proregion of cathepsin L is required for proper folding, stability, and ER exit. Arch Biochem Biophys 311:19–27
90. Coulombe R, Grochulski P, Sivaraman J, Menard R, Mort JS, Cygler M (1996) Structure of human procathepsin L reveals the molecular basis of inhibition by the prosegment. EMBO J 15:5492–5503
91. Turk B, Dolenc I, Lenarcic B, Krizaj I, Turk V, Bieth JG, Bjork I (1999) Acidic pH as a physiological regulator of human cathepsin L activity. Eur J Biochem 259:926–932
92. Nakagawa T, Roth W, Wong P, Nelson A, Farr A, Deussing J, Villadangos JA, Ploegh H, Peters C, Rudensky AY (1998) Cathepsin L: critical role in Ii degradation and CD4 T cell selection in the thymus. Science 280:450–453
93. Shi GP, Villadangos JA, Dranoff G, Small C, Gu L, Haley KJ, Riese R, Ploegh HL, Chapman HA (1999) Cathepsin S required for normal MHC class II peptide loading and germinal center development. Immunity 10:197–206
94. Manoury B, Hewitt EW, Morrice N, Dando PM, Barrett AJ, Watts C (1998) An asparaginyl endopeptidase processes a microbial antigen for class II MHC presentation. Nature 396:695–699
95. Antoniou AN, Blackwood SL, Mazzeo D, Watts C (2000) Control of antigen presentation by a single protease cleavage site. Immunity 12:391–398
96. Manoury B, Mazzeo D, Fugger L, Viner N, Ponsford M, Streeter H, Mazza G, Wraith DC, Watts C (2002) Destructive processing by asparagine endopeptidase limits presentation of a dominant T cell epitope in MBP. Nat Immunol 3:169–174
97. Manoury B, Mazzeo D, Li DN, Billson J, Loak K, Benaroch P, Watts C (2003) Asparagine endopeptidase can initiate the removal of the MHC class II invariant chain chaperone. Immunity 18:489–498
98. Bevec T, Stoka V, Pungercic G, Dolenc I, Turk V (1996) Major histocompatibility complex class II-associated p41 invariant chain fragment is a strong inhibitor of lysosomal cathepsin L. J Exp Med 183:1331–1338
99. Zavasnik-Bergant V, Schweiger A, Bevec T, Golouh R, Turk V, Kos J (2004) Inhibitory p41 isoform of invariant chain and its potential target enzymes cathepsins L and H in distinct populations of macrophages in human lymph nodes. Immunology 112:378–385
100. Storm van's Gravesande K, Layne MD, Ye Q, Le L, Baron RM, Perrella MA, Santambrogio L, Silverman ES, Riese RJ (2002) IFN regulatory factor-1 regulates IFN-gamma-dependent cathepsin S expression. J Immunol 168:4488–4494
101. Thibodeau J, Bourgeois-Daigneault MC, Lapointe R (2012) Targeting the MHC Class II antigen presentation pathway in cancer immunotherapy. Oncoimmunology 1:908–916
102. Durrant LG, Ballantyne KC, Armitage NC, Robins RA, Marksman R, Hardcastle JD, Baldwin RW (1987) Quantitation of MHC antigen expression on colorectal tumours and its association with tumour progression. Br J Cancer 56:425–432
103. Tabibzadeh SS, Sivarajah A, Carpenter D, Ohlsson-Wilhelm BM, Satyaswaroop PG (1990) Modulation of HLA-DR expression in epithelial cells by interleukin 1 and estradiol-17 beta. J Clin Endocrinol Metab 71:740–747
104. Dazzi F, D'Andrea E, Biasi G, De Silvestro G, Gaidano G, Schena M, Tison T, Vianello F, Girolami A, Caligaris-Cappio F (1995) Failure of B cells of chronic lymphocytic leukemia in presenting soluble and alloantigens. Clin Immunol Immunopathol 75:26–32

105. Degener T, Momburg F, Moller P (1988) Differential expression of HLA-DR, HLA-DP, HLA-DQ and associated invariant chain (Ii) in normal colorectal mucosa, adenoma and carcinoma. Virchows Arch A Pathol Anat Histopathol 412:315–322
106. Thompson JA, Srivastava MK, Bosch JJ, Clements VK, Ksander BR, Ostrand-Rosenberg S (2008) The absence of invariant chain in MHC II cancer vaccines enhances the activation of tumor-reactive type 1 CD4+ T lymphocytes. Cancer Immunol Immunother 57:389–398
107. Frolich D, Blassfeld D, Reiter K, Giesecke C, Daridon C, Mei HE, Burmester GR, Goldenberg DM, Salama A, Dorner T (2012) The anti-CD74 humanized monoclonal antibody, milatuzumab, which targets the invariant chain of MHC II complexes, alters B-cell proliferation, migration, and adhesion molecule expression. Arthritis Res Ther 14:R54
108. Werle B, Staib A, Julke B, Ebert W, Zladoidsky P, Sekirnik A, Kos J, Spiess E (1999) Fluorometric microassays for the determination of cathepsin L and cathepsin S activities in tissue extracts. Biol Chem 380:1109–1116
109. Kos J, Sekirnik A, Kopitar G, Cimerman N, Kayser K, Stremmer A, Fiehn W, Werle B (2001) Cathepsin S in tumours, regional lymph nodes and sera of patients with lung cancer: relation to prognosis. Br J Cancer 85:1193–1200
110. Turk V, Turk B, Guncar G, Turk D, Kos J (2002) Lysosomal cathepsins: structure, role in antigen processing and presentation, and cancer. Adv Enzyme Regul 42:285–303
111. Felix K, Gaida MM (2016) Neutrophil-derived proteases in the microenvironment of pancreatic cancer—active players in tumor progression. Int J Biol Sci 12(3):302–313
112. Hajjar E, Broemstrup T, Kantari C, Witko-Sarsat V, Reuter N (2010) Structures of human proteinase 3 and neutrophil elastase–so similar yet so different. FEBS J277:2238–2254
113. Sun Z, Yang P (2004) Role of imbalance between neutrophil elastase and alpha 1-antitrypsin in cancer development and progression. Lancet Oncol 5:182–190
114. Gaida MM, Steffen TG, Gunther F, Tschaharganeh DF, Felix K, Bergmann F, Schirmacher P, Hansch GM (2012) Polymorphonuclear neutrophils promote dyshesion of tumor cells and elastase-mediated degradation of E-cadherin in pancreatic tumors. Eur J Immunol 42:3369–3380
115. Tan GJ, Peng ZK, Lu JP, Tang FQ (2013) Cathepsins mediate tumor metastasis. World J Biol Chem 4:91–101
116. Wiedow O, Meyer-Hoffert U (2005) Neutrophil serine proteases: potential key regulators of cell signalling during inflammation. J Intern Med 257:319–328
117. Padrines M, Wolf M, Walz A, Baggiolini M (1994) Interleukin-8 processing by neutrophil elastase, cathepsin G and proteinase-3. FEBS Lett 352:231–235
118. Chertov O, Ueda H, Xu LL, Tani K, Murphy WJ, Wang JM, Howard OM, Sayers TJ, Oppenheim JJ (1997) Identification of human neutrophil-derived cathepsin G and azurocidin/CAP37 as chemoattractants for mononuclear cells and neutrophils. J Exp Med 186:739–747
119. Valkovskaya N, Kayed H, Felix K, Hartmann D, Giese NA, Osinsky SP, Friess H, Kleeff J (2007) ADAM8 expression is associated with increased invasiveness and reduced patient survival in pancreatic cancer. J Cell Mol Med 11:1162–1174
120. Stocker W, Bode W (1995) Structural features of a superfamily of zinc-endopeptidases: the metzincins. Curr Opin Struct Biol 5:383–390
121. Krampert M, Kuenzle S, Thai SN, Lee N, Iruela-Arispe ML, Werner S (2005) ADAMTS1 proteinase is up-regulated in wounded skin and regulates migration of fibroblasts and endothelial cells. J Biol Chem 280:23844–23852
122. Decock J, Obermajer N, Vozelj S, Hendrickx W, Paridaens R, Kos J (2008) Cathepsin B, cathepsin H, cathepsin X and cystatin C in sera of patients with early-stage and inflammatory breast cancer. Int J Biol Markers 23(3):161–168
123. Jevnikar Z, Obermajer N, Bogyo M, Kos J (2008) The role of cathepsin X in the migration and invasiveness of T lymphocytes. J Cell Sci 121(16):2652–2661
124. Lauffenburger DA, Horwitz AF (1996) Cell migration: a physically integrated molecular process. Cell 84(3):359–369

125. Kos J, Jevnikar Z, Obermajer N (2009) The role of cathepsin X in cell signaling. Cell Adh Migr 3(2):164–166
126. Bauer S, Groh V, Wu J (1999) Activation of NK cells and T cells by NKG2D, a receptor for stress-inducible MICA. Science 285:727–729
127. Waldhauer I, Goehlsdorf D, Gieseke F, Weinschenk T, Wittenbrink M, Ludwig A, Stevanovic S, Rammensee HG, Steinle A (2008) Tumor-associated MICA Is Shed by ADAM proteases. Cancer Res 68(15):6368–6376
128. Ding Y, Xu D, Feng G, Bushell A, Muschel RJ, Wood KJ (2009) Mesenchymal stem cells prevent the rejection of fully allogenic islet grafts by the immunosuppressive activity of matrix metalloproteinase-2 and −9. Diabetes 58:1797–1806
129. Lee BK, Kim MJ, Jang HS, Lee HR, Ahn KM, Lee JH, Choung PH (2008) A high concentration of MMP-2/gelatinase A and MMP-9/gelatinase B reduce NK cell-mediated cytotoxicity against an oral squamous cell carcinoma cell line. In Vivo 22:593–597
130. Meisel R, Zibert A, Laryea M, Gobel U, Daubener W, Dilloo D (2004) Human bone marrow stromal cells inhibit allogeneic T-cell responses by indoleamine 2,3-dioxygenase-mediated tryptophan degradation. Blood 103:4619–4621
131. Sharma S, Yang SC, Zhu L, Reckamp K, Gardner B, Baratelli F, Huang M, Batra RK, Dubinett SM (2005) Tumor cyclooxygenase-2/prostaglandin E2-dependent promotion of FOXP3 expression and CD4+ CD25+ T regulatory cell activities in lung cancer. Cancer Res 65:5211–5220
132. Ma X, Kundu N, Rifat S, Walser T, Fulton AM (2006) Prostaglandin E receptor EP4 antagonism inhibits breast cancer metastasis. Cancer Res 66:2923–2927
133. Kundu N, Ma X, Holt D, Goloubeva O, Ostrand-Rosenberg S, Fulton AM (2009) Antagonism of the prostaglandin E receptor EP4 inhibits metastasis and enhances NK function. Breast Cancer Res Treat 117:235–242
134. Peng YP, Zhang JJ, Liang WB, Tu M, Lu ZP, Wei JS, Jiang KR, Gao WT, Wu JL, Xu ZK, Miao Y, Zhu Y (2014) Elevation of MMP-9 and IDO induced by pancreatic cancer cells mediates natural killer cell dysfunction. BMC Cancer 14:738
135. Liu Q, Sun Y, Rihn S, Nolting A, Tsoukas PN, Jost S, Cohen K, Walker B, Alter G (2009) Matrix metalloprotease inhibitors restore impaired NK cell-mediated antibody-dependent cellular cytotoxicity in human immunodeficiency virus type 1 infection. J Virol 83 (17):8705–8712
136. Park Y, Jin HS, Aki D et al (2014) The ubiquitin system in immuneregulation. Adv Immunol 124:17–66
137. Malynn BA, Ma A (2010) Ubiquitin makes its mark on immune regulation. Immunity 33:843–852
138. Van Loosdregt J, Fleskens V, Fu J et al (2013) Foxp3 by the deubiquitinase USP7 increases Treg-cell-suppressive capacity. Immunity 39:259–271
139. Shen M, Schmitt S, Buac D et al (2013) Targeting the ubiquitin-proteasome system for cancer therapy. Expert Opin Ther Targets 17:1091–1108
140. Harhaj EW, Dixit VM (2011) Deubiquitinases in the regulation of NF-κB signaling. Cell Res 21:22–39
141. Coornaert B, Baens M, Heyninck K et al (2008) T cell antigen receptor stimulation induces MALT1 paracaspase-mediated cleavage of the NF-kappaB inhibitor A20. Nat Immunol 9:263–271
142. Düwel M, Welteke V, Oeckinghaus A et al (2009) A20 negatively regulates T cell receptor signaling to NF-kappaB by cleaving Malt1 ubiquitin chains. J Immunol 182:7718–7728
143. Ferch U, Kloo B, Gewies A et al (2009) Inhibition of MALT1 protease activity is selectively toxic for activated B cell-like diffuse large B cell lymphoma cells. J Exp Med 206: 2313–2320
144. Compagno M, Lim WK, Grunn A et al (2009) Mutations of multiple genes cause deregulation of NF-kappaB in diffuse large B-cell lymphoma. Nature 459:717–721

145. Schmitz R, Hansmann ML, Bohle V et al (2009) TNFAIP3 (A20) is a tumor suppressor gene in Hodgkin lymphoma and primary mediastinal B cell lymphoma. J Exp Med 206:981–989
146. Goedegebuure P, Mitchem JB, Porembka MR et al (2011) Myeloid-derived suppressor cells: general characteristics and relevance to clinical management of pancreatic cancer. Curr Cancer Drug Targets 11:734–751
147. Yu J, Du W, Yan F, Wang Y et al (2013) Myeloid-derived suppressor cells suppress antitumor immune responses through IDO expression and correlate with lymph node metastasis in patients with breast cancer. J Immunol 190:3783–3797
148. Shao B, Wei X, Luo M et al (2015) Inhibition of A20 expression in tumor microenvironment exerts anti-tumor effect through inducing myeloid-derived suppressor cells apoptosis. Sci Rep 5:16437
149. Stromnes IM, Blattman JN, Tan X et al (2010) Abrogating Cbl-b in effector CD8 (+) T cells improves the efficacy of adoptive therapy of leukemia in mice. J Clin Invest 120:3722–3734

Protease, an Advance Therapeutic Target in Cancer

Animesh Chowdhury, Donatella Romaniello, Soma Ghosh, Tapati Chakraborti, Soumitra Roy and Sajal Chakraborti

Abstract

Proteases are known to be associated with cancer development because of their aptitude to degrade extracellular matrices, which enables invasion and metastasis. Recent studies have demonstrated that a variety of substrates are the main target of these important enzymes and favour all steps of tumour evolution. An extensive number of reports have been available which shows a positive correlation between the activity of several proteases and tumour progression suggesting the usefulness of protease inhibitors as anticancer drugs. Nowadays, the cure for metastatic diseases is still a utopia and many efforts are focused on finding new sensitive biomarkers for a precise diagnostic as well as prognostic and therapy. In this scenario, understanding of the proteases, how they are involved in early to the end point in cancer progression is needed. In this review, we will focus on the role of proteases as prognostic and therapeutic targets in various types of cancers.

Keywords

Protease · Metalloprotease · Serine protease · Cystein protease
Aspartate protease · Cancer

A. Chowdhury (✉) · D. Romaniello · S. Ghosh
Department of Biological Regulation, Weizmann Institute of Science, Rehovot 76100, Israel
e-mail: animeshchowdhury9@gmail.com

T. Chakraborti · S. Chakraborti
Department of Biochemistry and Biophysics, University of Kalyani, Kalyani, West Bengal, India

S. Roy
Department of Physiology & Cell Biology, Faculty of Health Sciences, Goldman Medical School, Ben Gurion University of the Negev, Beer Sheva, Israel

S. Chakraborti and N.S. Dhalla (eds.), *Pathophysiological Aspects of Proteases*,
DOI 10.1007/978-981-10-6141-7_13

1 Introduction

Cancer is a major burden of disease with millions getting affected worldwide each year. It is associated with abnormal cell growth that has potential to invade and/or spread to other parts of the body. The cancer cells grow and divide to create more cells which results in tumour formation. However, not all tumours are cancerous and are referred as benign tumours. The benign tumours are not malignant and so neither do they invade nearby tissue nor spread to other parts of the body. The molecular mechanisms of the complex interplay between the tumour cells and its microenvironment play a critical role in the formation of tumorigenesis leading to cancer. At a stage of cancer invasion and metastasis, interstitial connective tissue and basement membranes get degraded and molecular events that take place during tumorigenesis and signalling molecules related with the progression of cancer development are being studied over the past decades.

Proteolysis is one of the vital processes among the biological reactions. It is an irreversible regulatory mechanism and facilitated by the selectively cleavage of specific substrate (s). Proteolytic activity has been attributed to a class of enzymes called proteases. These enzymes are of wide variety and play a significant role in biological and pathological processes. In addition, multimeric and multicatalytic proteases are known to degrade multiple intracellular proteins, called proteasomes, essential for biological processes [1]. Proteases in normal cells are very worth in executing crucial biological processes; however, recent studies proved the fact that these enzymes are also essential in the development of tumour growth and metastasis [2]. The recent accessibility of the genome sequence of different organisms makes it easy for the recognition of the entire protease repertoire, which has been defined as degradome. The human degradome is made up of at least 569 proteases that are divided into five broadly classified groups: metalloproteases (194 in number), serine proteases (176 in number), cysteine (150 in number), threonine proteases (28 in number), and aspartic proteases (21 in number) [3]. Serine, cysteine, and threonine proteases are associated with covalent catalysis. In contrast, metalloproteases and aspartic proteases are involved in non-covalent catalysis [4]. Serine, cysteine, and threonine act directly as nucleophiles that attack an amide carbonyl C, whereas aspartate, glutamate, and metalloproteases activate a water molecule that then acts as a nucleophile. These enzymes are also classified into exopeptidases and endopeptidases depending upon the position of the peptide bond in a protein they cleave. Exopeptidases truncate one or several amino acids from either the N- or the C-terminus of a peptide, whereas endopeptidases cleave an internal peptide bond. A positive correlation between the aggressiveness of tumour and the secretion of various proteases has been found, yet the secretion of some specific proteases in tumour cells make prognosis complicated. The ability to degrade extracellular matrices and proteins, proteases are strongly linked with cancer progression including invasion and metastasis. Nevertheless, proteases are not completely expressed by cancer cells and in several cases tumour cells induce the expression of proteolytic enzymes in non-neoplastic neighbouring cells, taking over their activity to favour tumour development [5]. The aim of this article is to critically review the current status of proteases as prognostic markers in cancers mainly focusing on metallo, cysteine, serine, and aspartic proteases. The present review will give the reader a strong milieu in the field of cancer research associated with proteases and an outlook of how this field can advance further.

2 Metalloproteases in Cancer Development

The extracellular matrix (ECM) is composed of highly variable and dynamic components that regulate cell behaviour. ECM is a major component of the local microenvironment or niche of cancer cells that plays important roles in cancer development. It is a complex network of macromolecules with distinctive physical, biochemical, and biomechanical properties and tightly controlled during embryonic development and organ homeostasis; however, the ECM is usually deregulated and becomes messed up during cancer progression. Deregulated ECM metabolism affects cancer progression by directly promoting cellular transformation and metastasis. ECM remodelling (synthesis and degradation) is mainly controlled by matrix metalloproteases (MMPs), which are zinc-dependent endopeptidases and synthesized as inactive form and activated during pathophysiological condition and also secreted from the cells. To date, 28 members of MMP have been identified and 23 of them are found in humans. MMPs are classified according to their structure and/or ECM substrate specificity [6]. Collagenases, including MMP-1, MMP-8, MMP-13, and MMP-18 (Xenopus) can degrade native collagen-like collagens I, II, and III. MMPs also cleave other ECM and non-ECM molecules. Although stromelysins including stromelysin-1 (MMP-3), stromelysin-2 (MMP-10) and stromelysin-3 (MMP-11) share structural similarity with the collagenases but are not able to cleave native collagen, gelatinases such as gelatinase-A (MMP-2) and gelatinase-B (MMP-9) are well known to cleave mostly denatured collagens (gelatins) as well as type IV collagen in basement membranes. They have 3 structural repeats of a type II fibronectin domain inserted in the catalytic domain, which bind to gelatin, collagens, and laminin. Matrilysins, which lack the hemopexin domain, include matrilysin-1 (MMP-7) and matrilysin-2 (MMP-26, endometase). They mainly degrade ECM constituents such as laminin and entactin. Membrane-Type MMPs (MT-MMPs) are entrenched in the plasma membrane of the cells via type I or II domains or glycosylphosphatidylinositol (GPI) anchors. Their family includes MT1-MMP (MMP-14), MT2-MMP (MMP-15), MT3-MMP (MMP-16), and MT4-MMP (MMP-24), and the GPI-anchored proteins MT5- and MT6-MMP (MMP-17 and MMP-25). MT1-MMP can cleave type I, type II, and type III collagen, and other components of ECM and can activate proMMP-2 to MMP-2. Other MMPs such as MMP-11, MMP-12, MMP-19, MMP-20, MMP-22, MMP-23, and MMP-28 do not conform easily to this classification system.

Regulation of MMPs' activity is a complex process including three different levels of activation: (a) regulation of MMPs gene expression; (b) regulation of MMP enzyme activity by cystein switch mechanism; and (c) inhibition of MMPs by its endogenous inhibitors, called tissue inhibitor of metalloproteases (TIMPs). High expression of a variety of MMPs has been observed in almost all type of human cancer and correlates with advanced stage of invasion and metastasis [7–11]. Among the MMPs, MMP-2 and MMP-9 have been implicated as the most important prognostic factor in various types of cancer model [12]. A previous study examined the correlation between MMP-2 and breast tumour progression in parallel with the clinicopathological parameters such as tumour size, histological type, lymph node status [13]. Immunoreactivity of MMP-2 was noticed in the cytoplasm in both tumour and tumour stromal cells and significantly associated with tumour size [13]. Another study suggested that MMP-2 overexpression is present in the malignant breast tumours induced by paracrine stimulation of soluble factors [14]. In the same manner, also MMP-9 expression and activity were investigated in 81 malignant breast tumours and correlated with known pathological parameters in which MMP-9 was seen to be overexpressed in node-positive tumours and the preoperative blood serum of patients [15]. Using immunoperoxidase staining, the prognostic value of MMPs was investigated in a case study of 210 breast cancer tissues: mainly MMP-9 staining was detected in cancer cells and slightly in surrounding stromal cells; however, MMP-9 expression was not noticed in normal breast tissue [16]. Thus, to find out the optimal treatment, the overexpression of MMP-2 and MMP-9 in breast cancer could be useful as a prognosis marker to subdivide node-negative breast cancer patients. Besides these studies, the impact of targeting the expression of MT1-MMP in breast cancer and its clinical relevance was also investigated. Jiang et al. showed that MT1-MMP staining was present in both membrane and cytoplasm in breast cancer cells [17]. Tumour cells exhibited stronger staining in comparison with normal mammary epithelial cells showing slightly higher levels of the MT1-MMP transcript compared with normal tissues [17].

In a recent work, cytoplasmic and stromal expression of MMP-2 was investigated in 39 (54.2%) lung cancer patients and the overexpression of MMP-2 was correlated with tumour size, lymph node and distant metastasis [18]. Other studies also correlated the high expression of MMP-2 and MMP-9 with lung cancer invasion, metastasis, and progression [19–23].

Recent studies have also demonstrated the importance of MMP-2 and MMP-9 in ovarian cancer progression and metastasis. A new study specified the role of MMP-2 in ovarian cancer progression presenting that knockdown of C3G suppressed cell invasion, intravasation, and extravasation in parallel with the reduction of Rap1 (Ras-proximate-1 or Ras-related protein 1) activity and secretion of MMP-2 and MMP-9 [24]. Therefore, C3G-mediated activation of Rap1 could direct the tumour pattern of human ovarian cancer by promoting the secretion of MMP-2 and MMP-9 [24]. Hilary et al., using organotypic 3D culture of the omentum, showed increased MMP-2 proteolytic activity during adherence to ovarian cancer cells [25]. The activated MMP-2 further cleaves the matrix proteins fibronectin, vitronectin, and collagen-I into smaller fragments, which then promotes cancer cell adhesion and invasion by binding to their cognate integrin receptors [25]. Research has shown that in vivo inhibition of MMP-2 using a siRNA or a blocking antibody before adhesion decreases the number of metastasis and tumour weight in a significant manner in a xenograft mouse model [25]. Another novel study proposed that inhibitor of DNA binding/differentiation 1 (Id1) could enhance endothelial progenitor cell (EPC) angiogenesis in ovarian cancer involving PI3 K/Akt and NF-κB/MMP-2 signalling pathways [26]. Therefore, Id1 and its downstream effectors MMP-2 could be possible targets for the treatment of ovarian cancer due to their role in angiogenesis.

For colorectal cancer (CRC), several research studies have revealed an interrelation between increased MMP-2 and MMP-9 expression and their outcome. Compared to without lymph node metastasis, an increased level of plasma MMP-2 expression has been noticed in lymph node-positive patients with CRC [15, 27]. Many studies have suggested the effectiveness of serum MMPs as markers for CRC invasion, as they show higher expression levels of MMP-2 and MMP-9 proteins in the sera of the patients compared to normal controls. These results have greater diagnostic sensitivity than two other biomarkers currently used in clinical practice, CEA and CA19-9 [16, 28]. Another study observed the interaction of β1-integrins with MMP-2 in colon cancer cells. Here, the MMP-2 was highly expressed in invasive colon cancer cells leading to CRC progression [17, 29]. Considering other MMPs, MMP-1 genetic polymorphisms are also linked with increased CRC susceptibility [30]. Also, possible role for MMP-13 too as a prognostic marker for CRC was evaluated where it indicated that there was increased MMP-13 expression with advanced cancer stage, leading to nearly eightfold increased risk of post-operative relapse compared to those without MMP-13 overexpression [31, 32].

3 Cysteine Proteases in Cancer Development

Like MMPs, other proteases, e.g. cathepsin cysteine proteases are also required for ECM degradation. Cathepsins of the cysteine protease family are mainly localized in lysosomes and endosomes, and digest intracellular or endocytosed proteins. Among the members of the cysteine protease enzyme family, the catalytic site of CA-clan papain-like cysteine proteases possesses cysteine (Cys), histidine (His), and aspartate (Asp) residues and is highly conserved [33]. Cysteine proteases are acceptably validated targets for the treatment of many human diseases. In cancer, several cysteine proteases endow with specific targets for enzyme inhibitors, especially cathepsins and calpains. The active-site cysteine offers specificity to design many inhibitors over other families of proteases, such as aspartate and serine; however, the strategy for inhibition often (a) employs covalent enzyme modification and (b) achieves selectivity within families of cysteine proteases. The cysteine proteases can be subgrouped into several families involving interleukin-1-converting enzyme (ICE), the calpain family, and the papain family [34, 35].

The papain group of the cysteine proteases family are most abundant among all cysteine proteases. This family consists of papain, related plant proteases, and lysosomal cathepsins B, C, F, H, K, L, O, S, V, W, and X [35, 36]. Cathepsins N and T have also been included in the family; albeit a detailed characterization is still waiting to be explained [35]. Nearly all of the cathepsins are endopeptidases [35, 37], but cathepsins B and H could be functional as a dipeptidyl carboxypeptidase [38] and as an aminopeptidase [39], respectively. Cathepsin C is an amino dipeptidase [35, 36] and cathepsin X is a carboxy-mono or carboxy-dipeptidase [40]. Turk et al. disclosed the activation process of cathepsins [36], and according to their concept, cathepsins are produced as preproenzymes and after removal of the prepeptide during the passage to the endoplasmic reticulum, procathepsin is formed. Finally, the active cathepsin is synthesized after proteolytic removal of the propeptide in the acidic environments of late endosomes or lysosomes. This final proteolytic process is mediated by the involvement of several proteases, such as pepsin, neutrophil elastase. Most cathepsins like cathepsins B, F, H, K, L, and V are usually active optimally in acidic environments and are merely weakly active at neutral pH. In contrast, cathepsin S is generally active at neutral pH optimally [36].

There are numerous studies published which shows a correlation between cysteine protease and invasion, proliferation, or metastasis of tumour cells. Cysteinyl cathepsins were invented to be associated with the regulation of cell proliferation involving signal transduction pathway. Inhibition of cathepsins by a synthetic broad-spectrum cysteine protease inhibitor has been shown to markedly reduce tumour cell proliferation in a mouse model of pancreatic islet cells [41]. Gocheva et al. demonstrated that lack of cathepsins B or L reduced tumour cell proliferation [42]. Although cathepsin X has been found to suppress proliferation of mononuclear cells by activation of Macrophage-1 antigen (MAC-1) (CD11b/CD18), yet cathepsin X induces the proliferation of T lymphocytes by activation of lymphocyte function-associated antigen 1 (LFA-1) (CD11a/CD18) [43]. A recent study has shown that the decline of Cux1 progression by cathepsin L deletion results in the accumulation of Cux1, downregulation of p21/p27 leading to induce cell proliferation [44].

Early reports indicated a correlation between cathepsin B and cancer implying that cathepsin B is released from malignant human breast tumour explants and is found in the blood serum of patients with neoplastic vaginal lesions. Cathepsin B has also been shown to be associated with the dissolution and remodelling of connective tissue and basement membrane in the processes of tumour growth, invasion, and metastasis [45–47]. Cathepsin B among all of the cysteine protease was also identified as the most efficient lysosomal protease to be associated with breast cancer [48]. A recent study established that while cathepsins V and D expression levels were noticed to be connected with breast cancer metastasis, the expression levels of cathepsins B and D were associated with poor disease-free survival in breast cancer patients [49]. In addition, univariate analysis displayed that the expression of cathepsin B enzyme and metastasis to the bone in breast cancer model were also associated with poor disease-free survival [49]. Moreover, Cathepsin B was recognized as a potential prognostic and therapeutic marker for human lung squamous cell carcinoma and non-small cell lung cancer (NSCLC) [50, 51]. One previous study investigated the role of cathepsin B contributing in the mechanisms of invasion by ovarian cancer [52]. This study showed that expression of cathepsin B is closely related to the invasion of ovarian cancer and suppression of cathepsin B activity using cathepsin B inhibitors markedly blocked cancer cells growth [52]. Kawasaki et al. demonstrated that the level of cathepsins D and B was also associated with tumour growth and invasion in oral squamous cell carcinoma [53]. While cathepsins B and L have been seen more frequently overexpressed in chronic atrophic gastritis with dysplasia, only cathepsin B has been significantly overexpressed in laryngeal carcinoma [54]. Additionally, contribution in regulation of angiogenesis revealed another distinct role of cathepsin in tumour progression [47, 55]. Considering another cysteine proteases, a very recent report implicated the role of cathepsin K in breast cancer cells indicating that cathepsin K induces platelet dysfunction involving cell-signalling cascade [56]. This study demonstrated that platelets were aggregated by cathepsin K but not by other cysteine cathepsins in a dose-dependent manner and PAR-3 and PAR-4 are the main targets of cathepsin K in this scenario. Moreover, studying co-culture experiments, this study showed that platelets activated by cathepsin K induce the upregulation of SHH (Sonic hedgehog), PTHrP (Parathyroid hormone-related protein), OPN (Osteopontin), and TGFβ in epithelial-mesenchymal-like cells from patients with luminal B breast cancer and concluded how cathepsin K induces platelet dysfunction with the involvement of cellular signalling affected by cathepsin K in breast cancer cells [56]. Duong et al. confirmed that increase in cathepsin K protein and mRNA levels is associated with the induction of human breast cancer primary and metastatic tumour growth [57]. These data evidenced the role of cathepsin K in breast cancer skeletal growth and metastasis and inhibition of cathepsin K activity using pharmacological or genetic inhibitors may stand for a novel oral therapy for the treatment of metastatic breast cancer [57]. Chen and Platt [58], using multiplex zymography, verified that cathepsin K was unique compared to cathepsins L and S and the level of cathepsin K was significantly higher for all cancers (lung, cervical, breast) even at the earlier stage. A large number of studies indicated the role of cathepsin K in osteoclast-mediated bone degradation, which has been assumed to be produced by cancer cells that metastasize to bone where cathepsin K functions in the proteolytic pathways leading to cancer cell invasion. Highly selective and potent cathepsin K inhibitors have been recently developed and shown to be positive antiresorptive agents in the treatment of osteoporosis. Furthermore, preclinical studies revealed that inhibition of cathepsin K using specific inhibitors reduced the chance of breast cancer-induced osteolysis and skeletal tumour burden. The possible reason in the depletion of skeletal tumour burden is recommended to be due to the antiresorptive activity of cathepsin K inhibitors that leading to deprive cancer cells of bone-derived growth factors which are essentially required for tumour growth [59]. The cysteine protease cathepsin L is also often thought to act as a tumour promoter by enhancing tumour progression and metastasis. This happens due to the increase of cathepsin L activity in various tumour tissues. A recent study demonstrated that cathepsin L expression is linked with the progression of breast cancer invasion and metastasis [60].

Cathepsin L has also been observed to be involved in prostate cancer and bone metastasis, and cathepsin L inactivation has been found to inhibit prostate cancer cell dissemination in a preclinical bone metastasis model [61].

Beside cathepsin, calpains are also cysteine proteases which are calcium-dependent [62, 63]. Two major isoforms of calpain, μ-calpain or calpain-1, which entails micromolar Ca^{2+} concentrations for activity, and m-calpain or calpain-2, which requires millimolar Ca^{2+} concentrations, are ubiquitously expressed; however, other tissue-specific forms of calpains are also exist. For example, calpain3 (CAPN3) is a skeletal muscle-specific protease, but its expression also appears transiently in the human embryonic heart [64]. Both μ- and m-calpains form heterodimers with a common regulatory subunit, calpain 4/CAPNS1 (calpain small-1). Binding of Ca^{2+} to μ- or m-calpain leading to the release of constraints that are imposed by domain interactions results an activation process [65]. On the other side, in the absence of cytosolic calcium flux, calpains are also activated through the direct phosphorylation at serine 50 by extracellular signal-regulated kinases (ERK). Calpains (specifically the ubiquitous μ- and m-calpain) play the important role in numerous physiological and pathological phenomena. Several studies indicated the importance of calpains in cancer development and progression, including cell transformation, migration and tumour invasion, apoptosis/survival, as well as angiogenesis [66]. Therefore, calpains could be considered as a potential anticancer targets.

A large number of data have been demonstrated indicating the role played by the two ubiquitous calpains (μ- and m) during malignant transformation. Carragher et al. first revealed that the balance of calpain system was modified by the oncoprotein v-Src [67]. They showed that v-Src induces an increase in the expression of m-calpain and also the degradation of its endogenous inhibitor calpastatin leading to a significant increase in the calpain activity [67]. They also suggested that the m-calpain activity is also significantly augmented by several other oncoproteins, such as v-Myc, v-Jun, k-Ras, and v-Fos during the cell transformation [68]. The increased calpain activity is responsible for the anchorage-independent growth of the v-Src-transformed cells. Niapour et al. also demonstrated that c-Myc stimulates calpain activity through the suppression of calpastatin expression and promotes the transformation of cells [69]. These different studies strongly demonstrate that calpains could be major effectors of malignant transformation.

A recent study indicated the role of calpain in breast cancer cell invasion and metastasis [70]. This study revealed that altering calpain activity, ezrin controls focal adhesion (FA) and invadopodia dynamics; two key processes essentially entailed for efficient invasion of cancer cells to induce breast cancer cell invasion. This study also demonstrated that ezrin-depleted cells show reduced calpain-mediated cleavage of the FA and invadopodia-associated proteins such as talin, focal adhesion kinase (FAK), and cortactin and also reduced calpain-1-specific membrane localization, suggesting a requirement for calpain-dependent ezrin in breast cancer invasion and metastasis [70]. Another recent study revealed the correlation between the expression of calpain-2 and breast cancer-specific overall survival in both pragmatically defined basal-like and triple-negative phenotype subgroups. Furthermore, the importance of calpain-2 expression was verified in a separate, defined cohort of invasive breast cancer patients. This provides strong evidence for the prognostic significance of calpain-2 expression [71]. However, Lau et al. recently established that calpains are responsible for the apoptosis in human small cell lung cancer [72]. They showed that capsaicin, which is a known agonist of the TRPV receptor that induces apoptosis in human SCLC cells, was mediated via the TRPV receptor family member: TRPV6. Inhibition of TRPV6 receptor using genetic inhibitor eliminated the capsaicin apoptotic activity in SCLC cells. Immunostaining and ELISA study confirmed that TRPV6 receptor was expressed significantly higher in human small cell lung cancer (SCLC) tissues (from patients) and SCLC cell lines but almost absent in normal lung tissues. This study also indicated that pro-apoptotic activity of capsaicin was mediated by the intracellular calcium-dependent calpain-mediated pathway. Capsaicin treatment to human SCLC cells increased the activity of both ubiquitous calpain (μ and m) by threefold relative to untreated SCLC cells, and the calpain is downstream of the TRPV6 receptor in capsaicin-induced SCLC cells [72]. In contrast, a recent past study suggested that calpains are overactivated in alveolar rhabdomyosarcoma (ARMS) [73]. This study showed that compared to non-malignant myoblasts, peroxiredoxins (Prx) family member, Prx IV is overexpressed by five times in ARMS cells. Prxs are known to be overexpressed and associated with the progression of several tumours. Moreover, the inhibition of calpains using pharmacological inhibitors resulted in a decrease in Prx IV abundance. Thus, calpains could be accepted as the tumour phenotype of ARMS cells especially through Prx IV regulation and, thus, might represent a potential therapeutic target to inhibit progression of ARMS tumour [73]. Considering lung cancer, two studies have shown that the migration of the tumour cells induced by nicotine and NNK (Nicotine-derived nitrosamine ketone, formed by the nitrosation of nicotine) was associated with the activation of both μ- and m-calpains [74, 75]. These two cigarette smoke components induce an increase in the phosphorylation of the two ubiquitous calpains (μ and m), which promote invasion and migration. However, the signalling pathways responsible for the activation of these enzymes appear different for the two components of cigarette smoke. While the phosphorylation of calpains induced by nicotine was observed to be mediated by PKC iota, the NNK-induced calpain phosphorylation is dependent on ERK/MAPK pathway [74, 75]. Another study has highlighted the association of m-calpain with the invasion and migration in lung cancer cells. According to this study, fibronectin stimulates the invasion and migration of lung cancer cells through the activation of FAK-ERK/MAPK signalling pathway, resulting augmentation of m-calpain activity [76]. The overexpression of calpain-1 and calpain-2 is also connected with the progression of ovarian cancer. Storr et al. in their study indicated that the expression of m-calpain is closely associated with response to platinum-based chemotherapy, progression-free and overall survival in ovarian cancer [77]. Taking into account the various cellular functions of m-calpain, unusual overexpression and high activity could also play an important role in prostate cancer progression including migration, invasion, and metastasis. It is realistic that m-calpain may be acceptable as a target to restrict tumour progression. In fact, inhibition or downregulation of m-calpain by its pharmacological or genetic inhibitors significantly reduces migration and invasion of DU-145 prostate cancer cells in vitro and in vivo [78]. Furthermore, m-calpain results in cancer cell alteration towards androgen-independent growth and cancer cell proliferation during androgen deprivation treatment. This is due to the cleavage of androgen receptor (AR) resulting in a truncated, functional AR without ligand-binding domain in androgen-sensitive prostate cancer cells [79]. Prolonged reduction in androgen levels promotes overexpression of m-calpain and increased its activity, leading to an increase in the fragmental cleavage of AR and filamin A (FlnA). Enhanced expression of m-calpain followed by FlnA may result in the development of an aggressive phenotype of prostate cancer. Therefore, it is assumed that a drug combination, wherein androgen deprivation with m-calpain inhibition, could result in new therapeutic strategy to prevent prostate cancer [80]

On the other hand, Lal et al. demonstrated that m-calpain expression is required for the metastasis of glioblastoma cells within the dynamic microenvironment of the brain, in a zebrafish model and thus m-calpain could be an important therapeutic target to ameliorate the progression of brain tumorigenesis [81]. Calpains are also appearing to be strongly implicated in angiogenesis through the regulation of migration and survival of endothelial cells [82]. A previous study suggested that calpains are essential for the growth factor-stimulated migration of microvascular endothelial cells highlighting particularly the role of m-calpain during angiogenesis. Indeed, VEGF induces the migration of endothelial cells by augmenting an increase of m-calpain expression and activity to enable tail retraction [83]. Thus, inhibition of calpains with pharmacological or genetic (siRNA) inhibitors or by overexpression of calpastatin could prevent VEGF-induced angiogenesis.

4 Serine Proteases in Cancer Development

The serine proteases are the most flexible and widely studied proteolytic enzymes till date, and trypsin among the all serine proteases grips the most valuable functions responsible for digestion, blood coagulation, fibrinolysis, development, fertilization, apoptosis, and immunity. Serine proteases are grouped into 13 clans and 40 families. The family name stems from the nucleophilic serine (Ser) in the enzyme's active site, which attacks the carbonyl moiety of the substrate peptide bond forming an acyl-enzyme intermediate [84]. Catalytic triad of Asp, His, and Ser residues is the major object for nucleophilicity of the catalytic Ser and commonly referred to as the charge relay system [85]. An inactive trypsin-like serine protease zymogen precursor is mainly activated by the proteolytic processing involving cleavage of the proprotein precursor between residues 15 and 16 (chymotrypsinogen numbering) [86]. The nascent N-terminus induces conformational change in the enzyme through the formation of an ion-pair with the highly conserved D194 that organizes both the oxy anion hole and substrate-binding site. On the other hand, proteases like tissue-type plasminogen activator possesses Lys at position 156 that connects D194 with an ion-pair that confers a catalytically competent fold without proteolytic cleavage at residue 15 [87].

Kallikreins are a subgroup of the serine protease enzyme family. The human kallikrein family includes 15 kallikrein genes. Kallikreins are expressed differentially in almost all tissues, including prostate, breast, ovary, and testis, and are regulated by steroid hormones in cancer cell lines. There is growing evidence connecting kallikreins and cancer. Prostate-specific antigen (PSA: hK3) and human glandular kallikrein (hK2) are widely accepted tumour prognosis markers for prostate cancer nowadays and are expressed highly in the prostate. hK6, hK10, and hK11 are also promising new serum biomarkers for ovarian and prostate cancer diagnosis and prognosis. Some other kallikreins are differentially expressed in various endocrine-related malignancies, and they have strong predictive value. Kallikrein became ideal markers for prostatic diseases due to their restricted tissue expression and secretion into biological fluids. Several studies presented detailed explanation on hK2 and hK3 as a biomarker in the progression of cancer [88]. Furthermore, hK3, mainly a promising marker for prostate cancer diagnosis and prognosis, recently accepted as a biomarker for breast cancer prognosis [89–91]. hK4 is also another candidate showing role in inducing prostate cancer and acts as a prognostic biomarker for prostate cancer [92].

Human kallikreins possibly contributes in many stages of the metastatic cascade by advancing tumour cell detachment and by facilitating invasion through multiple ECM barriers, independently or together with other extracellular proteases, and, potentially, by contributing to the metastatic spread of prostate cancer cells to bone. In line with this, one study showed that the invasion of MDA-MB-231 breast cancer cells into Matrigel was suppressed by a synthetic hK1 inhibitor [93] whereas other study demonstrated that invasion of LNCaP prostate cancer cells (androgen-sensitive human prostate adenocarcinoma cells) through Matrigel was shown to be reduced by hK3-neutralizing antibodies [94, 95]. Several lines of evidence suggested the role of kallikreins in the formation of osteoblastic (bone-forming) metastases that occur in ~90% of prostate cancer cases. A previous study demonstrated that hK3 is an important mediator of prostate cancer cell–bone endothelium interactions. An hK3 antibody treatment prevented prostate cancer cell adhesion to BMECs and knockdown of KLK3 mRNA by RNA interference reduced in the production of hK3 in prostate cancer cells leading to lower adhesive ability [96]. Kallikreins could be crucial to induce angiogenesis also. Several in vitro studies proved that kallikreins are responsible for angiogenesis by directly and/or indirectly disrupting ECM barriers. It has been reported previously that kallikreins, such as hK2, hK3, hK6, hK7, and hK14, directly catalyse the hydrolysis of a distinct and partially overlapping set of ECM proteins, which might promote endothelial-cell as well as tumour cell migration and invasion [97–102]. Several hKs might also indirectly facilitate ECM degradation. For example, hK1 indirectly stimulates the activation of pro-MMP2 and pro-MMP9, which are type IV collagenases that degrade collagen IV and other constituents of basement membranes [103, 104]. Alternatively, hK2 and hK4 also activates the uPA–uPAR system leading to the degradation of a wide spectrum of ECM components through plasmin, accompanied by the activation of sequestered pro-angiogenic growth factors, e.g. vascular endothelial growth factor (VEGF) and pro-MMPs [105, 106]. On the other hand, hK3 might directly induce angiogenesis through the activation of the pro-angiogenic growth factor TGFβ [107]. Moreover, hK1 is exceedingly expressed in angiogenic endothelial cells [108] and involves in a proteolytic cascade-regulating tumour angiogenesis. hKs has also been suggested to play a critical role to induce tumour cells growth through the activation of PAR [109].

Urokinase plasminogen activator (uPA) is a serine protease that shows a crucial role in cellular migration, tissue remodelling, and cancer metastasis. A number of groups have suggested uPA as a marker of disease outcome in node-negative breast cancer patients [110–112]. However, it is also prognostic in node-positive patients, premenopausal patients, post-menopausal patients, and the ER-positive subgroup [113, 114]. uPA has also been shown to be a prognostic marker in cancers other than that of the breast. A study with 76 completely resected gastric cancers using univariate analysis indicated that high levels of uPA were significantly associated with decreased survival [115]. Although using multivariate analysis with variables such as tumour size, nodal status, the presence of distant metastases, grade, and PAI-1, it has been suggested that uPA was not an independent prognostic marker. However, if PAI-l was removed from the calculation, uPA became an independent factor. Several preliminary studies also suggested that uPA is an effective factor of disease outcome in patients with bladder, lung, cervix, and ovary cancer [116–119]. However, in these cancers, uPA by itself has not yet been shown to act as an independent prognostic factor.

5 Aspartate Proteases in Cancer Development

Aspartic proteases include several important enzymes, such as pepsin, chymosin, renin, cathepsin D, and these are isolated from numerous fungi. Among the aspartic proteases, cathepsin–D (Cath-D) has a great importance in pathophysiology. Cath-D is ubiquitously distributed in lysosomes and its main function is to degrade proteins in lysosomes at an acidic pH [120]. Besides its classical role as a major protein-degrading enzyme in lysosomes and phagosomes, Cath-D can also activate precursors of biologically active proteins in pre-lysosomal compartments of specialized cells [121]. However, Cath-D has been recognized as an important prognosis marker in cancer development due to its role in cancer progression and development. The direct role of Cath-D in cancer metastasis was first demonstrated in rat tumour cells in which Cath-D overexpression increased the metastatic potential in vivo [122, 123]. Several reports demonstrated that Cath-D induces cancer cell proliferation, fibroblast outgrowth, angiogenesis, and metastasis [124–126]. Berchem et al. [127] revealed that Cath-D is responsible for the stimulation of metastasis which has a positive effect on cell proliferation, favouring the growth of micrometastases, rather than increasing the invasive potential in a rat tumour model. Its expression has also been observed to be increased and secreted at high levels by human epithelial breast cancer cells [124, 128, 129], suggesting Cath-D could be a therapeutic marker of breast cancer [130, 131].

6 Conclusion and Future Direction

Proteases are among the first group of molecules causally involved in metastasis which act as prognostic markers in cancer. Numerous studies around cancer research involving proteases made us to be thoughtful that proteases are fuel in the cancer research and worth therapeutic targets for drug design. The evidence discussed above illustrates that overexpression of proteases is synonymous with tumour progression and poor clinical prognosis. Some of the proteases could positively regulate several aspects of cancer and are summarized in Table 1. The functional exploration of the proteolytic systems that are associated with cancer has led to a change in the way we view in this complex field. We have described

Table 1 Proteases in different types of cancers

Family	Name	Location	Cancer
MMPs	General	Extracellular	Most
	MMP-1, MMP-8, MMP-13		Breast
	MMP-2, MMP-9		Breast, colorectal, lung, malignant gliomas, ovarian
	MMP-14	Membrane	Breast
Cysteine protease	General	Intracellular, lysosome	Most
	Cathepsin K	Extracellular	Breast, bone
	Cathepsin B	Extracellular and pericellular under pathological conditions	Breast, cervix, colon, colorectal, gastric, head and neck, liver, lung, melanoma, ovarian, pancreatic, prostate, thyroid
	Cathepsin L	Lysosome	Breast, colorectal, lung, prostate, ovarian
	Calpains	Intracellular, non-lysosome	Breast, lung
Serine protease	uPA, uPAR	Membrane, pericellular	Cervical, colorectal, gastric, prostate, lung, breast
	General	Intracellular, secreted	Most
	hK1		Possibly in ovarian, breast, and prostate
	hK2		Prostate and breast cancer
	PSA (hK 3)		Prostate, ovarian, breast
	hK6		Ovarian cancer
	hK10		Colon, ovarian, pancreatic, head and neck
	hK11		Ovarian and prostate cancer
	hK15		Ovarian, prostate
Aspartate protease	Cathepsin E	Endosomal structures, ER, Golgi	Cervical, gastric, lung, pancreas, breast
	Cathepsin D	Lysosome	Breast, colorectal, ovarian, prostate, lung

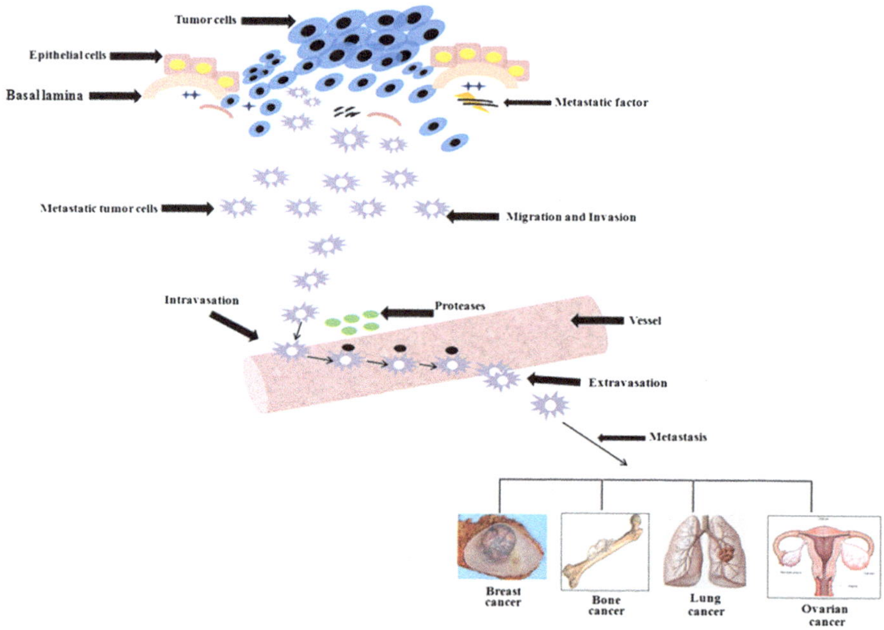

Fig. 1 Role of proteases in different types of cancers

tumour-associated proteases as non-specific and late-acting pro-metastatic enzymes and discussed as key proteins involved in early stages of cancer. Finally, we depicted the role of these pro-tumorigenic enzymes in various types of cancer metastasis. There are many reports published before this into clinical benefits for cancer patients but, hopefully, this attempt would help to ascertain the building of a conceptual framework regarding the antiprotease therapeutic drug and also facilitate to design anticancerous drug in our war on cancer. Figure 1 represents the involvement of proteases in tumour progression and metastasis leading to various types of cancer.

References

1. Lopez-Otin C, Hunter T (2010) The regulatory crosstalk between kinases and proteases in cancer. Nat Rev Cancer 10:278–292
2. Yang Y, Hao Hong H, Yin Zhang Y, Weibo CW (2009) Molecular Imaging of proteases in cancer. Cancer Growth Metastasis 2:13–27
3. Lopez-Otin C, Matrisian LM (2007) Emerging roles of proteases in tumour suppression. Nat Rev Cancer 7:800–808
4. Turk B (2006) Targeting proteases: successes, failures and future prospects. Nat Rev Drug Discov 5:785–799
5. Zucker S, Cao J, Chen WT (2000) Critical appraisal of the use of matrix metalloproteinase inhibitors in cancer treatment. Oncogene 19:6642–6650

6. Raffetto JD, Khalil RA (2008) Matrix metalloproteinases and their inhibitors in vascular remodeling and vascular disease. Biochem Pharmacol 75:346–359
7. Coussens LM, Fingleton B, Matrisian LM (2002) Matrix metalloproteinase inhibitors and cancer: trials and tribulations. Science 295:2387–2392
8. Overall CM, López-Otín C (2002) Strategies for MMP inhibition in cancer: innovations for the post-trial era. Nat Rev Cancer 2:657–672
9. Lockhart AC, Braun RD, Yu D, Ross JR, Dewhirst MW, Humphrey JS, Thompson S, Williams KM, Klitzman B, Yuan F, Grichnik JM, Proia AD, Conway DA, Hurwitz HI (2003) Reduction of wound angiogenesis in patients treated with BMS-275291, a broad spectrum matrix metalloproteinase inhibitor. Clin Cancer Res 9:586–593
10. Woessner JF Jr (1991) Matrix metalloproteinases and their inhibitors in connective tissue remodeling. FASEB J 5:2145–2154
11. Chambers AF, Matrisian LM (1997) Changing views of the role of matrix metalloproteinases in metastasis. J Natl Cancer Inst 89:1260–1270
12. Egeblad M, Werb Z (2002) New functions for the matrix metalloproteinases in cancer progression. Nat Rev Cancer 2:161–174
13. Nakopoulou L, Tsirmpa I, Alexandrou P, Louvrou A, Ampela C, Markaki S, Davaris PS (2003) MMP-2 protein in invasive breast cancer and the impact of MMP-2/TIMP-2 phenotype on overall survival. Breast Cancer Res Treat 77:145–155
14. Singer CF, Kronsteiner N, Marton E, Kubista M, Cullen KJ, Hirtenlehner K, Seifert M, Kubista E (2002) MMP-2 and MMP-9 expression in breast cancer-derived human fibroblasts is differentially regulated by stromal-epithelial interactions. Breast Cancer Res Treat 72:69–77
15. Nanda DP, Sil H, Moulik S, Biswas J, Mandal SS, Chatterjee A (2013) Matrix metalloproteinase-9 as a potential tumor marker in breast cancer. J Environ Pathol Toxicol Oncol 32:115–129
16. Scorilas A, Karameris A, Arnogiannaki N, Ardavanis A, Bassilopoulos P, Trangas T, Talieri M (2001) Overexpression of matrix-metalloproteinase-9 in human breast cancer: a potential favourable indicator in node-negative patients. Br J Cancer 84:1488–1496
17. Jiang WG, Davies G, Martin TA, Parr C, Watkins G, Mason MD, Mansel RE (2006) Expression of membrane type-1 matrix metalloproteinase, MT1-MMP in human breast cancer and its impact on invasiveness of breast cancer cells. Int J Mol Med 17:583–590
18. Osman NM, Osman WM (2016) SDF-1 and MMP2 cross talk in cancer cells and tumor microenvironment in non-small cell lung cancer. Egypt J Chest Dis Tubercul 65:517–525
19. Guo CB, Wang S, Deng C, Zhang DL, Wang FL, Jin XQ (2007) Relationship between matrix metalloproteinase 2 and lung cancer progression. Mol Diagn Ther 11:183–192
20. Ali-Labib R, Louka ML, Galal IH, Tarek M (2014) Evaluation of matrix metalloproteinase-2 in lung cancer. Proteomics Clin Appl 8:251–257
21. Schveigert D, Cicenas S, Bruzas S, Samalavicius NE, Gudleviciene Z, Didziapetriene J (2013) The value of MMP-9 for breast and non-small cell lung cancer patients' survival. Adv Med Sci 58:73–82
22. Zheng S, Chang Y, Hodges KB, Sun Y, Ma X, Xue Y, Williamson SR, Lopez-Beltran A, Montironi R, Cheng L (2010) Expression of KISS1 and MMP-9 in non-small cell lung cancer and their relations to metastasis and survival. Anticancer Res 30:713–718
23. Hrabec E, Strek M, Nowak D, Hrabec Z (2001) Elevated level of circulating matrix metalloproteinase-9 in patients with lung cancer. Respir Med 95:1–4
24. Che YL, Luo SJ, Li G, Cheng M, Gao YM, Li XM, Dai JM, He H, Wang J, Peng HJ, Zhang Y, Li WY, Wang H, Liu B, Linghu H (2015) The C3G/Rap1 pathway promotes secretion of MMP-2 and MMP-9 and is involved in serous ovarian cancer metastasis. Cancer Lett 359:241–249
25. Kenny HA, Lengyel E (2009) MMP-2 functions as an early response protein in ovarian cancer metastasis. Cell Cycle 8:683–688

26. Su Y, Gao L, Teng L, Wang Y, Cui J, Peng S, Fu S (2013) Id1 enhances human ovarian cancer endothelial progenitor cell angiogenesis via PI3 K/Akt and NF-κB/MMP-2 signaling pathways. J Transl Med 11:132
27. Langenskiold M, Holmdahl L, Falk P, Ivarsson ML (2005) Increased plasma mmp-2 protein expression in lymph node-positive patients with colorectal cancer. Int J Colorectal Dis 20:245–252
28. Dragutinovic VV, Radonjic NV, Petronijevic ND, Tatic SB, Dimitrijevic IB, Radovanovic NS, Krivokapic ZV (2011) Matrix metalloproteinase-2 (mmp-2) and -9 (mmp-9) in preoperative serum as independent prognostic markers in patients with colorectal cancer. Mol Cell Biochem 355:173–178
29. Kryczka J, Stasiak M, Dziki L, Mik M, Dziki A, Cierniewski C (2012) Matrix metalloproteinase-2 cleavage of the beta1 integrin ectodomain facilitates colon cancer cell motility. J Biol Chem 287:36556–36566
30. Decock J, Paridaens R, Ye S (2008) Genetic polymorphisms of matrix metalloproteinases in lung, breast and colorectal cancer. Clin Genet 73:197–211
31. Leeman MF, McKay JA, Murray GI (2002) Matrix metalloproteinase 13 activity is associated with poor prognosis in colorectal cancer. J Clin Pathol 55:758–762
32. Huang MY, Chang HJ, Chung FY, Yang MJ, Yang YH, Wang JY, Lin SR (2010) Mmp13 is a potential prognostic marker for colorectal cancer. Oncol Rep 24:1241–1247
33. Lecaille F, Kaleta J, Brömme D (2002) Human and parasitic papain-like cysteine proteases: their role in physiology and pathology and recent developments ininhibitor design. Chem Rev 102:4459–4488
34. Chapman HA, Riese RJ, Shi GP (1997) Emerging roles for cysteine proteases in human biology. Annu Rev Physiol 59:63–88
35. Turk B, Turk V, Turk D (1997) Structural and functional aspects of papain-like cysteine proteinases and their protein inhibitors. Biol Chem 378:141–150
36. Turk B, Turk D, Turk V (2000) Lysosomal cysteine proteases: more than scavengers. Biochim Biophys Acta 1477:98–111
37. Turk V, Turk B, Turk D (2001) Lysosomal cysteine proteases: facts and opportunities. EMBO J 20:4629–4633
38. Aronson NN Jr, Barrett AJ (1978) The specificity of cathepsin B. Hydrolysis of glucagon at the C-terminus by a peptidyldipeptidase mechanism. Biochem J 17:759–765
39. Koga H, Mori N, Yamada H, Nishimura Y, Tokuda K, Kato K, Imoto T (1992) Endo- and aminopeptidase activities of rat cathepsin H. Chem Pharm Bull (Tokyo) 40:965–970
40. Klemencic I, Carmona AK, Cezari MH, Juliano MA, Juliano L, Guncar G, Turk D, Krizaj I, Turk V, Turk B (2000) Biochemical characterization of human cathepsin X revealed that the enzyme is an exopeptidase, acting as carboxymonopeptidase or carboxydipeptidase. Eur J Biochem 267:5404–5412
41. Joyce JA, Baruch A, Chehade K, Meyer-Morse N, Giraudo E, Tsai FY, Greenbaum DC, Hager JH, Bogyo M, Hanahan D (2004) Cathepsin cysteine proteases are effectors of invasive growth and angiogenesis during multistage tumorigenesis. Cancer Cell 5:443–453
42. Gocheva V, Zeng W, Ke D, Klimstra D, Reinheckel T, Peters C, Hanahan D, Joyce JA (2006) Distinct roles for cysteine cathepsin genes in multistage tumorigenesis. Genes Dev 20:543–556
43. Obermajer N, Repnik U, Jevnikar Z, Turk B, Kreft M, Kos J (2008) Cysteine protease cathepsin X modulates immune response via activation of β2 integrins. Immunology 124:76–88
44. Alcalay NI, Sharma M, Vassmer D, Chapman B, Paul B, Zhou J, Brantley JG, Wallace DP, Maser RL, Vanden Heuvel GB (2008) Acceleration of polycystic kidney disease progression in cpk mice carrying a deletion in the homeodomain protein Cux1. Am J Physiol Renal Physiol 295:1725–1734
45. Schmitt M, Jänicke F, Graeff H (1992) Protease matrix degradation and tumor cell spread. Fibrinolysis 6:1–17

46. Sloane BF, Moin K, Lah TT (1994) Lysosomal enzymes and their endogenous inhibitors in neoplasia. In: Pretlow TG, Pretlow TP (eds) Biochemical and molecular aspects of selected cancers. Academic, New York, pp 411–466
47. Joyce JA, Hanahan D (2004) Multiple roles for cysteine cathepsins in cancer. Cell Cycle 3:1516–1619
48. Poole AR, Tiltman KJ, Recklies AD, Stoker TAM (1978) Differences in secretion of the proteinases cathepsin B at the edges of human breast carcinomas and fibroadenomas. Nature 273:545–547
49. Sun T, Jiang D, Zhang L, Su Q, Mao W, Jiang C (2016) Expression profile of cathepsins indicates the potential of cathepsins B and D as prognostic factors in breast cancer patients. Oncol Lett 11:575–583
50. Gong F, Peng X, Luo C, Shen G, Zhao C, Zou L, Li L, Sang Y, Zhao Y, Zhao X (2013) Cathepsin B as a potential prognostic and therapeutic marker for human lung squamous cell carcinoma. Mol Cancer 12:125
51. Kayser K, Richter N, Hufnagl P, Kayser G, Kos J, Werle B (2003) Expression, proliferation activity and clinical significance of cathepsin B and cathepsin L in operated lung cancer. Anticancer Res 23:2767–2772
52. Nishikawa H, Ozaki Y, Nakanishi T, Blomgren K, Tada T, Arakawa A, Suzumori K (2004) The role of cathepsin B and cystatin C in the mechanisms of invasion by ovarian cancer. Gynecol Oncol 92:881–886
53. Kawasaki G, Kato Y, Mizuno A (2002) Cathepsin expression in oral squamous cell carcinoma: relationship with clinicopathologic factors. Oral Surg Oral Med Oral Pathol Oral Radiol Endod 93:446–454
54. Macabeo-Ong M, Shiboski CH, Silverman S, Ginzinger DG, Dekker N, Wong DTW, Jordan RCK (2003) Quantitative analysis of cathepsin L mRNA and protein expression during oral cancer progression. Oral Oncol 39:638–647
55. Joyce JA, Baruch A, Chehade K, Meyer-Morse N, Giraudo E (2004) Cathepsin cysteine proteases are effectors of invasive growth and angiogenesis during multistage tumorigenesis. Cancer Cell 5:409–410
56. Andrade SS, Gouvea IE, Silva MC, Castro ED, de Paula CA, Okamoto D, Oliveira L, Peres GB, Ottaiano T, Facina G, Nazário AC, Campos AH, Paredes-Gamero EJ, Juliano M, da Silva ID, Oliva ML, Girão MJ (2016) Cathepsin K induces platelet dysfunction and affects cell signaling in breast cancer—molecularly distinct behavior of cathepsin K in breast cancer. BMC Cancer 16:173
57. Duong LT, Wesolowski GA, Leung P, Oballa R, Pickarski M (2014) Efficacy of a cathepsin K inhibitor in a preclinical model for prevention and treatment of breast cancer bone metastasis. Mol Cancer Ther 13:2898–2909
58. Chen B, Platt MO (2011) Multiplex zymography captures stage-specific activity profiles of cathepsins K, L, and S in human breast, lung, and cervical cancer. J Transl Med 9:109
59. Le Gall C, Bonnelye E, Clézardin P (2008) Cathepsin K inhibitors as treatment of bone metastasis. Curr Opin Support Palliat Care 2:218–222
60. Tholen M, Wolanski J, Stolze B, Chiabudini M, Gajda M, Bronsert P, Stickeler E, Rospert S, Reinheckel T (2015) Stress-resistant translation of cathepsin L mRNA in breast cancer progression. J Biol Chem 290:15758–15769
61. Sudhan DR, Pampo C, Rice L, Siemann DW (2016) Cathepsin L inactivation leads to multimodal inhibition of prostate cancer cell dissemination in a preclinical bone metastasis model. Int J Cancer 138:2665–2677
62. Goll DE, Thompson VF, Li H, Wei W, Cong J (2003) The calpain system. Physiol Rev 83:731–801
63. Zatz M, Starling A (2005) Calpains and disease. N Engl J Med 352:2413–2423
64. Fougerousse F, Anderson LV, Delezoide AL, Suel L, Durand M, Beckmann JS (2000) Calpain3 expression during human cardiogenesis. Neuromuscul Disord 10:251–256

65. Letavernier E, Zafrani L, Perez J, Letavernier B, Haymann JP, Baud L (2012) The role of calpains in myocardial remodelling and heart failure. Cardiovasc Res 96:38–45
66. Leloup L, Wells A (2011) Calpains as potential anti-cancer targets. Expert Opin Ther Targets 15:309–323
67. Carragher NO, Frame MC (2002) Calpain: a role in cell transformation and migration. Int J Biochem Cell Biol 34:1539–1543
68. Carragher NO, Fonseca BD, Frame MC (2004) Calpain activity is generally elevated during transformation but has oncogene-specific biological functions. Neoplasia 6:53–73
69. Niapour M, Yu Y, Berger SA (2008) Regulation of calpain activity by c-Myc through calpastatin and promotion of transformation in c-Myc-negative cells by calpastatin suppression. J Biol Chem 283:21371–21381
70. Hoskin V, Szeto A, Ghaffari A, Greer PA, Côté GP, Elliott BE (2015) Ezrin regulates focal adhesion and invadopodia dynamics by altering calpain activity to promote breast cancer cell invasion. Mol Biol Cell 26:3464–3479
71. Storr SJ, Lee KW, Woolston CM, Safuan S, Green AR, Macmillan RD, Benhasouna A, Parr T, Ellis IO, Martin SG (2012) Calpain system protein expression in basal-like and triple-negative invasive breast cancer. Ann Oncol 23:2289–2296
72. Lau JK, Brown KC, Dom AM, Witte TR, Thornhill BA, Crabtree CM, Perry HE, Brown JM, Ball JG, Creel RG, Damron CL, Rollyson WD, Stevenson CD, Hardman WE, Valentovic MA, Carpenter AB, Dasgupta P (2014) Capsaicin induces apoptosis in human small cell lung cancer via the TRPV6 receptor and the calpain pathway. Apoptosis 19:1190–1201
73. Roumes H, Pires-Alves A, Gonthier-Maurin L, Dargelos E, Cottin P (2010) Investigation of peroxiredoxin IV as a calpain-regulated pathway in cancer. Anticancer Res 30:5085–5089
74. Xu L, Deng X (2004) Tobacco-specific nitrosamine 4-(methylnitrosamino)-1-(3-pyridyl)-1-butanone induces phosphorylation of mu- and m-calpain in association with increased secretion, cell migration, and invasion. J Biol Chem 279:53683–53690
75. Xu L, Deng X (2006) Protein kinase Ciota promotes nicotine-induced migration and invasion of cancer cells via phosphorylation of micro- and m-calpains. J Biol Chem 281:4457–4466
76. Meng XN, Jin Y, Yu Y, Bai J, Liu GY, Zhu J, Zhao YZ, Wang Z, Chen F, Lee KY, Fu SB (2009) Characterisation of fibronectin-mediated FAK signalling pathways in lung cancer cell migration and invasion. Br J Cancer 101:327–334
77. Storr SJ, Safuan S, Woolston CM, Abdel-Fatah T, Deen S, Chan SY, Martin SG (2012) Calpain-2 expression is associated with response to platinum based chemotherapy, progression-free and overall survival in ovarian cancer. J Cell Mol Med 16:2422–2428
78. Mamoune A, Luo JH, Lauffenburger DA, Wells A (2003) Calpain-2 as a target for limiting prostate cancer invasion. Cancer Res 63:4632–4640
79. Libertini SJ, Tepper CG, Rodriguez V, Asmuth DM, Kung HJ, Mudryj M (2007) Evidence for calpain-mediated androgen receptor cleavage as a mechanism for androgen independence. Cancer Res 67:9001–9005
80. Liu T, Mendes DE, Berkman CE (2014) Prolonged androgen deprivation leads to overexpression of calpain 2: implications for prostate cancer progression. Int J Oncol 44:467–472
81. Lal S, La Du J, Tanguay RL, Greenwood JA (2012) Calpain 2 is required for the invasion of glioblastoma cells in the zebrafish brain microenvironment. J Neurosci Res 90:769–781
82. Bodnar RJ, Yates CC, Wells A (2006) IP-10 blocks vascular endothelial growth factor-induced endothelial cell motility and tube formation via inhibition of calpain. Circ Res 98:617–625
83. Su Y, Cui Z, Li Z, Block ER (2006) Calpain-2 regulation of VEGF-mediated angiogenesis. FASEB J 20:1443–1451
84. Hedstrom L (2002) Serine protease mechanism and specificity. Chem Rev 102:4501–4524

85. Blow DM, Birktoft JJ, Hartley BS (1969) Role of a buried acid group in the mechanism of action of chymotrypsin. Nature 221:337–340
86. Neurath H, Dixon GH (1957) Structure and activation of trypsinogen and chymotrypsinogen. Fed Proc 16:791–801
87. Renatus M, Engh RA, Stubbs MT, Huber R, Fischer S, Kohnert U, Bode W (1997) Lysine 156 promotes the anomalous pro-enzyme activity of tPA: X-ray crystal structure of single-chain human tPA. EMBO J 16:4797–4805
88. Rittenhouse HG, Finlay JA, Mikolajczyk SD, Partin AW (1998) Human Kallikrein 2 (hK2) and prostate-specific antigen (PSA): two closely related, but distinct, kallikreins in the prostate. Crit Rev Clin Lab Sci 35:275–368
89. Black MH, Diamandis EP (2000) The diagnostic and prognostic utility of prostate specific antigen for diseases of the breast. Breast Cancer Res Treat 59:1–14
90. Yu H, Giai M, Diamandis EP, Katsaros D, Sutherland DJ, Levesque MA, Roagna R, Ponzone R, Sismondi P (1995) Prostate-specific antigen is a new favorable prognostic indicator for women with breast cancer. Cancer Res 55:2104–2110
91. Black MH, Giai M, Ponzone R, Sismondi P, Yu H, Diamandis EP (2000) Serum total and free prostate-specific antigen for breast cancer diagnosis in women. Clin Cancer Res 6:467–473
92. Lisle JE, Mertens-Walker I, Stephens CR, Stansfield SH, Clements JA, Herington AC, Stephenson SA (2015) Murine, but not human, ephrin-B2 can be efficiently cleaved by the serine protease kallikrein-4: implications for xenograft models of human prostate cancer. Exp Cell Res 333:136–146
93. Wolf WC, Evans DM, Chao L, Chao JA (2001) synthetic tissue kallikrein inhibitor suppresses cancer cell invasiveness. Am J Pathol 159:1797–1805
94. Webber MM, Waghray A, Bello D (1995) Prostate-specific antigen, a serine protease, facilitates human prostate cancer cell invasion. Clin Cancer Res 1:1089–1094
95. Ishii K, Otsuka T, Iguchi K, Usui S, Yamamoto H, Sugimura Y, Yoshikawa K, Hayward SW, Hirano K (2004) Evidence that the prostate-specific antigen (PSA)/Zn^{2+} axis may play a role in human prostate cancer cell invasion. Cancer Lett 207:79–87
96. Romanov VI, Whyard T, Adler HL, Waltzer WC, Zucker S (2004) Prostate cancer cell adhesion to bone marrow endothelium: the role of prostate-specific antigen. Cancer Res 64:2083–2089
97. Cloutier SM, Chagas JR, Mach JP, Gygi CM, Leisinger HJ, Deperthes D (2002) Substrate specificity of human kallikrein 2 (hK2) as determined by phage display technology. Eur J Biochem 269:2747–2754
98. Deperthes D, Frenette G, Brillard-Bourdet M, Bourgeois L, Gauthier F, Tremblay RR, Dubé JY (1996) Potential involvement of kallikrein hK2 in the hydrolysis of the human seminal vesicle proteins after ejaculation. J Androl 17:659–665
99. Lilja H (1985) A kallikrein-like serine protease in prostatic fluid cleaves the predominant seminal vesicle protein. J Clin Invest 76:1899–1903
100. Watt KW, Lee PJ, M'Timkulu T, Chan WP, Loor R (1986) Human prostate-specific antigen: structural and functional similarity with serine proteases. Proc Natl Acad Sci USA 83:3166–3170
101. Bernett MJ, Blaber SI, Scarisbrick IA, Dhanarajan P, Thompson SM, Blaber M (2002) Crystal structure and biochemical characterization of human kallikrein 6 reveals that a trypsin-like kallikrein is expressed in the central nervous system. J Biol Chem 277:24562–24570
102. Magklara A, Mellati AA, Wasney GA, Little SP, Sotiropoulou G, Becker GW, Diamandis EP (2003) Characterization of the enzymatic activity of human kallikrein 6: autoactivation, substrate specificity, and regulation by inhibitors. Biochem Biophys Res Commun 307:948–955
103. Tschesche H, Michaelis J, Kohnert U, Fedrowitz J, Oberhoff R (1989) Tissue kallikrein effectively activates latent matrix degrading metalloenzymes. Adv Exp Med Biol 247:545–548

104. Menashi S, Fridman R, Desrivieres S, Lu H, Legrand Y, Soria C (1994) Regulation of 92-kDa gelatinase B activity in the extracellular matrix by tissue kallikrein. Ann NY Acad Sci 732:466–468
105. Frenette G, Tremblay RR, Lazure C, Dube JY (1997) Prostatic kallikrein hK2, but not prostate-specific antigen (hK3), activates single-chain urokinase-type plasminogen activator. Int J Cancer 71:897–899
106. Takayama TK, McMullen BA, Nelson PS, Matsumura M, Fujikawa K (2001) Characterization of hK4 (prostase), a prostate-specific serine protease: activation of the precursor of prostate specific antigen (pro-PSA) and single-chain urokinase-type plasminogen activator and degradation of prostatic acid phosphatase. Biochemistry 40:15341–15348
107. Killian CS, Corral DA, Kawinski E, Constantine RI (1993) Mitogenic response of osteoblast cells to prostatespecific antigen suggests an activation of latent TGF-β and a proteolytic modulation of cell adhesion receptors. Biochem Biophys Res Commun 192:940–947
108. Emanueli C, Minasi A, Zacheo A, Chao J, Chao L, Salis MB, Straino S, Tozzi MG, Smith R, Gaspa L, Bianchini G, Stillo F, Capogrossi MC, Madeddu P (2001) Local delivery of human tissue kallikrein gene accelerates spontaneous angiogenesis in mouse model of hindlimb ischemia. Circulation 103:125–132
109. Jin E, Fujiwara M, Pan X, Ghazizadeh M, Arai S, Ohaki Y, Kajiwara K, Takemura T, Kawanami O (2003) Protease-activated receptor (PAR)-1 and PAR-2 participate in the cell growth of alveolar capillary endothelium in primary lung adenocarcinomas. Cancer 97:703–713
110. Janicke F, Schmitt M, Hafter R, Holrieder A, Babic R, Ulm K, Gossner W, Graeff H (1990) Urokinase-type plasminogen activator (u-PA) antigen is a predictor of early relapse in breast cancer. Fibrinolysis 4:69–78
111. Foekens JA, Schmitt M, van Putten WL, Peters HA, Bontenbal M, Jänicke F, Klijn JG (1992) Prognostic value of urokinase-type plasminogen activator in 671 primary breast cancer patients. Cancer Res 52:6101–6105
112. Janicke F, Schmitt M, Pache L, Ulm K, Harbeck N, Hofler H, Graeff H (1993) Urokinase plasminogen activator (uPA) and its inhibitor PAl- 1 are strong and independent prognostic factor in nodenegative breast cancer. Breast Cancer Res Treat 24:195–208
113. Duffy MJ, Reilly D, Nugent A, McDermott E, Faul C, Fennelly JJ, O'Higgins N (1992) Evaluation of proteolytic enzymes implicated in cancer metastasis as prognostic markers in breast cancer. Is J Med Sci 161:49
114. Duffy MJ, Reilly D, McDermott E, O'Higgins N, Fennelly JJ, Andreasen PA (1994) Urokinase plasminogen activator as a prognostic marker in different subgroups of patients with breast cancer. Cancer 74:2276–2280
115. Nekarda H, Siewert J, Schmitt M, Ulm K (1994) Tumor-associated proteolytic factors uPA and PAl-I and survival in totally resected gastric cancer. Lancet 343:117
116. Hasui Y, Marutsuka K, Suzumiya J, Kitada S, Osada Y, Sumiyoshi A (1992) The content of urokinase-type plasminogen activator antigen as a prognostic factor in urinary bladder cancer. Int J Cancer 50:871–873
117. Oka T, Ishida T, Nishino T, Sugimachi K (1991) Immunohistochemical evidence of urokinase plasminogen activator in primary and metastatic tumors of pulmonary carcinoma. Cancer Res 51:3522–3525
118. Kobayashi H, Fujishiro S, Terao T (1994) Impact of urokinase-type plasminogen activator and its inhibitor type I on prognosis in cervical cancer of the uterus. Cancer Res 54:6539–6548
119. Kuhn W, Pache L, Schmalfeldt S, Dettmar P, Schmitt M, Janicke F, Graeff H (1994) Urokinase (uPA) and PAl-I predict survival in advanced ovarian cancer patients (FIGO III) after radical surgery and platinum based chemotherapy. Gynecol Oncol 55:401–409
120. Barrett AJ, Cathepsin D (1970) Purification of isoenzymes from human and chicken liver. Biochem J 117:601–607

121. Diment S, Martin KJ, Stahl PD (1989) Cleavage of parathyroid hormone in macrophage endosomes illustrates a novel pathway for intracellular processing of proteins. J Biol Chem 264:13403–13406
122. Liaudet E, Garcia M, Rochefort H (1994) Cathepsin D maturation and its stimulatory effect on metastasis are prevented by addition of KDEL retention signal. Oncogene 9:1145–1154
123. Liaudet-Coopman E, Mélanie B, Danielle D, Marcel G, Glondu-Lassis M, Laurent-Matha V, Christine P, Henri R, Françoise V (2006) Cathepsin D: newly discovered functions of a long-standing aspartic protease in cancer and apoptosis. Cancer Lett 237:167–179
124. Vashishta A, Ohri SS, Proctor M, Fusek M, Vetvicka V (2007) Ribozyme- targeting procathepsin D and its effect on invasion and growth of breast cancer cells: an implication in breast cancer therapy. Int J Oncol 30:1223–1230
125. Hu L, Roth JM, Brooks P, Luty J, Karpatkin S (2008) Thrombin up-regulates cathepsin D which enhances angiogenesis, growth, and metastasis. Cancer Res 68:4666–4673
126. Ohri SS, Vashishta A, Proctor M, Fusek M, Vetvicka V (2008) The propeptide of cathepsin D increases proliferation, invasion and metastasis of breast cancer cells. Int J Oncol 32:491–498
127. Berchem GJ, Glondu M, Gleizes M, Brouillet JP, Garcia M, Liaudet CE (2002) Cathepsin-D affects multiple steps of tumor progression: proliferation, angiogenesis and apoptosis. Oncogene 51:5951–5955
128. Nicotra G, Castino R, Follo C, Peracchio C, Valente G (2010) The dilemma: does tissue expression of cathepsin D reflect tumor malignancy? The question: does the assay truly mirror cathepsin D mis-function in the tumor. Cancer Biomark 7:47–64
129. Radisky ES, Radisky DC (2010) Matrix metalloproteinase-induced epithelial-mesenchymal transition in breast cancer. J Mammary Gland Biol Neoplasia 15:201–212
130. Westley BR, May FE (1999) Prognostic value of cathepsin D in breast cancer. Br J Cancer 79:189–190
131. Rodriguez J, Vazquez J, Corte MD, Lamelas M, Bongera M (2005) Clinical significance of cathepsin D concentration in tumor cytosol of primary breast cancer. Int J Biol Markers 20:103–111

Part II
Role of Proteases in Different Diseases

Matrix Metalloproteinases in Parasitic Infections

Fabrizio Bruschi and Barbara Pinto

Abstract
Matrix metalloproteinases (MMPs) constitute a wide family of more than 20 distinct endopeptidases, either secreted or membrane-bound. They are involved in many physiological (embryogenesis, precursor or stem cell mobilization, tissue remodeling during wound healing, etc.) as well as pathological (inflammation, tumor progression and metastasis in cancer, vascular pathology, etc.) situations. These proteinases have been considered in the past exclusively for their degradation ability of molecules of extracellular matrix (ECM) (e.g., collagen, laminin, fibronectin) as well as that to release hidden epitopes from the ECM. In more recent years, it has been completely clarified that these enzymes are also involved in the immune response, acting on cytokines, hormones, and chemokines. Among many others, the so-called gelatinases, MMP-2, and MMP-9 are produced by neutrophils, macrophages, and monocytes. When infection is associated with leukocyte influx into specific organs, immunopathology and following tissue damage may occur, facilitated by gelatinases. The focus of this chapter will be on the participation of MMPs and in particular of gelatinases in either protozoan or helminth infections. A well-studied model is represented by cerebral malaria, for example, where MMPs play a crucial role in the pathogenesis of such disease. Also trypanosomosis and toxoplasmosis will be considered for protozoan infections, as well as neurocysticercosis, angiostrongyloidosis and trichinellosis, for helminth infections.

F. Bruschi (✉) · B. Pinto
Department of Translational Research, N.T.M.S., School of Medicine, University of Pisa, Via Roma, 55, 56126 Pisa, Italy
e-mail: fabrizio.bruschi@med.unipi.it

S. Chakraborti and N.S. Dhalla (eds.), *Pathophysiological Aspects of Proteases*,
DOI 10.1007/978-981-10-6141-7_14

Keywords
Matrix metalloproteinases · Tissue inhibitor of metalloproteinases
Malaria · Trypanosomosis · Toxoplasmosis · Neurocysticercosis
Angiostrongyloidosis · Trichinellosis

1 Introduction

Matrix metalloproteinases (MMPs) consist of a large family of proteolytic multi-domain calcium/zinc dependent endopeptidases, encoded by 24 distinct genes in human [1]. These proteinases can break down most of the components of the extracellular matrix (ECM), and even non-matrix proteins [2].

In this chapter, we will focus our attention on the role carried out by these enzymes in the inflammatory response which occurs during infections caused by several parasites, in particular by protozoa or helminths, with the aim to show that many mechanisms involved are similar in the different situations.

Before reporting on the data accumulated in the literature during the last years, dealing with the different parasitic infections, we think it necessary to briefly discuss about the structure and function of the extracellular matrix, which represents the main target of the endopeptidases, and then to proceed with the description of the different MMPs currently known.

2 The Matrix

The extracellular matrix (ECM) is a complex non-cellular component surrounding cells in connective tissues that provide them biochemical and structural support. It exists in two main forms, the interstitial extracellular matrix, in which collagen is the most abundant fibrous protein, and the epithelial-cell associated basal membrane (BM), the latter representing a thin fibrous network that underlies epithelial layers and separates cells from the connective tissue [3, 4]. In most tissues, the ECM is made up of three major types of interwoven proteins, i.e., structural fibrous proteins (collagen, elastin) glycoproteins (fibrillin, fibronectin, tenascins, and laminins) [5], and several classes of proteoglycans (such as aggrecans) [6, 7]. These consists of a "core protein" containing one or more large glysoaminoglycan (GAG) chains (chondrohitin sulfate, heparan sulfate and keratan sulfate) with the exception of hyaluronic acid.

ECM not only joins the cells together providing mechanical adhesion and protection, but also regulates cell–cell interaction in both normal and pathological conditions including cell differentiation, angiogenesis, and apoptosis [8] and influences their survival, development, shape, polarity, and behavior [7]. It may further provide a support for cell–cell and cell–matrix cohesion in tissues. This adhesion is regulated by adhesive molecules (CAM), such as those acting, for example in wound healing [4, 9]. Cell-ECM adhesion is mediated by ECM receptors mainly integrins, discoidin domain receptors, and syndecans [10]. The ECM is also involved in the transmission of extracellular signals to cells, thus influencing cell survival and proliferation, but also affecting cell differentiation and death [7, 11–13].

In addition to proteins, the major components of the ECM are water, and polysaccharides. However, each tissue has a specific ECM which may differ in composition and topology, and that is produced during tissue development [10]. Its amount varies depending on tissues. Indeed, it is scant in the muscle and nervous tissue, whereas it is abounding in blood and in cartilage and bone [14] tissue [15]. Components of the ECM are synthesized intracellularly by fibroblasts, chondrocytes, and osteoblasts, and only successively secreted into the ECM by exocytosis [16]. Once secreted, they gather with the matrix.

The ECM is a highly dynamic entity that undergoes continuous remodeling, through the action of enzymes or non-enzymatic molecules [10] during development, morphogenesis, tissue repair, and remodeling. Proteolysis certainly constitutes one of the main processes leading to structural and biochemical changes in the ECM [17]. Many different classes of proteolytic enzymes are involved in ECM degradation. The proteolytic system present in human tissues is quite complex, since more than 500 genes coding for proteases or *protease-like* molecules are present in the human genome. Within this heterogeneous enzyme system, the major enzymes degrading the ECM are represented by serine-proteases, cysteine-proteases as well as members of the family of MMPs [18].

3 The MMP Family

MMPs are members of a large superfamily of Ca^{2+}-dependent endopeptidases, called Metzincins, strictly related, containing an active site Zn^{2+} [2]. Currently, the MMP family consists of a group of at least 25 related, but distinct, soluble and membrane-bound proteases, 24 of which are present in mammals [19]. They are produced by a wide range of cell types, in particular inflammatory cells, and are secreted into the extracellular space in an inactive form, called zymogen or proenzyme, or pro-MMP that is activated by proteolytic cleavage [20, 21]. Fibroblasts [22], monocytes [23], neutrophils and macrophages [24, 25], T cells [26], smooth muscle cells [27], astrocytes [28], glial cells [29], eosinophils [30–32], and endothelial cells [33] can synthesize MMPs, which after their synthesis, can be accumulated into inflammatory cell granules. More frequently, they are secreted and found anchored to the cell surface or tethered to other proteins on the cell surface or within the ECM [34].

These proteinases are evolutionarily conserved and rigorously regulated [35, 36]. As shown in Fig. 1, all members of this MMPs share a primary highly conserved organization called "minimal domain" consisting of three common major functional regions: an amino-terminal (N-terminal) hydrophobic signal sequence called pre-domain (Pre) or signal peptide (SP), a propeptide domain (Pro) containing a thiol group (–SH) and a furin-cleavage site (Fu), the catalytic domain containing the zinc-binding site (Zn^{2+}) (Fig. 1).

As previously stated, the majority of MMPs have an amino-terminal signal peptide containing 18–30 residues rich in hydrophobic amino acids [37]. The signal peptide manages their synthesis to the endoplasmic reticulum [38], inside which the signal is removed, and to their secretion into the extracellular space [39].

The cleavable propeptide domains (about 80 amino acids) contain a highly conserved consensus sequence PRCXXPD next to the C-terminal end of these segments.

Prodomain is essential for maintaining enzymatic latency through a mechanism identified as "cysteine switch," where the unpaired cysteine in a highly conserved amino acid sequence (Pro-Arg-Cys-Gly-X-Pro-Asp) forms a bridge with the catalytic zinc [39]. The cleavage of the propeptide domain produces a conformational change that modifies the proenzyme to its active form.

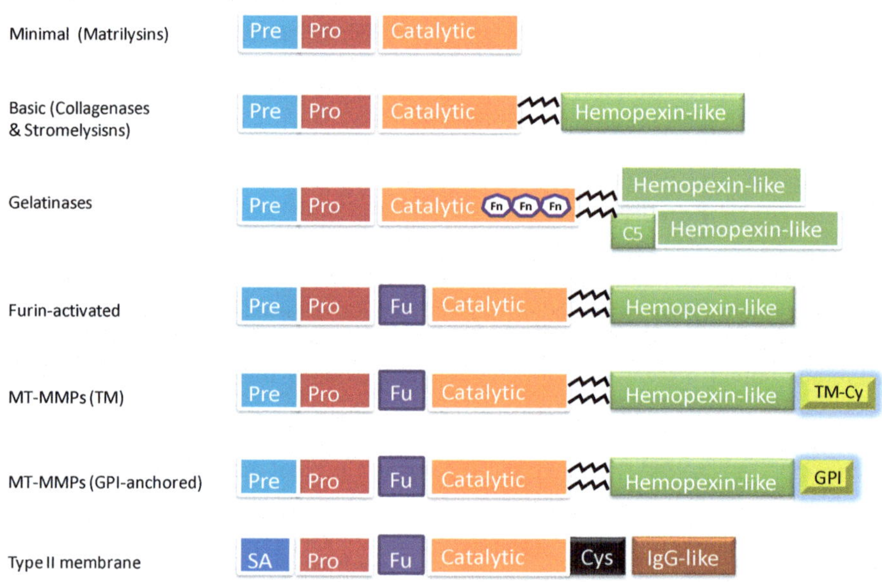

Fig. 1 Domain structure of the MMP family. The mammalian MMP family encompasses 25 members identified by different numbers or depending on their matrix substrates. The MMP motifs are indicated. *C5* type V collagen-like domain; *Col* collagenase-like protein; *Cs* cytosolic; *Cys* cysteine array; *Fn* fibronectin repeat; *Fr* furin-cleavage site; *Pro* prodomain; *SH* thiol group; *SP* signal peptide; *Zn* zinc. From Parks et al. [113], modified

The furin recognition site is a domain containing about 9 amino acids that are present in many MMPs including all membrane-bound MMPs. Its consensus sequence RXKR between their procatalytic and catalytic domains leads to intracellular activation cleavage by furin, a serine proteinase belonging to the convertase family [37].

The catalytic domain is characterized by a 3-histidine zinc-binding motif (HEXXHXXGXXH) to which the active site Zn^{2+} is bound [19] and a conserved methionine-turn motif following the active site. Indeed, this domain comprises two zinc ions and one or more calcium ions coordinated to different residues [40]. One Zn^{2+} ion (namely, the catalytic zinc) is embedded into the active site and is needed for the proteolytic activity of the MMPs. The other Zn^{2+} ion (also known as structural zinc) and the Ca^{++} ion are located in the catalytic domain [40]. The three histidine residues bound to the catalytic Zn^{2+} are conserved among all the MMPs.

Overall, the 3D conformation of the MMPs catalytic domain is similar. Therefore, substrate specificity is not simply explained by the structural features within the enzymes themselves [19].

With the exception of MMP-7, MMP-23, and MMP-26, a carboxy C-terminal hemopexin domain (Hpx) is present, linked to a short proline-rich hinge region, which likely has a critical role in substrate binding specificity [41–43] and in interactions of these enzymes with their tissue inhibitors [44]. The Hpx region (about 200 residues) is composed of four repeats with a Cys residue at either end that resemble plasma hemopexin and contain a disulfide bond (S–S) between the first and the last subdomain [39]. In matrilysins, this domain is absent, probably because of a deletion rather than of an evolutionary origin prior to the addition of Hpx [37, 45].

Other MMP motifs consist of a gelatin-binding fibronectin-like domain intercalated within their catalytic domain (gelatinases MMP-2 and MMP-9), a serine-, threonine-, and proline-rich collagen type V-like (C5) domain (MMP-9), a C-terminal transmembrane (TM) domain, a glycosylphosphatidylinositol (GPI) anchor, an N-terminal signal anchor (SA) and, in some cases, a cytoplasmic (Cs) domain [36, 46, 47].

One major distinction of MMPs is between secreted and membrane-anchored proteinases. MMPs may be anchored to the cell surface by their TM (MMP-14, MMP-15, MMP-16, and MMP-24), GPI (MMP-17 and MMP-25), or SA (MMP-23) motif. The TM domains and the GPI anchors are attached to the Hpx-like domain by a short linker peptide [42].

The MMPs have been further divided into several clades, at least five or six subgroups, according to substrate specificity and amino acid sequence homology (Table 1): collagenases, gelatinases, stromelysins, matrilysins, membrane-type MMPs and other MMPs [48, 49] (Table 2).

The smaller MMPs actually known are the matrilysins 1 and 2 (MMP-7, MMP-26) [50, 51]. These enzymes break down several ECM components, including gelatin and fibronectin [52, 53]. As previously stated, matrilysins do not contain the C-terminal Hpx-like domain and are consequently called the "minimal-domain MMPs" [43, 54].

Table 1 MMP family subgroups

Common name	MMP	Chromosomal location (human)	M.W. (kDa)	Collagen substrates	Some additional substrates*
Collagenases					
Collagenase-1	MMP-1	11q22-q23	55/45	I, II, III, VII, VIII, X	Aggrecan, gelatin
Collagenase-2	MMP-8	11q21-q22	75/58	I, II, III, VII, VIII, X	Aggrecan, gelatin, fibronectin
Collagenase-3	MP-13	11q22.3	60/48	I, II, III, IV, IX, X, XIV	
Collagenase-4	MMP-18	(*Xenopus*)	70/53		Aggrecan, gelatin, fibronectin
Gelatinases					
Gelatinasi A	MMP-2	16q13	72/66	I, II, III, IV, VII, X	Gelatin, fibronectin, fibrillin
Gelatinasi B	MMP-9	20q11.2-q13.1	92/86	IV, V	Gelatin, elastin, fibrillin
Stromelysins					
Stromelysin-1	MMP-3	11q23	57/45	II, III, IV, V, IX, X, XI	Gelatin, plasminogen
Stromelysin-2	MMP-10	11q22.3-q23	57/44	IV	Laminin, fibronectin elastin,
Stromelysin-3	MMP-11	22q11.2	51/44	IV	Fibronectin, laminin, aggrecan
Matrilysins					
Matrilysin-1	MMP-7	11q21-q22	28/19	IV	Fibronectin, laminin, gelatin
Matrilysin-2	MMP-26	11p-15	28/19	IV	Fibrinogen, fibronectin, gelatin
Metalloelastase	MMP-12	11q22.2-q22.3	54/45	IV	Elastin, fibronectin, latent TNF
MT-MMP Tm-type I					
MT1-MMP	MMP-14	14q11-q12	66/56	I, II, III	Gelatin, fibronectin, laminin
MT2-MMP	MMP-15	15q13-q21	72/60		Gelatin, fibronectin, laminin
MT3-MMP	MMP-16	8q21	64/52	III	Gelatin, fibronectin, laminin
MT5-MMP	MMP-24	20q11.2	–/52		Gelatin, fibronectin, laminin
GPI-anchored					Fibrinogen, fibrin
MT4-MMP	MMP-17	12q24.3	57/63		Fibrin, gelatin
MT6-MMP	MMP-25	16p13.3		IV	Fibronectin Gelatin, laminin

(continued)

Table 1 (continued)

Common name	MMP	Chromosomal location (human)	M.W. (kDa)	Collagen substrates	Some additional substrates*
Other MMPs					
	MMP-19	12q14	54/45	IV	Aggrecan, elastin, fibrillin
Enamelysin	MMP-20	11q22.3	54/22		Gelatin
	MMP-21	ND	70/53		Aggrecan
CA-MMP	MMP-23	1p36.3			Aggrecan
	MMP-27	11q24			Gelatin, casein, fibronectin
Epylisin	MMP-28	17q21.1	56/45		Casein

ND not determined. The list of substrates is not exhaustive

Table 2 Functional relevance of USPs in signaling pathways

Signaling pathway	Function	USPs
p53	Activators	USP7, USP10, USP24, USP29, USP42
	Repressors	USP2, USP4, USP5, USP7, USP15, USP22
NF-κB	Activators	USP2, USP6, USP7, USP17, USP21, USP25
	Repressors	CYLD, USP4, USP7, USP10, USP11, USP15, USP18, USP21, USP25, USP34
RTK	Activators	USP2, USP8, USP18
	Repressors	USP8
Wnt	Activators	USP4, USP5, USP14, USP34, USP47
	Repressors	CYLD, USP4, USP15
TGF-β	Activators	USP4, USP9X, USP11, USP15
	Repressors	CYLD

Collagenases can generally cleave the interstitial collagens I, II, and III and to digest other ECM as well as non-ECM proteins [2, 54].

The gelatinase group, (MMP-2 and MMP-9), mainly digests gelatin, the denatured form of collagen [2, 54]. Gelatinases contain three repeats homologous to the type II module of fibronectin incorporated into the catalytic domain that has been shown to be involved in binding to denatured collagen or gelatin [49]. This domain, known as the gelatin-binding domain or fibronectin type II-like domain, is unique to the gelatinases.

ECM components such as collagen IV and fibronectin are digested by the stromelysins MMP-3 and MMP-10. Another stromelysin (MMP-11 or stromelysin-3) has a different sequence and substrate specificity compared to MMP-3 and MMP-10. For this reason, some authors consider MMP-11 in the heterogeneous subgroup [2, 55].

MMP-12, or human macrophage elastase, is another member of the stromelysin subgroup initially found in alveolar macrophages of cigarette smokers [56].

Six isoforms of the membrane-type MMPs (MT-MMP) are currently known. They can digest ECM proteins such as gelatin, fibronectin, and laminin [57]. Differently from secreted MMPs, the MT-MMPs are membrane associated with cytoplasmic domains important in cellular signaling [58]. MT-MMPs in mammals include four type I transmembrane proteins (MT-1, MT-2, MT-3, MT-5-MMP) and two glycophosphatidylinositol-(GPI-) anchored proteins (MT4-MMP and MT6-MMP) [59, 60]. Moreover, most MT-MMPs activate pro-MMP-2 [2, 61]. Besides the highly conserved MMP functional domains, the MT-MMPs have also additional insertion sequences (IS) controlling the insertion of these enzymes into the cell membrane and conferring unique functional roles [58].

Finally, a heterogeneous subgroup includes the remaining MMPs which are characterized by different substrate specificity, amino acid sequence or domain organization [1, 65, 77]. To this subgroup belong MMP-19, MMP-20, MMP-21, MMP-23, MMP-27, and MMP-28, all able to cleave elastin and aggrecan [2, 62, 63].

4 The Role of MMPs in Normal and Pathological Conditions

MMPs have been considered for a long time as the principal enzymes responsible for the turnover and degradation of extracellular matrix [19]. Indeed, MMPs participate in remodeling and degradation of ECM and basement membranes (BM) occurring throughout the life in a number of physiological processes such as embryogenesis, cell proliferation, migration and differentiation, ovulation, mammary gland involution, as well as in epidermal wound healing, and tissue repair after injury such as myocardial infarction [48, 60, 64]. Some members of this family, including several MT-MMPs, likely play a role in homeostasis since they are expressed in resting tissues or in the regulation of several functions including cell survival, angiogenesis, inflammation, and signaling [65, 66], as well as in neuronal physiology of the adult brain [67].

However, recent findings have clearly demonstrated that MMPs also act on a wide range of non-matrix extracellular proteins, such as cytokines, chemokines, several receptors and membrane proteins, as well as antimicrobial peptides [19] and can also degrade proteins inside the cell, in the cytoplasm, mitochondria, and nucleus [68, 69]. Their targets include other proteinases, proteinase inhibitors, clotting factors, chemotactic molecules, latent growth factors, growth factor-binding proteins, cell surface receptors, cell–cell adhesion molecules, and theoretically all structural extracellular matrix proteins [38].

The gelatinases are type IV collagenases which specifically degrade type IV collagen, a component of the basal membrane [64]. These proteinases have been associated by many authors with inflammatory diseases. Indeed, MMP-9, which is present in very low levels in brain, is markedly up-regulated during an inflammatory response [70].

It is well established that MMP-9 plays a crucial role in chronic inflammation when neutrophil accumulation occurs [71]. MMP-9 is also present in neutrophil extracellular traps, which contain also a mixture of nucleic acids and histones, once released by these cells after activation by infectious agents such as for example *Candida albicans* [72] or *Schistosoma japonicum* [73].

Increase of MMP expression and activity is implicated in a number of acute and chronic pathological conditions, such as arthritis [74, 75], cardiovascular disease including acute myocardial infarction [76], chronic heart failure [77], chronic obstructive pulmonary disease [78–80], asthma [81], inflammatory bowel disease [42, 67] neuroinflammation [65], diabetes [82], tumor growth and metastasis [83–85], multiple sclerosis [86, 87] systemic sclerosis [47], Alzheimer's disease [88], brain injury and disease [89–91]. MMPs are also important in the etiopathogenesis of endometriosis, being involved in the ECM invasion of the endometriotic cells in the ectopic site [39].

Cell types express distinct patterns of MMPs under different stimuli [19]. Similarly, the same proteinase expressed by each cell type may act on different substrates and likely impacts different processes.

5 Regulation of MMP Activity

Several authors have drawn recently their attention on the regulation of MMPs [21, 38, 65]. In vivo, the function of MMPs is controlled at various levels: biosynthesis which include both gene transcription and protein translation, zymogen activation, compartmentalization (ie., the intracellular and pericellular distribution of enzymes), secretion, cell surface expression, activation and inhibition by protein inhibitors, ECM or cell surface localization, oxidative modification.

Physiologically, the expression of many MMPs is accurately and mostly regulated during the transcription of the gene and promoter elements, by specific signals [92]. In normal tissues, MMP expression is usually low, but during tissue remodeling, for example, their synthesis is rapidly induced. Transcription of MMPs is up- or down-regulated by various molecules such as growth factors (EGF, HGF, and TGF-ß), cytokines like interleukins (IL-1, IL-4, and IL-6), or tumor necrosis factor alpha (TNF-α), and chemokines [1, 38, 42, 93, 94].

Post-translational modifications also can occur in fact, activation of pro-MMP precursor zymogens and acetylation [38, 93, 94], represent another level of MMP regulation. As previously stated, most MMPs are generally produced by cells as pre-proenzymes in a latent state and then activated either pericellularly or outside the cell [48]. In contrast, the MT-MMP isoforms can be activated just inside the cell by furin. The signal peptide is cleaved during translation, and pro-MMPs are then obtained [21]. Activation of zymogens is an essential regulatory step of MMP activation and activity. The interaction between the thiol group of a conserved cysteine residue (Cys^{73}) in the prodomain and the Zn^{2+} of the catalytic site maintains the latent state of the pro-MMPs [42, 95]. Proenzymes are transformed to active enzymes by disruption of this interaction, a process known as the cysteine-switch mechanism [96], achievable by two main mechanisms: proteolytic cleavage of the prodomain, including autolysis, or by allosteric change of the thiol group present in the cysteine [42, 61]. A relevant side of the cysteine-switch mechanism activation is that the only step which is really crucial is the disruption of the interaction between the Zn^{2+} and the thiol group. The prodomain can remain attached, and the MMP can be active. Indeed, the thiol-Zn^{2+} interaction can also be disrupted by non-proteolytic mechanisms such as allosteric perturbation by non-substrate macromolecules like integrins and glysaminoglycans (GAGs) or in conditions reproduced in the laboratory using organomurcials, such as APMA, or SDS [19, 92]. In a gelatin zymography, for example, SDS in the running buffer is able to disrupt the Zn^{2+}-thiol binding thus making fully active the proenzyme. The reader is addressed for more details to reviews focusing on regulation of MMPs activity [19, 61, 92].

The extracelluar proteolytic activation of the proenzyme is regulated by several steps which are involving also other MMPs as well as serine proteinases like plasmin [38, 59]. Glycosilation may provide an additional level of control [97].

The intracellular and pericellular distribution of enzymes (compartmentalization) is another noteworthy mechanism for regulating the specificity of MMP proteolytic activity. Indeed, both secreted and membrane-bound MMPs are localized inside the cell to, the nucleus, cytosol, and organelles [39].

After being activated, MMPs are further regulated by endogenous inhibitors, autodegradation, and selective endocytosis. MMP-2, MMP-9, and MMP-13, for example, are internalized by a low density lipoprotein receptor-related protein (LRP) pathway [98]. Control over MMP activity is exerted by specific endogenous inhibitors, among them α2-macroglobulin and tissue inhibitors of MMPs (TIMPs) are the most important [42, 65].

TIMPs constitute a family of proteins which are secreted and are able to inhibit MMP activity outside the cells [48]. In mammals, four TIMPs have been identified at a gene level, namely TIMP-1, TIMP-2, TIMP-3 and TIMP-4 [99]. Inside normal cells, TIMPs rigidly regulate MMPs [49, 100]. TIMPs are synthesized in many tissues and by different cell populations. Their expression is regulated during several processes like, for example, development and tissue remodeling [99]. TIMPs inhibit almost all MMPs which have been evaluated. However, they differ in their affinity for specific MMPs, but their binding does not always mean inhibition [42]. Their amino-terminal domain is required for the inhibitory ability since this domain binds to the active site of the MMPs [49], whereas the other domain (C-terminal) interacts with the Hpx-like domain present in the proenzyme [60]. The balance between MMPs and their corresponding TIMPs is a key factor in normal and pathological tissue remodeling [49, 101]. TIMPs are also involved in MMP-independent TIMP regulation of cell behavior [102]. Furthermore, some TIMPs also may act as anti-angiogenetic [103] and anti-apoptotic factors [104].

A further control of MMPs regulation is exerted by the availability and affinity of substrates, but also by compartmentalization of the proteinases. Reviews of this topic were recently published [61, 92].

6 MMPs and Protozoan Infections

6.1 Malaria

Malaria is one of the major public health problems worldwide; in fact, it causes 300–500 million clinical cases and about 1 million fatalities annually. Five *Plasmodium* spp. may cause malaria in humans, but the most pathogenic species is certainly *Plasmodium falciparum*, particularly in Sub-Saharan Africa [105], although also *Plasmodium vivax* is responsible of a significant burden of morbidity and associated mortality [106]. Among the different pictures characterizing severe malaria, one of those potentially fatal is cerebral malaria (CM). Three mechanisms are envisaged as responsible for this complication: (i) the accumulation of red cells in the capillaries, facilitated by adhesion of infected erythrocytes to endothelial cell surface; (ii) the increase of Blood Brain Barrier (BBB) permeability; (iii) as a consequence of the latter point, the passage of toxic products or proinflammatory molecules like TNF-α, interleukin (IL)-1 and (IL)-6, but also chemokines and other molecules into the brain tissue [36, 107, 108]. Proinflammatory cytokines, like those produced by the Th1 cells activated by the parasite, up-regulate adhesion molecules present on the surface of the endothelial cells. This results in the erythrocyte accumulation in the different organs [107, 109].

The same above mechanisms are responsible also for tissue damage affecting other organs (lungs, kidney, and placenta in pregnant women) during severe malaria [107, 110, 111].

Different strategies are used by the parasite to evade the host immune response, for instance, *Plasmodium* spp. codes and produce antigenic proteins, underlying variation, localized on the surface of the red cells. That is the case of the *P. falciparum* erythrocyte membrane protein-1 (PfEMP-1) which allows the parasite binding to the molecules localized on the endothelial cell membrane [107].

Between the blood and the nervous tissue or fluid spaces, there are different barrier layers which regulate molecular exchange. The most important of these is represented by the BBB, constituted by the endothelium which is located between blood and brain interstitial fluid. This barrier is permeable in a selective way and regulates ion and nutrient transport into the brain, representing a filter between the CNS and the blood and limiting the passage of molecules from the blood circulation to the nervous tissue.

The blood vessels in the CNS are lined by specialized endothelial cells which are strictly joined by tight junctions and surrounded by a basal lamina along with astrocyte end-foot protuberances [112]. The choroid plexus epithelium between blood and the ventricular CSF and the arachnoid epithelium between blood and the subarachnoid CSF represent other barriers [112].

When BBB integrity is lost for any reason the transit of potentially toxic molecules into the nervous tissue is facilitated, leading to possible tissue damage. During CM, vascular structure alteration may damage seriously the BBB, and this process is enhanced by the inflammatory cascade previously cited [110].

MMPs, produced by infiltrating leukocytes or by cells of the CNS, mediate in many neurological diseases the disruption of the BBB [113] by cleaving matrix proteins which play a fundamental role in maintaining the integrity of BBB as well as in neuronal survival [114]. Experimental studies show that when MMPs are injected intraparenchymally a rupture of the BBB and the subsequent permeabilization of capillaries is obtained. Even MMP inhibitors, like TIMP-2, are involved in this process, by reducing the degradation of extracellular matrix, caused for example, by type IV collagenase, resulting in protection of the BBB [115].

At least two mechanisms can be envisaged to explain the role of either MMPs or TIMPs in CM pathogenesis: (i) modification of BBB structure and (ii) their participation as effectors and regulators of the immune response, since MMPs can cleave different molecules which participate in such response, like cytokines, chemokines, etc. [36]. TIMPs are family of proteins which are involved in many functions such as the turnover of extracellular matrix, the tissue remodeling, etc. [116].

In vitro, studies have clearly shown that human monocytes incubated with free hemozoin (Hz), a catabolic product of heme component of hemoglobin or Hz-containing trophozoites and endothelial cells, increase MMP-9 expression, production, and activity [36, 117]. The same does not occur for MMP-2 [118].

Data on the possible pathogenetic role of MMPs in CM, occurring in humans, derived from autopsy studies or from evaluations in blood or serum from malaria patients (either uncomplicated or severe) are not conclusive and consistent [36, 119].

In the brains of patients with fatal CM, pro-angiogenic uPAR resulted particularly concentrated [120], as well as MMP-1 and VEGF [121] in Durck granulomas, but anti-angiogenic endostatin/collagen XVIII was not observed in ring hemorrhage areas [122]. All these data suggest that the pro-MMP-9 underlies activation in those areas where parasite sequestration and vascular damage are more evident [36].

Activation of the human MMP-9 gene was observed using microarray analysis carried out on blood from Kenyan children affected by severe *P. falciparum* malaria, with a contemporary increase in neutrophils [123].

However, serum MMP-9 levels were not modified in Ghanaian and Gabonese children, independently on the severity of clinical picture [119, 124]. In the study carried out in Gabon, TIMP-2 levels resulted even lower in malaria patients, compared to healthy people. Vice versa, TIMP-1 resulted associated to the severity of malaria. As regards serum MMP-8 levels, these resulted increased in Gabonese children independently on the severity of disease [124], but not in Nigerian children suffering of uncomplicated malaria [125].

According to all these data, it is possible to conclude that TIMP-1 and MMP-8 levels may preconize the severity of malaria. In particular, TIMP-1 seems to prevent further the damage induced by MMPs by blocking the enzymatic activity of MMP-9, and to a lesser extent, that of MMP-8.

As in other inflammatory conditions involving the brain, such as the experimental model of lipopolysaccharide (LPS)-injured brain or in multiple sclerosis [126, 127], MMP-9 might also be involved significantly in CM pathogenesis, by facilitating the BBB permeability and infiltration of leukocytes. In particular, this enzyme might be induced by the catabolic product of hemoglobin, the hemozoin, in the endothelials, as observed by in vitro studies. Furthermore, this molecule, a side product of parasite invasion of red cells, increases the expression of MMP-1, MMP-3, and also of TIMP-2 [128].

In malaria, caused by *P. falciparum*, according to the analysis of post-mortem brain samples, MMP-1 resulted accumulated in astrocytes of the BBB and in macrophages/microglial cells, which are present in Dürck's granulomas [121], represented by microglial-astroglial nodules surrounding the damaged vessels.

From all these data, we can argue that in *P. falciparum* malaria, particularly when complicated by CM, there is an unbalance between MMP and TIMP, in favor of the former. However, the conclusive demonstration of the role played by the MMPs in disrupting the BBB during CM has to be obtained, and we need more research to highlight this issue [113].

7 Artesunate and MMPs

During *Plasmodium* infection, the blood levels of TNF-α increase because of the activation of monocyte-macrophages, caused either by infected erythrocytes or parasite molecules, such as the malaria pigment hemozoin. As previously said, this induces an up-regulation of MMP-9 either in monocytes or macrophages with a following proteolytical effect not only on the ECM but also on precursors of inflammatory cytokines, amplifying the inflammatory response [36].

Artemisinin and its derivatives such as artemisone, artesunate, and dihydroartemisinin (DHA) can immunomodulate the host response, in fact, they inhibit the mRNA synthesis and the production of MMP-9 in human monocyte-like cells, after stimulation with hemozoin or TNF-α. Artesunate, artemisone, and DHA were capable to antagonize MMP-9 secretion as well as its expression, stimulated by hemozoin, up to 50%.

The derivatives significantly down-regulated both TNF-α-induced MMP-9 secretion and mRNA levels of the enzyme, in a more active way than hemozoin itself. Both hemozoin and TNF-α increased the transcription dependent on NF-kB by 11 and 7.7 fold, respectively. Artesunate, artemisone, and DHA inhibited the NF-kB driven transcription, induced by hemozoin, by 28, 34, and 49%, respectively. In a similar manner, the derivatives, differently from artemisinin, prevented TNF-α-induced NF-kB driven transcription by 47–51%. Furthermore, contrary to artemisinin, its derivatives such as artemisone, artesunate, and DHA, modulated MMP-9, and also other genes which depend on NF-kB, like TNF-α.

Artemisinins may turn off the inflammatory response of monocytes in vivo. Thus, the beneficial clinical effects of artemisinins for the treatment of malaria would include not only the direct anti-parasitic activities but also the ability to attenuate the inflammatory response, thus reducing the risk of progression to the more severe form of the disease, like the CM [129].

8 The Role of MMPs in the Malaria Vector

A link between MMP and vector competence in mosquito-*Plasmodium* interactions has been observed recently. In fact MMPs resulted involved in the process of tissue invasion within the mosquitoes by malaria stages (ookinetes and sporozoites). In the genome of the malaria vector *Anopheles gambiae* there are three genes encoding MMPs, among them the *A. gambiae*, MMP-1 (*Ag*MMP-1) gene has resulted expressed during the blood digestion, midgut epithelium invasion by *Plasmodium* ookinetes, and the production of oocysts. The two isoforms of *Ag*MMP-1 derive from alternative splicing: the secreted (S-MMP-1), which is associated with hemocytes, and the membrane-type MMPs (MT-MMP-1). This latter is particularly abundant in the cell attachment sites of the midgut epithelium. During ookinete midgut invasion, in particular, MT-MMP-1 resulted increased in expression, in zymogen maturation, and subcellular redistribution. All these modifications suggest an implication of such protease in the midgut epithelial healing which follows ookinete invasion.

Furthermore, during oocyst development, using the RNA interference-mediated silencing of the *Ag*MMP-1 gene, a post-invasion protective function of this gene was observed [130].

9 African Trypanosomosis

This disease is caused by *Trypanosoma* (*T. brucei*), of which two species affect humans (*Trypanosoma brucei gambiense* and *Trypanosoma brucei rhodesiense*) and represents one of the most important infection, caused by parasites, in Sub-Saharan Africa: the human African trypanosomiasis (HAT), also called the human sleeping sickness, and Nagana which affects the cattle. In humans, the disease can progress in two stages: the hemolymphatic one and later on the meningo-encephalitic stage which occurs when parasites succeed in crossing the BBB and invading the CNS [131].

After introduction during the blood meal by the tsetse fly bite, metacyclic trypomastigotes underlie morphological and behavioral modifications in the circulation from which only later they can migrate through the BBB [132]. After passing through the choroid plexus, the parasites enter into the CSF, to finally reach the nervous tissue. At that time, it is possible to observe an increased expression of receptors, present on the host endothelium, like ICAM-1. Then parasites accumulate in the brain, and this is followed by dramatic leukocyte arrival and activation of astrocytes and microglia, leading to chronic encephalopathy, possibly fatal, when not treated [133].

The mechanisms which cause the disruption of BBB, during African trypanosomosis, are not very well clarified; for example [134], the parasite can transit through the barrier using a cysteine proteinase, partially in a Ca^{++}-dependent way. However, the passage of inflammatory cells and trypomastigotes through the basal lamina into the brain during the meningo-encephalitic stage of HAT may have mechanisms involving certainly the MMPs, in both cases.

It was therefore elucidated that MMP-9 and ICAM-1 represent reliable staging markers for *T. b. gambiense* HAT, and that MMPs as well as ICAMs, alone or in combination, can predict the meningo-encephalitic stage of HAT [135]. Both MMP-2 and MMP-9 create a localized temporary opening of the *glia limitans* (a component of the BBB) by selectively cleaving the β-dystroglycan subunit which anchors the astrocyte end-feet to the parenchymal membrane [136]. In this way, the gelatinases MMP-2 and MMP-9 allow the penetration of inflammatory cells through the outer parenchymal basal membrane into the brain.

MMP-3, MMP-8, and MMP-12, in the brain of *T. b. brucei*-infected mice, at thirty days post-infection, underwent an up-regulation of mRNA expression, concomitantly with a massive accumulation in parasites and leukocytes. On the contrary, the levels of MMP-1b, MMP-2, MMP-7, MMP-9, MMP-11, MMP-13, MMP-14 and MMP-19 as well as TIMP-1 and TIMP-2 mRNA were unmodified. MMP-10 mRNA expression resulted even undetectable [137].

A neutral *T. brucei*-MMP with a marked proteolytic activity on either gelatin or casein, and with a molecular mass of approximately 40 kDa, was partially characterized from the biochemical point of view. Classical inhibitors of MMPs resulted able to inhibit its activity [138].

The MMPs present in the *T. brucei brucei* extract depend strictly on the surrounding environment, with a stability of the proteolytic activity at pH7 but not pH5 or 9 and at 37 °C, as regards the temperature [138].

The levels of MMP-2, MMP-9, ICAM-1, VCAM-1, and E-selectin in the CSF of patients infected by *T. b. gambiense* in the two stages of the disease.

Furthermore, by the evaluation of ICAM-1 and MMP-9, alone or in combination, it was possible to distinguish between stage 1 and stage 2 patients with HAT with very high sensitivity and specificity, and a better performance compared to that obtained with other known markers such as the chemokine CXCL10 [135].

In conclusion, also in trypanosomosis, as it is well established in CM, increased MMP levels play a crucial role in facilitating the passage not only of the parasites, but also of leukocytes, through the BBB into the nervous tissue, where they may induce pathological modifications.

10 Toxoplasmosis

Toxoplasma gondii is an obligate intracellular parasite able to infect virtually any homeoterm animal species [139].

In humans, *Toxoplasma* is widespread and in individuals with an immature or suppressed immune response the infection can cause severe disease, an example is represented by toxoplasmic encephalitis [140].

In *Toxoplasma* encephalitis in mice, CD4+ and CD8+ T cells are mainly recruited to the brain, where they can prevent the reactivation of latent infections [141]. This process is mediated by the Th1-dependent IFN-γ which stimulates anti-parasitic effector mechanisms and regulates chemokine expression and the consequent leukocyte recruitment [142].

Different molecules such as IL-1, IL-23, TNF-α, and COX-2 stimulate, during toxoplasmosis, MMP production in the brain [143].

For example, an increase in CD4+ and CD8+ T cells producing both MMP-8 and MMP-10, has been observed in the brain tissue of *T. gondii*-infected mice [144].

Furthermore, in invading T cells as well as in CNS-resident astrocytes, during infection, TIMP-1 is expressed [143]. In wild type mice, changes in tissue morphology and signs of astrocyte activation occur, contrary to what happens in infected TIMP-1 KO mice where an increase in CD4+ T cells along with a significantly reduced parasite burden in the brain was observed, differently from the peripheral amount of parasites. This was not accompanied by any substantial pathological change in the brain according to histological analyses.

The up-regulation of TIMP-1 during infection might inhibit the pathogen clearance by reducing lymphocyte penetration into the CNS, driven by MMPs. The resulting inhibition of MMPs by an increased expression of TIMP-1 may represent an evasion mechanism of the parasite; in this way, it would try to control the arrival of immune cells or a host response to down-regulate immune-mediated damage.

When the TIMP-1 gene is disrupted in infected mice, in fact, perivascular inflammation is reduced or even absent, but a higher number of CD4+ T cells infiltrating the brain parenchyma is observed, probably reflecting a failure of control of the degradation processes of the basal lamina, carried out by MMPs. This further suggests that these proteinases are essential for the parasite clearance from the brain.

In congenital toxoplasmosis, the parasite arrives to the fetus after crossing the placental barrier.

MMP-12 might participate to elastin degradation which occurs when toxoplasmosis affects pregnant women where serum levels of this proteinase resulted increased along with elastin degradation products [144].

Furthermore, in *T. gondii*-infected pregnant women, it has been recently observed that *T. gondii* induces an increase in MMP-2 and MMP-9 serum and umbilical cord levels.

By co-immunoprecipitation analyses, it was possible in fact to show that MMP-2 and MMP-9 can interact with fibronectin in pregnant women, where the fibronectin monomer of 220 kDa resulted higher than in healthy controls. It is possible to envisage that MMP-2 and MMP-9 may be involved in extracellular matrix degradation and placental barrier dysfunction, which facilitates *T. gondii* transmission to the fetus [145].

11 MMPs and Helminth Infections

11.1 Neurocysticercosis

Neurocysticercosis (NCC) is the infection of the CNS caused by the larval stage of *Taenia solium*, which is acquired after parasite egg ingestion (NCC without taeniasis), or when the parasite cycle is completed, after raw or poorly cooked pork consumption (NCC with taeniasis) [146]. NCC is the most frequent cause of acquired epilepsy globally [147].

The clinical picture of NCC is characterized more frequently by active seizures. The disease is slowly progressive, and multiple factors are involved in determining the severity of the symptoms; among them, the degree of inflammatory reaction in the brain tissue is certainly relevant. It may also happen frequently; however, that NCC patients could remain asymptomatic for long periods, but we do not know the reason [148].

According to studies carried out in experimental animals, MMP expression plays a crucial role in the differential breakdown of the BBB, where distinct populations of immune cells play a role, and different MMPs can cleave cytokines, chemokines, and adhesion molecules [149]. In NCC patients, serum levels of MMP-2 and MMP-9 correlated with clinical manifestations of the disease (presence of seizures).

Mean serum MMP-2 levels were higher independently of the presence of symptoms in NCC cases in comparison with healthy controls, with no difference between the two patient groups.

On the contrary, MMP-9 serum levels were significantly increased in symptomatic NCC patients in comparison with asymptomatic NCC cases or healthy individuals [150].

Recently, higher levels of MMP-9 were found to correlate with epilepsy [151, 152], suggesting that in symptomatic NCC patients the seizures might be facilitated by increased levels of this gelatinase.

In a study conducted in India, some Toll-like receptor (TLR-4) gene polymorphisms in patients with solitary cysticercus granuloma, which is the most frequently observed type of NCC in that country, resulted particularly frequent, suggesting a possible association to the susceptibility to infection of individuals with such genetic traits [153].

The TLR-4 gene polymorphisms Asp/Gly and Thr/Ile, Gly (Asp/Gly plus Gly/Gly) and Ile (Thr/Ile plus Ile/Ile) resulted significantly associated with calcified neurocysticercosis. Vice versa, Gly/Gly and Ile/Ile genotypes were not significantly associated with either group. Furthermore, the levels of MMP-9 resulted higher in calcified neurocysticercosis, particularly when calcifications were multiple.

The heterozygous forms Asp/Gly and Thr/Ile, on the contrary, were significantly associated with seizure recurrence. The Gly (Asp/Gly plus Gly/Gly) and Ile (Thr/Ile plus Ile/Ile) genotypes were also significantly associated with seizure recurrence, to which also serum MMP-9 levels were significantly associated. The Authors suggested that The TLR-4 gene abnormalities may facilitate the inflammatory response around calcified neurocysticercosis leading to an increase in perilesional edema and following risk of seizures [154].

12 Nematode Infections

MMPs have been described in several parasitic nematodes (*Brugia malayi*, *Toxocara canis*, *Strongyloides stercoralis*, *Nippostrongylus brasiliensis*, *Dirofilaria immitis*, *Trichuris suis*, *Ancylostoma caninum*, *Caenorhabditis elegans*, and *Gnathostoma spinigerum*), but their sole in the life cycle is not always well clarified [155].

It is supposed that in parasitic nematodes, as in the case of hookworms, TIMPs play crucial roles in the host-parasite relations, which include the invasion process and establishment in the vertebrate animal hosts [116].

By a large-scale investigation of TIMP proteins in several human parasites (nematodes such as *Necator americanus*, *Ascaris suum*, but also the liver flukes *Clonorchis sinensis* and *Opisthorchis viverrini*, as well as the schistosome blood flukes), 15 protein sequences with a high homology to known eukaryotic TIMPs were predicted from the sequence data available for parasitic helminths and subjected to in-depth bioinformatic analyses [116].

A very well-studied nematode model in vivo as regards MMP involvement in the mechanisms responsible for organ damage (mainly the brain) mediated by MMPs is infection caused by *Angiostrongylus cantonensis*, a parasitic nematode which causes an eosinophilic meningitis which can occur mainly in the southeast Asia [156].

This rat lung worm is obligate to migrate inside the cerebral parenchyma in its hosts [157]. When non-permissive hosts such as humans and mice are infected with this parasite, the worms pass to the brain and mature into young adults; however, contrary to rats, they fail to reach maturity within the heart and lungs [157]. This nematode has a particular tropism for the nervous tissue; in fact, it requires a sejour in the CNS of mammalian hosts to grow [158].

The infective L_3 larvae orally infect the final host and by the general circulation arrive to the CNS, here they molt twice to transform in immature adults and enter the subarachnoid space. In a permissive host (like rats), immature adults migrate from the brain to the lungs, where they fulfill the maturation process [157].

In non-permissive hosts, however, the immature adults remain in the CNS of the host, causing eosinophilic meningitis or meningoencephalitis [159]. What we know about this nematode infection derives most from experimental studies in rodents. In fact, mice infected with *A. cantonensis* undergo eosinophilic meningitis, with a peak at around three weeks of eosinophilia in the CSF [160, 161]. The presence of MMP-9 in CSF was revealed 10 days after inoculation (a.i.) with a peak between day 15 and 25 a.i.

The MMP-9 is localized within inflammatory cells such as eosinophils and macrophages, accumulated in the subarachnoid space of experimentally infected mice, according to immunohistochemical results, suggesting that infiltrating leukocytes are the relevant source of MMP-9 in this type of meningitis [162].

The increase of MMP-9 activity was paralleled by that, even more rapid, of CSF eosinophils and the inflammatory reaction of the subarachnoid space. Differently from MMP-9, MMP-2 activity did not change during infection [162].

In *A. cantonensis*-infected patients, affected by, eosinophilic meningitis, the amounts of MMP-2, MMP-9, and TIMP-1 resulted significantly increased in the CSF, compared to healthy controls, whereas TIMP-4 levels resulted significantly lower. MMP-9 in CSF increases in patients in parallel with CSF leukocyte counts and CSF/serum albumin ratio (QAlb) values. When patients are treated with mebendazole and dexamethasone, during recovery from eosinophilic meningitis, a gradual decrease in MMP-9 levels, QAlb and TIMP-1, and a contemporary increase in those of TIMP-4, was observed. During the acute stage of infection, TIMP-4 levels were lower, even when MMP-9 and TIMP-1 have already decreased, until eosinophil meningitis was not definitely declined. This suggests that in these patients TIMP-4 has an important role in regulating the proteolytic-derived BBB damage, [155]. These results in humans further supported what we knew from experimental studies.

The activity of MMP-9 in the CSF of patients was not completely suppressed, since simultaneously TIMP-4 decreases, leading to BBB dysfunction, as indicated by the higher QAlb values detected in these patients. Changes in the cytokine milieu have a strong impact on modulation of TIMP expression and may be responsible for modifications of the levels of these proteins (MMPs/TIMPs) in this kind of meningitis [155].

Also, MMP-12 along with its substrate, elastin (present in the meningeal vessel of the subarachnoid space), is involved in the inflammatory response. MMP-12/TIMP-1 ratio, in fact, is significantly increased in the CSF of *A. cantonensis*-infected mice from day 10 post-infection (p.i.) and reached the peak on days 20 and 25 p.i. The production of MMP-12 is correlated with several parameters (elastin degradation, number of eosinophils, blood–CSF barrier permeability, and pathological modifications in the subarachnoid space).

When the antihelmintic albendazole is associated with doxycycline (used in this case as a non-selective MMP inhibitor), the amounts of MMP-12, elastin, and Evans blue accumulation in the CSF in mice affected by meningitis were significantly reduced, indicating that MMP-12 participates to the degradation of elastin [163].

The invasion by the parasite of the host meninges or nervous parenchima occurs with a concomitant increase of molecules such as tissue-type plasminogen activator (tPA), urokinase-type PA (uPA), MMP-9, and MMP-12 in the CSF [163–165]. As a result of the action of proteolytic enzyme increase, the brain barrier is disrupted, facilitating the eosinophil infiltration, as well as larval diffusion.

During *A. cantonensis* infection, leukocytes can pass the blood–CSF barrier, then entering the subarachnoid space by crossing the wall of meningeal vessels, formed by endothelial cells and elastic tissue which is degraded by proteinases such as MMP-12, released by infiltrating leukocytes [163].

After MMP-9 and -12 activation, the BBB becomes more permeable because tight junction proteins are destroyed. It has been shown that MMP-9 through the NF-κB/MMP-9 signaling pathway in angiostrongyloidosis mediates the degradation of claudin-5, followed by blood–CSF barrier impairment in the brain [166] (Fig. 2).

13 MMPs and Trichinellosis

Trichinellosis is a foodborne parasitic disease caused by nematodes of the genus *Trichinella*. We presently distinguish eight species in this genus (*Trichinella spiralis*, *Trichinella nativa*, *Trichinella britovi*, *Trichinella murrelli*, *Trichinella patagoniensis*, *Trichinella pseudospiralis*, *Trichinella papuae*, and *Trichinella zimbabwensis*) and three genotypes distributed in different geographical areas [167]. The first above five species may be grouped in the clade of encapsulating species, whereas the three remaining, in that of non-encapsulating species [167].

Infection is acquired by ingestion of raw or undercooked meat from infected (i.e., containing the L_1 infective larva) mammals, birds, and reptiles. This larva, inside the nurse cell, may be surrounded or not by a collagen capsule (see above), is released by gastric juice action in the stomach and after arriving to the intestine molts in adult male or female worms. After mating, the females, localized inside the epithelial intestinal cells, begin to shed the newborn larvae which enter the blood and lymph circulation to arrive to the final destination, the skeletal muscle cell, the only tissue where they can survive and develop [168]. Encapsulated species such as *T. spiralis* and non-encapsulated species like *T. pseudospiralis* behave differently in the host-parasite relationship as regards, for example the inflammatory response at muscle level which is stronger in the former [169].

In experimental murine infection with the nematode *Trichinella* spp. MMP-9 and MMP-2 underlie significant level changes during infection [170].

In mice infected with *T. spiralis*, the total MMP-9 serum level increased 6 days p.i., whereas that of total MMP-2 started to increase later. A similar result was obtained in mice infected with *T. pseudospiralis*, but the level of MMP-9 resulted

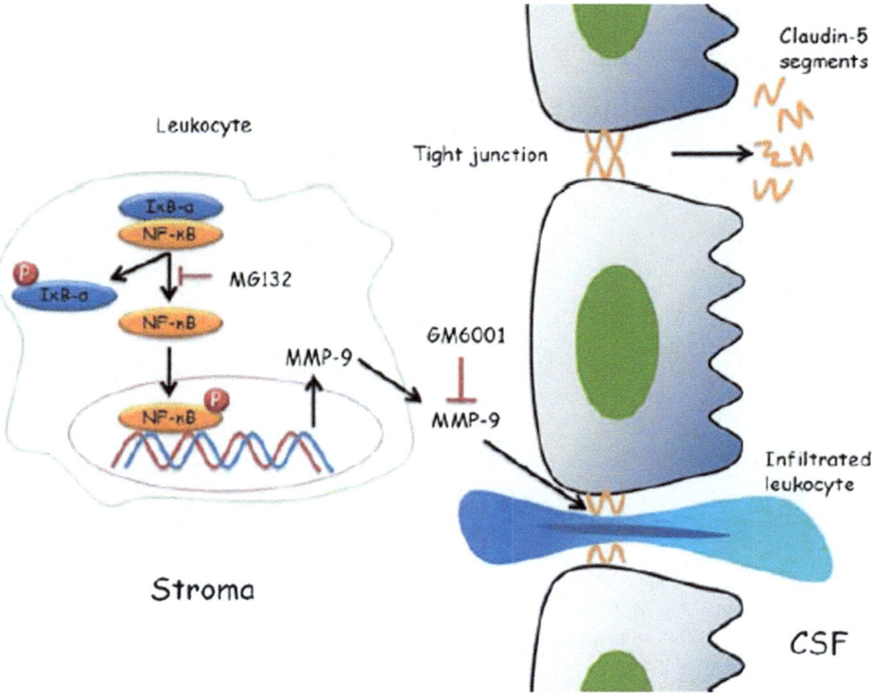

Fig. 2 Mechanisms of matrix metalloproteinase (MMP)-9 leading to claudin-5 degradation via the NF-κB pathway. The activation of NF-κB up-regulates MMP-9 production in *Angiostrongylus cantonensis*-induced leukocytes. Blocking of MMP-9 activity can reduce claudin-5 degradation and blood–CSF barrier permeability during angiostrongyloidosis meningoencephalitis. Therefore, MMP-9 is suggested to cause claudin-5 degradation and that it promotes leukocyte infiltration into the CSF by the paracellular route during *A. cantonensis* infection in the mouse choroid plexus. From Lai [76] with permission

lower than that found in *T. spiralis*-infected mice. As regards MMP-2 levels, significant differences were also observed between the two experimental groups of mice (infected either with *T. spiralis* or *T. pseudospiralis*).

Furthermore, serum TIMP-1 mean levels in *T. spiralis* or *T. pseudospiralis*-infected animals resulted significantly higher than in control mice at 21 days p.i. However, the TIMP-1 level was significantly higher in *T. spiralis* infected than in *T. pseudospiralis* infected animals, only at two weeks of infection. The kinetics of TIMP-1 paralleled that of TNF-α. On the contrary, IL-1β did not change during the whole period of observation in both groups of animals infected with the two *Trichinella* species.

In conclusion, MMP-9 and MMP-2 have resulted reliable markers of inflammation in both *T. spiralis* and *T. pseudospiralis* infections [170].

Serum levels of these MMPs were also analyzed in patients affected by trichinellosis during an outbreak caused by *T. britovi*. A representative zymography carried out in patient sera is shown in Fig. 3.

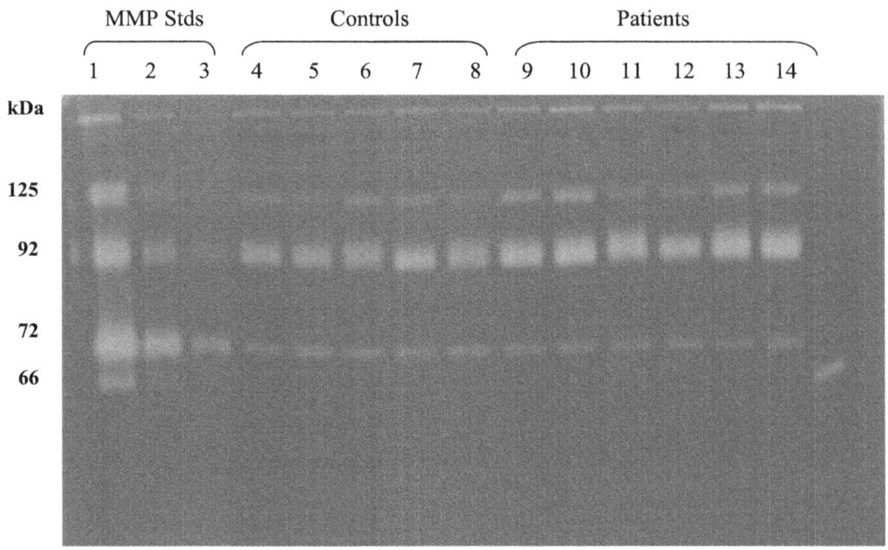

Fig. 3 Zymographic analysis of sera from *T. britovi*-infected patients and healthy individuals. Lane 1–3 and lane 15, Molecular standards; lane 4–8, healthy controls; lane 9–14 *T. britovi*-infected patients. The number on the left show the molecular weights of respectively: MMP-9/NGAL complex (125-kDa), Pro-MMP-9 (92-kDa), Pro-MMP-2 (72-kDa), Active-MMP-2 (66-kDa)

Gelatinolytic activity, corresponding to pro-MMP-9 resulted significantly increased in most of the patients in comparison with the healthy group. The same did not occur for pro-MMP-2 activity. The zymographic analysis of the gels showed the presence in serum samples of gelatinase bands at approximately 125, 92, and 72-kDa, corresponding to the MMP-9/Neutrophil gelatinase-associated lipocalin (NGAL) complex and proenzyme forms of MMP-9 and MMP-2, respectively. MMP-9/NGAL serum level was significantly augmented in patients, compared to healthy controls.

The MMP-9 levels resulted increased also by ELISA in the patient sera (Fig. 4a), with a significant correlation with zymographic data, as shown in Fig. 4b.

Furthermore, they resulted higher in more affected patients (suffering diarrhea, facial edemas, and myalgia). On the contrary, no difference was observed in patients with or without eosinophilia and muscle enzyme increase. In the light of these results, MMP-9 might be considered as a marker of inflammation in *T. britovi* patients, differently from MMP-2 which did not result significantly changed in patient sera, in comparison with controls (Fig. 4).

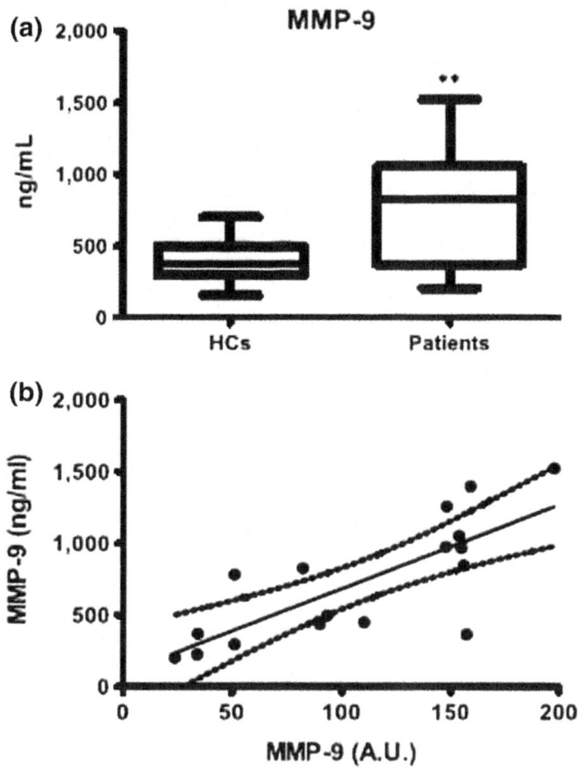

Fig. 4 **a** MMP-9 serum levels in trichinellosis patients. ** indicates the significant difference compared to healthy controls ($p = 0.0037$). **b** in a simple regression analysis, MMP-9 activity and MMP-9 serum levels were highly significantly related to each other ($r = 0.77$, $p < 0.001$) in patients. Solid line indicates the linear regression line. Dot lines indicate the 95% Confidence Intervals. MMP-9 levels are expressed as concentration (ng/ml) in ELISA test and as arbitrary units of gelatinolytic activity (A.U.) × 106 in the gelatin zymography. From Bruschi et al. [18] with permission

14 Concluding Remarks

Parasitic infections (caused by protozoa such as *Plasmodium*, African *Trypanosoma*, *T. gondii* or helminths such as *T. solium*, *A. cantonensis*, *Trichinella*), have in common an increase in levels of several MMPs, which can be induced either directly or indirectly by regulating cytokine levels, therefore, with an imbalance between such enzymes and TIMPs. The result of this unbalance is the amplified inflammation mediated damage which may occur in several organs.

References

1. Abbott NJ, Rönnbäck L, Hansson E (2006) Astrocyte-endothelial interactions at the blood-brain barrier. Nat Rev Neurosci 7:41–53
2. Affara NI, Andreu P, Coussens LM (2009) Delineating protease functions during cancer development. Methods Mol Biol 539:1–32
3. Agrawal S, Anderson P, Durbeej M, van Rooijen N, Ivars F, Opdenakker G, Sorokin LM (2006) Dystroglycan is selectively cleaved at the parenchymal basement membrane at sites of leukocyte extravasation in experimental autoimmune encephalomyelitis. J Exp Med 203:1007–1019
4. Alvarez JI, Teale JM (2008) Multiple expression of matrix metalloproteinases in murine neurocysticercosis: implications for leukocyte migration through multiple central nervous system barriers. Brain Res 1214:145–158
5. Armah H, Dodoo AK, Wiredu EK, Stiles JK, Adjei AA, Gyasi RK, Tettey Y (2005) High-level cerebellar expression of cytokines and adhesion molecules in fatal, paediatric, cerebral malaria. Ann Trop Med Parasitol 99:629–647
6. Armah HB, Wilson NO, Sarfo BY, Powell MD, Bond VC, Anderson W et al (2007) Cerebrospinal fluid and serum biomarkers of cerebral malaria mortality in Ghanaian children. Malar J 6:147
7. Avolio C, Ruggieri M, Giuliani F, Liuzzi GM, Leante R, Riccio P, Livrea P, Trojano M (2003) Serum MMP-2 and MMP-9 are elevated in different multiple sclerosis subtypes. J Neuroimmunol 136:46–53
8. Baird JK (2013) Evidence and implications of mortality associated with acute *Plasmodium vivax* malaria. Clin Microbiol Rev 26(1):36–57
9. Bendeck MP, Irvin C, Reidy M, Smith L, Mulholland D, Horton M, Giachelli CM (2000) Smooth muscle cell matrix metalloproteinase production is stimulated via alpha(v) beta(3) integrin. Arterioscler Thromb Vasc Biol 20(6):1467–1472
10. Bode W, Gomis-Rüth FX, Stöckler W (1993) Astacins, serralysins, snake venom and matrix metalloproteinases exhibit identical zinc-binding environments (HEXXHXXGXXH and Met-turn) and topologies and should be grouped into a common family, the “metzincins”. FEBS Lett 331:134–140
11. Bode W, Maskos K (2003) Structural basis of the matrix metalloproteinases and their physiological inhibitors, the tissue inhibitors of metalloproteinases. Biol Chem 384:863–872
12. Borden P, Heller RA (1997) Transcriptional control of matrix metalloproteinases and the tissue inhibitors of matrix metalloproteinases. Crit Rev Eukaryot Gene Expr 7(1–2):159–178
13. Brew K, Dinakarpandian D, Nagase H (2000) Tissue inhibitors of metalloproteinases: evolution, structure and function. Biochim Biophys Acta 1477:267–283
14. Briasoulis A, Tousoulis D, Papageorgiou N, Kampoli AM, Androulakis E, Antoniades C, Tsiamis E, Latsios G, Stefanadis C (2012) Novel therapeutic approaches targeting matrix metalloproteinases in cardiovascular disease. Curr Top Med Chem 12:1214–1221
15. Bruschi F, Chiumiento L (2011) Trichinella inflammatory myopathy: host or parasite strategy? Parasit Vec 23(4):42
16. Bruschi F, Dupouy-Camet J (2014) Trichinellosis. In: Bruschi F (ed) Helminth infections and their impact on global Public Health. Springer, Wien, pp 229–273
17. Bruschi F, Bianchi C, Fornaro M, Naccarato G, Menicagli M, Gomez-Morales MA, Pozio E, Pinto B (2014) Matrix metalloproteinase (MMP)-2 and MMP-9 as inflammation markers of *Trichinella spiralis* and Trichinella pseudospiralis infections in mice. Parasite Immunol 36 (10):540–549
18. Bruschi F, D’Amato C, Piaggi S, Bianchi C, Castagna B, Paolicchi A, Pinto B (2016) Matrix metalloproteinase (MMP)-9: a reliable marker for inflammation in early human trichinellosis. Vet Parasitol 14:S0304-4017(16)30105-4

19. Cantacessi C, Hofmann A, Pickering D, Navarro S, Mitreva M, Loukas A (2013) TIMPs of parasitic helminths—a large-scale analysis of high-throughput sequence datasets. Parasit Vectors 6:156
20. Cauwe B, Opdenakker G (2010) Intracellular substrate cleavage: a novel dimension in the biochemistry, biology and pathology of matrix metalloproteinases. Crit Rev Biochem Mol Biol 45(5):351–423
21. Chen KM, Liu JY, Lai SC, Hsu LS, Lee HH (2006) Association of plasminogen activators and matrix metalloproteinase-9 proteolytic cascade with blood-CNS barrier damage of angiostrongyliasis. Int J Exp Pathol 87:113–119
22. Chiu PS, Lai SC (2013) Matrix metalloproteinase-9 leads to claudin-5 degradation via the NF-κB pathway in BALB/c mice with eosinophilic meningoencephalitis caused by Angiostrongylus cantonensis. PLoS ONE 8:e53370
23. Chou PH, Lai SC (2011) Elevated concentrations of matrix metalloproteinase-12 and elastin degradation products in the sera of pregnant women infected with *Toxoplasma gondii*. Annals of Tropical Med Parasitol 105:225–231
24. Chuah C, Jones MK, Burke ML, McManus DP, Owen HC, Gobert GN (2014) Defining a pro-inflammatory neutrophil phenotype in response to schistosome eggs. Cell Microbiol 16:1666–1677
25. Clark RT, Nance JP, Noor S, Wilson EH (2010) T cell production of matrix metalloproteases and inhibition of parasite clearance by TIMP-1 during chronic toxoplasma infection in the brain. ASN Neurol 3:1–12
26. Coussens LM, Fingleton B, Matrisian LM (2002) Matrix metalloproteinase inhibitors and cancer: trials and tribulations. Science 295(5564):2387–2392
27. Creemers EE, Cleutjens JP, Smits JF, Daemen MJ (2001) Matrix metalloproteinase inhibition after myocardial infarction: a new approach to prevent heart failure? Circ Res 89:201–210
28. Darcy F, Santoro F (1994) Toxoplasmosis. In: Kierszenbaum F (ed) Parasitic Infections and the Immune System. Academic Press, Waltham, MA, USA, pp 163–201
29. Das S, Mandal M, Chakraborti T, Mandal A, Chakraborti S (2003) Structure and evolutionary aspects of matrix metalloproteinases: a brief overview. Mol Cell Biochem 253:31–40
30. Davis GE, Donald R, Senger DR (2005) Endothelial extracellular matrix: biosynthesis, remodeling, and functions during vascular morphogenesis and neovessel stabilization. Circ Res 97:1093–1107
31. Deininger MH, Winkler S, Kremsner PG, Meyermann R, Schluesener HJ (2003) Angiogenic proteins in brains of patients who died with cerebral malaria. J Neuroimmunol 142:101–111
32. Deininger MH, Fimmen B, Kremsner PG, Meyermann R, Schluesener HJ (2002) Accumulation of endostatin/collagenXVIII in brains of patients who died with cerebral malaria. J Neuroimmunol 131:216–221
33. de Sousa KP, Atouguia J, Silva MS (2010) Partial biochemical characterization of a metalloproteinase from the bloodstream forms of *Trypanosoma brucei brucei* parasites. Protein J 29(4):283–289
34. Dietmann A, Helbok R, Lackner P, Issifou S, Lell B, Matsiegui PB, Reindl M, Schmutzhard E, Kremsner PG (2008) Matrix metalloproteinases and their tissue inhibitors (TIMPs) in Plasmodium falciparum malaria: serum levels of TIMP-1 are associated with disease severity. J Infect Dis 197:1614–1620
35. Dzwonek J, Rylski M, Kaczmarek L (2004) Matrix metalloproteinases and their endogenous inhibitors in neuronal physiology of the adult brain. FEBS Lett 567:129–135
36. Egeblad M, Werb Z (2002) New functions for the matrix metalloproteinases in cancer progression. Nat Rev Cancer 2:161–174
37. Enanga B, Burchmore R, Stewart ML, Barrett MP (2002) Sleeping sickness and the brain. Cell Mol Life Sci 59:845–858

38. Fanjul-Fernández M, Folgueras AR, Cabrera S, López-Otín C (1803) Matrix metalloproteinases: evolution, gene regulation and functional analysis in mouse models. Biochim Biophys Acta 1:3–19
39. Fauser S, Deininger MH, Kremsner PG, Magdolen V, Luther T, Meyermann R, Schluesener HJ (2000) Lesion associated expression of urokinase-type plasminogen activator receptor (uPAR, CD87) in human cerebral malaria. J Neuroimmunol 111:234–240
40. Fata JE, Leco KJ, Voura EB, Yu HY, Waterhouse P, Murphy G, Moorehead RA, Khokha R (2001) Accelerated apoptosis in the Timp-3-deficient mammary gland. J Clin Invest 108:831–841
41. Ferrer E, Garate T (2014) Taeniasis and cysticercosis. In: Bruschi F (ed) Helminth infections and their impact on global public health. Springer, Wien, pp 202–227
42. Fillmore HL, VanMeter TE, Broaddus WC (2001) Membrane-type matrix metalloproteinases (MT-MMPs): expression and function during glioma invasion. J Neurooncol 53:187–202
43. Ford CC, Rosenberg GA (1999) Matrix metalloproteinases and neuroinflammation in multiple sclerosis. In: Proteases in the brain. Proteases in biology and disease, vol 3, pp 351–371
44. Frantz C, Stewart KM, Weaver VM (2010) The extracellular matrix at a glance. J Cell Sci 123(Pt 24):4195–4200
45. García HH, Gonzalez AE, Evans CA, Gilman RH Cysticercosis Working Group in Peru (2003) *Taenia solium* cysticercosis. Lancet 362:547–556
46. Gattazzo F, Urciuolo A, Bonaldo P (1840) Extracellular matrix: a dynamic microenvironment for stem cell niche. Biochim Biophys Acta 8:2506–2519
47. Gazzinelli R, Xu Y, Hieny S, Cheever A, Sher A (1992) Simultaneous depletion of CD4+ and CD8+T lymphocytes is required to reactivate chronic infection with *Toxoplasma gondii*. J Immunol 149:175–180
48. Gentili C, Cancedda R (2009) Cartilage and bone extracellular matrix. Curr Pharm Des 15:1334–1348
49. Geurts N, Opdenakker G, Van den Steen PE (2012) Matrix metalloproteinases as therapeutic targets in protozoan parasitic infections. Pharmacol Ther 133:257–279
50. Ghajar CM, George SC, Putnam AJ (2008) Matrix metalloproteinase control of capillary morphogenesis. Crit Rev Eukaryot Gene Expr 18:251–278
51. Gharagozlian S, Svennevig K, Bangstad HJ, Winberg JO, Kolset SO (2009) Matrix metalloproteinases in subjects with type 1 diabetes. BMC Clin Pathol 16(9):7
52. Gottschall PE, Yu X (1995) Cytokines regulate gelatinase A and B (matrix metalloproteinase 2 and 9) activity in cultured rat astrocytes. J Neurochem 64:1513–1520
53. Goulielmaki E, Sidén-Kiamos I, Loukeris TG (2014) Functional characterization of *Anopheles* matrix metalloprotease 1 reveals its Aagonistic role during sporogonic development of malaria parasites. Infect Immun 82(11):4865–4877
54. Grab DJ, Kennedy PG (2008) Traversal of human and animal trypanosomes across the blood–brain barrier. J Neurovirol 14:344–351
55. Griffiths MJ, Shafi MJ, Popper SJ, Hemingway CA, Kortok MM, Wathen A, Rockett KA, Mott R, Levin M, Newton CR, Marsh K, Relman DA, Kwiatkowski DP (2005) Genomewide analysis of the host response to malaria in Kenyan children. J Infect Dis 191:1599–1611
56. Gupta SP (2012) Matrix metalloproteinases inhibitors: specificity of binding and structure-activity relationships. In: Gupta SP (ed) Experientia supplements, vol 103. Springer
57. Hadler-Olsen E, Fadnes B, Sylte I, Uhlin-Hansen L, Winberg JO (2011) Regulation of matrix metalloproteinase activity in health and disease. FEBS J 278(1):28–45
58. Hainard A, Tiberti N, Robin X, Ngoyi DM, Matovu E, Enyaru JC, Müller M, Turck N, Ndung'u JM, Lejon V, Sanchez JC (2011) Matrix metalloproteinase-9 and intercellular adhesion molecule 1 are powerful staging markers for human African trypanosomiasis. Trop Med Int Health 16(1):119–126

59. Halper J, Kjaer M (2014) Basic components of connective tissues and extracellular matrix: elastin, fibrillin, fibulins, fibrinogen, fibronectin, laminin, tenascins and thrombospondins. M Adv Exp Med Biol 802:31–47
60. Herron GS, Werb Z, Dwyer K, Banda MJ (1986) Secretion of metalloproteinases by stimulated capillary endothelial cells. In Production of pro collagenase and pro stromelysin exceeds expression of proteolytic activity. J Biol Chem 261:2810–2813
61. Heuser K, Hoddevik EH, Taubøll E, Gjerstad L, Indahl U, Kaczmarek L, Berg PR, Lien S, Nagelhus EA, Ottersen OP (2010) Temporal lobe epilepsy and matrix metalloproteinase 9: a tempting relation but negative genetic association. Seizure 19:335–338
62. Hijova E (2005) Matrix metalloproteinases: their biological functions and clinical implication. Bratisl Lek Listy 106:127–132
63. Hong Z, Lin YM, Qin X, Peng JL (2012) Serum MMP-9 is elevated in children with asthma. Mol Med Rep 5:462–464
64. Hou RF, Tu WC, Lee HH, Chen KM, Chou HL, Lai SC (2004) Elevation of plasminogen activators in cerebrospinal fluid of mice with eosinophilic meningitis caused by *Angiostrongylus cantonensis*. Int J Parasitol 34:1355–1364
65. Hsu WY, Chen JY, Chien CT, Chi CS, Han NT (1990) Eosinophilic meningitis caused by Angiostrongylus cantonensis. Pediatr Infect Dis J 9:443–445
66. Ingber DE (2006) Mechanical control of tissue morphogenesis during embryological development. Int J Dev Biol 50:255–266
67. Iyer RP, Patterson NL, Fields GB, Lindsey ML (2012) The history of matrix metalloproteinases: milestones, myths, and misperceptions. Am J Physiol Heart Circ Physiol 303(8): H919–H930
68. Jiang Y, Goldberg ID, Shi YE (2002) Complex roles of tissue inhibitors of metalloproteinases in cancer. Oncogene 21:2245–2252
69. Jones CB, Sane DC, Herrington DM (2003) Matrix metalloproteinases: a review of their structure and role in acute coronary syndrome. Cardiovasc Res 59:812–823
70. Kim YS, Joh TH (2012) Matrix metalloproteinases, new insights into the understanding of neurodegenerative disorders. Biomol Ther (Seoul) 20(2):133–143
71. Klein T, Bischoff R (2011) Physiology and pathophysiology of matrix metalloproteases. Amino Acids 41(2):271–290
72. Kotra LP, Zhang L, Fridman R, Orlando R, Mobashery S (2002) N-Glycosylation pattern of the zymogenic form of human matrix metalloproteinase-9. Bioorg Chem 30:356–370
73. Kristensson K, Nygård M, Bertini G, Bentivoglio M (2010) African trypanosome infections of the nervous system: parasite entry and effects on sleep and synaptic functions. Prog Neurobiol 91:152–171
74. Kumar V, Fausto N, Abbas A (2004) Robbins and Cotran: pathologic basis of disease, 7th edn. Elsevier, Philadelphia
75. Lachuriya G, Garg RK, Jain A, Malhotra HS, Singh AK, Jain B, Kumar N, Verma R, Sharma PK (2016) Toll-like receptor-4 polymorphisms and serum matrix metalloproteinase-9 in newly diagnosed patients with calcified neurocysticercosis and seizures. Medicine (Baltimore) 95(17):e3288
76. Lai SH (2014) Angiostrongyloidosis. In: Bruschi F (ed) Helminth infections and their impact on global public health. Springer, Wien, pp 461–477
77. Lee HH, Chou HL, Chen KM, Lai SC (2004) Association of matrix metalloproteinase-9 in eosinophilic meningitis of BALB/c mice caused by *Angiostrongylus cantonensis*. Parasitol Res 94:321–328
78. Leppert D, Lindberg RL, Kappos L, Leib SL (2001) Matrix metalloproteinases: multifunctional effectors of inflammation in multiple sclerosis and bacterial meningitis. Brain Res Rev 36:249–257
79. Leppert D, Waubant E, Galardy R, Bunnett NW, Hauser SL (1995) T cell gelatinases mediate basement membrane transmigration in vitro. J Immunol 154:4379–4389

80. Lin J, Lindsay MP (2008) MMP roles in the initiation and progression of cardiac remodeling leading to congestive heart failure. In: Lagente V, Boichot E (eds) Matrix metalloproteinases in tissue remodelling and inflammation, pp 99–122
81. Lin CQ, Bissell MJ (1993) Multi-faceted regulation of cell differentiation by extracellular matrix. FASEB J 7:737–743
82. Lindner D, Zietsch C, Becher PM, Schulze K, Schultheiss HP, Tschöpe C, Westermann D (2012) Differential expression of matrix metalloproteases in human fibroblasts with different origins. Biochem Res Int 2012:875742
83. Lohi J, Wilson CL, Roby JD, Parks WC (2001) Epilysin, a novel human matrix metalloproteinase (MMP-28) expressed in testis and keratinocytes and in response to injury. J Biol Chem 276:10134–10144
84. Lukes A, Mun-Bryce S, Lukes M, Rosenberg GA (1999) Extracellular matrix degradation by metalloproteinases and central nervous system diseases. Mol Neurobiol 19:267–284
85. Lynch JP 3rd (2003) Idiopathic pulmonary fibrosis. Series: Lung biology in health and disease, p 185
86. Magenta D, Sangiovanni E, Basilico N, Haynes RK, Parapini S, Colombo E, Bosisio E, Taramelli D, Dell'Agli M (2014) Inhibition of metalloproteinase-9 secretion and gene expression by artemisinin derivatives. Acta Trop 140:77–83
87. Maleski M, Hockfield S (1997) 9 Glial cells assemble hyaluronan-based pericellular matrices in vitro. Glia 20:193–202
88. Mannello F, Medda V (2012) Nuclear localization of matrix metalloproteinases. Prog Histochem Cytochem. 47(1):27–58
89. Masocha W, Rottenberg ME, Kristensson K (2006) Minocycline impedes African trypanosome invasion of the brain in a murine model. Antimicrob Agents Chemother 50:1798–1804
90. Massova I, Kotra LP, Fridman R, Mobashery S (1998) Matrix metalloproteinases: structures, evolution, and diversification. FASEB J 12:1075–1095
91. Matthews KR, Gull K (1994) Cycles within cycles: the interplay between differentiation and cell division in Trypanosoma brucei. Parasitol Today 10:473–476
92. McLean AR, Ataide R, Simpson JA, Beeson JG, Fowkes FJ (2015) Malaria and immunity during pregnancy and postpartum: a tale of two species. Parasitology 142(8):999–1015
93. Miller LH, Dror I, Baruch KM, Ogobara KD (2002) The pathogenic basis of malaria. Nature 415:673–679
94. Mocchegiani E, Giacconi R, Costarelli L (2011) Metalloproteases/anti-metalloproteases imbalance in chronic obstructive pulmonary disease: genetic factors and treatment implications. Curr Opin Pulm Med 17:S11–S19
95. Mosher DF, Adams JB (2012) Adhesion-modulating/matricellular ECM protein families: a structural, functional and evolutionary appraisal. Matrix Biol 31:155–161
96. Mott JD, Werb Z (2004) Regulation of matrix biology by matrix metalloproteinases. Curr Opin Cell Biol 16:558–564
97. Mun-Bryce S, Rosenberg GA (1998) Gelatinase B modulates selective opening of the blood–brain barrier during inflammation. Am J Physiol 274:R1203–R1211
98. Muroski ME, Roycik MD, Newcomer RG, Van den Steen PE, Opdenakker G, Monroe HR, Sahab ZJ, Sang QX (2008) Matrix metalloproteinase-9/gelatinase B is a putative therapeutic target of chronic obstructive pulmonary disease and multiple sclerosis. Curr Pharm Biotechnol 9:34–46
99. Murphy G (2011) Tissue inhibitors of metalloproteinases. Murphy Genome Biol 12:1–7. http://genomebiology.com/2011/12/11/233
100. Murphy G, Nagase H (2011) Localizing matrix metalloproteinase activities in the pericellular environment. FEBS J 278(1):2–15
101. Nagase H, Visse R, Murphy G (2006) Structure and function of matrix metalloproteinases and TIMPs. Cardiovascular Res 69:562–573
102. Nagase H, Woessner JF Jr (1999) Matrix metalloproteinases. J Biol Chem 274:21491–21494

103. Nielsen BS, Timshel S, Kjeldsen L, Sehested M, Pyke C, Borregaard N, Dano K (1996) 92 kDa type IV collagenase (MMP-9) is expressed in neutrophils and macrophages but not in malignant epithelial cells in human colon cancer. Int J Cancer 65:57–62
104. Nishimura K, Hung T (1997) Current views on geographic distribution and modes of infection of neurohelminthic diseases. J Neurol Sci 145:5–14
105. Noguera R, Nieto OA, Tadeo I, Fariñas F, Alvaro T (2012) Extracellular matrix, biotensegrity and tumor microenvironment. An update and overview. Histol Histopathol 27:693–705
106. Noone C, Parkinson M, Dowling DJ, Aldridge A, Kirwan P, Molloy SF et al (2013) Plasma cytokines, chemokines and cellular immune responses in pre-school Nigerian children infected with *Plasmodium falciparum*. Malar J 12:5
107. Ohno I, Ohtani H, Nitta Y, Suzuki J, Hoshi H, Honma M, Isoyama S, Tanno Y, Tamura G, Yamauchi K, Nagura H, Shirato K (1997) Eosinophils as a source of matrix metalloproteinase-9 in asthmatic airway inflammation. Am J Respir Cell Mol Biol 16 (3):212–219
108. Oikonomidi S, Kostikas K, Tsilioni I, Tanou K, Gourgoulianis KI, Kiropoulos TS (2009) Matrix metalloproteinases in respiratory diseases: from pathogenesis to potential clinical implications. Curr Med Chem 16:1214–1228
109. Okada S, Kita H, George TJ, Gleich GJ, Leiferman KM (1997) Migration of eosinophils through basement membrane components in vitro: role of matrix metalloproteinase-9. Am J Respir Cell Mol Biol 17:519–528
110. Overall CM (2002) Molecular determinants of metalloproteinase substrate specificity: matrix metalloproteinase substrate binding domains, modules, and exosites. Mol Biotechnol 22:51–86
111. Page-McCaw A, Ewald AJ, Werb Z (2007) Matrix metalloproteinases and the regulation of tissue remodelling. Nat Rev Mol Cell Biol 8:221–233
112. Park HI, Turk BE, Gerkema FE, Cantley LC, Sang QX (2002) Peptide substrate specificities and protein cleavage sites of human endometase/matrilysin-2/matrix metalloproteinase-26. J Biol Chem 277(38):35168–35175
113. Parks WC, Wilson CL, Lòpez-Boado YS (2004) Matrix metalloproteinases as modulators of inflammation and innate immunity. Nat Rev 4:617–628
114. Parks WC, Mecham RP (eds) (1998) Matrix metalloproteinases. Academic press, London
115. Peng WJ, Yan JW, Wan YN, Wang BX, Tao JH, Yang GJ, Pan HF, Wang J (2012) Matrix metalloproteinases: a review of their structure and role in systemic sclerosis. J Clin Immunol 32(6):1409–1414
116. Plopper G (2007) The extracellular matrix and cell adhesion. In: Lewin B, Cassimeris L, Lingappa V, Plopper G, Sudbury MA (eds), Cells. Jones and Bartlett
117. Polimeni M, Prato M (2014) Host matrix metalloproteinases in cerebral malaria: new kids on the block against blood-brain barrier integrity? Fluids Barriers CNS 11(1):1–24
118. Pozio E, Zarlenga D (2013) New pieces of the *Trichinella* puzzle. Int J Parasitol 43:983–997
119. Prato M (2011) Malarial pigment does not induce MMP-2 and TIMP-2 protein release by human monocytes. Asian Pac J Trop Med 4:756
120. Prato M, D'Alessandro S, Van den Steen PE, Opdenakker G, Arese P, Taramelli D, Basilico M (2011) Natural haemozoin modulates matrix metalloproteinases and induces morphological changes in human microvascular endothelium. Cell Microbiol 13:1275–1285
121. Prato M, Giribaldi G (2011) Matrix metalloproteinase-9 and haemozoin: wedding rings for human host and *Plasmodium falciparum* parasite in complicated malaria. J Trop Med 2011:628435
122. Puxeddu I, Ribatti D, Crivellato E, Levi-Schaffer F (2005) Mast cells and eosinophils: a novel link between inflammation and angiogenesis in allergic diseases. J Allergy Clin Immunol 116(3):531–536
123. Ra HJ, Parks WC (2007) Control of matrix metalloproteinase catalytic activity. Matrix Biol 26(8):587–596

124. Raffetto JD, Khalil RA (2008) Matrix metalloproteinases and their inhibitors in vascular remodeling and vascular disease. Biochem Pharmacol 75(2):346–359
125. Ravi A, Pallavi G, Sitaraman SV (2007) Matrix metalloproteinases in inflammatory bowel disease: boon or a bane? Inflamm Bowel Dis 13:97–107
126. Rosenberg GA, Kornfeld M, Estrada E, Kelley RO, Liotta LA, Stetler-Stevenson WG (1992) TIMP-2 reduces proteolytic opening of blood–brain barrier by type IV collagenase. Brain Res 576:203–207
127. Roycik MD, Fang X, Sang QX (2009) A fresh prospect of extracellular matrix hydrolytic enzymes and their substrates. Curr Pharm Des 15:1295–1308
128. Sasaki O, Sugaya H, Ishida K, Yoshimura K (1993) Ablation of eosinophils with anti-IL-5 antibody enhances the survival of intracranial worms of *Angiostrongylus cantonensis* in the mouse. Parasite Immunol 15:349–354
129. Shapiro SD, Kobayashi DK, Ley TJ (1993) Cloning and characterization of a unique elastolytic metalloproteinase produced by human alveolar macrophages. J Biol Chem 268:23824–23829
130. Siefert SA, Sarkar R (2012) Matrix metalloproteinases in vascular physiology and disease. Vascular 20:210–216
131. Shikani HJ, Freeman BD, Lisanti MP, Weiss LM, Tanowitz HB, Desruisseaux MS (2012) Cerebral malaria: we have come a long way. Am J Pathol 181:1484–1492
132. Sciutto E, Fragoso G, Fleury A, Laclette JP, Sotelo J, Aluja A, Vargas L, Larralde C (2000) *Taenia solium* disease in humans and pigs: an ancient parasitosis disease rooted in developing countries and emerging as a major health problem of global dimensions. Microbes Infect 2:1875–1890
133. Singh A, Garg RK, Jain A, Malhotra HS, Prakash S, Verma R, Sharma PK (2015) Toll like receptor-4 gene polymorphisms in patients with solitary cysticercus granuloma. J Neurol Sci 355(1–2):180–185
134. Springman EB, Angleton EL, Birkedal-Hansen H, Van Wart HE (1990) Multiple modes of activation of latent human fibroblast collagenase: evidence for the role of a Cys73 active-site zinc complex in latency and a “cysteine switch” mechanism for activation. Proc Natl Acad Sci 87:364–368
135. Srivastava PK, Dastidar SG, Ray A (2007) Chronic obstructive pulmonary disease: role of matrix metalloproteases and future challenges of drug therapy. Expert Opin Investig Drugs 16:1069–1078
136. Steffens B, Hakkinen L, Larjava H (2001) Proteolytic events of wound-healing-coordinated interactions among matrix metalloproteinanses (MMPs), integrins, and extracellular matrix molecules. Crit Rev Oral Biol Med 12:373–398
137. Sterlicht MD, Werb Z (2001) How matrix metalloproteinases regulate cell behaviour. Ann. Rev Cell Dev Biol 17:463–516
138. Strack A, Asensio VC, Campbell IL, Schluter D, Deckert M (2002) Chemokines are differentially expressed by astrocytes, microglia and inflammatory leukocytes in Toxoplasma encephalitis and critically regulated by interferon-gamma. Acta Neuropathol 103:458–468
139. Stracke OJ, Fosang JA, Last K, Mercuri AF, Pendas MA, Llano E, Perris R, Di Cesare EP (2000) Matrix metalloproteinases 19 and 20 cleave aggrecan and cartilage oligomeric matrix protein (COMP). FEBS Lett 478:52–56
140. Sugaya H, Yoshimura K (1998) T-cell-dependent eosinophilia in the cerebrospinal fluid of the mouse infected with Angiostrongylus cantonensis. Parasite Immunol 10:127–138
141. Tetlow LC, Adlam DJ, Woolley DE (2001) Matrix metalloproteinase and proinflammatory cytokine production by chondrocytes of human osteoarthritic cartilage: associations with degenerative changes. Arthritis Rheum 44:585–594
142. Tocchi A, Parks WC (2013) Functional interactions between matrix metalloproteinases and glycosaminoglycans. FEBS J 280(10):2332–2341
143. Tsai HC, Chung LY, Chen ER, Liu YC, Lee SSJ, Chen YS, Sy CL, Wann SR, Yen CM (2008) Association of matrix metalloproteinase-9 and tissue inhibitors of

metalloproteinase-4 in cerebrospinal fluid with blood-brain barrier dysfunction in patients with eosinophilic meningitis caused by *Angiostrongylus cantonensis*. Am J Trop Med Hyg 78:20–27

144. Urban CF, Reichard U, Brinkmann V, Zychlinsky A (2006) Neutrophil extracellular traps capture and kill *Candida albicans* yeast and hyphal forms. Cell Microbiol 8:668–676
145. Uría JA, López-Otín C (2000) Matrilysin-2, a new matrix metalloproteinase expressed in human tumors and showing the minimal domain organization required for secretion, latency, and activity. Cancer Res 60:4745–4751
146. van der Heyde HC, Nolan J, Combes V, Gramaglia I, Grau GE (2006) A unified hypothesis for the genesis of cerebral malaria: sequestration, inflammation and hemostasis leading to microcirculatory dysfunction. Trends Parasitol 22:503–508
147. Van Hove I, Lemmens K, Van de Velde S, Verslegers M, Moons L (2012) Matrix metalloproteinase-3 in the central nervous system: a look on the bright side. J Neurochem 123(2):203–216
148. Van Wart HE, Birkedal-Hansen H (1990) The cysteine switch: a principle of regulation of metalloproteinase activity with potential applicability to the entire matrix metalloproteinase gene family. Proc Natl Acad Sci USA 87:5578–5582
149. Verma A, Prasad KN, Nyati KK, Singh SK, Singh AK, Paliwal VK, Gupta RK (2011) Association of MMP-2 and MMP-9 with clinical outcome of neurocysticercosis. Parasitol 138:1423–1428
150. Visse R, Nagase H (2003) Matrix metalloproteinases and tissue inhibitors of metalloproteinases: structure, function, and biochemistry. Circ Res 92:827–839
151. Wang MF, Lai SC (2013) Fibronectin degradation by MMP-2/MMP-9 in the serum of pregnant women and umbilical cord with *Toxoplasma gondii* infection. J Obstet Gynaecol 33(4):370–374
152. Wang Q, Wu ZD, Wei J, Owen RL, Lun ZR (2012) Human Angiostrongylus cantonensis: an update. Eur J Clin Microbiol Infect Dis 31:389–395
153. Wang XX, Tan MS, Yu JT, Tan L (2014) Matrix metalloproteinases and their multiple roles in Alzheimer's disease. Biomed Res Int 2014:908636
154. Watt FM, Fujiwara H (2011) Cell–extracellular matrix interactions in normal and diseased skin. Cold Spring Harb Perspect Biol 3(4):a005124
155. Wei PC, Tsai CH, Chiu PS, Lai SC (2011) Matrix metalloproteinase-12 leads to elastin degradation in BALB/c mice with eosinophilic meningitis caused by *Angiostrongylus cantonensis*. Int J Parasitol 41:1175–1183
156. Welgus HG, Campbell EJ, Cury JD, Eisen AZ, Senior RM, Wilhelm SM, Goldberg GI (1990) Neutral metalloproteinases produced by human mononuclear phagocytes. Enzyme profile, regulation, and expression during cellular development. J Clin Invest 86:1496–1502
157. Wells GM, Catlin G, Cossins JA, Mangan M, Ward GA, Miller KM, Clements JM (1996) Quantitation of matrix metalloproteinases in cultured rat astrocytes using the polymerase chain reaction with a multi-competitor cDNA standard. Glia 18:332–340
158. Werb Z, Sympson CJ, Alexander CM, Thomasset N, Lund LR, MacAuley A, Ashkenas J, Bissell MJ (1996) Extracellular matrix remodelling and the regulation of epithelial-stromal interactions during differentiation and involution. Kidney Int Suppl 54:S68–S74
159. Wiese S, Karus M, Faissner A (2012) Astrocytes as a source for extracellular matrix molecules and cytokines. Front Pharmacol 3:120. doi:10.3389/fphar.2012.00120
160. Wong SY, Remington JS (1993) Biology of *Toxoplasma gondii*. AIDS 7:299–316
161. World Health Organization (2015) (WHO): World Malaria Report. World Health, Organization, Geneva
162. Xu X, Jackson PL, Tanner S, Hardison MT, Abdul Roda M, Blalock JE, Gaggar A (2011) A self-propagating matrix metalloprotease-9 (MMP-9) dependent cycle of chronic neutrophilic inflammation. PLoS ONE 6:e15781
163. Yan C, Boyd DD (2007) Regulation of matrix metalloproteinase gene expression. J Cell Physiol 211(1):19–26

164. Yang Z, Strickland DK, Bornstein P (2001) Extracellular MMP-2 levels are regulated by the low-density lipoprotein-related scavenger receptor and thrombospondin 2. J Biol Chem 276:8403–8408
165. Yin P, Yang L, Zhou HY, Sun RP (2011) Matrix metalloproteinase-9 may be a potential therapeutic target in epilepsy. Med Hypotheses 76:184–186
166. Yong VW (2005) Metalloproteinases: mediators of pathology and regeneration in the CNS. Nat Rev Neurosci 6:931–944
167. Yong VW, Power C, Forsyth P, Edwards DR (2001) Metalloproteinases in biology and pathology of the nervous system. Nature Rev 2:502–511
168. Yoshida W, Uzuki M, Nishida J, Shimamura T, Sawai T (2009) Examination of in vivo gelatinolytic activity in rheumatoid arthritis synovial tissue using newly developed in situ zymography and image analyzer. Clin Exp Rheumatol 27:587–593
169. Zucker S, Pei D, Cao J, Lopez-Otin C (2003) Membrane type-matrix metalloproteinases (MT-MMP). Curr Top Dev Biol 54:1–74
170. Zucker S, Vacirca J (2004) Role of matrix metalloproteinases (MMPs) in colorectal cancer. Cancer Metastasis Rev 23:101–117

Roles of *Candida albicans* Aspartic Proteases in Host-Pathogen Interactions

Mariusz Gogol, Oliwia Bochenska, Marcin Zawrotniak, Justyna Karkowska-Kuleta, Dorota Zajac and Maria Rapala-Kozik

Abstract

Candida albicans—a common opportunistic fungal pathogen of humans—causes serious, disseminated invasive infections (candidiases) executed due to the action of several groups of virulence factors. One of the most critical is a family of secreted aspartic proteases involved in the destruction of host proteins and tissues. This chapter aims to characterize biochemical and structural properties of these enzymes that determine their functions and summarize their specific roles in the development and propagation of fungal infections. Candidal aspartic proteases deregulate the host biochemical homeostasis, by impairing the major proteolytic cascades such as the blood coagulation, the kallikrein-kinin system, and the complement system, by unleashing the activity of host proteases due to the degradation of specific endogenous inhibitors and by the inactivation of antimicrobial peptides and proteins produced by host cells. The degradation of important host proteins influences the fungal adhesion to the host cell surfaces, promotes the subsequent tissue damages, and enables the further dissemination of the pathogen. Confirmed multiple roles of candidal aspartic proteases in the host-pathogen interactions during candidiasis qualify these enzymes as promising potential targets for novel antifungal therapies.

M. Gogol · M. Zawrotniak · J. Karkowska-Kuleta · M. Rapala-Kozik (✉)
Department of Comparative Biochemistry and Bioanalytics, Faculty of Biochemistry, Biophysics and Biotechnology, Jagiellonian University,
Gronostajowa 7, 30-387 Krakow, Poland
e-mail: maria.rapala-kozik@uj.edu.pl

M. Gogol · O. Bochenska · D. Zajac
Department of Analytical Biochemistry, Faculty of Biochemistry, Biophysics and Biotechnology, Gronostajowa 7, 30-387 Krakow, Poland

S. Chakraborti and N.S. Dhalla (eds.), *Pathophysiological Aspects of Proteases*,
DOI 10.1007/978-981-10-6141-7_15

Keywords
Candida albicans · Fungal infection · Aspartic proteases (Saps) · Protein degradation · Host cell interaction · Antibacterial peptides · Kinins

1 Introduction

The current progress in the development of advanced health technologies, rescuing critically ill patients, has inevitably led to a considerable increase in the number of immunocompromised individuals who are particularly vulnerable to infections caused by opportunistic microorganisms including pathogenic fungi [1]. The catheterization and application of parenteral nutrition, hematologic malignancy, surgical treatments, including those related to cancer therapy and organ transplantation, HIV infection, the use of immunosuppressive therapies and broad-spectrum antibiotic treatment, as well as inherited immunodeficiency, old age and prematurity, are the main risk factors contributing to the development of opportunistic fungal infections [2–5].

The *Candida* spp. yeasts belongs to the most commonly identified fungal opportunistic pathogens of humans, responsible for serious, disseminated invasive infections [6, 7]. These fungi are also considered to be some of the major pathogens responsible for nosocomial bloodstream infections associated with high mortality rates in the range of 40–70% of infected patients [1, 2, 8]. *Candida albicans*—the most prevalent species from the *Candida* genus—is a part of normal human microbiota as a commensal microorganism that colonizes the skin and mucous membranes of the oral cavity, gastrointestinal tract, or genitourinary system [9, 10]. The genital or oral carriage of these yeasts is reported to be present in 20–65% of healthy individuals. However, when the delicate balance between microbial colonizers and the host is disturbed, *Candida* yeasts can cause annoying and painful superficial infections including thrush, oral candidosis, or candidal vulvovaginitis that can affect considerably large number of individuals colonized by these fungi [11–14].

The significant changes in the global distribution of particular *Candida* species and diversified prevalence among different groups of patients have been noticed over the last few decades [3, 15]. Despite the fact that *C. albicans* is still the major infectious agent from the genus *Candida* responsible for approximately 50% of all candidiases worldwide, other, so-called non-albicans *Candida* species have been emerging as fungal pathogens of humans, attributed to an increasing share in the overall number of candidal infections [16, 17]. This group includes mainly four species—*C. glabrata*, *C. tropicalis*, *C. parapsilosis* and *C. krusei*—that greatly differ in terms of the mechanisms of their pathogenicity [18].

Candida yeasts display a broad range of virulence attributes (Fig. 1) that allow them to successfully colonize and invade the host organism. During the infection, several mechanical, physical, and chemical protective barriers and biochemical defenses have to be affected by pathogens [19].

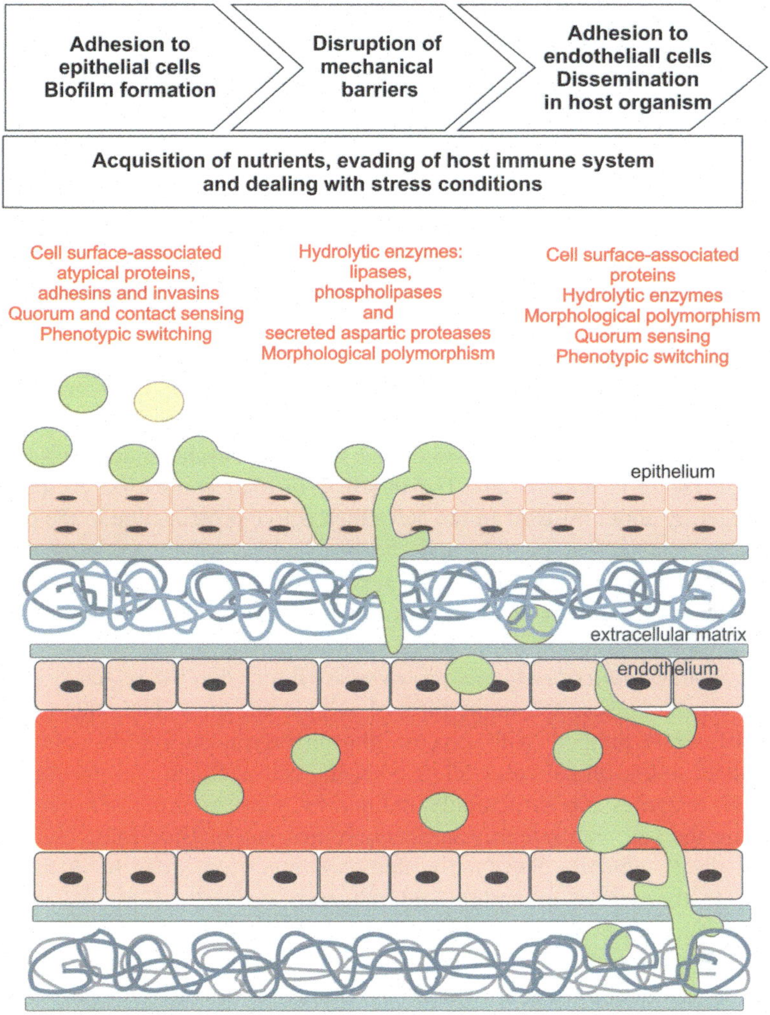

Fig. 1 Involvement of *C. albicans* virulence factors and traits related to its pathogenicity in the sequential stages of interaction with the human host during infection (based on [20, 21])

The candidal virulence factors primarily include two large groups of proteins: (i) an abundant set of cell surface-exposed proteins (adhesins and invasins) [22–24] and (ii) secreted hydrolytic enzymes such as lipases, phospholipases, and aspartic proteases [25–27]. The presence of such molecules, in a combination with other virulence-related fungal traits like a morphological polymorphism—i.e., the ability of the fungus to grow as unicellular, ovoid yeast-like forms or as filamentous forms (true hyphae or pseudohyphae)—the contact- and quorum-sensing, the phenotypic switching ("white" → "opaque") and the biofilm formation greatly facilitates the initiation of the first contact of fungal cells with the host, followed by further dissemination within the human body and the subsequent development of infection [28]. All of these features and abilities contributing to the fungal pathogenicity work together to successfully combat or evade the host immune system and take control of the processes involved in maintaining physiological homeostasis of the host. In particular, the role of proteolytic enzymes in this phenomenon can hardly be overestimated [21, 26].

2 The Family of *C. albicans* Secreted Aspartic Proteases (Saps)

The proteolytic activity of *C. albicans* was first described in 1965 [29] and has more recently been assigned to 10 secreted aspartic proteases (Saps). The open reading frames (ORFs) for their genes, located on five different chromosomes, vary between 1173 and 1764 bp in length [26]. The products of their expression are preproenzymes longer by about 60–200 amino acids than active proteins due to the presence of the N-terminal signal peptide and the propeptide, which are proteolytically removed by the signal peptidase and Kex2 protease, respectively, during the classical secretory pathway in order to form the final products with a molecular mass within the range of 35–50 kDa [30, 31].

Although all *C. albicans* aspartic proteases are directed to the secretory pathway, only Sap1–Sap8 are secreted in the form of soluble enzymes. The other two, Sap9 and Sap10, are equipped with glycosylphosphatidylinositol (GPI) anchor that attaches them to the fungal cell wall or both, the cell wall and cell membrane [25, 32]. These enzymes are structurally similar to yapsins, proteases involved in maintaining the cell wall integrity in *Saccharomyces cerevisiae* [32].

Within the *C. albicans* Sap family, three separate groups can be distinguished according to the degree of amino acid sequence similarity [25]. Among them, Sap4–Sap6 and Sap1–Sap3 represent the highest similarity degree (Fig. 2). The expression of these proteases seems to be dependent on the specific morphological form of the yeast, because Sap1–Sap3 are expressed primarily by the yeast-like forms, whereas Sap4–Sap6 are characteristic for hyphal forms. Saps that belong to these groups have a high sequence similarity to Sap8 which constitutes the third group together with Sap7 that shares only 20–27% sequence identity with other proteases and with Sap9 and Sap10 which are slightly closer homologs [30, 33].

3 Biochemical Properties of *C. albicans* Aspartic Proteases

So far, for four out of ten Saps of *C. albicans*, the crystal structures have been solved, including Sap1 in an inhibitor-free form [34], the complexes of Sap2 with inhibitor A70450 [35, 36] and benzamidine [37], Sap3 in the free form and bound with pepstatin A [38], and the complex of Sap5 with pepstatin A [34]. The latter structure

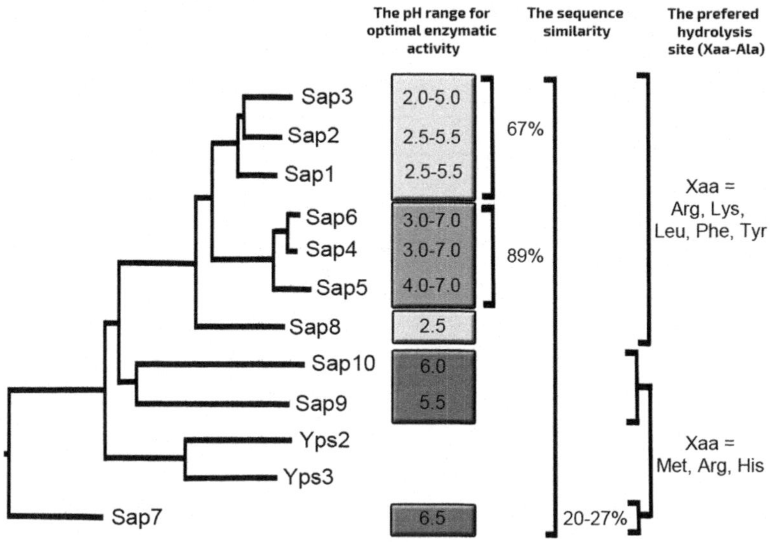

Fig. 2 Dendrogram presenting the amino acid sequence similarity between the members of *C. albicans* aspartic protease family (Sap1–Sap10). A close relationship of Saps to *S. cerevisiae* yapsins (Yps) is also emphasized. The optimal pH for Sap action and the preferential hydrolysis sites are also specified [25, 30]

is presented in Fig. 3. All structurally characterized Sap isoenzymes show similar features that classify them as pepsin-like aspartic proteases. These kidney-shaped bilobed globular proteins predominantly consisting of β-sheets are clearly divided into N-terminal and C-terminal domains, each providing one catalytic Asp residue, belonging to highly conserved regions in aspartic proteases with the motifs Asp-Ser-Gly or Asp-Thr-Gly and the disulfide bridge. Another conserved region that overlaps the active center is a β-hairpin loop, commonly known as the active site flap, which contains a catalytically essential Tyr residue. The disulfide bridges tie together the N-terminal and C-terminal entrance loops. The presence of the N-terminal entrance loop, consisting of 11 amino acids and specified as the second site flap [39], distinguishes Saps from the other aspartic proteases [34, 38, 40].

In addition, the structural and conformational differences allow the Sap isoenzyme structures to be divided into distinct subgroups. The most significant differences between Sap1–Sap3 and Sap5 regard the substrate binding site pockets, which have different characteristics, shapes, and sizes. On one hand, the S3 and S4 pockets, relatively large and with negative polar character in Sap1–Sap3, in Sap5 have a reduced size and the polar character is turned highly positive due to the substitutions of Leu297 (Sap1–Sap3), Asp299 (Sap2), Asp120 (Sap2, Sap3), and Gly299 (Sap1, Sap3) residues with Arg residues. On the other hand, the S2 pocket is enlarged in Sap5 because of the replacement of Asn131 (Sap1, Sap2) with Gly. The S1 and S2 substrate binding pockets in the central region of the enzymes show

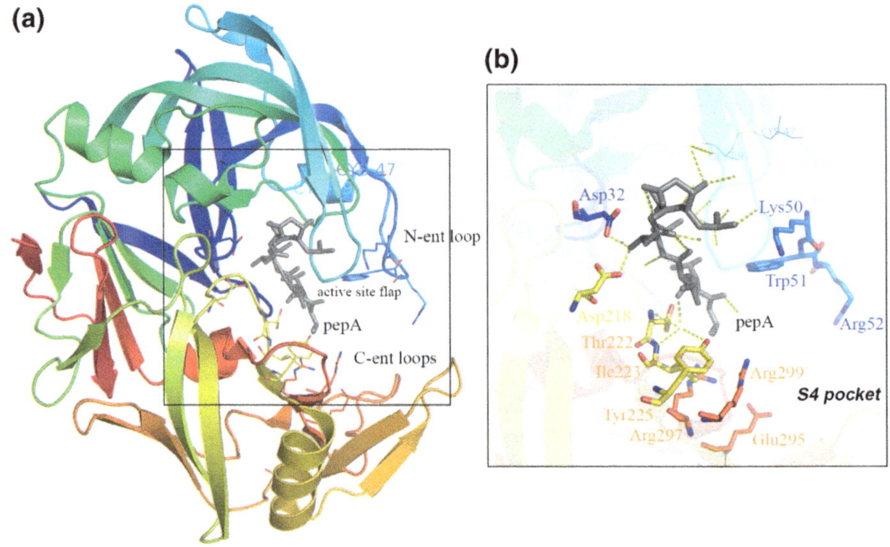

Fig. 3 Structure of the complex of Sap5 with pepstatin A (pepA). **a** The overall structure of the Sap molecule, showing its kidney-shaped appearance and the predominance of β-sheets. The loops tied together by the disulfide bridges are specified as the N-terminal entrance loop (cysteine residues 47 and 59 showed as blue sticks) (N-ent loop) and the C-terminal entrance loops (cysteine residues 256 and 294) (C-ent loops). **b** A close-up view of the active site with marked conserved Asp32 and Asp218 residues and the disulfide bridge (Cys47 and Cys 59), Lys50, Trp51, and Arg52 residues forming the N-ent loop directed to the active site cleft and the residues Thr222, Ile223, Tyr225, Glu295, Arg 297, and Arg299 involved in substrate binding in the S4 pocket. Yellow dashed lines indicate hydrogen bonds between pepA and residues that form the substrate binding pocket. The figure was made with Pymol [41], PDB ID:2QZX

only a few differences, of which the formation of an extra cavity within the S1 pocket in Sap5 due to the substitution of Arg195 and Glu193 with Thr195 and Lys193 is the most important. Another significant difference is a narrowed entrance to the active site cleft in Sap5 relative to Sap1–Sap3 because the N-terminal loop in the former has larger residues, such as Lys50, Trp51, and Arg52, pointing down into the substrate binding cleft. Although the active site is strongly negative in all isoenzymes, the overall electrostatic charge of Sap5 is positive while Sap1, Sap2 and Sap3 are negatively charged. The difference in overall electrostatic charge of the molecule can underlie a rise of the optimal pH for Sap5 activity, compared to Sap1–Sap3 [34].

The pH dependence of Sap enzymatic activity was characterized for recombinant proteins [30, 42–44] as well as for Sap1–Sap3 purified from *C. albicans* culture supernatants [45] and determined using resorufin-labeled casein [30, 42], FRETS-25Ala library [30], bovine hemoglobin [43], bovine serum albumin [45], and a peptide, histatin 5 (His5) [44]. The analysis performed in the broad range of pH between 2.0 and 7.5 indicated that most of Saps displayed the optimum for

Table 1 Biochemical properties of recombinant Saps [30]

Enzyme	Sap1	Sap2	Sap3	Sap4	Sap5	Sap6	Sap7	Sap8	Sap9	Sap10
Molecular mass (kDa)	36	36	37	37	37	37	47	35	53	45
Optimal pH for hydrolytic activity	5	4	3	5	5	5	6.5	2.5	5.5	6
pH range for activity	2.5–6.5	2.5–5.5	2–5	2.5–7	4.5–6.5	2.5–6.5	4–7.5	2–6.5	2.5–7	3–7
Pepstatin A inhibition	+	+	+	+	+	+	–	+	+	+
N-glycosylation	–	–	–	+	–	+	+	+	+	+

proteolytic activities at pH 3.0–5.0 (Table 1), a feature typical for aspartic proteases; however, slightly variable results were obtained by different research groups. The pH optimal for Sap4–Sap6, Sap7, Sap9–Sap10 is less acidic, and the enzymes are still active at neutral pH. Sap3 and Sap8 differ from other Saps in showing substantial activity at pH 2.0. Interestingly, Sap8 shows the lowest pH optimum of 2.5, whereas Sap7, which is the less related with other Saps, shows the highest (6.5). The activity of all Saps, except Sap7, is inhibited by the classic aspartic protease inhibitor, pepstatin A. It has been suggested that pepstatin A insensitivity is due to the presence of Met242 and Thr467 residues which restrict the accessibility of pepstatin A to the binding site [46]. The biochemical characteristics of Sap isozymes are summarized in Table 1 [30].

Substrate specificities of all ten Sap isoenzymes (briefly summarized in Fig. 2) were determined by using FRETS-25Xaa libraries [30] or distinct peptide substrates [32, 43, 47]. A study of the substrate specificities at the P1 and P1′ sites for Sap1–Sap3 and Sap6 [43] showed that P1′ specificities are generally broader than those observed for P1. In general, Sap1–Sap6 and Sap8 have a broad substrate specificity and, like other aspartic proteases, prefer to hydrolyze peptide bonds after hydrophobic residues such as Leu, Phe, and Tyr, but also after positively charged residues such as Arg and Lys. In contrast, Sap7, Sap9, and Sap10 have narrower substrate specificities and prefer at the P1 site residues such as Met, Arg, and His Sap9 and Sap10 perform hydrolysis after dibasic (LysArg, LysLys) or monobasic (Lys, Arg) residues [32, 47], and almost all Saps hydrolyze peptide bonds before Ala [30]. Based on the similarities and differences in substrate specificities, Sap isozymes can be categorized into three groups [30]. Group 1 comprises Sap7 and Sap10, and most notably differs from other Saps in terms of substrate specificity, which here is narrowest. Sap4–Sap6 are categorized into group 2, and group 3 comprises Sap1–Sap3 and Sap8–Sap9, with a very similar, broad substrate specificity.

4 Functions of Saps in *C. albicans* Virulence

C. albicans exploits its proteolytic enzymes for host tissue invasion and inactivation of the host's immune defense, to establish fungal infection. These factors are mandatory for the degradation of tissue barriers and acquiring nutrition at different host niches [48]. The adaptation to the host environment and the propagation of infections as well as further dissemination demand the involvement of the proteases in, for instance, combating the host immune cells like neutrophils or mononuclear phagocytes [49] as well as inactivating proteins of the complement system [50]. The roles of Saps during host infection, in terms of interactions of these pathogen proteases with major cellular, proteinaceous and peptide targets of the host, and the further consequences of these interactions, are briefly summarized in Fig. 4, and will be discussed in detail in the following subsections.

4.1 Expression of *C. albicans* Aspartic Protease Genes During Candidal Infection

As the individual members of *C. albicans* Sap family play diverse roles during both commensal and pathogenic interactions with the host, their genes are differentially expressed at various body sites, depending on the type, phase, and site of the infection [25, 51–53]. *SAP1–SAP3* gene expression was detected in both yeast and hyphal cells. *SAP1–SAP3* genes, predominantly expressed during mucosal infections [51, 54], were suggested to be significant for infection process in general, whereas the gene encoding Sap4, which belongs to the Sap4–Sap6 subfamily, was expressed in the hyphal phase during the adhesion to and penetration of epithelial cells [33, 52]. This Sap isoenzyme was suggested to be essential for the development of systemic infections and to be involved in avoiding the immune response [55]. In vivo, analysis of the expression of *SAP1–SAP8* genes in oral candidiasis [25, 51, 52] showed the highest frequency of *SAP2* expression both in colonized and infected patients. During oral infections, the expression of *SAP1*, *SAP4*, *SAP7*, and *SAP8* genes was also detected at a significant level [25]. In the oral reconstituted human epithelium (RHE) model, *SAP1* and *SAP3* expression was detected, followed by *SAP2* and *SAP8* expression [57]. In contrast, in the RHE model of vaginal candidiasis, *SAP2* and *SAP9* expression preceded *SAP1* expression [58]. Nevertheless, in both oral and vaginal RHE models, *SAP1–SAP3* and *SAP9* contributed to tissue damage [25, 58, 59]. In addition, an extensive expression of *SAP1–SAP3* and *SAP7–SAP8* genes was observed in a model that mimicked bloodstream infections [55, 60]. An investigation of *SAP7* expression during *C. albicans* adhesion to the intestinal human cells indicated a meaningful role of Sap7 during the initial adaptation of *C. albicans* to intestinal tract which decreased over time [61]. Sap8 correlates mostly with oral or vaginal infections, with the expression of its gene detected in a RHE model together with *SAP1–SAP3* [33]. *SAP9* was one of the most expressed genes in vivo during human mucosal infections and in

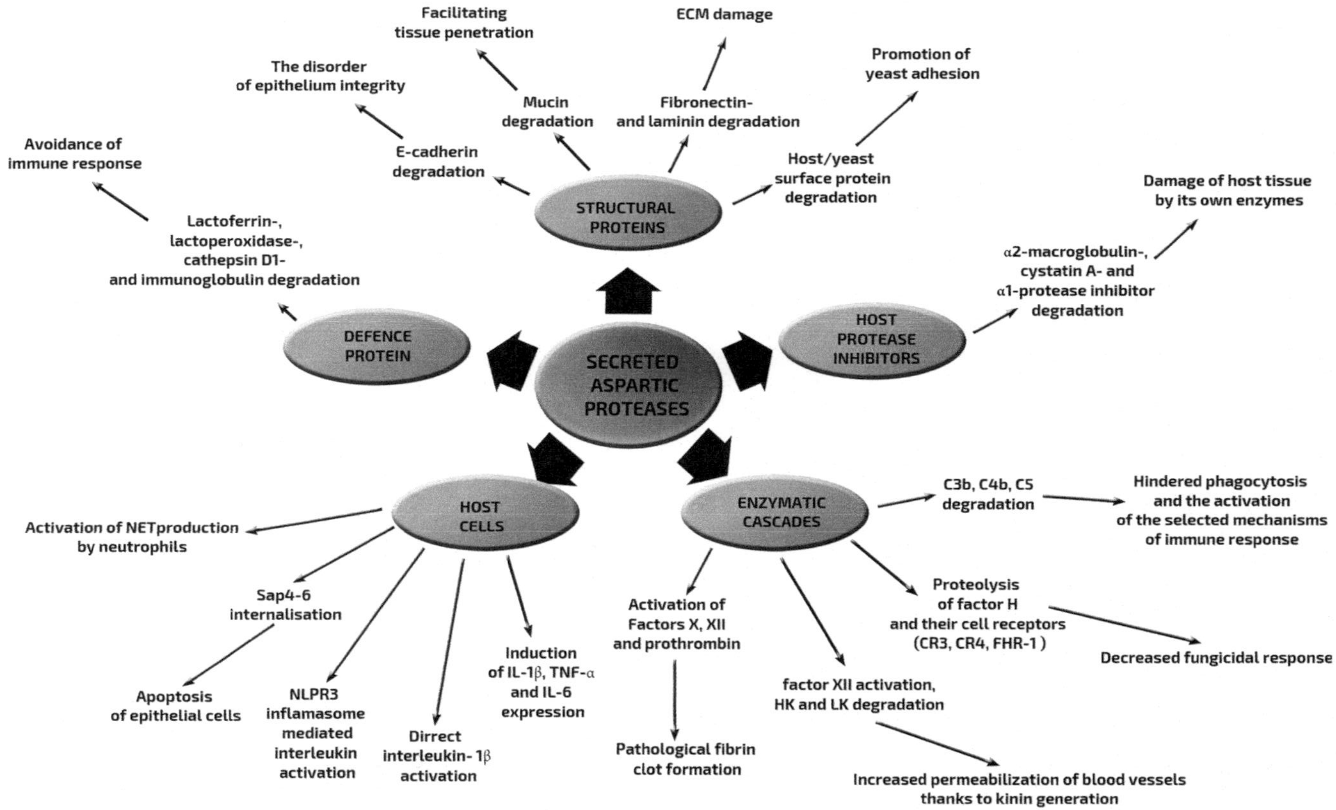

Fig. 4 Roles of *C. albicans* Saps in the course of candidal infections: interactions of Saps with major targets in the host, with further consequences for the development and progress of infection

Table 2 Expression of Sap-encoding genes in humans and in animal models

Model	Infection	Proteases involved	References
Mouse and guinea pig	Systemic infections	Sap4–Sap6 are involved in the progression of systemic infection	[62]
Mouse and guinea pig	Disseminated infection	Sap1–Sap3 presumably play an important role during disseminated candidiasis	[63]
Mouse	Gastrointestinal infection	Expression of *SAP1–SAP3* is lower than that of *SAP4–SAP6*	[64]
Mouse	Disseminated candidiasis	*SAP1*, *SAP2*, *SAP4*, *SAP5*, *SAP6*, and *SAP9* are the most commonly expressed Sap-encoding genes within 72 h after infection	[65]
Mouse	Disseminated candidiasis	Sap1–Sap6 do not play a significant role in the murine model of disseminated candidiasis	[66]
Mouse	Keratitis infection model	*SAP6* is associated with corneal pathogenicity and appears to be associated with morphological transformation into invasive hyphae	[67]
Rat	Vaginal candidiasis	Proteases are actively secreted during rat vaginitis and localized in the cell wall of *C. albicans* during infection	[68]
Human	Oral and cutaneous candidiasis	Expression of *SAP1–SAP3* is higher than *SAP4–SAP6* in oral and cutaneous candidiasis	[52]
Human	Vaginal candidiasis	Expression of *SAP2*, *SAP9*, and *SAP10* occurs in the early stages of infection. Transcripts of *SAP1*, *SAP4*, and *SAP5* appear after 12 h, while *SAP6*- and *SAP7*-transcripts were found in the late stages of infection (after 24 h); *SAP1–SAP2* play a key role during infection	[58]
Human	Oral candidiasis	At least one of *SAP1-SAP3* genes is expressed during oral candidiasis	[59]
Human	Oral and vaginal candidiasis	*SAP5* and *SAP9* are the most highly expressed Sap-encoding genes in patients with oral and vaginal candidiasis	[54]
Human	Oral candidiasis	The presence of *SAP1*-, *SAP3*- and *SAP6*-encoding genes in the case of the 29-year-old female patient suffering from acute oral candidiasis and *SAP2* gene obtained from an HIV-infected patient from a lesion of chronic oral candidosis (the clinical specimens)	[57]

oral RHE models [54] and detected in both infectious and commensal forms of *C. albicans*. The Sap9 enzyme was suggested to play a role in the cell wall integrity, as well as the efficiency of yeast cells in contact with the mucosal surface of the host [32]. The clinical specimens of patients suffering from acute oral candidosis and from a lesion of chronic oral candidosis showed the expression of *SAP1*, *SAP2*, *SAP3*, and *SAP6* genes [57]. The data regarding the expression of Sap-encoding genes, involving human samples and animal models, are summarized in Table 2.

4.2 Degradation of Main Functional Proteins of the Host

A broad spectrum of host substrates hydrolyzed by Saps, identified in early studies, included lactoferrin, lactoperoxidase, cathepsin D, albumin, hemoglobin, and the extracellular matrix components such as keratin, collagen, and vimentin [25, 48]. Sap2, produced in high amounts by yeast-like forms of *C. albicans,* was shown to contribute to the damage of mucin, the main component of protective layer for the mucous membrane [69]. The degradation of mucin can facilitate not only the penetration of mucous barrier enabling the further invasion of tissues, but can also provide carbon and nitrogen required for fungal growth [70].

The activity of Saps also affects the structural integrity of the epithelium. During the contact of *C. albicans* with the human oral mucosa, *C. albicans* uses its proteolytic potential to degrade E-cadherin in epithelial adherent junctions. Studies with protease mutant strains indicated that Sap5 is the major enzyme responsible for this process [71].

The invasion of host tissue can also be facilitated due to the degradation of proteinaceous components of subendothelial extracellular matrix, mainly laminin and fibronectin. Their proteolysis could presumably be important in the process of bloodstream penetration by the yeast [72]. Evidence suggested that strains that possessed a higher proteolytic activity strongly adhered to epidermal keratinocytes [73], corneocytes [74], and cells of the oral mucosa [75]. Although the role of Sap activity in this process was partly confirmed with the use of the specific inhibitor, pepstatin A, the mechanism of this effect was not fully elucidated. It was proposed that the increased adhesion to host cells resulted from the Sap-dependent degradation of certain host surface proteins which revealed additional potential binding sites for *C. albicans* [56].

Presumably, Sap9 and Sap10 can also participate in the regulation of adhesion due to the degradation of chitin synthases that participate in the process of cell wall formation, and a number of yeast surface proteins such as yeast-form cell wall protein 1 (Ywp1), agglutinin-like sequence protein 2 (Als2), and the protein repressed during hyphae development 3 (Rhd3) [32, 47].

4.3 The Interaction of Saps with the Complement System and the Antibodies

After invading host tissues, *C. albicans* encounters the innate immune system that acts against the pathogen through numerous antimicrobial peptides, the complement system, and specialized immune proteins and cells [50]. *C. albicans* aspartic proteases can prevent or modulate the functionality of immunoglobulins, in particular immunoglobulin A, resistant to the majority of bacterial proteases, which influences the attachment of *C. albic*ans to buccal epithelial cells [76, 77].

The components of the complement system involved in pathogen removal were also shown to be targets of Saps [78, 79]. Sap1, Sap2, and Sap3 degraded complement proteins C3b, C4b, and C5, preventing both phagocytosis and the final formation of the terminal complement complex (TCC) and thus the activation of the selected mechanisms of the immune response [78]. Furthermore, Sap2 was able to degrade factor H, a complement system controller whose binding to the surface of *C. albicans* cells can increase the fungicidal response through bridging yeast and host immune cells. The same protease could also degrade CR3 and CR4, and FHR-1 receptor involved in the recognition of pathogens by the cells of the immune system [80].

4.4 Propelling the Host Proteolytic Cascades: The Kinin Production and Clot Formation

During fungal cell proliferation and further dissemination, Saps are able to degrade important components of proteolytic cascades involved in maintaining the biochemical homeostasis of the host organism, including proteins that comprise the contact system (i.e., the surface-activated kinin-generating system in plasma) and the blood clotting pathways [78, 80–83].

Primarily aimed at effectively defending against microbial infections, the activation of the contact system that results in the production of kinins can also, to some extent, be beneficial for pathogens [84]. The kinins are vasoactive peptides [85] and play an important role of inflammation mediators [86]. The increased permeabilization of blood vessels caused by kinins can not only support the migration of immune cells but can also increase the availability of nutrients for the pathogens and the possibility of disseminating the infection [87]. The contact system consists of two initially inactive serine proteases—factor XII and plasma prekallikrein—and a non-enzymatic protein, high-molecular-mass kininogen (HK). The factor XII is activated on a contact with negatively charged cell surfaces and subsequently activates prekallikrein. The active kallikrein releases a kinin—the nonapeptide bradykinin—from HK [88]. Kinins can also be released from the low-molecular-mass kininogen (LK), which does not belong to the contact system, but serves as the substrate for tissue kallikrein [89].

During the infection, *C. albicans* can trigger the release of kinins indirectly, by the activation of factor XII [90] or by direct action of Saps on HK molecule [81]. It was reported that mixtures of proteases secreted into a growth medium by both morphological forms of *C. albicans* released kinins from both HK and LK [81]. Studies with purified Sap2 confirmed its ability to release kinins from kininogens and also clearly indicated that LK is a protein much more susceptible to the direct proteolysis than HK [91]. The detailed analysis of LK degradation by all Saps showed that the majority of proteases, except Sap7, are able to produce kinins [92]. Sap3 released Met-Lys-bradykinin—a peptide able to exert the kinin-like biological effects through the activation of cellular kinin receptors—to a very high yield. Other Saps (except Sap9) could produce small amounts of the same kinin and bradykinin but primarily generated peptides comprising a kinin sequence extended at N- and/or C-termini. These peptides could be further processed by the action of Sap9, resulting in an efficient production of Met-Lys-bradykinin. The cooperative degradation of kininogens by several Saps can be exploited by *C. albicans* to produce an optimal amount of kinins at the site of the infection. However, Sap9 alone released des-Arg^1-bradykinin, a peptide devoid of kinin-like biological activity [92].

The activation of coagulation factor XII by Saps can also initiate the process of blood clotting; however, this protein is not the only component of the cascade that is susceptible to a limited proteolysis by Saps. The process of fibrin clot formation can also be assisted by the Sap-dependent activation of factor X and prothrombin. A contribution of Saps to the development of the fibrin clot may underlie the septic coagulation and an insufficient peripheral circulation during infections [83, 93].

4.5 Deregulation of Host Proteases by Degradation of Their Inhibitors

Saps are able to degrade proteinaceous inhibitors that control the activity of main proteases of the host. Sap2 was shown to cleave α2-macroglobulin [94], cystatin A [95], and α1-protease inhibitor (A1PI), the latter also being susceptible to the action of Sap1, Sap3, Sap 4, and Sap9 [96]. By impairing the enzyme-inhibitor balance through the inactivation of the inhibitor molecule, Saps can indirectly contribute to the destruction of host tissues, supporting the process of infection. For instance, it was shown that the proteolytic cleavage of A1PI may assist in damaging epithelial and endothelial cells, caused by neutrophil extracellular traps (NETs). NETs are defense structures composed of DNA and microbicidal molecules released from neutrophil after a contact with a variety of microorganisms and molecules, which appear in response to infection, such as interleukin 8 (IL-8) [97]. Pathogens located within the NETs are exposed to direct contact with microbicidal molecules including neutrophil elastase (NE). The degradation of A1PI contributes to unrestrained NE activity, which might lead to the damage of host tissues, thereby supporting the colonization of the host by *C. albicans* cells that survived despite the killing properties of NETs. Furthermore, the degraded inhibitor had a reduced ability to complexation of IL-8 which can contribute to the increased influx of neutrophils to the site of infection and the progression of NET formation [96].

4.6 Interaction with Host Cells

The first host cells that come into contact with *C. albicans* but play an important function in preventing the fungal invasion are epithelial cells. Saps probably assist the adhesion of fungal cells to epithelium, acting as a ligand for the host cell receptors or modifying the surface components of both types of cells to allow for better interaction [25]. The E-cadherin cleavage by Saps resulted in epithelial integrity destruction and an increase of monolayer permeability [71, 98].

Sap4-Sap6 have the ability to bind to epithelial surface integrins. This interaction, mediated by Sap amino acid motif RGD/KGD, enables the internalization of

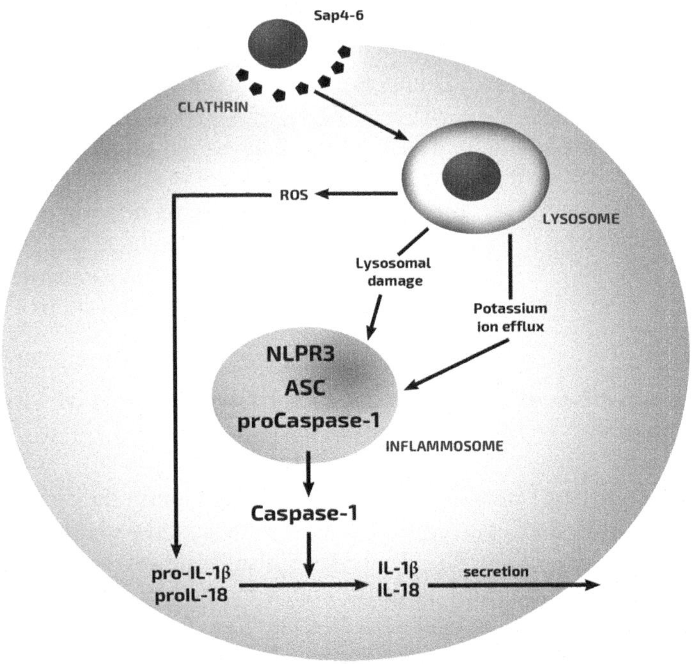

Fig. 5 NLRP3 inflammasome activation by Sap2 and Sap6 that leads to IL-1β and IL-18 production [103 modified]

proteases to endosomes and lysosomes by a still unknown mechanism. The location of these proteases in acidic interior of lysosomes leads to their activation, resulting in a partial permeabilization of the lysosome membrane and a subsequent caspase activation [99]. This kind of apoptotic pathway has been well established and occurs in many pathological conditions [100].

Proteases of *C. albicans* have the ability to induce cytokine expression by host cells. An increase in the production of interleukin-1β (IL-1β) and tumor necrosis factor (TNF-α) was observed in the macrophages after stimulation with Sap1–Sap3 and Sap6. Additionally, all these proteases except Sap3 increased the secretion of IL-6. The induction of cytokine secretion appeared to be regulated by the Akt/NF-κB activation and was independent of the Sap enzymatic activity [101]. It was also shown that aspartic proteases expressed by both morphological forms of *C. albicans* were able to produce IL-1β by direct, limited proteolysis of its precursor [102].

The formation of mature interleukin requires the involvement of apoptosis-associated speck-like protein (ASC), pro-caspase-1, and activated NLRP3 inflammasome, which is a pro-inflammatory complex presented in the monocytes, monocyte-derived macrophages, and dendritic cells (Fig. 5). It was shown that Sap2 and Sap6 were able to stimulate this inflammatory process, undergoing an internalization via the clathrin-dependent mechanism [103].

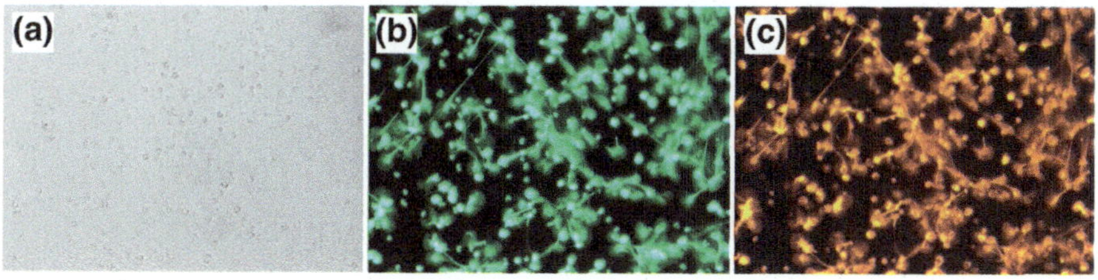

Fig. 6 Sap treatment of neutrophils (**a**) activated the release of NETs with antifungal properties. The visualization of NETs was performed with detection of DNA (Sytox staining, **b**) and histones (staining with antibody against histone 3, **c**). Based on the authors' unpublished data

At the place of infection, the fungal pathogens are faced with locally presented or attracted phagocytes, mainly neutrophils and macrophages. Regardless of their efficiency, *C. albicans* cells have developed several mechanisms to evade phagocyte control, with the engagement of proteolytic enzymes (see the reviews: [104, 105]).

Neutrophils, equipped with many receptors sensitive to fungal pathogen-associated molecular patterns (PAMP), are the main immune cells involved in the host defense against *C. albicans.* These phagocytes migrate to the place of the fungal infection across Sap concentration gradient, in a dose-dependent manner [106]. The chemoattractant properties of those proteases, confirmed in a study of *Candida* mutant strains with selective deletion of Sap genes, were attributed to Sap9 [107]. The influx of neutrophils to the place of infection was also confirmed in vivo, in a mice model, in response to Sap2 [108].

Saps can also influence the ROS production by the activation of neutrophil oxidative burst, as was deduced from an observed decrease of ROS formation during the contact of neutrophils with *C. albicans* mutant strains with deleted Sap9-encoding gene [107]. Both processes seemed to depend on Sap proteolytic activity, because enzymatically inactive Saps did not cause neutrophil activation [108]. A further stimulation of neutrophils with Sap9 also triggered the apoptosis of these cells [107]. On the other hand, independently of their enzymatic activity, Saps induce interleukin release by epithelial cells. Interleukins, especially IL-8, are chemoattracting agents for neutrophils; therefore, inactive Saps can also indirectly modulate the influx of neutrophils to the infection foci [109].

The formation of NETs, composed of decondensed chromatin incrusted with a subset of granular proteins, is a relatively recently recognized mechanism used by neutrophils to capture and kill the microbial pathogens outside of the phagocyte cells [110]. We recently found *SAP* genes to be overexpressed in the yeast during the contact with NET-forming neutrophils [111]. On the other hand, the *SAP*-encoded proteins triggered the NET release in a dose-dependent manner, engaging two different mechanisms, which depended on fungal morphology and correlated with preference to secreted Sap type (Fig. 6). For Sap1–Sap2 and Sap8–Sap10, we observed the release of NETs in a ROS-dependent way. Sap5 and Sap7 acted similarly, but with lower efficiency. A ROS-independent mechanism of NETosis was observed for neutrophils treated with Sap4 and Sap6, for which the NADPH oxidase inhibition only partially lowered the NET release (unpublished data).

4.7 Sap Impact on the Action of Antimicrobial Peptides Produced by the Host

Besides degrading a vast number of the host proteins, the arsenal of ten *C. albicans* Saps effectively degrades and neutralizes some of human antimicrobial peptides (AMPs) which are key components of the innate immune system of the host, and represent a first, primitive line of defense against attacks of a wide range of microorganisms and are often called natural antibiotics.

It has been shown that *C. albicans* cells use Saps to hydrolytically inactivate His5, a histidine-rich cationic salivary component that possesses potent antimicrobial activities, in particular against *C. albicans* [112]. The investigation of four *C. albicans* aspartic proteases—Sap2, Sap5, Sap9, and Sap10—showed that His5 is effectively cleaved by all these enzymes except Sap5 and that the main protease responsible for His5 degradation is Sap9. A recent study expanded the known characteristic of His5 degradation on all ten *C. albicans* proteases using peptide chemistry methods [44] and indicated that seven Sap family members (Sap1–Sap4 and Sap7–Sap9, used as recombinant proteins) could rapidly degrade this salivary AMP under conditions corresponding to the oral cavity environment.

Human cathelicidin LL-37 was another human AMP shown to be prone to Sap-dependent degradation and inactivation [111]. This cationic α-helical AMP with antimicrobial and immunomodulatory properties [113], constitutively expressed in epithelial cells and the cells of the innate immune system such as human neutrophils, was cleaved into multiple products by six Sap enzymes, Sap1–Sap4 and Sap8–Sap9. The progress of degradation deprived this peptide from its fungicidal activity, thus enabling the pathogen to survive and propagate despite presence of AMP.

Nonetheless, at the initial stages of Sap treatment of two peptides, LL-37 and His5, truncated derivatives—LL-25, LL8-37, His-21, His-17, and His-13—that still possessed some antifungal activity were produced and, therefore, the body's first line of defense against the infection was initially sustained [44, 111]. However, the LL-25 peptide, despite possessing antifungal properties, was devoid of the immunomodulatory properties of full-length LL-37, i.e., did not affect the generation of ROS by neutrophils, lowered the chemoattractant activity toward neutrophils by significantly decreased calcium flux and IL-8 production after neutrophil stimulation, and also lost the function of an inhibitor of neutrophil apoptosis [111].

It was recently demonstrated [114] that Saps can degrade and inactivate two antimicrobial peptides—designated NAT26 and HKH20—that can potentially be excised from human kininogens by host proteases such as NE [115, 116]. The NAT26, a helical and positively charged peptide responsible for the antimicrobial properties of the domain 3 of LK and HK, was effectively cleaved by all Saps except Sap10. In contrast, the HKH20, a histidine- and lysine-rich, positively

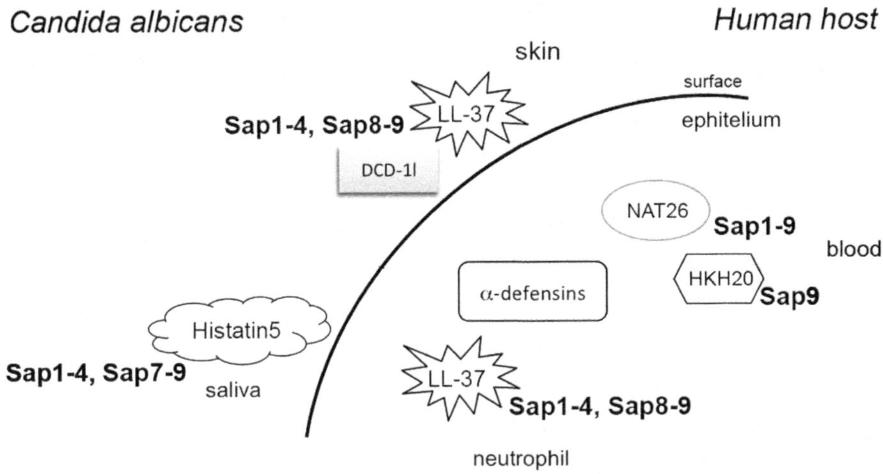

Fig. 7 Scheme presenting the places where AMP are prevalent in the human cells/tissues, with the indication of *C. albicans* aspartic proteases capable of their degradation

charged peptide derived from domain 5 of HK, was completely cleaved only by Sap9.

A susceptibility of human AMPs to Sap action is schematically summarized in Fig. 7.

A detailed kinetic analysis of Sap-catalyzed degradation of LL-37 [111], His5 [44], and NAT26 and HKH20 [114] revealed that, despite the high hydrolytic activity of these proteases in an acidic environment, most of them were able to process the peptides over a broad pH range, with the highest activity at a neutral pH for Sap3, Sap4, and Sap9. These findings suggest that proteolytic processing of AMP can possibly proceed in the various niches of the body where *C. albicans* reside.

4.8 Involvement of Saps in the Formation and Resistance of Polymicrobial Biofilm

On mucosal or artificial surfaces, *C. albicans* forms three-dimensional polymicrobial communities with extracellular matrix layers also containing host immune cells. Such a complex biofilm is highly resistant to the host's immune activity and antifungal drugs [117].

The role of Saps in biofilm formation is poorly recognized. In the simple artificial models, it was demonstrated that *C. albicans* biofilm secreted more Saps than the planktonic counterparts [118]. In the oral mucosal epithelia model coinfected with *C. albicans* and *Streptococcus oralis*, the microorganisms synergized to activate the host enzyme, calpain 1, involved in the cleavage of epithelial junction proteins and increased fungal invasion but the fungal protease activity was not required for this affect [119]. In the study of a biofilm, formed by *C. albicans* and oral streptococci, the cell wall-associated Sap9 was found to be required for the control of hyphal filamentation of *C. albicans*, and for the regulation of mixed species biofilm formation [120].

5 Secreted Aspartic Proteases of Other *Candida* Species and Their Roles in Candidal Infections

Apart from *C. albicans,* three non-albicans *Candida* species possess in their genome the genes that encode secreted aspartic proteases: *C. parapsilosis* (*SAPP1–SAPP3*), *C. tropicalis* (*SAPT1–SAPT4*), and *C. dubliniensis* (*SAPCD1–SAPCD4, SAPCD7–SAPCD10*).

C. parapsilosis possesses three *SAPP* genes (*SAPP1–SAPP3*) and two *SAPP2* homologs that demonstrate 91.5% amino acid sequence identity. To date, only the products of *SAPP1* and *SAPP2* genes have been isolated and purified [121, 122]. Both Sapp1 and Sapp2 are extracellular enzymes; however, Sapp1 was also reported to occur in a cell wall-attached form [123]. They have an identical molecular mass of 37 kDa, and their amino acid sequence is 53% identical [124]. While the secretion of Sapp2 does not depend on the type of nitrogen source, the production of Sapp1 is induced by the presence of protein in the growth medium [121]. The optimal pH for Sapp1 and Sapp2 hydrolytic activity is in the acidic pH range, and Sapp1 possesses a broad substrate specificity [121, 125].

Currently, the tertiary structures of Sapp1 and of one of the Sapp2 homologs are known [122, 125]. During the maturation of proenzymes, they can be activated autocatalytically or by the action of endoproteinase Kex2, and only removed peptides possess potential glycosylation sites [124, 126].

It was reported that *C. parapsilosis* aspartic proteases demonstrated the ability to degrade several host proteins, i.e., Sapp1 hydrolyzed IgA and activated prothrombin and coagulation factor X, while Sapp2 degraded keratin and trypsinogen [121, 127]; both proteases had the ability to generate biologically active kinins from human kininogens [81, 128].

Sapt1 is the only secreted aspartic protease of the Sapt family that could be successfully isolated from *C. tropicalis* cultures. The amino acid sequence similarities between particular Sapts do not exceed 63% but they are closely related to the enzyme produced by *C. albicans,* with Sapt1 showing a large similarity to Sap8, while Sapt4 showed similarities to the Sap1–Sap3 subfamily [129]. A tertiary structure of Sapt1 is already known [130]. The production of this enzyme can be induced by the presence of an exogenous protein, such as bovine serum albumin, in the culturing medium [131]. *C. tropicalis* proteases might be directly involved in fungal invasion, with the destruction of the host tissues and degradation of the host proteins, e.g., of the human kininogens with generating of kinin-related peptides [81], and their activity may be also important for fungal cell adhesion and development of the disseminated candidiasis in patients with leukemia and neutropenia

[132]. However, the expression of the *SAPT* genes during the colonization of the oral epithelium was not strictly correlated with the invasion [133].

C. tropicalis also produces tropiase, a protease that does not belong to the Sapt family. In contrast to the previously described enzymes, tropiase demonstrates the proteolytic activity in a broader pH range of 7–9 and is stable at pH between 3 and 12 and at high temperatures [134]. Tropiase is involved in the degradation of casein, keratin, and collagen. Moreover, this enzyme also hydrolyzes α and β chains of fibrinogen, but without demonstrated clotting activity and fibrin formation. Interestingly, the purified tropiase possesses hemorrhagic and capillary permeability-increasing activities which may highly contribute to the development of candidiasis [134, 135].

Another non-albicans *Candida* species, *C. glabrata,* which is more closely related to baker's yeast *S. cerevisiae* than to other pathogenic species of *Candida genus,* possesses in its genome at least 11 *YPS* genes which are similar in structure to *SAP* genes [136]. The *YPS*-encoded proteins (yapsins) constitute a family of aspartic proteinases with a GPI-anchor, involved in the maintenance of cell wall integrity and cell–cell interactions. *C. glabrata* yapsins are considered to have a strong structural similarity to Sap9 and Sap10 [32].

6 Fungal Aspartic Proteases as Drug Targets—Future Application

From the perspective of Saps' contribution to the fungal pathogenicity, they are ideal drug targets, especially in the context of increasing resistance of *C. albicans* strains against the commonly used antifungal agents. The inactivation of Saps could successfully stop the infections on different levels correlated with the facets of the infection process. They can be the targets for the process of tissue barrier degradation, destruction of the host's defense molecules, acquiring nutrients for pathogen propagation, the adhesion, and biofilm formation on the host tissues or abiotic surfaces [39, 137–140].

However, due to their wide substrate specificity and broad range of pH operation [30], it is difficult to find a universal inhibitor, working with high efficiency. Pepstatin A, the most popular inhibitor of aspartic proteases like pepsin and cathepsin D, is also effective toward *C. albicans* Saps as was presented in many in vitro studies [141]. In the model of human oral candidiasis, it was shown that the inhibition of Saps with pepstatin A can influence the fungal adhesion and invasion, associated with a reduction of tissue damages [142]. However, its inhibition profile toward Saps is not universal. The activity of Sap9 and Sap10, GPI-anchored to the fungal cell membrane, was only partially blocked by pepstatin A [47], and Sap7 was proven to be insensitive to pepstatin A [46]. Some researchers also suggested that the effectiveness of pepstatin A action can be restricted to the used models or conditions [143]. Moreover, the possible therapeutic application of pepstatin A in mice model failed due to its metabolism in the liver and rapid clearance from the blood [144].

An interesting opportunity has opened for finding Saps inhibitors since an observation of declined *Candida* infections in AIDS patients after the application of antiretroviral therapy that included HIV protease inhibitors (HIV PIs). Since *Candida* Saps and HIV proteinase belong to the same class of aspartic proteases, it was postulated that HIV PIs can be also effective against Saps in vivo [36, 145]. Among analyzed compounds, ritonavir was the most potent inhibitor of Sap2, the main candidal protease active in fungal infection, while saquinavir, indinavir, and nelfinavir inhibited Sap2 activity with lower efficiency [72, 146]. The inhibitory activity of these compounds toward Sap1–Sap3 and the attenuation of *Candida* cell adhesion were observed in RHE model of oral candidiasis and in a model of experimental rat vaginitis [147]. However, it is not clear whether the modulation of *Candida* adhesion to epithelial cells, observed in vitro, results only from proteolytic activity of Saps, that can affect some proteinaceous targets on the surface of epithelial cells, allowing for better adherence of fungus to them. The current results also pointed out a possible interruption of some specific interactions of aspartic proteases, not connected with their enzymatic activity, with the host proteins located on the surface of these cells. Such drugs, blocking the adhesion of *Candida* cells, would be particularly attractive. However, this issue was addressed by only one report for oral candidiasis [148].

Currently used antifungal drugs, which take advantage of their inhibitory activity toward *Candida* aspartic proteases, do not satisfy the medical standards, due to their potency, pharmacokinetic properties, and increased toxicity at higher concentration [149]. Therefore, new sources of Sap inhibitors are being sought including programmed, variable domain antibodies, produced against Sap2, whose protective properties were observed in experimental rat vaginal candidiasis [150]. Many peptidomimetic inhibitors have been developed that are derived from the structure of pepstatin A, which inhibited proteases of different *Candida* species [148]. Also the molecular modeling and new key structural information about Saps' active centers have been adopted for a design and synthesis of new Sap inhibitors [151–153]. Research was also focused on natural sources of inhibitors, including the extracts of plants [154], bacteria [155], and marine organisms [156] or artificial materials such as triangular gold nanoparticles [157].

The current progressive increase in fungal resistance to the available drugs, a noticeable shift of the infection profiles toward non-albicans *Candida* species, as well as the problems with toxicity and delivery of existing drugs to the place of infection should prompt the search for various fungal targets. At this point, candidal aspartic proteases, with their broad spectrum of engagement in physiological and pathological processes, seem to be among the best candidates.

Acknowledgements This work was supported in part by the National Science Centre of Poland (grant no. 571 UMO-2012/05/B/NZ1/00003 awarded to M.R.-K). Faculty of Biochemistry, Biophysics, and Biotechnology of Jagiellonian University is a partner of the Leading National Research Center (KNOW) supported by the Ministry of Science and Higher Education.

References

1. Yapar N (2014) Epidemiology and risk factors for invasive candidiasis. Ther Clin Risk Manag 10:95–105
2. Pfaller MA, Diekema DJ (2007) Epidemiology of invasive candidiasis: a persistent public health problem. Clin Microbiol Rev 20:133–163
3. Horn DL, Neofytos D, Anaissie EJ et al (2009) Epidemiology and outcomes of candidemia in 2019 patients: data from the prospective antifungal therapy alliance registry. Clin Infect Dis 48:1695–1703
4. Pana ZD, Farmaki E, Roilides E (2014) Host genetics and opportunistic fungal infections. Clin Microbiol Infect 20:1254–1264
5. Eggimann P, Que YA, Revelly JP, Pagani JL (2015) Preventing invasive *Candida* infections. Where could we do better? J Hosp Infect 89:302–308
6. Perlroth J, Choi B, Spellberg B (2007) Nosocomial fungal infections: epidemiology, diagnosis, and treatment. Med Mycol 45:321–346
7. Kullberg BJ, Arendrup MC (2016) Invasive candidiasis. N Engl J Med 374:794–795
8. Gudlaugsson O, Gillespie S, Lee K et al (2003) Attributable mortality of nosocomial candidemia, revisited. Clin Infect Dis 37:1172–1177
9. Soll DR, Galask R, Schmid J et al (1991) Genetic dissimilarity of commensal strains of *Candida* spp. carried in different anatomical locations of the same healthy women. J Clin Microbiol 29:1702–1710
10. Cannon RD, Chaffin WL (1999) Oral colonization by *Candida albicans*. Crit Rev Oral Biol Med 10:359–383
11. Akpan A, Morgan R (2002) Oral candidiasis. Postgrad Med J 78:455–459
12. Achkar JM, Fries BC (2010) *Candida* infections of the genitourinary tract. Clin Microbiol Rev 23:253–273
13. Mendling W, Brasch J, Cornely OA et al (2015) Guideline: vulvovaginal candidosis (AWMF 015/072), S2k (excluding chronic mucocutaneous candidosis). Mycoses S1:1–15
14. Patil S, Rao RS, Majumdar B, Anil S (2015) Clinical appearance of oral *Candida* infection and therapeutic strategies. Front Microbiol 6:1391
15. Lewis RE (2009) Overview of the changing epidemiology of candidemia. Curr Med Res Opin 25:1732–1740
16. Diekema D, Arbefeville S, Boyken L et al (2012) The changing epidemiology of healthcare-associated candidemia over three decades. Diagn Microbiol Infect Dis 73:45–48
17. Arendrup MC (2013) *Candida* and candidaemia. Susceptibility and epidemiology. Dan Med J 60:B4698
18. Krcmery V, Barnes AJ (2002) Non-albicans *Candida* spp. causing fungaemia: pathogenicity and antifungal resistance. J Hosp Infect 50:243–260
19. Karkowska-Kuleta J, Rapala-Kozik M, Kozik A (2009) Fungi pathogenic to humans: molecular bases of virulence of *Candida albicans*, *Cryptococcus neoformans* and *Aspergillus fumigatus*. Acta Biochim Pol 56:211–224
20. Naglik JR, Moyes DL, Wächtler B, Hube B (2011) *Candida albicans* interactions with epithelial cells and mucosal immunity. Microbes Infect 13:963–976
21. Polke M, Hube B, Jacobsen ID (2015) *Candida* survival strategies. Adv Appl Microbiol 91:139–235
22. Liu Y, Filler SG (2011) *Candida albicans* Als3, a multifunctional adhesin and invasin. Eukaryot Cell 10:168–173
23. Karkowska-Kuleta J, Kozik A (2014) Moonlighting proteins as virulence factors of pathogenic fungi, parasitic protozoa and multicellular parasites. Mol Oral Microbiol 29:270–283
24. Karkowska-Kuleta J, Kozik A (2015) Cell wall proteome of pathogenic fungi. Acta Biochim Pol 62:339–351

25. Naglik JR, Challacombe SJ, Hube B (2003) *Candida albicans* secreted aspartyl proteinases in virulence and pathogenesis. Microbiol Mol Biol Rev 67:400–428
26. Schaller M, Borelli C, Korting HC, Hube B (2005) Hydrolytic enzymes as virulence factors of *Candida albicans*. Mycoses 48:365–377
27. Hruskova-Heidingsfeldova O (2008) Secreted proteins of *Candida albicans*. Front Biosci 13:7227–7242
28. Höfs S, Mogavero S, Hube B (2016) Interaction of *Candida albicans* with host cells: virulence factors, host defense, escape strategies, and the microbiota. J Microbiol 54: 149–169
29. Staib F (1965) Serum-proteins as nitrogen source for yeastlike fungi. Sabouraudia 4: 187–193
30. Aoki W, Kitahara N, Miura N et al (2011) Comprehensive characterization of secreted aspartic proteases encoded by a virulence gene family in *Candida albicans*. J Biochem 150:431–438
31. Dos Santos ALS (2010) HIV aspartyl protease inhibitors as promising compounds against *Candida albicans*. World J Biol Chem 1:21–30
32. Albrecht A, Felk A, Pichova I (2006) Glycosylphosphatidylinositol-anchored proteases of *Candida albicans* target proteins necessary for both cellular processes and host-pathogen interactions. J Biol Chem 281:688–694
33. Silva NC, Nery JM, Dias ALT (2014) Aspartic proteinases of *Candida* spp.: role in pathogenicity and antifungal resistance. Mycoses 57:1–11
34. Borelli C, Ruge E, Lee JH et al (2008) X-ray structures of Sap1 and Sap5: structural comparison of the secreted aspartic proteinases from *Candida albicans*. Proteins 72:1308–1319
35. Cutfield SM, Dodson EJ, Anderson BF et al (1995) The crystal structure of a major secreted aspartic proteinase from *Candida albicans* in complexes with two inhibitors. Structure 3:1261–1271
36. Abad-Zapatero C, Goldman R, Muchmore SW et al (1996) Structure of a secreted aspartic protease from *C. albicans* complexed with a potent inhibitor: implications for the design of antifungal agents. Protein Sci 5:640–652
37. Behnen J, Koster H, Neudert G et al (2012) Experimental and computational active site mapping as a starting point to fragment-based lead discovery. Chem Med Chem 7:248–261
38. Borelli C, Ruge E, Schaller M et al (2007) The crystal structure of the secreted aspartic proteinase 3 from *Candida albicans* and its complex with pepstatin A. Proteins 68:738–748
39. Stewart K, Abad-Zapatero C (2001) *Candida* proteases and their inhibition: prospects for antifungal therapy. Curr Med Chem 8:941–948
40. Monod M, Staib P, Borelli C (2013) Candidapepsin. In: Handbook of proteolytic enzymes, vol 1, pp 159–166
41. Delano WL (2006) The PyMol molecular graphics system. Delano Scientific LLC, San Carlos
42. Borg-von Zepelin M, Beggah S, Boggian K et al (1998) The expression of the secreted aspartyl proteinases Sap4 to Sap6 from *Candida albicans* in murine macrophages. Mol Microbiol 28:543–554
43. Koelsch G, Tang J, Loy JA et al (2000) Enzymic characteristics of secreted aspartic proteases of *Candida albicans*. Biochim Biophys Acta 1480:117–131
44. Bochenska O, Rapala-Kozik M, Wolak N et al (2016) The action of ten secreted aspartic proteases of pathogenic yeast *Candida albicans* on major human salivary antimicrobial peptide, histatin 5. Act Biochi Pol 63:1–8
45. Smolenski G, Sullivan PA, Cutfield SM, Cutfield JF (1997) Analysis of secreted aspartic proteinases from *Candida albicans*: purification and characterization of individual Sap1, Sap2 and Sap3 isoenzymes. Microbiology 143:349–356
46. Aoki W, Kitahara N, Miura N et al (2012) *Candida albicans* possesses Sap7 as a pepstatin A-insensitive secreted aspartic protease. PLoS ONE 7:1–9

47. Schild L, Heyken A, de Groot PWJ et al (2011) Proteolytic cleavage of covalently linked cell wall proteins by *Candida albicans* Sap9 and Sap10. Eukaryot Cell 10:98–109
48. Hube B (1998) Possible role of secreted proteinases in *Candida albicans* infections. Rev Iberoam Micol 15:65–68
49. Cheng SC, Joosten LA, Kullberg BJ et al (2012) Interplay between *Candida albicans* and the mammalian innate host defense. Infect Immun 80:1304–1313
50. Zipfel PF, Hallström T, Riesbeck K (2013) Human complement control and complement evasion by pathogenic microbes–tipping the balance. Mol Immunol 56:152–160
51. Naglik JR, Newport G, White TC et al (1999) In vivo analysis of secreted aspartyl proteinase expression in human oral candidiasis. Infect Immun 67:2482–2490
52. Schaller M, Januschke E, Schackert C et al (2001) Different isoforms of secreted aspartyl proteinases (Sap) are expressed by *Candida albicans* during oral and cutaneous candidosis *in vivo*. J Med Microbiol 50:743–747
53. Staniszewska M, Siennicka K, Pilat J et al (2012) Role of aspartic proteinases in *Candida albicans* virulence. Part II: Expression of SAP1-10 aspartic proteinase during *Candida albicans* infections *in vivo*. Post Mikrobiol 51:137–142
54. Naglik JR, Moyes D, Makwana J et al (2008) Quantitative expression of the *Candida albicans* secreted aspartyl proteinase gene family in human oral and vaginal candidiasis. Microbiology 154:3266–3280
55. Staniszewska M, Bondarczyk MM, Siennicka K et al (2012) In vitro study of secreted aspartyl proteinases Sap1 to Sap3 and Sap4 to Sap6 expression in *Candida albicans* pleomorphic forms. Pol J Microbiol 61:247–256
56. Naglik JR, Rodgers C, Shirlaw PJ et al (2003) Differential expression of *Candida albicans* secreted aspartyl proteinase and phospholipase B genes in humans correlates with active oral and vaginal infections. J Infect Dis 188:469–479
57. Schaller M, Korting HC, Schafer W et al (1998) Investigations on the regulation of secreted aspartyl proteases in a model of oral candidiasis in vivo. Mycoses 41:69–73
58. Schaller M, Bein M, Korting HC et al (2003) The secreted aspartyl proteinases Sap1 and Sap2 cause tissue damage in an in vitro model of vaginal candidiasis based on reconstituted human vaginal epithelium. Infect Immun 71:3227–3234
59. Schaller M, Hube B, Ollert MW et al (1999) In vivo expression and localization of *Candida albicans* secreted aspartyl proteinases during oral candidiasis in HIV-infected patients. J Invest Dermatol 112:383–386
60. Staniszewska M, Bondaryk M, Malewski T, Kurzatkowski W (2014) Quantitative expression of *Candida albicans* aspartyl proteinase genes SAP7, SAP8, SAP9, SAP10 in human serum in vitro. Pol J Microbiol 63:15–20
61. Staniszewska M, Bondaryk M, Zukowski K, Chudy M (2015) Role of SAP7-10 and morphological regulators (EFG1, CPH1) in *Candida albicans* hypha formation and adhesion to colorectal carcinoma Caco-2. Pol J Microbiol 64:203–210
62. Sanglard D, Hube B, Monod M et al (1997) A triple deletion of the secreted aspartyl proteinase genes SAP4, SAP5, and SAP6 of *Candida albicans* causes attenuated virulence. Infect Immun 65:3539–3546
63. Hube B, Sanglard D, Odds FC et al (1997) Disruption of each of the secreted aspartyl proteinase genes SAP1, SAP2 and SAP3 of *Candida albicans* attenuates virulence. Infect Immun 65:3529–3538
64. Kretschmar M, Felk A, Staib P et al (2002) Individual acid aspartic proteinases (Saps) 1–6 of *Candida albicans* are not essential for invasion and colonization of the gastrointestinal tract in mice. Microb Pathog 32:61–70
65. Felk A, Kretschmar M, Albrecht A et al (2002) *Candida albicans* hyphal formation and the expression of the Efg1-regulated proteinases Sap4 to Sap6 are required for the invasion of parenchymal organs. Infect Immun 70:3689–3700

66. Correia A, Lermann U, Teixeira L et al (2010) Limited role of secreted aspartyl proteinases Sap1 to Sap6 in *Candida albicans* virulence and host immune response in murine hematogenously disseminated candidiasis. Infect Immun 78:4839–4849
67. Jackson BE, Wilhelmus KR, Hube B (2007) The role of secreted aspartyl proteinases in *Candida albicans* keratitis. Invest Ophthalmol Vis Sci 48:3559–3565
68. Stringaro A, Crateri P, Pellegrini G et al (1997) Ultrastructural localization of the secretory aspartyl proteinase in *Candida albicans* cell wall in vitro and in experimentally infected rat vagina. Mycopathologia 137:95–105
69. Hube B, Monod M, Schofield DA et al (1994) Expression of seven members of the gene family encoding secretory aspartyl proteinases in *Candida albicans*. Mol Microbiol 14: 87–99
70. Colina AR, Aumont F, Deslauriers N et al (1996) Evidence for degradation of gastrointestinal mucin by *Candida albicans* secretory aspartyl proteinase. Infect Immun 64:4514–4519
71. Villar CC, Kashleva H, Nobile CJ et al (2007) Mucosal tissue invasion by *Candida albicans* is associated with E-cadherin degradation, mediated by transcription factor Rim101p and protease Sap5p. Infect Immun 75:2126–2135
72. Morschhäuser J, Virkola R, Korhonen TK, Hacker J (1997) Degradation of human subendothelial extracellular matrix by proteinase-secreting *Candida albicans*. FEMS Microbiol Lett 153:349–355
73. Ollert MW, Söhnchen R, Korting HC et al (1993) Mechanisms of adherence of *Candida albicans* to cultured human epidermal keratinocytes. Infect Immun 61:4560–4568
74. Ray TL, Payne CD (1988) Scanning electron microscopy of epidermal adherence and cavitation in murine candidiasis: a role for *Candida* acid proteinase. Infect Immun 56:1942–1949
75. Borg M, Rüchel R (1988) Expression of extracellular acid proteinase by proteolytic *Candida* spp. during experimental infection of oral mucosa. Infect Immun 56:626–631
76. Rüchel R (1986) Cleavage of immunoglobulins by pathogenic yeasts of the genus *Candida*. Microbiol Sci 3:316–319
77. Marcotte H, Lavoie MC (1998) Oral microbial ecology and the role of salivary immunoglobulin A. Microbiol Mol Biol Rev 62:71–109
78. Gropp K, Schild L, Schindler S et al (2009) The yeast *Candida albicans* evades human complement attack by secretion of aspartic proteases. Mol Immunol 47:465–475
79. Luo S, Skerka C, Kurzai O, Zipfel PF (2013) Complement and innate immune evasion strategies of the human pathogenic fungus *Candida albicans*. Mol Immunol 56:161–169
80. Svoboda E, Schneider AE, Sándor N et al (2015) Secreted aspartic protease 2 of *Candida albicans* inactivates factor H and the macrophage factor H-receptors CR3 (CD11b/CD18) and CR4 (CD11c/CD18). Immunol Lett 168:13–21
81. Rapala-Kozik M, Karkowska-Kuleta J, Ryzanowska A et al (2010) Degradation of human kininogens with the release of kinin peptides by extracellular proteinases of *Candida* spp. Biol Chem 391:823–830
82. Rüchel R (1983) On the renin-like activity of *Candida* proteinases and activation of blood coagulation in vitro. Zentralbl Bakteriol Mikrobiol Hyg A 255:368–379
83. Kaminishi H, Hamatake H, Cho T et al (1994) Activation of blood clotting factors by microbial proteinases. FEMS Microbiol Lett 121:327–332
84. Frick IM, Björck L, Herwald H (2007) The dual role of the contact system in bacterial infectious disease. Thromb Haemost 98:497–502
85. Cockcroft JR, Chowienczyk PJ, Brett SE, Ritter JM (1994) Effect of NG-monomethyl-L-arginine on kinin-induced vasodilation in the human forearm. Br J Clin Pharmacol 38:307–310
86. Golias C, Charalabopoulos A, Stagikas D el al (2007) The kinin system-bradykinin: biological effects and clinical implications. Multiple role of the kinin system-bradykinin. Hippokratia 11:124–128

87. Imamura T, Tanase S, Szmyd G et al (2005) Induction of vascular leakage through release of bradykinin and a novel kinin by cysteine proteinases from *Staphylococcus aureus*. J Exp Med 201:1669–1676
88. Wu Y (2015) Contact pathway of coagulation and inflammation. Thromb J 13:17
89. Lalmanach G, Naudin C, Lecaille F, Fritz H (2010) Kininogens: more than cysteine protease inhibitors and kinin precursors. Biochimie 92:1568–1579
90. Kaminishi H, Tanaka M, Cho T et al (1990) Activation of the plasma kallikrein-kinin system by *Candida albicans* proteinase. Infect Immun 58:2139–2143
91. Bras G, Bochenska O, Rapala-Kozik M et al (2012) Extracellular aspartic protease SAP2 of *Candida albican*s yeast cleaves human kininogens and releases proinflammatory peptides, Met-Lys-bradykinin and des-Arg(9)-Met-Lys-bradykinin. Biol Chem 393:829–839
92. Kozik A, Gogol M, Bochenska O et al (2015) Kinin release from human kininogen by 10 aspartic proteases produced by pathogenic yeast *Candida albicans*. BMC Microbiol 15:60
93. Rüchel R (1983) On the role of proteinases from *Candida albicans* in the pathogenesis of acronecrosis. Zentralbl Bakteriol Mikrobiol Hyg A 255:524–536
94. Kaminishi H, Miyaguchi H, Tamaki T et al (1995) Degradation of humoral host defense by *Candida albicans* proteinase. Infect Immun 63:984–988
95. Tsushima H, Mine H, Kawakami Y et al (1994) *Candida albicans* aspartic proteinase cleaves and inactivates human epidermal cysteine proteinase inhibitor, cystatin A. Microbiology 140:167–171
96. Gogol M, Ostrowska D, Klaga K et al (2016) Inactivation of α1-proteinase inhibitor by *Candida albicans* aspartic proteases favors the epithelial and endothelial cell colonization in the presence of neutrophil extracellular traps. Acta Biochim Pol 63:1163
97. Zawrotniak M, Rapala-Kozik M (2013) Neutrophil extracellular traps (NETs)—formation and implications. Acta Biochim Pol 60:277–284
98. Moyes DL, Richardson JP, Naglik JR (2015) *Candida albicans*-epithelial interactions and pathogenicity mechanisms: scratching the surface. Virulence 6:338–346
99. Wu H, Downs D, Ghosh K et al (2013) *Candida albicans* secreted aspartic proteases 4–6 induce apoptosis of epithelial cells by a novel Trojan horse mechanism. FASEB J 27:2132–2144
100. Johansson AC, Appelqvist H, Nilsson C et al (2010) Regulation of apoptosis-associated lysosomal membrane permeabilization. Apoptosis 15:527–540
101. Pietrella D, Rachini A, Pandey N et al (2010) The inflammatory response induced by aspartic proteases of *Candida albicans* is independent of proteolytic activity. Infect Immun 78:4754–4762
102. Beauséjour A, Grenier D, Goulet JP, Deslauriers N (1998) Proteolytic activation of the interleukin-1beta precursor by *Candida albicans*. Infect Immun 66:676–681
103. Pietrella D, Pandey N, Gabrielli E et al (2013) Secreted aspartic proteases of *Candida albicans* activate the NLRP3 inflammasome. Eur J Immunol 43:679–692
104. Jiménez-López C, Lorenz MC (2013) Fungal immune evasion in a model host-pathogen interaction: *candida albicans* versus macrophages. PLoS Pathog 9(11):e1003741
105. Cheng SC, Sprong T, Joosten LA et al (2012) Complement plays a central role in *Candida albicans*-induced cytokine production by human PBMCs. Eur J Immunol 42:993–1004
106. Ran Y, Iwabuchi K, Yamazaki M et al (2013) Secreted aspartic proteinase from *Candida albicans* acts as a chemoattractant for peripheral neutrophils. J Dermatol Sci 72:191–193
107. Hornbach A, Heyken A, Schild L et al (2009) The glycosylphosphatidylinositol-anchored protease Sap9 modulates the interaction of *Candida albicans* with human neutrophils. Infect Immun 77:5216–5224
108. Pericolini E, Gabrielli E, Amacker M et al (2015) Secretory aspartyl proteinases cause vaginitis and can mediate vaginitis caused by *Candida albicans* in mice. MBio 6:e00724–15
109. Gabrielli E, Sabbatini S, Roselletti E et al (2016) In vivo induction of neutrophil chemotaxis by secretory aspartyl proteinases of *Candida albicans*. Virulence 29:1–7

110. Brinkmann V, Reichard U, Goosmann C et al (2004) Neutrophil extracellular traps kill bacteria. Science 303:1532–1535
111. Rapala-Kozik M, Bochenska O, Zawrotniak M et al (2015) Inactivation of the antifungal and immunomodulatory properties of human cathelicidin LL-37 by aspartic proteases produced by the pathogenic yeast *Candida albicans*. Infect Immun 83:2518–2530
112. Meiller TF, Hube B, Schild L et al (2009) A novel immune evasion strategy of *Candida albicans*: proteolytic cleavage of a salivary antimicrobial peptide. PLoS ONE 4:e5039
113. Vandamme D, Landuyt B, Luyten W, Schoofs L (2012) A comprehensive summary of LL-37, the factotum human cathelicidin peptide. Cell Immunol 280:22–35
114. Bochenska O, Rapala-Kozik M, Wolak N et al (2015) Inactivation of human kininogen-derived antimicrobial peptides by secreted aspartic proteases produced by the pathogenic yeast *Candida albicans*. Biol Chem 396:1369–1375
115. Frick IM, Akesson P, Herwald H et al (2006) The contact system—A novel branch of innate immunity generating antibacterial peptides. EMBO J 25:5569–5578
116. Nordahl EA, Rydengård V, Mörgelin M, Schmidtchen A (2005) Domain 5 of high molecular weight kininogen is antibacterial. J Biol Chem 280:34832–34839
117. Ganguly S, Mitchell AP (2011) Mucosal biofilm of *Candida albicans*. Curr Opin Microbiol 14:380–385
118. Mendes A, Mores AU, Carvalho AP et al (2007) *Candida albicans* biofilms produce more secreted aspartyl protease than the planktonic cells. Biol Pharm Bull 30:1813–1815
119. Xu H, Sobue T, Bertolini M, Thompson A, Dongari-Bagtzoglou A (2016) *Streptocuccus oralis* and *Candida albicans* synergistically activate calpain to degrade E-cadherin from oral epithelial junctions. J Infect Dis 13:pii:jiw201
120. Dutton LC, Jenkinson HF, Lamont RJ, Nobbs AH (2016) Role of *Candida albicans* secreted aspartyl protease Sap9 in interkingdom biofilm formation. Pathog Dis 74:pii:ftw005
121. Hrusková-Heidingsfeldová O, Dostál J, Majer F et al (2009) Two aspartic proteinases secreted by the pathogenic yeast *Candida parapsilosis* differ in expression pattern and catalytic properties. Biol Chem 390:259–268
122. Dostál J, Pecina A, Hrusková-Heidingsfeldová O et al (2015) Atomic resolution crystal structure of Sapp2p, a secreted aspartic protease from *Candida parapsilosis*. Acta Cryst D 71:2494–2504
123. Vinterová Z, Sanda M, Dostál J et al (2011) Evidence for the presence of proteolytically active secreted aspartic proteinase 1 of *Candida parapsilosis* in the cell wall. Protein Sci 20:2004–2012
124. Hrušková-Heidingsfeldová O, Dostál J, Hamal P et al (2001) Enzymological characterization of secreted proteinases from *Candida parapsilosis* and *Candida lusitaniae*. Collect Czech Chem Commun 66:1707–1719
125. Dostál J, Brynda J, Hrusková-Heidingsfeldová O et al (2009) The crystal structure of the secreted aspartic protease 1 from *Candida parapsilosis* in complex with pepstatin A. J Struct Biol 167:145–152
126. Horváth P, Nosanchuk JD, Hamari Z et al (2012) The identification of gene duplication and the role of secreted aspartyl proteinase 1 in *Candida parapsilosis* virulence. J Infect Dis 205:923–933
127. Merkerová M, Dostál J, Hradilek M et al (2006) Cloning and characterization of Sapp2p, the second aspartic proteinase isoenzyme from *Candida parapsilosis*. FEMS Yeast Res 6:1018–1026
128. Bras G, Bochenska O, Rapala-Kozik M et al (2013) Release of biologically active kinin peptides, Met-Lys-bradykinin and Leu-Met-Lys-bradykinin from human kininogens by two major secreted aspartic proteases of *Candida parapsilosis*. Peptides 48:114–123
129. Parra-Ortega B, Cruz-Torres H, Villa-Tanaca L, Hernández-Rodríguez C (2009) Phylogeny and evolution of the aspartyl protease family from clinically relevant *Candida* species. Mem Inst Oswaldo Cruz 104:505–512

130. Symersky J, Monod M, Foundling SI (1997) High-resolution structure of the extracellular aspartic proteinase from *Candida tropicalis* yeast. Biochemistry 36:12700–12710
131. Zaugg C, Borg-Von Zepelin M, Reichard U et al (2001) Secreted aspartic proteinase family of *Candida tropicalis*. Infect Immun 69:405–412
132. Kontoyiannis D, Vaziri I, Hanna H et al (2001) Risk factors for *Candida tropicalis* fungemia in patients with cancer. Clin Infect Dis 33:1676–1681
133. Silva S, Negri M, Henriques M et al (2010) Silicone colonization by non-*Candida albicans Candida* species in the presence of urine. J Med Microbiol 59:747–754
134. Okumura Y, Inoue N, Nikai T (2007) Isolation and characterization of a novel acid proteinase, tropiase, from *Candida tropicalis IFO 0589*. Nihon Ishinkin Gakkai Zasshi 48:19–25
135. Chen YV, Rosli R, Fong SH et al (2012) Histopathological characteristics of experimental *Candida tropicalis* induced acute systemic candidiasis in BALB/c Mice. Int J Zool Res 1:12–22
136. Kaur R, Ma B, Cormack BP (2007) A family of glycosylphosphatidylinositol-linked aspartyl proteases is required for virulence of *Candida glabrata*. Proc Natl AcadSci USA 104:7628–7633
137. Nguyen JT, Hamada Y, Kimura T, Kiso Y (2008) Design of potent aspartic protease inhibitors to treat various diseases. Arch Pharm 341:523–535
138. Braga-Silva LA, Santos ALS (2011) Aspartic protease inhibitors as potential anti-*Candida albicans* drugs: impacts on fungal biology, virulence and pathogenesis. Curr Med Chem 18:2401–2419
139. Santos ALS (2011) Aspartic proteases of human pathogenic fungi are prospective targets for the generation of novel and effective antifungal inhibitors. Curr Enz Inhib 7:96–118
140. Bondaryk M, Kurzątkowski W, Staniszewska M (2013) Antifungal agents commonly used in the superficial and mucosal candidiasis treatment: mode of action and resistance development. Postępy Dermatol Alergol 30:293–301
141. Kuriyama T, Williams DW, Lewis MA (2003) In vitro secreted aspartyl proteinase activity of *Candida albicans* isolated from oral diseases and healthy oral cavities. Oral Microbiol Immunol 18:405–407
142. Schaller M, Schäfer W, Korting HC, Hube B (1998) Differential expression of secreted aspartyl proteinases in a model of human oral candidiosis and in patient samples from oral cavity. Mol Microbiol 29:605–615
143. Lermann U, Morschhäuser J (2008) Secreted aspartic proteases are not required for invasion of reconstituted human epithelia by *Candida albicans*. Microbiol 154:3281–3295
144. Rüchel R, Ritter B, Schaffrinski M (1990) Modulation of experimental systemic murine candidosis by intravenous pepstatin. Zentralbl Bakteriol Mikrobiol Hyg 273:391–403
145. Cauda R, Tacconelli M, Tumbarello M et al (1999) Role of protease inhibitors in preventing recurrent oral candidosis in patients with HIV infection: a prospective case-control study. J Acquir Immun Defic Syndr 21:20–25
146. Borg-Von Zeppelin M, Meyer I, Thomssen R et al (1999) HIV-protease inhibitors reduce cell adherence of *Candida albicans* strains by inhibition of yeast secreted aspartic proteases. J Investig Dermatol 113:747–751
147. Cassone A, De Bernardis F, Torosantucci A et al (1999) In vitro and in vivo anticandidal activity of human immunodeficiency vírus protease inhibitors. J Infect Dis 180:448
148. Pichova I, Pavlickova L, Dostal J et al (2001) Secreted aspartic proteases of *Candida albicans*, *Candida tropicalis*, *Candida parapsilosis* and *Candida lusitaneae*. Inhibition with peptidomimetic inhibitors. Eur J Biochem 268:2669–2677
149. Santos A, Braga-Silva L (2013) Aspartic protease inhibitors: effective drugs against the human fungal pathogen *Candida albicans*. Mini Rev Med Chem 13:155–162
150. De Bernardis F, Liu H, O'Mahony R et al (2007) Human domain antibodies against virulence traits of *candida albicans* inhibits fungus adherence to vaginal epithelium and protect against experimental vaginal candidiasis. J Infect Dis 195:149–157

151. Fear G, Komarnytsky S, Raskin I (2007) Protease inhibitors and their peptidomimetic derivatives as potential drugs. Pharmacol Ther 113:354–368
152. Cadicamo C, Mortier J, Wolber G et al (2013) Design, synthesis, inhibition studies, and molecular modeling of pepstatin analogs addressing different secreted aspartic proteases of *Candida albicans*. Biochem Pharmacol 85:881–887
153. Zielinska P, Staniszewska M, Bondaryk M et al (2015) Design and studies o multiple mechanism of *anti-Candida* activity of new potent-Trp-rich peptide dendrimers. Eur J Med Chem 105:106–119
154. Höfling JF, Mardegan RC, Anibal PC et al (2011) Evaluation of antifungal activity of medicinal plant extracts against oral *Candida albicans* and proteinases. Mycopathologia 172:117–124
155. Sato T, Nagai K, Shibazaki M et al (1994) Novel aspartyl protease inhibitors, YF-0200R-A and B. J Antibiot (Tokyo) 47:566–570
156. Christopeit T, Øverbø K, Danielson H, Nilsen IW (2013) Efficient screening of marine, FRET extracts for protease inhibitors by combining fret based activity assays and surface plasmon resonance spectroscopy based binding assays. Mar Drugs 11:4279–4293
157. Hajjar FHE, Jebali A, Hekmatimoghaddam S (2015) The inhibition of *Candida albicans* secreted aspartyl proteinase by triangular gold nanoparticles. Nanomedicine J 2:54–59

Pseudomonas aeruginosa and Its Arsenal of Proteases: Weapons to Battle the Host

Anna Clara M. Galdino, Marta H. Branquinha, André L.S. Santos and Lívia Viganor

Abstract

Pseudomonas aeruginosa is a ubiquitous and opportunistic human pathogen that represents a critical problem to the clinician due to the increased number of resistant strains isolated from hospital settings. In addition, there is a great variety of pathologies associated with this versatile Gram-negative bacterium. *P. aeruginosa* cells are able to produce an incredible arsenal of virulence factors, especially secreted molecules that act singly or together to ensure the establishment, maintenance, and persistence of a successful infection in susceptible hosts. In this context, pseudomonal proteases' roles are highlighted due to their ability to cleave key host proteinaceous substrates as well as to modulate several biological processes, for example, escaping and modulating the host immune responses in the bacterial own favor. Proteases secreted by *P. aeruginosa* include elastase A (LasA), elastase B (LasB), alkaline protease (AP), protease IV (PIV), *Pseudomonas* small protease (PASP), large protease A (LepA), MucD, and *P. aeruginosa* aminopeptidase (PAAP). In the present review, we discuss the role of each of these relevant proteases produced by

A.C.M. Galdino · A.L.S. Santos
Programa de Pós-Graduação em Bioquímica (PPGBq), Instituto de Química (IQ), Universidade Federal do Rio de Janeiro (UFRJ), Rio de Janeiro, Brazil

A.C.M. Galdino · M.H. Branquinha · A.L.S. Santos (✉) · L. Viganor
Departamento de Microbiologia Geral, Instituto de Microbiologia Paulo de Góes (IMPG), Centro de Ciências da Saúde (CCS), Universidade Federal do Rio de Janeiro, Rio de Janeiro, Brazil
e-mail: andre@micro.ufrj.br

L. Viganor (✉)
The Centre for Biomimetic & Therapeutic Research, Focas Research Institute, Dublin Institute of Technology, Camden Row, Dublin 8, Dublin, Ireland
e-mail: liviaviganor@gmail.com

S. Chakraborti and N.S. Dhalla (eds.), *Pathophysiological Aspects of Proteases*,
DOI 10.1007/978-981-10-6141-7_16

P. aeruginosa taking into consideration their main biological functions in the bacterium–host interaction that favors the establishment of the infectious process.

Keywords
Pseudomonas aeruginosa · Proteases · Virulence factors

1 Introduction

Pseudomonads are bacteria well known for their metabolic versatility and widespread spatiotemporal distribution [1]. One of the most important species of pseudomonads is, with no doubt, *Pseudomonas aeruginosa*, which is a fascinating ubiquitous Gram-negative bacterium with rod shape measuring 0.5–0.8 μm × 1.5–3.0 μm (Fig. 1a) [1, 2]. *P. aeruginosa* presents the following metabolic features: non-fermentative, catalase positive, oxidase positive, ammonia producer, and usually aerobic, but it also can grow in an anaerobic environment if nitrate, citrate, and arginine are available [3]. The production of 2-aminoacetophenone by the bacterial cells generates the fruity grape-like odor that is characteristic of this pseudomonad species. On blood agar plates, colonies of *P. aeruginosa* often display beta-hemolysis and a greenish metallic sheen due to the production of pigments [2]. The characteristic that most distinguishes *P. aeruginosa* from the other pseudomonads, and from the other species of Gram-negative non-fermenting bacteria, is its ability to produce pyocyanin, a blue-green phenazine pigment that gives the green color to the bacterial colony (Fig. 1b) and also to the pus observed in *P. aeruginosa-infected* tissues. This pigment and several others, such as pyochelin

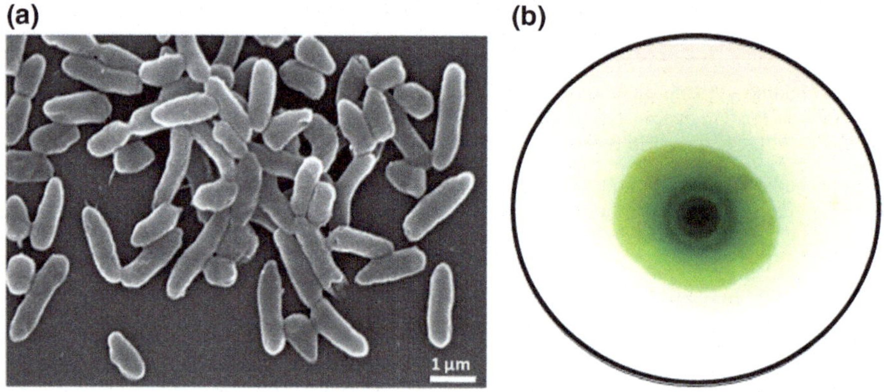

Fig. 1 Scanning electron microscopy (**a**), showing the characteristic bacterial rod shape, and colony morphology (**b**), evidencing the pyocyanin pigment, of *Pseudomonas aeruginosa*

(purple-cyan), pyoverdin (yellow, green and fluorescent), pyomelanin (light-brown), and pyorubin (red-brown), are secondary metabolites of *P. aeruginosa*, which play an important role in bacterial nutrition, such as iron acquisition and pathogenesis [2, 3]. Almost all *P. aeruginosa* strains are motile due to the presence of a single polar flagellum that facilitates the locomotion and colonization of a wide range of environmental niches [2]. This microorganism can grow within the temperature range from 4 to 42 °C in terrestrial (soil) and aquatic habitats (polluted, salt, and freshwater) as well as on the surface of animate hosts (insects, plants, animals, and humans) and inanimate surroundings, mainly in the hospital environment (distilled water, disinfectants, sinks, medical devices, and equipment), being an important causative agent of nosocomial infections, particularly in intensive care units (ICUs) [1–4]. One of the interesting characteristics of *P. aeruginosa* is its pan-genome, which presents a larger genetic repertoire than the human genome. This intriguing feature explains the broad metabolic capabilities of *P. aeruginosa* and its distribution and adaptability in diverse environments [5].

P. aeruginosa is one of the most important bacterial species for public health considerations due to its high resistance to different classes of antibiotics and its capability to cause serious health care-associated as well as nosocomial infections [6, 7]. Results reported from an International Nosocomial Infection Control Consortium (INICC) surveillance study, performed between 2007 and 2012, in Latin America, Asia, Africa, and Europe, in which prospective data were collected from 605,310 patients hospitalized in 503 ICUs, displayed frequencies of 42.8% of *Pseudomonas* isolates resistant to amikacin and 42.4% to imipenem [8]. In the USA, an estimated 51,000 health care-associated *P. aeruginosa* infections occur each year, in which more than 6,000 (13%) of these are multidrug-resistant and 400 deaths per year are attributed to these infections [9]. The analyses based on data extracted from the Public Health England (PHE) voluntary surveillance database in the period 2008–2012 showed that 92% of *Pseudomonas* spp. isolates identified from bacteremia in 3,457 reports were *P. aeruginosa* [10]. In Brazil, the National Health Surveillance Agency (ANVISA), through the National Monitoring Microbial Resistance Network Health Services (RM Network), published a report that shows the main etiologic agents and the resistance phenotypes responsible for causing primary bloodstream infections associated with the use of central venous catheter in adult patients interned at ICUs from Brazilian hospitals between January and December 2013. According to that study, 18,233 notifications were reported, of which 1,850 (10.1%) were caused by *P. aeruginosa*, being the fifth pathogen most often reported as the etiologic agent. The resistance rate to the carbapenems reached 37.4% (692 *P. aeruginosa* isolates) [11]. Additionally, the Infectious Diseases Society of America has highlighted *P. aeruginosa* as part of a faction of antibiotic-resistant bacteria, called 'the ESKAPE pathogens'—*Enterococcus faecium*, *Staphylococcus aureus*, *Klebsiella pneumoniae*, *Acinetobacter baumannii*, *Pseudomonas aeruginosa,* and *Enterobacter* spp., capable of 'escaping' the bactericidal action of antibiotics and mutually representing new paradigms in pathogenesis, transmission, and resistance [12].

P. aeruginosa is extensively resistant to multiple drugs and is increasingly resistant to most available antibiotics, being a great emergency problem in the hospital settings worldwide [13]. Interestingly, *P. aeruginosa* has evolved over time in its ability to find new ways to be resistant to different classes of chemical compounds as well as to build strategies to exchange genetic materials, allowing that other bacteria also become drug-resistant [5]. Generally, resistance usually occurs due to a combination of factors acting synergistically: (i) *P. aeruginosa* is intrinsically resistant to antimicrobial agents due to its outer membrane/cell envelope composition that reduces the permeability of several drugs; and (ii) *P. aeruginosa* expresses a powerful repertoire of resistance mechanisms that can be developed through mutations in the genomic content that regulates resistance genes, and also acquired from other organisms via plasmids, transposons, or bacteriophages [14].

As a major opportunistic pathogen for humans, *P. aeruginosa* causes a plenty variety of acute and chronic infections and presents significant levels of morbidity and mortality [15, 16]. *P. aeruginosa* typically infects through airways, wounds, urinary tract, ear canal, via ocular and implanted medical devices (e.g., catheters or ventilators). Thereby, it is the main cause of eschars, conjunctivitis, keratitis, corneal ulcer, osteomyelitis, otitis, urinary infections, surgical site infections, bloodstream infections in ICUs and hospital-acquired pneumonia in immunocompromised individuals, mainly in patients with severe burn wounds, AIDS, lung cancer, chronic obstructive pulmonary disease, bronchiectasis, and cystic fibrosis [16–18].

It is known that Gram-negative bacteria are common causes of a huge diversity of infections including, intra-abdominal infections (IAIs), urinary tract infections (UTIs), ventilator-associated pneumonia (VAP), and bacteremia [19]. In particular, *P. aeruginosa* is one of the most important pathogens in the hospital setting, being responsible for 27% of all pathogens and 70% of all Gram-negative bacteria causing health care-associated infections in the USA, and it is the most common Gram-negative organism causing VAP and the second most common organism causing catheter-associated UTIs [7, 19]. The Centers for Disease Control and Prevention found that *P. aeruginosa* totalized 7.1% of health care-associated infection in the USA in 2011, being the second most common cause of pneumonia in hospital settings and the third most common Gram-negative bacterium to cause bloodstream infections [20]. *P. aeruginosa* is also a major cause of concern in the cystic fibrosis setting, being the most common pathogen isolated from cystic fibrosis sputum, and approximately 70% of adult cystic fibrosis patients are chronically colonized by this microorganism [21, 22].

The pathogenic potential of *P. aeruginosa* is not only due to its metabolic/genetic versatility and both intrinsic and acquired antibiotic resistance. Its ability to form biofilm and to produce an arsenal of virulence attributes, including cell-associated determinants (e.g., lipopolysaccharide, pili, and flagellum) and soluble secreted factors (e.g., extracellular polysaccharides, exotoxins, pigments, and proteases), is very important for the survival and adaptation of this pathogen in distinct environments [17, 22, 23].

2 *Pseudomonas aeruginosa*: Establishing and Maintaining an Infection

In order to establish an infection, *P. aeruginosa* cells count on a suite of virulence factors (Fig. 2) [17, 24]. These factors act together not only causing injuries on the host epithelial cell lining but also inducing dysfunctions in physiology and function, such as host cell shape, membrane permeability, and protein synthesis, as well as manipulating/overcoming host defenses, down-modulating the immune responses and preventing *P. aeruginosa* endocytosis and obstructing clearance mechanisms, thereby allowing this microbe to persist in cells/tissues and to establish an infection in the host [25, 26]. The virulence of *P. aeruginosa* is mediated by multiple mechanisms, but the major contributor is the production of extracellular proteases. In general, these enzymes regulate multiple cellular and physiological processes

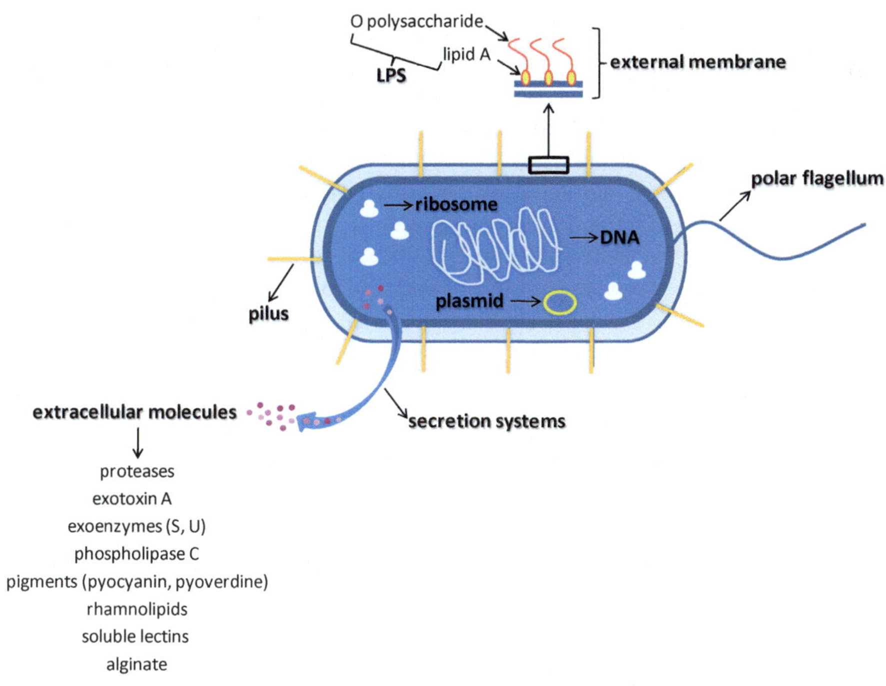

Fig. 2 Virulence factors expressed/produced by *P. aeruginosa* cells: (i) lipopolysaccharide (LPS) that induces cytokine production, (ii) pili that help bacterial adherence to the respiratory epithelial cells, (iii) flagellum that participates in mobility, adherence, and internalization events, (iv) extracellularly released molecules like proteases (responsible for the cleavage of key host proteins), exotoxin A (inhibition of host protein synthesis), exoenzyme S (induces cytotoxic effect), exoenzyme U (antiphagocytic effect), phospholipase C (cleavage of membrane phospholipids), pigments (many biological effects, like pyocyanin that induces free radicals in host cells), rhamnolipids (detergent action), soluble lectins (inhibition of beating of lung cells), and alginate (phagocytosis inhibition, antifungal action, and host immune responses)

and are essential to the success of the infection. They degrade a wide array of host proteins, impairing host defenses and destroying physical barriers that normally prevent attachment and penetration of the bacteria [26–28].

3 Proteolytic Enzymes Produced by *Pseudomonas aeruginosa*

P. aeruginosa is able to extracellularly release different kinds of proteases (Fig. 3), which together are responsible for invasion and destruction of host tissues. Because of the relevant roles played by proteases on the physiopathology of *P. aeruginosa,* it has been shown that the majority of environmental and clinical strains of *P. aeruginosa* exhibited proteolytic activity, particularly elastase activity [29–31]. According to Stover and co-workers [32], approximately 3% of the whole *P. aeruginosa* genome is composed by open reading frames that encode proteases [32]. Thus, the high genomic variability allows the bacterium to adapt its virulence arsenal machinery to support the variations of environment conditions, and for that, protease production in *P. aeruginosa* can vary greatly (Fig. 4) [32].

The expression of extracellular proteolytic enzymes in *P. aeruginosa* is directly influenced by environmental factors and changes in the physicochemical properties of culture medium (e.g., nutrients, temperature, pH, and aeration), which significantly modulate the production of these crucial virulence factors [26, 33]. In addition, the amount of protease produced depends on the cell cycle moment (e.g., lag, exponential, or stationary growth phase) and on the growing lifestyle (e.g., planktonic or biofilm). For instance, the total protease production (Fig. 5a) as well as the specific elastase secretion increases along the first 48 h of in vitro cultivation

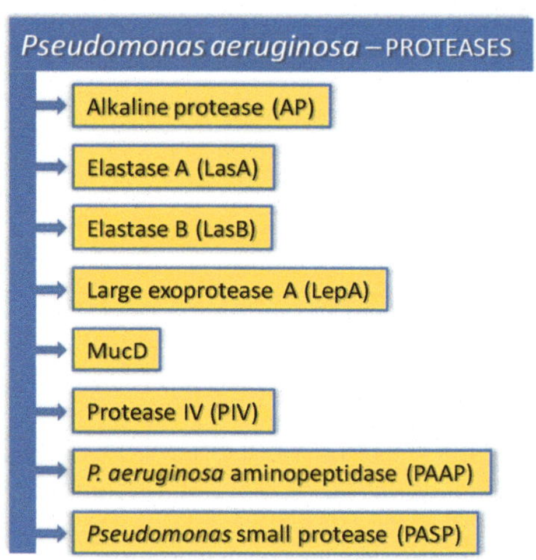

Fig. 3 Proteases secreted by *P. aeruginosa* cells

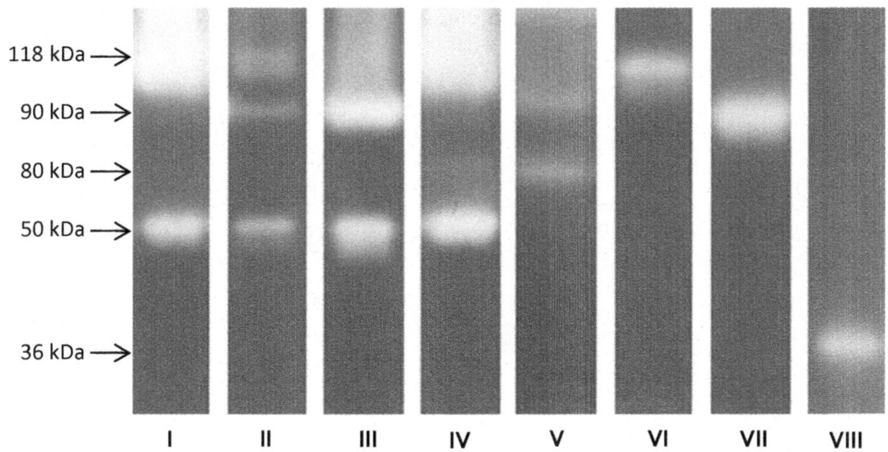

Fig. 4 Production of extracellular proteases in clinical isolates of *P. aeruginosa* recovered from different anatomical sites. The proteolytic profiles were characterized by sodium dodecyl sulfate-containing polyacrylamide gel electrophoresis (SDS-PAGE) containing 0.1% gelatin as the protein substrate. Profile I—118 + 50 kDa; Profile II—118 + 90 + 50 kDa; Profile III—90 + 50 kDa; Profile IV—118 + 80 + 50 kDa; Profile V—90 + 80 kDa; Profile VI—118 kDa; Profile VII—90 kDa, and Profile VIII—36 kDa

of *P. aeruginosa* planktonic cells (Fig. 5b). Further, according to Hastie and co-workers [34], after 85 h of bacterial growth, the elastase production dropped off.

3.1 Elastase B

One of the best proteases characterized in *Pseudomonas* is elastase B (LasB), also known as pseudolysin. This 33-kDa enzyme belongs to the M4 thermolysin-like family of neutral, Zn-dependent metallo-endopeptidases (Fig. 6). This enzyme is encoded by *lasB* gene as a pre-pro-protein, containing at the N-terminal region a signal peptide of 23 amino acids that transport the enzyme through the inner membrane to periplasmic place by bacterial secretory system [35].

The first and the most studied substrate of elastase B is bovine and human elastin [36–38]. Some reports correlate the elastinolytic activity of elastase B to *Pseudomonas* infections in cystic fibrosis patients [39–43]. Histological studies have detected altered elastin fibers in lung alveoli of cystic fibrosis patients on autopsy, indicating a probable elastase activity on cystic fibrosis lung [39]. In addition, the elastase activity is associated with vascular inflammation during *P. aeruginosa* infection, since the disorganization of elastin fiber in vascular tissue caused by protease degradation was observed [44]. Previously, our group analyzed the production of virulence attributes in 96 clinical strains of *P. aeruginosa* recovered from patients attended at hospitals located in three States of Brazil (Espírito Santo, Minas Gerais, and Rio de Janeiro), and it was shown that all bacterial strains exhibited a

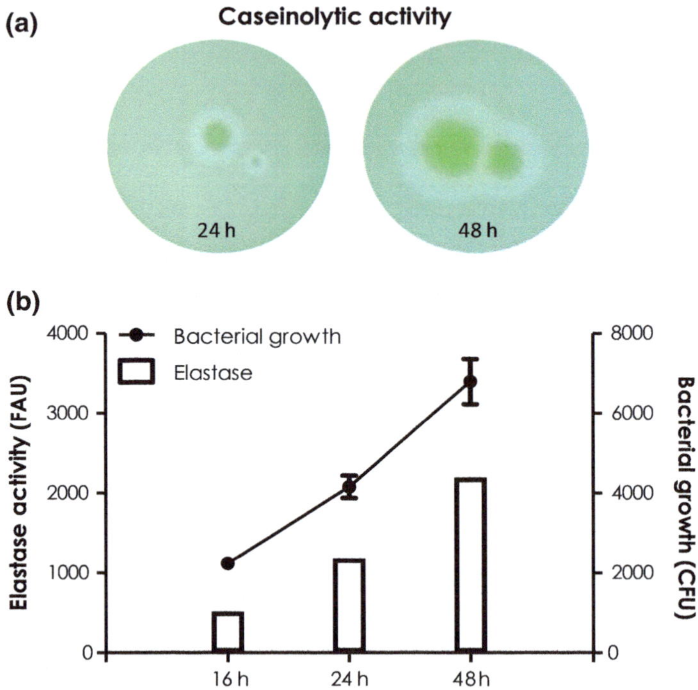

Fig. 5 Protease detection in *P. aeruginosa*. **a** Total extracellular protease production was analyzed by the degradation of casein (1%) incorporated into Luria Bertani agar medium up to 48 h at 37 °C. **b** The elastase activity was measured in the cell-free culture supernatant obtained from *P. aeruginosa* cells grown in tryptic soy broth up to 48 h at 37 °C, using the fluorogenic peptide substrate Abz-Ala-Gly-Leu-Ala-*p*-Nitro-Benzyl-Amide. Results were expressed as fluorescence arbitrary units (FAU). In parallel, the number of bacterial cells along each time point was evaluated by plating cells onto agar medium and expressed as colony-forming units (CFU)

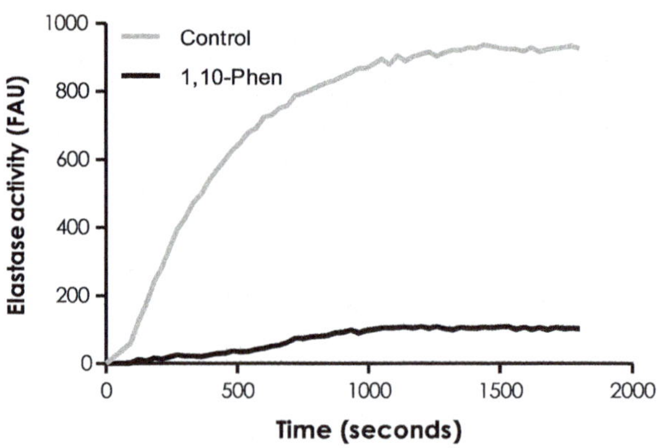

Fig. 6 Elastase of *P. aeruginosa* is a typical zinc-metalloprotease. The purified elastase B is able to cleave the fluorogenic peptide substrate Abz-Ala-Gly-Leu-Ala-*p*-Nitro-Benzyl-Amide along the time. Conversely, 1,10-phenanthroline (1,10-Phen), a metalloprotease inhibitor, at 10 μM was able to block the substrate cleavage. FAU, fluorescence arbitrary units

homogeneous elastase activity, with an average of 1069.28 ± 213.95 fluorogenic arbitrary units (FAU) with no correlation with the original anatomical site of isolation [16]. On the other hand, *P. aeruginosa* strains recovered from trachea, urinary tract, and wounds of patients attended at University Medical Center/Texas Tech Health Sciences Center were able to produce different amounts of elastase [45]. Woods and co-workers [46] showed that Canadian *P. aeruginosa* strains isolated from acute lung infections showed the highest production of elastase (0.053 ± 0.021 mg/ml) compared with elastase activity of strains isolated from burns, wounds, cystic fibrosis lung, and blood.

LasB is also able to cleave other host extracellular matrix proteins, such as collagen type III and IV. Interestingly, after subcutaneous injection of purified elastase B into mice, an intense degradation of basement membranes was observed, and elastase B was responsible for severe hemorrhage and tissue damage [47]. Several studies have demonstrated that LasB-associated epithelial disruption is mediated by the attack to intracellular tight junctions and cytoskeleton reorganization via inhibition of protein kinase C and activation of EGFR, ERK1/2 and NFκB, urokinase, and protease-activated receptor 2 (PAR-2) [48–53]. Elastase B can also interfere with the host bacterial clearance by degrading several components of innate and adaptive immune defenses, including tumor necrosis factor-α (TNF-α), interferon-γ (IFN-γ) and interleukin-2 (IL-2), monocyte chemotactic protein-1 (MCP-1), and epithelial neutrophil activating protein-78 (ENA-78) [52–57]. In addition, it was shown that elastase B was efficient in the inactivation of key components of the complement system such as fluid-phase and cell-bound C1 and C3 and fluid-phase C5, C8, and C9 [44]. This multifunctional enzyme is also able to cleave surfactant protein A and D (SP-A and SP-D), also known as collectin. SP-A and SP-D are synthesized by alveolar type II epithelial cells and are responsible for the recognition and binding to oligosaccharides present on the cell surface of many bacteria to be phagocytized by host macrophages [58]. Previously, Meyer and co-workers [59] have reported that a decrease on the SP-A and SP-D levels in bronchoalveolar lavage (BAL) was observed in the lung of cystic fibrosis individuals. Also, SP-D knockout mice were more sensible to *P. aeruginosa* corneal infections when compared to wild-type animals, and only the wild-type mice recovered completely of the infection [60]. Based on this, elastase B was suggested to be responsible for the SP-D degradation in the eye [25, 26]. Furthermore, pseudomonal elastase can interact with host adaptive immune system by degrading immunoglobulins [61–63]. Bainbrigde and Flick [61] showed that elastase B was able to cleave IgG molecules recovered from cystic fibrosis patients and the degradation products bound to IgG-receptors of human neutrophils, thereby inhibiting the opsonization of bacterial invaders. Lomholt and Kilian [63] reported the IgA degradation in tears from patients infected with *P. aeruginosa.* They also observed that isogenic mutants of *P. aeruginosa* knockout to either elastase or alkaline protease were not able to completely inhibit the IgA degradation, indicating that several proteases were working in concert to cleave IgA.

Furthermore, elastase B plays a key role in the differentiation of pseudomonal biofilms. Tielen and co-workers [64] showed that strains that overexpress *lasB* gene were not able to form robust biofilms, and they observed the formation of few microcolonies after 72 h of contact with glass surface. Those authors also assigned that *lasb*-overexpressed strain shifted the composition of its extracellular polymeric substances, reducing the alginate content as well as enhancing the rhamnolipids concentration [64]. However, Yu and co-workers [65] demonstrated that elastase B is crucial for biofilm formation. They observed that Δ*lasB* mutant decreased the biofilm formation through down-regulation of rhamnolipids synthesis.

3.2 Elastase A

Another extracellular protease produced by *P. aeruginosa* is elastase A (LasA), a metalloprotease that belongs to the subgroup A of M23 family of staphylolytic or β-lytic zinc metallo-endopeptidases. LasA is codified as an elastase A pre-pro-protein with molecular mass of 40 kDa [66, 67]. After its synthesis in intracellular bacterial environment, LasA is secreted via type II secretion machinery and when it is secreted to the extracellular space, LasA is immediately converted to its mature and active form of 27 kDa due to the cleavage by other pseudomonal-secreted endopeptidases, such as LasB, LysC, and protease IV [68, 69].

Elastase A is also called as staphylolysin, because it is able to cleave the pentaglycine bonds in the peptidoglycan of *Staphylococcus aureus* [70]. As well, LasA degrades several glycine-rich synthetic peptides [71]. However, LasA exhibited a limited elastinolytic activity [72]. Kessler and co-workers [71] showed that LasA prefers cleaving Gly–Ala peptide bonds within the Gly-Gly-Ala sequences surrounded by apolar sequences. Such sequences are uncommon in elastin, resulting in low elastinolytic activity [26, 73]. Besides its own intrinsic elastinolytic activity, LasA enhances significantly the elastinolytic activity of other proteases, including LasB in *P. aeruginosa*, but also human leukocyte elastase and human neutrophil elastase [74, 75]. Moreover, LasA is responsible for inducing shedding of the host cell surface proteoglycan syndecan-1 (co-receptor proteins), which has been shown to be important for *P. aeruginosa* survival [25, 26].

3.3 Alkaline Protease

Another pseudomonal protein shown to be important for phagocytic evasion is alkaline protease (AprA), which is also known as aeruginolysin. Alkaline protease is a 50-kDa zinc-metalloprotease, member of subfamily B of the M10 peptidase family and metzincin superfamily. AprA, encoded by *aprA* gene, has a C-terminal secretion signal located within the last 50 amino acid residues necessary to be translocated and secreted by AprD, APrE, and AprF membrane proteins, which form the bacterial type I secretory machinery [35].

It was reported that alkaline protease is able to degrade a large number of host proteins, including fibronectin and laminin, important components of basal lamina and endothelium. Therefore, alkaline protease develops an important function in invasion and hemorrhagic tissue necrosis in infections caused by *P. aeruginosa* [76]. Furthermore, this protease was found in many isolates of *P. aeruginosa* recovered from different human anatomical sites with especial elevated expression in clinical isolates from eyes, gastrointestinal tract, and mucoid wounds exacerbated in cystic fibrosis patients [25, 61]. AprA is important to bacterial escape from the host immunological defenses, degrading complement proteins (C1q, C2, and C3) and cytokines (IFN-γ, TNF-α and IL-6) [76]. Also, alkaline protease and elastase B are able to inhibit chemotaxis of neutrophils and block efficiently the phagocytosis, which gives the pathogen an advantage in escaping from phagocyte cells that are one of the first lines of host defense mechanisms [25, 31, 77, 78]. Moreover, alkaline protease is able to inhibit flagellin recognition by TLR5 due to the degradation of free flagellin monomers, helping *P. aeruginosa* cells to avoid the immune detection [79]. This enzyme has also been shown to aid *P. aeruginosa* survival in iron limitations conditions during human infections by cleaving transferrin that increase the siderophore-mediated iron uptake [80]. Gupta and co-workers [81] also reported that treatment of mouse corneal tissue with alkaline protease (50 ng) increases the binding of *P. aeruginosa* to the epithelial surface.

3.4 Protease IV

P. aeruginosa secretes a serine-type protease designated as protease IV (PIV) or lysyl endopeptidase (PrpL), a 26-kDa protease belonging to the chymotrypsin family S1 that has been demonstrated to be an important virulence factor in the rabbit cornea, but is found in clinical isolates recovered from all the anatomical sites analyzed [35, 82]. Its catalytic domain is formed by the triad His_{72}, Asp_{122}, and Ser_{198}. Moreover, it was demonstrated that the residue Ser_{197} adjacent to Ser_{198} is critical to the catalytic activity [83]. Protease IV is encoded by *piv* gene (PA4175), with a full length of 48 kDa, which is initially expressed in the cytoplasm in a pre-pro-enzyme form and then processed to the 26-kDa mature protease after its secretion into the extracellular milieu [83].

PIV participates in the tissue invasion/damage processes and hemorrhagic events due to the cleavage of fibrinogen. It is well known that fibrinogen is required after vascular damage, but the degradation of fibrinogen by PIV leads to hemorrhage during *P. aeruginosa* infection [84]. PIV is also important to evade host immune defenses because it is able to degrade plasminogen, immunoglobulin, C1q and C3, and host antimicrobial peptide LL-37 [25, 68]. Furthermore, Malloy and co-workers [82] observed that PIV degrades the surfactant proteins, SP-A, SP-D, and SP-B, by a time- and dose-depended way in cell-free bronchoalveolar lavage fluid. Those authors reported that degradation of SPs by protease IV reduced the association among bacteria and alveolar macrophage. Interestingly, the incubation of pulmonary surfactant with pseudomonal protease IV reduced the ability of the surfactant to diminish the superficial tension within the lung [82]. Protease IV has been shown to be an iron-regulated protein, suggesting that its expression is regulated irrespective of *quorum sensing* system, which is distinct from other pseudomonal proteases [69]. Protease IV has also been correlated to ring abscess lesions present in pseudomonal keratitis [68]. Corroborating this finding, Engel and co-workers [85] showed that protease IV-deficient mutants exhibited lower ocular virulence in rabbits when intrastromally infected.

3.5 Pseudomonas Small Protease

P. aeruginosa small protease (PASP) is described as a 18.5-kDa secreted zinc-dependent leucine aminopeptidase. *PASP* gene has been found in a large number of *P. aeruginosa* clinical strains, but its higher expression is found during the ocular infection [86]. Previous reports showed that PASP is found only in the bacterial supernatant culture. According to Tang and co-workers [86], the sequence of *PASP* gene appears to have a signal peptide consistent with that needed for type II secretion system.

Direct inoculation of purified PASP into the rabbit cornea causes severe ocular pathology, including epithelial erosion and ulcer in stroma, edema, and neutrophil infiltration into the corneal stroma [87]. PASP has also been demonstrated to cleavage host proteins required for maintaining structure of cornea, such as collagens, fibrinogen (but not fibrin), complement C3, and antimicrobial peptide LL-37. Studies of PASP, coupled with those of PIV, strongly support the hypothesis that *Pseudomonas* proteases play a major role in keratitis [87].

3.6 Large Exoprotease A

Large exoprotease A (LepA) is an exoprotease with molecular mass of ~100 kDa produced by *P. aeruginosa*. LepA, as well as thrombin and trypsin, cleaves human protease-activated receptors (PARs) 1, 2, and 4 in order to activate the critical transcription factor NF-κB, which is associated with host inflammatory and immune responses [49, 88].

3.7 MucD

MucD was reported to be a serine endoprotease that is localized within the periplasmic space. Data suggest that MucD induced a significant reduction on the levels of IL-1β, neutrophil-chemoattractant chemokines KC, and macrophage-inflammatory protein-2 (MIP-2) in the early stages of bacterial infection as well as it inhibited the recruitment of polymorphonuclear (PMN) cells into the cornea. Furthermore, a decrease in PMN cells recruited to infection site favored the establishment of infection by *P. aeruginosa*. MucD may be secreted to the extracellular space, interfering with the biological functions of cytokines and chemokines, but further investigation is needed to understand the mechanisms underlying the role of MucD in keratitis [89, 90].

3.8 Aminopeptidase

The *P. aeruginosa* aminopeptidase (PAAP) or leucine aminopeptidase has been speculated as complementary enzyme to the activity of other endopeptidases. PAAP has an important function in bacterial physiology; it acts releasing free amino acids/small peptides from protein fragments produced by the others *P. aeruginosa* endopeptidases, thereby providing low molecular mass nutrients that can be taken up by the bacterium, which in turn may promote bacterial growth and proliferation [26].

4 Conclusions

P. aeruginosa is a metabolically versatile bacterium that can cause a wide range of severe opportunistic infections in hospitalized patients. To cause this huge variety of infections, *P. aeruginosa* has an arsenal of proteases that are involved in critical events of bacterial pathogenicity and virulence, which are important for survival in the host, tissue invasion, and evasion of host immune defenses. Therefore, this review has highlighted the importance of each pseudomonal protease in bacterial physiology and/or in infectious events. In this context, inhibitors able to block the proteases produced by *P. aeruginosa* cells would represent a new drug class quite promising to combat this widespread bacterial pathogen.

Acknowledgements This study was supported by grants from the following Brazilian Agencies: Coordenação de Aperfeiçoamento de Pessoal de Nível Superior (CAPES), Conselho Nacional de Desenvolvimento Científico e Tecnológico (CNPq), and Fundação de Amparo à Pesquisa no Estado do Rio de Janeiro (FAPERJ).

References

1. Tümmler B, Wiehlmann L, Klockgether J et al (2014) Advances in understanding *Pseudomonas*. F1000 Prime 6:9
2. Jayaseelan S, Ramaswamy D, Dharmaraj S (2014) Pyocyanin: production, applications, challenges and new insights. World J Microbiol Biotechnol 30:1159–1168
3. Vasil ML (1986) *Pseudomonas aeruginosa*: biology, mechanisms of virulence, epidemiology. J Pediatr 108:800–805
4. Wiehlmann L, Wagner G, Cramer N et al (2007) Population structure of *Pseudomonas aeruginosa*. PNAS 104:8101–8106
5. Ghodhbane H, Elaidi S, Sabatier JM et al (2015) Bacteriocins active against multi-resistant gram negative bacteria implicated in nosocomial infections. Infect Disord Drug Targets 15:2–12
6. Bartram J, Cotruvo J, Exner M, Fricker C, Glasmacher A (2003) WHO—World Health Organization. Heterotrophic plate counts and drinking-water safety. IWA Publishing, London. ISBN: 1 84339 025 6
7. Nielsen SL (2015) The incidence and prognosis of patients with bacteremia. Dan Med J 62: B5128
8. Rosenthal VD, Maki DG, Mehta Y, Leblebicioglu H et al (2014) International nosocomial Infection Control Consortium. International Nosocomial Infection Control Consortium

(INICC) report, data summary of 43 countries for 2007–2012. Device-associated module. Am J Infect Control 42:942–956
9. Centers for Disease Control and Prevention. Healthcare-associated Infections (HAIs). *Pseudomonas aeruginosa* in Healthcare Settings. Available in: http://www.cdc.gov/hai/organisms/pseudomonas.html. Accessed on 17 June 2016
10. Public Health England (2012) *Pseudomonas aeruginosa*: guidance, data and analysis—voluntary surveillance of *Pseudomonas* spp. and *Stenotrophomonas* spp. causing bacteraemia in England, Wales and Northern Ireland. Available in: https://www.gov.uk/government/publications/pseudomonas-spp-and-stenotrophomonas-spp-voluntary-surveillance-2012. Accessed on 17 June 2016
11. ANVISA—Agência Nacional de Vigilância Sanitária. Boletim Informativo Segurança do Paciente e Qualidade em Serviços de Saúde—Ano V nº 09|Dezembro de 2014. Available in: http://portal.anvisa.gov.br. Accessed on 17 June 2016
12. Pendleton JN, Gorman SP, Gilmore BF (2013) Clinical relevance of the ESKAPE pathogens. Expert Rev Anti Infect Ther 11:297–308
13. Buhl M, Peter S, Willmann M (2013) Prevalence and risk factors associated with colonization and infection of extensively drug-resistant *Pseudomonas aeruginosa*: a systemic review. Expert Rev Anti-infect Ther 13:1159–1170
14. El Zowalaty ME, Al Thani AA, Webster TJ et al (2015) *Pseudomonas aeruginosa*: arsenal of resistance mechanisms, decades of changing resistance profiles, and future antimicrobial therapies. Future Microbiol 10:1683–1706
15. Bentzmann S, Plésiat P (2011) The *Pseudomonas aeruginosa* opportunistic pathogen and human infections. Environ Microbiol 13:1655–1665
16. Silva LV, Galdino ACM, Nunes APF et al (2014) Virulence attributes in Brazilian clinical isolates of *Pseudomonas aeruginosa*. Int J Med Microbiol 304:990–1000
17. Balasubramanian D, Schneper L, Kumari H, Mathee K (2013) A dynamic and intricate regulatory network determines *Pseudomonas aeruginosa* virulence. Nucleic Acids Res 41:1–20
18. Savoia D (2014) New perspectives in the management of *Pseudomonas aeruginosa* infections. Future Microbiol 9:917–928
19. Kaye KS, Pogue JM (2015) Infections caused by resistant Gram-negative bacteria: epidemiology and management. Pharmacotherapy 35:949–962
20. McCarthy K (2015) *Pseudomonas aeruginosa*: evolution of antimicrobial resistance and implications for therapy. Semin Respir Crit Care Med 36:44–55
21. Sousa AM, Pereira MO (2014) *Pseudomonas aeruginosa* diversification during infection development in cystic fibrosis lungs—a review. Pathogens 3:680–703
22. Oliver A, Mulet X, López-Causapé C, Juan C (2015) The increasing threat of *Pseudomonas aeruginosa* high-risk clones. Drug Resist Update 22:41–59
23. Kung VL, Ozer EA, Hauser AR (2010) The accessory genome of *Pseudomonas aeruginosa*. Microbiol Mol Biol 74:621–664
24. Crousilles A, Maunders E, Bartlett S, Fan C et al (2015) Which microbial factors really are important in *Pseudomonas aeruginosa* infections? Future Microbiol 10:1825–1836
25. Ballok AE, O'Toole GA (2013) Pouring salt on a wound: *Pseudomonas aeruginosa* virulence factors alter Na^+ and Cl^- flux in the lung. J Bacteriol 195:4013–4019
26. Kessler E, Safrin M (2014) Elastinolytic and proteolytic enzymes. In *Pseudomonas* methods and protocols. Methods Mol Biol 1149:135–169
27. McCarty SM, Cochrane CA, Clegg PD, Percival SL (2012) The role of endogenous and exogenous enzymes in chronic wounds: a focus on the implications of aberrant levels of both host and bacterial proteases in wound healing. Wound Repair Regen 20:125–136
28. Gellatly SL, Hancock REW (2013) *Pseudomonas aeruginosa*: new insights into pathogenesis and host defenses. Pathog Dis 67:159–173
29. Schmidtchen A, Wolff H, Hansson C (2001) Differential proteinase expression by *Pseudomonas aeruginosa* derived from chronic leg ulcers. Acta Derm Venereol 81:406–409

30. Tingpej P, Smith L, Rose B et al (2007) Phenotypic characterization of clonal and nonclonal *Pseudomonas aeruginosa* strains isolated from lungs of adults with cystic fibrosis. J Clin Microbiol 45:1697–1704
31. Thibodeau PH, Butterworth MB (2013) Proteases, cystic fibrosis and the epithelial sodium channel (ENaC). Cell Tissue Res 351:309–323
32. Stover CK, Pham XQ, Erwin AL et al (2000) Complete genome sequence of *Pseudomonas aeruginosa* PAO1, an opportunistic pathogen. Nature 406:959–964
33. Marquart ME, Dajcs JJ, Caballero AR et al (2005) Calcium and magnesium enhance the production of *Pseudomonas aeruginosa* protease IV, a corneal virulence factor. Med Microbiol Immunol 194:39–45
34. Hastie AT, Hingley ST, Kueppers F (1983) Protease production by *Pseudomonas aeruginosa* isolates from patients with cystic fibrosis. Infect Immun 40:506–513
35. Hoge R, Pelzer A, Rosenau F, Wilhelm S (2010) Weapons of a pathogen: proteases and their role in virulence of *Pseudomonas aeruginosa*. In: Méndez-Vilas A (ed) Current research, technology and education topics in applied microbiology and microbial biotechnology. Formatex Research Center, Badajoz, pp. 383–395
36. Hamdaoui A, Wund-Bisseret F, Bieth JG (1987) Fast solubilization of human lung elastin by *Pseudomonas aeruginosa* elastase. Am Rev Respir Dis 135:860–863
37. Saulnier JM, Curtil FM, Duclos MC, Wallach JM (1989) Elastolytic activity of *Pseudomonas aeruginosa* elastase. Biochim Biophys Acta 995:285–290
38. Yang J, Zhao HL, Ran YL et al (2015) Mechanistic insights into elastin degradation by pseudolysin, the major virulence factor of the opportunistic pathogen *Pseudomonas aeruginosa*. Sci Rep 9936
39. Bruce MC, Poncz L, Klinger JD et al (1985) Biochemical and pathologic evidence for proteolytic destruction of lung connective tissue in cystic fibrosis. Am Rev Respir Dis 132:529–535
40. Erickson DL, Endersby R, Kirkham A et al (2002) *Pseudomonas aeruginosa* quorum-sensing systems may control virulence factor expression in the lungs of patients with cystic fibrosis. Infect Immun 70:1783–1790
41. Kosorok MR, Zeng L, West SE et al (2001) Acceleration of lung disease in children with cystic fibrosis after *Pseudomonas aeruginosa* acquisition. Pediatr Pulmonol 32:277–287
42. Voynow JA, Fischer BM, Zheng S (2008) Proteases and cystic fibrosis. Int J Biochem Cell Biol 40:1238–1245
43. Van't Wout EF, van Schadewijk A, van Boxtel R et al (2015) Virulence factors of *Pseudomonas aeruginosa* induce both the unfolded protein and integrated stress responses in airway epithelial cells. PLoS Pathog 11:e1004946
44. Schultz DR, Miller KD (1974) Elastase of *Pseudomonas aeruginosa*: inactivation of complement components and complement-derived chemotactic and phagocytic factors. Infect Immun 10:128–135
45. Hamood A, Griswold G, Colmer J (1996) Characterization of elastase-deficient clinical isolates of *Pseudomonas aeruginosa*. Infect Immun 64:3154–3160
46. Woods DE, Schaffer MS, Rabin HR et al (1988) Phenotypic comparison of *Pseudomonas aeruginosa* strains isolated from a variety of clinical sites. J Bacteriol 170:4309–4314
47. Komori Y, Nonogaki T, Nikai T (2001) Hemorrhagic activity and muscle damaging effect of *Pseudomonas aeruginosa* metalloproteinase (elastase). Toxicon 39:1327–1332
48. Bentzmann S, Polette M, Zahm JM et al (2000) *Pseudomonas aeruginosa* virulence factors delay airway epithelial wound repair by altering the actin cytoskeleton and inducing overactivation of epithelial matrix metalloproteinase-2. Lab Invest 80:209–219
49. Kida Y, Higashimoto Y, Inoue H et al (2008) A novel secreted protease from *Pseudomonas aeruginosa* activates NF-kappaB through protease-activated receptors. Cell Microbiol 10:491–504
50. Clark CA, Thomas LK, Azghani AO (2011) Inhibition of protein kinase C attenuates *Pseudomonas aeruginosa* elastase-induced epithelial barrier disruption. Am J Respir Cell Mol Biol 45:1263–1271

51. Cosgrove S, Chotirmall SH, Greene CM et al (2011) Pulmonary proteases in the cystic fibrosis lung induce interleukin 8 expression from bronchial epithelial cells via a heme/meprin/epidermal growth factor receptor/Toll-like receptor pathway. J Biol Chem 286:692–704
52. Kuang Z, Hao Y, Walling BE et al (2011) *Pseudomonas aeruginosa* elastase provides an escape from phagocytosis by degrading the pulmonary surfactant protein-A. PLoS ONE 6: e27091
53. Nomura K, Obata K, Keira T et al (2014) *Pseudomonas aeruginosa* elastase causes transient disruption of tight junctions and downregulation of PAR-2 in human nasal epithelial cells. Respir Res 18:15–21
54. Parmely M, Gale A, Clabaugh M et al (1990) Proteolytic inactivation of cytokines by *Pseudomonas aeruginosa*. Infect Immun 58:3009–3014
55. Horvat RT, Clabaugh M, Duval-Jobe C, Parmely MJ (1989) Inactivation of human gamma interferon by *Pseudomonas aeruginosa* proteases: elastase augments the effects of alkaline protease despite the presence of alpha 2-macroglobulin. Infect Immun 57:1668–1674
56. Theander TG, Kharazmi A, Pedersen BK et al (1988) Inhibition of human lymphocyte proliferation and cleavage of interleukin-2 by *Pseudomonas aeruginosa* proteases. Infect Immun 56:1673–1677
57. Leidal KG, Munson KL, Johnson MC et al (2003) Metalloproteases from *Pseudomonas aeruginosa* degrade human RANTES, MCP-1, and ENA-78. J Interferon Cytokine Res 23:307–318
58. Mariencheck WI, Alcorn JF, Palmer SM (2003) *Pseudomonas aeruginosa* elastase degrades surfactant proteins A and D. Am J Respir Cell Mol Biol 28:528–537
59. Meyer KC, Sharma R, Brown M et al (2000) Function and composition of pulmonary surfactant and surfactant-derived fatty acid profiles are altered in young adults with cystic fibrosis. Chest 118:164–174
60. McCormick CC, Hobden JA, Balzli CL et al (2007) Surfactant protein D in *Pseudomonas aeruginosa* keratitis. Ocular Immun Inflam 15:371–379
61. Bainbridge T, Fick RB (1989) Functional importance of cystic fibrosis immunoglobulin G fragments generated by *Pseudomonas aeruginosa* elastase. J Lab Clin Med 114:728–733
62. Heck LW, Alarcon PG, Kulhavy RM et al (1990) Degradation of IgA proteins by *Pseudomonas aeruginosa* elastase. J Immunol 144:2253–2257
63. Lomholt JA, Kilian M (2008) Degradation of uniquely glycosylated secretory immunoglobulin A in tears from patients with *Pseudomonas aeruginosa* keratitis. Invest Ophthalmol Vis Sci 49:1944–4939
64. Tielen P, Rosenau F, Wilhelm S et al (2010) Extracellular enzymes affect biofilm formation of mucoid *Pseudomonas aeruginosa*. Microbiology 156:2239–2252
65. Yu H, He X, Xie W et al (2014) Elastase LasB of *Pseudomonas aeruginosa* promotes biofilm formation partly through rhamnolipid-mediated regulation. Can J Microbiol 60:227–235
66. Schad PA, Iglewski BH (1988) Nucleotide sequence and expression in Escherichia coli of the *Pseudomonas aeruginosa* lasA gene. J Bacteriol 170:2784–2789
67. Kessler E, Safrin M, Gustin JK et al (1998) Elastase and the LasA protease of *Pseudomonas aeruginosa* are secreted with their propeptides. J Biol Chem 273:30225–30231
68. Engel LS, Hill JM, Caballero AR (1998) Protease IV, a unique extracellular protease and virulence factor from *Pseudomonas aeruginosa*. J Biol Chem 273:16792–16797
69. Wilderman PJ, Vasil AI, Johnson Z (2001) Characterization of an endoprotease (prpl) encoded by a pvds-regulated gene in *Pseudomonas aeruginosa*. Infect Immun 69:5385–5394
70. Barequet IS, Bourla N, Pessach YN et al (2012) Staphylolysin is an effective therapeutic agent for *Staphylococcus aureus* experimental keratitis. Graefes Arch Clin Exp Ophthalmol 250:223–229
71. Kessler E, Safrin M, Abrams WR, Rosenbloom J, Ohman DE (1997) Inhibitors and specificity of *Pseudomonas aeruginosa* LasA. J Biol Chem 272:9884–9889

72. Kessler E, Safrin M, Blumberg S, Ohman DE (2004) A continuous spectrophotometric assay for *Pseudomonas aeruginosa* LasA prote- ase (staphylolysin) using a two-stage enzymatic reaction. Anal Biochem 328:225–232
73. Vessillier S, Delolme F, Bernillon J, Saulnier J, Wallach J (2001) Hydrolysis of glycine-containing elastin pentapeptides by LasA, a metalloelastase from *Pseudomonas aeruginosa*. Eur J Biochem 268:1049–1057
74. Peters JE, Galloway DR (1990) Purification and characterization of an active fragment of the LasA protein from *Pseudomonas aeruginosa*: enhancement of elastase activity. J Bacteriol 172:2236–2240
75. Peters JE, Park SJ, Darzins A et al (1992) Further studies on *Pseudomonas aeruginosa* LasA: analysis of specificity. Mol Microbiol 6:1155–1162
76. Laarman AJ, Bardoel BW, Ruyken M et al (2012) *Pseudomonas aeruginosa* alkaline protease blocks complement activation via the classical and lectin pathways. J Immunol 188:386–393
77. Kharazmi A, Hoiby N, Doring G, Valerius NH (1984) *Pseudomonas aeruginosa* exoproteases inhibit human neutrophil chemiluminescence. Infect Immun 44:587–591
78. Hong YQ, Ghebrehiwet B (1992) Effect of *Pseudomonas aeruginosa* elastase and alkaline protease on serum complement and isolated components C1q and C3. Clin Immunol Immunopathol 62:133–138
79. Bardoel BW, van Kessel KP, van Strijp JA, Milder FJ (2012) Inhibition of *Pseudomonas aeruginosa* virulence: characterization of the AprA-AprI interface and species selectivity. J Mol Biol 415:573–583
80. Kim SJ, Park RY, Kang SM (2006) *Pseudomonas aeruginosa* alkaline protease can facilitate siderophore-mediated iron-uptake via the proteolytic cleavage of transferrins. Biol Pharm Bull 29:2295–22300
81. Gupta SK, Masinick SA, Hobden JA et al (1996) Bacterial proteases and adherence of *Pseudomonas aeruginosa* to mouse cornea. Exp Eye Res 62:641–650
82. Malloy JL1, Veldhuizen RA, Thibodeaux BA et al (2005) *Pseudomonas aeruginosa* protease IV degrades surfactant proteins and inhibits surfactant host defense and biophysical functions. Am J Physiol Lung Cell Mol Physiol 288:409–418
83. Traidej M, Caballero AR, Marquart ME et al (2003) Molecular analysis of *Pseudomonas aeruginosa* protease IV expressed in *Pseudomonas putida*. Invest Ophthalmol Vis Sci 44:190–196
84. Matsumoto K (2004) Role of bacterial proteases in pseudomonal and serratial keratitis. Biol Chem 385:1007–1016
85. Engel LS, Hobden JA, Moreau JM et al (1997) Pseudomonas deficient in protease IV has significantly reduced corneal virulence. Invest Ophthalmol Vis Sci 38:1535–1542
86. Tang A, Marquart ME, Fratkin JD et al (2009) Properties of PASP: a *Pseudomonas* protease capable of mediating corneal erosions. Invest Ophthalmol Vis Sci 50:3794–3801
87. Tang A, Caballero AR, Marquart ME, O'callaghan RJ (2013) *Pseudomonas aeruginosa* small protease (PASP), a keratitis virulence factor. Invest Ophthalmol Vis Sci 54:2821–2828
88. Kida Y, Shimizu T, Kuwano K (2011) Cooperation between LepA and PlcH contributes to the in vivo virulence and growth of *Pseudomonas aeruginosa* in mice. Infect Immun 79:211–219
89. Mochizuki Y, Suzuki T, Oka N, Zhang Y et al (2014) *Pseudomonas aeruginosa* MucD protease mediates keratitis by inhibiting neutrophil recruitment and promoting bacterial survival. Invest Ophthalmol Vis Sci 55:240–246
90. Okuda J, Hayashi N, Tanabe S et al (2011) Degradation of interleukin 8 by the serine protease MucD of Pseudomonas aeruginosa. Infect Chemother 17:782–792

Serine Proteases as Metabolic Regulators in Yeast

Alexis N. Campetelli and Noelia E. Monesterolo

Abstract

Serine proteases are enzymes that break peptidic bonds in proteins, being serine the nucleophilic amino acid at the enzyme's active site. These proteases are found ubiquitously in both eukaryotes and prokaryotes. Based on their structure, serine proteases may be classified into two super families: chymotrypsin-like (trypsin-like) or subtilisin-like. This chapter focuses on aspects related to the molecular structure, subcellular localization, mechanism of action, and physiological role, among others, of Kexin, Ynm3p, Prb1p, Ssy5p, Lpx1p, and Pcp1p, the best characterized serine proteases found in yeast, taking as a reference the almost fully characterized yeast model *Saccharomyces cerevisiae.* Table 1 summarizes the general information about the proteases mentioned above, and such general information is illustrated in Fig. 1.

1 Introduction

Serine proteases are widely distributed in nature and found in all kingdoms of cellular life as well as many viral genomes. Above 30% of all known protein degrading enzymes are serine proteases, highlighting its importance in evolution. They are grouped into 13 major clans and 40 families. This way to classify segregates proteases into clans based on their action mechanism and families on evolutionary basis. The family name has its origin based in the nucleophilic Ser in the enzyme active site. This serine attacks the carbonyl group of the substrate

A.N. Campetelli (✉) · N.E. Monesterolo
Departmento de Biología Molecular, Facultad de Ciencias Exactas,
Físico-Químicas y Naturales, Univesidad Nacional de Río Cuarto, Río Cuarto, Argentina
e-mail: acampetelli@exa.unrc.edu.ar

S. Chakraborti and N.S. Dhalla (eds.), *Pathophysiological Aspects of Proteases*,
DOI 10.1007/978-981-10-6141-7_17

Table 1 General information about serine proteases that will be discussed throughout this chapter

Standard name	Pcp1p	Ssy5p	Prb1p	Kex2p	Ynm3p	Lpx1p
Systematic name	YGR101w	YJL156c	YEL060c	YNL238w	YNL123w	YOR084w
Enzyme class N°	E.C. 3.4.21.105	E.C. 3.4.21.–	E.C. 3.4.21.48	E.C. 3.4.21.61	E.C. 3.4.21.–	E.C. 3.1.1.– E.C. 3.4.21.–?
Aliases	Rhomboid protease Rbd1, MDM37	RAA3, SPS-sensor component serine protease SSY5	proteinase B (PrB) CVT1	Kexin QDS1, VMA4, yscF, SRB1	Ynm3 or Nma111 (nuclear mediator of apoptosis)	Peroxisomal matrix-localized lipase
Description						
Length (a.a.)	346	699	635	814	997	387
Mol. weight (Da)	38839.9	77518.4	69606.1	89957.0	110860.3	43730.5
Isoelectric point	10.78	6.42	6.36	4.59	5.80	8.29
Location	Mitochondrial inner membrane	Plasma membrane	Vacuoles	Secretory pathway	Nucleus	Peroxisomal/cytosol
Tipe of serine protease	Rhomboid	Chymotrypsin-like	Subtilisin-like	Subtilisin-like	Chymotrypsin-like	?
Substrate	Mgm1p, Ccp1p	Stp1p, Stp2p	Histone H3	α factor, Kex2p, etc. (S. cerevisiae), Sap2 (C. albicans), Lip2 (Y. lipolítica)	Bir1p	p-nitrophenyl butyrate DPG, Glycerophosphocholine, Tubulin

(continued)

Table 1 (continued)

Standard name	Pcp1p	Ssy5p	Prb1p	Kex2p	Ynm3p	Lpx1p
Function	Pcp1 converts L-Mgm1 into S-Mgm. S-Mgm is crucial for wild-type mitochondrial morphology and maintenance of mitochondrial DNA Pcp1 participates in the second processing step, yielding the mature Ccp1, protein involved in degradation of reactive oxygen species	Ssy5p is responsible for the endoproteolytic processing of transcription factors Stp1 and Stp2, allowing them to enter the nucleus and activate the transcription of amino acid permease genes	PrB activity increases significantly and participates in degradation events. Prb1 cleaves histone H3 tail, under nutritional stress and during the sporulation process	Participates in several cellular processes including mating pheromone processing, hyphal development, cell wall remodeling, and activation of preprohormons, neuropeptides, and extracellular enzymes	Ynm3 exhibits its proapoptotic activity because of Bir1p degradation (Bir1p acts as an inhibitor of apoptosis) Regulator of fatty acid metabolism Chaperone activity in vitro	Shows acyl esterase, lipase, and phospholipase A activity toward PNB, DPG and BPC Lpx1p hydrolyzes acetilated tubulin, inducing AcTub/H^+-ATPase complex dissociation, which results in H^+-ATPase activation
Reference	Urban et al. [75], Herlan et al. [80]	Forsberg et al. [85], Poulsen et al. [87], Pfirrmann et al. [91]	Moehle et al. [62], Pérez et al. [72], Xue et al. [65]	Rockwell et al. [18, 33]	Padmanabhan, et al. [45], Belanger et al. [36], Walter et al. [47], Fahrenkrog et al. [44]	Thoms et al. [50, 58], Campetelli et al. [51]

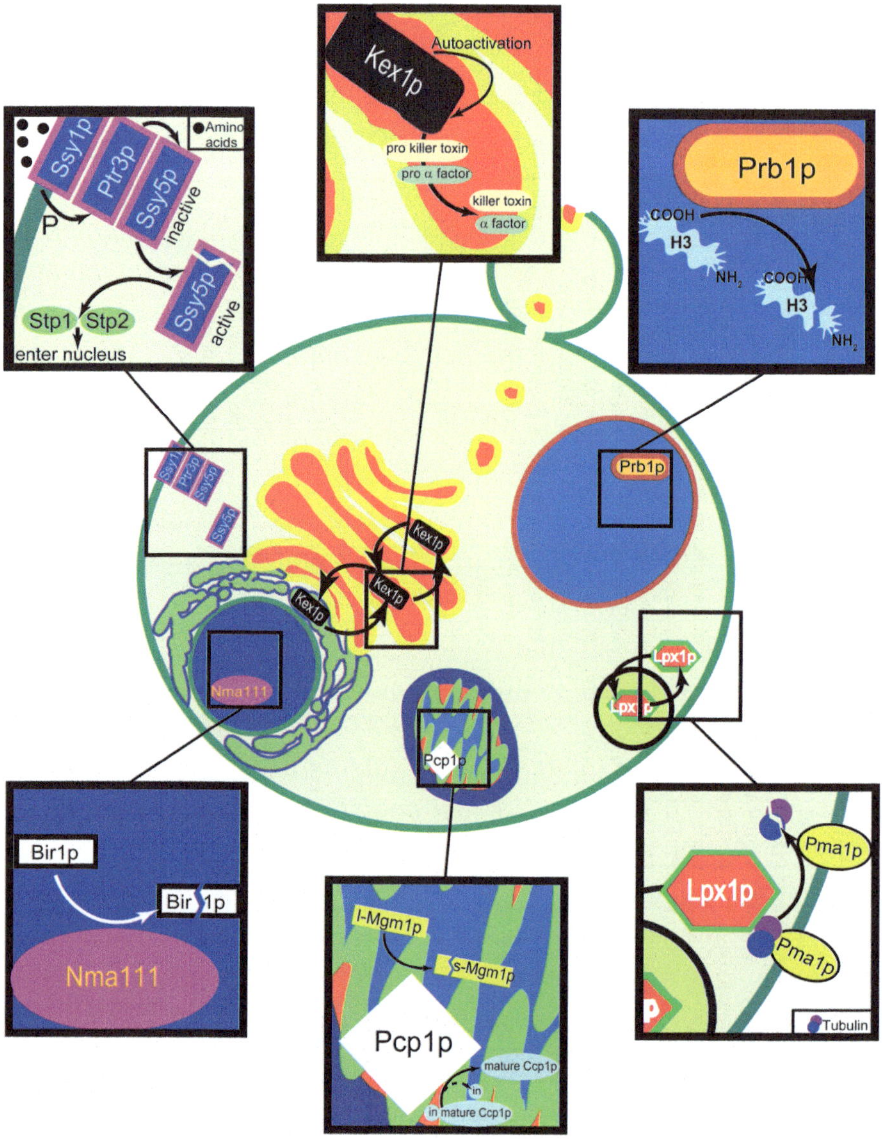

Fig. 1 Several metabolics and cell signalling pathways are regulated by serine proteases in yeasts

peptide bond to form an acyl-enzyme intermediate [1]. Nucleophilicity of this serine depends generally on a catalytic sequence of Asp, His, and Ser residues, commonly termed as the charge relay system [2]. The success of the catalytic break depends on this triad, which is found in four distinct evolutionary pathways. Serine proteases are normally endoproteases, and their substrates are generally cleaved close to the middle of their primary sequence. However, some families of proteases that remove one or more amino acids from the ends of target polypeptide chains have been described (Table 1 and Fig. 1).

In the yeast *S. cerevisiae*, a group of serine proteases composed by Kexin, Ynm3p, Prb1p, Ssy5p, Lpx1p, and Pcp1p, play key roles in the cell cycle progress. The aim of this chapter is to review the aspects related to the molecular structure, subcellular localization, mechanism of action, and physiological role of these serine proteases.

2 Kexin

Kex2p was first described in *S. cerevisiae*. Wickner and colleagues identified this as a genetic locus for killer toxin expression [3], but it was immediately related to the production of α-factor, the mating pheromone secreted by MATα haploid cells [4]. Further structural characterization of these two effectors showed that both are excised from larger precursors by specific endoproteolysis [5]. Later, it was showed that Kex2 encodes a Ca^{2+}-dependent serine protease, with a single transmembrane domain, followed by a cytosolic tail which is responsible for Kex2p partitioning at the late compartments secretory pathway of yeasts.

2.1 Homologs

While Kex2p has its respective homologs in other fungi (krp1 in *Schizosaccharomyces pombe* [6], KEX1 in *Kluyveromyces lactis* [7], XPR6 in *Yarrowia lipolytica* [8], and Kex2 in *Candida albicans*, *Candida glabrata* and *Aspergillus niger* [9–11]), and other organisms like *Drosophila melanogaster*, the mammalian homologs are numerous, including the prohormone convertases PC1/3 and PC2, which clivates proinsulin, prohormones, and neuropeptides. These enzymes lack the transmembrane domain and are mainly expressed in neuroendocrine cells and localized at the regulated secretory pathway. Another Kex2p-related mammalian homolog is furin, which is found more ubiquitously, and has a transmembrane domain that, as in the case of Kex2p, allows it to cycle between late compartments of the constitutive secretory pathway. Other homologs in mammals include PC4, PC5/6, PC7, and PACE4, with some of them displaying several isoforms.

In mammalian, the family members are classified into two distinguishable subfamilies, according to whether they harbor a transmembrane domain or not, and therefore, its delivery to the constitutive or regulated secretory pathway.

2.2 Subcellular Localization

These two arms of this family are needed in significantly different localizations. Kex2p, as well as furin, display a fast cycling between different compartments in the secretory pathway. There, they are temporally exposed to their target substrates, which are mixed with an excess of incorrect substrates, on a timescale of minutes. In the other hand, PC1/3 and PC2 localize into the dense-core secretory granule, where most precursors predominate, during a longer period of time.

2.3 Molecular Structure

Characteristics of the primary structures of Kexin-related enzymes are visibly conserved. In general, most of them harbor an N-terminal signal sequence responsible for entering the secretory pathway. This N-terminal sequence is followed by a prodomain, which once the protein reach maturity is autocatalytically removed [12, 13], a catalytic domain (with the catalytic triad playing a pivotal role) [14], and an exclusive domain called the P-domain in PC-type serine proteases [15] —which is necessary for substrate recognition and activity. PCs harbor C-terminal domains that are important for correct subcellular localization since they carry sorting signals. In these enzymes, the single transmembrane domain is followed by a cytosolic tail containing several signaling sequences that allow them to cycle between the trans-Golgi network and endosomal compartments [16]. Structural studies have shown that regions C-terminal to the P-domain could work as flexible harms in vivo. The catalytic site of Kex2p and furin display similar fold and topology as those of the other subclass, subtilisin, and proteinase K; however, some of the loops at the core surface differ in length, and some α-helix and β-sheet are slightly shifted. These differences build the platform that interacts with the P-domain and alter the substrate-binding pocket. Other conserved residue is the well-known “oxyanion hole” Asn, which, at least in subtilisin, has been proposed as a stabilizer of the transition state.

2.4 Action Mechanism

In general, kexins cleaves amide or ester substrates, and the action mechanism is described by acyl transfer reactions carried out by the classical serine protease mechanism, which uses the catalytic triad of serine, histidine, and aspartate. In the case of proteases of the subtilase superfamily, a fourth conserved catalytic residue, the oxyanion hole Asn is also an active residue in the action mechanism. But let explain more exactly the role of these group of residues on the action mechanism. The Ser provide the hydroxyl nucleophile for attack of the scissile bond; in the other hand, His acts as the catalytic base for de-protonation of the nucleophilic Ser, and finally, Asp stabilizes the assembling of positive charge on the protonated His [17]. During the nucleophilic attack on the scissile bond, the negative charge of the carbonyl oxygen is stabilized by the “oxyanion hole,” which further contributes a pair of hydrogen bond donors to the carbonyl oxygen atom. Once a substrate binds a serine protease, the Ser attacks the scissile bond, giving rise to the first step of the reaction, the acylation. Then, the separated C-terminal fragment of the substrate is released, while the nucleophilic Ser remains covalently linked as an ester to the remaining N-terminal fragment, being the second step of the reaction. Finally, a solvent water molecule hydrolyzates this acyl-enzyme intermediate (deacylation), thereby releasing the protease-N-terminal fragment complex.

Unlike degradative proteases, processing proteases and therefore kexins must be exceptionally accurate and very efficient, because they are often only exposed to their substrates for extremely short time periods.

How these proteases exploit the serine protease mechanism to achieve the required efficiency and specificity can be partially explained, at least partially, on the specificity of P sites. In the case of P4, either positive charge or a large aliphatic side chain is sufficient to satisfy the enzyme–substrate interaction [18]. Kex2 recognizes basic residues like Arg, Lys, and ornithine at P2 [19] with no differences in catalytic parameters. Consequently, the contribution of the aliphatic region of the Lys or Arg to catalysis is poor at this position. However, when these interactions are omitted by replacement of Arg or Lys by acidic or aromatic residues, larger defects are observed. At P1, Kex2 displays high specificity for Arg. Substitution by Lys or ornithine at this site resulted in a serious *kcat/KM* defect of at least 70-fold despite the positive charge is conserved with both residues, demonstrating the overriding importance of electrostatic interaction with Arg at the P1 position as a determinant of Kex2 specificity.

In the structure of Kex2p [20], three Ca^{2+} ions can be found. However, in subtilisin, only two Ca^{2+} ions are found. While two of these ions occupy sites equivalent both in kexins and subtilisins, the third Ca^{2+} in the Kex2p structure is absent in the crystal structures of any subtilisin or proteinase K, albeit the third Ca^{2+} binding site is conserved. In kexins, this bound Ca^{2+} is placed at the base of the S1 pocket, where interacts with two conserved Asp, one conserved Glu and some water molecules. A key role in substrate recognition has been conferred to this ion because one of the conserved Asp residues contacted also directly contacts the side chain of the substrate P1 Arg.

2.5 Physiological Role

In yeast, kexins display a broad range of activities including the processing of cell wall building and repairing proteins [16], key proteins for the aerial hyphae development [21], lipases [22], α-type mating prepropheromones and killer toxins [23], polysaccharide-degrading enzymes [24, 25], zymogens of extracellular proteinases, and themselves as well [8, 11, 26].

While in *S. cerevisiae*, kex2-null mutants are viable, some conditional morphological abnormalities are observed [27]. Some of the found defects include the following: poor vacuolar pH control [28], cold-sensitive growth [27], impaired meiosis, and a probable impairment of the RNA polymerase II complex formation [29]. In the opportunistic fungal pathogen *C. albicans*, Kex2p participates in pathways involved in the expression of virulence factors such as the secretion of proteinases and hyphal formation. When the growth conditions allow hyphae production, the double mutant kex2/kex2 cells become larger than the wild-type, multiple buds and nuclei are often observed, abnormal hyphae are produced, and secreted aspartyl proteinase (SAP) are misprocessed, leading to diminished secretion of the enzyme [11]. In a mouse model of systemic infection, the *C. albicans* double mutant *kex2/kex2* is markedly impaired, lacking its ability to invade the kidney, which is one of the main virulence factors in this condition. Also, the null mutants lose the ability to produce hyphae either when grown embedded in agar, when ingested by macrophages, or when disseminated in mice [30].

2.6 Inhibitors

As mentioned above, Kex2p requires Ca^{2+} ions for activity, and thus is inhibited by chelators such as EDTA. It can be additionally inhibited by reducing agents such as DTT and p-chloromercuribenzoate [31], by peptidyl chloromethyl ketones and short peptides with the same sequence than the recognition site of the substrate [32], by the serine protease inhibitor DFP and provably by PMSF, and by heavy metals such as Cu^{2+} and Zn^{2+} that could be competing with Ca^{2+} ions. On the other hand, K^+ and other monovalent metal cations have been shown to activate Kex2, probably due to a specific allosteric interaction [33].

An extra system to control the activity of proteases is carried out at the level of compartmentalization. Many proteases localize in degradative compartments such as the yeast vacuole. In the case of protease leakage from these organelles, some cytosolic proteins act as inhibitors, presumably as insurance against massive proteolysis. Furthermore, some sequences that can be found naturally, including the prodomains of the proteases themselves, are heavy inhibitors.

3 Ynm3p

Ynm3p or Nma111 (from **n**uclear **m**ediator of **a**poptosis) is a member of evolutionarily conserved apoptosis-regulatory proteins. In *S. cerevisiae,* it is an HtrA (high-temperature requirement A)-like serine protease of about 111 KDa. In this yeast, Ynm3p break Bir1p, the only known IAP (**i**nhibitor-of-**a**poptosis **p**rotein) in yeast. Ynm3p induces apoptosis in a serine protease-dependent manner and exhibits molecular chaperone function under high temperature stress, and its activity occurs exclusively in the nucleus.

3.1 Homologs

Homologs of HtrA/DegP are ubiquitously disseminated in all kingdoms. Normally, these proteins are constituted by two conserved core domains, a chymotrypsin-like protease domain, and at least one PDZ domain found always in the C-terminal side.

While most of the protease-chaperone systems require ATP, the HtrA represents the first well-known protein quality control factor that does not use this energetic molecule [34, 35].

3.2 Subcellular Localization

Under steady-state conditions, Ynm3p is a nuclear protein. It was shown, by a heterokaryon assay, that Ynm3p cannot shuttle between the nucleus and the cytoplasm, remaining in the nucleus, when apoptosis is induced by H_2O_2.

The nuclear localization of Ynm3p is given by a nuclear localization signal (NLS) localized in the N-terminal end of the mature protein. The protein Kap95p, which recognizes this NLS sequence, has been proposed as the nuclear import receptor for Ynm3p [36].

In mammalian cells, HtrA2 is mostly localized in the mitochondria and translocated to the cytosol during apoptosis. Once in the cytosol, it interacts with IAP, which allows the activation of caspases [37–39]. In the case of yeast, although Ynm3p is mainly nuclear, a subpopulation remains associated with mitochondria suggesting a putative role in mitochondrial homeostasis during aging, since Ynm3p null mutants displayed poor oxidative growth during prolonged incubation.

3.3 Molecular Structure

As mentioned above, an exclusive feature of the HtrA protein family is presence of two conserved domains. A catalytic serine protease domain followed by one or two PDZ domains. PDZ domains are very usual protein interaction motives normally involved in signaling pathways and structurally consist of small globular structures [34, 40–42]. In *S. cerevisiae*, Ynm3p harbors an internal duplication of the HtrA-like sequence. While an active domain retaining the complete catalytic triad is placed in the N-terminal end, the C-terminal end harbors the other incomplete serine protease site, which is inactive [34, 40, 43].

3.4 Action Mechanism

The yeast Ynm3p null mutant can grow normally at elevated temperatures and apoptotic signals as ROS accumulation and chromatin condensation are absent [44]. In the other hand, overexpression of Ynm3p induces apoptosis. When the Ser235 residue is replaced by cysteine (S235C mutant), a notorious reduction of proapoptotic properties is observed. Surprisingly, the other main function of this protein, it is its chaperone activity, would depend on Ser236 [45].

In mammalians, the homolog Omi/HtrA2 induces apoptosis by binding and degrading cellular IAPs [37, 39]. The hallmark characteristic of IAPs is the presence of BIR (baculovirus IAP repeat) domains [46]. In *S. cervisiae*, only one BIR protein has been determined, Bir1p, which is a substrate for Ynm3p, and therefore the only IAP in this organism. Recombinant purified Bir1p interacts physically with Ynm3p in vitro, and this interaction depends on the N-terminal HtrA repeat [47]. How Bir1p is degraded by Ynm3p is not clear; however, it was structurally demonstrated in the *E. coli* HtrA homolog DegP that substrate binding to the PDZ domain of DegP results in oligomerization of hexameric DegP giving as a results an active enzyme. In HtrA proteases, the specific loop 3 (L3) that works as a conserved molecular switch senses the activation signal and transmits this information to the active site which acts as canonical serine protease [48].

3.5 Physiological Role

The main role of Ynm3p in yeast is related to induction of apoptosis. Bir1p is a substrate of Ynm3p and prevents apoptosis. Its degradation in hands of Ynm3p, instead, results in apoptosis induction. Since both proteins localize in the nucleus, the process takes place in this organelle.

By two hybrid analyses, it could be shown that Ynm3p interacts with the acyl-CoA synthetase Faa1p and Faa4p, and therefore regulating their activities. Ynm3p null mutants display increased import of fatty acids, misregulated expression of Δ^9-acyl-CoA desaturase, and accumulation of free fatty acids and triglycerides. So, an important role of Ynm3p in fatty acid metabolism must be considered [49].

Ynm3 exhibits general chaperone activity in vitro. Ynm3p inhibited the aggregation of citrate synthase (which is not degraded by Ynm3p) in in vitro thermal induced aggregation assay. This inhibition was achieved in the absence of ATP, suggesting a no conventional chaperone activity. This additional function would involve the PDZ domain, since recombinant proteins lacking these domains lose their ability to confer thermos protection [45].

4 Lpx1p

The green fluorescent protein (GFP) tagged form of Lpx1p was found in peroxisomes in *S. cerevisiae*. Its localization was confirmed by sedimentation assays [50]. Its transport into the peroxisome depends on the QKL motif present in its PTS1 (peroxisomal targeting signal) domain.

The first functions attributed to Lpx1p were in vitro acyl hydrolase and phospholipase A activities. However, it was shown later that Lpx1p would participate in events that lead to Pma1p (the yeast plasma membrane proton pump) activation by hydrolyzing its natural inhibitor tubulin [51].

4.1 Homologs

Its global structure has many similarities to E-2AMS hydrolase from *Mesorhizobium loti*, an α/β hydrolase from *Novosphingobium aromaticivorans* and of *E. coli* YbfF. The function of the *Novosphingobium* protein is unknown. *E. coli* YbfF [52, 53] has been described as an esterase with activity toward p-nitrophenyl butyrate [54].

4.2 Subcellular Localization

The PTS1 domain is the responsible of the delivery of most matrix proteins to the peroxisome. Pex5p, the peroxisomal receptor which recognizes the PTS1 localized at the C-terminus of Lpx1p, is involved in the peroxisomal delivery of Lpx1p [55]. However, some remaining Lpx1p is observed in the cytosol, and other yeast proteins that contain the QKL motif present in the PTS1 are not targeted to peroxisomes, like the elongation factor EFB1 [56] or RPT4, an ATPase of the 19S regulatory particle of the 26S proteasome required for spindle pole body duplication which is mainly nuclear [57].

4.3 Molecular Structure

The molecular mass of Lpx1p is about 44 kDa. Its structure is composed of a catalytic triad with an unusual location of the acid residue after strand β6 of the canonical α/β-hydrolase fold. The active site is covered by a four-helix cap domain, and the interface between the α/β-hydrolase core and this domain would be the potential substrate-binding site. This interface would also form the tunnel connecting with the protein interior and widens into a cavity. Other two tunnels that connect the protein surface with the active site have been found, and a role in substrate accessibility was given. Lpx1p is a homodimer that results from the binding between the cap domain of one monomer and the C-terminal helix of the other, giving as a result a complex of about 6000 $Å^2$. The tunnel and the cavity are located at the interface between the α/β hydrolase core fold and the cap. The N-terminal loop of the Lpx1p cap domain (residues 174–190) has been proposed as lid that would regulate substrate access to the active site, because it displays high flexibility [58].

4.4 Action Mechanism

In Lpx1p, residues Ser145, Glu169, and His323 located after strands β5, β6, and β8, respectively, conform the typical catalytic triad. In Lpx1p, Ser295 occupies the position after strand β7 where the acidic residue of the triad is generally located.

Some authors suggest that due to the presence of this additional serine, the term catalytic triad should be changed to "catalytic tetrad" [53, 58].

Regarding its enzymatic activity, Lpx1p hydrolyzed the test substrate p-nitrophenyl butyrate (PNB) with a K_M of 6.3 μM and Vmax of 0.17 μmol s^{-1} suggesting esterase activity. Lpx1p also exerts lipase activity toward DPG of 5.6 pmol h^{-1} μg^{-1} and phospholipase A activity of 7.9 pmol h^{-1} μg^{-1} when the BODIPY dye-labeled glycerophosphocholine (BPC) is used as substrate [50].

There is also biochemical and genetic evidence indicating that Lpx1p would exert serine protease activity on tubulin; however, the protease activity of Lpx1p was not reported in vitro with purified Lpx1p and tubulin [51].

4.5 Physiological Role

Oleic acid strongly induced Lpx1p expression, which is also regulated by stress-associated transcription factors [59, 60]. In Lpx1p null mutants, peroxisomes display abnormal morphology, being mainly vesiculated. This drastic phenotype is similar to the peroxisomal morphology found in some peroxisomal disorders [50].

In *S. cerevisiae*, the plasma membrane H^+-ATPase (Pma1p) is inhibited by acetylated tubulin. Glucose, the natural activator of Pma1p, dissociates tubulin from the pump, restoring activity [61]. It was demonstrated that glucose-induced dissociation of the Pma1p-tubulin complex happens when a membrane tubulin-specific protease is activated. This protease has similar biochemical characteristics than Lpx1p. Membrane tubulin degradation was not observed in Lpx1p null mutants treated with glucose, the Pma1p-tubulin complex remained assembled, and no changes in Pma1p activity were observed. The proposed mechanism of H^+-ATPase activation by glucose involves a decrease in the cytosolic pH, generated by tubulin inhibition of Pma1p, a concomitant fall in intracellular pH that results in Lpx1p activation that hydrolyzes AcTub, accelerating the process of the AcTub/H^+-ATPase complex dissociation and the activation of the enzyme. All these data postulate Lpx1p as an endogenous regulator of intracellular pH in *S. cerevisiae* [51].

5 Prb1p

The Prb1p (also proteinase B or PrB) is a soluble serine protease of the subtilisin family that resides in the *S. cerevisiae* vacuoles. To reach the mature form, Prb1p is proteolytically processed four times. Prb1p activity is mostly detected under nutritional stress and during the sporulation process. As many serine proteases, it is completely inhibited by phenylmethylsulphonylfluoride (PMSF) and diisopropylphosphofluoride (DPF). Prb1p is responsible of the degradation of the histone H3 N-terminal tail in *S. cerevisiae*.

5.1 Homologs

After maturation, Prb1p is a 33 kD glycoprotein whose primary sequence is highly homolog to subtilisins, proteinase K, and thermitase. This serine protease harbors a free cysteine residue near the active site, and this residue is conserved in proteinase K and thermitase, but not in subtilisins.

The mouse embryonic stem cells lysosomal protein Cathepsin L also cuts the histone H3 N-terminus, and this activity has been associated with gene expression regulation during mouse embryogenesis. Other studies reported Glutamate dehydrogenase (GDH) as another histone H3-specific protease in chicken liver, and its role has been connected to gene expression during aging.

The alkaline proteases of *Aspergillus* species, such as protease B of *C. albicans* and thermomycolin from *Malbranchea pulchella,* display also functional homology with Prb1p. Protease B of *C. albicans* catalyzes a subtilisin-like cleavage of the oxidized β chain of insulin [62].

5.2 Subcellular Localization

The vacuole has been proposed as the lysosome for yeast. In *S. cerevisiae,* a variety of hydrolases were found in vacuole, among them two endopeptidases, the proteinase A and B, and five exopeptidases, carboxypeptidases Y and S, aminopeptidases I and Co, as well as dipeptidyl aminopeptidase V [63].

5.3 Molecular Structure

The PRB1 gene encodes a 73 kDa protein precursor (preproPrb1) that needs four proteolytic cleavage steps for maturation. First, the signal peptide (SS) is removed; later, in autocatalytic manner, a fragment of 260 amino acids (P1) localized at the N-Terminus is removed. The third cleavage, catalyzed by Protease A, eliminates a small region named P2 in the C-terminus. Finally, the segment P3 (about 6 KDa, in the C-Terminus) is auto catalytically removed to give the 31 kDa mature enzyme (mPrb1) [64].

5.4 Action Mechanism

In vitro experiments made by Xue et al. [65] with purified Prb1p from yeast demonstrated that the cleavage site in H3 is in the N-terminus between Lys23 and Ala24. Nevertheless, the authors mention the possibility that Prb1p might be required for the activity of yet another unidentified nuclear serine protease which cleaves H3 at Ala21/Ser22 [65].

5.5 Physiological Role

Prb1p participates in the degradation events that occur under nutritional stress and during the sporulation process [62, 65]. The proteolytic cleavage in the histone H3 N-terminus is a modification conserved in *S. cerevisiae*. The eukaryotic nucleosome contains two copies each of the core histones H2A, H2B, H3, and H4 [66]. These histones are highly conserved in evolution and suffer several modifications, both at the N-terminus that extends from their globular cores and at the globular domains themselves.

Many modifications in the histone H3 N-terminal tail are involved in gene regulation and silencing of transcription [67], in the assembly and remodeling of heterochromatin [68, 69], replication [70], nucleosome assembly [71], both in sporulation and stationary phase.

When yeast cells are grown exponentially in YEPD, a commonly used rich medium, the levels of protease B activity are low, because glucose is a repressor of PRB1 gene expression. Prb1p activity is regulated not only at the transcriptional level, but some posttranslational events have been also stablished [72].

5.6 Inhibitors

PrB1, like most serine proteases, is completely inhibited by the subtilisin-like protease inhibitors PMSF, DPF, $HgCl_2$, chymostatin, and antipain. However, under growth conditions where PRB1 expression is induced, higher concentration of these inhibitors is necessary. A potent natural endogenous inhibitor is I2B, which is encoded by the PBI2 gene and is localized in the cytoplasm [73].

6 Pcp1p Rhomboid Protease

Pcp1p rhomboid protease is a mitochondrial serine protease. Its main function is the processing of several mitochondrial proteins and the maintenance of mitochondrial DNA and mitochondrial morphology. It belongs to the superfamily of intramembrane rhomboid-peptidase GlpG [74].

6.1 Homologs

Rhomboid proteins reside in the internal membrane of mitochondria. Its serine proteases activity is involved in epidermal growth factor receptor (EGFR) activation in *Drosophila* [75]. Rhomboids serine proteases are conserved throughout evolution [76] and are present even in eukaryotes with no EGFRs, suggesting additional roles. At least five active rhomboids have been found throughout the mammalian genome, including the presenilin-associated rhomboid-like (PARL) which also localize at the inner mitochondrial membrane [77]. PCP1 gene homologs have been found in *Kluyveromyces lactis*, and *Eremothecium gossypii.*

6.2 Subcellular Localization

Rbd1p–GFP colocalized with the yeast mitochondrial protein porin, indicating its mitochondrial localization. More precisely, Rbd1p localizes in the mitochondrial inner membrane as an integral membrane protein with six predicted transmembrane domains (TMDs) [78].

6.3 Molecular Structure

The PCP1 gene from *S. cerevisiae* encodes a protein of 38.8 kDa. Its primary structure contains a mitochondrial targeting sequence, six or seven transmembrane helices, a weak ABC3 transport family pattern, and a rhomboid-GlpG motif as predicted by the program MitoProt II (MIPS) [74].

During log-phase growth, wild-type yeast cells display tubular mitochondrial structures around the cell cortex; however, the mitochondria of Rbd1 null mutants appear as small fragments and aggregated masses throughout the cell and nucleoid structures, which represent mitochondrial DNA are not observed upon DAPI staining [79].

6.4 Action Mechanism

Pcp1p is an endoprotease whose catalytic mechanism involves the common catalytic triad found in most serine proteases, consisting of a serine nucleophile that is activated by a proton relay from an acidic residue and a basic residue which is usually histidine.

6.5 Physiological Role

S. cerevisiae has two rhomboids: Rbd1p and Rbd2p. Rbd1p main substrates are cytochrome c peroxidase (Ccp1p); and Mgm1p (a dynamin-like GTPase).

I. Removal of a Mitochondrial Signal Sequence

The yeast protein cytochrome c peroxidase (Ccp1p) is involved in ROS (reactive oxygen species) degradation. This protein is imported from the nucleus into the mitochondrial intermembrane space. It is synthesized as an immature precursor with an N-terminal pre-sequence, which is proteolytically removed in two steps during the delivery of the protein. The mAAA protease (which functions in the degradation of incorrectly folded or unassembled mitochondrial proteins) was identified as essential for the first processing step. The second step is carried out by the PCP1 gene product, giving the mature Ccp1 protein [74].

II. Processing of Proteins for Mitochondrial Morphology and Mitochondrial DNA Preservation

The structure of mitochondria is very dynamic, and this dynamism is mainly controlled by fusion and fission processes. In addition to regulating these processes, Mgm1p also remodels the mitochondrial inner membrane. Mgm1p exists in two forms of different lengths; the large isoform (l-Mgm1) harbors an N-terminal transmembrane domain that is absent in the short isoform (s-Mgm1). Both isoforms are necessary for complete functioning. The conversion of l-Mgm1 into s-Mgm1 depends on Pcp1p. Therefore, the processing of l-Mgm1 by Pcp1 is crucial for wild-type mitochondrial morphology and maintenance of mitochondrial DNA [80].

6.6 Inhibitors

Like other serine proteases, Prb1p is inhibited by different inhibitors, among which are synthetic chemical inhibitors and natural inhibitors. There is a family of natural inhibitors called "serpins" (abbreviation of serine proteases inhibitors), which form a covalent bond with serine proteases, thereby inhibiting their function, and also artificial small molecules that are irreversible inhibitors called AEBSF and PMSF [81].

7 Ssy5p

SPS complex is formed by Ssy1p, Ptr3p, and Ssy5p and is localized at the yeast plasma membrane. This system involved in amino acid metabolism detects the presence of amino acids in the extracellular medium. Upon amino acid sensing, this system triggers metabolic signals that regulate the functional expression of a battery of amino acid-related enzymes that includes amino acid transporters and amino acid metabolizing enzymes [82].

7.1 Homologs

Ssy5p shares structural homology with a lytic protease from *Lysobacter enzymogenes*. The Ssy5p protease domain sequence was modeled using PHYRE [83], and this region was predicted to have predominantly α structure, while the N-terminal proregion displayed a high helical content [84]. The regions where more homology was detected include the domains that flank the Ser and Asp of the catalytic triad.

7.2 Subcellular Localization

Ssy1p, Ptr3p, and Ssy5p are sequentially ordered at the plasma membrane. There, they adopt different conformations and modifications that will depend on the amino acids in the extracellular side and the presence of the other two partners [85].

7.3 Molecular Structure

We will focus on Ssy5 which acts as a chymotrypsin-like serine proteases. It is expressed as a zymogen that by autocatalysis cleaves the peptide bond between residues 381 and 382, yielding a C-terminal catalytic (Cat) domain that contains the conserved catalytic triad. The deletion of Ser640 results in loss of function, confirming its triad membership [86, 87].

7.4 Action Mechanism

The first event is the sensing of amino acids in the extracellular medium by the primary amino acid receptor Ssy1p [88], which triggers the signal resulting in hyperphosphorylation of Ptr3p [89]. As a result of this hyperphosphorylation, Ssy5 becomes activated and cleaves Stp1 and Stp2, which now are allowed to enter the nucleus which activates the transcription of amino acid permease genes [90].

As mentioned before, Ssy5p is synthesized with a large N-terminal prodomain. After auto processing, the N-terminal prodomain and Cat domain remain associated. The prodomain displays a potent inhibitory effect against the Cat domain yielding a yet inactive protease that will be only activated via Ssy1p receptor, ensuring its metabolic function [91]. The prodomain removal is mediated by a conserved phosphodegron. Upon amino acid induction, the phosphodegron is hyperphosphorylated by casein kinase I. As a consequence, the prodomain becomes polyubiquitylated by the action of Skp1/Cullin/Grr1 E3 ubiquitin ligase complex (SCFGrr1). Finally, the polyubiquitylated prodomain is irreparably degraded by the 26S proteasome [82].

7.5 Physiological Role

In yeast, the transcription factors that regulate the expression of genes related to amino acid transport and metabolism are Stp1p and Stp2p. These transcription factors are synthetized as inactive precursors and localize at the cytoplasm. When the Ssy1p-Ptr3p-Ssy5p complex sense amino acids in the extracellular milieu, the activated Ssy5p processes Stp1p and Stp2p and activates them. The active transcription factors can now move to the nucleus where they induce the expression of amino acid-related genes.

8 Conclusions

In summary, serine proteases play several important roles for normal yeast cell function.

S. cerevisiae mating is a key process that takes place when two haploid cells of different mating type (a and α) are close enough to form a diploid a/α cell through shmoo formation. Diploid cells are able to enter sporulation when complete nutrient deprivation occurs, allowing cells to wait "alive" for better growing conditions. Kex2p is crucial for mating, since it maturates the α-factor, one of the mating pheromones. Kexins also participate in the maturation of enzymes involved in hyphae production. *C. albicans* ability to form hyphae has been proposed as a virulence factor, as these structures are often observed invading tissues.

In mammalians, HtrA proteins function as potential modulators of apoptosis and chemotherapy-induced cytotoxicity. The yeast homolog Ynm3p shares some of these functions as its proapoptotic role has been probed. The expression level of some HtrA serine proteases have been documented to be low in some specific cancer types, making them an attractive candidate for future study [92]. It has been postulated that HtrA serine proteases potentially interact with caspase-mediated cell death pathways, postulating them as potential targets of therapeutic treatment. Future studies unraveling the connection between caspases and HtrA serine proteases will be indispensable to fully explore the therapeutic potential of this novel serine protease pathway. Similarly, further screenings to identify the substrates of these proteases will also be essential to better understand their function and for the successful design of therapeutic drugs.

Similar to kexins, Prb1p participates in cellular events related to nutrient deprivation and following sporulation, degrading key proteins in the vacuolar environment. In the case of Ssy5p, the sensing of extracellular aminoacids by the plasma membrane Ssy1p-Ptr3p-Ssy5p sensor results in the cleavage of the transcription factors Stp1p and Stp2p, which are activated and translocated to the nucleus, where induction of genes related to the metabolism of aminoacids is launched.

Mitochondrial morphology and the processing of reactive oxygen species (ROS) that take place in this organelle depend on the activity of rhomboid proteases. It is noteworthy that these two important cellular processes rely on the activity of rhomboid proteases.

In summary, all the serine proteases described in this chapter participate in cellular events of crucial relevance, ranging from sporulation and apoptosis induction to cell breathing and nutrient processing and sensing. All these physiological processes may be accurately regulated in order to maintain normal cell function. Acquiring more in-depth knowledge about these enzymes will be of invaluable importance, since addressing questions that are still unanswered will open up new possibilities to understand pathophysiological processes where these enzymes are involved and to design new drugs and therapies.

References

1. Hedstrom L (2002) Serine protease mechanism and specificity. Chem Rev 102(12):4501–4524. cr000033x [pii]
2. Blow DM, Birktoft JJ, Hartley BS (1969) Role of a buried acid group in the mechanism of action of chymotrypsin. Nature 221(5178):337–340
3. Wickner RB, Leibowitz MJ (1976) Two chromosomal genes required for killing expression in killer strains of *Saccharomyces cerevisiae*. Genetics 82(3):429–442
4. Leibowitz MJ, Wickner RB (1976) A chromosomal gene required for killer plasmid expression, mating, and spore maturation in *Saccharomyces cerevisiae*. Proc Natl Acad Sci USA 73(6):2061–2065
5. Fuller RS, Sterne RE, Thorner J (1988) Enzymes required for yeast prohormone processing. Annu Rev Physiol 50:345–362. doi:10.1146/annurev.ph.50.030188.002021
6. Davey J, Davis K, Imai Y, Yamamoto M, Matthews G (1994) Isolation and characterization of krp, a dibasic endopeptidase required for cell viability in the fission yeast *Schizosaccharomyces pombe*. EMBO J 13(24):5910–5921
7. Tanguy-Rougeau C, Wesolowski-Louvel M, Fukuhara H (1988) The *Kluyveromyces lactis* KEX1 gene encodes a subtilisin-type serine proteinase. FEBS Lett 234(2):464–470. doi:10.1016/0014-5793(88)80139-X
8. Enderlin CS, Ogrydziak DM (1994) Cloning, nucleotide sequence and functions of XPR6, which codes for a dibasic processing endoprotease from the yeast *Yarrowia lipolytica*. Yeast 10(1):67–79. doi:10.1002/yea.320100107
9. Bader O, Schaller M, Klein S, Kukula J, Haack K, Muhlschlegel F, Korting HC, Schafer W, Hube B (2001) The KEX2 gene of *Candida glabrata* is required for cell surface integrity. Mol Microbiol 41(6):1431–1444. doi:2614 [pii]
10. Jalving R, van de Vondervoort PJ, Visser J, Schaap PJ (2000) Characterization of the kexin-like maturase of *Aspergillus niger*. Appl Environ Microbiol 66(1):363–368
11. Newport G, Agabian N (1997) KEX2 influences *Candida albicans* proteinase secretion and hyphal formation. J Biol Chem 272(46):28954–28961
12. Anderson ED, Molloy SS, Jean F, Fei H, Shimamura S, Thomas G (2002) The ordered and compartment-specfific autoproteolytic removal of the furin intramolecular chaperone is required for enzyme activation. J Biol Chem 277(15):12879–12890. doi:10.1074/jbc.M108740200; M108740200 [pii]
13. Wilcox CA, Fuller RS (1991) Posttranslational processing of the prohormone-cleaving Kex2 protease in the *Saccharomyces cerevisiae* secretory pathway. J Cell Biol 115(2):297–307
14. Siezen RJ, Leunissen JA (1997) Subtilases: the superfamily of subtilisin-like serine proteases. Protein Sci 6(3):501–523. doi:10.1002/pro.5560060301
15. Gluschankof P, Fuller RS (1994) A C-terminal domain conserved in precursor processing proteases is required for intramolecular N-terminal maturation of pro-Kex2 protease. EMBO J 13(10):2280–2288
16. Molloy SS, Anderson ED, Jean F, Thomas G (1999) Bi-cycling the furin pathway: from TGN localization to pathogen activation and embryogenesis. Trends Cell Biol 9(1):28–35. S0962-8924(98)01382-8
17. Perona JJ, Craik CS (1995) Structural basis of substrate specificity in the serine proteases. Protein Sci 4(3):337–360. doi:10.1002/pro.5560040301
18. Rockwell NC, Wang GT, Krafft GA, Fuller RS (1997) Internally consistent libraries of fluorogenic substrates demonstrate that Kex2 protease specificity is generated by multiple mechanisms. Biochemistry 36(7):1912–1917. doi:10.1021/bi961779l; bi961779l [pii]
19. Brenner C, Fuller RS (1992) Structural and enzymatic characterization of a purified prohormone-processing enzyme: secreted, soluble Kex2 protease. Proc Natl Acad Sci USA 89(3):922–926
20. Holyoak T, Wilson MA, Fenn TD, Kettner CA, Petsko GA, Fuller RS, Ringe D (2003) 2.4 A resolution crystal structure of the prototypical hormone-processing protease Kex2 in complex

with an Ala-Lys-Arg boronic acid inhibitor. Biochemistry 42(22):6709–6718. doi:10.1021/bi034434t
21. Wosten HA, Bohlmann R, Eckerskorn C, Lottspeich F, Bolker M, Kahmann R (1996) A novel class of small amphipathic peptides affect aerial hyphal growth and surface hydrophobicity in *Ustilago maydis*. EMBO J 15(16):4274–4281
22. Pignede G, Wang H, Fudalej F, Gaillardin C, Seman M, Nicaud JM (2000) Characterization of an extracellular lipase encoded by LIP2 in *Yarrowia lipolytica*. J Bacteriol 182(10):2802–2810
23. Julius D, Blair L, Brake A, Sprague G, Thorner J (1983) Yeast alpha factor is processed from a larger precursor polypeptide: the essential role of a membrane-bound dipeptidyl aminopeptidase. Cell 32(3):839–852. doi:10.1016/0092-8674(83)90070-3
24. Goller SP, Schoisswohl D, Baron M, Parriche M, Kubicek CP (1998) Role of endoproteolytic dibasic proprotein processing in maturation of secretory proteins in *Trichoderma reesei*. Appl Environ Microbiol 64(9):3202–3208
25. Iguchi K, Hirano H, Kishida M, Kawasaki H, Sakai T (1997) Cloning of a protopectinase gene of Trichosporon penicillatum and its expression in *Saccharomyces cerevisiae*. Microbiology 143(Pt 5):1657–1664. doi:10.1099/00221287-143-5-1657
26. Germain D, Dumas F, Vernet T, Bourbonnais Y, Thomas DY, Boileau G (1992) The pro-region of the Kex2 endoprotease of Saccharomyces cerevisiae is removed by self-processing. FEBS Lett 299(3):283–286. doi:10.1016/0014-5793(92)80132-Z
27. Komano H, Fuller RS (1995) Shared functions in vivo of a glycosyl-phosphatidylinositol-linked aspartyl protease, Mkc7, and the proprotein processing protease Kex2 in yeast. Proc Natl Acad Sci USA 92(23):10752–10756
28. Oluwatosin YE, Kane PM (1998) Mutations in the yeast KEX2 gene cause a Vma(-)-like phenotype: a possible role for the Kex2 endoprotease in vacuolar acidification. Mol Cell Biol 18(3):1534–1543
29. Martin C, Young RA (1989) KEX2 mutations suppress RNA polymerase II mutants and alter the temperature range of yeast cell growth. Mol Cell Biol 9(6):2341–2349
30. Newport G, Kuo A, Flattery A, Gill C, Blake JJ, Kurtz MB, Abruzzo GK, Agabian N (2003) Inactivation of Kex2p diminishes the virulence of Candida albicans. J Biol Chem 278 (3):1713–1720. doi:10.1074/jbc.M209713200; M209713200 [pii]
31. Fuller RS, Brake A, Thorner J (1989) Yeast prohormone processing enzyme (KEX2 gene product) is a Ca^{2+}-dependent serine protease. Proc Natl Acad Sci USA 86(5):1434–1438
32. Angliker H, Wikstrom P, Shaw E, Brenner C, Fuller RS (1993) The synthesis of inhibitors for processing proteinases and their action on the Kex2 proteinase of yeast. Biochem J 293(Pt 1):75–81
33. Rockwell NC, Fuller RS (2002) Specific modulation of Kex2/furin family proteases by potassium. J Biol Chem 277(20):17531–17537. doi:10.1074/jbc.M111909200; M111909200 [pii]
34. Pallen MJ, Wren BW (1997) The HtrA family of serine proteases. Mol Microbiol 26(2):209–221
35. Spiess C, Beil A, Ehrmann M (1999) A temperature-dependent switch from chaperone to protease in a widely conserved heat shock protein. Cell 97(3):339–347. doi:10.1016/S0092-8674(00)80743-6
36. Belanger KD, Walter D, Henderson TA, Yelton AL, O'Brien TG, Belanger KG, Geier SJ, Fahrenkrog B (2009) Nuclear localisation is crucial for the proapoptotic activity of the HtrA-like serine protease Nma111p. J Cell Sci 122(21):3931–3941. doi:10.1242/jcs.056887; jcs.056887 [pii]
37. Martins LM, Iaccarino I, Tenev T, Gschmeissner S, Totty NF, Lemoine NR, Savopoulos J, Gray CW, Creasy CL, Dingwall C, Downward J (2002) The serine protease Omi/HtrA2 regulates apoptosis by binding XIAP through a reaper-like motif. J Biol Chem 277(1):439–444. doi:10.1074/jbc.M109784200; M109784200 [pii]

38. Suzuki Y, Imai Y, Nakayama H, Takahashi K, Takio K, Takahashi R (2001) A serine protease, HtrA2, is released from the mitochondria and interacts with XIAP, inducing cell death. Mol Cell 8(3):613–621. doi:10.1016/S1097-2765(01)00341-0
39. Verhagen AM, Silke J, Ekert PG, Pakusch M, Kaufmann H, Connolly LM, Day CL, Tikoo A, Burke R, Wrobel C, Moritz RL, Simpson RJ, Vaux DL (2002) HtrA2 promotes cell death through its serine protease activity and its ability to antagonize inhibitor of apoptosis proteins. J Biol Chem 277(1):445–454. doi:10.1074/jbc.M109891200; M109891200 [pii]
40. Clausen T, Kaiser M, Huber R, Ehrmann M (2011) HTRA proteases: regulated proteolysis in protein quality control. Nat Rev Mol Cell Biol 12(3):152–162. doi:10.1038/nrm3065; nrm3065 [pii]
41. Harris BZ, Lim WA (2001) Mechanism and role of PDZ domains in signaling complex assembly. J Cell Sci 114(Pt 18):3219–3231
42. Vande Walle L, Lamkanfi M, Vandenabeele P (2008) The mitochondrial serine protease HtrA2/Omi: An overview. Cell Death Differ 15(3):453–460. doi:10.1038/sj.cdd.4402291; 4402291 [pii]
43. Ponting CP (1997) Evidence for PDZ domains in bacteria, yeast, and plants. Protein Sci 6 (2):464–468. doi:10.1002/pro.5560060225
44. Fahrenkrog B, Sauder U, Aebi U (2004) The *S. cerevisiae* HtrA-like protein Nma111p is a nuclear serine protease that mediates yeast apoptosis. J Cell Sci 117(1):115–126. doi:10.1242/jcs.00848; 117/1/115 [pii]
45. Padmanabhan N, Fichtner L, Dickmanns A, Ficner R, Schulz JB, Braus GH (2009) The yeast HtrA orthologue Ynm3 is a protease with chaperone activity that aids survival under heat stress. Mol Biol Cell 20(1):68–77. doi:10.1091/mbc.E08-02-0178; E08-02-0178 [pii]
46. Vaux DL, Silke J (2005) IAPs, RINGs and ubiquitylation. Nat Rev Mol Cell Biol 6(4):287–297. doi:10.1038/nrm1621; nrm1621 [pii]
47. Walter D, Wissing S, Madeo F, Fahrenkrog B (2006) The inhibitor-of-apoptosis protein Bir1p protects against apoptosis in *S. cerevisiae* and is a substrate for the yeast homologue of Omi/HtrA2. J Cell Sci 119 (9):1843–1851. doi:10.1242/jcs.02902; jcs.02902 [pii]
48. Krojer T, Sawa J, Huber R, Clausen T (2010) HtrA proteases have a conserved activation mechanism that can be triggered by distinct molecular cues. Nat Struct Mol Biol 17(7):844–852. doi:10.1038/nsmb.1840; nsmb.1840 [pii]
49. Tong F, Black PN, Bivins L, Quackenbush S, Ctrnacta V, DiRusso CC (2006) Direct interaction of *Saccharomyces cerevisiae* Faa1p with the Omi/HtrA protease orthologue Ynm3p alters lipid homeostasis. Mol Genet Genomics 275(4):330–343. doi:10.1007/s00438-005-0089-1
50. Thoms S, Debelyy MO, Nau K, Meyer HE, Erdmann R (2008) Lpx1p is a peroxisomal lipase required for normal peroxisome morphology. FEBS J 275(3):504–514. doi:10.1111/j.1742-4658.2007.06217.x; EJB6217 [pii]
51. Campetelli AN, Monesterolo NE, Previtali G, Santander VS, Amaiden MR, Arce CA, Valdez-Taubas J, Casale CH (2013) Activation of H(+)-ATPase by glucose in *Saccharomyces cerevisiae* involves a membrane serine protease. Biochim Biophys Acta 1830(6):3593–3603. doi:10.1016/j.bbagen.2013.03.012; S0304-4165(13)00090-1 [pii]
52. McCulloch KM, Mukherjee T, Begley TP, Ealick SE (2010) Structure determination and characterization of the vitamin B6 degradative enzyme (E)-2-(acetamidomethylene)succinate hydrolase. Biochemistry 49(6):1226–1235. doi:10.1021/bi901812p
53. Park SY, Lee SH, Lee J, Nishi K, Kim YS, Jung CH, Kim JS (2008) High-resolution structure of ybfF from Escherichia coli K12: a unique substrate-binding crevice generated by domain arrangement. J Mol Biol 376(5):1426–1437. doi:10.1016/j.jmb.2007.12.062; S0022-2836(07)01711-1 [pii]
54. Kim JH, Jang KS, Yang YH, Kim YG, Lee JH, Oh MK, Kim BG, Lee CS (2008) Rapid functional identification of putative genes based on the combined in vitro protein synthesis with mass spectrometry: a tool for functional genomics. Anal Biochem 375(1):11–17. doi:10.1016/j.ab.2008.01.007; S0003-2697(08)00011-0 [pii]

55. Thoms S, Erdmann R (2006) Peroxisomal matrix protein receptor ubiquitination and recycling. Biochim Biophys Acta 1763 (12):1620–1628. doi:10.1016/j.bbamcr.2006.08.046; S0167-4889(06)00249-7 [pii]
56. Hiraga K, Suzuki K, Tsuchiya E, Miyakawa T (1993) Cloning and characterization of the elongation factor EF-1 beta homologue of *Saccharomyces cerevisiae*. EF-1 beta is essential for growth. FEBS Lett 316 (2):165–169. doi:10.1016/0014-5793(93)81208-H
57. McDonald HB, Byers B (1997) A proteasome cap subunit required for spindle pole body duplication in yeast. J Cell Biol 137(3):539–553
58. Thoms S, Hofhuis J, Thoing C, Gartner J, Niemann HH (2011) The unusual extended C-terminal helix of the peroxisomal alpha/beta-hydrolase Lpx1 is involved in dimer contacts but dispensable for dimerization. J Struct Biol 175(3):362–371. doi:10.1016/j.jsb.2011.06.008; S1047-8477(11)00189-4 [pii]
59. Lucau-Danila A, Delaveau T, Lelandais G, Devaux F, Jacq C (2003) Competitive promoter occupancy by two yeast paralogous transcription factors controlling the multidrug resistance phenomenon. J Biol Chem 278(52):52641–52650. doi:10.1074/jbc.M309580200; M309580200 [pii]
60. Smith JJ, Marelli M, Christmas RH, Vizeacoumar FJ, Dilworth DJ, Ideker T, Galitski T, Dimitrov K, Rachubinski RA, Aitchison JD (2002) Transcriptome profiling to identify genes involved in peroxisome assembly and function. J Cell Biol 158(2):259–271. doi:10.1083/jcb.200204059; jcb.200204059 [pii]
61. Campetelli AN, Previtali G, Arce CA, Barra HS, Casale CH (2005) Activation of the plasma membrane H-ATPase of *Saccharomyces cerevisiae* by glucose is mediated by dissociation of the H(+)-ATPase-acetylated tubulin complex. FEBS J 272(22):5742–5752. doi:10.1111/j.1742-4658.2005.04959.x; EJB4959 [pii]
62. Moehle CM, Tizard R, Lemmon SK, Smart J, Jones EW (1987) Protease B of the lysosomelike vacuole of the yeast *Saccharomyces cerevisiae* is homologous to the subtilisin family of serine proteases. Mol Cell Biol 7(12):4390–4399
63. Teichert U, Mechler B, Muller H, Wolf DH (1989) Lysosomal (vacuolar) proteinases of yeast are essential catalysts for protein degradation, differentiation, and cell survival. J Biol Chem 264(27):16037–16045
64. Mark KG, Meza-Gutierrez F, Johnson JR, Newton BW, Krogan NJ, Toczyski DP (2015) Prb1 protease activity is required for its recognition by the F-box protein Saf1. Biochemistry 54 (29):4423–4426. doi:10.1021/acs.biochem.5b00504
65. Xue Y, Vashisht AA, Tan Y, Su T, Wohlschlegel JA (2014) PRB1 is required for clipping of the histone H3N terminal tail in *Saccharomyces cerevisiae*. PLoS ONE 9(2):e90496. doi:10.1371/journal.pone.0090496; PONE-D-13-50140 [pii]
66. Luger K, Mader AW, Richmond RK, Sargent DF, Richmond TJ (1997) Crystal structure of the nucleosome core particle at 2.8 A resolution. Nature 389(6648):251–260. doi:10.1038/38444
67. Mann RK, Grunstein M (1992) Histone H3N-terminal mutations allow hyperactivation of the yeast GAL1 gene in vivo. EMBO J 11(9):3297–3306
68. Chatterjee N, Sinha D, Lemma-Dechassa M, Tan S, Shogren-Knaak MA, Bartholomew B (2011) Histone H3 tail acetylation modulates ATP-dependent remodeling through multiple mechanisms. Nucleic Acids Res 39(19):8378–8391. doi:10.1093/nar/gkr535; gkr535 [pii]
69. Sperling AS, Grunstein M (2009) Histone H3N-terminus regulates higher order structure of yeast heterochromatin. Proc Natl Acad Sci USA 106(32):13153–13159. doi:10.1073/pnas.0906866106; 0906866106 [pii]
70. Vogelauer M, Rubbi L, Lucas I, Brewer BJ, Grunstein M (2002) Histone acetylation regulates the time of replication origin firing. Mol Cell 10(5):1223–1233. doi:S1097276502007025
71. Burgess RJ, Zhou H, Han J, Zhang Z (2010) A role for Gcn5 in replication-coupled nucleosome assembly. Mol Cell 37 (4):469–480. doi:10.1016/j.molcel.2010.01.020; S1097-2765(10)00071-7 [pii]

72. Perez J, Gomez A, Roncero C (2010) Upregulation of the PRB1 gene in the *Saccharomyces cerevisiae* rim101Delta mutant produces proteolytic artefacts that differentially affect some proteins. Yeast 27(8):575–581. doi:10.1002/yea.1776
73. Slusarewicz P, Xu Z, Seefeld K, Haas A, Wickner WT (1997) I2B is a small cytosolic protein that participates in vacuole fusion. Proc Natl Acad Sci USA 94(11):5582–5587
74. Esser K, Tursun B, Ingenhoven M, Michaelis G, Pratje E (2002) A novel two-step mechanism for removal of a mitochondrial signal sequence involves the mAAA complex and the putative rhomboid protease Pcp1. J Mol Biol 323(5):835–843. doi:S0022283602010008
75. Urban S, Lee JR, Freeman M (2001) Drosophila rhomboid-1 defines a family of putative intramembrane serine proteases. Cell 107(2):173–182. doi:S0092-8674(01)00525-6
76. Urban S, Schlieper D, Freeman M (2002) Conservation of intramembrane proteolytic activity and substrate specificity in prokaryotic and eukaryotic rhomboids. Curr Biol 12(17):1507–1512. doi:S0960982202010928
77. Sekine S, Kanamaru Y, Koike M, Nishihara A, Okada M, Kinoshita H, Kamiyama M, Maruyama J, Uchiyama Y, Ishihara N, Takeda K, Ichijo H (2012) Rhomboid protease PARL mediates the mitochondrial membrane potential loss-induced cleavage of PGAM5. J Biol Chem 287(41):34635–34645. doi:10.1074/jbc.M112.357509; M112.357509 [pii]
78. Dimmer KS, Fritz S, Fuchs F, Messerschmitt M, Weinbach N, Neupert W, Westermann B (2002) Genetic basis of mitochondrial function and morphology in *Saccharomyces cerevisiae*. Mol Biol Cell 13(3):847–853. doi:10.1091/mbc.01-12-0588
79. McQuibban GA, Saurya S, Freeman M (2003) Mitochondrial membrane remodelling regulated by a conserved rhomboid protease. Nature 423 (6939):537–541. doi:10.1038/nature01633; nature01633 [pii]
80. Herlan M, Vogel F, Bornhovd C, Neupert W, Reichert AS (2003) Processing of Mgm1 by the rhomboid-type protease Pcp1 is required for maintenance of mitochondrial morphology and of mitochondrial DNA. J Biol Chem 278(30):27781–27788. doi:10.1074/jbc.M211311200; M211311200 [pii]
81. Breugelmans B, Simonet G, van Hoef V, Van Soest S, Vanden Broeck J (2009) Pacifastin-related peptides: Structural and functional characteristics of a family of serine peptidase inhibitors. Peptides 30(3):622–632. doi:10.1016/j.peptides.2008.07.026; S0196-9781(08)00324-0 [pii]
82. Omnus DJ, Pfirrmann T, Andreasson C, Ljungdahl PO (2011) A phosphodegron controls nutrient-induced proteasomal activation of the signaling protease Ssy5. Mol Biol Cell 22 (15):2754–2765. doi:10.1091/mbc.E11-04-0282; mbc.E11-04-0282 [pii]
83. Kelley LA, MacCallum RM, Sternberg MJ (2000) Enhanced genome annotation using structural profiles in the program 3D-PSSM. J Mol Biol 299(2):499–520. doi:10.1006/jmbi.2000.3741; jmbi.2000.3741 [pii]
84. Fuhrmann CN, Kelch BA, Ota N, Agard DA (2004) The 0.83 A resolution crystal structure of alpha-lytic protease reveals the detailed structure of the active site and identifies a source of conformational strain. J Mol Biol 338(5):999–1013. doi:10.1016/j.jmb.2004.03.018; S0022283604003080 [pii]
85. Forsberg H, Ljungdahl PO (2001) Genetic and biochemical analysis of the yeast plasma membrane Ssy1p-Ptr3p-Ssy5p sensor of extracellular amino acids. Mol Cell Biol 21(3):814–826. doi:10.1128/MCB.21.3.814-826.2001
86. Abdel-Sater F, El Bakkoury M, Urrestarazu A, Vissers S, Andre B (2004) Amino acid signaling in yeast: casein kinase I and the Ssy5 endoprotease are key determinants of endoproteolytic activation of the membrane-bound Stp1 transcription factor. Mol Cell Biol 24 (22):9771–9785. doi:10.1128/MCB.24.22.9771-9785.2004; 24/22/9771 [pii]
87. Poulsen P, Lo Leggio L, Kielland-Brandt MC (2006) Mapping of an internal protease cleavage site in the Ssy5p component of the amino acid sensor of *Saccharomyces cerevisiae* and functional characterization of the resulting pro- and protease domains by gain-of-function genetics. Eukaryot Cell 5(3):601–608. doi:10.1128/EC.5.3.601-608.2006; 5/3/601 [pii]

88. Wu B, Ottow K, Poulsen P, Gaber RF, Albers E, Kielland-Brandt MC (2006) Competitive intra- and extracellular nutrient sensing by the transporter homologue Ssy1p. J Cell Biol 173 (3):327–331. doi:10.1083/jcb.200602089; jcb.200602089 [pii]
89. Liu Z, Thornton J, Spirek M, Butow RA (2008) Activation of the SPS amino acid-sensing pathway in *Saccharomyces cerevisiae* correlates with the phosphorylation state of a sensor component, Ptr3. Mol Cell Biol 28(2):551–563. doi: 10.1128/MCB.00929-07; MCB.00929-07 [pii]
90. Andreasson C, Ljungdahl PO (2002) Receptor-mediated endoproteolytic activation of two transcription factors in yeast. Genes Dev 16(24):3158–3172. doi:10.1101/gad.239202
91. Pfirrmann T, Heessen S, Omnus DJ, Andreasson C, Ljungdahl PO (2010) The prodomain of Ssy5 protease controls receptor-activated proteolysis of transcription factor Stp1. Mol Cell Biol 30(13):3299–3309. doi:10.1128/MCB.00323-10; MCB.00323-10 [pii]
92. Chien J, Campioni M, Shridhar V, Baldi A (2009) HtrA serine proteases as potential therapeutic targets in cancer. Curr Cancer Drug Targets 9(4):451–468

Caspases: Regulatory Mechanisms and Their Implications in Pathogenesis and Therapeutics

Aasna L. Parui and Kakoli Bose

Abstract

Caspases (**c**ysteine-**asp**artic prote**ases**) belong to a family of endoproteases that regulate the cellular networks governing distinct physiological functions, which mainly include apoptotic cell death and inflammation. Few decades of research have generated a plethora of information on human caspases, their homologues, substrates and role in apoptosis. Programmed cell death (*aka* apoptosis) is an energy-dependent natural physiological process that is adapted by multicellular organisms for maintenance of tissue homeostasis. It involves a complex network of cellular proteases primarily caspases, which regulate selective removal of ageing and unwanted cells, thereby maintaining a precise balance between cell survival and death. Disruption of this stasis often leads to various pathophysiological conditions including inflammation, neurodegenerative disorders and cancer. Thus, being promoter of apoptosis, several caspases are prime targets of numerous research endeavours that aim to encounter and impede the advancement of these diseases. The main objective of this chapter is to provide a comprehensive overview of the structural and mechanistic aspects of caspases as well as the pathways involved in their activation and regulation. Besides, it also elaborates on the role of this protease family in different diseases and the current therapeutic strategies that are being devised to modulate their functions with desired characteristics.

Keywords

Caspase · Programmed cell death · Proteases · Regulation
Cancer · Neurodegenerative diseases · Therapeutics

A.L. Parui · K. Bose (✉)
Integrated Biophysics and Structural Biology Lab,
Advanced Centre for Treatment, Research and Education in Cancer (ACTREC),
Tata Memorial Centre, Navi Mumbai 410210, Maharashtra, India
e-mail: kbose@actrec.gov.in

S. Chakraborti and N.S. Dhalla (eds.), *Pathophysiological Aspects of Proteases*,
DOI 10.1007/978-981-10-6141-7_18

1 Introduction

Apoptosis is an evolutionarily conserved phenomenon of **p**rogrammed **c**ell **d**eath (PCD) that is observed in a wide variety of multicellular biological systems. It is primarily needed for the selective removal of ageing, unwanted and impaired cells. It is also essentially involved in chemical-induced cell death and maintenance of several processes such as normal cell turnover, immune system, embryonic development, metamorphosis and hormone-dependent atrophy [1]. During apoptosis, a series of distinct morphological and biochemical alterations lead to the disintegration of cellular components into smaller apoptotic bodies. Thus, in contrast to necrosis, a process wherein cell death results into an inflammatory response, apoptosis is a more controlled process where the contents are strictly maintained within the cell membrane [2]. Deregulated apoptosis has been implicated in many human diseases, including neurodegenerative disorders such as Alzheimer's and Huntington's disease, ischaemic damage, autoimmune disorders and several forms of cancer [3, 4]. Therefore, tight regulation of PCD is critical for proper maintenance of cellular homeostasis, i.e. fine balance between cell proliferation and death.

Since its first revelation by Carl Vogt way back in the early nineteenth century, a plethora of research endeavours on this vital cellular process improved much of our understanding towards the complexities of the biochemical pathways involved [5]. This includes the interaction and crosstalk of critical molecules with myriads of other cellular components to form specialized machinery responsible for this form of cell suicide. The central component of this machinery is an intricate proteolytic system involving a family of proteases. These proteases upon activation via a specific proapoptotic signal result in the cleavage of several essential cellular proteins such as cytoskeletal proteins, DNA repair enzymes and nuclear proteins. The apoptotic machinery is majorly regulated by a class of cysteine proteases called Caspases (**C**ysteinyl **asp**artate-specific prote**ase**), although several other less prominent proteases such as granzymes, calpains, cathepsins and HtrAs are also involved [6]. Well-studied proteolytic systems reveal an intriguing fact that proteolytic reactions are specific and irreversible unlike post-translational modifications [7]. Hence, these enzymes are found at the apex of irreversible processes such as development, cell cycle and, most importantly, cell death.

Apart from being mostly synthesized as precursors called zymogens that have little catalytic activity, proteases are known to regulate their own activation through positive and negative feedbacks. In case of caspases, the active protease can directly or indirectly activate its own precursor and result in an exponential rate of activation that ensures the protease accomplishes its goal more efficiently. The effective concentration of active proteases is also accompanied by the presence of endogenous inhibitors that regulate this crucial process by establishing a significant concentration threshold [7]. This prevents the consequences of an undesirable event of spontaneous cell death upon enzyme activation. Since caspases are the key proteases involved in apoptosis, deregulation of this intricate process can lead to a variety of human pathologies. Therefore, understanding caspase regulation becomes extremely important so as to rationally manipulate apoptosis for therapeutic gain.

In this chapter, we attempt to describe the properties of caspases, the cellular pathways controlling their activation and their modes of regulation, and also discuss the potential therapeutic strategies involving caspase modulation.

2 The Mediators of Apoptosis: Caspases

For a considerably long period of time, apoptosis has remained a descriptive cell biology phenomenon, until Sulston and Horvitz initiated cell lineage study in 1977 and elaborated more on programmed cell death in the nematode *Caenorhabditis elegans* (*C. elegans*). The group demonstrated that 131 out of 1090 somatic cells of *C. elegans* invariantly underwent the cell death process during normal development [8, 9]. This information was consequently followed up by genetic and molecular studies, wherein the discovery of cell lineage-dependent apoptosis paved the way for genetic characterization of its molecular machinery in *C. elegans* [10, 11]. During this investigation, a cysteine protease CED-3 (**Ce**ll **D**eath Abnormality-**3**) that is required for cell death in *C. elegans* was found to be related to mammalian ICE (**I**nterleukin-1β–**C**onverting **E**nzyme or caspase-1). The central components of the apoptotic machinery in *C. elegans* additionally included CED-4 and CED-9. In mammals, the homologues of CED-9, CED-4 and CED-3 were found to be present in the Bcl-2 (**B C**ell **l**ymphoma 2) family, Apaf-1/NOD-like (**N**ucleotide-binding **O**ligomerization **D**omain-like) receptor family and caspase family, respectively [12–15]. Since the mammalian homologues of CED-3, CED-4 and CED-9 have been found to regulate apoptosis, it is likely that the cellular machinery that modulates programmed cell death in *C. elegans* shares its evolutionary origin with that of mammalian apoptosis. This seminal finding, together with the observation that overexpression of ICE also induces apoptosis, suggested that it possibly plays a key role in promoting cell death. Thus, programmed cell death in *C. elegans* represents a primordial type of apoptosis that implicated the role of caspases in this vital cellular process. Expression of a caspase inhibitor protein, CrmA (**C**ytokine **r**esponse **m**odifier gene product encoded by cowpox virus), in sensory neurons prevented neuronal cell death induced by trophic factor deprivation [16]. This also provided the first evidence for understanding the functional role of caspases in neuronal cell death. Finally, caspase-3 was shown to be a critical mediator of the endogenous apoptotic pathway in mammalian cells [17], and in this way, the journey towards understanding caspases continued.

2.1 General Features of Caspases

The caspases comprise a family of cysteine proteases that use a cysteine residue as the catalytic nucleophile and share an impeccable specificity for cleaving their target peptide substrates after specific aspartate residues. This concerted action of caspases is essentially responsible for apoptosis; however, a subgroup of the caspase family is also involved in inflammation, where they act as procytokine activators [18]. Further identification of caspases in *C. elegans*, Drosophila and other vertebrate species demonstrated the evolutionary conservation of the death machinery. Till date, 14 distinct mammalian caspases have been identified, of which 11 caspases are from humans: caspases-1–10 and caspase-14 [19]. Caspase-11 and Caspase-12 are murine enzymes that are homologues of human caspase-4 and caspase-5, while caspase-13 was later found to represent a bovine homologue of caspase-4. Thus, the number of caspases discovered increased over the phylogenetic time, with four caspases being identified in *C. elegans*, seven in Drosophila, as well as 11 in mice and humans [20].

Overall, the caspases share similarities in a number of distinct features that include their amino acid sequence, structure and substrate specificity [21]. They are generally expressed as proenzymes (30–50 kD) that contain three domains: an amino (NH_2) terminal domain, a large subunit p20 ($\sim$20 kD) and a small subunit p10 ($\sim$10 kD). In some cases, these subunits are separated by a linker region that has unknown function. The NH_2-terminal domain is found to be highly variable in sequence as well as length and is involved in regulating the activation of proenzyme. Moreover, all these domains are derived from the proenzyme by cleavage at caspase consensus sites. This implies that the caspases can be activated either autocatalytically or in a cascade by enzymes with similar specificity. Generally, the activation is brought about by proteolytic processing between domains, which is then followed by heterodimerization caused due to the association of the large and small subunits. Crystal structures of two active caspases (caspase-1 and caspase-3) further reveal the presence of two heterodimers that associate to form a tetramer with two independent catalytic sites [22–24]. The large and small subunits are closely associated with each catalytic domain, and both these subunits contribute residues necessary for substrate binding and catalysis (described in Sects. 3.1 and 3.2).

As per research reports, caspases have an unusual and absolute requirement for cleavage after aspartic acid and hence, belong to C14 family of the clan CD of cysteine proteases. However, the substrate specificities of the individual caspases are distinct, being determined by the residues present in the pockets of P2, P3 and P4 of the substrate. Proteases are generally known to possess the substrate-binding pockets S1–S2–S3 ... S*n* (where '*n*' represents the number of substrate-binding sites). Amino acids on the amino- and carboxy-terminal of the scissile bond in the substrate are called the P and P′ sites, respectively. For example, if S2, S1 and S1′ are the three well-defined substrate-binding sites, which involve both the main and side chain contacts between substrate and enzyme residues, P2, P1 and P1′ become the corresponding substrate residues that bind to the enzyme's respective subsites. Apart from that, recognition of an active-site pentapeptide QACXG (where X is R, Q or G) is also found to be necessary for efficient catalysis [1]. This preferred pentapeptide recognition motif, that lies NH_2-terminal to the cleavage site, differs significantly among caspases, and hence, varied diversity exists in the biological functions of caspases [25]. Since not all proteins containing the optimal tetrapeptide or pentapeptide sequence are cleaved, the tertiary structural elements can be said to influence the event of substrate recognition.

The '*oxyanion hole*' in caspases is known to be important for polarizing and stabilizing the scissile P1 carbonyl group of the substrate. It is a pocket in the active site of an enzyme that hydrogen bonds the carbonyl oxygen of P1 residue of the substrate and stabilizes the negative charge on a deprotonated oxygen or alkoxide in the transition state. Thus, it lowers the activation energy required for the reaction and in turn promotes catalysis [26]. The concept of an oxyanion hole was first proposed in trypsin-like serine proteases. In caspases, it is generally formed by the backbone nitrogens of a strictly conserved Gly and the catalytic dyad Cys residue [27]. Along with a His residue, the catalytic Cys is primarily involved in forming the crucial catalytic dyad. Thus, in general, the active-site residues in caspases are in the order His and Cys forming the catalytic dyad. Both active-site residues are preceded by blocks of hydrophobic residues. The catalytic residues are found in the motifs His-Gly and Ala-Cys. These motifs are conserved throughout clan CD. For example, in caspase-1, amino acids such as Arg, Gln, Arg and Ser are known to be involved in shaping the P1 carboxylate-binding pocket. These amino acids are together found to be conserved in all the other caspases, except for a conservative substitution of Thr for Ser in caspase-8. Thus, this justifies the unconditional requirement of an Asp in the P1 position. However, the residues forming the P2–P4 binding pocket are not found to be well conserved, and therefore, they may be involved in determining the substrate specificities of different caspases (Table 1). In consensus with this, the cleavage of proteins by caspases is not only found to be specific, but also highly efficient ($k_{cat}/K_m > 10^6\ M^{-1}\ s^{-1}$) [7]. This strict specificity governs the fact that apoptosis is not accompanied by indiscriminate protein digestion; rather, a selective set of proteins is cleaved usually at a single site and in a coordinated manner leading to programmed cell death.

2.2 Classification of Caspases

Despite multiple rubrics that have been used to classify caspases, the tight structure–function correlations existing within the caspase family mainly govern the classification. Depending upon the two majorly known functions, the caspases are grouped into proapoptotic and proinflammatory subfamilies. Proapoptotic caspases (caspase-2, caspase-3, caspase-6, caspase-7, caspase-8, caspase-9 and caspase-10) are generally known to be involved in mediating cell death signalling transduction, whereas proinflammatory caspases (caspase-1, caspase-4, caspase-5, caspase-11 and caspase-12) are important in maturation of cytokines and promoting inflammation. Among the inflammatory caspases, mouse caspase-11 was found to be a

Table 1 Human caspases and their substrate specificities

Group	Human caspases	Alternative names	UniProtKB accession identifier	Substrate specificities P4–P3–P2–P1↓	P4 Site Preference
I	Caspase-1	ICE	P29466	W-E-H-D	Large P4
	Caspase-4	ICH-2, ICE(rel)-II, Mih1, Protease TX	P49662	(W/L)-E-H-D	
	Caspase-5	ICH-3, ICE(rel)-III, Mih1, Protease TY	P51878	(W/L)-E-H-D	
II	Caspase-2	ICH-1, NEDD2	P42575	D-E-H-D	Small charged P4
	Caspase-3	Apopain, CPP-32, Protein Yama, SCA-1	P42574	D-E-V-D	
	Caspase-7	CMH-1, ICE-LAP3, Mch-3	P55210	D-E-V-D	
III	Caspase-6	Mch-2	P55212	V-E-H-D	Intermediate P4
	Caspase-8	CAP4, FLICE, MACH, Mch-5	Q14790	L-E-T-D	
	Caspase-9	APAF-3, ICE-LAP6, Mch-6	P55211	L-E-H-D	
	Caspase-10	FLICE2, Mch-4, ICE-like apoptotic protease 4	Q92851	(I/L/V) E-X-D	–
–	Caspase-12	–	Q6UXS9	–	–
–	Caspase-14	–	P31944	–	–

The table represents the preferred tetrapeptide substrate sequences (P4–P1) for various human caspases that have been obtained from a positional-scanning combinatorial substrate library. On the basis of the size of S4 subsite and P4 residue, the caspases can be divided into three groups exhibiting distinct cleavage specificity. The terms P1–P4 and S4 have been described in the text

homologue of human caspase-4, while caspase-13 is a bovine ortholog of human caspase-4. In humans, caspase-12 is generally truncated due to the presence of a premature stop codon, but in African descendants, a read-through mutation causes expression of the full-length protein and increases the risk of sepsis [28]. Overall, this form of caspase classification is incomprehensive, as there are growing evidences that indicate the role of caspases in multiple cellular processes other than just being proapoptotic or proinflammatory (as described below). Moreover, apoptosis can also be induced through the activation of 'proinflammatory' caspases.

An alternative method of caspase classification is based on their positions in the apoptotic signalling cascades. Herein, they are categorized as initiator caspases (caspase-1, caspase-2, caspase-4, caspase-5, caspase-8, caspase-9, caspase-10, caspase-11 and caspase-12) and effector/executioner caspases (caspase-3, caspase-6 and caspase-7). This form of classification also categorizes caspases according to the length of their recruitment domains (or prodomains), which are placed at their N-termini. The initiator caspases that act upstream in cell death pathways share long, structurally similar prodomains. These prodomains comprise one of the two characteristic protein–protein interaction modules: DED (**D**eath **E**ffector **D**omain) or CARD (**C**aspase **A**ctivation and **R**ecruitment **D**omain) domains. The prodomain is separated from the catalytic domain by a flexible interdomain linker that is cleaved during the activation and maturation process.

Although poor solution behaviour and tendency to aggregate have limited the scope of structural studies on procaspases, recent advances in structural biology and protein engineering have helped circumvent the problem. As a result, few procaspases and several prodomain structures are currently available that revealed a classical hallmark structure of a **D**eath-fold **D**omain (DD) superfamily exhibited by these caspases [29, 30]. Generally, the group of proenzymes exists as stable monomers in the cell until they are activated by dimerization. Once dimerized, the initiator caspases achieve sufficient activity to autoprocess and subsequently cleave between their prodomain and intersubunit linker. The prodomains also allow initiator caspases to bind to the activation complexes by interacting with other adaptor proteins. Functionally, the initiator caspase-2 and caspase-9 are involved in the intrinsic pathway that is activated either by mitochondrial damage, cytotoxic stress, chemotherapeutic drugs or through certain developmental signals. On the other hand, the initiator caspase-8 and caspase-10 are activated by the extrinsic pathway, upon ligand binding to a death receptor at the membrane, which belongs to TNFR (**T**umour **N**ecrosis **F**actor **R**eceptor) superfamily (as described in Sects. 4.1 and 4.2). In contrast to the initiator caspases, the effector caspases do not require death scaffolds for dimer formation [31, 32]. These caspases possess short prodomains that are typically processed and activated by upstream caspases, thus performing the downstream execution steps of apoptosis by cleaving multiple cellular substrates. As the executioner caspases exist in the cell as preformed but inactive homodimers that must be processed by initiator caspases, their inappropriate activation is usually prevented. Upon activation, a single executioner caspase can cleave and activate other executioner caspases, resulting in an accelerated feedback loop of caspase activation. However, the described upstream and downstream relationship is not absolute and may only exist transitorily during very early phases of apoptosis [33]. The catalytic processing and different prodomain types in caspases have been illustrated in Fig. 1.

The protease families can also be classified according to their substrate preferences, despite the absolute requirement for an aspartate residue at the P1 site of the caspase substrates. A positional-scanning synthetic combinatorial library approach

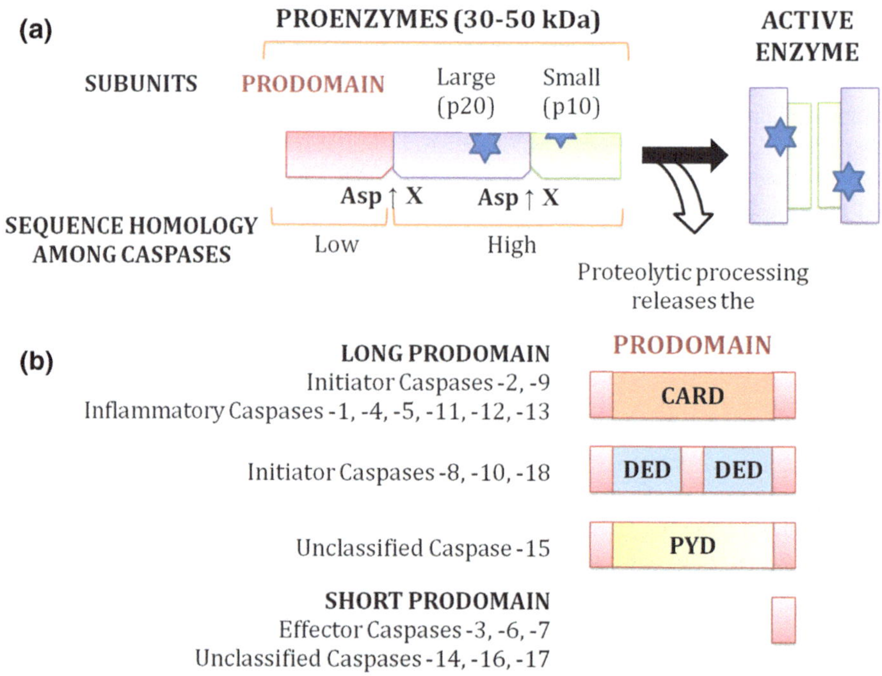

Fig. 1 Domain organization of the caspase protein family. **a** Caspase zymogen (proenzyme) comprises NH_2-terminal prodomain as well as large and small subunits separated by short linker region that harbours the cleavage site (represented by Asp↑X). Proteolytic processing releases the prodomain, and subsequent association of the large and small subunits leads to heterodimerization. Two such heterodimers come together to form the mature active caspase that possesses two independent catalytic sites (indicated by *star*). **b** Based on the size and length of prodomains, procaspases are functionally categorized into inflammatory, initiator and effector caspases

was employed to delineate the cleavage specificities of different caspases in vitro using the general structure of peptide-aminomethylcoumarin (Ac-X-X-X-Asp-AMC) substrates [34]. With this approach, the preferred tetrapeptide substrate sequences (P4–P1) for caspases-1–11 were determined [25] (Table 1), and based on that, these caspases were separated into three classes. The first group (class I) includes caspase-1, caspase-4 and caspase-5, which prefer bulky, hydrophobic side chains in the P4 substrate position, best described by the sequence —WEHD. The second class (class II) consists of caspase-2, caspase-3, caspase-7 and CED-3 that show a preference for DEXD (where X is V or H). Finally, the third class (class III), comprising caspase-6, caspase-8 and caspase-9, displays preference for the optimal sequence (L/V)EXD (where X is V, T or H) (Table 1). The P4 site (i.e. fourth amino acid N-terminal to the cleavage site) has also been found to be a critical determinant of substrate specificity. Those in Group II require Asp for efficient catalysis, while the Group III caspases although have broader specificity at P4 position prefer those with larger aliphatic side chains [25] (Table 1). The optimal cleavage site of caspase-11 has been determined as (I/L/V/P)EHD [35], while substrate recognition sequence for caspase-12 and caspase-14 is yet be to determined [36]. Overall, this form of substrate specificity classification is also in coherence with the aforementioned functional subdivisions of the family as shown in Fig. 1. The substrate preferences match well with the sites found in the known in vivo substrates, such as IL-1β and IL-18, laminin A, PARP and DNA-PK [25]. The processing sites between the large and small subunits in all initiator caspases resemble their own substrate preference, which explains the self-activation ability of initiator caspases. Furthermore, the processing sites in effector caspases fit the preferences of group II caspases, and therefore, they act as downstream activation targets of the initiator caspases. Interestingly, while the presence of a CARD domain in caspase-2 allows it to play the initiator role, both its substrate preference and the cleavage site between the large and small subunits are found to be strikingly similar to that of the executioner caspases (DXXD). These characteristics suggest that caspase-2 is a proximal responder to autoactivation in apoptosis signalling under certain conditions [37–39]. Apart from that, it can also function as an executioner caspase by cleaving cellular targets [19]. An interesting amount of data has been generated through the substrate specificity studies that may further help understand the differences in the physiological roles of individual caspases. However, since these data are derived using a short synthetic substrate in vitro, they need to be interpreted cautiously before being considered conclusive. So far, more than 700 substrates for mammalian caspases have been reported and compiled into a searchable online database known as 'CASBAH' (http://bioinf.gen.tcd.ie/casbah/) [40].

The other caspases that still remain unclassified include caspase-14, whose expression is known to be restricted to epidermal keratinocytes and is specifically involved in differentiation [41]. Similar to the effector caspases, caspase-14 possesses a short prodomain with no adaptor regions. The list of unclassified caspases further includes caspase-15, caspase-16 and caspase-17. Caspase-15 contains a **Py**rin-like region in its pro**d**omain (PYD), like those found in zebrafish counterparts Caspy and Caspy2 [42]. It is also known to be expressed in several other mammalian species including pigs, dogs and cattle [43]. Like caspase-14, caspase-16 and caspase-17 also contain a short prodomain with no adaptor regions [44]. Caspase-16 has been conserved in marsupials and placental mammals, including humans, while caspase-17, which is most similar to caspase-3, has been conserved among fish, frog, chicken, lizard and the platypus but is absent from marsupials and placental mammals. On the contrary, caspase-18 is found in opossums as well as in chickens and is known to contain two tandem DED regions in its prodomain like caspase-8 and caspase-10. It is most likely thought to act as an initiator apoptotic caspase [44]. Further studies on these newly identified caspases will help understand their physiological roles as well as complementarities and redundancy in this protein family.

3 Caspase Structure, Specificity and Catalysis

With the structural determination of caspase-1, elucidation of the structures of several other caspases, their inhibitors and caspase–inhibitor complexes came to the forefront. This focused endeavour provided tremendous information on caspase structure, mode of activation, specificity and regulation. In the following section, we discuss the major characteristics of the '*caspase fold*' and its significance in the various cellular processes.

3.1 Structure of Active Caspases

All caspase structures exhibit the fundamental catalytic domain that comprises a large (17–20 kDa; p20) subunit and a small (10–12 kDa; p10) subunit, which are tightly packed together into a compact ellipsoid of approximate dimensions of 25 Å × 50 Å × 30 Å [45]. A twofold crystallographic axis relates the pairs of catalytic domains, each of which arises from a single procaspase molecule.

During activation and/or maturation, each of the catalytic domains is cleaved between the large (p20) and small (p10) subunits, which in turn interact intimately with each other to form a heterodimer (p20–p10). The secondary structure of these mature caspases is composed of a slightly twisted, but mostly parallel, six core β-strands. These strands are further sandwiched between two layers of α-helices, with two main helices on one face of the protein and three helices on the other. Thus, while the large subunit comprises the first four core β-strands and helices 1–3, the last two core β-strands and helices 4–5 form the small subunit. The quaternary structures of caspases are finally formed when two such heterodimers associate with each other through antiparallel alignment of their C-terminal β6 strands, generating a continuous 12-stranded β-sheet [46]. Therefore, the active form of a caspase is generally referred to as '*homodimers of heterodimers*' [22, 23], where two active sites are positioned at opposite ends of each molecule (Fig. 2). However, this topology is quite intriguingly different for caspase-9, where only one '*active centre*' is formed [19, 47].

Generally, the '*active centre*' of caspases is exclusively formed by flexible loops (L1–L4) [19], where these loops are involved in shaping the substrate-binding groove. While loop L1 constitutes one side of the groove, L4 represents the other side. Loop L3, along with a β-hairpin, is collectively referred to as L3 and is located at the base of the groove. Loop L2, which harbours the catalytic residue cysteine, is positioned at one end of the groove with its side chain primed for binding and catalysis [48]. The catalytic histidine (another catalytic dyad residue) is part of a loop extending from the C-terminal end of β3. Loop L1 and a portion of L2 constitute a part of the large subunit, whereas L3 and L4 come from its smaller counterpart (Fig. 2; only L3 in case of caspase-3). Among all the caspases, the size and topology of the L1 and L3 loops are found to be highly conserved, whereas the L2 and L4 loops display a greater degree of variation.

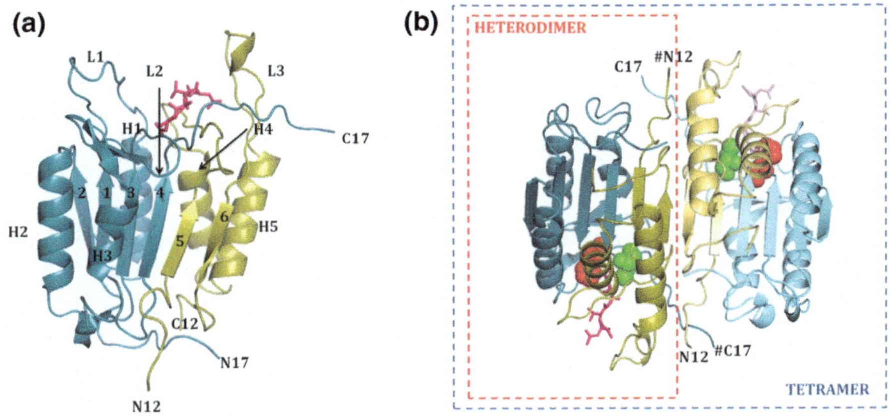

Fig. 2 Crystal structure of mature caspase-3 in complex with a tetrapeptide aldehyde inhibitor (PDB ID: 1CP3). **a** Cartoon representing mature caspase-3 heterodimer comprising the large subunit (p17) and the small subunit (p12). The 17 kDa large subunit (*deep teal*) comprises the first four core β-strands and helices 1–3 (present on one face of the protein), while the last two cores β-strands and helices 4–5 (present on the other face) form the 12 kDa small subunit (*olive*). The N- and C-termini of the large (p17) and small (p12) subunits are represented as N17-C17 and N12-C12, respectively. The β-strands are numbered from 1 to 6, the α-helices are represented by *upper-case letters* (H1–H5), and the loops are shown as L1–L3. **b** Quaternary structure of the active form of caspase-3. Cartoon representing each heterodimer (shown as *A* and *B*) comprising large and small subunits undergoes processing at the intersubunit linker (indicated by *dotted arrow*). Upon activation, two such heterodimers (*dark and light shades,* respectively) associate to form an active tetramer (*blue dotted box* comprising molecules *A* and *B*), with a continuous 12-stranded β-sheet that is sandwiched by α-helices. The #N17 and #C12 represent the N- and C-termini of one of the heterodimers (*dark-coloured dimer*). His-121 and Cys-163 of the catalytic dyad of caspase-3 are shown in *red* and *green,* respectively. In the tetramer, the two active sites are positioned at opposite ends of each molecule. The bound inhibitor Ac-DVAD-fmk is shown as stick model (*pink*). The figures are generated using PyMOL (DeLano Scientific, USA)

Typically, a caspase is dimeric and the intersubunit linker needs to be essentially cleaved for its activation. For effector caspases, equilibrium favours the inactive dimer, while for the activation of initiator caspases, dimerization seems to be the main challenge to overcome. Moreover, since the initiator caspases have longer intersubunit linkers than effector caspases, their cleavage in effector caspases is not necessary for activation. It would rather stabilize the active conformation of these caspases. In general, cleavage of the intersubunit linker occurs first soon after dimerization and is further followed by cleavage of the prodomain. Prior to cleavage, the intersubunit linker from one monomer occupies the dimer interface. Upon activation, cleavage of the intersubunit linker, i.e. loop L2, liberates the C terminus of large subunit (referred to as L2′) and the N terminus of small subunit (Fig. 2). The C-terminal portion of the linker L2′ now vacates the central cavity and rotates about 180° towards the active site. Herein, L2′ forms contact with L2, L3 and L4 from the opposite monomer and develop the so-called ‘*loop bundle contacts*’, which are known to stabilize the active site. This movement of L2′ out of the dimer interface allows L3 to slide in towards the interface and form the substrate-binding pocket. Rotation of a key arginine on L2 from a solvent-exposed position into the interface allows its neighbouring residue, the catalytic cysteine, to assume its proper position for catalysis. Thus, for the construction of substrate-binding groove of the catalytic domain, a total of five loops seem to be important: loops L1 to L4 that come from one monomer and loop L2′ that is a part of the other monomer.

As mentioned earlier in the text, in addition to the catalytic domain, inflammatory and initiator caspases carry at least one or two copies of prodomains. These death domains have a globular architecture, wherein six amphipathic α-helices are arranged in an antiparallel α-helical bundle, with helices α1–α5 forming a conserved α-helical Greek key topology [45]. Despite the lack of obvious sequence similarity, these prodomains are found to share a common fold. However, few structural changes are also observed such as variation in lengths of α-helices and surface distribution of charged as well as hydrophobic residues. While the structural similarity strongly suggests a common evolutionary origin for the prodomains [49, 50], these subtle structural differences define their functional specificity. The primary function of prodomains is the recruitment of caspases to adaptor proteins and subsequent formation of intricate higher-order oligomers that are prerequisites for caspase activation in vivo. The mechanism for the same has been further elaborated in Sects. 4.1 and 4.2.

3.2 Active-Site Architecture

As described earlier, the binding pockets for P4–P3–P2–P1 positions in the substrate are known as S4–S3–S2–S1 subsites, respectively. Since caspases exhibit a very stringent specificity for P1 residue, their S1 pockets are found to be almost identical. S1 pocket is a deep, highly basic groove that is constructed of side chains of strictly conserved residues Arg (placed in the β1–α1 region of the L1 loop), Gln (positioned at the beginning of L2 loop) and Arg (placed at the end of L3 loop). In accordance to this, enzymes that exhibit open α/β structures generally have their active sites placed at positions within or close to the C-terminal end of central β-sheet in which loops originating in two adjacent β-strands lead to the opposite helical layers [45]. This general rule is shown to be valid for caspases, wherein the catalytic dyad residue His is located in the loop that connects strand β3 to the 'front' helix α3, while the neighbouring β1 strand is followed by the 'back' helix α1. Here, 'back' and 'front' refer to the standard caspase orientation. The loop comprising the β4–β5 linker contains the second catalytic dyad residue Cys. It protrudes from strand β4 (terminating in Gln) towards the front helical layer, while the one coming from the adjacent strand β5 leads to the back helix α4 (Fig. 2). Finally, the β5–α4 loop contains Arg (placed at the end of L3 loop) that is involved in substrate fixation. These series of interactions and coordinated movements lead to the effective interaction of S1 pocket residues with the catalytic dyad residues.

The S1 pocket is ideally designed to accommodate an aspartate side chain, and therefore, replacement of a glutamate residue at P1 leads to up to fourfold reduction in catalytic efficiency for cleavage of peptides [51]. Notably, mouse caspase-12 is a less catalytically efficient caspase, as it contains Lys at the position equivalent to the residue Arg [45]. S2 pocket is the only site to show a considerable alteration during substrate binding. In caspase-3 and caspase-7, it is formed by side chains of aromatic residues Tyr, Trp and Phe and can favourably accommodate small aliphatic residues such as Ala and Val. However, due to the substitution of Tyr → Val (Ala in caspase-2), the subsite S2 is found to be larger in inflammatory and initiator caspases [45]. This Val forms a large hydrophobic S2 pocket that is structurally tuned to accommodate residues with bulkier side chains such as His or Thr as P2 residues. Besides having a role in the S1 pocket, the Arg residue placed at the end of L3 loop also mediates main-chain–main-chain hydrogen bond interaction, through its guanidinium group, with the carboxylate of a P3 glutamate residue [45]. This justifies the preference for the glutamate residue in small peptide substrates and inhibitors [25, 52, 53]. Moreover, tighter binding of inhibitors containing a glutamic acid residue at position P3 is promoted by adjacent basic residues of the S3 pocket like Arg (placed in the $\beta 1$–$\alpha 1$ region of the L1 loop) [45]. Although S1 and S3 subsites exhibit similar character, the chemical nature and geometry of S4 pocket do not appear to be conserved among all caspases [54]. This provides a rationale for the distinct macromolecular specificities and hence functions of the different caspase subclasses.

In the inflammatory caspase-1 as well as in caspase-4 and caspase-5, the S4 subsite is found to be an extended, shallow hydrophobic depression that prefers to accommodate indole moiety of large aromatic side chains such as Trp residue [34]. On the contrary, caspase-2, caspase-3 and caspase-7 possess a narrow pocket containing several side chains that serve to stabilize the aspartic acid residue present in P4 position. In caspase-3, this P4 aspartic acid seems to be required for efficient hydrolysis, as the replacement of this residue with any other amino acid results in >100-fold decrease in k_{cat}/K_m [54]. The S4 subsites of caspase-8 and caspase-9 occupy intermediate positions and show preference for small hydrophobic Val or Leu residue [33]. However, caspase-8 is shown to equally tolerate both small hydrophobic [55] and acidic residues like Ile and Asp, respectively, at P4 position [56]. An occupied S5 pocket is an exceptional and additional requirement in caspase-2 so as to bring about efficient substrate cleavage. The presence of a P5 residue confers 35-fold increase in catalytic efficiency, and this may be due to a better burial of the P4 aspartate [57]. This extended specificity has also been well understood from the high-resolution structure of caspase-2 in complex with acetyl-Leu-Asp-Glu-Ser-Asp-CHO [57]. The P5 Leu side chain occupies a small groove formed by two residues of the hydrophobic S5 pocket, Pro and the phenyl moiety of Tyr. This explains the preferential accommodation of small hydrophobic residues at this position in caspase-2. The primed subsites (P1′–P4′), however, seem to be less restrictive than the S1–S4 pockets and have been recently characterized crystallographically [58]. Using fluorescent peptidyl substrates, an apparent degree of discrimination for small residues (Gly, Ala, Ser) at position P1′ has been observed in several caspases. The large aromatic side chains (Phe/Tyr) were also found to be well tolerated, whereas polar residues or proline were prohibited [51]. Thus, these investigations verified and extended the observations done on the P1′ substrate specificity of caspase-1 and caspase-4 [52, 59].

In summary, a consensus caspase substrate sequence can be defined as (W/Y/V/L/D)-E-X-D↓-Φ with P4-P3-P2-P1↓-P1′. Herein, X represents any amino acid, ↓ represents the scissile bond, and Φ stands for a small uncharged residue. Knowledge of this inherent specificity that is exhibited by caspases has led many investigators to construct peptide-based substrate reporters for measuring caspase activity, both in academic and in commercial settings. Such reporter-based substrates typically constitute a tetrapeptide sequence matching a preferred caspase consensus coupled to a fluorogenic probe at P1 position. The probe is released upon cleavage and produces a fluorescent signal. Although these methods are useful to individually characterize purified caspases in vitro, they cannot distinguish between caspases in a complex milieu, as significant overlaps also exist in the specificity of caspases.

3.3 Proposed Catalytic Mechanism

Experimental investigations using time-resolved crystallography or any detailed theoretical calculations of substrate cleavage have not yet been reported for caspases. However, the similarity of chemical groups involved in catalysis and crystal structures of active caspases bound to substrate-like inhibitors have provided valuable information [45]. It has been found that caspases have a course of catalytic reaction that is similar to what is generally observed in serine proteases and other families of cysteine proteases. As described earlier, caspases contain a catalytic dyad comprising Cys and His. On the basis of the accepted catalytic mechanism for cysteine proteases, a mechanism of catalysis has been proposed for caspases [60]. Herein, the catalytic histidine first abstracts a proton from the nucleophilic catalytic cysteine. The anionic sulphur group of this deprotonated Cys now attacks the carbonyl oxygen of P1 residue of the substrate and forms a covalent tetrahedral intermediate with the peptide substrate. This intermediate state is stabilized by imidazole moiety of His, which in turn protonates α-amino moiety of the peptide leaving group and triggers release of the C-terminal part of the peptide. The deprotonated Nδ atom of His now abstracts a proton from water molecule that further attacks the thioester carbonyl, resulting in the formation of a second tetrahedral intermediate. This is followed by the cleavage of the covalent bond formed between carbonyl carbon and sulphur of Cys. Finally, hydrolysis frees the N-terminal part of the peptide and reprotonates catalytic histidine, and this regenerated enzyme can now start up a new proteolytic cycle. The proposed mechanism thus practically reverses roles of the '*classic oxyanion hole*' and His [45].

However, there arise some controversies with the proposed mechanism. Unlike most proteases, a 6–7 Å distance between the catalytic dyad residues is found to be larger in caspases and this makes the occurrence of a direct hydrogen transfer slightly unlikely [26]. Molecular dynamics simulation studies show that these residues cannot exist as a charged pair prior to catalysis [61]. This clearly indicates that at such an unusual Nδ1/Nε2–Sγ distance, His cannot accept a thiol proton of Cys. It also implies that Cys may not be pre-polarized, but the nucleophile may develop along the reaction coordinate [26]. The step takes place in unity with the pH optima of all caspases lying in a narrow range of 6.8–7.4 [62]. Further, several line of evidences raise an intriguing question on the presence of a third catalytic residue in caspases. Many investigators, including Wilson and colleagues, proposed a possible participation of the carbonyl oxygen of Arg residue to be an important element of the caspase-1 catalytic machinery [23, 24, 55]. Herein, the oxygen atom accepts a hydrogen bond from the Nε atom of His and might either affect the basicity of imidazole moiety and/or orientate it similarly to the side chains of catalytic Asn/Asp residues in clan CA cysteine peptidases and the carboxylate of Asp in trypsin-like proteases. However, the participation of this third catalytic residue in caspase-mediated catalysis still remains as a disputed topic [45]. Many other investigators, including Miscione and co-workers as well as Sulpizi and his group, have also proposed reaction schemes that differ from the classical one in some or the other step [27, 61]. Thus, rigorous quantum mechanics or molecular dynamics calculations along with structural and mutagenesis studies are required to establish a universal caspase catalytic mechanism.

4 Cellular Pathways Controlling Caspase Activation

There arises a need to control active caspases via stringent mechanisms as they are responsible for causing rapid cell death. These mechanisms primarily include synthesis of caspases as inactive zymogens, involvement of highly evolved upstream regulatory pathways that control their activation, and presence of endogenous inhibitors. The apoptotic signalling pathways that lead to the processing of caspase zymogens are subdivided into two major categories: extrinsic and intrinsic pathways of apoptosis.

4.1 The Extrinsic Pathway of Apoptosis

As stated earlier, this pathway includes the cell surface sensor-directed apoptotic signals involving the **D**eath-**I**nducing **S**ignalling **C**omplex (DISC) and is usually exemplified by the signalling of the death receptor family. Belonging to a family of type I transmembrane proteins, the **D**eath **R**eceptors (DRs) are characterized by the presence of multiple cysteine-rich repeats in their extracellular region that is required for recognition of their specific ligands (Table 2). Few examples include TNF-R1 (**T**umour **N**ecrosis **F**actor **R**eceptor **1**), Fas (also known as CD95 or APO-1), DR3, DR4 (TRAIL-R1), DR5 (TRAIL-R2) and DR6. These DRs are also known to have the protein–protein interaction module known as **D**eath **D**omain (DD)

Table 2 Extrinsic pathway molecules and their functions

Group	Protein	Acronym	Alternate names
Receptors	Fatty acid synthetase receptor	FasR	Fas receptor, TNFRSF6, APT1, CD95
	Tumor necrosis factor receptor 1	TNFR1	TNF receptor, TNFRSF1A, p55 TNFR, CD120a
	Death receptor 3	DR3	TNFRSF12, Apo3, WSL-1, TRAMP, LARD, DDR3
	Death receptor 4	DR4	TNFRSF10A, TRAILR1, APO2, KILLER
	Death receptor 5	DR5	TNFRS10B, TRAIL-R2, TRICK2, CHE1
Extracellular ligands	Fatty acid synthetase ligand	FasL	Fas ligand, TNFSF6, Apo1, apoptosis antigen ligand 1, CD95L, CD178, APT1LG1
	Tumor necrosis factor alpha	TNF-α	TNF ligand, TNFA
	Apo2 ligand	Apo2L	TRAIL/TNFSF10
Adaptor proteins	Fas-associated death domain	FADD	MORT1
	TNF receptor-associated death domain	TRADD	TNFRSF1A associated via death domain
	Receptor-interacting protein	RIP	RIPK1
Decoy protein	FLICE-inhibitory protein	c-FLIP	Casper, I-FLICE, FLAME-1, CLARP, MRIT
Initiator protein	Cysteinyl aspartic acid protease-8	Caspase-8	MACH-1, MCH5, ICE-LAP6, Mch6, Apaf-3

Adapted from [63]

in their cytoplasmic tails. Similar to the prodomains in caspases, these DDs belong to the Death Domain superfamily and are involved in initiating assembly of large macromolecular complexes through a series of homotypic interactions.

When the cell surface sensors detect extracellular danger signals through ligand binding, clustering of the receptors recruits a DD-containing adaptor molecule, FADD (**F**as-**A**ssociated protein with **D**eath **D**omain) through DD–DD interactions (Fig. 3). For Fas and TRAIL (**T**umour necrosis factor-**R**elated **A**poptosis-**I**nducing **L**igand) receptors, the adaptor protein FADD is recruited to DISC through its C-terminal DD. TRAIL-mediated extrinsic pathway involves five TRAIL receptors, TRAIL R1-R4 and osteoprotegerin (OPG) [63]. The first two promote cell death in different tumour cells, while TRAIL R3 and TRAIL R4 are devoid of cytoplasmic death domains (DD) and act as '*decoy*' proteins with significantly reduced proapoptotic activities. OPG, however, with a lower binding affinity for TRAIL 1 and 2 ligands has limited effect at physiological temperature [64]. Generally, the binding of the ligand induces receptor multimerization, and this exposes another death-fold domain (DED) in FADD. The exposed DEDs of FADD in turn interact with the N-terminal DEDs of monomeric procaspase-8 molecules (Fig. 3a). The complex, so formed at the cytoplasmic tail of engaged DR, now represents a caspase activation platform called DISC, and it may include multiple adaptor molecules. Further, recruitment and oligomerization of caspase-8 in DISC cause its autocatalytic cleavage in between the large and small subunits as well as between large subunit and prodomain. The first cleavage stabilizes the active dimer that is required for homodimer activity in this pathway, while the second cleavage releases it from DISC [65]. The formation of DISC increases the local concentration of monomeric initiator caspase zymogens through proximity induced by the adaptor molecules [66]. The active caspase-8, so formed, further cleaves and activates the downstream or executioner procaspases, mainly procaspase-3 and procaspase-7. The activated caspase-3 cleaves **X**-linked **I**nhibitor of **A**poptosis **P**rotein (XIAP), which is the most widely expressed endogenous inhibitor that plays an important role in regulating cell survival (Fig. 3a). This event also creates a positive feedback loop of self-activation, thereby synergistically enhancing the cell death process [67, 68]. Apart from sharing similarities in their physiological role, active caspase-7 differs in significant ways from caspase-3. It specifically cleaves several substrates that are not cleaved by caspase-3, and its activation requires caspase-1 inflammasomes under inflammatory conditions [69]. Subsequently, these effector caspases bring about proteolytic cleavage of important cellular proteins, thus culminating into apoptosis [70, 71]. TRAIL is also capable of inducing apoptosis through an intrinsic pathway, which gets activated when mitochondrial proapoptotic molecules Smac/DIABLO or HtrA2 restore caspase-3 activity by relieving inhibitory effect of XIAP [72, 73]. Smac/DIABLO (**S**econd **m**itochondria-derived **a**ctivator of **c**aspases/**D**irect **IAP**-**B**inding protein with **Lo**w pI) and HtrA2 (**H**igh-**t**emperature **r**equirement protease **A2**) both are mitochondrial molecules that are released into the cytosol upon apoptotic induction. In addition to this, TRAIL weakly induces NF-κB pathway with the help of adaptor proteins Rip and TRAF-2 [74].

In case of TNFR1 (**T**umour **N**ecrosis **F**actor **R**eceptor-1) receptor, the signalling pathways engaged downstream are more complex and its activation does not necessarily lead to cell death but may potentially lead to different outcomes (Fig. 3b). Herein, a different DD-containing adaptor TRADD (**TNFR**-**A**ssociated **D**eath **D**omain) is first recruited upon the interaction of ligand TNFα with TNFR1 and induces TNFR1 ligation. Subsequently, this leads to the formation of two distinct signalling complexes [75].

(1) *Complex I formation*: Since TRADD does not directly bind to or activates caspase-8, the ligand instead acts as a membrane-bound scaffold that recruits additional signalling molecules such as a serine/threonine-protein kinase RIP1 (**R**eceptor-**I**nteracting **P**rotein-1) and ubiquitin ligases, TRAF2 and cIAP1/2 (**c**ellular **I**nhibitor of **A**poptosis **P**rotein) (Fig. 3b). This event leads to the formation of a signalling platform known as Complex I. The assembly and non-degradative ubiquitylation of its components by molecules such as cIAPs and LUBAC (**L**inear **Ub**iquitin **A**ssembly **C**omplex) culminate in the activation

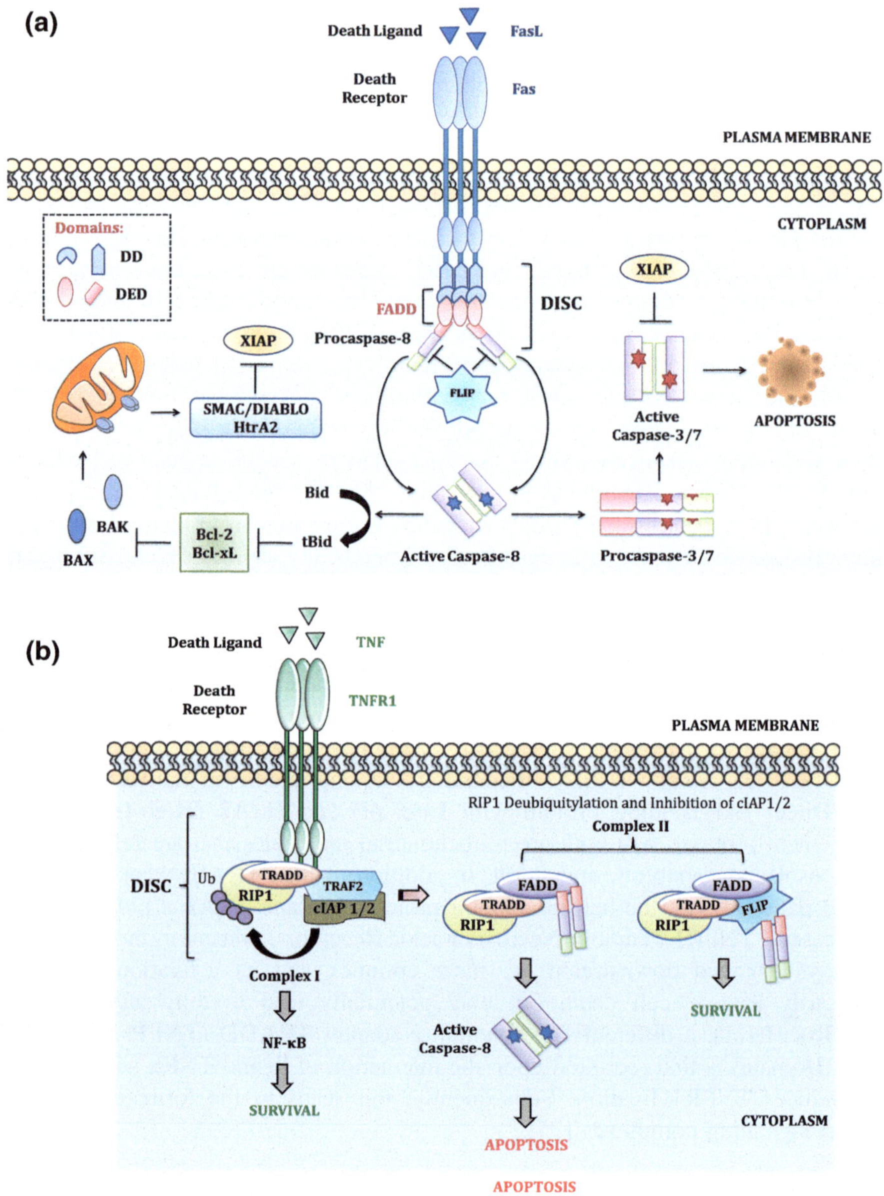
(a)
Death Ligand
FasL
Death Receptor
Fas
PLASMA MEMBRANE
CYTOPLASM
Domains:
DD
DED
FADD
Procaspase-8
DISC
XIAP
FLIP
SMAC/DIABLO
HtrA2
XIAP
Active Caspase-3/7
APOPTOSIS
Bid
tBid
BAK
BAX
Bcl-2
Bcl-xL
Active Caspase-8
Procaspase-3/7
(b)
Death Ligand
TNF
Death Receptor
TNFR1
PLASMA MEMBRANE
RIP1 Deubiquitylation and Inhibition of cIAP1/2
Complex II
DISC
Ub
TRADD
RIP1
TRAF2
cIAP 1/2
FADD
TRADD
RIP1
FADD
TRADD
RIP1
FLIP
Complex I
NF-κB
SURVIVAL
Active Caspase-8
APOPTOSIS
SURVIVAL
CYTOPLASM
APOPTOSIS

◀ **Fig. 3** Receptor-mediated extrinsic cell death pathway. The death receptor signalling pathway is triggered when death ligands such as FasL, TNF or TRAIL engage with their cognate receptors—Fas, TNFR, or DR5, respectively. **a** Ligation initiates the recruitment of adaptor proteins, FADD or TRADD (through DD–DD interactions), towards Fas and TRAIL receptors, respectively. This further recruits zymogenic initiator procaspase-8 (through DED–DED interactions) and forms the caspase activation platform called DISC. Activation of caspase-8 promotes maturation of executioner caspase-3, caspase-6 and caspase-7 that further bring about substrate proteolysis and cell death. On the other hand, it crosstalks with the mitochondrial cell death pathway by cleaving the proapoptotic Bid protein. The truncated Bid (tBid) inhibits the antiapoptotic Bcl-2 and Bcl-xL proteins, which prevent the oligomerization of proapoptotic Bax and Bak proteins. Smac/DIABLO and HtrA2 are the mitochondrial proteins that act as XIAP antagonists and promote apoptosis. **b** The TNFR receptor ligation with TNF-α leads to further recruitment of the adaptor protein, TRADD, and the kinase, RIP1. The death-inducing complex (DISC), so formed, further attracts additional partners such as the ubiquitin ligases, TRAF2 and cIAP1/2, which catalyse the non-degradative ubiquitylation of RIP1. Furthermore, it stabilizes a prosurvival and proinflammatory signalling platform called complex I. However, inhibition of cIAP1/2 or activation of deubiquitylases destabilizes complex I and allows subsequent release of TRADD and RIP1 into the cytosol. These proteins, in turn, recruit FADD as well as caspase-8 and promote caspase activation. Expression of FLIP inhibits the activity of both complexes, as it heterodimerizes with caspase-8 and simultaneously blocks apoptosis

of NF-κB signalling pathway followed by an inflammatory response rather than causing cell death [76].

(2) *Complex II formation*: TNFR1 ligation can lead to cell death only when ubiquitylation of RIP1 is compromised by inhibition of ubiquitin ligases cIAP1/2 or through direct de-ubiquitylation of RIP1. TRADD and RIP1 are then released from TNFR1 to form a series of dynamic cytosolic signalling platforms that is collectively known as complex II. This complex leads to apoptosis when cytosolic TRADD recruits FADD (via DD–DD interaction), which in turn binds to and activates caspase-8 similarly to the DISC (Fig. 3b) [77, 78]. However, when activation of NF-κB induces the expression of caspase-8-like protein FLIP (**F**LICE-**L**ike **I**nhibitory **P**rotein) [79], apoptosis does not proceed upon TNFR1 ligation. FLIP is a molecule that inhibits caspase-8 activation. It possesses a domain structure similar to that of caspase-8 and preferentially binds to FADD (via DED–DED interactions) as well as a monomer of caspase-8. Thus, the caspase-8–FLIP heterodimer is catalytically active [80, 81], but does not promote apoptosis probably due to rapid degradation of the complex [82]. The FLIP expression and assembly of complex II under conditions of active NF-κB signalling therefore lead to a survival outcome.

This overall balance of complexes I and II ultimately decides the fate of life or death; however, the process still remains elusive.

The outcome of DR-mediated activation of caspase-8 also depends on cell types, which are categorized on the basis of their dependency on the mitochondria for the induction of Fas/TRAIL death receptor-mediated apoptosis. In certain cell types, known as type I cells, it initiates apoptosis directly by cleaving and thereby activating downstream executioner caspase-3 and caspase-7 [83]. The apoptotic signal amplification through the existence of such caspase cascades becomes essential, as initiator caspases are unable to complete the execution of cell death on their own. In addition, such pathways may enable cells to maintain a stronger control over the cellular suicide mechanism at multiple points. In type II cells, where apoptotic signalling is initiated by an insufficient amount of active caspase-8 or downstream caspases, death receptor signalling must be amplified by a mitochondrial amplification step to induce efficient cell death [84]. In this case, caspase-8 cleaves cytosolic BH3-only proapoptotic Bcl-2 family member Bid, and the truncated Bid (tBid) then translocates to mitochondria, thus promoting the mitochondrial cell death pathway (described in Sect. 4.2.1) [85, 86]. The event subsequently promotes **M**itochondrial **O**uter **M**embrane **P**ermeabilization (MOMP), which results in the release of intracellular **I**nhibitor of **A**poptosis **Pro**teins (IAPs) antagonists—Smac and HtrA2. The IAPs are known to block executioner caspase function unless they are suppressed by proteins released from mitochondria. Smac and HtrA2 neutralize XIAP and allow apoptosis to proceed. Through the knockout studies in mice, thymocytes were reported to be type I, while liver cells were type II cells [87], and these cells differ well in their IAPs content [88, 89].

4.2 Intracellular Sensor-Mediated Pathway (Intrinsic Pathway)

Being another major category of apoptotic signalling, this pathway is activated by stimuli such as DNA damage and cytotoxic drugs which act inside the cell. It is very similar to type II cell death receptor mechanism in terms of its requirement for signal amplification through mitochondrial damage. However, since cells possess multiple means of targeting mitochondria for the release of apoptogenic factors, this category could be further subdivided into following discrete signalling pathways that influence the mechanism of caspase activation in their distinct ways.

4.2.1 Mitochondrial Pathway Activation Through Apoptosome

This form of intrinsic pathway depends on various factors released from the mitochondria upon its activation by a number of stimuli. These include growth factor deprivation, cytoskeletal disruption and build up of unfolded proteins, DNA damage and so forth. The apoptotic stimuli consequently trigger **M**itochondrial **O**uter **M**embrane **P**ermeabilization (MOMP), a phenomenon that induces release of all soluble proteins present in the mitochondrial intermembrane space into the cytosol. MOMP is tightly regulated by myriads of Bcl-2 family proteins. These include proapoptotic effector proteins, Bax and Bak, that are necessary and sufficient for MOMP; the antiapoptotic Bcl-2, Bcl-xL and Mcl1 proteins that block MOMP; and the BH3 (**B**cL-2 **H**omology domain)-only proteins, Bid, Bim, Bad and Noxa, which activate the proapoptotic effectors and/or neutralize antiapoptotic Bcl-2 proteins [90]. The activation of one or more members of the BH3-only protein family above a certain critical threshold helps in overcoming the inhibitory effect of antiapoptotic Bcl-2 family. On the other hand, proapoptotic effectors—Bax (**B**cl-2 **a**ssociated **X** protein) and BAK (**B**cl-2 homologous **A**ntagonist **K**iller)—are directly responsible for the loss of mitochondrial outer membrane integrity (Fig. 4). Upon activation, they form large oligomers that insert into the mitochondrial outer membrane, disrupt it [91–94] and initiate the release of cytochrome c [95]. Apart from cytochrome c, the other key players of this intrinsic pathway include Apaf-1 (**Ap**optotic **p**rotease-**a**ctivating **f**actor-1) and initiator caspase, caspase-9. Both of these proteins exist in a resting cell as cytosolic, inactive monomers. Apaf-1 contains an N-terminal CARD, an expanded nucleotide-binding domain NACHT and a C-terminal WD40 repeat domain (Trp-Asp sequence repeating 40 times). The binding of cytochrome c to WD40 domain of Apaf-1 monomer triggers hydrolysis of the Apaf-1 cofactor, dATP (deoxy-adenosine triphosphate) to dADP (deoxy-adenosine diphosphate) [96]. A conformational change subsequent to this interaction exposes a nucleotide-binding site in the NACHT domain of Apaf-1 where dATP binds. This induces a second conformational change in Apaf-1 that exposes both its oligomerization and CARD domains. Seven such activated Apaf-1 monomers then assemble into an oligomeric complex that contains the CARDs in the centre. These CARDs together then recruit and interact with the prodomain (CARD) of caspase-9 (Fig. 4) [97, 98]. The wheel-shaped multimeric protein complex, so formed, is termed as '*apoptosome*', which comprises Apaf-1, cytochrome c as well as the cofactor dATP/ATP. At the centre of the apoptosome, interaction of the exposed CARDs brings inactive caspase-9 monomers into close proximity for activation and autoprocessing [99]. The lifetime of an active apoptosome is directly proportional to the amount of unprocessed caspase-9 present [100], since caspase-9 sustains the catalytic activity in this bound state [26, 101, 102]. This activated initiator protease further processes other downstream caspases, caspase-3 and caspase-7, to carry out apoptosis (Fig. 4).

4.2.2 Caspase-2 Activation Through PIDDosome

Being one of the earliest caspases to be discovered, caspase-2 is the most highly conserved initiator caspase with a long prodomain. Upon treatment with chemotherapeutic drugs, caspase-2-deficient germ cells and oocytes have been found to become resistant to cell death [103], which implicates it to be involved in the process of oocyte apoptosis [104] and degenerating neurons [105–107].

In response to apoptotic stresses like heat shock, cytoskeletal disruption, metabolic perturbation and DNA damage, inactive procaspase-2 monomers are known to induce oligomerization and get activated by induced proximity (mechanism described in Sect. 5.2) [108]. Procaspase-2 oligomerization is primarily mediated by the adapter protein RAIDD (**RI**P-**A**ssociated **I**CH-1 homologous protein with a **D**eath **D**omain). RAIDD was originally identified as a protein associated with RIP, the DD-containing Ser/Thr kinase involved in TNF signalling. It has both an N-terminal CARD and a C-terminal DD and acts as a bridge between caspase-2 and the adapter molecule, PIDD (**p**53-**I**nduced **D**eath **D**omain) (Fig. 5a). PIDD is a death domain-containing protein that is involved in p53-dependent apoptosis. The

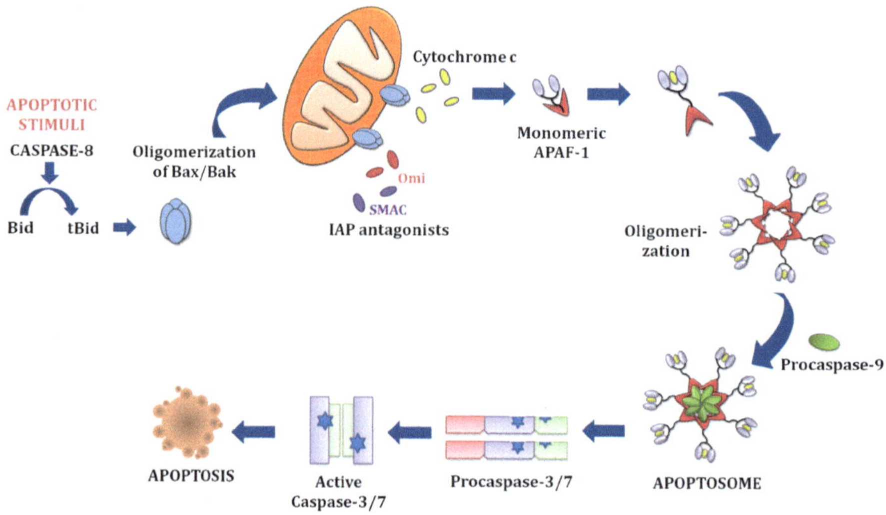

Fig. 4 Mitochondria-mediated intrinsic cell death pathway. Activation of caspase-8 through the extrinsic pathway promotes the truncation of proapoptotic molecule, Bid. The truncated version of this protein (tBid) induces oligomerization of proapoptotic effector proteins such as Bax and Bak. The large oligomers, so formed, then get inserted into the mitochondrial outer membrane, and the subsequent membrane permeabilization induces the efflux of cytochrome c and other antiapoptotic proteins such as HtrA2 and Smac/DIABLO. The released cytochrome c binds to the inactive monomer of Apaf-1, which induces a conformational change and promotes oligomerization of the latter to further form a heptameric wheel-like structure called '*apoptosome*'. This complex later recruits and activates initiator procaspase-9 through the interaction of exposed CARDs. Activated caspase-9 then propagates a proteolytic cascade by cleaving effector caspase-3 and caspase-7 to trigger apoptosis

activation platform is formed when RAIDD binds to procaspase-2 via CARD–CARD interaction that is being regulated through phosphorylation of the caspase (elaborated in Sect. 7.5). Procaspase-2 bound RAIDD molecules further form a complex with the adapter protein PIDD through adapter-mediated homotypic interaction.

PIDD is a 100 kDa protein that is composed of seven N-terminal LRRs (**Leu**cine-**R**ich **R**epeats), two ZU-5 (**ZO**-1 and **U**nc**5**-like) domains, a putative oligomerization domain UPA (uncharacterized protein domain in **U**NC5, **P**IDD and **A**nkyrin) and a C-terminal death domain (DD). It contains two intein regions that promote its autoproteolytic processing in cells to generate three fragments. PIDD-N (48 kDa) contains the N-terminal LRRs and the proximal ZU-5 domain; PIDD-C (51 kDa) comprises the distal ZU-5 domain, UPA and DD; and PIDD-CC (37 kDa) lacks the ZU-5 domain but retains an intact UPA and DD [109]. The first cleavage produces PIDD-C that forms a complex with RIP1 and NEMO (a component of NF-κB-activating kinase complex), which in turn leads to NF-κB activation [109, 110]. The second cleavage generates PIDD-CC that binds to RAIDD via DD–DD interaction [109, 111]. The PIDD–RAIDD–procaspase-2 large protein complex, so

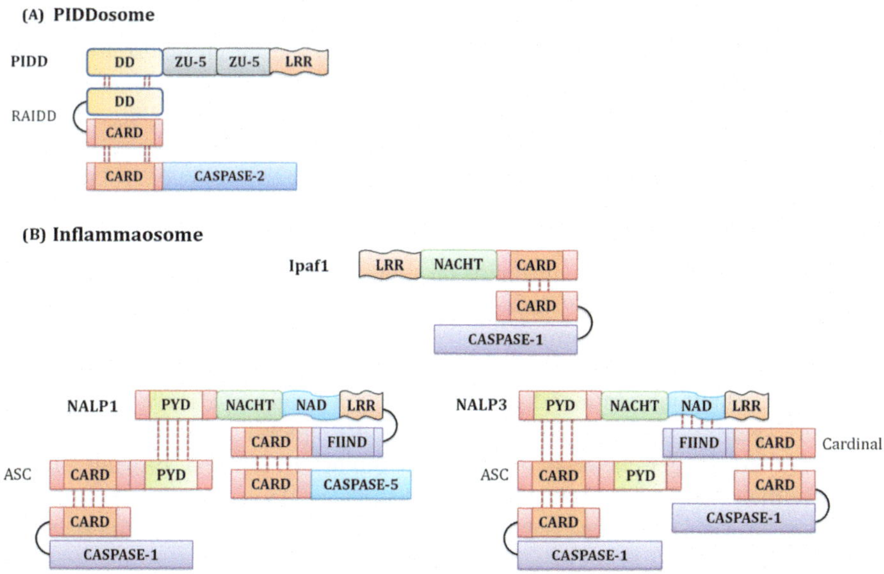

Fig. 5 Caspase activation through PIDDosome and inflammasome. **a** PIDDosome structure. **b** Inflammasome structure. The *red dotted lines* indicate the interaction between the domains of different protein molecules involved in forming the PIDDosome and inflammasome. The acronyms for the domains involved have been described in the text

formed, has been shown to mediate caspase-2 activation in response to genotoxic stimuli and is together termed as PIDDosome. The crystal structure of PIDDosome has revealed the presence of multiple PIDD and RAIDD subunits [112, 113], and the structure resembles the CD95–FADD complex that is involved in procaspase-8 activation.

Caspase-2 activation can also occur through a mechanism that involves p53-dependent CD95 upregulation and recruitment of caspase-8 to the DISC complex. This coordination of caspase-2 and caspase-8 cleaves Bid and subsequently activates mitochondrial apoptosis [114, 115]. However, caspase-2 is also involved in other divergent functions such as protection against DNA damage [116] or cancer development [117], and its activation has been found to occur even in the absence of PIDD in some cases [118, 119].

4.2.3 Proinflammatory Caspase Activation Through Inflammasome

Apart from being proapoptotic factors, several caspases also act as critical mediators of innate immune responses as mentioned earlier. These proinflammatory caspases are known to regulate both cytokine processing and inflammatory ignalling. In humans, caspase-1, caspase-4, caspase-5, and caspase-12 are included in the inflammatory subset, whereas the same function is served by caspase-1, caspase-11, and caspase-12 in mice. The genes encoding the human proinflammatory caspases are located in close proximity on chromosome 11q22, while the murine proinflammatory caspases are clustered on chromosome 9A1 in the following order from the telomere: caspase-1, caspase-11 and caspase-12. The organization of the syntenic region in humans is similar, except that the location of murine caspase-11 gene is replaced by the genes of human caspase-5 and caspase-4 [120]. Similar to their proapoptotic counterparts, these caspases have large pro-domains containing CARDs and are produced as inactive zymogens in resting cells. The procaspases are activated only after cellular stimulation through infectious agents (e.g. viruses, bacteria and fungi) or inert substances (e.g. uric acid crystals, calcium phosphate crystals, alum and asbestos) [121]. These stimuli induce inflammation through the formation of a cytosolic caspase activation platform known as inflammasome.

The external stimuli are initially recognized by a family of **P**attern-**R**ecognition **R**eceptors (PRRs), including TLRs (**T**oll-**L**ike **R**eceptors) and NLRs (**N**ucleotide-binding domain, **L**eucine-rich repeat-containing **R**eceptors) proteins. PRRs recognize the conserved PAMPs (**P**athogen **A**ssociated **M**olecular **P**atterns), which are unique to each pathogen, and therefore, are essential molecular structures for pathogen survival and to regulate innate immune responses. In a resting cell, the NLRs exist as monomers and are held in an inactive conformation. Different NLR-driven inflammasomes contain several other NLR members [33], which mainly include three subfamilies of proteins:

(1) *IPAF (or NLRC4) subfamily*: IPAF contains domains like CARD, NACHT and the C-terminal LRR domain. It associates specifically with CARD of caspase-1 through CARD–CARD interaction, while the NACHT domain induces oligomerization and promotes proximity of the caspases. The LRR domain is probably involved in ligand sensing (Fig. 5b).
(2) *NAIP subfamily*: This NLR subfamily of proteins shares highest sequence similarity with IPAF in its NACHT and LRR domains. However, NAIP harbours three N-terminal-BIR domains (**B**aculovirus **I**nhibitor of apoptosis **R**epeats) instead of CARD, and the protein was originally proposed to interact with IPAF. This suggests that it may be a part of the same caspase-1-activating complex [122, 123] that includes both IPAF and NAIP.
(3) *NALP subfamily*: NLRs like NALP1, NALP2 and NALP3 were shown to be the central scaffold of inflammasomes. These proteins harbour NACHT and LRR like those present in IPAF and NAIP, but they are additionally characterized by an N-terminal **Py**rin **D**omain (PYD). The PYD of NALPs is known to interact with the N-terminal PYD of the adapter protein ASC (**A**poptosis associated **S**peck-like protein containing **C**ARD). ASC also contains a C-terminal CARD that binds and further recruits the inactive inflammatory procaspase, typically procaspase-1. The inflammasome, so formed, may also further recruit caspases such as caspase-5 via the C-terminal CARD of NALP1 or another caspase-1 molecule via the C-terminal CARD domain of protein CARDINAL that is another component of the NALP2/3 inflammasome (Fig. 5b).

Towards the end of the pathway, activated caspase-1 processes proinflammatory cytokines such as pro-IL-1β, pro-IL-18 and pro-IL-33 that eventually leads to inflammation. The repertoire of caspase-1-activating molecular machines thus seems to be potentially very complex, as there exists a differential requirement of IPAF, ASC and NALP3 in sensing diverse stimuli activating caspase-1 [124, 125].

Besides the aforementioned caspase-dependent pathways, multiple non-caspase proteases are also shown to process and activate caspases directly. For example, upon exposure to endoplasmic reticulum stress, the Ca^{2+}-activated protease, m-calpain can process caspase-12 [126], whereas the lysosomal protease cathepsin B was shown to process caspase-1 and caspase-11 in vitro [127, 128]. The serine protease granzyme B directly activates caspase-3 and is critical for the killing of virally infected cells by cytotoxic lymphocytes (CTL) [129].

In conclusion, cells possess a vast array of caspase activation cascades that respond differently and specifically to a variety of cytotoxic and inflammatory stimuli.

5 Molecular Basis of Caspase Activation

Apart from a few special cases, caspase processing is usually mediated by caspases themselves. This was supported by early observations made in bacteria, wherein simple overexpression of wild-type caspases, but not active-site mutant zymogens, resulted in spontaneous activation of these active caspases [62, 130]. Likewise, in the highly purified samples of caspase-1, cleavage between its large and small subunits was assisted with a concomitant increase in activity [131–133]. The concept that the processing sites of individual initiator caspases often conform to their own substrate specificities also provided a mechanistic explanation for their autoactivation [25]. This observation in turn correlated with the presence and importance of residual activity in caspase zymogens [133]. Initiator caspases such as caspase-8 and caspase-9 are known to be quite active as zymogens, whereas the executioner caspase-3 is the one that acts as a 'true' zymogen, since it is much less active [66]. Thus, such differences in caspase zymogen activity highlight the ability of initiator caspases to serve as proximal responders to proapoptotic signals due to their high propensity towards autoactivation. It also necessitates the activation of executioner caspases by their predecessors.

5.1 Activation of Effector Caspases: Transactivation Mechanism

The process of executioner caspase activation is generally termed as '*transactivation*', since it is mediated *in trans* by another heterologous caspase. As demonstrated through in vitro experiments, the upstream or apical initiator caspase-8 and caspase-10 that possess long prodomains can efficiently cleave and activate executioner procaspase-3 and procaspase-7, which either have or lack short prodomains. These downstream caspases further serve in amplifying the death signal and cause cell demise. As discussed in the mitochondrial intrinsic pathway, caspase-9 cleaves and activates caspase-3 and caspase-7. Activated caspase-3 further cleaves caspase-2 and caspase-6 as well as provides a feed-forward amplification loop by cleaving procaspase-9 [134]. Caspase-7 has no significant role in sensitivity to intrinsic cell death. In contrast, caspase-3 can inhibit ROS production and is the effector caspase necessary for efficient cell killing [135]. According to a model proposed for the maturation of procaspase-3 dimer, it is suggested that the zymogen exists as a dimer in vivo and the removal of the N-terminal propeptide is neither required nor beneficial for its activation [32]. Thus, in contrast to the caspases containing long prodomains, the propeptide of caspase-3 does not appear to be involved in dimerization. However, a research work suggested that apart from propeptide cleavage, additional unknown structural determinants may also play a critical role in the activation process of caspases such as caspase-6 [136].

On the basis of biochemical and structural characterization, active and zymogenic forms of caspase-7 were found to be homodimers [137, 138]. Upon activation, the overall structure of the enzyme undergoes very minor changes, except for the rearrangement of crucial catalytic loops L2, L3 and L4, which undergo dramatic conformational transitions. In procaspase-7, these three critical loops stay away from each other in a conformation prohibitive to substrate binding and this provides a molecular mechanism that explains its zymogenicity [19]. During the activation process, cleavage at Asp rescues L2′ loop from the neighbouring small subunits in caspase-7 homodimer and allows it to form a tight 'active bundle' with L2 and L4 loops. This initial cleavage in turn produces 90° rotation of L2 and makes catalytic Cys accessible to the solvent. The subsequent processing steps thus become dispensable for the full activation of caspases. However, this activation mechanism just primes caspase-7 for substrate binding. The comparison of the active caspase-7 with its inhibitor-bound structure suggests that the substrate-binding groove in the former is still not properly formed, as L2′ loop remains in the closed conformation [137]. Upon substrate/inhibitor binding, the formation of the '*active bundle*' is completed and it triggers 180° flipping of L2′ loop. This subsequently results in its upward movement towards L1–L4, which completes the active bundle formation. Therefore, binding of the substrate by caspases is a dynamic 'induced-fit' process [48]. This information signifies that caspase processing and substrate binding lead to continuous sequence of conformational changes that eventually guide the 'loop bundle' formation and proper arrangement of the active centre [137, 138]. This unique mechanism thus emphasizes the importance of dimerization for executioner caspase activity.

5.2 Activation of Initiator Caspases: Allosteric Regulation and Induced Proximity Mechanism

On the basis of all crystal structures available for caspases, it was believed that they are activated by proteolytic cleavage within their linker region [134]. However, the initiator caspase-8 and caspase-9 are unique among all the other caspases, as studies reveal that proteolytic cleavage is neither sufficient nor necessary for their activation [26, 47, 101]. In contrast to the zymogens of effector caspase-3 and caspase-7 that are already present as dimers in their latent forms, the initiator caspases reside as monomers and require dimerization to assume an active conformation. The precise detail of initiator caspase activation, however, still remains elusive. Since there is no proteolytic enzyme upstream of these caspases, the zymogens tend to exhibit high intrinsic residual activities and low zymogenicity values (i.e. ratio of activity of the cleaved to the uncleaved form). The zymogenicity for caspase-8 and caspase-9 were found to be of at most 100 or 10, respectively, in contrast to the values of >10,000 for caspase-3 [134]. This also raises an intriguing question as to how such activity is regulated in living cells.

This intrinsic zymogenicity feature of initiator caspases is very much similar to the mechanistically unrelated serine proteases of the blood clotting and fibrinolytic cascades. However, the initiator caspases vary from other serine proteases due to the cofactor induced oligomerization that result in the activation of these upstream proteases [134]. Unlike effector caspases, cleavage at the intersubunit linker in initiator molecules is dispensable for its activity; however, this structural rearrangement might provide stability to the protease. This oligomerization is adequate enough to trigger the processing of inflammatory and initiator procaspases, thus establishing the central postulate of the induced proximity model.

The *Induced Proximity Model* was first proposed in 1998, which states that the initiator caspases autoprocess themselves when brought into close proximity of each other [66]. Being the most upstream caspase in the extrinsic pathway, activation of caspase-8 is brought about by the adaptor-mediated multimerization. Even though this zymogen possesses significant activity, subsequent proteolytic cleavage is still required for its full activation [139] and an '*induced proximity*' mechanism has been suggested for the same. In this model, FADD-mediated trimerization increases the local concentration of caspase-8 zymogen and brings about intermolecular processing of the protease [19]. The model has been further refined by Salvesen and colleagues to a '*proximity-induced dimerization*' model. Herein, biochemical studies show that dimerization of initiator caspases drives their activation. Thus, internal proteolysis is a secondary event that does not activate these apical caspases but results in partial stabilization of activated dimers [140]. Moreover, studies of FLIP-mediated caspase-8 activation have provided additional inputs on the allosteric aspect of this model [80, 141]. As described earlier, FLIP is a close homologue of caspase-8 and is catalytically inactive. It generally exists in two splice variant forms: short FLIP(S) and long FLIP(L) [142]. The former is a homologue of caspase-8 prodomain and can directly block caspase-8 recruitment into the DISC [143]. However, the long form contains a pseudo-catalytic domain that lacks the active-site Cys. This form has been known to induce a reordering of the catalytic centre of caspase-8 in a manner similar to dimerization-driven caspase-9 activation [80]. Also, the L2 loop of this protease might be long and flexible that supports its autocleavage and results in the generation of a partially processed, but fully active, heterodimer of FLIP(L) and full-length caspase-8. Thus, the activation of caspase-8 may include both allosteric and processing-based mechanisms.

Caspase-9 is unique among all the other caspases. The X-ray structural studies reveal the active site of monomer to be in the inactive conformation, and its activation is driven by a dimerization-dependent conformational change. Upon activation, the interaction surface of the neighbouring monomer provides an acceptor site for the activation loop and results into a rearrangement in the substrate-binding pocket and reorientation of the catalytic Cys, thus forming a functional active site. Interestingly, caspase-9 dimer possesses only one active site, but the significance of this finding is currently unknown [47]. Allosteric regulation of this protease is served by the apoptosome complex (as earlier described in Sect. 4.2.1), which promotes its homodimerization and activation [144]. Therefore, the apoptosome complex can be considered as a holoenzyme, in which the caspase is 1000-fold more active than its monomeric form [101].

In contrast to caspase-8, L2 loop of caspase-9 might be extended enough to eliminate the need for proteolytic activation completely [47]. Therefore, it exhibits a much higher level of catalytic activity in the apoptosome through '*induced conformation model*'. This model proposes that the apoptosome may directly activate monomeric caspase-9 through allosteric modification, as described earlier. Alternatively, the apoptosome may assemble dimeric caspase-9 into a higher-order complex that results in the modification of active-site conformation for its enhanced activity. This form of allosteric mechanism may also contribute to the activation of caspase-1 and caspase-5 in '*inflammasome*' complex involving NALP1 [145].

6 Caspase Cross-Activation

As previously described, the established activation mechanisms precisely explain the initiator–executioner dichotomy of caspase family. However, this segmentation primarily applies only to the initial stages of apoptotic signalling. A complex network of caspase cross-activation supposedly exists in the downstream steps of apoptosis [19]. Recombinant caspase-1, caspase-2, caspase-3, caspase-6, caspase-7 and caspase-8 are known to process numerous in vitro-translated caspase zymogens [146]. For instance, caspase pair caspase-6/caspase-3, caspase-6/caspase-7 and caspase-8/caspase-3 form reciprocal feedback loops. Caspase-8 has been shown to process all caspases efficiently, whereas caspase-1 and caspase-11 are able to activate executioner caspases, but not vice versa. It has been demonstrated that caspase-2 can be activated by a host of caspases, but it failed to process any tested caspase.

As executioner caspases are also capable of processing the normally upstream activator molecules in vitro, these findings largely highlight the complexity of caspase transactivation patterns. An intricate network of one such caspase transactivation event has been ably demonstrated in a cell-free apoptosis system [147]. The addition of cytochrome c to post-nuclear Jurkat (JK) cell lysates resulted in the initiation of caspase-9/caspase-3 partnership, with caspase-3 subsequently processing caspase-2 and caspase-6. These, in turn, activate apical caspase-8 and caspase-10. On the other hand, besides processing caspase-3, caspase-9 also cleaves caspase-7. Caspase-3 simultaneously forms a feedback loop to cleave caspase-9. The primary function of the latter event is to remove binding site for XIAP from the L20 loop of caspase-9, thus changing its direction towards inhibition [148].

These in vitro observations briefly elaborate the potential complexity of putative downstream caspase cross-activation networks induced inside cells. Although upstream activation may initially involve particular caspases, there also resides the probability that a vast array of additional family members may eventually get activated. Such transactivation processes blur one thin line of distinction existing between initiator and executioner caspases. This may serve to expand the repertoire of accessible caspase substrates and stimulate diverse downstream signalling processes, for example, cytokine processing versus apoptosis [19]. Thus, in the event of inhibition or malfunctioning of a specific pathway, the transactivation and cross-activation processes can subsequently provide multiple redundant pathways that ensure smooth execution of apoptotic programme. Subsequently, it can also accelerate the same programme through massive caspase activation, trigger specific apoptotic execution subroutines and generate sustained signalling through positive feedback loops.

7 Regulation of Caspases

Since proteolysis is an irreversible process, activation of most of the constitutively expressed caspases is tightly regulated in multiple cell types. The activation mechanisms described above represent by far the most important ways to regulate caspase activity. Caspases with long prodomains play a regulatory role in apoptosis by activating those effector caspases with shorter prodomains, as they may undergo structural rearrangements required for their autoactivation. However, to prevent unwanted physiological responses, cells may also employ additional regulatory mechanisms that have been well established and are elaborated in this section.

7.1 Transcriptional Regulation of Caspases

The existence of an endogenous suicide mechanism was first deduced from the fact that new protein synthesis is required in certain apoptotic paradigms [149]. Although this is not applicable in most cases, subsequent analysis of the murine caspase-11 demonstrated the importance of this regulatory mechanism and its critical role in controlling the caspase activity before or during an apoptotic stimulus. Usually, the basal expression level of caspase-11 in healthy mice and resting cells is found to be very low. However, in vivo treatment with lipopolysaccharide (LPS) or other pathological stimuli led to a remarkable upregulation of caspase-11 expression [35, 150]. The transcription of caspase-5, which is the putative human caspase-11 homologue, has also been found to be upregulated upon LPS treatment [151]. On the translational front, mRNA synthesis, protein expression and catalytic activation of caspase-5 were highly regulated in response to various proinflammatory stimuli (such as LPS), ATP and endoplasmic reticulum (ER) stress inducers [152]. The caspase-2 transcriptional regulation appears to be coordinated with transcriptional induction of its activators PIDD and RAIDD upon DNA damage [153]. A bit counter-intuitive study showed a p53/p21-dependent pathway for downregulating caspase-2 mRNA expression occurs in resting cells as well as in response to DNA damage [154] and thus reduces unwanted extraneous cell death.

Caspase-2 is known to have two isoforms: the proapoptotic long isoform, caspase-2L, and the antiapoptotic short isoform, caspase-2S. Through RNAi-mediated loss-of-function approach, it has been demonstrated that the RNA-binding protein HuR negatively regulates both caspase-2 mRNA and the protein levels [155]. Attenuation of HuR sensitizes adenocarcinoma cells by increasing the translation of caspase-2L and enhancing intrinsic apoptosis. On the other hand, the nucleotide excision repair factor XPC (genetic deficiency of which increases predisposition to cancer) is known to enhance DNA damage-induced apoptosis by downregulating caspase-2S [156]. In case of caspase-9, the alternative splicing results in two different caspase-9 transcripts that differ by four exons. The splice variant caspase-9a is found to be proapoptotic, while the shorter isoform, caspase-9b, is antiapoptotic [157, 158]. However, this splicing mechanism has been found to be defective in **N**on-**S**mall-**C**ell **L**ung **C**arcinoma (NSCLC), a type of epithelial lung cancer that accounts for about 85% of all lung cancers [159]. Moreover, caspase-9a to caspase-9b ratio increases with K-Ras^{12V} overexpression as caspase-9a promotes apoptosis, while **E**pidermal **G**rowth **F**actor **R**eceptor (EGFR) overexpression lowers this ratio. K-Ras is known to be a downstream component of EGFR signalling network and associates the growth-promoting signals from cell surface to the nucleus. On the other hand, the EGFR signalling network regulates cancer cell proliferation, apoptosis and tumour-induced neo-angiogenesis.

RfCasp10, a homologue of caspase-10 from black rockfish (*Sebastes schlegelii*), has been found to be upregulated upon pathogenic stimuli elicited by *Streptococcus iniae* and mediates apoptosis in immune-relevant tissues such as blood, liver and spleen [160]. Caspase-10 mRNA level is also negatively regulated in Jurkat cells expressing HIV tat gene when compared with control-treated Jurkat cells [161]. The tat protein (HIV-1-trans-activating protein) is a nuclear transcriptional activator of viral gene expression that enhances the efficiency of viral transcription. The reduction of caspase-10 mRNA, coupled with a concomitant increase in the expression of c-FLIP, is sufficient enough to enhance the resistance of

Tat-expressing T cells to TRAIL-induced apoptosis. Elaborating on caspase-14, it is known to be transcriptionally induced during the terminal differentiation of keratinocytes [162].

Deregulation of oncogenic E2F (E2 factor) transcription factor by adenovirus E2A oncogene overexpression or through loss of the tumour suppressor Rb gene results in the accumulation of zymogen forms of multiple caspases through a direct transcriptional mechanism [163]. The E2F proteins belong to a family of transcription factors that are downstream effectors of the retinoblastoma (RB) protein pathway and are believed to play a critical role in cell division control. Increase in zymogen concentration, thus, appears to potentiate p53-mediated apoptotic signals, and this ably explains why oncogene-expressing cells exhibit an increased sensitivity to multiple apoptotic stimuli. This area of research has now been exploited for anticancer therapy. The adapter protein, Apaf-1, has also been found to be transcriptionally regulated by E2F1 and p53 [164].

7.2 Metabolic Regulation of Caspases

Numerous signalling pathways, linking cell survival and metabolism, have been shown to directly influence the activation of caspase-dependent cell death [165] and regulate apoptosis. For example, Akt (protein kinase B) is a serine/threonine-specific protein kinase that plays a crucial role in regulating multiple cellular processes such as metabolism, cell proliferation, apoptosis, transcription, cell migration and angiogenesis. Cell survival mediated by this oncogenic protein is dependent on glucose metabolism to inhibit Bax activation and cytochrome c release [33]. The addition of glucose-6-phosphate in Xenopus oocyte system has been shown to delay the activation of caspases. NADPH (**N**icotinamide **A**denine **D**inucleotide **P**hosphate), which gets generated through the pentose phosphate pathway, is important in inhibiting caspase-2 activation through the inhibitory phosphorylation at Ser-135 present in the CARD domain of the zymogen. Phosphorylation is also mediated by calcium/calmodulin-dependent protein kinase II (CaMKII) [104], which allows the binding of 14-3-3ζ that prevents caspase-2 association with the PIDDosome [166].

The genome-wide screen for genes required for apoptosis revealed multiple genes directly involved in cellular metabolism of Drosophila cells. These included the genes encoding **c**itrate **s**ynthase (CS) and 3-ketoacyl-acyl carrier protein synthase, which protects against apoptosis that is induced by the removal of DIAP1, a direct inhibitor of Drosophila caspases [167]. Further studies reveal that the cellular NADPH levels also modulate the Drosophila initiator caspase, Dronc, through Ser-130 phosphorylation. Depletion of NADPH removes this inhibitory phosphorylation and results in the activation of Dronc, which is then followed by cell death [168]. Further studies, thus, become essential in identifying the precise metabolic changes mediated by the up-/downregulation of the constitutively expressed genes that safeguard cells against apoptosis.

7.3 Nitrosylation

Nitric oxide (NO) is a short-lived gaseous free radical that regulates apoptosis through **c**yclic **G**uanosine **M**ono**p**hosphate (cGMP)-dependent and cGMP-independent pathways. NO donors are known to directly inhibit caspase activity through active-site nitrosylation [169], as the catalytic cysteine of caspases is very active and susceptible to such modifications. This mechanism has been implicated in the NO-mediated inhibition of apoptosis or inflammatory cytokine production.

Denitrosylation of caspase-3 occurs in cells following treatment with Fas ligand, and this increases the caspase activity with subsequent cell death in response to Fas. It has also been shown that only a subset of caspase-3 gets localized to the mitochondria and is nitrosylated [170]. Moreover, the primary adhesion fibroblasts isolated from patients showed diminished apoptotic responsiveness upon this nitrosylation event [171]. The crucial role of nitrosylation in regulating caspase-3 has been well established in other models, including treatment of cardiac muscle cells with pharmacological NO donor, S-Nitroso-*N*-acetyl-DL-penicillamine (SNAP) and doxorubicin [164]. Conversely, it has been demonstrated that a single cysteine in thioredoxin (Trx), a class of small redox proteins, is capable of bringing about a targeted, reversible trans-nitrosylation reaction with active-site cysteine of caspase-3 in response to intracellular NO levels [172].

Both in vitro and in vivo studies have shown that abnormal S-nitrosylation contributes to the pathogenesis of Parkinson's disease (PD) by impairing anticaspase-3 and thus, the antiapoptotic function of endogenous caspase inhibitor, XIAP [173, 174]. Procaspase-9 is also known to induce its cleavage by trans-nitrosylating XIAP via Trx system [175]. NO induces S-nitrosylation of caspase-8, which in turn reduces the sensitivity of hepatocytes towards TNF-α/Actinomycin D-induced apoptosis [176]. While carrying out in vivo studies in rat livers treated with TNF-α and D-galactosamine, the liver-specific NO donor 'V-PYRRO/NO' was found to block caspase-8 activity, Bid cleavage and mitochondrial cytochrome c release [176]. Caspase-9 has also been shown to be modified by nitrosylation when intracellular NO levels increase [177], and this event preferentially occurs in the mitochondrial compartment [170]. On the other hand, treatment of cell lines with SNAP does not block mitochondrial cytochrome c release but reduces caspase-9 activation in a cell-free system.

Cerebral ischaemia is a condition, wherein there is insufficient blood flow to the brain. It is promoted by the activation of neuronal nitric oxide synthase (nNOS) that significantly increases S-nitrosylation of Fas. S-nitrosoglutathione (GSNO) pre-treatment has been known to protect against this condition, as the compound inhibits post-ischaemic nNOS activation and NO release. This subsequently disrupts the activation of proapoptotic caspase-8, Bid as well as caspase-9 and caspase-3 [178]. Thus, it downregulates Fas S-nitrosylation and the consequent Fas signalling cascade. On the contrary, KRGE (**K**orean **R**ed **G**inseng **E**xtract) is another traditional herbal medicine that is known to prevent endothelium dysfunction in the cardiovascular system, increase nitric oxide (NO) production via Akt-dependent activation of endothelial NO synthase (eNOS) and thus, inhibit caspase-9/caspase-3 activities [179]. IL-15 is also known to differentially regulate caspase activity and influence susceptibility to T cell death. It was been reported that this interleukin molecule, with reduced glycolysis, increases the reactive nitrogen as well as oxygen species and promotes T cell survival via S-nitrosylation-mediated inhibition of caspase-3 [180]. Thus, the outcome of the NO generation might ultimately depend on the level of its production in the cellular context.

7.4 Oxidation

Similar to nitrosylation, excessive generation of reactive oxygen species (ROS) may also lead to inhibition of caspases. In the intrinsic mitochondrial apoptotic pathway, it is the oxidized cytochrome c that is capable of inducing the caspase cascade, and therefore, it is suggested that the redox state of cytochrome c might regulate apoptosis initiation by controlling the cellular ROS level [181]. XIAP is also known to regulate apoptosis by modulating the intracellular ROS levels and by enhancing the expression of antioxidant genes such as SOD-1/-2 and Txn2 [182]. Melatonin, a pineal hormone that regulates circadian and seasonal rhythms, promotes human platelet apoptosis by significantly increasing the generation of intracellular ROS and Ca^{2+}, thus facilitating events that lead to mitochondrial damage [183]. The NOD-like receptors (NLRs), NOD2 and the downstream adaptor protein, RIP2, are known to regulate ROS production. Increase in the intracellular ROS levels amplifies caspase-11 expression and non-canonical NLRP3 (**N**ucleotide-binding domain, **L**eucine-**R**ich-containing family, **P**yrin domain-containing-3) inflammasome activation by enhancing the c-Jun N-terminal kinase (JNK) signalling pathway [184]. This inflammasome complex plays a critical role in the pathogenesis of various acute or chronic inflammatory diseases.

Recent in vivo studies have revealed that caspase-2 confers protection against oxidative stress [185]. Stimulation of IL-1β secretion from the macrophages increases the activity of xanthine oxidase (XO). This enzyme is known to increase ROS generation and brings about NLRP3 inflammasome activation [186]. On the other hand, carbon monoxide and the transcription activator Nrf2 (**N**uclear factor E2-**r**elated **f**actor-2) have been shown to negatively regulate this inflammasome activation event and prevented subsequent mitochondrial dysfunction by inhibiting caspase-1 activation and secretion of IL-1β [187, 188]. Although ROS possesses high intrinsic toxicity, inhibition of caspases by such stimuli might not result in the alleviation of death. On the contrary, it may lead to a diversion from apoptosis to necrosis that is a non-specific caspase-independent death caused by overwhelming stress [189, 190]. For example, CHOP (**C/EBP** **Ho**mologous **P**rotein) is a transcription factor that is activated at multiple levels during various extracellular or endoplasmic reticulum (ER) stresses and plays a critical role in ER stress-induced apoptosis. LGH00168 is a novel activator of the CHOP promoter that is found to decrease the antiapoptotic Bcl-2 expression. However, it does not induce apoptosis but rather leads to RIP1-dependent necroptosis via caspase-8 inhibition and mito-ROS production [191].

7.5 Phosphorylation

Phosphorylation is an extensively utilized cellular regulatory mechanism that is best in bringing about rapid and reversible modulation of multiple signalling networks. As activation of caspases often initiates an irreversible process, direct phosphorylation-dependent caspase modulation might not be appropriate. However, phosphorylation plays an important role in regulating various other steps in apoptosis signalling. In response to prosurvival stimuli through growth factor stimulation, Akt and ERK (**E**xtracellular signal **R**egulated **K**inase) or MAPK (**M**itogen-**A**ctivated **P**rotein **K**inase) have been shown to phosphorylate caspase-9 at Ser-196 or Thr-125 residue, respectively [192, 193]. These phosphorylation events lead to inhibition of caspase-9 activity and therefore prevent apoptosis. Later studies have demonstrated that during hyperosmotic stress, phosphorylation on Ser-144 of caspase-9 by a serine/threonine kinase, PKC ζ (**P**rotein **K**inase **C zeta**), causes inhibition of procaspase activation and holds back the intrinsic apoptotic pathway [194]. The kinase is known to be involved in a variety of cellular processes such as proliferation, differentiation and secretion. Moreover, during mitosis, caspase-9 is phosphorylated at Thr-125 by the mitosis-promoting kinase complex, CDK1–cyclin-B1 (cell cycle regulator **C**yclin-**d**ependent **k**inase 1 (CDK1) bound to cyclin B1) [195]. During cell cycle, this process is thought to dampen the threshold for intrinsic apoptosis signals, especially upon prolonged mitotic arrest. However, this phosphorylation event does not affect binding of caspase-9 to Apaf-1, and its mechanism still remains obscure.

It has also been demonstrated that when NADPH levels decrease in cells, the enzyme PP1 (**P**rotein **P**hosphatase 1) dephosphorylates caspase-2 and permits its activation by the PIDDosome [166]. CDK1–cyclin-B1 also phosphorylates caspase-2 [196] in the interdomain region between the apparently large and small subunits of active caspases. This event directly inhibits the generation of mature, stable enzyme and thus, suppresses apoptosis. Conversely, mimicking the CDK-2 phosphorylation of full-length Bcl-xL (an antiapoptotic member of Bcl-2 protein family), at Ser-73, results into its cleavage and caspase activation [197]. Recently, it has been demonstrated that in tumour cells exhibiting high expression of Plk-3 (a **P**olo-**l**ike **k**inase that acts as an important regulator of cell cycle progression), ligand stimulation of the death receptor CD95 brings about increase in activity of Plk-3. This event further leads to stimulation of extrinsic death pathway through Thr-273 phosphorylation of caspase-8 by the kinase [198].

7.6 Ubiquitination and Degradation

Since activated caspases are short-lived species inside cells and exhibit a more dynamic turnover than the inactive zymogens, another mechanism of caspase regulation involves degradation via the proteasomal pathway [199]. Apparently, it has been suggested that proteins responsible for the rapid removal of caspases include IAPs, and therefore, an increasing attention has been paid in recent years to understand the ubiquitination/ubiquitylation of caspases by IAPs. As later elaborated in Sect. 7.7.2, many IAPs are known to contain RING domain and ubiquitin-associated domains that are involved in ubiquitin ligation of caspases [200, 201]. For example, XIAP contains a carboxy-terminal RING domain critical for its E3 ubiquitin ligase activity. E3 ubiquitin ligases are essential components of the proteasomal degradation process. They are known to recruit an E2 ubiquitin-conjugating enzyme and assist or directly catalyse the transfer of ubiquitin from E2 to its required protein substrate. XIAP is shown to polyubiquitylate the large subunit of active caspase-9 in vitro through its E3 ubiquitin ligase activity, but it cannot perform the same with the inactive procaspase-9 [164]. Conversely, GIDE (**G**rowth **I**nhibition and **D**eath **E**3 Ligase) is another mitochondrially located E3 ubiquitin ligase that induces apoptosis via activating caspases and inhibiting XIAP function [202]. Hence, additional studies become requisite to fully understand the channel by which XIAP-mediated caspase-9 ubiquitination can be controlled, as well as the significance of this activity in controlling apoptotic progression.

IAPs other than XIAP, such as cIAP1 and cIAP2, are known to monoubiquitylate caspase-3 in vitro [203]. The physiological relevance of cIAP2 has been demonstrated in neutrophil granulocytes, wherein induction of these IAPs by endotoxins has been found to accelerate the degradation of activated caspase-3 and subsequently reduce apoptosis during sepsis [204]. A recent study highlights the role of cIAP1 in ubiquitination of an intermediate processed form of caspase-3, which subsequently leads to proteasome-dependent degradation of the effector caspase as well as increases resistance to TRAIL-induced apoptosis [205]. The study is supported by a work that demonstrates the stabilization of cleaved caspase-3 (active subunits) and enhancement of apoptosis upon treatment of cells with the proteasome inhibitor, lactacystin [206]. Another research work has established that XIAP can polyubiquitylate active caspase-3 (not procaspase-3) and lead to its proteasome-dependent degradation [207]. However, further validation of these observations in animal models is required for harnessing their therapeutic potential [164]. Similar to caspase-3, it has been reported that cIAP1 can only ubiquitylate the mature protease, rather than the partially processed form as seen in caspase-3 [205].

In case of caspase-8, a cullin3 (CUL3)-based E3 ligase performs the polyubiquitination event and enhances the enzymatic activity of this apical protease [208]. After recruitment of caspase-8 to the DISC, this ubiquitination allows binding of active caspase-8 to p62. p62, being a poly-Ub-binding protein (Fig. 6), facilitates an association between CUL3-modified caspase-8 and similar other complexes in the cell that together lead to the formation of an aggregate of active

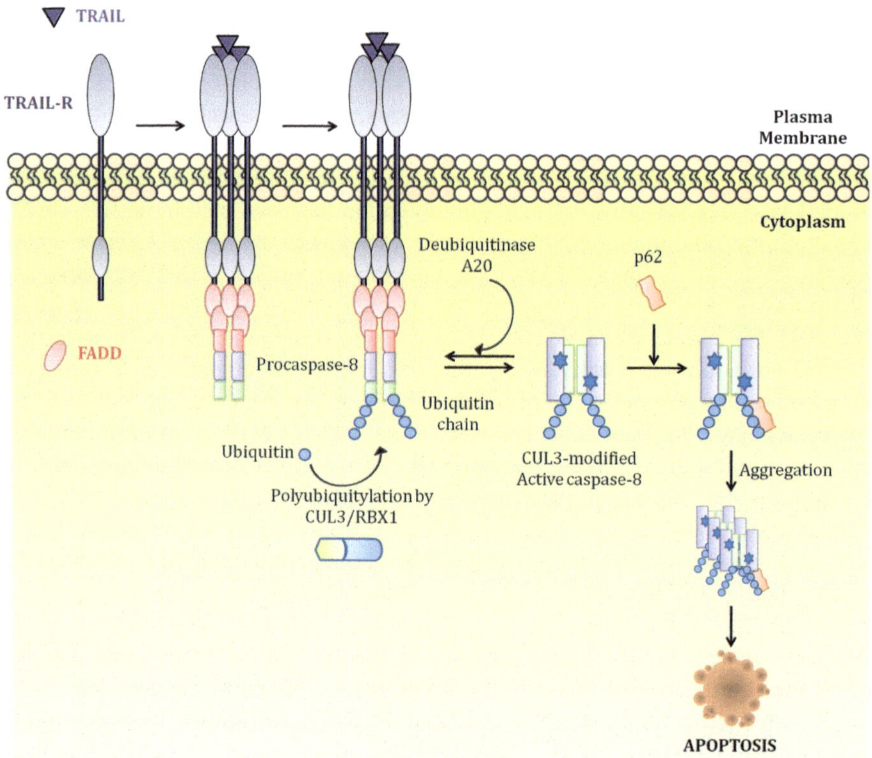

Fig. 6 Regulation of caspases through ubiquitination. Initiation of the extrinsic apoptotic pathway via the death receptor, TRAIL-R, occurs upon binding of TRAIL ligand and subsequent receptor multimerization at the plasma membrane. This leads to the formation of DISC and further activation of procaspase-8 through autoprocessing and dimerization. Upon activation, a cullin3 (CUL3)-based E3 ligase promotes ubiquitination of caspase-8, which could be reversed by deubiquitinase, A20. The ubiquitination process further leads to the interaction of active caspase-8 with a poly-Ub-binding protein, p62. Interaction of the CUL3-modified caspase-8 with p62 results into aggregation of this complex in the cytosol. These aggregated foci are supposed to have enhanced caspase-8 activity due to stabilization of the active dimer and thus trigger apoptosis

caspase-8 and p62. Interestingly, unlike other caspases, caspase-8 activity appears to be enhanced within these aggregated foci due to an increased stability of cleaved caspase-8. However, the deubiquitinase A20 has been reported to reverse the CUL3-mediated caspase-8 polyubiquitination [208].

7.7 Protein–Protein Interactions

Since programmed cell death involves a network of proteins that work in a coordinated manner, protein–protein interactions involving caspases play pivotal roles in regulating these proteases and maintenance of cellular homeostasis. Caspases exhibit the presence of well-conserved recruitment domains such as CARD or DED that mediate their homotypic interactions with adaptor proteins or other proteins of the apoptotic pathway often forming multiprotein complexes. Mature caspases also interact with substrate and inhibitor molecules through different binding sites as well. Therefore, the nature of these protein–protein interactions impacts the apoptotic pathway either positively or in a negative manner.

As discussed in the above subsections, there exist a variety of proteins that directly or indirectly bring about specific modifications in the caspases and promote their activation. For example, the serine protease granzyme B has been shown to catalyse activation of several effector caspases in vitro, thus leading to the assumption that this is one of its primary roles in cytotoxic lymphocyte-mediated cell death [209]. Moreover, proteins such as CAS (**C**ellular **A**poptosis **S**usceptibility protein) and Hsp70 (**H**eat **s**hock **p**rotein 70) act as mediators of a tumour suppressor protein PHAPI and together enhance caspase-9 activation. This is accomplished by facilitating the nucleotide exchange on Apaf-1, which is the backbone of caspase-9-activating apoptosome [210]. Thus, in addition to the protein–protein interactions, the levels of other (non-protein) cellular components, such as nucleotides and ions, also appear to play critical roles in regulating certain steps in caspase activation and promoting apoptosis.

On the contrary, the ability of inhibitors to mimic the binding site of a good substrate and take advantage of the substrate-binding cleft makes them suitable candidates in regulating complex proteolytic processes. In the apoptotic systems, these inhibitors may establish thresholds that determine the concentration of active cell death proteases required to initiate cell disassembly. Thus, it also prevents the consequences of accidental or spontaneous proenzyme activation and is known to confine the activity of these enzymes to specific cellular locations. The direct caspase inhibitors are some of the first proteins to be discovered that regulate caspase activity. These viral caspase inhibitors include CrmA (**C**ytokine **r**esponse **m**odifier A) from the cowpox virus and p35/p49 proteins produced by baculoviruses. Both of these are active-site-directed suicide inhibitors and do not appear to have direct cellular counterparts.

CrmA is a member of the serpin family and is a potent inhibitor ($K_i < 1$ nM) of some active initiator and inflammatory caspases such as caspase-1, caspase-8, and caspase-10. Expression of CrmA prevents the activation of both the executioner caspases (caspase-3 and caspase-7) as observed in a model of Fas-induced apoptosis [211]. On the other hand, the baculovirus protein p35 has no known homologues, and its selectivity for caspases is not clearly defined. Thus, these viral inhibitors are known to achieve rapid inhibition of caspases in a relatively non-selective manner, as a consequence of their requirement to present substrate-like sequences to the caspase.

Apart from these, there exist many proteins that affect caspase activity directly or indirectly by modulating the activation platform of these caspases. The inhibitory mechanism employed by decoy molecules involves proteins that are structurally related to caspase prodomains and would therefore compete for the same adaptors within activation platforms. For example, caspase-1-related CARD-only decoy proteins such as COP (**C**ARD-**O**nly **P**rotein), INCA (**In**hibitory **CARD**) and ICEBERG (a novel inhibitor of IL-1β generation) bind to caspase-1 prodomain via CARD–CARD interactions and prevent its recruitment to inflammasomes [212]. A few examples of significant proteins that interact directly with caspases and alter either their enzymatic activities are briefly elaborated below.

7.7.1 FLIP

The FLIP (**F**ADD-like **I**CE inhibitory **P**rotein) family of proteins was first discovered in a viral context, while its mammalian and other homologues were discovered concurrently [213]. As earlier mentioned, these proteins are similar to procaspase-8 in their primary sequence but lack essential catalytic residues. This structural identity helps them compete with procaspase-8 for binding to its cofactor, FADD, and prevent caspase activation. The long form of the mammalian cellular FLIP (c-FLIP(L)) is usually recruited to the DISC, where it is partially processed, while a portion of this protein is still retained at the DISC. This interaction with the caspase activation platform inhibits any subsequent recruitment of caspase-8 [164]. However, c-FLIP(L) induces a conformation of procaspase-8 that allows partial but not complete proteolytic processing, whereas the shorter splice variant, c-FLIP(S), can completely prevent partial processing of DISC-bound caspase-8 [214].

Interestingly, the FLIP proteins have also been shown to function as activators of the apoptotic process [80, 141]. In this case, c-FLIP(L) appears to heterodimerize with caspase-8, thus activating the caspase dimer. This hypothesis has been proven with the use of kosmotropes that induce the formation of FLIP(L)/caspase-8 heterodimers. In contrast to the proapoptotic caspase-8 homodimer, these heterodimers have been shown to have a lower kinetic barrier to activation [215] and can promote cell survival [216, 217]. Thus, it is suggested that it may be the local concentration of c-FLIP at the DISC that ultimately determines its role as an inhibitor or activator of caspase-8.

7.7.2 IAPs

The inhibitor of apoptosis proteins (IAPs) includes a conserved family of proteins that contain at least one **B**aculovirus **I**AP **R**epeat (BIR) domain that is a novel zinc-binding fold of about 70 amino acids. As discussed in Sect. 7.6, these proteins may also contain a RING domain that confers E3 ubiquitin ligase function [218]. Mammalian IAPs, mainly **X**-linked **IAP** (XIAP) and **c**ellular **IAPs** (cIAPs) are the best-characterized vertebrate members of the IAP family, which are able to convert the survival signal into a cell death-inducing signal or vice versa. For example, XIAP has been shown to be involved in the potent as well as selective inhibition of caspase-3 and caspase-7 in vitro. This multidomain protein efficiently and selectively inhibits the executioner caspases by binding to their active sites via its BIR2 domain. It also inhibits the initiator caspase-9 by preventing or reversing dimerization via its BIR3 domain [199]. These domains of XIAP combine two specific but relatively weak interactions with their target caspases for achieving specific tight inhibition via a two-site mechanism, and therefore, they are mechanistically distinct from the viral inhibitors mentioned earlier in this section [219].

The inhibitory action of XIAP can be antagonized by mitochondrial proapoptotic proteins Smac/DIABLO and HtrA2, which are released during an apoptotic stimulus along with cytochrome c [220–222]. As a direct inhibitor of these caspases, XIAP has been detected on the apoptosome [102, 223]. XIAP inhibits caspase-9 by a mechanism where the third BIR binds to the processed amino terminus of monomeric caspase-9 and prevents its dimerization [224]. A recent study in melanoma cell lines revealed caspase-mediated cleavage of XIAP via a positive feedback loop. This feedback event reduces caspase inhibition and XIAP levels through proteasomal degradation [225].

Other IAPs related to XIAP include cIAP1, cIAP2, ILP-2, ML-IAP as well as survivin; however, these do not directly inhibit caspases [219]. Like XIAP, cIAP1 and cIAP2 have three BIRs and a RING domain. Although they are incapable of inhibiting caspases under physiological circumstances, they are known to cause an indirect abrogation of caspase activity. This is accomplished by binding to IAP antagonists such as Smac/Diablo, thus allowing more XIAP molecules to be free to inhibit caspases [224]. As discussed in the previous subsections, they may also influence signalling via NF-κB and MAPK, or target caspases for ubiquitylation and proteasomal degradation. cIAP1 also specifically blocks apoptosis downstream of cytochrome c release by binding to and inhibiting active caspase-9 within the apoptosome [226]. In addition, XIAP and DIAP proteins have been shown to act as direct stoichiometric inhibitors of caspases. Apart from that, these IAPs also bring about specific modifications or interact with other proteins and indirectly assist in the regulation of caspase activity. For example, human oncoembryonic protein AFP (**A**lpha-**F**eto**p**rotein) is known to induce apoptosis in tumour cells by positively regulating cytochrome c/dATP-mediated apoptosome complex formation. It accelerates the release of active caspase-9 and caspase-3 as well as displaces cIAP-2 from the apoptosome [227]. Through abrogation of the inhibitory signalling, AFP affects XIAP–caspase interaction and rescues caspase-3 from inhibition [228]. Thus, IAPs are key players that are majorly involved in regulating caspase activity and in determining the cellular fate.

8 Caspase Regulation in Non-apoptotic Cellular Processes

Apart from their indispensable role in apoptosis, several active caspases have been found to be involved in various non-apoptotic cellular functions such as inflammation, protein secretion and differentiation. In the absence of a cell death signal, caspase activity may exhibit deadly consequences due to the dearth of a fine-tuned regulatory process. Therefore, tight control of caspases becomes particularly critical under circumstances in which the proteases are activated for short, well-defined periods of time or at isolated subcellular locations. For example, in hippocampal neurons, the transient activation of caspase-3 has been shown to play a role in synaptic LTD (**L**ong-**t**erm **D**epression: an activity-dependent reduction in the efficacy of neuronal synapses) and the internalization of the glutamate receptor, AMPA (α-**a**mino-3-hydroxy-5-**m**ethyl-4-isoxazole-**p**ropionic **a**cid receptor) [229]. In fact, treatment with NMDA (**N**-**m**ethyl-**D**-**a**spartate), a specific agonist at the NMDA receptor that mimics the action of glutamate, facilitates mitochondrial release of cytochrome c and caspase-9 activation, which are subsequently necessary for caspase-3 activation.

Further studies reveal that overexpression of a non-cleavable Akt1 (Protein kinase B) in hippocampal slice cultures inhibited LTD. Thus, the cleavage and inactivation of Akt1 is considered as a part of the link between caspase-3 activity and LTD [164]. Recent studies have shown the involvement of XIAP in regulating LTD and learning rate. Herein, XIAP deletion causes an increase in the caspase-3 activity within central neurons. This enhances AMPA receptor internalization, sharply increases LTD and thus, plays an important physiological role [230]. Another study using **O**lfactory **S**ensory **N**eurons (OSNs) has demonstrated the non-apoptotic role for Apaf-1 and caspase-9 signalling in development [231]. This non-apoptotic caspase activity leads to cleavage of the membrane-bound protein, Semaphorin 7A, which is critical for appropriate formation of axonal projections [232]. A variety of problems exist in mice lacking Apaf-1 or caspase-9 expression, and these include immature as well as erroneously routed OSN axons. However, Apaf-1 and caspase-9 do not impact the generation of OSNs, but rather affect specific formation of axonal projections. Thus, apart from being involved in cell death processes, caspases perform a plethora of cellular functions that are as diverse as signal transduction, cytoskeletal remodelling, and are also known to direct the behaviour of cells nearby the apoptotic cell through the caspase-dependent secretion of paracrine signalling factors [233].

9 Caspases in Pathophysiology of Diseases

Cells adopt different ways to regulate the normal functioning of its intricate apoptotic machinery by maintaining a precise balance between different agonists and antagonists. A slight imbalance in this homeostatic process can, therefore, have detrimental effects on the normal functioning of a healthy cell. Moreover, caspase activation can be considered as a double-edged sword, wherein diseases either associated with or that emanate from its low or high activity might vary from neurodegenerative disorders to cancer. Defective caspase activation that leads to inadequate cell death can promote tumourigenesis, whereas tremendous caspase activity and the resulting excessive cell death can initiate neurodegenerative processes. Caspases are also known to be involved in inflammation, where insufficient activation can lead to increased susceptibility towards infection, while hyperactivation promotes inflammatory conditions. In that light, restoring normal apoptosis in these different diseased conditions seems to be one of the major challenges of current biomedical research.

9.1 Caspases in Cancer

Cancer, the disease that involves rapid creation of abnormal cells, is a leading cause of death worldwide, and according to WHO (**W**orld **H**ealth **O**rganization), it accounted for 8.2 million deaths alone in 2012. To encounter cellular events such as mutations, altered gene expressions, variable occurrence of several apoptosis-related genes such as p53 tumour suppressor gene, alteration in the epigenetic profile and mRNA instability, etc., there exist several inherent sophisticated mechanisms that recognize and repair DNA mutations before they can become oncogenic. However, since evasion of apoptosis is one of the crucial hallmarks of human cancers and caspases are known to be the primary molecules that execute this cell death modality, deregulation of caspase activity and the pathways in which they participate might promote tumourigenesis. Also, through genetic and inhibitor studies, it has been demonstrated that the inactivation of individual caspases is not usually sufficient to either prevent progression of caspase cascade or disrupt the alternative non-apoptotic mechanisms. The malignant cells, instead, seem to gain a survival advantage by inactivating signalling mediators upstream of caspase activation [234].

In correlation to this, many cancers have shown to exhibit either reduced expression of proapoptotic caspases [235], or specific inactivating mutations. The absence of the inflammatory caspase-1 caused an increase in the proliferation of colonic epithelial cells in early-stage tumours, whereas it reduced apoptosis at the advanced stage [236]. The prevalence of mutations in another inflammatory caspase, caspase-5, was found to be highest in microsatellite instability (MSI)-positive gastric carcinomas [108]. Caspase-6 mutations have been found in 2% of 150 human cancers of colonic or gastric origin [237], and it is also suggested that decreased expression of this protease might assist in gastric cancer development [238].

Somatic mutations in caspase-7 have been found to affect the executioner function and result in pathogenesis of some human solid cancers, including colon, oesophageal and head/neck carcinomas. These tumour-derived caspase-7 mutants caused reduced apoptosis upon their overexpression in 293T human kidney cells [239]. Wild-type caspase-8 is known to act as a tumour suppressor in childhood neuroblastomas by exhibiting amplification of the proto-oncogene, N-myc [240]. N-myc is a nuclear transcription factor that is highly expressed in the foetal brain, where it is critical for normal brain development and its overexpression can lead to tumourigenesis. Inactivating caspase-8 mutations leads to loss of caspase-8 function and renders neuroblastoma cell lines to become resistant to death receptor-induced apoptosis [108]. In a study of 180 human colorectal tumours, the somatic caspase-8 mutations were detected in 5% of invasive carcinomas. Three of these mutations acted in a dominant-negative fashion and caused decrease in the caspase-8-mediated apoptosis [241]. In a similar case study of hepatocellular carcinomas (HCCs), a single somatic caspase-8 frameshift mutation resulted in a two-base-pair deletion and caused premature termination of translation as well as subsequent loss of caspase-8 function [242]. On the other hand, inactivating caspase-10 mutations have been found in **n**on-**H**odgkin's **l**ymphoma (NHL) samples [243], while rare caspase-10 mutations have also been detected in T cell acute lymphoblastic leukaemia, multiple myeloma [244] as well as in colon, breast, lung, hepatocellular [245] and gastric cancers [246].

Certain caspase polymorphisms also seem to affect caspase abundance or activity and therefore, have variable tumorigenic outcome. Polymorphism is a type of discontinuous genetic variation that results from evolutionary processes and leads to the occurrence of several different forms or types of individuals among the members of a single species. **S**ingle **N**ucleotide **P**olymorphisms (SNPs) are one of the most common types of genetic variation in a single nucleotide that occurs at a specific position in the genome, once in every 300 nucleotides on an average. Inheritance of such polymorphisms includes examples such as D302H variant in caspase-8 which substitutes histidine for an aspartic acid residue. In addition, a six-nucleotide deletion (-652 6N del) polymorphism has been found to be associated with a significantly reduced overall risk of breast cancer [247, 248] and other cancers [249]. Here, these alterations code the mutated caspase-8 protein which supposedly enhances its proapoptotic effects and thus prevents tumour cell persistence, although this remains to be proven under in vivo conditions [108].

Significant alterations of caspase-8, caspase-9 or caspase-1 have also been found to be associated with NHL [250]. In a study including 36 apoptosis pathway genes, germline variation in the caspase-9 gene as well as alterations of caspase-9 at SNP levels has been shown to be associated with NHL risk [251]. In lung cancer patients, there exist various polymorphisms that significantly decreased the risk of lung cancer. For example, as compared to the other haplotypes, increased promoter activity of the GC haplotype enhances caspase-9 expression, which is perhaps responsible for granting the protective effect [252]. A haplotype usually refers to the genetic make-up of an organism that is inherited together from a single parent because of the genetic linkage, and therefore, it can also refer to the inheritance of a cluster of SNPs. Thus, if a haplotype is associated with a certain disease, one can identify patterns of genetic variation and recognize the gene/s responsible for causing the disease.

Many studies have analysed the role that caspase-3 gene alterations play in promoting tumourigenesis. Compared with the GG genotype, the TT polymorphic variant was found to increase the risk of squamous cell carcinomas of the head and neck (SCCHN) in certain subgroups that included younger (<56 year) subjects, males and non-smokers [253]. In a stretch of DNA sequence in the chromosome, the description of a sequence change starts with a number referring to the first nucleotide affected (e.g. 76A > T), substitutions are designated by a '>' character, and the letters indicate the nucleotides involved in the substitution (A to T substitution). In lung cancer patients bearing at least one allele with a −928A > G, 77G > A or 17532A > C polymorphism, a significantly decreased risk of lung cancer as compared with those homozygous for the wild-type caspase-3 allele was observed [254]. Similarly, certain caspase-3 variants were found to reduce multiple myeloma risk and NHL [255, 256]. Of the 35 selected caspase-7 SNPs in the endometrial cancer patients, four were found to be in high linkage disequilibrium and caused an increased risk, while two SNPs were associated with reduced risk. The two other caspase-7 SNPs caused an increased risk as compared to individuals who are homozygous for the major caspase-7 alleles. Thus, in all, the polymorphisms existing in the crucial caspases and their haplotypes distinctly define an individual's genetic susceptibility towards cancer development.

9.2 Caspases in Inflammatory Diseases

A wide variety of inflammatory and autoimmune diseases have been attributed to the production of IL-1 and subsequent activation of caspase-1 [257]. Gout and pseudogout are the two most common crystal-induced arthropathies and autoinflammatory disorders whose pathogenesis has been linked to inflammatory responses in joints that are activated by the deposition of crystals. **M**ono**s**odium **U**rate (MSU) crystals are deposited in case of gout, while calcium pyrophosphate dihydrate crystals are present in pseudogout. Apparently, the deposition of crystals is mediated by the NLRP3 inflammasome [108]. Moreover, mutations in NLRP3 component of the inflammasome cause three rare inherited autoinflammatory diseases that are collectively known as **C**ryopyrin-**A**ssociated **P**eriodic **S**yndromes (CAPS). These mutations bring about alterations in the NLRP3-encoded protein, cryopyrin, and result into hyperactive NLRP3 inflammasome activity along with elevated IL-1β levels [258, 259].

Excessive caspase-1 activity has also been implicated to play a role in type 2 diabetes (T2D), as the elevated IL-1β levels are known to be involved in T2D development [260] that ultimately contributes to insulin resistance [261]. Familial Mediterranean fever (FMF) is another autoinflammatory disease that arises from missense mutations affecting the carboxy-terminal domain of pyrin, which in turn influences its interaction with procaspase-1 [262]. The C-terminal domain of pyrin is necessary and sufficient for the interaction, and binding is reduced by these FMF-associated mutations. On the other hand, a rare autosomal-dominant genetic disorder PAPA (**P**yogenic sterile **A**rthritis, **P**yoderma gangrenosum and **A**cne syndrome) is caused by mutations in the CD2-binding protein 1 (CD2BP1) gene. These mutations result in increased binding of CD2BP1 to pyrin and consecutively reduce its ability to inhibit inappropriate inflammasome activation [263].

Non-sterol isoprenoid deficiency, especially geranylgeranyl groups, is characteristic of an autosomal recessive disorder known as Mevalonate kinase deficiency (MKD) [264]. It is an autosomal recessive metabolic disorder that disrupts the biosynthesis of cholesterol and isoprenoids. It is characterized by recurrent episodes of fever, which usually begin during infancy. This isoprenoid deficiency is known to induce PI3K pathway-dependent procaspase-1 activation that leads to increased IL-1β production [265]. Thus, abnormality in the functioning and homeostasis maintenance by caspase-1 and IL-1β levels can considerably lead to severe inflammatory diseases.

9.3 Caspases in Other Diseases

Apart from playing a significant role in various physiological processes, caspases are also involved in the pathophysiology of several degenerative disorders and autoimmune diseases. From studies of post-mortem human brain samples and animal models, caspase activation has been demonstrated to play a key role in multiple neurodegenerative diseases like Alzheimer's disease (AD), Parkinson's disease (PD), Huntington's disease (HD) and **A**myotrophic **L**ateral **S**clerosis (ALS) [266]. Even though apoptosis is not the solitary player of cell demise in these disorders, it acts as a key component in partnership with other detrimental events such as inflammation, mitochondrial dysfunction, oxidative stress or protein misfolding and aggregation.

Rohn and co-workers have reported caspase-mediated cleavage of β-amyloid precursor protein (APP) [267]. In a murine AD model, caspase activation associated with this disease onset occurred earlier than the induction of neuronal apoptosis [268]. Likewise, caspase activation has also been noted before the development of neurofibrillary tangles of Tau in the brain of tau transgenic mice [269]. A nucleotide substitution from G to A in a particular SNP placed in the 5′ untranslated region of caspase-3 influenced the susceptibility towards Kawasaki disease (KD). Kawasaki disease is a rare childhood illness that brings about inflammation in medium-sized blood vessels throughout the body and is largely seen in children under the age of five years. The described substitution eliminates the binding of the NFAT (**N**uclear **F**actor of **A**ctivated **T** cells) transcription factor to the DNA sequence surrounding the SNP and alters caspase-3 expression in immune effector cells [270]. **A**utoimmune **L**ympho**p**roliferative **S**yndrome (ALPS) is another diseased condition that exhibits dominant mutations in CD95, CD95L or caspase-10 [108] and is thought to be caused by insufficient apoptosis of autoreactive T cells during negative thymic selection [271].

10 Targeting Caspases for Treatment of Various Diseases

As described in the previous section, deregulation of apoptosis can either cause damage to normal tissues through excessive apoptosis or may contribute to tumourigenesis, wherein apoptosis is prevented. The mechanistic significance of either inhibiting caspases to prevent excessive apoptosis in several diseases including neurodegeneration or reinstating apoptotic signalling towards selective targeting of malignant cells immensely exploits the caspase family of death-inducing molecules as a powerful therapeutic platform.

10.1 Caspase Activation and Treatment of Cancer

Chemotherapeutic drugs and irradiation are toxic to both normal and cancerous cells, and the damage is then translated through multiple steps of caspase activation, which ultimately leads to cell death. However, the therapy fails when these steps are compromised, and therefore, employment of alternative therapeutic strategies become essential. Within the last decade, several attempts have been made to develop molecules that directly activate caspases for use in cancer therapy. Justicidin A, a novel derivative obtained from the plant *J. procumbens*, is known to induce apoptosis in hepatocellular carcinoma through the activation of both intrinsic and extrinsic apoptosis pathways [272]. For example, through the activation of caspase-8 in the intrinsic pathway, justicidin A increases intracellular Bid that modulates $\Delta\psi_m$ (change in mitochondrial membrane potential) and causes the release of cytochrome c, but in turn it also decreases antiapoptotic Bcl-xL. Subsequently, the release of proapoptotic Smac/DIABLO and HtrA2 from the mitochondria activates caspase-9 and caspase-3 that initiate the intrinsic pathway. 6′-Hydroxy justicidin A (JR6) is another derivative that shares similarity with the lead antitumour drug. Podophyllotoxin has been shown to have promising apoptosis-induction efficacy in human bladder cancer cells [273]. Retinoic acid (RA) and its derivatives such as 13-cis RA, 9-cis RA and 4HPR (fenretinide) are also known to induce paracrine release of TRAIL in promyelocytic leukaemia cells as well as upregulate caspase-8 expression in neuroblastoma, medulloblastoma and small cell lung cancer cell lines [274]. This induced expression in turn sensitizes these cancer cells to TNF-α and other chemotherapeutics [275].

Since loss of caspase expression has also laid the grounds for many human cancers including neuroblastomas and small cell carcinomas, reinstating the expression and activation of caspases, especially caspase-8, might be of significant therapeutic advantage. This would also be particularly helpful in tumours caused due to gene dosage effect or hypermethylation of caspase-8 promoters. For example, 5-aza-2′ deoxycytidine (decitabine) is a cytosine nucleoside analogue that promotes genome-wide demethylation by inhibiting DNA methyl transferase covalent binding, which upregulates the expression of transactivators [276]. Therefore, rather than mediating caspase-8 promoter-specific demethylation, the compound increases caspase-8 promoter availability by allowing the binding of SP1 and ETS-like transcription factors [277, 278]. These transcription factors bind with high affinity to GC-rich motifs and to DNA sites with a central GGA(A/T) DNA sequence, respectively, and thus, regulate a variety of processes such as cell growth, apoptosis, differentiation and immune responses. The compound has been demonstrated to restore caspase-8 in a number of cancers including breast cancer, neuroblastoma, medulloblastoma, Ewing's sarcoma, lung carcinoma and gastric carcinoma [279].

Histone deacetylase inhibitors have also been known to assure adequate acetylation and transcriptional availability of caspase-3 promoter. These inhibitors have been demonstrated to activate apoptosis and increase sensitization to TRAIL-induced, radiation-induced and chemotherapy-induced apoptosis in prostate cancer, lung cancer, Ewing's sarcoma and medulloblastoma [279]. Similar to caspase-8 expression, approximately 30% of melanoma and lung cancer cell lines harbour inactivating mutations in the IFN-γ signalling pathway [280]. Therefore, it has been found that IFN-γ ably sensitizes the caspase-8-deficient neuroblastoma, medulloblastoma and Ewing's sarcoma cells by inducing apoptosis through death receptor ligand, cytotoxic drugs and radiotherapy in vitro [281, 282].

One of the challenging strategies involves selective activation of death receptor complexes that are directly linked to initiator caspases. High-throughput screening (HTS) studies have identified a series of molecules that include a-(trichloromethyl)-4-pyridineethanol (PETCM), gambonic acid and its derivative MX-2060, which can efficiently activate the executioner caspase-3 in vitro [283–285], and induce apoptosis in cancer cell lines. **P**rocaspase-**A**ctivating **C**ompound-1 (PAC-1) is another promising caspase-3 activator that contains a zinc-chelating motif [286] and activates the proenzyme through this motif [287]. However, the mechanism put forth by the research group seems to be controversial. Another independent group has strongly counter suggested that PAC-1 cannot directly activate executioner caspases, but uses an indirect and less effective activation mechanism instead [288]; however, further investigation is required to shed light into this ambiguity. As shown in recent in vivo canine studies, a '*next-generation*' compound (S-PAC-1) has been proved to be efficacious in inducing partial tumour regression [289].

Alternatively, active research is also underway in identifying compounds that activate caspases indirectly, which are currently under clinical trials. Some of these agents are known to block endogenous caspase inhibitors such as the Bcl-2 and IAP proteins [290], while some act as analogues of the endogenous IAP inhibitor Smac/DIABLO [291]. In addition, there are activators and antibodies that engage death receptors such as Fas, TNFR and TRAIL [292]. Survivin, the smallest protein in the IAP family, is found to be the fourth most upregulated gene in several cancers as compared to normal tissue [293]. Its increased expression is associated with loss of apoptosis, enhanced survival, angiogenesis [294] and resistance to therapy [295]. Therapeutic targeting of survivin is currently in various clinical trials, including the use of antisense oligonucleotides. LY2181308 is one such antisense molecule that induces caspase-3 proteolytic activity, cell cycle arrest and inhibition of cytokinesis in human tumour cells.

Various research laboratories and pharmaceutical companies have also attempted to develop recombinant forms of TRAIL or TRAIL receptor agonists (e.g. receptor-specific mAb) that are under various stages of clinical trials [296]. Apart from these, caspase activators (both peptidomimetics and small molecules) are being designed to specifically target cancer cells. Inducible caspases such as caspase-9 under the control of prostate-specific promoter specifically target prostate cancer cells [297]. Gene therapeutic targeting of caspases to cancer cells has been another popular approach in combating tumourigenesis [298].

It is important to note that the apoptosis-inducing agents for cancer treatment also encounter the same limitations of delivery and specificity as traditional chemotherapeutics. Exceptions such as certain caspase activators that inhibit anti-apoptotic molecules like Bcl-2 seem to have an enhanced therapeutic index when used to treat cancer cells that rely mainly on antiapoptotic proteins to stave off cell death [299]. However, a lot more needs to be done in this area of research to bring caspases in the forefront of cancer therapeutics. Substantial research endeavours using interdisciplinary approach have been currently set forth to rationally understand the structural and catalytic properties of active caspases and thus, consecutively manipulate the intricate apoptotic machinery for therapeutic interventions.

10.2 Caspase Therapy in Inflammatory Diseases

Advances in understanding the role of caspases in promoting various forms of inflammatory diseases have resulted in the development of various therapeutic targets. An inducible caspase-9 '*safety switch*' agent has been currently under phase I/II clinical trial and is used for the elimination of autoreactive lymphocytes in **g**raft-**v**ersus-**h**ost **d**isease (GVHD). The agent consists of a truncated caspase-9 protein that lacks the CARD domain and is fused to a forkhead protein-binding sequence. When used as a therapy, this caspase-9 safety switch protein is virally transduced into allo-depleted T cells, which are then administered to patients who have received a T cell depleted stem cell transplant [300]. The transduced protein agent then dimerizes in the presence of a particular small molecule, activates the hydrolytic function of enzyme and in turn, triggers apoptosis [301]. Since caspase-1 activation and subsequent generation of active IL-1β have also been found to be responsible for many inflammatory disorders, clinical trials are currently under way to assess the efficacy of antagonizing inhibitors and related molecules. These agents can antagonize either the generation or function of IL-1β or its receptor (IL-1R). For example, IL-1β antagonists have been found to be effective in the treatment of type 2 diabetes [302], confirming the important role of the NLRP3 inflammasome containing caspase-1 as a sensor of metabolic stress [303]. Patients with gout, pseudogout or pulmonary silicosis have also shown great improvement after treatment with such antagonists [108].

10.3 Caspase Inhibition and Prevention of Cell Death in Neurodegeneration

As described earlier, excessive apoptosis has been found to be responsible for several serious pathologies including neurodegenerative diseases, ischaemia-reperfusion injury, graft-versus-host disease and autoimmune disorders. Though there have been less striking therapeutic effects obtained by inhibiting caspase activity than those achieved through caspase activation, there still exist several instances, where caspase inhibition has ameliorated the symptoms of these diseased conditions that are affected by inappropriate apoptotic cell death. For example, preventing apoptosis through p35 expression prevents blindness in Drosophila mutants with retinal degeneration, which indicates that inhibition of caspases can functionally rescue cells from death [304].

A recent clinical trial has examined the therapeutic potential of a caspase inhibitor used for curing chronic hepatitis virus C infection, a condition that is accompanied by detrimental hepatocyte apoptosis [108]. Similarly, peptidyl caspase inhibitors have been found to be effective in animal models for stroke, myocardial ischaemia-reperfusion injury, liver disease and traumatic brain injury. Caspase inhibitors have also shown promising results in mouse models of neurodegenerative diseases such as Alzheimer's disease [305]. However, to prevent the risk of autoimmune disease or tumour progression to occur by chance, tissue-specific delivery of selective inhibitors becomes essential. Also, there are difficulties involved in generating small molecule, non-peptide inhibitors of proteolytic enzymes that would be selective, stable and would penetrate membranes effectively. Despite these, there has been an appreciable amount of work done on cysteine protease inhibition that provides a lead for the identification of several classes of potent, reversible and irreversible caspase inhibitors [306].

10.4 Current Scenario and Future Perspectives

Although a huge deluge of information on apoptosis has been made available from the current literature, translation of that knowledge into therapeutics still remains mostly unrealized due to several relevant concerns. For example, application of caspase inhibitors has been limited by their side effects, ineffectiveness due to redundancy in the apoptotic pathway and lack of an appropriate animal model system. However, a precise amalgamation of different aspects of various disease management strategies, including early detection, diagnosis and combination therapy, might circumvent the problem and will also provide a wider applicability of caspase mimetics/modulators in modern clinics.

References

1. Cohen GM (1997) Caspases: the executioners of apoptosis. Biochem J 326(Pt 1):1–16
2. Raff M (1998) Cell suicide for beginners. Nature 396(6707):119–122. doi:10.1038/24055
3. Thompson CB (1995) Apoptosis in the pathogenesis and treatment of disease. Science 267 (5203):1456–1462
4. Nicholson DW (1996) ICE/CED3-like proteases as therapeutic targets for the control of inappropriate apoptosis. Nat Biotechnol 14(3):297–301. doi:10.1038/nbt0396-297
5. Clarke PG, Clarke S (1996) Nineteenth century research on naturally occurring cell death and related phenomena. Anat Embryol 193(2):81–99
6. Philchenkov A (2004) Caspases: potential targets for regulating cell death. J Cell Mol Med 8 (4):432–444
7. Thornberry NA, Lazebnik Y (1998) Caspases: enemies within. Science 281(5381):1312–1316
8. Sulston JE, Horvitz HR (1977) Post-embryonic cell lineages of the nematode, *Caenorhabditis elegans*. Dev Biol 56(1):110–156
9. Sulston JE, Schierenberg E, White JG, Thomson JN (1983) The embryonic cell lineage of the nematode *Caenorhabditis elegans*. Dev Biol 100(1):64–119

10. Ellis HM, Horvitz HR (1986) Genetic control of programmed cell death in the nematode *C. elegans*. Cell 44(6):817–829
11. Hengartner MO, Ellis RE, Horvitz HR (1992) Caenorhabditis elegans gene ced-9 protects cells from programmed cell death. Nature 356(6369):494–499. doi:10.1038/356494a0
12. Miura M, Zhu H, Rotello R, Hartwieg EA, Yuan J (1993) Induction of apoptosis in fibroblasts by IL-1 beta-converting enzyme, a mammalian homolog of the *C. elegans* cell death gene ced-3. Cell 75(4):653–660
13. Zou H, Henzel WJ, Liu X, Lutschg A, Wang X (1997) Apaf-1, a human protein homologous to *C. elegans* CED-4, participates in cytochrome c-dependent activation of caspase-3. Cell 90(3):405–413
14. Gross A, McDonnell JM, Korsmeyer SJ (1999) BCL-2 family members and the mitochondria in apoptosis. Genes Dev 13(15):1899–1911
15. Fritz JH, Ferrero RL, Philpott DJ, Girardin SE (2006) Nod-like proteins in immunity, inflammation and disease. Nat Immunol 7(12):1250–1257. doi:10.1038/ni1412
16. Gagliardini V, Fernandez PA, Lee RK, Drexler HC, Rotello RJ, Fishman MC, Yuan J (1994) Prevention of vertebrate neuronal death by the crmA gene. Science 263(5148):826–828
17. Nicholson DW, Ali A, Thornberry NA, Vaillancourt JP, Ding CK, Gallant M, Gareau Y, Griffin PR, Labelle M, Lazebnik YA et al (1995) Identification and inhibition of the ICE/CED-3 protease necessary for mammalian apoptosis. Nature 376(6535):37–43. doi:10.1038/376037a0
18. Tschopp J, Martinon F, Burns K (2003) NALPs: a novel protein family involved in inflammation. Nat Rev Mol Cell Biol 4(2):95–104. doi:10.1038/nrm1019
19. Degterev A, Boyce M, Yuan J (2003) A decade of caspases. Oncogene 22(53):8543–8567. doi:10.1038/sj.onc.1207107
20. Lamkanfi M, Declercq W, Kalai M, Saelens X, Vandenabeele P (2002) Alice in caspase land. A phylogenetic analysis of caspases from worm to man. Cell Death Differ 9(4):358–361. doi:10.1038/sj/cdd/4400989
21. Nicholson DW, Thornberry NA (1997) Caspases: killer proteases. Trends Biochem Sci 22 (8):299–306
22. Walker NP, Talanian RV, Brady KD, Dang LC, Bump NJ, Ferenz CR, Franklin S, Ghayur T, Hackett MC, Hammill LD et al (1994) Crystal structure of the cysteine protease interleukin-1 beta-converting enzyme: a (p20/p10)2 homodimer. Cell 78(2):343–352
23. Wilson KP, Black JA, Thomson JA, Kim EE, Griffith JP, Navia MA, Murcko MA, Chambers SP, Aldape RA, Raybuck SA et al (1994) Structure and mechanism of interleukin-1 beta converting enzyme. Nature 370(6487):270–275. doi:10.1038/370270a0
24. Rotonda J, Nicholson DW, Fazil KM, Gallant M, Gareau Y, Labelle M, Peterson EP, Rasper DM, Ruel R, Vaillancourt JP, Thornberry NA, Becker JW (1996) The three-dimensional structure of apopain/CPP32, a key mediator of apoptosis. Nat Struct Biol 3(7):619–625
25. Thornberry NA, Rano TA, Peterson EP, Rasper DM, Timkey T, Garcia-Calvo M, Houtzager VM, Nordstrom PA, Roy S, Vaillancourt JP, Chapman KT, Nicholson DW (1997) A combinatorial approach defines specificities of members of the caspase family and granzyme B. Functional relationships established for key mediators of apoptosis. J Biol Chem 272(29):17907–17911
26. Stennicke HR, Salvesen GS (1999) Catalytic properties of the caspases. Cell Death Differ 6 (11):1054–1059. doi:10.1038/sj.cdd.4400599
27. Miscione GP, Calvaresi M, Bottoni A (2010) Computational evidence for the catalytic mechanism of caspase-7. A DFT investigation. J Phys Chem B 114(13):4637–4645. doi:10.1021/jp908991z
28. Saleh M, Vaillancourt JP, Graham RK, Huyck M, Srinivasula SM, Alnemri ES, Steinberg MH, Nolan V, Baldwin CT, Hotchkiss RS, Buchman TG, Zehnbauer BA, Hayden MR, Farrer LA, Roy S, Nicholson DW (2004) Differential modulation of endotoxin

responsiveness by human caspase-12 polymorphisms. Nature 429(6987):75–79. doi:10.1038/nature02451
29. Kersse K, Verspurten J, Vanden Berghe T, Vandenabeele P (2011) The death-fold superfamily of homotypic interaction motifs. Trends Biochem Sci 36(10):541–552. doi:10.1016/j.tibs.2011.06.006
30. Vaughn DE, Rodriguez J, Lazebnik Y, Joshua-Tor L (1999) Crystal structure of Apaf-1 caspase recruitment domain: an alpha-helical Greek key fold for apoptotic signaling. J Mol Biol 293(3):439–447. doi:10.1006/jmbi.1999.3177
31. Milam SL, Clark AC (2009) Folding and assembly kinetics of procaspase-3. Protein Sci (A Publication of the Protein Society) 18(12):2500–2517. doi:10.1002/pro.259
32. Pop C, Chen YR, Smith B, Bose K, Bobay B, Tripathy A, Franzen S, Clark AC (2001) Removal of the pro-domain does not affect the conformation of the procaspase-3 dimer. Biochemistry 40(47):14224–14235
33. Li J, Yuan J (2008) Caspases in apoptosis and beyond. Oncogene 27(48):6194–6206. doi:10.1038/onc.2008.297
34. Rano TA, Timkey T, Peterson EP, Rotonda J, Nicholson DW, Becker JW, Chapman KT, Thornberry NA (1997) A combinatorial approach for determining protease specificities: application to interleukin-1beta converting enzyme (ICE). Chem Biol 4(2):149–155
35. Kang SJ, Wang S, Hara H, Peterson EP, Namura S, Amin-Hanjani S, Huang Z, Srinivasan A, Tomaselli KJ, Thornberry NA, Moskowitz MA, Yuan J (2000) Dual role of caspase-11 in mediating activation of caspase-1 and caspase-3 under pathological conditions. J cell Biol 149(3):613–622
36. Zhivotovsky B (2003) Caspases: the enzymes of death. Essays Biochem 39:25–40
37. Lassus P, Opitz-Araya X, Lazebnik Y (2002) Requirement for caspase-2 in stress-induced apoptosis before mitochondrial permeabilization. Science 297(5585):1352–1354. doi:10.1126/science.1074721
38. Paroni G, Henderson C, Schneider C, Brancolini C (2002) Caspase-2 can trigger cytochrome C release and apoptosis from the nucleus. J Biol Chem 277(17):15147–15161. doi:10.1074/jbc.M112338200
39. Robertson JD, Enoksson M, Suomela M, Zhivotovsky B, Orrenius S (2002) Caspase-2 acts upstream of mitochondria to promote cytochrome c release during etoposide-induced apoptosis. J Biol Chem 277(33):29803–29809. doi:10.1074/jbc.M204185200
40. Luthi AU, Martin SJ (2007) The CASBAH: a searchable database of caspase substrates. Cell Death Differ 14(4):641–650. doi:10.1038/sj.cdd.4402103
41. Rendl M, Ban J, Mrass P, Mayer C, Lengauer B, Eckhart L, Declerq W, Tschachler E (2002) Caspase-14 expression by epidermal keratinocytes is regulated by retinoids in a differentiation-associated manner. J Invest Dermatol 119(5):1150–1155. doi:10.1046/j.1523-1747.2002.19532.x
42. Masumoto J, Zhou W, Chen FF, Su F, Kuwada JY, Hidaka E, Katsuyama T, Sagara J, Taniguchi S, Ngo-Hazelett P, Postlethwait JH, Nunez G, Inohara N (2003) Caspy, a zebrafish caspase, activated by ASC oligomerization is required for pharyngeal arch development. J Biol Chem 278(6):4268–4276. doi:10.1074/jbc.M203944200
43. Eckhart L, Ballaun C, Uthman A, Kittel C, Stichenwirth M, Buchberger M, Fischer H, Sipos W, Tschachler E (2005) Identification and characterization of a novel mammalian caspase with proapoptotic activity. J Biol Chem 280(42):35077–35080. doi:10.1074/jbc.C500282200
44. Eckhart L, Ballaun C, Hermann M, VandeBerg JL, Sipos W, Uthman A, Fischer H, Tschachler E (2008) Identification of novel mammalian caspases reveals an important role of gene loss in shaping the human caspase repertoire. Mol Biol Evol 25(5):831–841. doi:10.1093/molbev/msn012
45. Fuentes-Prior P, Salvesen GS (2004) The protein structures that shape caspase activity, specificity, activation and inhibition. Biochem J 384(Pt 2):201–232. doi:10.1042/BJ20041142

46. Mittl PR, Di Marco S, Krebs JF, Bai X, Karanewsky DS, Priestle JP, Tomaselli KJ, Grutter MG (1997) Structure of recombinant human CPP32 in complex with the tetrapeptide acetyl-Asp-Val-Ala-Asp fluoromethyl ketone. J Biol Chem 272(10):6539–6547
47. Renatus M, Stennicke HR, Scott FL, Liddington RC, Salvesen GS (2001) Dimer formation drives the activation of the cell death protease caspase 9. Proc Natl Acad Sci USA 98 (25):14250–14255. doi:10.1073/pnas.231465798
48. Shi Y (2002) Mechanisms of caspase activation and inhibition during apoptosis. Mol Cell 9 (3):459–470
49. Hofmann K (1999) The modular nature of apoptotic signaling proteins. Cell Mol Life Sci (CMLS) 55(8–9):1113–1128
50. Fairbrother WJ, Gordon NC, Humke EW, O'Rourke KM, Starovasnik MA, Yin JP, Dixit VM (2001) The PYRIN domain: a member of the death domain-fold superfamily. Protein Sci (A Publication of the Protein Society) 10(9):1911–1918. doi:10.1110/ps.13801
51. Stennicke HR, Renatus M, Meldal M, Salvesen GS (2000) Internally quenched fluorescent peptide substrates disclose the subsite preferences of human caspases 1, 3, 6, 7 and 8. Biochem J 350(Pt 2):563–568
52. Talanian RV, Quinlan C, Trautz S, Hackett MC, Mankovich JA, Banach D, Ghayur T, Brady KD, Wong WW (1997) Substrate specificities of caspase family proteases. J Biol Chem 272(15):9677–9682
53. Garcia-Calvo M, Peterson EP, Leiting B, Ruel R, Nicholson DW, Thornberry NA (1998) Inhibition of human caspases by peptide-based and macromolecular inhibitors. J Biol Chem 273(49):32608–32613
54. Thornberry NA (1997) The caspase family of cysteine proteases. Br Med Bull 53(3):478–490
55. Watt W, Koeplinger KA, Mildner AM, Heinrikson RL, Tomasselli AG, Watenpaugh KD (1999) The atomic-resolution structure of human caspase-8, a key activator of apoptosis. Structure 7(9):1135–1143
56. Blanchard H, Donepudi M, Tschopp M, Kodandapani L, Wu JC, Grutter MG (2000) Caspase-8 specificity probed at subsite S(4): crystal structure of the caspase-8-Z-DEVD-cho complex. J Mol Biol 302(1):9–16. doi:10.1006/jmbi.2000.4041
57. Schweizer A, Briand C, Grutter MG (2003) Crystal structure of caspase-2, apical initiator of the intrinsic apoptotic pathway. J Biol Chem 278(43):42441–42447. doi:10.1074/jbc.M304895200
58. Becker JW, Rotonda J, Soisson SM, Aspiotis R, Bayly C, Francoeur S, Gallant M, Garcia-Calvo M, Giroux A, Grimm E, Han Y, McKay D, Nicholson DW, Peterson E, Renaud J, Roy S, Thornberry N, Zamboni R (2004) Reducing the peptidyl features of caspase-3 inhibitors: a structural analysis. J Med Chem 47(10):2466–2474. doi:10.1021/jm0305523
59. Sleath PR, Hendrickson RC, Kronheim SR, March CJ, Black RA (1990) Substrate specificity of the protease that processes human interleukin-1 beta. J Biol Chem 265 (24):14526–14528
60. Cade C, Clark C (2015) Caspases—key players in apoptosis. In: Bose K (ed) Proteases in apoptosis: pathways, protocols and translational advances. Springer, pp 31–51. doi:10.1007/978-3-319-19497-4
61. Sulpizi M, Rothlisberger U, Carloni P (2003) Molecular dynamics studies of caspase-3. Biophys J 84(4):2207–2215. doi:10.1016/S0006-3495(03)75026-7
62. Stennicke HR, Salvesen GS (1997) Biochemical characteristics of caspases-3, -6, -7, and -8. J Biol Chem 272(41):25719–25723
63. Singh N, Bose K (2015) Apoptosis: pathways, molecules and beyond. In: Bose K (ed) Proteases in apoptosis: pathways, protocols and translational advances. Springer, pp 1–30. doi:10.1007/978-3-319-19497-4
64. Ashkenazi A (2002) Targeting death and decoy receptors of the tumour-necrosis factor superfamily. Nat Rev Cancer 2(6):420–430. doi:10.1038/nrc821

65. Dickens LS, Powley IR, Hughes MA, MacFarlane M (2012) The 'complexities' of life and death: death receptor signalling platforms. Exp Cell Res 318(11):1269–1277. doi:10.1016/j.yexcr.2012.04.005
66. Salvesen GS, Dixit VM (1999) Caspase activation: the induced-proximity model. Proc Natl Acad Sci USA 96(20):10964–10967
67. Logue SE, Martin SJ (2008) Caspase activation cascades in apoptosis. Biochem Soc Trans 36(Pt 1):1–9. doi:10.1042/BST0360001
68. Ferreira KS, Kreutz C, Macnelly S, Neubert K, Haber A, Bogyo M, Timmer J, Borner C (2012) Caspase-3 feeds back on caspase-8, Bid and XIAP in type I Fas signaling in primary mouse hepatocytes. Apoptosis Int J Prog Cell Death 17(5):503–515. doi:10.1007/s10495-011-0691-0
69. Lamkanfi M, Kanneganti TD (2010) Caspase-7: a protease involved in apoptosis and inflammation. Int J Biochem Cell Biol 42(1):21–24. doi:10.1016/j.biocel.2009.09.013
70. Juo P, Kuo CJ, Yuan J, Blenis J (1998) Essential requirement for caspase-8/FLICE in the initiation of the Fas-induced apoptotic cascade. Curr Biol (CB) 8(18):1001–1008
71. Varfolomeev EE, Schuchmann M, Luria V, Chiannilkulchai N, Beckmann JS, Mett IL, Rebrikov D, Brodianski VM, Kemper OC, Kollet O, Lapidot T, Soffer D, Sobe T, Avraham KB, Goncharov T, Holtmann H, Lonai P, Wallach D (1998) Targeted disruption of the mouse caspase 8 gene ablates cell death induction by the TNF receptors, Fas/Apo1, and DR3 and is lethal prenatally. Immunity 9(2):267–276
72. Deng Y, Lin Y, Wu X (2002) TRAIL-induced apoptosis requires Bax-dependent mitochondrial release of Smac/DIABLO. Genes Dev 16(1):33–45. doi:10.1101/gad.949602
73. Verhagen AM, Silke J, Ekert PG, Pakusch M, Kaufmann H, Connolly LM, Day CL, Tikoo A, Burke R, Wrobel C, Moritz RL, Simpson RJ, Vaux DL (2002) HtrA2 promotes cell death through its serine protease activity and its ability to antagonize inhibitor of apoptosis proteins. J Biol Chem 277(1):445–454. doi:10.1074/jbc.M109891200
74. Cook AL, Frydenberg M, Haynes JM (2002) Protein kinase G activation of K(ATP) channels in human-cultured prostatic stromal cells. Cell Signal 14(12):1023–1029
75. Micheau O, Tschopp J (2003) Induction of TNF receptor I-mediated apoptosis via two sequential signaling complexes. Cell 114(2):181–190
76. Newton K, Dixit VM (2012) Signaling in innate immunity and inflammation. Cold Spring Harb Perspect Biol 4(3). doi:10.1101/cshperspect.a006049
77. Christofferson DE, Yuan J (2010) Necroptosis as an alternative form of programmed cell death. Curr Opin Cell Biol 22(2):263–268. doi:10.1016/j.ceb.2009.12.003
78. Declercq W, Vanden Berghe T, Vandenabeele P (2009) RIP kinases at the crossroads of cell death and survival. Cell 138(2):229–232. doi:10.1016/j.cell.2009.07.006
79. Micheau O, Lens S, Gaide O, Alevizopoulos K, Tschopp J (2001) NF-kappaB signals induce the expression of c-FLIP. Mol Cell Biol 21(16):5299–5305. doi:10.1128/MCB.21.16.5299-5305.2001
80. Micheau O, Thome M, Schneider P, Holler N, Tschopp J, Nicholson DW, Briand C, Grutter MG (2002) The long form of FLIP is an activator of caspase-8 at the Fas death-inducing signaling complex. J Biol Chem 277(47):45162–45171. doi:10.1074/jbc.M206882200
81. Pop C, Oberst A, Drag M, Van Raam BJ, Riedl SJ, Green DR, Salvesen GS (2011) FLIP(L) induces caspase 8 activity in the absence of interdomain caspase 8 cleavage and alters substrate specificity. Biochem J 433(3):447–457. doi:10.1042/BJ20101738
82. Geserick P, Hupe M, Moulin M, Wong WW, Feoktistova M, Kellert B, Gollnick H, Silke J, Leverkus M (2009) Cellular IAPs inhibit a cryptic CD95-induced cell death by limiting RIP1 kinase recruitment. J cell Biol 187(7):1037–1054. doi:10.1083/jcb.200904158
83. Scaffidi C, Fulda S, Srinivasan A, Friesen C, Li F, Tomaselli KJ, Debatin KM, Krammer PH, Peter ME (1998) Two CD95 (APO-1/Fas) signaling pathways. EMBO J 17(6):1675–1687. doi:10.1093/emboj/17.6.1675

84. Samraj AK, Keil E, Ueffing N, Schulze-Osthoff K, Schmitz I (2006) Loss of caspase-9 provides genetic evidence for the type I/II concept of CD95-mediated apoptosis. J Biol Chem 281(40):29652–29659. doi:10.1074/jbc.M603487200
85. Li H, Zhu H, Xu CJ, Yuan J (1998) Cleavage of BID by caspase 8 mediates the mitochondrial damage in the Fas pathway of apoptosis. Cell 94(4):491–501
86. Luo X, Budihardjo I, Zou H, Slaughter C, Wang X (1998) Bid, a Bcl2 interacting protein, mediates cytochrome c release from mitochondria in response to activation of cell surface death receptors. Cell 94(4):481–490
87. Ozoren N, El-Deiry WS (2002) Defining characteristics of types I and II apoptotic cells in response to TRAIL. Neoplasia 4(6):551–557. doi:10.1038/sj.neo.7900270
88. Jost PJ, Grabow S, Gray D, McKenzie MD, Nachbur U, Huang DC, Bouillet P, Thomas HE, Borner C, Silke J, Strasser A, Kaufmann T (2009) XIAP discriminates between type I and type II FAS-induced apoptosis. Nature 460(7258):1035–1039. doi:10.1038/nature08229
89. Spencer SL, Gaudet S, Albeck JG, Burke JM, Sorger PK (2009) Non-genetic origins of cell-to-cell variability in TRAIL-induced apoptosis. Nature 459(7245):428–432. doi:10.1038/nature08012
90. Green DR, Llambi F (2015) Cell death signaling. Cold Spring Harb Perspect Biol 7(12). doi:10.1101/cshperspect.a006080
91. Eskes R, Desagher S, Antonsson B, Martinou JC (2000) Bid induces the oligomerization and insertion of Bax into the outer mitochondrial membrane. Mol Cell Biol 20(3):929–935
92. Korsmeyer SJ, Wei MC, Saito M, Weiler S, Oh KJ, Schlesinger PH (2000) Pro-apoptotic cascade activates BID, which oligomerizes BAK or BAX into pores that result in the release of cytochrome c. Cell Death Differ 7(12):1166–1173. doi:10.1038/sj.cdd.4400783
93. Dewson G, Kratina T, Sim HW, Puthalakath H, Adams JM, Colman PM, Kluck RM (2008) To trigger apoptosis, Bak exposes its BH3 domain and homodimerizes via BH3:groove interactions. Mol Cell 30(3):369–380. doi:10.1016/j.molcel.2008.04.005
94. Dewson G, Kratina T, Czabotar P, Day CL, Adams JM, Kluck RM (2009) Bak activation for apoptosis involves oligomerization of dimers via their alpha6 helices. Mol Cell 36(4):696–703. doi:10.1016/j.molcel.2009.11.008
95. Chipuk JE, Green DR (2008) How do BCL-2 proteins induce mitochondrial outer membrane permeabilization? Trends Cell Biol 18(4):157–164. doi:10.1016/j.tcb.2008.01.007
96. Kim HE, Du F, Fang M, Wang X (2005) Formation of apoptosome is initiated by cytochrome c-induced dATP hydrolysis and subsequent nucleotide exchange on Apaf-1. Proc Natl Acad Sci U S A 102(49):17545–17550. doi:10.1073/pnas.0507900102
97. Qin H, Srinivasula SM, Wu G, Fernandes-Alnemri T, Alnemri ES, Shi Y (1999) Structural basis of procaspase-9 recruitment by the apoptotic protease-activating factor 1. Nature 399 (6736):549–557. doi:10.1038/21124
98. Acehan D, Jiang X, Morgan DG, Heuser JE, Wang X, Akey CW (2002) Three-dimensional structure of the apoptosome: implications for assembly, procaspase-9 binding, and activation. Mol Cell 9(2):423–432
99. Yu X, Acehan D, Menetret JF, Booth CR, Ludtke SJ, Riedl SJ, Shi Y, Wang X, Akey CW (2005) A structure of the human apoptosome at 12.8 A resolution provides insights into this cell death platform. Structure 13(11):1725–1735. doi:10.1016/j.str.2005.09.006
100. Malladi S, Challa-Malladi M, Fearnhead HO, Bratton SB (2009) The Apaf-1*procaspase-9 apoptosome complex functions as a proteolytic-based molecular timer. EMBO J 28 (13):1916–1925. doi:10.1038/emboj.2009.152
101. Rodriguez J, Lazebnik Y (1999) Caspase-9 and APAF-1 form an active holoenzyme. Genes Dev 13(24):3179–3184
102. Bratton SB, Walker G, Srinivasula SM, Sun XM, Butterworth M, Alnemri ES, Cohen GM (2001) Recruitment, activation and retention of caspases-9 and -3 by Apaf-1 apoptosome and associated XIAP complexes. EMBO J 20(5):998–1009. doi:10.1093/emboj/20.5.998

103. Bergeron L, Perez GI, Macdonald G, Shi L, Sun Y, Jurisicova A, Varmuza S, Latham KE, Flaws JA, Salter JC, Hara H, Moskowitz MA, Li E, Greenberg A, Tilly JL, Yuan J (1998) Defects in regulation of apoptosis in caspase-2-deficient mice. Genes Dev 12(9):1304–1314
104. Nutt LK, Margolis SS, Jensen M, Herman CE, Dunphy WG, Rathmell JC, Kornbluth S (2005) Metabolic regulation of oocyte cell death through the CaMKII-mediated phosphorylation of caspase-2. Cell 123(1):89–103. doi:10.1016/j.cell.2005.07.032
105. Troy CM, Stefanis L, Greene LA, Shelanski ML (1997) Nedd2 is required for apoptosis after trophic factor withdrawal, but not superoxide dismutase (SOD1) downregulation, in sympathetic neurons and PC12 cells. J Neurosci (Official Journal of the Society for Neuroscience) 17(6):1911–1918
106. Troy CM, Rabacchi SA, Friedman WJ, Frappier TF, Brown K, Shelanski ML (2000) Caspase-2 mediates neuronal cell death induced by beta-amyloid. J Neurosci (The Official Journal of the Society for Neuroscience) 20(4):1386–1392
107. Troy CM, Rabacchi SA, Hohl JB, Angelastro JM, Greene LA, Shelanski ML (2001) Death in the balance: alternative participation of the caspase-2 and -9 pathways in neuronal death induced by nerve growth factor deprivation. J Neurosci (The Official Journal of the Society for Neuroscience) 21(14):5007–5016
108. McIlwain DR, Berger T, Mak TW (2013) Caspase functions in cell death and disease. Cold Spring Harb Perspect Biol 5(4):a008656. doi:10.1101/cshperspect.a008656
109. Tinel A, Janssens S, Lippens S, Cuenin S, Logette E, Jaccard B, Quadroni M, Tschopp J (2007) Autoproteolysis of PIDD marks the bifurcation between pro-death caspase-2 and pro-survival NF-kappaB pathway. EMBO J 26(1):197–208. doi:10.1038/sj.emboj.7601473
110. Janssens S, Tinel A, Lippens S, Tschopp J (2005) PIDD mediates NF-kappaB activation in response to DNA damage. Cell 123(6):1079–1092. doi:10.1016/j.cell.2005.09.036
111. Tinel A, Tschopp J (2004) The PIDDosome, a protein complex implicated in activation of caspase-2 in response to genotoxic stress. Science 304(5672):843–846. doi:10.1126/science.1095432
112. Park HH, Wu H (2006) Crystal structure of RAIDD death domain implicates potential mechanism of PIDDosome assembly. J Mol Biol 357(2):358–364. doi:10.1016/j.jmb.2005.12.082
113. Park HH, Logette E, Raunser S, Cuenin S, Walz T, Tschopp J, Wu H (2007) Death domain assembly mechanism revealed by crystal structure of the oligomeric PIDDosome core complex. Cell 128(3):533–546. doi:10.1016/j.cell.2007.01.019
114. Sidi S, Sanda T, Kennedy RD, Hagen AT, Jette CA, Hoffmans R, Pascual J, Imamura S, Kishi S, Amatruda JF, Kanki JP, Green DR, D'Andrea AA, Look AT (2008) Chk1 suppresses a caspase-2 apoptotic response to DNA damage that bypasses p53, Bcl-2, and caspase-3. Cell 133(5):864–877. doi:10.1016/j.cell.2008.03.037
115. Olsson M, Vakifahmetoglu H, Abruzzo PM, Hogstrand K, Grandien A, Zhivotovsky B (2009) DISC-mediated activation of caspase-2 in DNA damage-induced apoptosis. Oncogene 28(18):1949–1959. doi:10.1038/onc.2009.36
116. Shi M, Vivian CJ, Lee KJ, Ge C, Morotomi-Yano K, Manzl C, Bock F, Sato S, Tomomori-Sato C, Zhu R, Haug JS, Swanson SK, Washburn MP, Chen DJ, Chen BP, Villunger A, Florens L, Du C (2009) DNA-PKcs-PIDDosome: a nuclear caspase-2-activating complex with role in G2/M checkpoint maintenance. Cell 136(3):508–520. doi:10.1016/j.cell.2008.12.021
117. Ho LH, Taylor R, Dorstyn L, Cakouros D, Bouillet P, Kumar S (2009) A tumor suppressor function for caspase-2. Proc Natl Acad Sci U S A 106(13):5336–5341. doi:10.1073/pnas.0811928106
118. Manzl C, Krumschnabel G, Bock F, Sohm B, Labi V, Baumgartner F, Logette E, Tschopp J, Villunger A (2009) Caspase-2 activation in the absence of PIDDosome formation. J Cell Biol 185(2):291–303. doi:10.1083/jcb.200811105

119. Ribe EM, Jean YY, Goldstein RL, Manzl C, Stefanis L, Villunger A, Troy CM (2012) Neuronal caspase 2 activity and function requires RAIDD, but not PIDD. Biochem J 444 (3):591–599. doi:10.1042/BJ20111588
120. Martinon F, Tschopp J (2004) Inflammatory caspases: linking an intracellular innate immune system to auto inflammatory diseases. Cell 117(5):561–574. doi:10.1016/j.cell.2004.05.004
121. Franchi L, Munoz-Planillo R, Nunez G (2012) Sensing and reacting to microbes through the inflammasomes. Nat Immunol 13(4):325–332. doi:10.1038/ni.2231
122. Zamboni DS, Kobayashi KS, Kohlsdorf T, Ogura Y, Long EM, Vance RE, Kuida K, Mariathasan S, Dixit VM, Flavell RA, Dietrich WF, Roy CR (2006) The Birc1e cytosolic pattern-recognition receptor contributes to the detection and control of legionella pneumophila infection. Nat Immunol 7(3):318–325. doi:10.1038/ni1305
123. Vinzing M, Eitel J, Lippmann J, Hocke AC, Zahlten J, Slevogt H, N'Guessan PD, Gunther S, Schmeck B, Hippenstiel S, Flieger A, Suttorp N, Opitz B (2008) NAIP and Ipaf control Legionella pneumophila replication in human cells. J Immunol 180(10):6808–6815
124. Mariathasan S, Newton K, Monack DM, Vucic D, French DM, Lee WP, Roose-Girma M, Erickson S, Dixit VM (2004) Differential activation of the inflammasome by caspase-1 adaptors ASC and Ipaf. Nature 430(6996):213–218. doi:10.1038/nature02664
125. Mariathasan S, Weiss DS, Newton K, McBride J, O'Rourke K, Roose-Girma M, Lee WP, Weinrauch Y, Monack DM, Dixit VM (2006) Cryopyrin activates the inflammasome in response to toxins and ATP. Nature 440(7081):228–232. doi:10.1038/nature04515
126. Nakagawa T, Yuan J (2000) Cross-talk between two cysteine protease families. Activation of caspase-12 by calpain in apoptosis. J Cell Biol 150(4):887–894
127. Schotte P, Van Criekinge W, Van de Craen M, Van Loo G, Desmedt M, Grooten J, Cornelissen M, De Ridder L, Vandekerckhove J, Fiers W, Vandenabeele P, Beyaert R (1998) Cathepsin B-mediated activation of the proinflammatory caspase-11. Biochem Biophys Res Commun 251(1):379–387. doi:10.1006/bbrc.1998.9425
128. Vancompernolle K, Van Herreweghe F, Pynaert G, Van de Craen M, De Vos K, Totty N, Sterling A, Fiers W, Vandenabeele P, Grooten J (1998) Atractyloside-induced release of cathepsin B, a protease with caspase-processing activity. FEBS Lett 438(3):150–158
129. Darmon AJ, Nicholson DW, Bleackley RC (1995) Activation of the apoptotic protease CPP32 by cytotoxic T-cell-derived granzyme B. Nature 377(6548):446–448. doi:10.1038/377446a0
130. Orth K, O'Rourke K, Salvesen GS, Dixit VM (1996) Molecular ordering of apoptotic mammalian CED-3/ICE-like proteases. J Biol Chem 271(35):20977–20980
131. Thornberry NA, Bull HG, Calaycay JR, Chapman KT, Howard AD, Kostura MJ, Miller DK, Molineaux SM, Weidner JR, Aunins J et al (1992) A novel heterodimeric cysteine protease is required for interleukin-1 beta processing in monocytes. Nature 356(6372):768–774. doi:10.1038/356768a0
132. Ramage P, Cheneval D, Chvei M, Graff P, Hemmig R, Heng R, Kocher HP, Mackenzie A, Memmert K, Revesz L et al (1995) Expression, refolding, and autocatalytic proteolytic processing of the interleukin-1 beta-converting enzyme precursor. J Biol Chem 270 (16):9378–9383
133. Yamin TT, Ayala JM, Miller DK (1996) Activation of the native 45-kDa precursor form of interleukin-1-converting enzyme. J Biol Chem 271(22):13273–13282
134. Jacobson M, McCarthy N (2002) Apoptosis. The molecular biology of programmed cell death. Oxford University Press, UK
135. Brentnall M, Rodriguez-Menocal L, De Guevara RL, Cepero E, Boise LH (2013) Caspase-9, caspase-3 and caspase-7 have distinct roles during intrinsic apoptosis. BMC Cell Biol 14:32. doi:10.1186/1471-2121-14-32
136. Cowling V, Downward J (2002) Caspase-6 is the direct activator of caspase-8 in the cytochrome c-induced apoptosis pathway: absolute requirement for removal of caspase-6 prodomain. Cell Death Differ 9(10):1046–1056. doi:10.1038/sj.cdd.4401065

137. Chai J, Wu Q, Shiozaki E, Srinivasula SM, Alnemri ES, Shi Y (2001) Crystal structure of a procaspase-7 zymogen: mechanisms of activation and substrate binding. Cell 107(3):399–407
138. Riedl SJ, Fuentes-Prior P, Renatus M, Kairies N, Krapp S, Huber R, Salvesen GS, Bode W (2001) Structural basis for the activation of human procaspase-7. Proc Natl Acad Sci U S A 98(26):14790–14795. doi:10.1073/pnas.221580098
139. Muzio M, Stockwell BR, Stennicke HR, Salvesen GS, Dixit VM (1998) An induced proximity model for caspase-8 activation. J Biol Chem 273(5):2926–2930
140. Boatright KM, Renatus M, Scott FL, Sperandio S, Shin H, Pedersen IM, Ricci JE, Edris WA, Sutherlin DP, Green DR, Salvesen GS (2003) A unified model for apical caspase activation. Mol Cell 11(2):529–541
141. Chang DW, Xing Z, Pan Y, Algeciras-Schimnich A, Barnhart BC, Yaish-Ohad S, Peter ME, Yang X (2002) c-FLIP(L) is a dual function regulator for caspase-8 activation and CD95-mediated apoptosis. EMBO J 21(14):3704–3714. doi:10.1093/emboj/cdf356
142. Irmler M, Thome M, Hahne M, Schneider P, Hofmann K, Steiner V, Bodmer JL, Schroter M, Burns K, Mattmann C, Rimoldi D, French LE, Tschopp J (1997) Inhibition of death receptor signals by cellular FLIP. Nature 388(6638):190–195. doi:10.1038/40657
143. Kirchhoff S, Muller WW, Li-Weber M, Krammer PH (2000) Up-regulation of c-FLIPshort and reduction of activation-induced cell death in CD28-costimulated human T cells. Eur J Immunol 30(10):2765–2774. doi:10.1002/1521-4141(200010)30:10<2765:AID-IMMU2765>3.0.CO;2-W
144. Shiozaki EN, Chai J, Shi Y (2002) Oligomerization and activation of caspase-9, induced by Apaf-1 CARD. Proc Natl Acad Sci U S A 99(7):4197–4202. doi:10.1073/pnas.072544399
145. Martinon F, Burns K, Tschopp J (2002) The inflammasome: a molecular platform triggering activation of inflammatory caspases and processing of proIL-beta. Mol Cell 10(2):417–426
146. Van de Craen M, Declercq W, Van den brande I, Fiers W, Vandenabeele P (1999) The proteolytic procaspase activation network: an in vitro analysis. Cell Death Differ 6 (11):1117–1124. doi:10.1038/sj.cdd.4400589
147. Slee EA, Harte MT, Kluck RM, Wolf BB, Casiano CA, Newmeyer DD, Wang HG, Reed JC, Nicholson DW, Alnemri ES, Green DR, Martin SJ (1999) Ordering the cytochrome c-initiated caspase cascade: hierarchical activation of caspases-2, -3, -6, -7, -8, and -10 in a caspase-9-dependent manner. J Cell Biol 144(2):281–292
148. Srinivasula SM, Hegde R, Saleh A, Datta P, Shiozaki E, Chai J, Lee RA, Robbins PD, Fernandes-Alnemri T, Shi Y, Alnemri ES (2001) A conserved XIAP-interaction motif in caspase-9 and Smac/DIABLO regulates caspase activity and apoptosis. Nature 410 (6824):112–116. doi:10.1038/35065125
149. Lockshin RA (1969) Programmed cell death. Activation of lysis by a mechanism involving the synthesis of protein. J Insect Physiol 15(9):1505–1516
150. Wang S, Miura M, Jung Y, Zhu H, Gagliardini V, Shi L, Greenberg AH, Yuan J (1996) Identification and characterization of Ich-3, a member of the interleukin-1beta converting enzyme (ICE)/Ced-3 family and an upstream regulator of ICE. J Biol Chem 271(34):20580–20587
151. Lin XY, Choi MS, Porter AG (2000) Expression analysis of the human caspase-1 subfamily reveals specific regulation of the CASP5 gene by lipopolysaccharide and interferon-gamma. J Biol Chem 275(51):39920–39926. doi:10.1074/jbc.M007255200
152. Bian ZM, Elner SG, Khanna H, Murga-Zamalloa CA, Patil S, Elner VM (2011) Expression and functional roles of caspase-5 in inflammatory responses of human retinal pigment epithelial cells. Invest Ophthalmol Vis Sci 52(12):8646–8656. doi:10.1167/iovs.11-7570
153. Krumschnabel G, Manzl C, Villunger A (2009) Caspase-2: killer, savior and safeguard–emerging versatile roles for an ill-defined caspase. Oncogene 28(35):3093–3096. doi:10.1038/onc.2009.173

154. Baptiste-Okoh N, Barsotti AM, Prives C (2008) Caspase 2 is both required for p53-mediated apoptosis and downregulated by p53 in a p21-dependent manner. Cell Cycle 7(9):1133–1138. doi:10.4161/cc.7.9.5805
155. Winkler C, Doller A, Imre G, Badawi A, Schmid T, Schulz S, Steinmeyer N, Pfeilschifter J, Rajalingam K, Eberhardt W (2014) Attenuation of the ELAV1-like protein HuR sensitizes adenocarcinoma cells to the intrinsic apoptotic pathway by increasing the translation of caspase-2L. Cell Death Dis 5:e1321. doi:10.1038/cddis.2014.279
156. Wang QE, Han C, Zhang B, Sabapathy K, Wani AA (2012) Nucleotide excision repair factor XPC enhances DNA damage-induced apoptosis by downregulating the antiapoptotic short isoform of caspase-2. Cancer Res 72(3):666–675. doi:10.1158/0008-5472.CAN-11-2774
157. Seol DW, Billiar TR (1999) A caspase-9 variant missing the catalytic site is an endogenous inhibitor of apoptosis. J Biol Chem 274(4):2072–2076
158. Srinivasula SM, Ahmad M, Guo Y, Zhan Y, Lazebnik Y, Fernandes-Alnemri T, Alnemri ES (1999) Identification of an endogenous dominant-negative short isoform of caspase-9 that can regulate apoptosis. Cancer Res 59(5):999–1002
159. Shultz JC, Goehe RW, Wijesinghe DS, Murudkar C, Hawkins AJ, Shay JW, Minna JD, Chalfant CE (2010) Alternative splicing of caspase 9 is modulated by the phosphoinositide 3-kinase/Akt pathway via phosphorylation of SRp30a. Cancer Res 70(22):9185–9196. doi:10.1158/0008-5472.CAN-10-1545
160. Elvitigala DA, Whang I, Jung HB, Lim BS, Nam BH, Lee J (2015) Molecular delineation of a caspase 10 homolog from black rockfish (Sebastes schlegelii) and its transcriptional regulation in response to pathogenic stress. Gene 570(2):288–294. doi:10.1016/j.gene.2015.05.068
161. Gibellini D, Re MC, Ponti C, Vitone F, Bon I, Fabbri G, Grazia Di Iasio M, Zauli G (2005) HIV-1 Tat protein concomitantly down-regulates apical caspase-10 and up-regulates c-FLIP in lymphoid T cells: a potential molecular mechanism to escape TRAIL cytotoxicity. J Cell Physiol 203(3):547–556. doi:10.1002/jcp.20252
162. Eckhart L, Ban J, Fischer H, Tschachler E (2000) Caspase-14: analysis of gene structure and mRNA expression during keratinocyte differentiation. Biochem Biophys Res Commun 277 (3):655–659. doi:10.1006/bbrc.2000.3698
163. Nahle Z, Polakoff J, Davuluri RV, McCurrach ME, Jacobson MD, Narita M, Zhang MQ, Lazebnik Y, Bar-Sagi D, Lowe SW (2002) Direct coupling of the cell cycle and cell death machinery by E2F. Nat Cell Biol 4(11):859–864. doi:10.1038/ncb868
164. Parrish AB, Freel CD, Kornbluth S (2013) Cellular mechanisms controlling caspase activation and function. Cold Spring Harbor Perspect Biol 5(6). doi:10.1101/cshperspect.a008672
165. Hammerman PS, Fox CJ, Thompson CB (2004) Beginnings of a signal-transduction pathway for bioenergetic control of cell survival. Trends Biochem Sci 29(11):586–592. doi:10.1016/j.tibs.2004.09.008
166. Nutt LK, Buchakjian MR, Gan E, Darbandi R, Yoon SY, Wu JQ, Miyamoto YJ, Gibbons JA, Andersen JL, Freel CD, Tang W, He C, Kurokawa M, Wang Y, Margolis SS, Fissore RA, Kornbluth S (2009) Metabolic control of oocyte apoptosis mediated by 14-3-3zeta-regulated dephosphorylation of caspase-2. Dev Cell 16(6):856–866. doi:10.1016/j.devcel.2009.04.005
167. Yi CH, Sogah DK, Boyce M, Degterev A, Christofferson DE, Yuan J (2007) A genome-wide RNAi screen reveals multiple regulators of caspase activation. J Cell Biol 179 (4):619–626. doi:10.1083/jcb.200708090
168. Yang CS, Thomenius MJ, Gan EC, Tang W, Freel CD, Merritt TJ, Nutt LK, Kornbluth S (2010) Metabolic regulation of drosophila apoptosis through inhibitory phosphorylation of Dronc. EMBO J 29(18):3196–3207. doi:10.1038/emboj.2010.191

169. Li J, Billiar TR, Talanian RV, Kim YM (1997) Nitric oxide reversibly inhibits seven members of the caspase family via S-nitrosylation. Biochem Biophys Res Commun 240 (2):419–424. doi:10.1006/bbrc.1997.7672
170. Mannick JB, Schonhoff C, Papeta N, Ghafourifar P, Szibor M, Fang K, Gaston B (2001) S-Nitrosylation of mitochondrial caspases. J Cell Biol 154(6):1111–1116. doi:10.1083/jcb.200104008
171. Jiang ZL, Fletcher NM, Diamond MP, Abu-Soud HM, Saed GM (2009) S-nitrosylation of caspase-3 is the mechanism by which adhesion fibroblasts manifest lower apoptosis. Wound Repair and Regeneration (Official Publication of the Wound Healing Society [and] the European Tissue Repair Society) 17(2):224–229. doi:10.1111/j.1524-475X.2009.00459.x
172. Mitchell DA, Marletta MA (2005) Thioredoxin catalyzes the S-nitrosation of the caspase-3 active site cysteine. Nat Chem Biol 1(3):154–158. doi:10.1038/nchembio720
173. Tsang AH, Lee YI, Ko HS, Savitt JM, Pletnikova O, Troncoso JC, Dawson VL, Dawson TM, Chung KK (2009) S-nitrosylation of XIAP compromises neuronal survival in Parkinson's disease. Proc Natl Acad Sci U S A 106(12):4900–4905. doi:10.1073/pnas.0810595106
174. Wu W, Wan OW, Chung KK (2015) S-nitrosylation of XIAP at Cys 213 of BIR2 domain impairs XIAP's anti-caspase 3 activity and anti-apoptotic function. Apoptosis Int J Prog Cell Death 20(4):491–499. doi:10.1007/s10495-015-1087-3
175. Zhang D, Zhao N, Ma B, Wang Y, Zhang G, Yan X, Hu S, Xu T (2016) Procaspase-9 induces its cleavage by transnitrosylating XIAP via the Thioredoxin system during cerebral ischemia-reperfusion in rats. Sci R 6:24203. doi:10.1038/srep24203
176. Kim YM, Kim TH, Chung HT, Talanian RV, Yin XM, Billiar TR (2000) Nitric oxide prevents tumor necrosis factor alpha-induced rat hepatocyte apoptosis by the interruption of mitochondrial apoptotic signaling through S-nitrosylation of caspase-8. Hepatology 32(4 Pt 1):770–778. doi:10.1053/jhep.2000.18291
177. Torok NJ, Higuchi H, Bronk S, Gores GJ (2002) Nitric oxide inhibits apoptosis downstream of cytochrome C release by nitrosylating caspase 9. Cancer Res 62(6):1648–1653
178. Yin XH, Yan JZ, Hou XY, Wu SL, Zhang GY (2013) Neuroprotection of S-nitrosoglutathione against ischemic injury by down-regulating Fas S-nitrosylation and downstream signaling. Neuroscience 248:290–298. doi:10.1016/j.neuroscience.2013.06.012
179. Kim YM, Kim JH, Kwon HM, Lee DH, Won MH, Kwon YG, Kim YM (2013) Korean Red Ginseng protects endothelial cells from serum-deprived apoptosis by regulating Bcl-2 family protein dynamics and caspase S-nitrosylation. J Ginseng Res 37(4):413–424. doi:10.5142/jgr.2013.37.413
180. Saligrama PT, Fortner KA, Secinaro MA, Collins CC, Russell JQ, Budd RC (2014) IL-15 maintains T-cell survival via S-nitrosylation-mediated inhibition of caspase-3. Cell Death Differ 21(6):904–914. doi:10.1038/cdd.2014.10
181. Li M, Wang AJ, Xu JX (2008) Redox state of cytochrome c regulates cellular ROS and caspase cascade in permeablized cell model. Protein Pept Lett 15(2):200–205
182. Resch U, Schichl YM, Sattler S, de Martin R (2008) XIAP regulates intracellular ROS by enhancing antioxidant gene expression. Biochem Biophys Res Commun 375(1):156–161. doi:10.1016/j.bbrc.2008.07.142
183. Girish KS, Paul M, Thushara RM, Hemshekhar M, Shanmuga Sundaram M, Rangappa KS, Kemparaju K (2013) Melatonin elevates apoptosis in human platelets via ROS mediated mitochondrial damage. Biochem Biophys Res Commun 438(1):198–204. doi:10.1016/j.bbrc.2013.07.053
184. Lupfer CR, Anand PK, Liu Z, Stokes KL, Vogel P, Lamkanfi M, Kanneganti TD (2014) Reactive oxygen species regulate caspase-11 expression and activation of the non-canonical NLRP3 inflammasome during enteric pathogen infection. PLoS Pathog 10(9):e1004410. doi:10.1371/journal.ppat.1004410
185. Shalini S, Puccini J, Wilson CH, Finnie J, Dorstyn L, Kumar S (2015) Caspase-2 protects against oxidative stress in vivo. Oncogene 34(38):4995–5002. doi:10.1038/onc.2014.413

186. Ives A, Nomura J, Martinon F, Roger T, LeRoy D, Miner JN, Simon G, Busso N, So A (2015) Xanthine oxidoreductase regulates macrophage IL1beta secretion upon NLRP3 inflammasome activation. Nat Commun 6:6555. doi:10.1038/ncomms7555
187. Jung SS, Moon JS, Xu JF, Ifedigbo E, Ryter SW, Choi AM, Nakahira K (2015) Carbon monoxide negatively regulates NLRP3 inflammasome activation in macrophages. Am J Physiol Lung Cell Mol Physiol 308(10):L1058–L1067. doi:10.1152/ajplung.00400.2014
188. Liu X, Zhang X, Ding Y, Zhou W, Tao L, Hu R (2016) Nuclear factor E2-related factor-2 (Nrf2) negatively regulates NLRP3 inflammasome activity by inhibiting reactive oxygen species (ROS)-induced NLRP3 priming. Antioxid Redox Sig. doi:10.1089/ars.2015.6615
189. Nicotera P, Leist M, Ferrando-May E (1999) Apoptosis and necrosis: different execution of the same death. Biochem Soc Symp 66:69–73
190. Samali A, Nordgren H, Zhivotovsky B, Peterson E, Orrenius S (1999) A comparative study of apoptosis and necrosis in HepG2 cells: oxidant-induced caspase inactivation leads to necrosis. Biochem Biophys Res Commun 255(1):6–11. doi:10.1006/bbrc.1998.0139
191. Ma YM, Peng YM, Zhu QH, Gao AH, Chao B, He QJ, Li J, Hu YH, Zhou YB (2016) Novel CHOP activator LGH00168 induces necroptosis in A549 human lung cancer cells via ROS-mediated ER stress and NF-kappaB inhibition. Acta Pharmacol Sin. doi:10.1038/aps.2016.61
192. Cardone MH, Roy N, Stennicke HR, Salvesen GS, Franke TF, Stanbridge E, Frisch S, Reed JC (1998) Regulation of cell death protease caspase-9 by phosphorylation. Science 282 (5392):1318–1321
193. Allan LA, Morrice N, Brady S, Magee G, Pathak S, Clarke PR (2003) Inhibition of caspase-9 through phosphorylation at Thr 125 by ERK MAPK. Nat Cell Biol 5(7):647–654. doi:10.1038/ncb1005
194. Brady SC, Allan LA, Clarke PR (2005) Regulation of caspase 9 through phosphorylation by protein kinase C zeta in response to hyperosmotic stress. Mol Cell Biol 25(23):10543–10555. doi:10.1128/MCB.25.23.10543-10555.2005
195. Allan LA, Clarke PR (2007) Phosphorylation of caspase-9 by CDK1/cyclin B1 protects mitotic cells against apoptosis. Mol Cell 26(2):301–310. doi:10.1016/j.molcel.2007.03.019
196. Andersen JL, Johnson CE, Freel CD, Parrish AB, Day JL, Buchakjian MR, Nutt LK, Thompson JW, Moseley MA, Kornbluth S (2009) Restraint of apoptosis during mitosis through interdomain phosphorylation of caspase-2. EMBO J 28(20):3216–3227. doi:10.1038/emboj.2009.253
197. Seng NS, Megyesi J, Tarcsafalvi A, Price PM (2016) Mimicking Cdk2 phosphorylation of Bcl-xL at Ser73 results in caspase activation and Bcl-xL cleavage. Cell Death Discov 2. doi:10.1038/cddiscovery.2016.1
198. Helmke C, Raab M, Rodel F, Matthess Y, Oellerich T, Mandal R, Sanhaji M, Urlaub H, Rodel C, Becker S, Strebhardt K (2016) Ligand stimulation of CD95 induces activation of Plk3 followed by phosphorylation of caspase-8. Cell Res. doi:10.1038/cr.2016.78
199. Pop C, Salvesen GS (2009) Human caspases: activation, specificity, and regulation. J Biol Chem 284(33):21777–21781. doi:10.1074/jbc.R800084200
200. Blankenship JW, Varfolomeev E, Goncharov T, Fedorova AV, Kirkpatrick DS, Izrael-Tomasevic A, Phu L, Arnott D, Aghajan M, Zobel K, Bazan JF, Fairbrother WJ, Deshayes K, Vucic D (2009) Ubiquitin binding modulates IAP antagonist-stimulated proteasomal degradation of c-IAP1 and c-IAP2(1). Biochem J 417(1):149–160. doi:10.1042/BJ20081885
201. Gyrd-Hansen M, Darding M, Miasari M, Santoro MM, Zender L, Xue W, Tenev T, da Fonseca PC, Zvelebil M, Bujnicki JM, Lowe S, Silke J, Meier P (2008) IAPs contain an evolutionarily conserved ubiquitin-binding domain that regulates NF-kappaB as well as cell survival and oncogenesis. Nat Cell Biol 10(11):1309–1317. doi:10.1038/ncb1789
202. Zhang B, Huang J, Li HL, Liu T, Wang YY, Waterman P, Mao AP, Xu LG, Zhai Z, Liu D, Marrack P, Shu HB (2008) GIDE is a mitochondrial E3 ubiquitin ligase that induces apoptosis and slows growth. Cell Res 18(9):900–910. doi:10.1038/cr.2008.75

203. Huang H, Joazeiro CA, Bonfoco E, Kamada S, Leverson JD, Hunter T (2000) The inhibitor of apoptosis, cIAP2, functions as a ubiquitin-protein ligase and promotes in vitro monoubiquitination of caspases 3 and 7. J Biol Chem 275(35):26661–26664. doi:10.1074/jbc.C000199200
204. Mica L, Harter L, Trentz O, Keel M (2004) Endotoxin reduces CD95-induced neutrophil apoptosis by cIAP-2-mediated caspase-3 degradation. J Am Coll Surg 199(4):595–602. doi:10.1016/j.jamcollsurg.2004.05.272
205. Choi WY, Jin CY, Han MH, Kim GY, Kim ND, Lee WH, Kim SK, Choi YH (2009) Sanguinarine sensitizes human gastric adenocarcinoma AGS cells to TRAIL-mediated apoptosis via down-regulation of AKT and activation of caspase-3. Anticancer Res 29 (11):4457–4465
206. Chen L, Smith L, Wang Z, Smith JB (2003) Preservation of caspase-3 subunits from degradation contributes to apoptosis evoked by lactacystin: any single lysine or lysine pair of the small subunit is sufficient for ubiquitination. Mol Pharmacol 64(2):334–345. doi:10.1124/mol.64.2.334
207. Suzuki Y, Nakabayashi Y, Takahashi R (2001) Ubiquitin-protein ligase activity of X-linked inhibitor of apoptosis protein promotes proteasomal degradation of caspase-3 and enhances its anti-apoptotic effect in Fas-induced cell death. Proc Natl Acad Sci U S A 98(15):8662–8667. doi:10.1073/pnas.161506698
208. Jin Z, Li Y, Pitti R, Lawrence D, Pham VC, Lill JR, Ashkenazi A (2009) Cullin3-based polyubiquitination and p62-dependent aggregation of caspase-8 mediate extrinsic apoptosis signaling. Cell 137(4):721–735. doi:10.1016/j.cell.2009.03.015
209. Adrain C, Murphy BM, Martin SJ (2005) Molecular ordering of the caspase activation cascade initiated by the cytotoxic T lymphocyte/natural killer (CTL/NK) protease granzyme B. J Biol Chem 280(6):4663–4673. doi:10.1074/jbc.M410915200
210. Kim HE, Jiang X, Du F, Wang X (2008) PHAPI, CAS, and Hsp 70 promote apoptosome formation by preventing Apaf-1 aggregation and enhancing nucleotide exchange on Apaf-1. Mol Cell 30(2):239–247. doi:10.1016/j.molcel.2008.03.014
211. Chinnaiyan AM, Orth K, O'Rourke K, Duan H, Poirier GG, Dixit VM (1996) Molecular ordering of the cell death pathway. Bcl-2 and Bcl-xL function upstream of the CED-3-like apoptotic proteases. J Biol Chem 271(9):4573–4576
212. Kersse K, Vanden Berghe T, Lamkanfi M, Vandenabeele P (2007) A phylogenetic and functional overview of inflammatory caspases and caspase-1-related CARD-only proteins. Biochem Soc Trans 35(Pt 6):1508–1511. doi:10.1042/BST0351508
213. Thome M, Schneider P, Hofmann K, Fickenscher H, Meinl E, Neipel F, Mattmann C, Burns K, Bodmer JL, Schroter M, Scaffidi C, Krammer PH, Peter ME, Tschopp J (1997) Viral FLICE-inhibitory proteins (FLIPs) prevent apoptosis induced by death receptors. Nature 386(6624):517–521. doi:10.1038/386517a0
214. Krueger A, Schmitz I, Baumann S, Krammer PH, Kirchhoff S (2001) Cellular FLICE-inhibitory protein splice variants inhibit different steps of caspase-8 activation at the CD95 death-inducing signaling complex. J Biol Chem 276(23):20633–20640. doi:10.1074/jbc.M101780200
215. Boatright KM, Deis C, Denault JB, Sutherlin DP, Salvesen GS (2004) Activation of caspases-8 and -10 by FLIP(L). Biochem J 382(Pt 2):651–657. doi:10.1042/BJ20040809
216. Oberst A, Dillon CP, Weinlich R, McCormick LL, Fitzgerald P, Pop C, Hakem R, Salvesen GS, Green DR (2011) Catalytic activity of the caspase-8-FLIP(L) complex inhibits RIPK3-dependent necrosis. Nature 471(7338):363–367. doi:10.1038/nature09852
217. van Raam BJ, Salvesen GS (2012) Proliferative versus apoptotic functions of caspase-8 Hetero or homo: the caspase-8 dimer controls cell fate. Biochimica et biophysica acta 1824 (1):113–122. doi:10.1016/j.bbapap.2011.06.005
218. Crook NE, Clem RJ, Miller LK (1993) An apoptosis-inhibiting baculovirus gene with a zinc finger-like motif. J Virol 67(4):2168–2174

219. Eckelman BP, Salvesen GS, Scott FL (2006) Human inhibitor of apoptosis proteins: why XIAP is the black sheep of the family. EMBO Rep 7(10):988–994. doi:10.1038/sj.embor.7400795
220. Chai J, Du C, Wu JW, Kyin S, Wang X, Shi Y (2000) Structural and biochemical basis of apoptotic activation by Smac/DIABLO. Nature 406(6798):855–862. doi:10.1038/35022514
221. Du C, Fang M, Li Y, Li L, Wang X (2000) Smac, a mitochondrial protein that promotes cytochrome c-dependent caspase activation by eliminating IAP inhibition. Cell 102(1):33–42
222. Singh N, D'Souza A, Cholleti A, Sastry GM, Bose K (2014) Dual regulatory switch confers tighter control on HtrA2 proteolytic activity. FEBS J 281(10):2456–2470. doi:10.1111/febs.12799
223. Bratton SB, Lewis J, Butterworth M, Duckett CS, Cohen GM (2002) XIAP inhibition of caspase-3 preserves its association with the Apaf-1 apoptosome and prevents CD95- and Bax-induced apoptosis. Cell Death Differ 9(9):881–892. doi:10.1038/sj.cdd.4401069
224. Callus BA, Vaux DL (2007) Caspase inhibitors: viral, cellular and chemical. Cell Death Differ 14(1):73–78. doi:10.1038/sj.cdd.4402034
225. Hornle M, Peters N, Thayaparasingham B, Vorsmann H, Kashkar H, Kulms D (2011) Caspase-3 cleaves XIAP in a positive feedback loop to sensitize melanoma cells to TRAIL-induced apoptosis. Oncogene 30(5):575–587. doi:10.1038/onc.2010.434
226. Burke SP, Smith L, Smith JB (2010) cIAP1 cooperatively inhibits procaspase-3 activation by the caspase-9 apoptosome. J Biol Chem 285(39):30061–30068. doi:10.1074/jbc.M110.125955
227. Semenkova L, Dudich E, Dudich I, Tokhtamisheva N, Tatulov E, Okruzhnov Y, Garcia-Foncillas J, Palop-Cubillo JA, Korpela T (2003) Alpha-fetoprotein positively regulates cytochrome c-mediated caspase activation and apoptosome complex formation. Euro J Biochem/FEBS 270(21):4388–4399
228. Dudich E, Semenkova L, Dudich I, Denesyuk A, Tatulov E, Korpela T (2006) Alpha-fetoprotein antagonizes X-linked inhibitor of apoptosis protein anticaspase activity and disrupts XIAP-caspase interaction. FEBS J 273(16):3837–3849. doi:10.1111/j.1742-4658.2006.05391.x
229. Li Z, Jo J, Jia JM, Lo SC, Whitcomb DJ, Jiao S, Cho K, Sheng M (2010) Caspase-3 activation via mitochondria is required for long-term depression and AMPA receptor internalization. Cell 141(5):859–871. doi:10.1016/j.cell.2010.03.053
230. Gibon J, Unsain N, Gamache K, Thomas RA, De Leon A, Johnstone A, Nader K, Seguela P, Barker PA (2016) The X-linked inhibitor of apoptosis regulates long-term depression and learning rate. FASEB J (Official Publication of the Federation of American Societies for Experimental Biology). doi:10.1096/fj.201600384R
231. Ohsawa S, Hamada S, Kuida K, Yoshida H, Igaki T, Miura M (2010) Maturation of the olfactory sensory neurons by Apaf-1/caspase-9-mediated caspase activity. Proc Natl Acad Sci U S A 107(30):13366–13371. doi:10.1073/pnas.0910488107
232. Pasterkamp RJ, Peschon JJ, Spriggs MK, Kolodkin AL (2003) Semaphorin 7A promotes axon outgrowth through integrins and MAPKs. Nature 424(6947):398–405. doi:10.1038/nature01790
233. Connolly PF, Jager R, Fearnhead HO (2014) New roles for old enzymes: killer caspases as the engine of cell behavior changes. Front Physiol 5:149. doi:10.3389/fphys.2014.00149
234. Olsson M, Zhivotovsky B (2011) Caspases and cancer. Cell Death Differ 18(9):1441–1449. doi:10.1038/cdd.2011.30
235. Philchenkov A, Zavelevich M, Kroczak TJ, Los M (2004) Caspases and cancer: mechanisms of inactivation and new treatment modalities. Exp Oncol 26(2):82–97
236. Hu B, Elinav E, Huber S, Booth CJ, Strowig T, Jin C, Eisenbarth SC, Flavell RA (2010) Inflammation-induced tumorigenesis in the colon is regulated by caspase-1 and NLRC4. Proc Natl Acad Sci U S A 107(50):21635–21640. doi:10.1073/pnas.1016814108

237. Lee JW, Kim MR, Soung YH, Nam SW, Kim SH, Lee JY, Yoo NJ, Lee SH (2006) Mutational analysis of the CASP6 gene in colorectal and gastric carcinomas. APMIS: acta pathologica, microbiologica, et immunologica Scandinavica 114(9):646–650. doi:10.1111/j.1600-0463.2006.apm_417.x
238. Yoo NJ, Lee JW, Kim YJ, Soung YH, Kim SY, Nam SW, Park WS, Lee JY, Lee SH (2004) Loss of caspase-2, -6 and -7 expression in gastric cancers. APMIS (acta pathologica, microbiologica, et immunologica Scandinavica) 112(6):330–335. doi:10.1111/j.1600-0463.2004.apm1120602.x
239. Soung YH, Lee JW, Kim HS, Park WS, Kim SY, Lee JH, Park JY, Cho YG, Kim CJ, Park YG, Nam SW, Jeong SW, Kim SH, Lee JY, Yoo NJ, Lee SH (2003) Inactivating mutations of CASPASE-7 gene in human cancers. Oncogene 22(39):8048–8052. doi:10.1038/sj.onc.1206727
240. Teitz T, Wei T, Valentine MB, Vanin EF, Grenet J, Valentine VA, Behm FG, Look AT, Lahti JM, Kidd VJ (2000) Caspase 8 is deleted or silenced preferentially in childhood neuroblastomas with amplification of MYCN. Nat Med 6(5):529–535. doi:10.1038/75007
241. Kim HS, Lee JW, Soung YH, Park WS, Kim SY, Lee JH, Park JY, Cho YG, Kim CJ, Jeong SW, Nam SW, Kim SH, Lee JY, Yoo NJ, Lee SH (2003) Inactivating mutations of caspase-8 gene in colorectal carcinomas. Gastroenterology 125(3):708–715
242. Soung YH, Lee JW, Kim SY, Sung YJ, Park WS, Nam SW, Kim SH, Lee JY, Yoo NJ, Lee SH (2005) Caspase-8 gene is frequently inactivated by the frameshift somatic mutation 1225_1226delTG in hepatocellular carcinomas. Oncogene 24(1):141–147. doi:10.1038/sj.onc.1208244
243. Shin MS, Kim HS, Kang CS, Park WS, Kim SY, Lee SN, Lee JH, Park JY, Jang JJ, Kim CW, Kim SH, Lee JY, Yoo NJ, Lee SH (2002) Inactivating mutations of CASP10 gene in non-Hodgkin lymphomas. Blood 99(11):4094–4099
244. Kim MS, Oh JE, Min CK, Lee S, Chung NG, Yoo NJ, Lee SH (2009) Mutational analysis of CASP10 gene in acute leukaemias and multiple myelomas. Pathology 41(5):484–487
245. Oh JE, Kim MS, Ahn CH, Kim SS, Han JY, Lee SH, Yoo NJ (2010) Mutational analysis of CASP10 gene in colon, breast, lung and hepatocellular carcinomas. Pathology 42(1):73–76. doi:10.3109/00313020903434371
246. Park WS, Lee JH, Shin MS, Park JY, Kim HS, Lee JH, Kim YS, Lee SN, Xiao W, Park CH, Lee SH, Yoo NJ, Lee JY (2002) Inactivating mutations of the caspase-10 gene in gastric cancer. Oncogene 21(18):2919–2925. doi:10.1038/sj.onc.1205394
247. MacPherson G, Healey CS, Teare MD, Balasubramanian SP, Reed MW, Pharoah PD, Ponder BA, Meuth M, Bhattacharyya NP, Cox A (2004) Association of a common variant of the CASP8 gene with reduced risk of breast cancer. J Natl Cancer Inst 96(24):1866–1869. doi:10.1093/jnci/dji001
248. Frank B, Bermejo JL, Hemminki K, Klaes R, Bugert P, Wappenschmidt B, Schmutzler RK, Burwinkel B (2005) Re: association of a common variant of the CASP8 gene with reduced risk of breast cancer. J Natl Cancer Inst 97(13):1012; author reply 1012–1013. doi:10.1093/jnci/dji178
249. Cacina C, Pence S, Turan S, Genc F, Ozdemir H, Kafadar A, Kaynar MY, Yaylim I (2015) Analysis of CASP8 D302H gene variants in patients with primary brain tumors. In vivo 29(5):601–604
250. Lan Q, Morton LM, Armstrong B, Hartge P, Menashe I, Zheng T, Purdue MP, Cerhan JR, Zhang Y, Grulich A, Cozen W, Yeager M, Holford TR, Vajdic CM, Davis S, Leaderer B, Kricker A, Schenk M, Zahm SH, Chatterjee N, Chanock SJ, Rothman N, Wang SS (2009) Genetic variation in caspase genes and risk of non-Hodgkin lymphoma: a pooled analysis of 3 population-based case-control studies. Blood 114(2):264–267. doi:10.1182/blood-2009-01-198697
251. Dogan A, Wang AH, Witzig TE, Call TG, Kay NE, Habermann TM, Slager SL, Cerhan JR (2010) Germline variation in apoptosis pathway genes and risk of non-Hodgkin's lymphoma. Cancer epidemiology, biomarkers & prevention: a publication of the American

Association for Cancer Research, cosponsored by the American Society of Preventive Oncology 19(11):2847–2858. doi:10.1158/1055-9965.EPI-10-0581
252. Park JY, Park JM, Jang JS, Choi JE, Kim KM, Cha SI, Kim CH, Kang YM, Lee WK, Kam S, Park RW, Kim IS, Lee JT, Jung TH (2006) Caspase 9 promoter polymorphisms and risk of primary lung cancer. Hum Mol Genet 15(12):1963–1971. doi:10.1093/hmg/ddl119
253. Chen K, Zhao H, Hu Z, Wang LE, Zhang W, Sturgis EM, Wei Q (2008) CASP3 polymorphisms and risk of squamous cell carcinoma of the head and neck. Clin Cancer Res (An Official Journal of the American Association for Cancer Research) 14(19):6343–6349. doi:10.1158/1078-0432.CCR-08-1198
254. Jang JS, Kim KM, Choi JE, Cha SI, Kim CH, Lee WK, Kam S, Jung TH, Park JY (2008) Identification of polymorphisms in the Caspase-3 gene and their association with lung cancer risk. Mol Carcinog 47(5):383–390. doi:10.1002/mc.20397
255. Hosgood HD 3rd, Baris D, Zhang Y, Zhu Y, Zheng T, Yeager M, Welch R, Zahm S, Chanock S, Rothman N, Lan Q (2008) Caspase polymorphisms and genetic susceptibility to multiple myeloma. Hematol Oncol 26(3):148–151. doi:10.1002/hon.852
256. Lan Q, Zheng T, Chanock S, Zhang Y, Shen M, Wang SS, Berndt SI, Zahm SH, Holford TR, Leaderer B, Yeager M, Welch R, Hosgood D, Boyle P, Rothman N (2007) Genetic variants in caspase genes and susceptibility to non-Hodgkin lymphoma. Carcinogenesis 28(4):823–827. doi:10.1093/carcin/bgl196
257. Gabay C, Lamacchia C, Palmer G (2010) IL-1 pathways in inflammation and human diseases. Nat Rev Rheumatol 6(4):232–241. doi:10.1038/nrrheum.2010.4
258. Brydges SD, Mueller JL, McGeough MD, Pena CA, Misaghi A, Gandhi C, Putnam CD, Boyle DL, Firestein GS, Horner AA, Soroosh P, Watford WT, O'Shea JJ, Kastner DL, Hoffman HM (2009) Inflammasome-mediated disease animal models reveal roles for innate but not adaptive immunity. Immunity 30(6):875–887. doi:10.1016/j.immuni.2009.05.005
259. Meng G, Zhang F, Fuss I, Kitani A, Strober W (2009) A mutation in the Nlrp3 gene causing inflammasome hyperactivation potentiates Th17 cell-dominant immune responses. Immunity 30(6):860–874. doi:10.1016/j.immuni.2009.04.012
260. Spranger J, Kroke A, Mohlig M, Hoffmann K, Bergmann MM, Ristow M, Boeing H, Pfeiffer AF (2003) Inflammatory cytokines and the risk to develop type 2 diabetes: results of the prospective population-based European prospective investigation into cancer and nutrition (EPIC)-potsdam study. Diabetes 52(3):812–817
261. Maedler K, Dharmadhikari G, Schumann DM, Storling J (2009) Interleukin-1 beta targeted therapy for type 2 diabetes. Expert Opin Biol Ther 9(9):1177–1188. doi:10.1517/14712590903136688
262. Papin S, Cuenin S, Agostini L, Martinon F, Werner S, Beer HD, Grutter C, Grutter M, Tschopp J (2007) The SPRY domain of Pyrin, mutated in familial mediterranean fever patients, interacts with inflammasome components and inhibits proIL-1beta processing. Cell Death Differ 14(8):1457–1466. doi:10.1038/sj.cdd.4402142
263. Shoham NG, Centola M, Mansfield E, Hull KM, Wood G, Wise CA, Kastner DL (2003) Pyrin binds the PSTPIP1/CD2BP1 protein, defining familial mediterranean fever and PAPA syndrome as disorders in the same pathway. Proc Natl Acad Sci U S A 100(23):13501–13506. doi:10.1073/pnas.2135380100
264. Mandey SH, Kuijk LM, Frenkel J, Waterham HR (2006) A role for geranylgeranylation in interleukin-1beta secretion. Arthritis Rheum 54(11):3690–3695. doi:10.1002/art.22194
265. Kuijk LM, Beekman JM, Koster J, Waterham HR, Frenkel J, Coffer PJ (2008) HMG-CoA reductase inhibition induces IL-1beta release through Rac1/PI3K/PKB-dependent caspase-1 activation. Blood 112(9):3563–3573. doi:10.1182/blood-2008-03-144667
266. Vila M, Przedborski S (2003) Targeting programmed cell death in neurodegenerative diseases. Nat Rev Neurosci 4(5):365–375. doi:10.1038/nrn1100
267. Rohn TT, Head E, Su JH, Anderson AJ, Bahr BA, Cotman CW, Cribbs DH (2001) Correlation between caspase activation and neurofibrillary tangle formation in Alzheimer's disease. Am J Pathol 158(1):189–198. doi:10.1016/S0002-9440(10)63957-0

268. D'Amelio M, Cavallucci V, Middei S, Marchetti C, Pacioni S, Ferri A, Diamantini A, De Zio D, Carrara P, Battistini L, Moreno S, Bacci A, Ammassari-Teule M, Marie H, Cecconi F (2011) Caspase-3 triggers early synaptic dysfunction in a mouse model of Alzheimer's disease. Nat Neurosci 14(1):69–76. doi:10.1038/nn.2709
269. de Calignon A, Fox LM, Pitstick R, Carlson GA, Bacskai BJ, Spires-Jones TL, Hyman BT (2010) Caspase activation precedes and leads to tangles. Nature 464(7292):1201–1204. doi:10.1038/nature08890
270. Onouchi Y, Ozaki K, Buns JC, Shimizu C, Hamada H, Honda T, Terai M, Honda A, Takeuchi T, Shibuta S, Suenaga T, Suzuki H, Higashi K, Yasukawa K, Suzuki Y, Sasago K, Kemmotsu Y, Takatsuki S, Saji T, Yoshikawa T, Nagai T, Hamamoto K, Kishi F, Ouchi K, Sato Y, Newburger JW, Baker AL, Shulman ST, Rowley AH, Yashiro M, Nakamura Y, Wakui K, Fukushima Y, Fujino A, Tsunoda T, Kawasaki T, Hata A, Nakamura Y, Tanaka T (2010) Common variants in CASP3 confer susceptibility to Kawasaki disease. Hum Mol Genet 19(14):2898–2906. doi:10.1093/hmg/ddq176
271. Fleisher TA (2008) The autoimmune lymphoproliferative syndrome: an experiment of nature involving lymphocyte apoptosis. Immunol Res 40(1):87–92. doi:10.1007/s12026-007-8001-1
272. Su CL, Huang LL, Huang LM, Lee JC, Lin CN, Won SJ (2006) Caspase-8 acts as a key upstream executor of mitochondria during justicidin a-induced apoptosis in human hepatoma cells. FEBS Lett 580(13):3185–3191. doi:10.1016/j.febslet.2006.04.085
273. He XL, Zhang P, Dong XZ, Yang MH, Chen SL, Bi MG (2012) JR6, a new compound isolated from justicia procumbens, induces apoptosis in human bladder cancer EJ cells through caspase-dependent pathway. J Ethnopharmacol 144(2):284–292. doi:10.1016/j.jep.2012.09.010
274. Ghobrial IM, Witzig TE, Adjei AA (2005) Targeting apoptosis pathways in cancer therapy. CA Cancer J Clin 55(3):178–194
275. Jiang M, Zhu K, Grenet J, Lahti JM (2008) Retinoic acid induces caspase-8 transcription via phospho-CREB and increases apoptotic responses to death stimuli in neuroblastoma cells. Biochem Biophys Acta 1783(6):1055–1067. doi:10.1016/j.bbamcr.2008.02.007
276. Banelli B, Casciano I, Croce M, Di Vinci A, Gelvi I, Pagnan G, Brignole C, Allemanni G, Ferrini S, Ponzoni M, Romani M (2002) Expression and methylation of CASP8 in neuroblastoma: identification of a promoter region. Nat Med 8(12):1333–1335; author reply 1335. doi:10.1038/nm1202-1333
277. Liedtke C, Zschemisch NH, Cohrs A, Roskams T, Borlak J, Manns MP, Trautwein C (2005) Silencing of caspase-8 in murine hepatocellular carcinomas is mediated via methylation of an essential promoter element. Gastroenterology 129(5):1602–1615. doi:10.1053/j.gastro.2005.08.007
278. Fulda S, Debatin KM (2006) 5-Aza-2'-deoxycytidine and IFN-gamma cooperate to sensitize for TRAIL-induced apoptosis by upregulating caspase-8. Oncogene 25(37):5125–5133. doi:10.1038/sj.onc.1209518
279. Hensley P, Mishra M, Kyprianou N (2013) Targeting caspases in cancer therapeutics. Biol Chem 394(7):831–843. doi:10.1515/hsz-2013-0128
280. Kaplan DH, Shankaran V, Dighe AS, Stockert E, Aguet M, Old LJ, Schreiber RD (1998) Demonstration of an interferon gamma-dependent tumor surveillance system in immunocompetent mice. Proc Natl Acad Sci U S A 95(13):7556–7561
281. Fulda S, Debatin KM (2002) IFNgamma sensitizes for apoptosis by upregulating caspase-8 expression through the Stat1 pathway. Oncogene 21(15):2295–2308. doi:10.1038/sj.onc.1205255
282. Tekautz TM, Zhu K, Grenet J, Kaushal D, Kidd VJ, Lahti JM (2006) Evaluation of IFN-gamma effects on apoptosis and gene expression in neuroblastoma–preclinical studies. Biochem Biophys Acta 1763(10):1000–1010. doi:10.1016/j.bbamcr.2006.06.014
283. Jiang X, Kim HE, Shu H, Zhao Y, Zhang H, Kofron J, Donnelly J, Burns D, Ng SC, Rosenberg S, Wang X (2003) Distinctive roles of PHAP proteins and prothymosin-alpha in a death regulatory pathway. Science 299(5604):223–226. doi:10.1126/science.1076807

284. Zhang HZ, Kasibhatla S, Wang Y, Herich J, Guastella J, Tseng B, Drewe J, Cai SX (2004) Discovery, characterization and SAR of gambonic acid as a potent apoptosis inducer by a HTS assay. Bioorg Med Chem 12(2):309–317
285. Fischer U, Schulze-Osthoff K (2005) New approaches and therapeutics targeting apoptosis in disease. Pharmacol Rev 57(2):187–215. doi:10.1124/pr.57.2.6
286. Putt KS, Chen GW, Pearson JM, Sandhorst JS, Hoagland MS, Kwon JT, Hwang SK, Jin H, Churchwell MI, Cho MH, Doerge DR, Helferich WG, Hergenrother PJ (2006) Small-molecule activation of procaspase-3 to caspase-3 as a personalized anticancer strategy. Nat Chem Biol 2(10):543–550. doi:10.1038/nchembio814
287. Peterson QP, Hsu DC, Goode DR, Novotny CJ, Totten RK, Hergenrother PJ (2009) Procaspase-3 activation as an anti-cancer strategy: structure-activity relationship of procaspase-activating compound 1 (PAC-1) and its cellular co-localization with caspase-3. J Med Chem 52(18):5721–5731. doi:10.1021/jm900722z
288. Denault JB, Drag M, Salvesen GS, Alves J, Heidt AB, Deveraux Q, Harris JL (2007) Small molecules not direct activators of caspases. Nat Chem Biol 3(9):519, author reply 520. doi:10.1038/nchembio0907-519
289. Peterson QP, Hsu DC, Novotny CJ, West DC, Kim D, Schmit JM, Dirikolu L, Hergenrother PJ, Fan TM (2010) Discovery and canine preclinical assessment of a nontoxic procaspase-3-activating compound. Cancer Res 70(18):7232–7241. doi:10.1158/0008-5472. CAN-10-0766
290. Vogler M, Dinsdale D, Dyer MJ, Cohen GM (2009) Bcl-2 inhibitors: small molecules with a big impact on cancer therapy. Cell Death Differ 16(3):360–367. doi:10.1038/cdd.2008.137
291. Chen DJ, Huerta S (2009) Smac mimetics as new cancer therapeutics. Anticancer Drugs 20 (8):646–658. doi:10.1097/CAD.0b013e32832ced78
292. Lu Y, Chen GQ (2011) Effector caspases and leukemia. Int J Cell Biol 2011:738301. doi:10. 1155/2011/738301
293. Velculescu VE, Madden SL, Zhang L, Lash AE, Yu J, Rago C, Lal A, Wang CJ, Beaudry GA, Ciriello KM, Cook BP, Dufault MR, Ferguson AT, Gao Y, He TC, Hermeking H, Hiraldo SK, Hwang PM, Lopez MA, Luderer HF, Mathews B, Petroziello JM, Polyak K, Zawel L, Kinzler KW et al (1999) Analysis of human transcriptomes. Nat Genet 23(4):387–388. doi:10.1038/70487
294. Kawasaki H, Altieri DC, Lu CD, Toyoda M, Tenjo T, Tanigawa N (1998) Inhibition of apoptosis by survivin predicts shorter survival rates in colorectal cancer. Cancer Res 58 (22):5071–5074
295. Church DN, Talbot DC (2012) Survivin in solid tumors: rationale for development of inhibitors. Curr Oncol R 14(2):120–128. doi:10.1007/s11912-012-0215-2
296. Amarante-Mendes GP, Griffith TS (2015) Therapeutic applications of TRAIL receptor agonists in cancer and beyond. Pharmacol Ther 155:117–131. doi:10.1016/j.pharmthera. 2015.09.001
297. Xie X, Zhao X, Liu Y, Zhang J, Matusik RJ, Slawin KM, Spencer DM (2001) Adenovirus-mediated tissue-targeted expression of a caspase-9-based artificial death switch for the treatment of prostate cancer. Cancer Res 61(18):6795–6804
298. Komata T, Kondo Y, Kanzawa T, Hirohata S, Koga S, Sumiyoshi H, Srinivasula SM, Barna BP, Germano IM, Takakura M, Inoue M, Alnemri ES, Shay JW, Kyo S, Kondo S (2001) Treatment of malignant glioma cells with the transfer of constitutively active caspase-6 using the human telomerase catalytic subunit (human telomerase reverse transcriptase) gene promoter. Cancer Res 61(15):5796–5802
299. Certo M, Del Gaizo Moore V, Nishino M, Wei G, Korsmeyer S, Armstrong SA, Letai A (2006) Mitochondria primed by death signals determine cellular addiction to antiapoptotic BCL-2 family members. Cancer Cell 9(5):351–365. doi:10.1016/j.ccr.2006.03.027
300. Tey SK, Dotti G, Rooney CM, Heslop HE, Brenner MK (2007) Inducible caspase 9 suicide gene to improve the safety of allodepleted T cells after haploidentical stem cell

transplantation. Biol Blood Marrow Transplant (Journal of the American Society for Blood and Marrow Transplantation) 13(8):913–924. doi:10.1016/j.bbmt.2007.04.005
301. Straathof KC, Pule MA, Yotnda P, Dotti G, Vanin EF, Brenner MK, Heslop HE, Spencer DM, Rooney CM (2005) An inducible caspase 9 safety switch for T-cell therapy. Blood 105(11):4247–4254. doi:10.1182/blood-2004-11-4564
302. Larsen CM, Faulenbach M, Vaag A, Ehses JA, Donath MY, Mandrup-Poulsen T (2009) Sustained effects of interleukin-1 receptor antagonist treatment in type 2 diabetes. Diabetes Care 32(9):1663–1668. doi:10.2337/dc09-0533
303. Schroder K, Tschopp J (2010) The inflammasomes. Cell 140(6):821–832. doi:10.1016/j.cell.2010.01.040
304. Davidson FF, Steller H (1998) Blocking apoptosis prevents blindness in Drsosophila retinal degeneration mutants. Nature 391(6667):587–591. doi:10.1038/35385
305. O'Brien RJ, Wong PC (2011) Amyloid precursor protein processing and Alzheimer's disease. Annu Rev Neurosci 34:185–204. doi:10.1146/annurev-neuro-061010-113613
306. Thornberry NA (1998) Caspases: key mediators of apoptosis. Chem Biol 5(5):R97–103

Gastric Pathology and Metalloproteinases

Snehasikta Swarnakar, Anirban Roy, Satyabrata Ghosh, Rajib Majumder and Sandip Paul

Abstract

The spectrum of gastric pathologies involves heterogeneity with respect to biochemical mechanisms and clinical outcome and is globally common. Each year, 5–6 million people worldwide are affected by gastric ulcer, gastric cancer and inflammatory bowel diseases, and mortality rate being >50% shows steep increase in incidence. Hence, understanding the underlying pathogenesis and better therapeutic strategies remain the major challenges in gastroenterology field. Current knowledge of gastric pathology reveals that extracellular proteases vastly influence functional irregularities of cells along with their responses to microenvironment. Based on studies on metalloproteinases and their inhibitors, it is well accepted about their important roles in physiological developmental processes as well as pathological conditions. From past several years of extensive research on matrix, metalloproteinases (MMPs) establish their critical role in several cellular functions including proliferation, apoptosis and angiogenesis. MMPs are a family of "molecular scissors" with ambivalent actions and ability to cleave extracellular matrix (ECM) proteins that in turn facilitate tissue remodelling. Approximately, 27 subtypes of MMPs are there having mutual interaction among each of them in gastrointestinal disorders. Functional overlap between the MMPs leads to non-specificity, which makes designing MMP inhibitors more difficult. Thus, specific MMP inhibitors would be promising therapeutic tool against inflammatory diseases including gastric diseases. This chapter illustrates the new insights into mechanism of MMP regulation in

S. Swarnakar (✉) · A. Roy · S. Ghosh · R. Majumder
Cancer Biology and Inflammatory Disorder Division, CSIR-Indian Institute of Chemical Biology, 4, Raja S. C. Mullick Road, Kolkata 700032, India
e-mail: snehasiktas@hotmail.com

S. Paul
Structural Biology and Bioinformatics Division, CSIR-Indian Institute of Chemical Biology, Kolkata 700032, India

S. Chakraborti and N.S. Dhalla (eds.), *Pathophysiological Aspects of Proteases*,
DOI 10.1007/978-981-10-6141-7_19

gastrointestinal inflammatory disorders encompassing clinical trials for MMP inhibitors and new therapeutic strategies by targeting specific MMP(s) to control gastrointestinal pathologies.

Keywords
Gastric ulcer · Cancer · Matrix metalloproteinase
Inhibitor · *Helicobacter pylori*

1 Introduction

Metalloproteinases degrade extracellular matrix (ECM) proteins and regulate both cell–cell and cell–ECM interactions, which influence cell differentiation, migration, proliferation and survival. They belong to metzincin group of proteases, characterized by the presence of zinc in the catalytic domain, that includes bacterial serralysins and astacins, adamalysins (a disintegrin and metalloproteinase domain or ADAMs) and matrixins (matrix metalloproteinases or MMPs) [1–3]. Metzincins use three histidine (H) residues to bind the zinc ion at their active site [4, 5]. A water molecule that is essential for hydrolysis of the peptide bond also coordinates with the metal ion as a fourth ligand in the active form of metallopeptidase. Members of this superfamily of enzymes are involved in diverse physiological processes as embryonic development, morphogenesis, bone formation, reproduction, cell adhesion and migration. Aberrant activities of metalloproteases have been implicated in various pathological conditions like arthritis, cancer, cardiovascular diseases, nephritis, central nervous system disorders and fibrosis [2, 6, 7]. There are ample literatures which state that ECM degradation plays pivotal role in gastrointestinal diseases, thus role of MMPs are evident [2, 3, 8]. Collectively, these enzymes are capable of degrading collagens, elastins, gelatin, matrix glycoproteins and proteoglycan as well as number of bioactive molecules.

According to the classification of proteases, based on their 3-D structure in the MEROPS database (http://merops.sanger.ac.uk), metallopeptidases may be classified into forty-six families. The families are further grouped into fourteen different clans based on metal ion binding motifs and 3-D structure similarities [9, 10]. MMPs are calcium-dependent, zinc containing endopeptidases [11]. The name is derived from consensus sequence and structural features, specifically a "HExxH" zinc-binding motif (zincin) and a C-terminal conserved methionine residue, which forms a conserved structure, called "met turn" [8]. MMP family comprises $\sim$27 member proteases characterized in humans, rodents and amphibians [12–14]. They were first described in vertebrates (1962), including humans, but are also found in invertebrates and plants. MMPs are secreted by a variety of connective tissues and pro-inflammatory cells including fibroblasts, osteoblasts, endothelial cells, macrophages, neutrophils and lymphocytes. These enzymes are expressed as zymogens, which are subsequently processed by other proteolytic enzymes (such as serine proteases, furin, plasmin and others) to generate the active forms through a cysteine switch mechanism.

MMPs are classified into collagenases, gelatinases, stromelysins and matrilysins depending on their specificity as depicted in Fig. 1. Another subclass of MMPs is membrane-type MMPs (MT-MMPs) that additionally contain a transmembrane and cytoplasmic domain [12]. The activities of most MMPs are very low or negligible in the normal steady-state tissues, and their expression is transcriptionally controlled by inflammatory cytokines, growth factors, hormones, cell–cell and cell–matrix interactions [15] (Fig. 2).

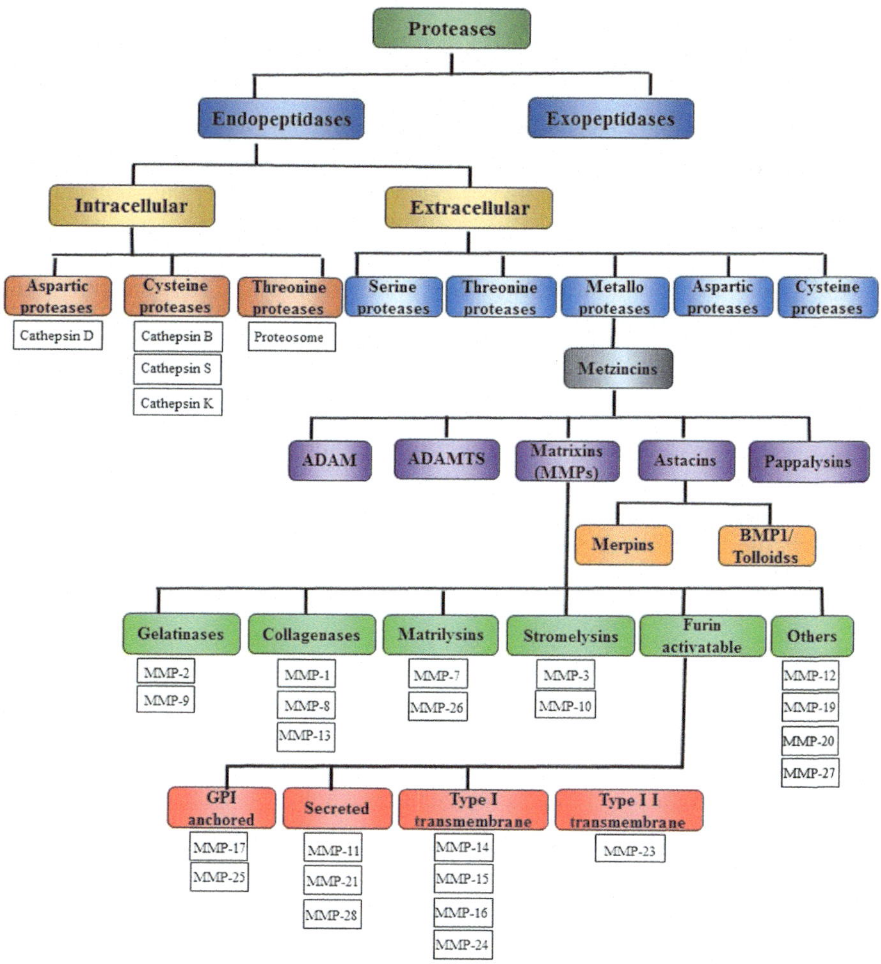

Fig. 1 Overview of the MMP family members and their evolutionary connection with other metzincin superfamily members

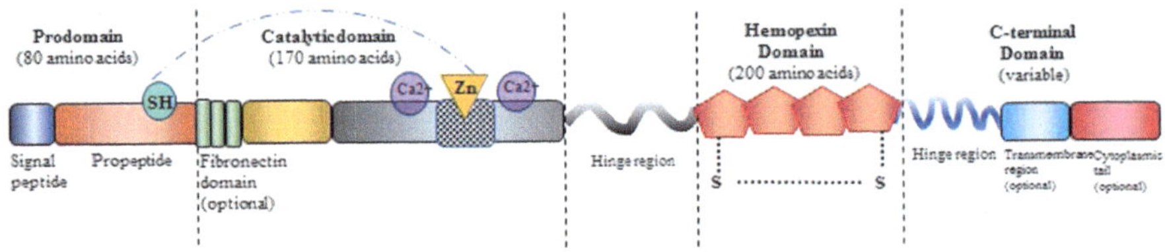

Fig. 2 Generalized domain structure and amino acid length of mammalian MMPs. MMPs contain a signal peptide followed by a propeptide which constitutes the pro-domain. It contains the conserved cysteine switch sequence, which makes a complex with the Zn^{2+} ion in the zymogen form of it. In case of only gelatinases (MMP-2 and MMP-9), the catalytic domain has a gelatin binding domain. Except for few MMPs (MMP-9, MMP-26, MMP-7), all other members contain a proline-rich hinge region followed by a hemopexin-like C-terminal domain, which helps in substrate recognition and its interaction with endogenous inhibitors. A major difference between secreted and cell surface anchored MMPs (MMP-14, MMP-15, MMP-16 and MMP-24) consists of intrinsic motif called transmembrane region and cytoplasmic tail

From the structural point of view, a typical MMP consists of approximately 80 amino acid long propeptide, about 170 amino acids catalytic metalloproteinase domain, followed by a linker peptide of variable length and a 200 amino acid long hemopexin (Hpx) domain (Fig. 3).

Among all the members of MMP family, MMP-7, MMP-26 and MMP-23 are the exceptions as they lack the Hpx domain along with the linker peptide, and MMP-23 has an additional cysteine-rich domain followed by an immunoglobulin-like domain after the metalloproteinase domain [16–19]. The signal peptide is removed during translation, and proMMPs are generated [20].

The activity of MMPs is very tightly regulated in the cell under normal physiological conditions. This regulation occurs at different levels; gene expression, proteolytic cleavage of the zymogens, transcription and inhibition of the active forms by various non-specific endogenous inhibitors such as α2-macroglobulin and specific tissue inhibitors of metalloproteinases (TIMPs) [1, 12, 13]. TIMPs inhibit active MMPs by forming 1:1 stoichiometric enzyme-inhibitor complexes leading to inhibition of their proteolytic activity [14, 15, 21]. TIMP-1, -2 and -4 are secreted, while TIMP-3 is sequestered to the ECM. The substrate specificity of TIMPs varies. A critical balance between MMPs and their endogenous inhibitors plays a pivotal role in vivo. Similar to MMPs, the proteolytic ADAM and ADAMTS family members are inhibited by specific TIMPs [18, 22, 23].

Reactive oxygen species (ROS) produced at the site of inflammation produced by activated neutrophils and macrophages has also a great influence on the function of MMPs [24]. These oxidants initially activate MMPs via oxidation of the pro-domain cysteine [25]. Eventually, MMPs may be inactivated by the enzyme myeloperoxidase secreted from inflammatory cells or by modification of catalytic domain amino acids by hypochlorous acid [26].

Detailed genetic and proteomic studies in experimental animals as well as in humans have provided insights into the involvement of MMPs in various disorders. The first human degenerative disease identified where MMPs were found to be

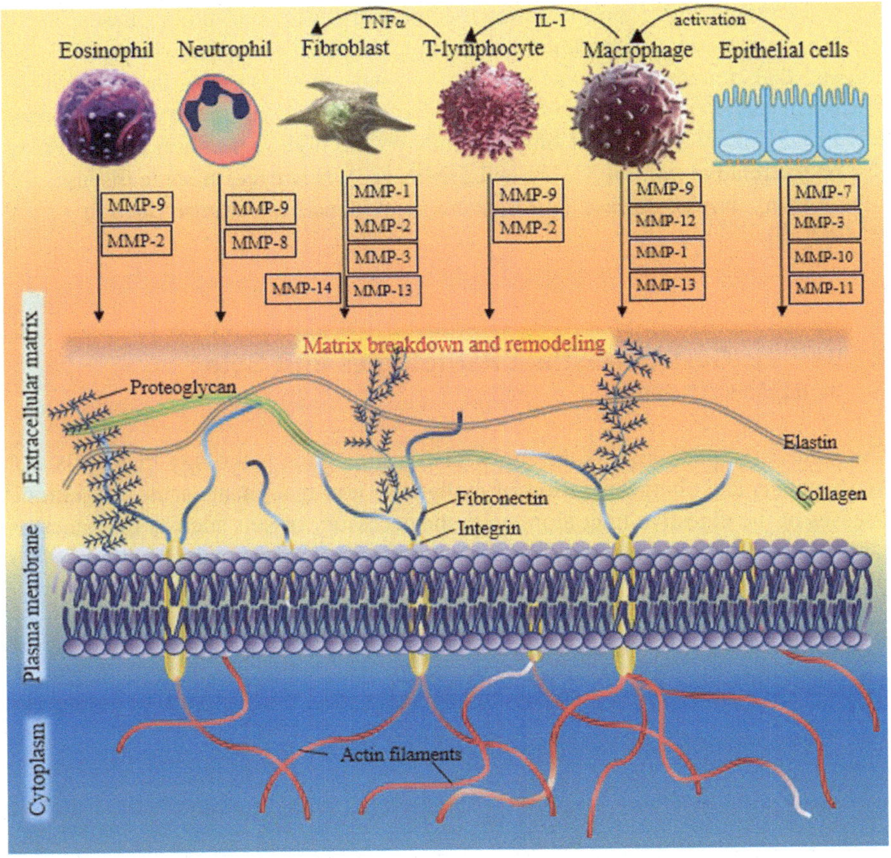

Fig. 3 Schematic diagram of intercellular signalling events that drive secretion of MMPs towards ECM during physiological and pathological conditions. Infiltrating immune cells secrete cytokines and MMPs simultaneously, which either activate other cells or degrade the ECM. These fragments of ECM components can mount the inflammatory response by a variety of events including immune cell chemotaxis, activation of receptors or chemokine ligands

linked was Sorsby's fundus dystrophy [27]. Stromelysin-1 knockout mice showed increased occurrence of collagen-induced arthritis. Several studies on MMP-null mice demonstrated impaired responses to pathological conditions. MMP-2, -7, -9 and -11 showed considerable [28, 29] influences on tumour progression and carcinogenesis in null mice. High expression levels of several MMPs have been correlated with tumour aggressiveness, stage and poor prognosis of various human cancers, but not always [29, 30]. MMPs are known to contribute to angiogenesis by degrading basement membranes, allowing for endothelial cell invasion, thus metastasis [25, 31, 32]. Abnormalities in ECM glycosaminoglycans and loss of glycosaminoglycans in epithelial basal lamina are detected in gastrointestinal inflammation like ulcerative colitis, peptic ulcers and Crohn disease [33, 34]. There is also increased expression of stromelysins, matrilysins and collagenases, which suggests a strong correlation among inflammation and tissue injury. Literatures also suggest that mucosal immune system triggers the response through MMP-dependent pathways [35]. The extent of damage in gastric tissues due to break-down of ECM by MMPs not only depends on the high expression of MMPs but also on the relative ratio of MMPs and TIMPs [23]. However, in some diseases like inflammatory bowel disease (IBD), there is evidence for overproduction of few MMPs [35].

2 Various Gastrointestinal Pathologies and Role of MMPs Therein

Gastrointestinal pathology is the subspecialty of surgical pathology that deals with the diagnosis and characterization of malignant, non-malignant, acute and chronic diseases of the digestive tract along with the accessory organs such as the pancreas, gallbladder, liver and intestine. MMPs play pivotal role in many gastrointestinal ailments like gastrointestinal mucositis, gastric ulcer, gastric cancer, colon cancer, pancreatic cancer, gallbladder cancer, hepatic cancer, colorectal cancer, etc. Detailed discussions have been provided in the following section.

2.1 Involvement of MMPs in Gastric Ulcer

Non-steroidal anti-inflammatory drugs (NSAIDs), corticosteroids, stress, alcohol consumption, smoking and family history are considered as risk factors in the pathogenesis of gastric ulcer [36]. Mucosal tissue injury may lead to gastric ulcer, which is triggered primarily by ischaemia, along with depletion of nutrient delivery [37]. There are many endogenous aggressive factors (gastric hydrochloric acid, pepsin, reactive free radicals and oxidants, leukotrienes, refluxed bile and endothelins) which actually counterbalanced by the protective factors like gastric mucosal barrier, bicarbonate, mucosal blood flow, surface active phospholipids, prostaglandins (PG), nitric oxide (NO) and antioxidants [38, 39].

MMPs, especially proMMP-2 (72-kDa gelatinase A) and proMMP-9 (92-kDa gelatinase B) as well as their active forms, are associated with gastric injury [40]. Although MMP-2 appears to be constitutively expressed by many cell types in culture, MMP-9 expression is induced during gastric ulcer development [41]. In addition, MMP-1 and 3 are also upregulated in gastric and duodenal ulcers [42]. Other studies reported that NSAIDs increase MMP-9 activity and suppress MMP-2 activity during gastric ulcer. In chronic conditions, MMP-2 activity also gets upregulated with MMP-9. This suggests that MMP-9 expression is crucial for the development of gastric ulcers, but MMP-2 may be involved in the turnover of gastric ECM. Menges et al. reported the upregulation of MMP-1 and MMP-9 during *Helicobacter pylori* (*H. pylori*) infection in cultured cells [43]. Acetic acid-induced experimental ulcer also showed the upregulation of MMP-9, but there were no significant change in the expression of MMP-2 [44]. In addition, infection can also influence the upregulation of MMP-1, -2, -3 and -7, but the mechanisms and pathways are not yet well understood. In contrast, in H_2O_2-mediated ulcers, MMP-2 activity and expression get downregulated [45]. Witzum et al. demonstrated that H_2O_2 alters the structure of MMP-2 by oxidation and catalytic domain inhibition [46, 47]. Singh et al. demonstrated that proMMP-9 along with pro and active MMP-2 gets upregulated in ethanol-induced gastric ulcer in experimental rats [48, 49]. Thus, the critical balance of MMP-9 and MMP-2 activities may be a determinant in the- progression as well as healing of gastric ulcer.

2.2 Role of MMPs in Gastric Cancer

Cancer is a multistage process, which requires various genetic and epigenetic changes in the tissue microenvironment. Alterations that occur during the malignant transformation are regulated by MMPs and their endogenous inhibitors TIMPs. Gastric cancer is the fourth most common cancer and the second leading cause of cancer-related death worldwide [50].

Different factors are having role and influence gastric cancer development and progression. Chronic inflammation or ulcer might be a linker for gastric cancer to many other types of malignancies for the future [51]. Studies of different surgical specimens showed chronic gastritis were more advanced in individuals with gastric cancer than in individuals with duodenal ulceration. It is now known that *H. pylori* is a major factor in both the induction of atrophic gastritis and histological progression to gastric cancer [52]. Literatures suggest that MMPs modulate the function of cytokines and chemokines and the consequences of functions of immunoregulatory cells during progression of gastrointestinal cancers.

Several knockout animal-based studies and case-control studies have confirmed that MMP-7 is an important member of MMP family that is upregulated in gastric carcinoma. Histology and immunohistochemistry revealed that it promotes tissue invasion and metastasis. Studies from our laboratory showed that single nucleotide polymorphism (SNP) in MMP-1 promoter at -519 A/G and MMP-3 promoter -375 C/G increases the risk of gastric cancer in Indian population [53]. Recently described downstream signalling molecules of MMP-7 include E-cadherin, Fas ligand and pro TNFα. E-cadherin is a cell adhesion molecule, which is responsible for the epithelial to mesenchymal (EMT) transition during cancer progression. Witty et al. reported that the colon cancer cells gained significant invasive potential when MMP-7 was transfected to them. Studies also emphasized the involvement of MMP-7 to the tumourigenicity and disease progression in malignant colorectal tumours [54]. Reverse transcription-polymerase chain reaction (RT-PCR) data revealed high expression of MMP-7 mRNA in the sentinel node lesions in patients with gastric carcinoma. In addition, SNP in MMP-7 promoter at -181 A/G increases the gastric cancer risk as reported from our laboratory [53, 55, 56].

In vitro studies on gastric cancer, cell lines demonstrated that the gene and protein levels of human epidermal growth factor 2 (HER 2) and MMP-9 are very tightly associated in the pathogenesis of gastric cancer [57]. Knocking down of HER 2 gene by shRNA significantly inhibited the invasion and metastasis of gastric cancer by downregulating the expression of MMP-9 while HER 2 overexpression again improved the MMP-9 transcription.

Yoo et al. found that signalling through sonic hedgehog pathway promotes the invasiveness of gastric tumours through activation of PI3 k/Akt pathway leading to EMT followed by MMP-9 activation [58]. Alakus et al. found that expression of MMP-2 was linked with the clinicopathological parameters in gastric cancer. High expression of MMP-2 from epithelial cells was associated with tumour stage and poor survival [59].

2.3 Specific Role of MMPs in Colorectal Cancer

Colorectal cancer (CRC) is a complex, multistage process, starts from neoplasia, followed by tissue invasion, vascular intra and extravasation and distant metastasis. The stromal cells of colon interact with the ECM and breakdown of ECM components is important for a cell to migrate from the primary site of tumour. Research on CRC has elucidated the role of distinct immune cells, cytokines and other immune mediators in virtually all steps of colon tumourigenesis, including initiation, promotion, progression and metastasis [60].

All groups of MMPs play role in the development as well as progression of CRC. The collagenases, i.e. MMP-1 and MMP-13 expressions were observed in the advanced stages of CRC with the lymph node involvement and poor prognosis. Huang et al. reported an approximately eightfold increased risk of post-operative recurrence in those patients who had MMP-13 overexpression [61]. There are studies on correlation of MMP-2 and -9 expressions with CRC and worse outcome. Patients having lymph node metastasis with CRC had an elevated level of plasma MMP-2 compared to the patients of early stage. Some reports also suggested serum MMPs as candidate biomarkers for CRC metastasis, as researchers have found higher ratio of expressions of MMP-2 and MMP-9 in CRC patients compared to normal subjects, and TGFβ is the key transcription factor responsible for MMP-9 expression [62, 63]. Elevated level of p38 gamma MAPK induces c-Jun synthesis, which in turn, increases the transcription of MMP-9, thus invasion in CRC [64]. TGFβ receptor kinase blocker was found effective to reduce MMP-9 expression and block CRC metastasis [65, 66]. In addition, MMP-7 was also found to activate proMMP-2 and proMMP-9 to promote lymph node metastasis, and upregulation of MMP-7 was found in $\sim$80% of advanced stage of CRC [64]. MMP-7 knockout mice models of CRC demonstrated decreased tumour burden and reduced colon cancer multiplicity. Interestingly, MMP-12, also called the metalloelastase, was found to be protective in CRC and its inhibition was lethal in experimental animal models. Higher expression of MMP-12 inhibits distant metastasis by downregulating VEGF expression and angiogenesis in CRC [64].

2.4 MMPs in Inflammatory Bowel Disease

Ulcerative colitis and Crohn's disease both together called as inflammatory bowel disease IBD, which can affect any segment of the gastrointestinal tract. From the epidemiological survey, it was seen that genetic predisposition for IBD might play a role towards development of malignancy from IBD [67]. Epidemiological studies estimated the occurrence of IBD is 1 in 1000 individuals in western countries, but the rate is rising globally because of the lifestyle and diet [68]. Histologically, the disease is characterized by presence of granulomas, fibrosis in the tissue space along with fistulae [69].

The levels of several MMPs including MMP-1, MMP-2, MMP-3, MMP-7, MMP-9, MMP-10, MMP-12, MMP-13 are modulated during IBD pathogenesis in the inflamed colon mucosa or serum [70, 71]. Gene expression profiling demonstrated the transcriptional upregulation of MMP-1 with the severity of the disease and is linked with hypoxia inducing factor-1. Most importantly, the critical ratio of MMP-1: TIMP-1 gets altered with the severity of the disease [70]. The secretory MMP-9 mucosal expression level as well as serum antigen level was found significantly higher in ulcerative colitis patients compared to healthy subjects [72]. In vivo gelatinases double knockout mice model showed resistance from DSS, TNBS and *Salmonella typhimurium*-induced colitis [73]. In addition, MMP-12$^{-/-}$ mice were protected from TNBS-induced colitis [74]. Beside gelatinases, stromelysins (MMP-3 and MMP-10) were also found upregulated in the inflamed areas of IBD patients. SiRNA mediated silencing of MMP-3 confers protection from DSS-induced colitis [75]. A study in New Zealand patient pool on SNP showed that genes of MMP-3, MMP-8, MMP-10 and MMP-14 were associated with IBD [76].

Accumulating data from several studies indicated that IL-17A and IL-17F can act as inducers for the secretion of MMP- 1 and -3 in subepithelial myofibroblasts and also promote the actions of IL-1 and TNF- on these MMPs via MAPK mediated pathway [77]. The disruption in the protease–antiprotease balance of MMP: TIMP may also promote fibrosis in the intestine during the disease progression. In humans, the fibrosis in the gut is inhibited by TGFβ/Smad pathway where MMPs are downregulated and TIMP expression gets upregulated. MMPs regulate both pro- and anti-angiogenic factors which may contribute to the pathogenesis of IBD or mucosal healing. While new mechanisms are emerging for the IBD pathogenesis, it is crucial to understand the scenario where MMPs play significant role in mucosal healing, ECM remodelling, regulation of angiogenesis or immune response during disease pathogenesis.

2.5 Implication of MMPs in Crohn's Disease

Crohn's disease is a chronic inflammatory disease of the digestive tract and also falls under IBD. It affects the end of the small intestine, i.e. the ileum, but it may also affect other parts of the gastrointestinal tract and the entire thickness of the intestinal wall [78]. Epidemiology states that Crohn's disease affects approximately

3 per 1000 individual in western countries, and it is less common in Asia and Africa [68].

MMPs have been strongly implicated in the tissue injury in Crohn's disease. Recently, elevated expressions of MMPs have been found in the inflamed tissues of patients having Crohn's disease [79], which implies that there is a role of MMPs in the increased proteolysis in the mucosa, ulceration followed by inflammation and fistula formation. MMP-9 gets upregulated in the inflamed tissues, and MMP-9 transcripts were found only in the highly inflamed regions of the tissues [80]. MMP-3 levels were also found elevated in mononuclear macrophage-like cells and fibroblasts in patients [81]. There were no significant differences in MMP-2 expression reported. Downregulation of TIMPs is also very significant as TIMP-1, TIMP-2 and TIMP-3 level goes down during acute stage of the disease [82]; thus, disrupts the protease–antiprotease homeostasis. In addition, high MMP-3 expression was consistently found in fistulae in patients suffering from Crohn's disease. Microarray analysis of the inflamed tissue lysates showed that MMP-3 transcripts and proteins were localised particularly in large mononuclear cells as well as macrophages. Although MMP-10 falls under the same stromelysin group of MMP family with MMP-3, transcripts as well as expression of MMP-10 were found negative. Moreover, SNP in the promoter region of MMP-3 gene in 5A/6A position confers higher rate of promoter activity and increases the susceptibility of the disease [83] (Table 1).

3 Role of Microbiome in Gastrointestinal Ailments

Human beings are inhabited by a complex array of microorganisms that interact with each other and with the host. They, as a whole, represent an integrated and functional ecosystem (microbiota) that have important role in human health and disease. The composition and diversity of the microbiota vary among different normal individuals [84]. Herein, the latest findings on gastrointestinal microbiota, in relation to their composition and prevalence in the presence or absence of *H. pylori* infection are highlighted. It was a notion that stomach is a sterile organ due to several innate defences including acid secretion, migrating motor complexes, enterosalivary circulation of nitrate. However, there is a significant influence of microorganisms on stomach and intestinal microenvironment, both in physiological and pathological conditions (Fig. 4).

3.1 *Helicobacter pylori* Mediated Gastric Ulcer and Involvement of MMPs

Nobel laureate Robin Warren and Barry Marshall discovered the role of *H. pylori* in gastric ulcers in the year 1982 and considerable literature documented the presence of many acid-resistant strains, such as *Streptococcus*, *Neisseria*, *Lactobacillus* etc.

Table 1 Involvement of different MMPs with their activating cytokine/chemokine in different gastrointestinal pathologies and expressing cell types

Disease	MMP involvement	Activating factor	Expressing cell
Gastric ulcer	MMP-1, MMP-2, MMP-3, MMP-7, MMP-10, MMP-12 MMP-14	Cytokine factor	Migrating epithelial cell, connective tissue
H. pylori induced gastritis	MMP-2 MMP-9	NF κβ, IL-12 TNR	Fibroblasts, keratinocytes, endothelial cells, monocytes and macrophages express
Gastric cancer	MMP-3 MMP-8 MMP-9	NF-κβ	Gastric epithelial cells
Colon cancer	MMP-1 MMP-2 MMP-7 MMP-9 MMP-13	TGF-β, SMAD	Colon epithelial cells
IBD	MMP-1, MMP-2 MMP-3, MMP-7 MMP-9, MMP-10 MMP-13	IL-37, IL-1, NF-κβ	Colonic epithelia
Crohn's disease	MMP-3 MMP-9	IL 12, TNF-α	Granulated epithelial cell

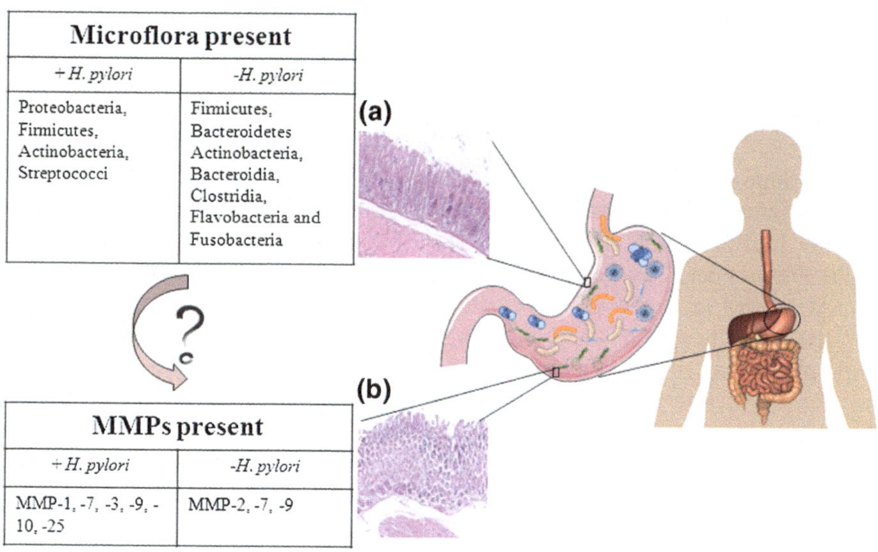

Fig. 4 Microbiome in human gut and MMPs expressed. Histology of the epithelium of stomach infested with pathogenic bacteria shows denudation and exfoliation of the layers in comparison with the control. Infiltration of inflammatory cells in the connective tissue is the benchmark of inflammation due to infection. The bacterial population and MMP expressed in presence and absence of *H. pylori* are shown in tables, but there is a gap in the information about the MMPs expressed in particular infection of the gastric system

[85–87]. *H. pylori* has been recognised as a Class I carcinogen [88]. There is enormous heterogeneity in the consequences of *H. pylori* infections. People acquire the infection early in life and are followed by a long quiescent phase when there is a chronic gastritis of variable intensity but with minimal symptoms [89]. Infection with *H. pylori* is not sufficient to induce gastric cancer, some other factors, i.e. bacterial and host cofactors are required to establish the disease [90, 91]. Only 10–15% of individuals infected with *H. pylori* develop ulcerative lesions in stomach, and the risk of gastric cancer is estimated to be approximately 1–3% [85, 92, 93].

Vitro studies suggest that *H. pylori* induce apoptosis of gastric epithelial cells and stimulate epithelial cells to secrete several chemoattractants [94]. Moreover, there is a marked increase in Th1-type cytokines, including IFN-γ, IL-12 and TNF-α in *H. pylori*-infected mucosa, all of which have been reported to be involved in tissue degradation in other systems [95, 96].

H. pylori infection induces the secretion of MMPs from a variety of gastric cells in vivo as well as in cultured cells, which in turn contribute to the pathogenesis of gastric ulcer and gastric cancer [97–99]. Gastric epithelial cells appear to be the major source of MMPs in *H. Pylori*-infected gastric tissues [100, 101]. A recent hospital-based study on gastric cancer patients with *H. pylori* infection revealed that the infection upregulated the expression of MMP-1 and MMP-10 [102, 103]. MMP-1 predominantly degrades the stroma, which is linked with invasion and metastasis [23]. In addition, microarray analysis of uninfected human gastric epithelial cell line (AGS) and *H. Pylori*-infected co-cultures demonstrated that along with MMP-10, several other MMPs, such as *MMP-1*, *MMP-7*, *MMP-25* genes were also upregulated [104, 105]. Among them, MMP-10 showed significant increase in expression in comparison to other MMPs [103].

3.2 Human Gastric Microbiota and Gastric Disorders

The gastrointestinal tract is the most populated organ in human body. Different sites of the GI tract are inhabited by different microbiota, including the stomach [106]. The very low pH value (median pH 1.4) of stomach makes it a very harsh and hostile environment for bacterial growth. Thus, the microbial colonization in stomach is very less (10^2–10^4 colony forming units (CFU)/g) compared to colon (10^{10}–10^{12} colony forming units (CFU)/g) [107]. The major constituent of human gastric microbiota is the Proteobacteria *H. pylori* although many other bacteria can also survive in this hostile environment, making it more diverse and complex. Along with the host physiology various other factors, including diet, *H. pylori* infection, enteral feeding, proton pump inhibitors, antibiotics and diseases shown to contribute in shaping the gastric microbiota [108]. With the advances of the DNA-based sequencing technologies, the culture-independent survey of human gastric microbiota is possible based on the analysis of a gastric biopsy sample. In 2013, Sheh and Fox summarized in their study that in stomach the most commonly found phyla are Proteobacteria, Firmicutes, Bacteroidetes, Actinobacteria and Fusobacteria. The most abundant phyla are Proteobacteria, Firmicutes and

Actinobacteria for *H. pylori* positive samples, while Firmicutes, Bacteroidetes and Actinobacteria are most abundant phyla for *H. pylori*-negative samples [109]. Interestingly, *H. pylori* being the most dominant bacteria comprise of 72–99% of sequencing reads in the stomach [110, 111]. Other commonly identified genera are of Streptococcus, Prevotella, Veillonella and Rothia species [112]. Surprisingly, the correlation was found only between the presence of Streptococci and peptic ulcer disease [113]. Another study compared gastric microbiota profile according to *H. pylori* status in chronic gastritis patients using high-throughput 16S rRNA sequencing. The microbiota of *H. Pylori*-negative patient sample was represented by the member of bacterial class alpha-, beta-, gamma-proteobacteria, bacilli, bacteroidia, clostridia, flavobacteria and fusobacteria [114].

The model proposed by Correa postulated that the chronic *H. pylori* infection of the gastric mucosa progresses through different stages like chronic active gastritis, intestinal metaplasia, dysplasia to subsequent development of gastric cancer [115]. Nonetheless, it has been clearly identified that *H. pylori* is the major risk factor in gastric cancer development [116, 117].

Other studies suggested that there is a gradual shift in the composition of gastric microbiota, which might plays a key role in the progression of pre-malignant lesions to gastric carcinoma. Additionally, gastric cancer samples showed decrease in the relative abundances in bacteria belonging to the phyla Proteobacteria, namely *Neisseria* spp, *Haemophilus* spp, *Bergeriella denitrificans*, *Epsilonproteo*bacteria and Helicobacteriaceae [114, 115] as well as Bacteroidetes (*Porphyromonas* spp and *Prevotella pallens*). The bacteria within the phyla Firmicutes are increased (Streptococcaceae, Lachnospiraceae and *Lactobacillus coleohominis*) [114, 115] or decreased *Streptococcus sinensis* [115].

4 Targeting MMPs as Therapeutic Strategy

Protease inhibitors are essential tools for the investigations of MMPs activities. They are useful not only for assessing the activity but also for inhibition of unwanted proteolysis in an experimental system. People started working on MMPs after proving its role on cancer stage, patient prognosis and death. Almost every pharmaceutical company started manufacturing MMP inhibitor (MMPI) to block MMP-mediated angiogenesis and metastasis. The programme started about 25 years ago and led to a number of small-molecule inhibitors in phase III clinical trials [25, 118, 119].

Chelating agents such as EDTA and 1,10-phenanthroline are routinely used in the laboratories to block MMP activities in in vitro experiments. Synthetic inhibitors commonly contain a chelating moiety, such as a carboxyl, a thiol, a phosphorous or a hydroxamic acid group. The chelating group is attached to a series of other groups that fit the specificity pocket of a particular metallopeptidase [120].

4.1 Endogenous MMP Inhibitors: TIMPs

TIMPs contain an N- and C-terminal domain of ~125 and 65 amino acids, respectively, with each containing three conserved disulfide bonds. The N-terminal domain folds as a separate unit and is capable of inhibiting MMPs. However, their range of activities is broader as it inhibits several disintegrin-metalloproteinases, namely ADAMs and ADAMTSs. Pathological conditions are associated with imbalanced MMP activities due to altered TIMP levels as important factor. Structural studies of TIMP-MMP complexes have allowed the generation of TIMP variants that selectively inhibit different groups of metalloproteinases. Engineering such variants is complicated by the fact that TIMPs can undergo changes in molecular dynamics induced by their interactions with MMPs. TIMPs are involved in cell growth and differentiation, cell migration, anti-angiogenesis, anti- and pro-apoptosis and synaptic plasticity [120, 121].

4.2 Antibody-Based Inhibition Targeting Catalytic Domain

Reports are available on the use of functional blocking antibodies, which have high potency for MMPs. Several functional blocking antibodies have been developed that selectively target the membrane-anchored MMPs. Combining a human antibody phage display library with automated selection and screening strategies resulted in the identification of a highly selective antibody-based MMP-14 inhibitor called DX-2400. It displayed anti-invasive, antitumour and anti-angiogenic properties and blocked MMP-14 mediated pro-MMP-2 processing [122]. To date, at least two monoclonal antibodies have been tested which bind to the catalytic domain, without interacting with the catalytic zinc. DX-2400 is a MMP-14 specific inhibitor, which binds to the catalytic domain with a K_i in the sub-nanomolar range.

Thus far, preclinical studies of DX-2400 indicated that the antibody is capable of inhibiting all of these activities while there was no measurable effect on other MMPs. In mouse studies, the drug was observed to decrease tumour burden significantly and decreased metastases in lung and liver. Further, DX-2400 was effective against HER2-positive xenografts both when used as a single agent or in combination with paclitaxel. This marks DX-2400 as an attractive candidate for patients diagnosed with triple-negative breast cancer, although clinical trials for this therapeutic have not yet been initiated [118, 122].

Other groups developed selective MMP-14 inhibitory antibodies that were successfully tested in vitro and in vivo. The neutralizing monoclonal antibody REGA-3G12 acts as a selective inhibitor of MMP-9 by binding against catalytic domain but not against the fibronectin or zinc-binding domains. A murine monoclonal antibody, termed REGA-3G12, has also been generated by hybridoma technology against the catalytic domain of human MMP-9. This inhibits MMP-9 without affecting activity of MMP-2, which shares high homology. Therefore, a therapeutic that can differentiate between these two highly similar enzymes may prove very useful [120]. In this regard, blocking antibodies were found to act on specific functions of the MMP rather than the general proteolytic activity. For example, 9E8 monoclonal antibody targets the MMP-2 activation capacity of MT1–MMP rather than the general proteolytic activity [118, 121].

The mechanism for TIMPs to inhibit MMPs can be utilized to develop different antibody-based strategies for effectively targeting the in vivo activity of MMP. A neutralizing antibody was developed based on the three-dimensional structure and amino acid sequence of MMP-13, which bind only to the active form of MMP-13. Monoclonal antibody against MMP-2 exhibited inhibition over MMP-2 activity but did not affect the structurally similar MMP-9 activity [121].

4.2.1 Hemopexin Domain Inhibitor and Small Molecule Inhibitor of MMPs

The interaction of inhibitor with the hemopexin-like domain prevents binding of endogenous partners that can promote angiogenesis or cancer cell migration. The hemopexin domain among different MMPs exhibits significantly less sequence and structural homology compared to the catalytic domain. This domain comprises a succession of four structurally similar hemopexin-like repeats to create a central funnel-like tunnel. Each hemopexin-like repeat is made up of four β-strands; the first three β-strands bear the highest homology across the MMP family whereas the b4 strands bear the least. As many as four structural ions have been found to be coordinated within this tunnel, and it has been proposed that these ions confer a stabilizing function for the whole domain. MMP-9 has only a sodium ion and displays a flexible architecture and considerable deviation from the structure of hemopexin domains reported for other MMPs. The first ion binding position, which is closest to the linker region connecting the hemopexin domain to the catalytic domain, is generally either a sodium or calcium ion [122].

In silico analysis of the MMP-14 hemopexin domain identified a druggable pocket-like site in the centre of the hemopexin structure. Binding of small molecule compounds in this site should, in theory, allosterically block dimerization. Compounds which bind to the hemopexin domains and prevent dimerization have been shown to significantly decrease tumour size, reduce MMP-mediated cell scattering/invasion, angiogenesis, and tumour metastasis both in vitro and in animal models. A subsequent docking study of small-molecule compounds led to identification of a compound which is selective for MMP-14 compared to MMP-2 and was not cytotoxic and did not affect catalytic activity (including MMP-14-mediated activation of MMP-2). Report is there for inhibition of hemopexin domain that was effective in attenuating cancer cell migration and in vivo-reduced tumour volume. In another study, an inhibitor was made with no proteolytic or cytotoxic effects while significantly decreased cancer cell migration and invasion and significantly decreased tumour size as well as the number of metastases in vivo [120, 123, 124].

Bimodal approach confers increased selectivity for MMP-2 as compared to individual subunit. A fusion protein has been designed which links the ten amino acid sequences of a MMP-2 selective inhibitory peptide (APP-IP, a β-amyloid precursor protein) to the N-terminus of TIMP-2. This macromolecular protein, which binds with a K_i in the sub-picomolar range, is designed to interact with both the active site and the hemopexin-like domain of MMP-2 [125–127].

In addition, small peptides have been used successfully to block dimer-induced functions of MMPs. Owing to induced intracellular cytoskeleton rearrangements necessary for processing of migration and invasion machinery, MMP-14 homodimerizes and also heterodimerizes with CD44. This includes proteolysis of proMMP-2 by MMP-14. A number of peptides generated that mimic the sequence of the residues of hemopexin-like domains required for dimerization were able to reduce tumourigenic effects in vitro and in vivo [128] (Table 2).

Table 2 Synthetic MMP inhibitors and their application in different diseases

Inhibitor name	MMPs inhibited	Disease model	Clinical trial	Reference
Batimastat (BB-94)	Broad-spectrum MMPs	Various tumours	Yet to be approved	[129]
Neovastat (AE-941)	MMPs 2, 9, 12; VEGFR-2	Renal cell carcinoma, non-small cell lymphoma	Phase III of clinical trial	[130]
Prinomastat (AG-3340)	MMPs 2, 3, 9, 13, and 14	Renal cell carcinoma	Phase III of clinical trial completed	[131]
Rebimastat (BMS-275291)	MMPs 1, 2, 8, 9, and 14	Advanced non-small cell lung cancer	Phase III of clinical trial	[132]
Marimastat (BB-2516)	Broad-spectrum MMPs	–	Development terminated due to poor performance in clinical trial	[133]
Ilomastat	MMPs (1–3, 8, 9)	–	–	[118]
Doxycycline hyclate (Dermostat, Periostat) [CollaGenex Pharmaceuticals]	Collagenase	Periodontal disorders	Launched	[127]
		Rosacea	Phase III	
		Acne	Phase II	
AZD 8955 [Astra Zeneca]	Collagenase	Osteoarthritis	Phase II	
PCK 3145 [Ambrilia Biopharma]	MMP9	Prostate cancer	Phase II	
Apratastat [Amgen/Wyeth]	MMP1, MMP9, MMP13, TACE	Rheumatoid arthritis	Phase II	
Incyclinide [CollaGenex Pharmaceuticals]	MMP2	Acne	Phase II	
		Brain cancer	Phase II	
		Kaposi's sarcoma	Phase II	
		Cancer metastases	Phase I	
		Solid tumours	Phase I	

(continued)

Table 2 (continued)

Inhibitor name	MMPs inhibited	Disease model	Clinical trial	Reference
ABT 518 [Abbott Laboratories]	Unknown	Solid tumours	Phase I	
MPC 2130 [Myriad Pharmaceuticals]	Unknown	Cancer	Phase I	
		Haematological malignancies	Phase I	
MMP12 inhibitor [Merck]	MMP12	Multiple sclerosis	Phase I	
AS111793 [Serono Pharmaceutical Research Institute (Geneva, Switzerland)]	MMP12	Reduces airway inflammation in mice exposed to cigarette smoke	–	[134]

5 Limitations/Challenges for MMPs Inhibition

Although several preclinical research that supported the importance of MMPs in cancer, all Phase III cancer trials using different inhibitors of MMPs are failed unfortunately. The major reason is lack of specificity of inhibitors and insufficient knowledge on the complexity of cancers [118]. The information of preclinical studies in the mouse models and the clinical trials in patients varied much that might be the reason behind the failure of the clinical trials as well as adverse effect on patients. The adverse effects were mostly due to their broad-spectrum inhibition of MMPs and the cross-inhibition among other family of proteins, e.g. ADAM (a disintegrin and metalloproteinase) and ADAMTS (ADAMs with thrombospondin motifs). In addition, several MMP inhibitors were neither metabolically stable, nor orally bio-available and toxic as well. In mammals, MMP genes are conserved indeed and are essential for normal functioning of the organism. It is worth mentioning that MMPs activities are not always harmful but becomes detrimental with its anti-targets actions in other physiological conditions. In contrast to other proteases (like caspases), most MMPs contain conserved amino acid sequences having high homology in the substrate binding domain which hinders the fabrication of specific substrate-based inhibitors. Interestingly, the evolution of many MMPs occurred by gene duplication in the mammalian genome, that leads to the formation of MMP genes clusters on particular chromosomes (for example, the chromosome-9 in the proximal mouse harbours ten MMP genes in less than 500 kb) and possess widespread homology in their amino acid sequence. As a consequence, translating in vitro research work with in vivo applications remains a difficult task. In vitro studies with active MMP and any protein are resulted in cleavage at particular sites. However, this cleavage might not essentially occur in an in vivo condition in physiological system. Scientists had attempted in knocking out many MMP-coding genes in mouse models (as in vivo systems) to investigate the consequences of the absence of these genes. However, not much has been evaluated in tissue-specific knockout mice as there is no obvious phenotypic abnormality in unstimulated conditions in most MMP-deficient mice (except for MMP-14 and MMP-20 deficient mice). Moreover, insufficient knowledge on the spatiotemporal activities of MMPs in pathological conditions adds to the unsuccessful attempts for clinical trial of MMP inhibitors.

6 Future Directions

A plethora of literature as well as supporting data revealed that MMPs play crucial roles in both physiological and pathological processes. They could be exploited as independent prognostic factors in gastrointestinal inflammation and malignancies. MMPs are associated with multiple diseases; hence they can be considered as drug targets to treat those diseases. A number of studies from knockout mice and in vitro cultured cells have shown that their involvement as integral part in acute as well as chronic inflammation. The major task for the future is to design specific MMP inhibitors and to elucidate the crosstalk among the members of MMP family. Newer activity-based imaging probes specific for MMPs will facilitate the elucidation of the structural role of inhibitors in gastrointestinal disorders. Although, clinical trials with the therapeutic MMP inhibitors encountered several challenges, studies in both in vivo and in vitro are in progress to target the specific MMP in gastrointestinal pathologies. Interaction between different transcription factors and different MMP promoters provides valuable insights into the mechanism of disease progression. Inhibition to specific MMP in gastrointestinal disorders and its effect at multiple cellular pathways become formidable task for therapeutic use. Although, monoclonal antibody-based therapy is promising for the prognosis and therapy of gastrointestinal cancers, however, its validation in experimental knockout animals and cancer models is prerequisite. Tailor-made therapies and drugs based on set of specific MMP in gastric disorders of different individual could be useful to develop good quality drug. Moreover, development of assay tool against a set of MMPs may lead to formulate commercially viable kits for early prognosis of gastrointestinal diseases using patient serum. Nonetheless, a role for MMPs in pathology of gastrointestinal tract could be related to tissue-specific expression and function of MMPs and be exploited as target for therapies.

Acknowledgements The financial assistance from CSIR network projects HUM (BSC 0119) and INDEPTH (BSC 0111) is thankfully acknowledged.

References

1. Verma RP, Hansch C (2007) Matrix metalloproteinases (MMPs): chemical-biological functions and (Q)SARs. Bioorg Med Chem 15:2223–2268
2. Lu P, Takai K, Weaver VM, Werb Z (2011) Extracellular matrix degradation and remodeling in development and disease. Cold Spring Harb Perspect Biol 3:a005058

3. Bonnans C, Chou J, Werb Z (2014) Remodelling the extracellular matrix in development and disease. Nat Rev Mol Cell Biol 15:786–801
4. Gomis-Rüth FX (2009) Catalytic domain architecture of metzincin metalloproteases. J Biol Chem 284:15353–15357
5. Lund J, Olsen OH, Sørensen ES, Stennicke HR, Petersen HH, Overgaard MT (2013) ADAMDEC1 is a metzincin metalloprotease with dampened proteolytic activity. J Biol Chem 288:21367–21375
6. Ikonomidou C (2014) Matrix metalloproteinases and epileptogenesis. Mol Cell Pediatr 1:6
7. Mizoguchi H, Yamada K (2013) Roles of matrix metalloproteinases and their targets in epileptogenesis and seizures. Clin Psychopharmacol Neurosci 11:45–52
8. Gong Y, Chippada-Venkata UD, Oh WK (2014) Roles of matrix metalloproteinases and their natural inhibitors in prostate cancer progression. Cancers 6:1298–1327
9. Massova I, Kotra LP, Fridman R, Maboshery S (1998) Matrix metalloproteinases: structures, evolution and diversification. FASEB J 12:1075–1095
10. Nagase H, Woessner JF Jr (1999) Matrix metalloproteinases. J Biol Chem 274:21491–21494
11. Sternlicht MD, Werb Z (2001) How matrix metalloproteinases regulate cell behavior. Annu Rev Cell Dev Biol 17:463–516
12. Harper E, Bloch KJ, Gross J (1971) The zymogen of tadpole collagenase. Biochemistry 10 (16):3035–3041
13. Ra H-J, Parks WC (2007) Control of matrix metalloproteinase catalytic activity. Matrix Biol 26:587–596
14. Löffek S, Schilling O, Franzke C-W (2011) Biological role of matrix metalloproteinases: a critical balance. Eur Respir J 38:191–208
15. Andrian E, Mostefaoui Y, Rouabhia M, Grenier D (2007) Regulation of matrix metalloproteinases and tissue inhibitors of matrix metalloproteinases by *Porphyromonas gingivalis* in an engineered human oral mucosa model. J Cell Physiol 211:56–62
16. Carvalho HF, Roque ACA, Iranzo O, Branco RJF (2015) Comparison of the internal dynamics of metalloproteases provides new insights on their function and evolution. PLoS ONE 10:e0138118
17. Gomis-Rüth FX (2003) Structural aspects of the metzincin clan of metalloendopeptidases. Mol Biotechnol 24:157–202
18. Tallant C, Marrero A, Gomis-Rüth FX (2010) Matrix metalloproteinases: fold and function of their catalytic domains. Biochimica Biophysica Acta Mol Cell Res 1803:20–28
19. Fridman R (2003) Surface association of secreted metalloproteinases. Curr Top Dev Biol Elsevier Sci. 54:75–100
20. Cerdà-Costa N, Gomis-Rüth FX (2014) Architecture and function of metallopeptidase catalytic domains. Protein Sci 23:123–144
21. Duan JX, Rapti M, Tsigkou A, Lee MH (2015) Expanding the activity of tissue inhibitors of metalloproteinase (TIMP)-1 against surface-anchored metalloproteinases by the replacement of its C-terminal domain: implications for anti-cancer effects. PLoS ONE 10(8):e0136384
22. Visse R, Nagase H (2003) Matrix metalloproteinases and tissue inhibitors of metalloproteinases structure, function, and biochemistry. Circ Res 92:827–839
23. Brew K, Nagase H (2010) The tissue inhibitors of metalloproteinases (TIMPs): an ancient family with structural and functional diversity. Biochim Biophys Acta 1803:55–71
24. Nita M, Grzybowski A (2016) The role of the reactive oxygen species and oxidative stress in the pathomechanism of the age-related ocular diseases and other pathologies of the anterior and posterior eye segments in adults. Oxid Med Cell Longev 3164734:1–23
25. Kessenbrock K, Plaks V, Werb Z (2010) Matrix metalloproteinases: regulators of the tumor microenvironment. Cell 141:52–67
26. Fu X, Kassim SY, Parks WC, Heinecke JW (2003) Hypochlorous acid generated by myeloperoxidase modifies adjacent tryptophan and glycine residues in the catalytic domain of matrix metalloproteinase-7 (matrilysin): an oxidative mechanism for restraining proteolytic activity during inflammation. J Biol Chem 278:28403–28409

27. Langton KP, McKie N, Smith BM, Brown NJ, Barker MD (2005) Sorsby's fundus dystrophy mutations impair turnover of TIMP-3 by retinal pigment epithelial cells. Hum Mol Genet 14(23):3579–3586
28. Page-McCaw A, Ewald AJ, Werb Z (2007) Matrix metalloproteinases and the regulation of tissue remodelling. Nat Rev Mol Cell Biol 8(3):221–233
29. Egeblad M, Werb Z (2002) New functions for matrix metalloproteinases in cancer progression. Nat Rev Cancer 2:161–174
30. Martin TA, Ye L, Slanders AJ, Lane J, Jiang WG (2013) Cancer invasion and metastasis: molecular and cellular perspective. In: Jandial R Metastatic cancer: clinical and biological perspectives. Landes Bioscience
31. Rundhaug JE (2003) Matrix metalloproteinases, angiogenesis, cancer. Clin Cancer Res 9:551–554
32. Sang QXA (1998) Complex role of matrix metalloproteinases in angiogenesis. Cell Res 8:171–177
33. Murch SH, MacDonald TT, Walker-Smith JA, Lionetti P, Levin M, Klein NJ (1993) Disruption of sulphated glycosaminoglycans in intestinal inflammation. Lancet 341:711–714
34. O'Sullivan S, Gilmer JF, Medina C (2015) Matrix metalloproteinases in inflammatory bowel disease: an update. Med Inflamm 964131:1–19
35. Shihab PK, Al-Roub A, Al-Ghanim M, Al-Mass A, Behbehani K, Ahmad R (2015) TLR2 and AP-1/NF-kappaB are involved in the regulation of MMP-9 elicited by heat killed *Listeria monocytogenes* in human monocytic THP-1 cells. J Inflamm 12:32–40
36. Hansen JM, Hallas J, Lauritsen JM, Bytzer P (1996) Non-steroidal anti-inflammatory drugs and ulcer complications: a risk factor analysis for clinical decision-making. Scand J Gastroenterol 31:126–130
37. Matsui H, Shimokawa O, Kanekon T, Nagano Y, Rai K, Hyodo I (2011) The pathophysiology of non-steroidal anti-inflammatory drug (NSAID) induced mucosal injuries in stomach and small intestine. J Clin Biochem Nutr 48(2):107–111
38. Rahal A, Kumar A, Singh V, Yadav B, Tiwari R, Chakraborty S, Dhama K (2014) Oxidative stress, prooxidants, and antioxidants: the interplay. BioMed Res Int ID 761264:19 p
39. Musumba C, Pritchard DM, Pirmohamed M (2009) Cellular and molecular mechanisms of NSAID-induced peptic ulcers. Aliment Pharmacol Ther 30(6):517–531
40. Frankowski H, Gu YH, Heo JH, Milner R, del Zoppo GJ (2012) Use of gel zymography to examine matrix metalloproteinase (gelatinase) expression in brain tissue or in primary glial cultures. Methods Mol Biol 814:221–233
41. Verma S, Kesh K, Ganguly N, Jana S, Swarnakar S (2014) Matrix metalloproteinases and gastrointestinal cancers: impacts of dietary antioxidants. World J Biol Chem 26:355–376
42. Wroblewski LE, Peek M, Wilson KT (2010) *Helicobacter pylori* and gastric cancer: factors that modulate disease risk. Clin Microbiol Rev 23:713–739
43. Cheng HC, Yang HB, Chang WL, Chen WY, Yeh YC, Sheu BS (2012) Expressions of MMPs and TIMP-1 in gastric ulcers may differentiate *H. pylori* infected from NSAID-related ulcers. Sci World J ID 539316:9
44. Cheng CL, Guo JS, Luk J, Koo MWL (2004) The healing effects of Centella and asiaticoside on acetic acid induced gastric ulcers in rats. Life Sci 74(18):2237–2249
45. Ganguly K, Kundu P, Banerjee A, Reiter RJ, Swarnakar S (2006) Hydrogen peroxide-mediated downregulation of matrix metalloproteinase-2 in indomethacin-induced acute gastric ulceration is blocked by melatonin and other antioxidants. Free Rad Biol Med 41:911–925
46. Witztum JL, Steinberg D (1991) Role of oxidized low density lipoprotein in atherogenesis. J Clin Invest 88:1785–1792
47. Kar S, Subbaram S, Carrico PM, Melendez JA (2010) Redox-control of matrix metalloproteinase-1: a critical link between free radicals, matrix remodeling and degenerative disease. Respir Physiol Neurobiol 31:299–306

48. Singh LP, Kundu P, Ganguly K, Mishra A, Swarnakar S (2007) Novel role of famotidine in downregulation of matrix metalloproteinase-9 during protection of ethanol-induced acute gastric ulcer. Free Rad Biol Med 43:289–299
49. Chakraborty S, Stalin S, Das N, Choudhury ST, Swarnakar Ghosh S S (2012) The use of nano-quercetin to arrest mitochondrial damage and MMP-9 upregulation during prevention of gastric inflammation induced by ethanol in rat. Biomaterials 33:2991–3001
50. Rahman R, Asombang AW, Ibdah JA (2014) Characteristics of gastric cancer in Asia. World J Gastroenterol 20:4483–4490
51. Fox JG, Wang TC (2007) Inflammation, atrophy, and gastric cancer. J Clin Invest 117:60–69
52. Correa P, Piazuelo MB (2011) *Helicobacter pylori* infection and gastric adenocarcinoma. US Gastroenterol Hepatol Rev 7(1):59–64
53. Dey S, Ghosh N, Saha D, Kesh K, Gupta A, Swarnakar S (2014) Matrix metalloproteinase-1 (MMP-1) promoter polymorphisms are well linked with lower stomach tumor formation in eastern Indian Population. PLoS ONE 9:e88040
54. Witty JP, McDonnell S, Newell KJ, Cannon P, Navre M, Tressler RJ, Matrisian LM (1994) Modulation of matrilysin levels in colon carcinoma cell lines affects tumorigenicity in vivo. Cancer Res 54:4805–4812
55. Dey S, Stalin S, Gupta A, Saha D, Kesh K, Swarnakar S (2012) Matrix metalloproteinase-3 gene promoter polymorphisms and their haplotypes are associated with gastric cancer risk in eastern Indian population. Mol Carcinog 51:E42–E53
56. Kesh K, Subramanian L, Ghosh N, Gupta V, Gupta A, Bhattacharya S, Mahapatra NR, Swarnakar S (2015) Association of MMP7-181A → G promoter polymorphism with gastric cancer risk. J Biol Chem 290:14391–14406
57. Shan YQ, Ying RC, Zhou CH, Zhu AK, Ye J, Zhu W et al (2015) MMP-9 is increased in the pathogenesis of gastric cancer by the mediation of HER2. Cancer Gene Ther 22:101–107
58. Yoo YA, Kang MH, Lee HJ, Kim BH, Park JK, Kim HK, Kim JS, Oh SC (2011) Sonic hedgehog pathway promotes metastasis and lymphangiogenesis via activation of Akt, EMT, and MMP-9 pathway in gastric cancer. Cancer Res 71:61–69
59. Alakus H, Grass AG, Hennecken JK, Bollschweiler E, Schulte C, Drebber U, Baldus SE, Metzger R, Hölscher AH, Mönig SP (2008) Clinicopathological significance of MMP-2 and its specific inhibitor TIMP-2 in gastric cancer. Histol Histopathol 23:917–923
60. Markman JL, Shiao SL (2015) Impact of the immune system and immunotherapy in colorectal cancer. J Gastrointest Oncol 6:208–223
61. Huang Z, Huang D, Ni S, Peng Z, Sheng W, Du X (2010) Plasma microRNAs are promising novel biomarkers for early detection of colorectal cancers. Int J Cancer 127(1):118–126
62. Fanjul-Fernández M, Folgueras AR, Cabrera S, López-Otín C (2010) Matrix metalloproteinases: evolution, gene regulation and functional analysis in mouse models. Biochim Biophys Acta Mol Cell Res 1803:3–19
63. Grivennikov SI (2013) Inflammation and colorectal cancer: colitis-associated neoplasia. Semin Immunopathol 35(2):229–244
64. Said AH, Raufman JP, Xie G (2014) The role of matrix metalloproteinases in colorectal cancer. Cancers 6(1):366–375
65. Cherukua HR, Mohamedalib A, Cantora DI, Tanc SH, Niced EC, Baker MS (2015) Transforming growth factor-b, MAPK and Wnt signaling interactions in colorectal cancer. EuPa Open Proteom 8:104–115
66. Iiizumi M, Liu W, Pai SK, Furuta E, Watabe K (2008) Drug development against metastasis-related genes and their pathways: a rationale for cancer therapy. Biochim Biophys Acta 1786(2):87–104
67. Zhiqin W, Palaniappan S, Ali R, Affendi R (2014) Inflammatory bowel disease-related colorectal cancer in the Asia-Pacific region: past, present, and future. Intest Res 12:194–204
68. M'Koma AE (2013) Inflammatory bowel disease: an expanding global health problem. Clin Med Insights Gastroenterol 6:33–47

69. Medina C, Radomski MW (2006) Role of matrix metalloproteinases in intestinal inflammation. J Pharmacol 318(3):933–938
70. O'Sullivan S, Gilmer JF, Medina C (2015) Matrix metalloproteinases in inflammatory bowel disease: an update. Mediators Inflamm ID 964131:19
71. Deban L, Correale C, Vetrano S, Malesci A, Danese S (2008) Multiple pathogenic roles of microvasculature in inflammatory bowel disease: a jack of all trades. Am J Pathol 172 (6):1457–1466
72. Lakatos G, Sipos F, Miheller P, Hritz I, Varga MZ, Juhasz M et al (2011) The behavior of matrix metalloproteinase-9 in lymphocytic colitis, collagenous colitis and ulcerative colitis. Pathol Oncol Rep 18(1):85–91
73. Shimoda M, Horiuchi K, Sasaki A et al (2016) Epithelial cell-derived a disintegrin and metalloproteinase-17 confers resistance to colonic inflammation through EGFR activation. EBioMedicine 5:114–124
74. Walter L, Harper C, Garg P (2013) Role of matrix metalloproteinases in inflammation/colitis-associated colon cancer. Immuno-Gastroenterol 2:22–28
75. Laroui H, Geem D, Xiao B et al (2014) Targeting intestinal inflammation with CD98 siRNA/PEI—loaded nanoparticles. Mol Ther 22(1):69–80
76. Godoy-Santos AL, Trevisan R, Fernandes TD, dos Santos MCLG (2011) Association of MMP-8 polymorphisms with tendinopathy of the primary posterior tibial tendon: a pilot study. Clinics 66(9):1641–1643
77. Li D-Q, Luo L, Chen Z, Kim H-S, Song XJ, Pflugfelder SC (2006) JNK and ERK MAP kinases mediate induction of IL-1β, TNF-α and IL-8 following hyperosmolar stress in human limbal epithelial cells. Exp Eye Res 82(4):588–596
78. Moon CM, Jung S-A, Kim S-E, Song HJ, Jung Y, Ye BD et al (2015) Clinical factors and disease course related to diagnostic delay in Korean Crohn's disease patients: results from the connect study. PLoS ONE 10(12):e0144390. doi:10.1371/journal.pone.0144390
79. Sullivan SO', Gilmer JF, Medina C (2015) Matrix metalloproteinases in inflammatory bowel disease: an update. Mediators Inflamm ID 964131:19 p
80. Pedersen G, Saermark T, Kirkegaard T, Brynskov J (2009) Spontaneous and cytokine induced expression and activity of matrix metalloproteinases in human colonic epithelium. Clin Exp Immunol 155(2):257–265
81. García MF, González-Reyes S, González LO et al (2010) Comparative study of the expression of metalloproteases and their inhibitors in different localizations within primary tumours and in metastatic lymph nodes of breast cancer. Int J Exp Pathol 91(4):324–334
82. Chang Y, Chiu Y, Cheng H et al (2015) Down-regulation of TIMP-1 inhibits cell migration, invasion, and metastatic colonization in lung adenocarcinoma. Tumor Biol 36:3957. doi:10.1007/s13277-015-3039-5
83. Pereira AC, Dias do Carmo E, Dias da Silva MA, Blumer Rosa LE (2012) Matrix metalloproteinase gene polymorphisms and oral cancer. J Clin Exp Dent 4(5):e297–e301
84. Jandhyala SM, Talukdar R, Subramanyam C, Vuyyuru H, Sasikala M, Reddy DN (2015) Role of the normal gut microbiota. World J Gastroenterol 21(29):8787–8803. doi:10.3748/wjg.v21.i29.8787
85. Kusters JG, van Vliet AHM, Kuipers EJ (2006) Pathogenesis of *Helicobacter pylori* infection. Clin Microbiol Rev 19(3):449–490
86. Nardone G, Compare D (2015) The human gastric microbiota: is it time to rethink the pathogenesis of stomach diseases? United Eur Gastroenterol J 3(3):255–260
87. Wang Z-K, Yang Y-S (2013) Upper gastrointestinal microbiota and digestive diseases. World J Gastroenterol 19(10):1541–1550
88. Wroblewski LE, Peek RM, Wilson KT (2010) *Helicobacter pylori* and gastric cancer: factors that modulate disease risk. Clin Microbiol Rev 23(4):713–739
89. Salih BA (2009) *Helicobacter pylori* infection in developing countries: the burden for how long? Saudi J Gastroenterol 15(3):201–207

90. Amsterdam KV, Van Vliet AHM, Kusters JG, Ende AVD (2006) Of microbe and man: determinants of *H. pylori* related diseases. Microbiol Rev 30(1):131–156
91. Fox JG, Wang TC (2007) Inflammation, atrophy and gastric cancer. J Clin Invest 117 (1):60–69
92. Fitzgerald RC, Caldas C (2004) Clinical implications of E-cadherin associated hereditary diffuse gastric cancer. Gut 53(6):775–778
93. Hu B, El Hajj N, Sittler S, Lammert N, Barnes R, Meloni-Ehrig A (2012) Gastric cancer: classification, histology and application of molecular pathology. J Gastrointest Oncol 3 (3):251–261
94. Alzahrani S, Lina TT, Gonzalez J, Pinchuk IV, Beswick EJ, Reyes VE (2014) Effect of *Helicobacter pylori* on gastric epithelial cells. World J Gastroenterol 20(36):12767–12780
95. Peek RM, Fiske C, Wilson KT (2010) Role of innate immunity in Helicobacter pylori-induced gastric malignancy. Physiol revi 90(3):831–858
96. White JR, Winter JA, Robinson K (2015) Differential inflammatory response to *Helicobacter pylori* infection: etiology and clinical outcomes. J Inflamm Res 8:137–147
97. Gööz M, Shaker M, Gööz P, Smolka AJ (2003) Interleukin 1β induces gastric epithelial cell matrix metalloproteinase secretion and activation during *Helicobacter pylori* infection. Gut 52(9):1250–1256
98. Pillinger MH, Marjanovic N, Kim SY, Lee YC, Scher JU, Roper J et al (2007) *Helicobacter pylori* stimulates gastric epithelial cell MMP-1 secretion via CagA-dependent and -independent ERK activation. J Biol Chem 282(26):18722–18731
99. Oliveira MJ, Costa AC, Costa AM, Henriques L, Suriano G, Atherton JC (2006) *Helicobacter pylori* induces gastric epithelial cell invasion in a c Met and type IV secretion system-dependent manner. J Biol Chem 281(46):34888–34896
100. Kundu P, De R, Pal I, Mukhopadhyay AK, Saha DR, Swarnakar S (2011) Curcumin alleviates matrix metalloproteinase-3 and -9 activities during eradication of *Helicobacter pylori* infection in cultured cells and mice. PLoS ONE 6(1):e16306
101. Stein M, Ruggiero P, Rappuoli R, Bagnoli F (2013) *Helicobacter pylori* CagA: from pathogenic mechanisms to its use as an anti-cancer vaccine. Front Immunol 4:328
102. Jiang H, Zhou Y, Liao Q, Ouyang H (2014) Helicobacter pylori infection promotes the invasion and metastasis of gastric cancer through increasing the expression of matrix metalloproteinase-1 and matrix metalloproteinase-10. Exp Ther Med 8(3):769–774
103. Costa AM, Ferreira RM, Pinto-Ribeiro I, Sougleri IS, Oliveira MJ, Carreto L et al (2016) Helicobacter pylori activates matrix metalloproteinase-10 in gastric epithelial cells via EGFR and ERK-mediated pathways. J Infect Dis 214(4)
104. Bebb JR, Letley DP, Thomas RJ, Aviles F, Collins HM, Watson SA et al (2003) Helicobacter pylori upregulates matrilysin (MMP-7) in epithelial cells in vivo and in vitro in a Cag dependent manner. Gut 52(10):1408–1413
105. Nam YH, Ryu E, Lee D, Shim HJ, Lee YC, Lee ST (2011) Cag-A phosphorylation dependent MMP-9 expression in gastric epithelial cells. Helicobacter 16(4):276–283
106. Dethlefsen L, Mcfall-Ngai M, Relman DA (2007) An ecological and evolutionary perspective on human-microbe mutualism and disease. Nature 449:811–818
107. Delgado S, Cabrera-Rubio R, Mira A, Suarez A, Mayo B (2013) Microbiological survey of the human gastric ecosystem using culturing an pyrosequencing methods. Microb Ecol 65:763–772
108. Wu WM, Yang YS, Peng LH (2014) Microbiota in the stomach: new insights. J Dig Dis 15:54–61
109. Sheh A, Fox JG (2013) The role of the gastrointestinal microbiome in Helicobacter pylori pathogenesis. Gut Microbes 4:505–531
110. Bik EM, Eckburg PB, Gill SR, Nelson KE, Purdom EA, Francois F, Perez-Perez E, Blaser MJ, Relman DA (2006) Molecular analysis of bacterial microbiota in human stomach. Proc Natl Acad Sci U S A 103:732–737

111. Andersson AF, Lindberg M, Jakobsson H, Bäckhed F, Nyrén P, Engstrand L (2008) Comparative analysis of human gut microbiota by barcoded pyrosequencing. PLoS ONE 3 (7):e2836
112. Khosravi Y, Dieye Y, Poh BH, Ng CG, Loke MF, Goh KL, Vadivelu J (2014) Culturable bacterial microbiota of the stomach of helicobacter pylori positive and negative gastric disease patients. Sci World J 2014:610421
113. Eun CS, Kim BK, Han DS, Kim SY, Kim KM, Choi BY, Song KS, Kim YS, Kim JF (2014) Differences in gastric mucosal microbiota profiling in patients with chronic gastritis, intestinal metaplasia, and gastric cancer using pyrosequencing methods. Helicobacter 19:407–416
114. Correa P (1992) Human gastric carcinogenesis: a multistep and multifactorial process—first American cancer society award lecture on cancer epidemiology and prevention. Cancer Res 52:6735–6740
115. Polk DB, Peek RM Jr (2010) Helicobacter pylori: gastric cancer and beyond. Nat Rev Caner 10:403–414
116. Uemura N, Okamoto S, Yamamoto S, Matsumura N, Yamaguchi S, Yamakido M, Taniyama K, Sasaki N, Schlemper RJ (2001) Helicobacter pylori infection and the development of gastric cancer. N Engl J Med 345:784–789
117. Aviles-Jimenez F, Vazquez-Jimenez F, Medrano-Guzman R, Mantilla A, Torres J (2014) Stomach microbiota composition varies between patients with non-atrophic gastritis and patients with intestinal type of gastric cancer. Sci Rep 4:4202
118. Vandenbroucke RE, Libert C (2014) Is there new hope for therapeutic matrix metalloproteinase inhibition? Nat Rev Drug Discov 13:904–927
119. Page-McCaw A, Ewald AJ, Werb Z (2007) Matrix metalloproteinases and the regulation of tissue remodelling. Nat Rev Mol Cell Biol 8:221–233
120. Cathcart J, Pulkoski-Gross A, Cao J (2015) Targeting matrix metalloproteinases in cancer: bringing new life to old ideas. Genes Dis 2:26–34
121. Devy L, Dransfield DT (2011) New strategies for the next generation of matrix-metalloproteinase inhibitors: selectively targeting membrane-anchored MMPs with therapeutic antibodies. Biochem Res Int 2011:1–11
122. Remacle AG, Golubkov VS, Shiryaev SA, Dahl R, Stebbins JL, Chernov AV, Cheltsov AV, Pellecchia M, Strongin AY (2012) Novel MT1-MMP small-molecule inhibitors based on insights into hemopexin domain function in tumor growth. Cancer Res 72:2339–2349
123. Coppola JM, Bhojani MS, Ross BD, Rehemtulla A (2008) A small-molecule furin inhibitor inhibits cancer cell motility and invasiveness. Neoplasia 10:363–370
124. Albini A, Tosetti F, Li VW, Noonan DM, Li WW (2012) Cancer prevention by targeting angiogenesis. Nat Rev Clin Oncol 9:498–509
125. Dormán G, Cseh S, Hajdú I, Barna L, Kónya D, Kupai K, Kovács L, Ferdinandy P (2010) Matrix metalloproteinase inhibitors: a critical appraisal of design principles and proposed therapeutic utility. Drugs 70:949–964
126. García-Pardo A, Opdenakker G (2015) Nonproteolytic functions of matrix metalloproteinases in pathology and insights for the development of novel therapeutic inhibitors. Metalloproteinases Med 2:19–28
127. Hu J, Van den Steen PE, Sang Q-XA, Opdenakker G (2007) Matrix metalloproteinase inhibitors as therapy for inflammatory and vascular diseases. Nat Rev Drug discov 6:480–498
128. Cathcart J, Pulkoski-Gross A, Cao J (2015) Targeting matrix metalloproteinases in cancer: bringing new life to old ideas. Genes Dis 2(1):26–34
129. Wojtowicz-Praga S, Low J, Marshall J, Ness E, Dickson R, Barter J, Sale M, McCann P, Moore J, Cole A (1996) Phase I trial of a novel matrix metalloproteinase inhibitor batimastat (BB-94) in patients with advanced cancer. Invest New Drugs 14:193–202
130. Lu C, Lee JJ, Komaki R et al (2010) Chemoradiotherapy with or without AE-941 in stage III non-small cell lung cancer: a randomized phase III trial. J Natl Cancer Inst 102(12):859–865

131. Bissett D, O'Byrne KJ, Von Pawel J, Gatzemeier U, Price A, Nicolson M, Mercier R, Mazabel E, Penning C, Zhang MH (2005) Phase III study of matrix metalloproteinase inhibitor prinomastat in non-small-cell lung cancer. J Clin Oncol 23:842–849
132. Leighl NB, Paz-Ares L, Douillard J-Y, Peschel C, Arnold A, Depierre A, Santoro A, Betticher DC, Gatzemeier U, Jassem J (2005) Randomized phase III study of matrix metalloproteinase inhibitor BMS-275291 in combination with paclitaxel and carboplatin in advanced non-small-cell lung cancer: National Cancer Institute of Canada-Clinical Trials Group Study BR. 18. J Clin Oncol 23:2831–2839
133. Sparano JA, Bernardo P, Stephenson P, Gradishar WJ, Ingle JN, Zucker S, Davidson NE (2004) Randomized phase III trial of marimastat versus placebo in patients with metastatic breast cancer who have responding or stable disease after first-line chemotherapy: Eastern Cooperative Oncology Group Trial E2196. J Clin Oncol 22:4683–4690
134. Le Quement C, Guenon I, Gillon JY, Valenca S, Cayron-Elizondo V, Lagente V, Boichot E (2008) The selective MMP-12 inhibitor, AS111793 reduces airway inflammation in mice exposed to cigarette smoke. Br J Pharmacol 154:1206–1215
135. Bonnans C, Chou J, Werb Z (2014) Remodelling the extracellular matrix in development and disease. Nat Rev Mol Cell Biol 15(12):786–801. doi:10.1038/nrm3904

Role of Proteases in Diabetes Mellitus

Shatadal Ghosh, Bhawna Pandey and Parames C. Sil

Abstract

Dipeptidyl peptidase-4 (DPP-4), a 110 kDa exopeptidase, selectively cleaves N-terminal dipeptides from a vast array of substrates. DPP-4 is expressed on the surface of many cell types and plays various important roles in diseases like cancer, inflammation, diabetes, obesity. In type 2 diabetes mellitus (T2DM), incretin hormones, glucagon-like peptide-1 (GLP-1), and glucose-dependent insulinotropic polypeptide (GIP) play major roles in the regulation of insulin secretion. Both GLP-1 and GIP are the substrates of DPP-4. That is why DPP-4 inhibitors have gained significantly increasing interest in treating T2DM recently. In addition to some general information on DPP-4, this chapter mainly describes its effects on relevant organs associated with T2DM and recent clinical trials. Besides, roles of some other proteases in diabetes mellitus have also been briefly discussed.

Keywords

Dipeptidyl peptidase-4 (DPP-4) · Incretin hormones
Glucagon-like peptide-1 (GLP-1) · Glucose-dependent insulinotropic polypeptide (GIP) · Clinical trials

S. Ghosh · B. Pandey · P.C. Sil (✉)
Division of Molecular Medicine, Bose Institute, P-1/12, CIT Scheme VII M, Kolkata 700054, West Bengal, India
e-mail: parames@jcbose.ac.in; parames_95@yahoo.co.in

S. Chakraborti and N.S. Dhalla (eds.), *Pathophysiological Aspects of Proteases*,
DOI 10.1007/978-981-10-6141-7_20

1 Introduction

Diabetes mellitus (DM) becomes one of the most common chronic, non-communicating diseases nowadays. DM is developed when insulin, responsible for the regulation of blood glucose, is either not produced, produced in insufficient amounts, or when the body cannot use insulin properly. Diabetes is mainly of three types: type 1 (T1DM), type 2 (T2DM), and gestational diabetes (GD). However, hyperglycaemia is common in all three types. If left untreated, sometimes even treated, DM generally accounts for long-term tissue damage throughout the body that explains the numerous complications associated with this disorder [1]. Acute complications include diabetic ketoacidosis (DKA) and nonketotic hyperosmolar state (NKHS). Vascular and nonvascular disorders are the main chronic symptoms [2–4]. In a global study in 2013, it is revealed that almost 382 million people are suffering from diabetes for a prevalence of 8.3%.

A major problem in the health care of diabetic patients is diabetic foot syndrome. Growth factors and proteases regulate the essential processes of normal wound healing [5, 6]. The diabetic foot syndrome occurs as a severe complication in the late stages of diabetes and is considered to be one of the most important complications of this disease. Diagnosis, classification, and proper therapy can help in reducing this problem. The rational development of treatments could be achieved only with the understanding of the cellular and molecular abnormalities that contribute to the diabetic foot syndrome.

As such to develop therapeutics and better understanding of the role of proteases, their detection becomes very important. Chromogenic and fluorometric substrates can be used for detection of protease activity. Immunoblotting or ELISA is also used for their detection. The least common methods for detection of these proteases are immunonephelometry, radial immunodiffusion, and immunoelectrophoresis. Also, degradomics is being used as an approach for screening the activity of proteases. The degradome represents the complete set of proteases expressed in certain medium at a specific time. Zymography and mass spectrometry or both are the approaches taken to serve the goal of degradomics.

Proteases not only play roles in catabolism of amino acids, but also perform vital roles involving the fate, functions, and interaction of proteins, leading to the regulation of different important cellular phenomenon including cellular signalling, transcription, and translation [7, 8]. Abnormalities in specific protease activity manifest in many pathologic conditions including DM. Proteases play vital roles in various biological processes including cell proliferation and differentiation, wound healing, tissue remodelling, angiogenesis, inflammation, immunity, autophagy, and cell death; all of which can be impaired, decreased, or augmented in DM [4, 6, 9–15]. So, these proteases might play a role in diabetic pathogenesis and associated complications. The present understanding, however, is limited due to poor data availability, although dipeptidyl peptidase-4 (DPP-4) inhibitors have been studied in some details [10, 11, 16–19]. DPP-4, also known as CD26, is a 110 kDa type II transmembrane glycoprotein [1–3]. Its inhibition has been shown to prolong biological activity of glucagon-like peptide-1 (GLP-1) and glucose-dependent insulinotropic polypeptide (GIP) which are incretins by nature and can stimulate a decrease in blood glucose levels. The approval of DPP-4 inhibitors for the treatment of type 2 diabetes mellitus (T2DM) has raised its importance in the pharmaceutical world. The role of proteases in diabetes also includes the association of single nucleotide polymorphisms (SNP)/genes with increased risk and protease profiling. Protease profiling, however, becomes important because of diagnostic and/or prognostic values.

The first part of this chapter basically deals with some general information about DPP-4 and its numerous biological functions, mainly protease activity, in regard to T2DM and its treatment. The other part describes an up-to-date understanding of the roles of DPP-4 and its inhibitors on the pathophysiology of T2DM. Some recent clinical studies are also discussed briefly.

2 Dipeptidyl Peptidase-4

A multifunctional protein, dipeptidyl peptidase-4 (DPP-4), shows a vast array of biological activities including: protease activity [1], binding with adenosine deaminase (ADA) [2] controlling viral entry through cell surface co-receptor [3], interaction with the extracellular matrix [4], and cell migration and proliferation [5]. The pleotropic DPP-4 action is further amplified by the variety of DPP-4 substrates that act as important biochemical messengers in many tissues.

2.1 Discovery, Genomic Organization, and Superfamily of related Enzymes

DPP-4 was discovered in 1966. This aminopeptidase was observed to have unique substrate characteristics [16]. In 1992 [20, 21], 70-kb human gene of DPP-4 was identified. This gene is situated on the long arm of chromosome 2 (2q24.3) comprising of 26 exons that encodes a 766-amino acid protein. The serine protease catalytic site is encoded by two exons (exons 21 and 22).

DPP-4 is a member of the serine peptidase/prolyl oligopeptidase gene family. This can be alternatively subclassified as fibroblast activation protein (FAP)/ seprase, the membrane-boundpeptidases, etc. The catalytic activity site within the C-terminal region of these related enzymes is highly conserved both in prokaryotes and eukaryotes. But the other sequences may differ significantly [22].

2.2 Molecular Function

DPP-4 transmits signals across cell membranes and interacts with other membrane proteins (Fig. 1). Interestingly, most of the protein is extracellular, including the

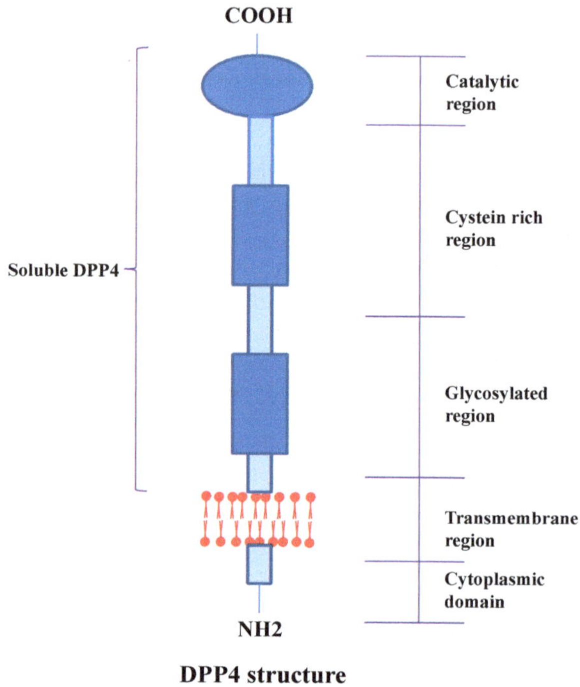

Fig. 1 DPP4 structure

C-terminal catalytic domain, a cysteine-rich area, and a large glycosylated region linked by a flexible region to the transmembrane segment. Only six N-terminal amino acids of DPP-4 are anchored into the cytoplasm. The active site was found to be Ser 630 which is flanked by the serine peptidase motif Gly-Trp-Ser630-Tyr-Gly-Gly-Tyr-Val [23, 24]. It was found that dimers are the predominant species [25]. Further studies confirm that monomeric form shows less activity and homodimerization significantly increases DPP-4 activity [26]. Crystal structure also demonstrated that two soluble homodimer can also form tetramers.

Membrane-bound DPP-4 can interact with several signalling molecules including ADA [27], CXCR4 receptor, caveolin-1 [28], the Na^+/H^+ exchanger, and the T cell antigen CD45 [29], and initiate intracellular signalling. The soluble form of DPP-4 (sDPP-4) was first identified in serum and saliva [30]. Later, it was also found in cerebrospinal and seminal fluid and bile. sDPP-4 can also activate intracellular signalling pathways and increases the proliferation of human lymphocytes.

2.3 Regulation of DPP-4 Expression

Regulation of DPP-4 expression is a complex phenomenon. It includes control at the level of both transcription and translation. The *DPP-4* gene lacks conventional TATAA or CCAAT promoter sequences rather contains a cytosine/guanine-rich promoter region [31]. Cell surface and intracellular DPP-4 expression is highly regulated though the control of DPP-4 shedding, which in turn increases the levels of the circulating DPP-4 (sDPP-4) form. Studies showed high activity of sDPP-4 in spleen, lung, thymus, and liver; and medium activity in the intestine, aorta as well as in bone marrow.

2.4 Post-translational Modifications of DPP-4

DPP-4 may go through N-glycosylation or N-terminal sialylatio. There are 8–11 potential N-glycosylation sites available on DPP-4 that contributes to its folding and stability [32] but not any other activity. N-terminal sialylation, on the other hand, facilitates trafficking of DPP-4 to the apical membrane [33].

2.5 DPP-4 Substrates

DPP-4 activity was first investigated in hydrolysis of dietary prolyl peptides. Numerous studies confirm that DPP-4 can cleave dozens of peptides (such as neuropeptides, chemokines, and regulatory peptides) containing a proline or alanine residue at position 2 from the amino-terminal region. Moreover, peptides having position 2 replaced with alternate residues (such as hydroxyproline, threonine, valine, dehydroproline > alanine > glycine, or leucine) at the penultimate position are also cleaved by DPP-4. This phenomenon indicates to a stereochemical preference [34].

The following are some important substrates for DPP-4 peptidase:

I. Physiological peptide substrates: These are intact peptides and/or DPP-4 cleavage products detected at different levels in experiments with wild type versus *DPP-4* knockout animals or human subjects treated with a selective DPP-4 inhibitor.
II. Pharmacological substrates: These are substrates shown to be cleaved at penultimate residue, predominantly in vitro.

2.5.1 Glucagon-Like Peptide-1 (GLP-1)

GLP-1 is secreted from L cells in the distal portion of the intestinal tract. It is derived from proglucagon after post-translational cleavage by prohormone convertase (PC)1/3. It is a neuropeptide and an incretin. There are two biologically active forms of GLP-1: GLP-1-(7-37) and GLP-1-(7-36) NH2. These are produced as a result of selective cleavage of proglucagon. Intact GLP-1 functions as an enhancer of glucose-stimulated insulin secretion. Moreover, it suppresses glucagon secretion, appetite, and gastric emptying [35]. DPP-4 cleavages this incretin and thereby destroys its above-said activities.

2.5.2 Glucagon-Like Peptide-2 (GLP-2)

GLP-2 is a 33-amino acid incretin co-secreted with GLP-1 from L cells in the distal bowel. HPLC analysis has detected both GLP-2 (1-33) and its DPP-4 cleavage product GLP-2 (3-33) in rat ileal [36]. A DPP-4 inhibitor (Val-Pyr) completely abolished formation of GLP-2 (3-33) in human plasma in vitro [37].

2.5.3 Glucose-Dependent Insulinotropic Polypeptide (GIP)

GIP is a 42-amino acid peptide. It is derived mainly within K cells in the duodenum and proximal jejunum from preproGIP after post-translational processing by PC1/3. GIP is also expressed in α-cells of the islet via PC2 processing. It stimulates insulin secretion. It is a physiological substrate for DPP-4 that cleaves GIP at specific site producing a dipeptide (Tyr-Ala) [35, 38].

Both GLP-1 and GIP increase the level of insulin. Moreover, they can also promote β-cell proliferation and suppress β-cell death [39, 40].Studies showed the blood glucose lowering effect of exogenously administered GLP-1 in many other organs like liver, adipose tissue, brain, and intestine [41]. It can also enhance glucose sensitivity of β-cell and induce glucose competence in previously unresponsive β-cells [42]. Besides, GLP-1 can also destroy hepatic glucose production [43]. In spite of many beneficial effects, GLP-1/GIP treatment is limited mainly because of its short half-life. The half-life of GLP-1 and GIP is less than 2 min and 7 min in human, respectively [39]. This is because DPP-4 acts on and thereby converts them into their inactive forms. This DPP-4 activity was first reported in early 1990s [17] and later confirmed in vivo (Fig. 2) [18].

Fig. 2 GLP-1 dependent effects of DPP-4

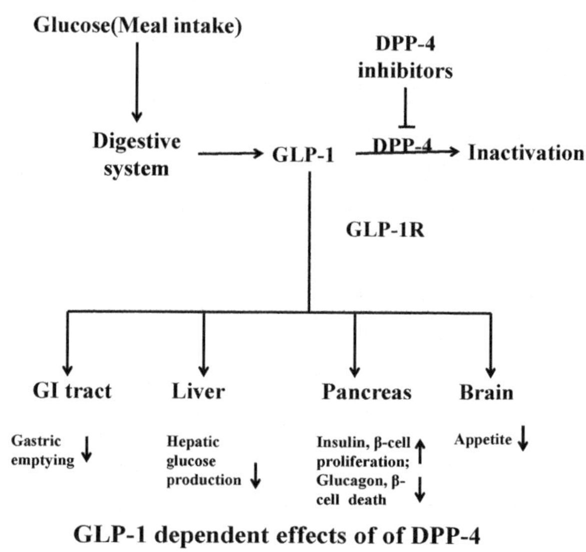

2.5.4 Glucagon

Glucagon is a peptide hormone produced by α-cells of the pancreas and stimulates hepatic glucose production. Mass spectrometry analysis revealed that incubation of purified porcine DPP-4 with glucagon in vitro resulted in the formation of glucagon (3-29) and glucagon (5-29). In human serum glucagon can also be metabolized to glucagon (3-29) ex vivo. However, Ile-thiazolidide, a DPP-4 inhibitor, prevents the metabolization [19]. In spite of this, the metabolization of glucagon in porcine plasma is not affected by another DPP-4 inhibitor valine-pyrrolidide [44]. Overall, the data available so far lead to the conclusion that glucagon is a pharmacological but not a physiological substrate.

2.5.5 Gastrin-Releasing Peptide (GRP)

GRP is a 27-amino acid peptide and a substrate for DPP-4 which produces GRP (3-27) and GRP (5-27) [45]. It has been observed that DPP-4 inhibitor valine-pyrrolidide when administered orally before infusion of GRP causes a 25% increase in the acute insulin response to i.v glucose in anesthetized mice [46].

2.5.6 Brain Natriuretic Peptide (BNP)

BNP is synthesized as a 134-amino acid precursor (preproBNP). It is then processed to a 108-residue proBNP (1-108) that is cleaved by furin or corin to BNP (1-32). BNP (1-32) promotes natriuresis through engagement of natriuretic peptide receptor type A. BNP (1-32) is cleaved by DPP-4 to a number of related metabolites. However, there are limited data on how DPP-4 inhibitors react on BNP [47].

2.5.7 Erythropoietin

Erythropoietin is a 166-amino acid protein. It has a proline at position 2 that makes it an ideal substrate for DPP-4. It was observed that hematopoiesis in response to radiation or chemotherapy was induced by genetic or pharmacological reduction of DPP-4 activity in mice. However, present understanding cannot confirm whether DPP-4 inhibition induces erythropoietin-dependent hematopoiesis in humans, and available data only indicate that erythropoietin may be a physiological DPP-4 substrate [50].

2.5.8 Eotaxin

Eotaxin is a chemokine, i.e. small peptides with chemoattractant properties. CC motif chemokine (CCL) 11, a 74-amino acid chemokine, is secreted by Th2 cells to attract eosinophils. CC chemokine ligand 11 (CCL11) is the member of all CC chemokine receptor 3 (CCR3) ligands and is shown to be selective for this receptor. It has been observed to be cleaved by DPP-4 into eotaxin (3-74). The potency for eosinophil chemotaxis is also reduced [51]. CCL11 also activates CCR5 but only in the micromolar range. Inactivating chemokine receptors in allergic disease models, it is tempting to study similar compounds derived from CCL11 as demonstrated for DP4-resistant n-nonanoyl (NNY)-CCL14 and for amino-oxypentane (AOP)-CCL5 [52].

3 DPP-4 Inhibitors and Diabetes

Incretin response is impaired in patients with type 2 diabetes, and DPP-4-mediated incretin degradation is thought to be responsible for that pathogenesis of type 2 diabetes. DPP-4 inhibitors are newly available oral drugs for type 2 diabetes used either solo or in combination with other oral medications (such as metformin, glitazones, and sulphonylurea) for type 2 diabetes [35]. DPP-4 inhibitors are used to inhibit the activity of DPP-4 and consequently to increase the half-life of incretins, which in turn lowers glucose level post-prandially [53].

It was observed that DPP-4 inhibitors are only moderately effective for the treatment of T2DM when used solo. It can only lower HbA1C level by 0.6–0.8% in patients with a starting level around 8% [54]. However, they are used increasingly now in clinic due to the safety and low side effect [55, 56]. Side effects of DPP-4 inhibitors in clinical trials include mainly respiratory tract and urinary tract infection as well as headache [57].

Currently, FDA or EU approved several DPP-4 inhibitors. These DPP-4 inhibitors may be divided into two classes depending on structure: DPP-4 dipeptide structures mimics and non-peptidomimetics. β-aminoacid-based sitagliptin, vildagliptin, and saxagliptin belong to the class one, whereas non-peptidomimetics inhibitors include alogliptin (modified pyrimidinedione) and xanthine-based linagliptin. Some of the DPP-4 inhibitors approved by FDA or EU are currently available in the market. FDA already approved sitagliptin (2006), alogliptin (2013), linagliptin (2011), and saxagliptin (2009).Vildagliptin was approved by EU in 2007. Both teneligliptin and anagliptin were approved in Japan in 2012. Apart from these, some other DPP-4 inhibitors like dutogliptin and gemigliptin are also going through clinical trials and awaiting FDA approval [58].

4 Impact of DPP-4 on T2DM-Relevant Organs and Associated Comorbidities

DPP-4 inhibitors maintain glucose metabolism by prolonging the effects of GLP and GIP-1, which in turn increases glucose-mediated insulin secretion and suppresses glucagon secretion [59]. Apart from the hypoglycaemic actions of DPP-4 inhibitors, newly acquired data indicate that incretin-based medications may also have an important role on the physiology of many organs. Thus, then next section of this chapter focuses on the role of DPP-4 action in the organs affected in T2DM and various complications associated with it (Fig. 3).

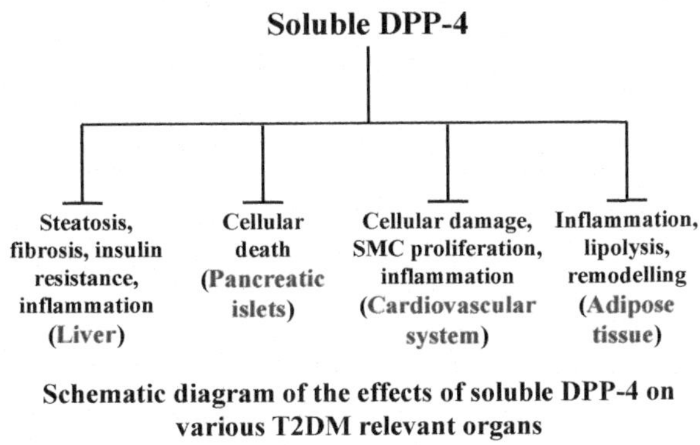

Fig. 3 Schematic diagram of the effects of soluble DPP-4 on various T2DM relevant organs

4.1 Adipose Tissue

Recent study showed that DPP-4 is highly expressed in human primary adipocytes and DPP-4 expression is increased in obese individuals [60, 61]. Visceral fat of obese persons showed the highest DPP-4 level. From this data, sDPP-4 may be taken as a novel adipokine released from primary human adipocytes. DPP-4 is also involved in adipose tissue lipolysis. It has been revealed that DPP-4-bound ADA, a monomeric enzyme catalyzing deamination of adenosine to inosine and ammonia [62, 63], has a 1000-fold greater activity than free ADA [64], thus reducing the antilipolytic effects of adenosine. In addition, DPP-4 takes part in adipose tissue remodelling and cell plasticity [65]. DPP-4 can also impair insulin signalling at the level of Akt in primary human adipocytes and induce inflammation [66].

4.2 Cardiovascular System

Many studies have shown the protective ability of DPP-4 inhibitors on the cardiovascular system. DPP-4 inhibitors have been shown to decrease ischaemia/reperfusion injury, reduce myocardial infarct size, and protect against myocardial ischaemia [67]. In SAVOR-TIMI53, EXAMINE, or VIVIDD, clinical trials conducted among cardiovascular patients failed to confirm cardioprotective effect of the used DPP-4 inhibitors [68–70]. Studies on GLP-1 receptor (GLP-1R) signalling indicated ameliorative cardiovascular effects GLP-1 itself. In fact, GLP-1R mRNA has been detected in the heart of rodents [71] and humans [72]. Pilot studies in patients with heart failure showed cardioprotection by GLP-1 infusion [73]. In another study, exenatide (GLP-1 analog) showed a significant (20%) reduction of CVD events compared with patients on other hypoglycaemic drugs [74].Apart from GLP-1, two other substrates of DPP-4, SDF-1α, and brain natriuretic peptide (BNP) showed cardioprotective effects. DPP-4 inhibitor sitagliptin increased circulating EPC levels and SDF-1α plasma concentrations [75] in T2DM patients. DPP-4 cleaves the physiologically active BNP (1–32) to BNP (3–

32) lowering plasma cGMP levels. Elevated BNP is a sensitive marker of heart failure and helps in cardiac remodelling and healing after acute MI [76, 77].

4.3 Liver

Diabetes and obesity are major risk factors for non-alcoholic fatty liver disease (NAFLd) [78] which is one of the main causes for chronic liver disease [79]. In the obese persons, elevated triglyceride degradation in adipose tissue causes an increased hepatic uptake of fatty acids leading to fat accumulation within the tissue. Moreover, reactive oxygen species (ROS), produced during lipid oxidation, also induce hepatocyte death. In human liver, DPP-4 was found only in hepatic acinar zones 2 and 3. This heterolobular distribution indicates that DPP-4 might be involved in the hepatic metabolism [80]. Moreover, mRNA level of DPP-4 has been observed to increase markedly in NAFLD livers compared to control [81]. Kaji et al. showed that sitagliptin significantly inhibits the development of liver fibrosis in rats by the suppression of collagen synthesis and hepatic stellate cell proliferation [82]. The mechanism of action of sitagliptin includes dephosphorylation of ERK1/2, p38, and Smad2/3 in the hepatic stellate cells. Moreover, hepatic steatosis was also prevented in several animal studies by DPP-4 inhibition [83]. Though clinical data are limited, some of them revealed the efficacy of DPP-4 inhibitors to improve the levels of liver transaminases and liver fat in diabetic patients [84, 85].

4.4 Pancreas

It has been shown that DPP-4 localization within the pancreatic islets differs between species [86, 87]. A recent study revealed DPP-4 activity in the conditioned medium of human islets indicating that DPP-4 is released from human islets as well [88]. Emerging in vitro and pre-clinical data demonstrated that DPP-4 inhibition may have beneficial effects on T2DM-induced β-cell dysfunction and apoptosis. Omar et al. showed the existence of DPP-4 in mouse and human islets. Their study also revealed that inhibition of islet DPP-4 activity directly stimulates insulin secretion GLP-1 dependently [87]. Moreover, DPP-4 inhibition has been shown to be associated with increased β-cell mass and function in various models of T2DM [89, 90]. These ameliorative effects, however, were revealed to be due to the transcriptional activation of pro-survival and anti-apoptotic genes, and the suppression of pro-apoptotic genes in β cells [91]. Clinical data of DPP-4 inhibitors on β cells are still limited. Both the short- and long-term treatments with vildagliptin increased β-cell function in humans [92, 93]. Interestingly, the SAVOR-TIMI53 trial, originally done to evaluate the cardiovascular safety of saxagliptin, revealed that DPP-4 inhibition may attenuate the progression of diabetes [94]. The exact mechanism how inhibition of DPP-4 activity increases insulin secretion and β-cell mass is still not fully clear. This is because prolonged GLP-1 activity or improved glycemic control could not explain the above-said phenomenon sufficiently.

5 Clinical Trials

A xanthine-based DPP-4 inhibitor, BI 1356, has reached phase III trials. This compound has shown potential to inhibit DPP-4 in vitro and in vivo. Pre-clinical trials confirmed that a once-daily administration of 5 mg to be efficient. This particular dose was also long lasting and without known side effects [95].

In an open label, multicentre phase I trial, the efficacy of DPP-4 inhibitor linagliptin was evaluated in African American patients with type 2 diabetes mellitus (T2DM). Forty-one patients were included in this study. 5 mg linagliptin, once daily for 7 days, was administered to them, followed by 7 days of out patient evaluation. It is to be mentioned that the exposure range and overall pharmacokinetic/pharmacodynamic profile of the drug in this study of that particular population with T2DM was very much comparable with that in other populations. Results showed that the geometric mean of DPP-4 inhibition at steady state was 84.2% at trough and 91.9% at maximum, and urinary excretion was low (0.5 and 4.4% of the dose excreted over 24 h, days 1 and 7) [96].

T2DM medication sometimes requires combination therapy. In this trial, scientists have assessed the effects of the DPP-4 inhibitor alogliptin (ALO) combined with the pioglitazone (PIO) on β-cell function and glycemic control in T2DM. Patients were treated with combined ALO at 25 mg and PIO at 30 mg daily or ALO 25 mg daily monotherapy or PBO. Results showed that short-term usage of ALO/PIO or ALO improved glycemic control in T2DM patients. β-cell function, however, was improved only by combined ALO/PIO treatment [61].

Gemigliptina is a selective DPP-4 inhibitor used to treat type 2 diabetes. In a combination therapy, the pharmacokinetics (PKs) of combination and gemigliptin, rosuvastatin, and irbesartan mono-therapies was evaluated. Randomized, open-label, 3-treatment, 6-sequence, 3-period, crossover studies were performed. Healthy male volunteers were given 50 mg gemigliptin alone, 20 mg rosuvastatin (part A) or 300 mg irbesartan alone (part B), with concomitant gemigliptin, and each drug was administered as part of once-daily, 7-day, repeated dosing regimens with a 14-day washout period. Results showed that gemigliptin have no effect on the PK properties of rosuvastatin or irbesartan and vice versa [97].

Another phase 2, multicentre, randomized, double-blind, placebo-controlled, 12-week dose-ranging study was conducted to evaluate the efficacy and safety of the DPP-IV inhibitor, PF-734200, in type 2 diabetic patients (men and women) with metformin as their sole diabetes medication. Results revealed that PF-734200 is safe and well tolerated at all the four doses. PF-734200 also significantly lowered HbA1c and 20 mg dose provided the most effective outcome in post-prandial glucose [98].

To determine efficacy of dutogliptin, a DPP-4 inhibitor in T2DM patients, a 12-week multicentre, randomized, double-blind, placebo-controlled trial in 423 patients was conducted. Results showed that dutogliptin treatment for 12 weeks improved glycaemic control in patients with T2DM who were primarily on a background medication of metformin. Results also showed well tolerability of dutogliptin at both the doses tested. The higher dose (400 mg), however, showed better outcome in terms of changes of HbA1c and FPG and subject number [99].

The aim of another study was to quantitatively evaluate the incretin effect after treatment with the DPP-4 inhibitor vildagliptin in patients with T2DM. Twenty-one patients (three women, 18 men) with T2DM primarily treated with metformin were studied in a two-period crossover design. They received 100 mg vildagliptin once daily or placebo for 13 days randomly. The incretin effect was measured on day 12 and 13. Results indicated that DPP-4 inhibition augmented insulin secretory responses both after oral glucose and during isoglycemic i.v. glucose infusions. But no net change in the incretin effect was observed [100].

The effect of GLP-1 in the human vasculature and how it is regulated by DPP-4 inhibition is still unknown. Besides, DPP-4 also degrades the vasodilator brain natriuretic peptide (BNP) to a less potent metabolite. This study was conducted with seventeen healthy subjects participated in this randomized, double-blinded, placebo-controlled crossover. Results showed that GLP-1 does not act as a direct vasodilator in humans, and sitagliptin does not regulate vascular function in healthy humans by degrading GLP-1 and BNP [101].

6 Some Other Proteases Associated with DM

DPP-4 is the most extensively studied proteases and has been developed as a major target for therapeutic purposes via the use of DPP inhibitors. However, a number of other proteases reported recently can serve as biomarkers as well as therapeutic targets for the treatment of DM. Strategies to lower the activity of calpain, cathepsin S, and neutral endopeptidase have been recently studied.

Cathepsin S (Cat S), a cysteine protease, is responsible for cleavage of invariant chain of MHC for proper antigen presentation. Increased activity of Cat S was reported in many pathological conditions. The adipose tissues of obese individuals over-express Cat S, and it has been found to promote neoplastic progression via matrix-derived angiogenic factors [102]. In addition to the cardiovascular and malignant diseases to which enhanced Cat S activity is expected to contribute, type 2 diabetes is one in which this enzyme is expected to play some important role as this disease is considered to be the most devastating comorbidities of obesity. Lafarge et al. [103] reported that anti-cathepsin therapy lowered blood glucose level to normal, comparable to that in cathepsin knockout mice. At present, Cat S inhibitors have been proposed for treatment of autoimmune diseases with the expectation that they could help to lower hepatic glucose output at risk for type 2 diabetes in obese individuals.

Neutral endopeptidase (NEP) is a protease and is reported to be involved in the degradation of vasoactive peptides (natriuretic peptides, calcitonin gene-related peptide, adrenomedullin, bradykinin as well as endothelin). Its levels increased in diabetic patients which facilitate the degradation of those vasoactive peptides. AVE7688, a vasopeptidase inhibitor, improves neurovascular and neural function in streptozotocin-induced type 1 diabetes animals as well as in other type 2 diabetes models [104].

Calpains, a superfamily of Ca^{2+}-activated proteases, are reported to be associated with various physiological and pathophysiological events including diabetes. Recently, increased activity of these proteases has been reported to be linked to reduced endothelium-derived nitric-oxide-mediated vasodilatation in diabetes [105]. Studies suggest that the calpain inhibitor, A-705253, *N*-(1-benzyl-2-carbamoyl-2-oxoethyl)-2-[E-2-(4-diethyl-aminomethylphenyl) ethen-1-yl]benzamide, improved penile nitrergic neurovascular function in streptozotocin-induced diabetic mice [106].

The majority of the serine protease inhibitors linked to diabetes constitute the family of serpins. The level of these proteins is found to be decreased in both T1D and T2D. Hyperglycaemia promotes pro-inflammatory conditions and results in impaired wound healing, a common secondary complication in diabetes. The release of pro-inflammatory cytokines activates the metalloproteases (MMPs). These MMPs have a role in wound repair, cell migration, and tissue remodelling. The use of protease inhibitors for therapeutic purposes for improved wound healing, especially in diabetic conditions, is prevalent.

Doxycycline, an antibiotic of tetracycline family, is an unusual metalloprotease inhibitor. It is reported to improve healing of chronic diabetic foot ulcers. Gelatin, a substrate for MMPs, can be used for healing of chronic wounds [106]. The temporal and spatial features of the activity of MMPs and TIMPs can be used as biomarkers for determining the progress of the diabetic conditions. Alterations in the level of metalloproteases (MMPs) have also been linked to diabetic conditions. MMP-2 is reported to increase in both T1D and T2D, diabetic nephropathy, retinopathy and lesions. MMP-8 is associated with diabetic nephropathy and retinopathy, whereas MMP-9 is associated with T1D, diabetic nephropathy, retinopathy, and macrovascular diseases. Metalloprotease aminopeptidase N is an important marker associated with diabetic nephropathy. Out of the four TIMPs, only TIMP1 and TIMP2 are linked with diabetes. Muller et al. [107] conducted a study in 16 patients with neuropathic diabetic foot ulcers. He proposed a correlation between high ratio of MMP-1/TIMP1 and good healing. Also, they reported that there is a decrease in the level of MMP-8 and MMP-9 in good healers. So, this study showed that the ratio of MMP-1/TIMP1 can be a predictor of healing, and thus their therapy will be designed accordingly. Also, those inhibitors of MMP-8 and MMP-9 can be used for therapeutics development for chronic wound healing purposes [107].

Caseiro et al. [108] performed protease profiling for type 1 diabetes (T1D) in serum, urine, and saliva. An increased activity of MMP-2 and MMP-9 was observed in serum. In urine and saliva, besides MMPs, increased activity of a number of proteases was found in type 1 diabetic patients with nephropathy and/or retinopathy like MMP-9/neutrophil gelatinase-associated lipocalin, aminopeptidase N, azurocidin, and kallikrein. This report suggests that these proteases found in biofluids can be potential biomarkers or screening targets for monitoring proteolytic activity in T1D patients [108].

Nowak et al. [109] has shown cathepsin D to be an important biomarker for insulin resistance (IR) in T2DM. Using a high throughput 92 protein as say, along with cathepsin D, they identified six other proteins as circulating IR markers in two large community cohorts [109].

7 Conclusion

After the approval of the first DPP-4 inhibitor on 17 October 2006, numerous patients with T2DM have been treated with DPP-4 inhibitors. Though DPP-4 inhibitors are generally well tolerated, adverse effects, especially in case of cardiovascular patients is not uncommon. The result of a recent study drives the scientists further to investigate the mechanism of action of DPP-4 inhibitor saxagliptin that has been shown to increase the rates of hospitalization for heart failure in few patients under trial. Moreover, many DPP-4 substrates are cleaved by multiple enzymes leading to renal or hepatic clearance. It challenges the dominant role of DPP-4 in maintaining peptide bioactivity. However, modern liquid chromatography mass spectrometry-based peptide profiling technology with the help of biochemistry could be a promising approach for revealing the clearer picture of DPP-4 action and would lead to a better and more accurate use of DPP-4 inhibitors pharmaceutically.

References

1. Brownlee M (2005) The pathobiology of diabetic complications. Unifying Mech 54:1615–1625
2. Rambhade S et al (2010) Diabetes mellitus—its complications, factors influencing complications and prevention—an overview. J Chem Pharm Res 2:2–7
3. Tripathi BK, Srivastava AK (2006) Diabetes mellitus: complications and therapeutics. Med Sci Monit Basic Res 12:RA130-RA147
4. Forbes JM, Cooper ME (2013) Mechanisms of diabetic complications. Physiol Rev 93:137–188
5. Lobmann R, Schultz G, Lehnert H (2005) Proteases and the diabetic foot syndrome: mechanisms and therapeutic implications. Diabetes Care 28:71–461
6. Falanga V et al (2005) Wound healing and its impairment in the diabetic foot. Lancet 366:43–1736
7. Lopez-Otin C, Bond JS (2008) Proteases: multifunctional enzymes in life and disease. J Biol Chem 283:7–30433
8. Lopez-Otin C, Overall CM (2002) Protease degradomics: a new challenge for proteomics. Nat Rev Mol Cell Biol 3:19–509
9. Spravchikov N et al (2001) Glucose Effects on Skin Keratinocytes. Implic Diabetes Skin Complicat 50:1627–1635

10. Balasubramanyam M, Rema M, Premanand C (2002) Biochemical and molecular mechanisms of diabetic. Curr Sci 83(12)
11. Hober D, Sauter P (2010) Pathogenesis of type 1 diabetes mellitus: interplay between enterovirus and host. Nat Rev Endocrinol 6:279–289
12. Bluestone JA, Herold K, Eisenbarth G (2010) Genetics, pathogenesis and clinical interventions in type[thinsp]1 diabetes. Nature 464:1293–1300
13. Association AD (2010) Diagnosis and classification of diabetes mellitus. Diabetes Care 33 (Supplement 1):S62–S69
14. Fierabracci A (2014) The putative role of proteolytic pathways in the pathogenesis of Type 1 diabetes mellitus: the 'autophagy' hypothesis. Med Hypotheses 82:553–557
15. Halban PA et al (2014) β-cell failure in type 2 diabetes: postulated mechanisms and prospects for prevention and treatment. Diabetes Care 37:1751–1758
16. Hopsu-Havu VK, Glenner GG (1966) A new dipeptide naphthylamidase hydrolyzing glycyl-prolyl-beta-naphthylamide. Histochemie 7:197–201
17. Mentlein R, Gallwitz B, Schmidt WE (1993) Dipeptidyl-peptidase IV hydrolyses gastric inhibitory polypeptide, glucagon-like peptide-1(7-36)amide, peptide histidine methionine and is responsible for their degradation in human serum. Eur J Biochem 214:35–829
18. Kieffer TJ, McIntosh CH, Pederson RA (1995) Degradation of glucose-dependent insulinotropic polypeptide and truncated glucagon-like peptide 1 in vitro and in vivo by dipeptidyl peptidase IV. Endocrinology 136:96–3585
19. Pospisilik JA et al (2001) Metabolism of glucagon by dipeptidyl peptidase IV (CD26). Regul Pept 96(3):41–133
20. Misumi Y et al (1992) Molecular cloning and sequence analysis of human dipeptidyl peptidase IV, a serine proteinase on the cell surface. Biochim Biophys Acta 1131:6–333
21. Tanaka T et al (1992) Cloning and functional expression of the T cell activation antigen CD26. J Immunol 149:6–481
22. Engel M et al (2003) The crystal structure of dipeptidyl peptidase IV (CD26) reveals its functional regulation and enzymatic mechanism. Proc Natl Acad Sci U S A 100:8–5063
23. Hiramatsu H et al (2003) Crystallization and preliminary X-ray study of human dipeptidyl peptidase IV (DPPIV). Acta Crystallogr D Biol Crystallogr 59:6–595
24. Rasmussen HB et al (2003) Crystal structure of human dipeptidyl peptidase IV/CD26 in complex with a substrate analog. Nat Struct Biol 19–25
25. Chien CH et al (2004) One site mutation disrupts dimer formation in human DPP-IV proteins. J Biol Chem 279:45–52338
26. Chien CH et al (2006) Identification of hydrophobic residues critical for DPP-IV dimerization. Biochemistry 45:12–7006
27. Abbott CA et al (1994) Genomic organization, exact localization, and tissue expression of the human CD26 (dipeptidyl peptidase IV) gene. Immunogenetics 40:8–331
28. Ohnuma K et al (2004) CD26 up-regulates expression of CD86 on antigen-presenting cells by means of caveolin-1. Proc Natl Acad Sci U S A 101:91–14186
29. Torimoto Y et al (1991) Coassociation of CD26 (dipeptidyl peptidase IV) with CD45 on the surface of human T lymphocytes. J Immunol 147:7–2514
30. Nagatsu I, Nagatsu T, Yamamoto T (1968) Hydrolysis of amino acid beta-naphthylamides by aminopeptidases in human parotid saliva and human serum. Experientia 24:8–347
31. Erickson RH et al (1999) Regulation of the gene for human dipeptidyl peptidase IV by hepatocyte nuclear factor 1 alpha. Biochem J 338:7–91
32. Fan H et al (1997) Domain-specific N-glycosylation of the membrane glycoprotein dipeptidylpeptidase IV (CD26) influences its subcellular trafficking, biological stability, enzyme activity and protein folding. Eur J Biochem 246:51–243
33. Delacour D et al (2003) 1-benzyl-2-acetamido-2-deoxy-alpha-D-galactopyranoside blocks the apical biosynthetic pathway in polarized HT-29 cells. J Biol Chem 278:809–37799
34. Tiruppathi C et al (1993) Genetic evidence for role of DPP IV in intestinal hydrolysis and assimilation of prolyl peptides. Am J Physiol 265:G9–G81

35. Campbell JE, Drucker DJ (2013) Pharmacology, physiology, and mechanisms of incretin hormone action. Cell Metab 17:37–819
36. Brubaker PL et al (1997) Circulating and tissue forms of the intestinal growth factor, glucagon-like peptide-2. Endocrinology 138:43–837
37. Hartmann B et al (2000) In vivo and in vitro degradation of glucagon-like peptide-2 in humans. J Clin Endocrinol Metab 85:8–2884
38. Deacon CF et al (2000) Degradation of endogenous and exogenous gastric inhibitory polypeptide in healthy and in type 2 diabetic subjects as revealed using a new assay for the intact peptide. J Clin Endocrinol Metab 85:81–3575
39. Baggio LL, Drucker DJ (2007) Biology of incretins: GLP-1 and GIP. Gastroenterology 132:57–2131
40. Kim SJ et al (2005) Glucose-dependent insulinotropic polypeptide (GIP) stimulation of pancreatic beta-cell survival is dependent upon phosphatidylinositol 3-kinase (PI3 K)/protein kinase B (PKB) signaling, inactivation of the forkhead transcription factor Foxo1, and down-regulation of bax expression. J Biol Chem 280:307–22297
41. Zander M et al (2002) Effect of 6-week course of glucagon-like peptide 1 on glycaemic control, insulin sensitivity, and beta-cell function in type 2 diabetes: a parallel-group study. Lancet 359:30–824
42. Holz GG, Kuhtreiber WM, Habener JF (1993) Pancreatic beta-cells are rendered glucose-competent by the insulinotropic hormone glucagon-like peptide-1(7-37). Nature 361:5–362
43. Larsson H, Holst JJ, Ahren B (1997) Glucagon-like peptide-1 reduces hepatic glucose production indirectly through insulin and glucagon in humans. Acta Physiol Scand 160:22–413
44. Deacon CF et al (2003) Differential regional metabolism of glucagon in anesthetized pigs. Am J Physiol Endocrinol Metab 285:E60–E552
45. Lambeir AM et al (2001) Kinetic study of the processing by dipeptidyl-peptidase IV/CD26 of neuropeptides involved in pancreatic insulin secretion. FEBS Lett 507:30–327
46. Ahren B, Hughes TE (2005) Inhibition of dipeptidyl peptidase-4 augments insulin secretion in response to exogenously administered glucagon-like peptide-1, glucose-dependent insulinotropic polypeptide, pituitary adenylate cyclase-activating polypeptide, and gastrin-releasing peptide in mice. Endocrinology 146:9–2055
47. Ghosh S, Banerjee S, Sil PC (2015) The beneficial role of curcumin on inflammation, diabetes and neurodegenerative disease: a recent update. Food Chem Toxicol 83:111–124
48. Ghosh S et al (2015) Curcumin protects rat liver from streptozotocin-induced diabetic pathophysiology by counteracting reactive oxygen species and inhibiting the activation of p53 and MAPKs mediated stress response pathways. Toxicol Rep 2:365–376
49. Brandt I et al (2006) Dipeptidyl-peptidase IV converts intact B-type natriuretic peptide into its des-SerPro form. Clin Chem 52:7–82
50. Broxmeyer HE et al (2012) Dipeptidylpeptidase 4 negatively regulates colony-stimulating factor activity and stress hematopoiesis. Nat Med 18:96–1786
51. Struyf S et al (1999) CD26/dipeptidyl-peptidase IV down-regulates the eosinophil chemotactic potency, but not the anti-HIV activity of human eotaxin by affecting its interaction with CC chemokine receptor 3. J Immunol 162:9–4903
52. Manns J et al (2007) The allergy-associated chemokine receptors CCR3 and CCR5 can be inactivated by the modified chemokine NNY-CCL11. Allergy 62:17–24
53. Drucker DJ, Nauck MA (2006) The incretin system: glucagon-like peptide-1 receptor agonists and dipeptidyl peptidase-4 inhibitors in type 2 diabetes. Lancet 368:705–1696
54. Inzucchi SE, McGuire DK (2008) New drugs for the treatment of diabetes: part II: Incretin-based therapy and beyond. Circulation 117:84–574
55. Raz I et al (2006) Efficacy and safety of the dipeptidyl peptidase-4 inhibitor sitagliptin as monotherapy in patients with type 2 diabetes mellitus. Diabetologia 49:71–2564

56. Lambeir AM, Scharpe S, De Meester I (2008) DPP-4 inhibitors for diabetes–what next? Biochem Pharmacol 76:43–1637
57. Jose T, Inzucchi SE (2012) Cardiovascular effects of the DPP-4 inhibitors. Diab Vasc Dis Res 9:16–109
58. Pattzi HM et al (2010) Dutogliptin, a selective DPP-4 inhibitor, improves glycaemic control in patients with type 2 diabetes: a 12-week, double-blind, randomized, placebo-controlled, multicentre trial. Diabetes Obes Metab 12:55–348
59. Pratley RE, Salsali A (2007) Inhibition of DPP-4: a new therapeutic approach for the treatment of type 2 diabetes. Curr Med Res Opin 23:919–931
60. Sell H et al (2013) Adipose Dipeptidyl peptidase-4 and obesity. Correlation with insulin resistance and depot-specific release from adipose tissue in vivo and in vitro 36:4083–4090
61. Lamers D et al (2011) Dipeptidyl peptidase 4 is a novel adipokine potentially linking obesity to the metabolic syndrome. Diabetes 60:1917–1925
62. Pacheco R et al (2005) CD26, adenosine deaminase, and adenosine receptors mediate costimulatory signals in the immunological synapse. Proc Natl Acad Sci USA 102:9583–9588
63. Schrader WP et al (1990) Characterization of the adenosine deaminase-adenosine deaminase complexing protein binding reaction. J Biol Chem 265:8–19312
64. Focosi D et al (2008) Conditioning response to granulocyte colony-stimulating factor via the dipeptidyl peptidase IV-adenosine deaminase complex. J Leukoc Biol 84:331–337
65. Lessard J et al (2015) Characterization of dedifferentiating human mature adipocytes from the visceral and subcutaneous fat compartments: fibroblast-activation protein alpha and dipeptidyl peptidase 4 as major components of matrix remodeling. PLoS ONE 10:e0122065
66. Rosmaninho-Salgado J et al (2012) Dipeptidyl-peptidase-IV by cleaving neuropeptide Y induces lipid accumulation and PPAR-γ expression. Peptides 37:49–54
67. Chinda K et al (2013) Cardioprotective effect of dipeptidyl peptidase-4 inhibitor during ischemia–reperfusion injury. Int J Cardiol 167:451–457
68. Bhatt DL, Cavender MA (2014) Do dipeptidyl peptidase-4 inhibitors increase the risk of heart failure? JACC Heart Fail 2:583–585
69. Scirica BM et al (2013) Saxagliptin and cardiovascular outcomes in patients with type 2 diabetes mellitus. N Engl J Med 369:1317–1326
70. White WB et al (2013) Alogliptin after acute coronary syndrome in patients with type 2 diabetes. N Engl J Med 369:1327–1335
71. Ban K et al (2008) Cardioprotective and vasodilatory actions of glucagon-like peptide 1 receptor are mediated through both glucagon-like peptide 1 receptor-dependent and–independent pathways. Circulation 117:2340–2350
72. Wei Y, Mojsov S (1996) Distribution of GLP-1 and PACAP receptors in human tissues. Acta Physiol Scand 157:355–357
73. Sokos GG et al (2006) Glucagon-like peptide-1 infusion improves left ventricular ejection fraction and functional status in patients with chronic heart failure. J Cardiac Fail 12:694–699
74. Best JH et al (2011) Risk of cardiovascular disease events in patients with type 2 diabetes prescribed the glucagon-like peptide 1 (GLP-1) receptor agonist exenatide twice daily or other glucose-lowering therapies. A retrospective analysis of the LifeLink database 34:90–95
75. Fadini GP et al (2010) The oral dipeptidyl peptidase-4 inhibitor sitagliptin increases circulating endothelial progenitor cells in patients with type 2 diabetes. Possible role of stromal-derived factor-1α. 33:1607–1609
76. Kuhn M (2012) Endothelial actions of atrial and B-type natriuretic peptides. Br J Pharmacol 166:522–531
77. Kawakami R et al (2004) Overexpression of brain natriuretic peptide facilitates neutrophil infiltration and cardiac matrix metalloproteinase-9 expression after acute myocardial infarction. Circulation 110:12–3306

78. Krawczyk M, Bonfrate L, Portincasa P (2010) Nonalcoholic fatty liver disease. Best Pract Res Clin Gastroenterol 24:695–708
79. Starley BQ, Calcagno CJ, Harrison SA (2010) Nonalcoholic fatty liver disease and hepatocellular carcinoma: a weighty connection. Hepatology 51:1820–1832
80. Itou M et al (2013) Dipeptidyl peptidase-4: a key player in chronic liver disease. World J Gastroenterol 19:306–2298
81. Miyazaki M et al (2012) Increased hepatic expression of dipeptidyl peptidase-4 in non-alcoholic fatty liver disease and its association with insulin resistance and glucose metabolism. Mol Med Rep 5:33–729
82. Kaji K et al (2014) Dipeptidyl peptidase-4 inhibitor attenuates hepatic fibrosis via suppression of activated hepatic stellate cell in rats. J Gastroenterol 49:481–491
83. Maiztegui B et al (2011) Sitagliptin prevents the development of metabolic and hormonal disturbances, increased β-cell apoptosis and liver steatosis induced by a fructose-rich diet in normal rats. Clin Sci 120:73–80
84. Itou M et al (2012) Dipeptidyl peptidase IV inhibitor improves insulin resistance and steatosis in a refractory nonalcoholic fatty liver disease patient: a case report. Case Rep Gastroenterol 6:538–544
85. Iwasaki T et al (2011) Sitagliptin as a novel treatment agent for non-alcoholic Fatty liver disease patients with type 2 diabetes mellitus. Hepatogastroenterology 58:5–2103
86. Liu L et al (2014) Dipeptidyl peptidase-4 (DPP-4): localization and activity in human and rodent islets. Biochem Biophys Res Commun 453:398–404
87. Omar BA et al (2014) Dipeptidyl peptidase 4 (DPP-4) is expressed in mouse and human islets and its activity is decreased in human islets from individuals with type 2 diabetes. Diabetologia 57:1876–1883
88. Shah P et al (2013) The DPP-4 inhibitor linagliptin restores β-cell function and survival in human isolated islets through GLP-1 stabilization. J Clin Endocrinol Metab 98:E1163–E1172
89. Duttaroy A et al (2011) The DPP-4 inhibitor vildagliptin increases pancreatic beta cell mass in neonatal rats. Eur J Pharmacol 650:703–707
90. Takeda Y et al (2012) Reduction of both beta cell death and alpha cell proliferation by dipeptidyl peptidase-4 inhibition in a streptozotocin-induced model of diabetes in mice. Diabetologia 55:404–412
91. Han SJ et al (2011) Effect of sitagliptin plus metformin on β-cell function, islet integrity and islet gene expression in Zucker diabetic fatty rats. Diabetes Res Clin Pract 92:213–222
92. Foley JE et al (2011) Beta cell function following 1 year vildagliptin or placebo treatment and after 12 week washout in drug-naive patients with type 2 diabetes and mild hyperglycaemia: a randomised controlled trial. Diabetologia 54:1985–1991
93. Mari A et al (2008) Characterization of the influence of vildagliptin on model-assessed β-cell function in patients with type 2 diabetes and mild hyperglycemia. J Clin Endocrinol Metab 93:103–109
94. Leibowitz G et al (2015) Impact of treatment with saxagliptin on glycaemic stability and β-cell function in the SAVOR-TIMI 53 study. Diabetes Obes Metab 17:487–494
95. Rungby J (2009) Inhibition of dipeptidyl peptidase 4 by BI-1356, a new drug for the treatment of beta-cell failure in type 2 diabetes. Expert Opin Investig Drugs 18:8–835
96. Friedrich C et al (2013) Pharmacokinetic and pharmacodynamic evaluation of linagliptin in African American patients with type 2 diabetes mellitus. Br J Clin Pharmacol 76:54–445
97. Van Raalte DH et al (2014) The effect of alogliptin and pioglitazone combination therapy on various aspects of beta-cell function in patients with recent-onset type 2 diabetes. Eur J Endocrinol 170:74–565
98. Choi HY et al (2015) Evaluation of the pharmacokinetics of the DPP-4 inhibitor gemigliptin when coadministered with rosuvastatin or irbesartan to healthy subjects. Curr Med Res Opin 31:41–229

99. Terra SG et al (2011) A dose-ranging study of the DPP-IV inhibitor PF-734200 added to metformin in subjects with type 2 diabetes. Exp Clin Endocrinol Diabetes 119:7–401
100. Vardarli I et al (2011) Inhibition of DPP-4 with vildagliptin improved insulin secretion in response to oral as well as "isoglycemic" intravenous glucose without numerically changing the incretin effect in patients with type 2 diabetes. J Clin Endocrinol Metab 96:54–945
101. Devin JK et al (2014) Dipeptidyl-peptidase 4 inhibition and the vascular effects of glucagon-like peptide-1 and brain natriuretic peptide in the human forearm. J Am Heart Assoc 3
102. Wang B, Sun J, Kitamoto S (2006) Cathepsin S controls angiogenesis and tumor growth via matrix-derived angiogenic factors. J Biol Chem 281:6020–6029
103. Lafarge JC et al (2014) Cathepsin S inhibition lowers blood glucose levels in mice. Diabetologia 57:1674–1683
104. Oltman CL et al (2009) Treatment of Zucker diabetic fatty rats with AVE7688 improves vascular and neural dysfunction. Diabetes Obes Metab 11:223–233
105. Hadi ARH et al (2007) Endothelial dysfunction in diabetes mellitus. Vascul Health Risk Manag 3:853–876
106. Nangle MR, Cotter MA, Cameron NE et al (2006) The calpain inhibitor, A-705253, corrects penile nitrergic nerve dysfunction in diabetic mice. Eur J Pharmacol 538(1–3):148–153
107. Caseiro Armando et al (2012) Protease profiling of different biofluids in type 1 diabetes mellitus. Clin Biochem 45:1613–1619
108. Nowak C et al (2016) Protein biomarkers for insulin resistance and type 2 diabetes risk in two large community cohorts. Diabetes 65:276–284
109. Muller M, Trocme C, Lardy B, Morel F, Halimi S, Benhamou PY (2008) Matrix metalloproteinases and diabetic foot ulcers: the ratio of MMP-1 to TIMP-1 is a predictor of wound healing. Diabetic Med 25(4):419–426

Role of Proteases in Regulating Cell Death Pathways

Thomas Divya, Sekar Vasudevan and Ganapasam Sudhandiran

Abstract

Cell death is a critical process involved during development, tissue homeostasis, and aging. Multiple forms of cell death exist such as apoptosis (type I cell death), necrosis, and autophagy (type II cell death). Recently, other selective forms of cell death such as pyroptosis, eryptosis, entosis, mitophagy, and oncosis are also reported. These cell death pathways collaborate with each other, and regulation of such mechanisms is crucial for maintaining cellular homeostasis. Interestingly, proteases are the one that mediate the cell death programs, and immense research is focused on elucidating the mechanisms through which protease regulates cell death program. In this chapter, we focus on various cell death pathways and how protease regulates these pathways.

Keywords

Apoptosis · Autophagy · Proteases · Necroptosis · Cell death

1 Introduction

Organisms live in an environment that constantly copes up with extrinsic and intrinsic damaging agents. In every second of our life, millions of new cells are born and millions of damaged cells die. Fate of a cell to death or survival is accomplished by a complex signaling network with several mediators. Cell death is a

T. Divya · S. Vasudevan · G. Sudhandiran (✉)
Cell Biology Laboratory, Department of Biochemistry, University of Madras, Guindy Campus, Chennai, India
e-mail: sudhandiran@yahoo.com

S. Chakraborti and N.S. Dhalla (eds.), *Pathophysiological Aspects of Proteases*,
DOI 10.1007/978-981-10-6141-7_21

decisive phenomenon that maintains tissue homeostasis and abolishes potentially damaged cells and organelles. Damages to the structural integrity of a cell lead to passive cell death. In another way, a cell itself can decide to die owing to more constrained biological processes. This passive type of cell death is uncontrolled which is harmful to the organism. In contrast, firmly coordinated active cell death supports to maintain homeostasis and a healthy cellular environment [1, 2]. The cellular proteolytic machinery includes numerous proteases localized in different intracellular compartments that play a crucial role in various cell death mechanisms [3]. In order to know how such proteases can regulate dismantling of a cell, one has to be aware of what protease signaling is in this context.

Initially, proteases which regulate a great variety of physiological processes were considered mainly as protein-degrading enzymes only. However, nowadays this view has totally changed, and proteases are treated as most important signaling molecules that control several critical processes including cell cycle, cell death, wound healing, protein and organelle recycling, and blood coagulation [4–6]. In addition with these crucial roles of proteases in maintaining normal cell functions, alterations in proteolytic cascade stimulate several pathological conditions such as cancer cell metastasis [7], neurodegenerative and neuroinflammatory diseases [8], inflammatory and cardiovascular diseases [9, 10]. The protease functions by cleaving and thereby modulating the properties of its substrate irreversibly [11]. Since the protease cleavage is an irreversible reaction, protease signaling is one of the effective methods for cell death, as once started the death mechanism cannot be reversed or prevented. One interesting fact about protease is that, for most proteases, it is uncertain how many physiologically significant substrates they have, how a given protease is activated within particular tissue, and how the characteristics of protease differ in disease conditions. The principal focus of this chapter is to discuss the potential role of proteases in regulating various cell death mechanisms which decide the fate of a cell under conditions of stress.

2 Protease Family

Five mechanistic classes of proteases are recognized according to the key catalytic group in the active site: cysteine (Cys), serine (Ser), threonine (Thr), aspartate (Asp), and metalloproteases. Yet another family of protease also has been identified, glutamic protease; the only type of protease that does not found in mammals. Cys, Ser, and Thr act directly as nucleophiles that attack an amide carbonyl C where histidines function as a base. Whereas, Asp and metalloproteases activate a water molecule that acts as a nucleophile which is non-covalent catalysis in which Asp residue and zinc serve as acids and bases [12, 13].

2.1 Cysteine Proteases

The major function of cysteine proteases is intracellular protein catabolism. In addition, they specifically function in the activation of selective signaling molecules such as interleukin, protein kinase C. Cysteine proteases also function in the degradation of extracellular protein. They are secreted by specific cell types under pathological conditions. Cathepsins are cysteine proteases localized in the lysosome, and calpains are the cysteine proteases localized in the cytosol [14, 15]. The catalytic thiol groups at the active site of cysteine proteases are highly oxidized, and hence, they are constrained to the reducing environment.

2.2 Serine Proteases

These are the group of proteases that closely associated with growth and differentiation. Serine protease activities are essential for normal physiological processes of the cell. However, abnormal regulation of serine proteases results in pathological conditions such as tumor development and metastasis [16].

2.3 Threonine Proteases

Threonine proteases are primarily involved in the process of polyubiquitination [17]. The major function of threonine proteases is intracellular protein turnover in cells, i.e., processing and degradation of major proteins that required for regulation of various cellular activities [18, 19].

2.4 Aspartate Proteases

Major function of aspartate proteases is the protein degradation especially in lysosomes and phagosomes. The classical example for aspartate protease is cathepsin D [20].

2.5 Metalloproteases

Matrix metalloproteases (MMPs) are a family of eighteen or more Zn-dependant endopeptidase family of enzymes. They specifically function in cleavage of the components of extracellular matrix [21]. These groups of enzymes are either secreted type or membrane bound type of enzymes.

3 Protease Signaling

To understand the role of protease in a biological process, the important task is to identify the protease's physiological substrates and its downstream effects. Identifying the downstream targets of protease cascade is challenging but very crucial as it is important in controlling various biological processes. There is significant difference between protease signaling and the other types of signaling cascades such as receptor signaling. Protease signaling is an irreversible pathway, and the signal is transmitted through the dismantling of substrates which results either activation or inactivation of the protein function [22]. Unlike other cellular signaling, the criterion for protein degradation by protease signaling is not necessary that the physiological substrate possesses the recognition sequence for a protease. On the other hand, the substrate has to be co-localized with the active protease in the same cellular compartment and so that the substrate can subsequently be processed [23]. Protease signaling can either be a simplest mechanism by which direct cleavage of the substrate leads to the conformational change to the protein, eg. processing of Interleukin-1β by cysteine proteases [24]. Alternatively, the protease signaling can be complex which involves sequential modification of the substrate. For example, the mechanism of protein degradation by ubiquitin/proteasome pathway involves two consecutive steps: (1) binding of various ubiquitin moieties to the protein substrate and (2) specific degradation of the targeted protein by proteasome complex [25].

4 Roles of Protease in Disease

In spite of the essential role of proteases in maintaining cell behavior and cellular turn over, the protease signaling may also play major role in disease progression. Studies have pointed out that the events in cancer progression, invasion and metastasis in which proteases are involved are not only late events but can occur early stages itself [26, 27]. Moreover, studies show that various other stages of cancer such as cell proliferation, apoptosis, angiogenesis are also protease dependant [28, 29]. Recently, mitochondrial proteases in the inner mitochondrial membrane have emerged as culprits in several human neurodegenerative diseases [30, 31]. Available reports suggest that coagulant and fibrinolytic proteases evoke pro-inflammatory and remodeling actions in inflammatory and fibro-proliferative diseases. Coagulants, plasminogen activators, plasmin, and plasmin-activated MMPs provoke cell-mediated responses via receptors (e.g., protease activated receptors) and co-receptors (e.g., integrins). Plasmin also indirectly contributes to inflammatory processes by growth factor receptor trans-activation [32–34]. Collectively, these reports indicate the importance of protease signaling in healthy as well as diseased conditions.

5 Proteases and Cell Death Signaling

Any external or internal triggers such as physical or chemical insult, hormonal activation, specific cell-derived signals that activate various cellular mediators can induce cell death [35]. Various types of cell deaths have been identified: necrosis, apoptosis, autophagy, and more specific death mechanisms such as mitophagy, pyroptosis, eryptosis, NETosis, autosis etc. In all the cell death mechanisms, proteases possess a key role by cleaving the protein substrates. Hence, the nature of substrate and the site of proteolytic cleavage are very much decisive in cell death signaling.

5.1 Protease Signaling in Apoptosis

Apoptosis is one of the most illustrated forms of programmed cell death which plays crucial role in embryogenesis, tissue homeostasis, and aging [36, 37]. The mechanism of apoptosis highly relies upon the regulated activation of proteases and hydrolases which rapidly cleave the cellular structures. The events of apoptosis that lead to the biological and morphological changes include shrinkage of the cell, plasma membrane blebbing, condensation of chromatin (pyknosis), nuclear fragmentation (karyorrhexis), and eventually cell death [38]. In concept, apoptosis occurs within the plasma membrane of the cell, it is an immunologically silent mechanism of cell death without affecting neighboring cells and eliciting any inflammatory response. Therefore, a network of controlled actions is needed to pack the cells for cleavage. This network of events absolutely requires the involvement of caspases [39, 40].

Caspases (cysteine-dependent aspartate-specific protease) are intracellular proteases that initiate a series of events to trigger cell death and an essential mediator of apoptosis. On the basis of the presence or absence of protein interaction domain at N-terminus, apoptotic caspases are majorly distinguished into two: initiator (caspase-8, -9, and -10) and effector (caspase-3, -6, and -7) caspases [41]. The importance of caspases in cell death mechanism is well established, and its function during apoptotic cell death was identified three decades ago itself [42]. There are two major pathways to initiate apoptotic cell death: mitochondrial-mediated intrinsic pathway and death receptor-mediated extrinsic pathway. Cellular stress such as cytotoxic drugs and DNA damage activates the intrinsic cell death pathway, and Bcl-2 family proteins are the major regulator of intrinsic pathway of cell death. The activation of Bcl-2 family proapoptotic mediators such as bcl-2 associated X (BAX) and bcl-2 homologous antagonist/killer (BAK) induces mitochondrial outer membrane permeabilization that results in release of cytochrome c to the cytosol. Cytochrome c then associated with Apaf-1 (apoptotic protease-activating factor 1) into a multimeric complex apoptosome which further activates caspase-9 [43]. Activation of extrinsic apoptosis pathway is mediated by death receptors and is initiated upon binding of ligands to tumor necrosis receptor super family (e.g., TNFR,

Fas, TRAIL) [44]. This activates the initiator caspase-8 or -10, through FAS-associated death domain protein (FADD) or TNFR-associated death domain protein (TRADD). Activation of initiator caspases by either intrinsic or extrinsic pathways mediates the activation of effector caspase-3, -6, and -7 [45]. The role of caspases is well known in apoptotic cell death. Several other proteases such as calpains and cathepsin are also involved in apoptotic machinery.

Calpains are Ca^{2+}-activated non-lysosomal cysteine proteases found in the mitochondria and cytosol. These cysteine proteases are crucial for a number of cellular processes such as cytoskeletal remodeling, attachment of proteins to the membrane, and several signaling pathways [46]. Since calpains are calcium-dependant proteases, loss of Ca^{2+} homeostasis deregulates its activity which results in tissue damage. Available reports provide evidences for the role of calpains in apoptotic cell death. When there is Ca^{2+} overload in endothelial cells, it induces the cleavage of mitochondrial calpain 1 that leads to accumulation of Ca^{2+} in mitochondria. Cleaved calpain 1 further cleaves Bcl-2 family protein BH3 interacting-domain (Bid), inducing cytochrome c release that leads to apoptosis [47]. Yet another protease calpain 10 specifically cleaves electron transport chain proteins that result in decreased mitochondrial respiration. Covington et al. have demonstrated that glucose-induced loss of calpain 10 leads to mitochondrial dysfunction and apoptosis in streptozotocin-induced diabetic rats with kidney dysfunction [48]. Caspase-12 and calpain-mediated extrinsic apoptotic cell death has been reported in HL-1 cardiomyocytes [49].

Cathepsins are being referred as ‘suicide bags’ because of their high degradation potential and of the presence of high level of hydrolases [50]. The role of lysosomal cathepsin was well known for a long time during autophagy and necrosis cell death, and it was totally unclear about their involvement in apoptosis. Two decades before, Deiss et al. have reported the possible role of cathepsin D in apoptosis and hence cathepsin D also considered as the proteases that positively mediates apoptosis [51]. Caspases can initiate apoptosis only upon its activation. In contrast, cathepsins that released to the cytosol following lysosomal membrane permeabilization are active and can initiate apoptotic machinery [52]. As discussed earlier, identification of substrate is an important task in protease signaling. For caspases, several hundred substrates are known [53]. However, very few substrates have been identified which mediates apoptosis. The key identified cathepsin substrate for apoptosis is Bid [54, 55]. Bid was initially known as substrate for caspase-8, the caspase which links intrinsic and extrinsic pathways of apoptosis [56]. In addition, Bcl-2, Bcl-XL, and Bax were also found to be as substrate of cathepsin D [57]. Cathepsin-mediated apoptosis is further confirmed as it can degrade X-chromosome-linked inhibitor of apoptosis (XIAP) protein which suggests that cathepsin mediates apoptosis downstream of mitochondria [58].

5.2 Relevance of Proteases in Autophagy

Autophagy, type II cell death, is a self-degradative process occurring in all eukaryotic cells which is crucial for maintaining tissue homeostasis. It is a well-organized cellular mechanism in response to external or internal stress [59]. The cellular components or damaged organelles of cytosol to be digested are sequestered into a double-membrane vesicle called autophagosomes. This vesicle fuses with lysosome, and the degradation of cellular components takes place at lysosome which contains hydrolases [60]. Autophagy is a cell survival mechanism which provides protection to the cells upon external or internal stress conditions such as starvation, endoplasmic reticulum (ER) stress, growth factor deprivation, and pathogen infection [61]. Multiple steps are required for the sequestration of cytoplasm through autophagy. These steps include induction of autophagy; packaging up of cargo to the vesicle; nucleation of autophagosome formation; expansion and completion of autophagosome; docking and fusion of the completed autophagosome with the lysosome; and breakdown of the damaged organelles and other cargos [62]. Several autophagy-related (Atg) genes are involved in autophagy. Five different complexes of Atg protein that function in autophagy have been identified: (1) the ULK1 complex, (2) Beclin-1/PI3KIII complex, (3) LC3/Atg8 lipid conjugation ubiquitin-like complex, (4) Atg5-Atg512-Atg16 complex, and (5) Atg9 protein and its cycling system [62].

The kinase complex mammalian target of rapamycin (mTOR) is a major checkpoint in autophagy signaling, which allows base-level formation of autophagosome; however, its inhibition triggers autophagy [63]. Impaired autophagic cell death has recently been identified in numerous pathological conditions including neurodegeneration, inflammation, and cancer [64–66]. Identifying the importance of autophagy in these pathological conditions considered autophagy as an emerging therapeutic target for the prevention of diseases through either its activation or inhibition. In this regard, protease signaling gains much attention recently in the regulation of autophagic cell death, as several proteolytic enzymes are recognized as indirect modulators of the autophagy signaling. Since autophagy is a lysosomal-degradative mechanism, a number of lysosomal proteases such as calpains, caspase, cathepsins are concerned to be participating in several stages of the autophagic pathway. Interestingly, yet another protease, the cysteine protease Atg4 also plays a crucial role in the execution of autophagic cell death.

The central role of Atg4 is well identified in the LC3/atg8 lipid conjugation complex, which is essential protein complex for the elongation and complete formation of autophagosome membrane. Lang et al. have reported for the first time the physical interaction of Atg4 with Atg8 in the yeast *Saccharomyces cerevisiae* [67]. During the formation of autophagosome, Atg8 covalently binds to phosphatidyl ethanolamine (PE). The function of protease Atg4 in the complex is that it removes C-terminal arginine of Atg8 and felicitates its binding with PE [68]. Additionally, Atg8 orthologs are differentially cleaved by the different Atg4 family members, with Atg4B representing the most potent of all of them in terms of substrate specificity. In fact, Atg4B is able to process a wide range of Atg8 orthologs [69], and the analysis of their structures has been very useful to understand the interaction between both enzyme and substrate [70]. Moreover, apart from their central role in the Atg8/LC3 lipid conjugation system, recent reports have pointed out a potential link between Atg4D and apoptosis, indicating a reputed role for Atg4D at the interface between autophagy and apoptosis [71, 72]. Dysfunctional autophagic responses with regard to Atg4 have been reported in several pathological conditions. The essential role of ATG4A in regulating the tumorigenicity of breast cancer stem cells has been reported in vivo [73]. The tumor suppressor role of Atg4C is reported in murine fibrosarcoma model [74]. These reports suggest the crucial role of this cysteine protease in regulating autophagic cell death.

The close connection of apoptosis and autophagy cell death is well illustrated. Available reports suggest that induction of either of these cell death mechanisms is depending on the cellular context. Blockade of apoptosis in cells results in the activation of autophagy, or inhibition of autophagy may trigger apoptotic cell death [75, 76]. Cysteine protease caspases are the major molecules that link apoptosis and autophagy. The role of proteases that links apoptosis and autophagy pathways is depicted in Fig. 1.

The proteases calpain and caspases can cleave human ATG proteins. hATG3 has been reported to be lysed by caspase-3, -6, and -8 [77]. Cho et al. have suggested that caspase-mediated cleavage of ATG6 links the apoptotic and autophagic signaling pathways and have reported Beclin-1-Atg 6 complex as caspase-3 target

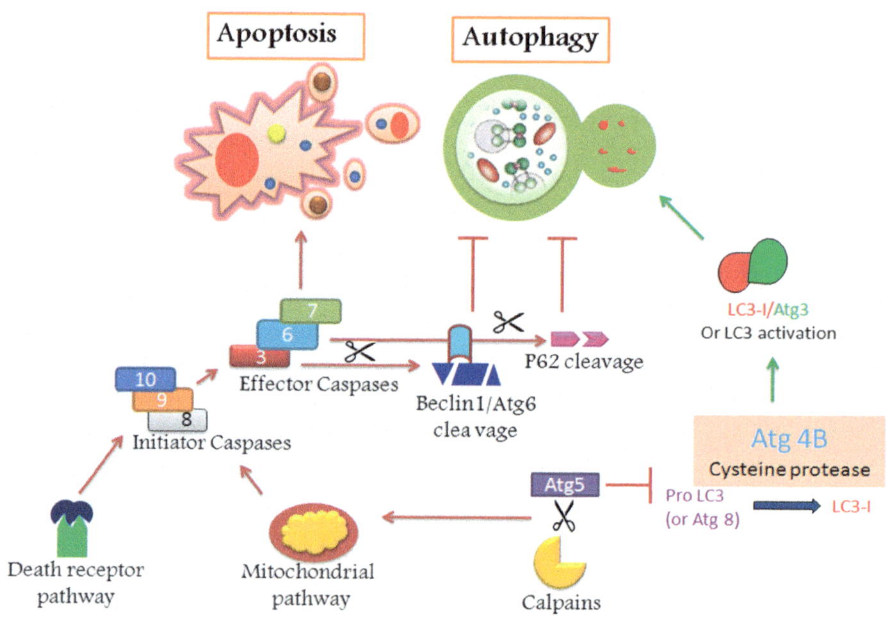

Fig. 1 Role of proteases in apoptosis and autophagic cell death. Mitochondrial and death receptor pathways initiate apoptosis through the activation of initiator and effector caspases. In addition, caspase-mediated cleavage of autophagy proteins Beclin1/Atg6 and p62 inhibits autophagy. The cysteine protease calpain mediates cleavage of Atg5 results in the liberation of a proapoptotic fragment of Atg5 that leads to mitochondrial outer membrane permeabilization and cell death. The cysteine protease Atg4B activates autophagy through the activation of LC3-mediated autophagosome formation

[78]. Caspase-6 and -8 cleave p62, an autophagy adaptor protein that leads to inhibition of autophagy [79]. Apart from caspases, cathepsins being major protease in the lysosome also function in the regulation of autophagy. Boland et al. have reported that inhibition of cathepsin-mediated proteolysis in Alzheimer's model results in accumulation of incompleted autophagosomes with partially degraded LC3II which suggests the importance of cathepsins along with caspases during autophagy [80]. In support to this view, a vital role of cathepsins in the degradation of damaged organelles and associated cargos within lysosome upon the fusion of autophagosomes with lysosome to form autolysosomes has been reported [81]. These reports suggest the autophagy regulatory function of cathepsins. Hence, it is obvious that protease signaling is significant in the regulation of autophagic cell death.

5.3 Proteases in Necrotic Cell Death

While apoptosis is a programmed type of cell death, necrosis, type III modality of cell death is not a controlled mechanism but is characterized by dysregulation of cellular activities and collapse of cellular environment under stress conditions. The morphological characteristic features of necrotic cell death include plasma membrane rupture, swelling of mitochondria, dissolution of the endoplasmic reticulum, and extensive vacuolation of the cytoplasm. During necrotic cell death, the damaged cellular contents are released into the intracellular space, and hence, it affects neighboring cells by inducing an inflammatory response [82]. The ligand–receptor interactions in necrotic cell death are highly regulated by the serine/threonine kinase receptor-interacting protein 1 (RIP1) which is calcium dependant. During necrosis, cytosolic calcium level significantly increases which typically lead to calcium overload in mitochondria that activates proteases and phospholipases [83]. Since calcium is the notorious factor in necrotic cell death, calcium-dependant proteases: calpains and cathepsins play major role in the regulation of necrosis. Because of the enormous destructive potential of cathepsins, massive lysosomal rupture leads to necrotic cell death. Jacobson et al. have reported cathepsin-mediated lysosome disruption and necrotic cell death in myeloid leukocytes [84]. Increased number of evidences show the interrelation between apoptosis and necrosis. For example, low concentration of an external stress insult induces apoptosis in cells, while the same stimuli at higher concentration induce necrosis in the same cells [85–87]. The necrotic cell death can be partially mediated by apoptotic cascade such as caspase 8/Bid strongly highlights the interplay of these two cell death mechanisms [88]. In necrotic cell death, the non-caspase protease cathepsin-induced morphological changes resemble the caspase-induced morphological changes in apoptosis [89]. However, the caspase-mediated cell death is highly regulated, while the cathepsin activity is not specific and is less precise, and hence, it mediates more disastrous necrotic cell death.

5.4 Protease in Necroptosis Signaling

The concept of uncontrolled and unregulated necrotic cell death has been changed nowadays, and evidences revealed the regulated mechanism of necrotic cell death. This programmed necrosis is termed as 'necroptosis' which has important functions in the organogenesis and in the maintenance of tissue homeostasis [90]. Tumor necrosis factor receptor 1 (TNFR1) initiates the necroptosis signaling in cells. However, TNF function in cell is dependent on cell type and its surroundings, and based on this, TNF administration can result in cell survival, apoptosis, or necroptosis [91]. In necroptosis signaling, the protease caspase 8 modulates the molecular events in response to TNFR1 activation. The RIP1, RIP3, and mixed lineage kinase domain-like (MLKL) proteins are the major regulators of necroptosis signaling. Under condition of blockade of apoptosis, necroptosis cell death serves as critical mechanism to remove damaged cells [92]. Necroptosis is also closely linked with apoptosis. Cysteine protease caspases interlink these two pathways. In response to external or internal stimuli, TNF induces a TRADD (Tumor necrosis factor receptor type 1-associated death domain)-dependant complex 1 by recruiting RIP1, TRADD, TRAFs (TNF-receptor-associated factors), etc. Formation of this complex ensures the activation of canonical nuclear factor-κB (NF-κB) pathway [93]. Role of caspases is significant here to decide whether to undergo apoptosis or necroptosis. The cleavage of RIP1 and RIP3 by caspase 8 initiates apoptosis. The inhibition of caspase activity leads to the auto-phosphorylation of RIP1 and RIP3 which recruit MLKL protein to form a complex called necrosome that initiates necroptosis [94]. Thus, RIP1 is the master switch protein that decides the fate of cell whether to undergo caspase-8-mediated apoptosis or MLKL-dependant necroptosis.

5.5 Proteases in Pyroptosis

Pyroptosis is a proteolytic cell death mechanism that first described in *Shigella flexneri*-infected macrophages [95]. It shares some of the features with apoptosis including DNA fragmentation and nuclear condensation. Pyroptosis is recognized as caspase-1-dependant modality of cell death [96]. In pyroptosis, caspases are directly activated via a caspase activation and recruitment domain (CARD)-containing inflammasome. Inflammatory response in cell induces caspase activation that results in initiation of cell death signaling. Caspases induce permeabilization of the plasma membrane that results in flushing of water and ions into the cell causing cell rupture and the release of cytosolic contents [97]. Along with caspase-1, caspase-11 is also gaining much attention in pyroptosis research [98, 99]. A research group recently reported that cleavage of pannexin-1 channel is essential for pyroptosis which is catalyzed by caspase 11 [100]. In contrast, yet another research group has reported that caspase-1 but not caspase-11 was required for inflammasome formation during pyroptosis in Legionella species as well as in mice [101].

5.6 Possible Role of Proteases in Eryptosis

The life span of circulating erythrocytes is normally 100–120 days [102]. How these cells are cleared? Prior to cellular senescence, erythrocytes may be removed by suicidal death, termed as eryptosis upon external stimuli or cell injury. Eryptosis is characterized by the similar morphological features of apoptosis as cell shrinkage and rupture of cell membrane. The breakdown of plasma membrane leads to the translocation of phosphatidylserine (PS) to the cell surface, a hallmark of suicidal cell death [103]. The translocation of PS to the cell surface is calcium dependent, and increased Ca^{2+} concentration inside the cell further facilitates the membrane rupture and translocation of PS [103]. Though erythrocytes express caspases and these caspases function in stimulating the expression of PS at outer membrane [104], the calcium-mediated membrane rupture is independent of caspase activity [105].

5.7 Proteases in ER-Stress-Mediated Cell Death

We have discussed the role of proteases in various modes of cell death. What exactly prompts a cell to die? Several direct and indirect factors cause cell death. In this context, endoplasmic reticulum (ER)-stress-mediated cell death is very much emerging and gaining attention of researchers recently. The preliminary observations of our laboratory demonstrate that ER stress plays an important role in inducing caspase-mediated apoptotic cell death in bleomycin-induced pulmonary fibrosis model. In this regard, it would be logic to discuss the potential role of protease in ER-stress-mediated cell death.

Since protein folding is taking place in ER, intracellular stress by a wide variety stressors disturbs the normal environment of the ER which leads to unfolded protein accumulation [106]. Normally, proper protein processing and folding occur in ER using molecular chaperones, lectins, and foldases [107]. However, any disturbance to normal protein folding results in ER-associated degradation (ERAD) pathways of mis-folded proteins [108]. Prolonged ER stress triggers apoptosis as a last choice of eukaryotes to dish out the dysfunctional cells [109]. As in other cell death mechanisms, caspases are the major proteases involved in ER-stress-mediated cell death. Available reports show that caspase-12 associates with activated inositol-requiring enzyme 1 (Ire1) which is a key mediator of ER stress pathway [110]. High concentration of calcium in ER under stress condition activates calpains which further stimulate the proteolytic activity of caspase-12 [111]. But caspase 12 is absent in most humans [112], and therefore, the relevance of caspase-12 in ER-stress-mediated apoptosis is questionable. In this point of view, human caspase-4 which is a close paralog of rodent caspase-12 may be assumed to participate in ER-stress-mediated cell death in humans [113]. However, further studies are warranted to validate the role of protease as therapeutic target in context of ER-stress-mediated cell death.

6 Conclusion

For long time, apoptosis was considered as the only mechanism for the removal of unwanted cells. However, it has become clear that along with apoptosis, various other cell death mechanisms such as autophagy, necroptosis, pyroptosis, NETosis, mitophagy play an important role in maintaining tissue homeostasis. These different modalities of cell death are being intensely studied in context of various diseases in order to elucidate the mechanism that how cell death contributes for disease progression. Since proteases play a decisive role in the regulation of cell death programs, elucidation of cell death pathways and the mechanisms through which they are regulated by proteases will lead to development of novel drugs for life-threatening diseases.

Acknowledgement The authors thank the support of University Grants Commission (UGC)-UPE Phase II for carrying out a part of work in calcium signaling.

References

1. Green DR, Llambi F (2015) Cell death signaling. Cold Spring Harb Perspect Biol 7(12): a006080
2. Galluzzi L, Lopez-Soto A, Kumar S, Kroemer G (2016) Caspases connect cell-death signaling to organismal homeostasis. Immunity 44(2):221–231
3. Ashkenazi A, Salvesen G (2014) Regulated cell death: signaling and mechanisms. Annu Rev Cell Dev Biol 30:337–356
4. Puente XS, Sanchez LM, Overall CM, Lopez-Otin C (2003) Human and mouse proteases: a comparative genomic approach. Nat Rev Genet 4(7):544–558
5. Davie EW, Ratnoff OD (1964) Waterfall sequence for intrinsic blood clotting. Science 145(1310):1312–3638
6. Macfarlane RG (1964) An enzyme cascade in the blood clotting mechanism, and its function as a biochemical amplifier. Nature 202:498–499
7. Bulteau AL, Bayot A (2011) Mitochondrial proteases and cancer. Biochim Biophys Acta 1807(6):595–601
8. Troy CM, Jean YY (2015) Caspases: therapeutic targets in neurologic disease. Neurotherapeutics 12(1):42–48
9. Qureshi N, Morrison DC, Reis J (2012) Proteasome protease mediated regulation of cytokine induction and inflammation. Biochim Biophys Acta 1823(11):2087–2093
10. Azevedo A, Prado AF, Antonio RC, Issa JP, Gerlach RF (2014) Matrix metalloproteinases are involved in cardiovascular diseases. Basic Clin Pharmacol Toxicol 115(4):301–314
11. Verdoes M, Verhelst SH (2016) Detection of protease activity in cells and animals. Biochim Biophys Acta 1864(1):130–142
12. Lopez-Otin C, Bond JS (2008) Proteases: multifunctional enzymes in life and disease. J Biol Chem 283(45):30433–30437
13. Scott CJ, Taggart CC (2010) Biologic protease inhibitors as novel therapeutic agents. Biochimie 92(11):1681–1688
14. Joyce JA, Hanahan D (2004) Multiple roles for cysteine cathepsins in cancer. Cell Cycle 3(12):619–1516
15. Jedeszko C, Sloane BF (2004) Cysteine cathepsins in human cancer. Biol Chem 385(11):1017–1027

16. Henneke I, Greschus S, Savai R, Korfei M, Markart P, Mahavadi P, Schermuly RT, Wygrecka M, Stürzebecher J, Seeger W, Günther A, Ruppert C (2010) Inhibition of urokinase activity reduces primary tumor growth and metastasis formation in a murine lung carcinoma model. Am J Respir Crit Care Med 181(6):611–619
17. Mitchell BS (2003) The proteasome–an emerging therapeutic target in cancer. N Engl J Med 348(26):2597–2598
18. Ciechanover A (1994) The ubiquitin-proteasome proteolytic pathway. Cell 79(1):13–21
19. Hochstrasser M (1995) Ubiquitin, proteasomes, and the regulation of intracellular protein degradation. Curr Opin Cell Biol 72(2):2215–2223
20. Barrett AJ (1970) Cathepsin D. Purification of isoenzymes from human and chicken liver. Biochem J 117(3):601–607
21. Li NG, Tang YP, Duan JA, Shi ZH (2014) Matrix metalloproteinase inhibitors: a patent review (2011–2013). Expert Opin Ther Pat 24(9):1039–1052
22. Turk B (2006) Targeting proteases: successes, failures and future prospects. Nat Rev Drug Discov 5(9):785–799
23. Turk B, Turk D, Turk V (2012) Protease signalling: the cutting edge. EMBO J 31(7):1630–1643
24. Thornberry NA, Bull HG, Calaycay JR, Chapman KT, Howard AD, Kostura MJ, Miller DK, Molineaux SM, Weidner JR, Aunins J et al (1992) A novel heterodimeric cysteine protease is required for interleukin-1 beta processing in monocytes. Nature 356(6372):768–774
25. Glickman MH, Ciechanover A (2002) The ubiquitin-proteasome proteolytic pathway: destruction for the sake of construction. Physiol Rev 82(2):373–428
26. Deryugina EI, Quigley JP (2006) Matrix metalloproteinases and tumor metastasis. Cancer Metastasis Rev 25(1):9–34
27. Gocheva V, Joyce JA (2007) Cysteine cathepsins and the cutting edge of cancer invasion. Cell Cycle 6(1):60–64
28. Duffy MJ (1996) Proteases as prognostic markers in cancer. Clin Cancer Res 2(4):613–618
29. Hu L, Roth JM, Brooks P, Luty J, Karpatkin S (2008) Thrombin up-regulates cathepsin D which enhances angiogenesis, growth, and metastasis. Cancer Res 68(12):4666–4673
30. Martinelli P, Rugarli E (2010) Emerging roles of mitochondrial proteases in neurodegeneration. Biochim Biophys Acta 1797(1):1–10
31. Tatsuta T, Augustin S, Nolden M, Friedrichs B, Langer T (2007) m-AAA protease-driven membrane dislocation allows intramembrane cleavage by rhomboid in mitochondria. EMBO J 26(2):325–335
32. Schuliga M (2015) The inflammatory actions of coagulant and fibrinolytic proteases in disease. Mediators Inflamm 437695
33. Xie Y, Gao K, Häkkinen L, Larjava HS (2009) Mice lacking beta6 integrin in skin show accelerated wound repair in dexamethasone impaired wound healing model. Wound Repair Regen 17(3):326–339
34. Florsheim E, Yu S, Bragatto I, Faustino L, Gomes E, Ramos RN, Barbuto JA, Medzhitov R, Russo M (2015) Integrated innate mechanisms involved in airway allergic inflammation to the serine protease subtilisin. J Immunol 194(10):4621–4630
35. Vicencio JM, Galluzzi L, Tajeddine N, Ortiz C, Criollo A, Tasdemir E, Morselli E, Ben Younes A, Maiuri MC, Lavandero S, Kroemer G (2007) Senescence, apoptosis or autophagy? When a damaged cell must decide its path–a mini-review. Gerontology 54 (2):92–99
36. Xiong S, Mu T, Wang G, Jiang X (2014) Mitochondria-mediated apoptosis in mammals. Protein Cell 5(10):737–749
37. Zhang Y, Herman B (2002) Ageing and apoptosis. Mech Ageing Dev 123(4):245–260
38. Saraste A, Pulkki K (2000) Morphologic and biochemical hallmarks of apoptosis. Cardiovasc Res 45(3):528–537
39. Hochreiter-Hufford A, Ravichandran KS (2013) Clearing the dead: apoptotic cell sensing, recognition, engulfment, and digestion. Cold Spring Harb Perspect Biol 5(1):a008748

40. Henson PM, Bratton DL (2013) Antiinflammatory effects of apoptotic cells. J Clin Invest 123(7):2773–2774
41. McIlwain DR, Berger T, Mak TW (2013) Caspase functions in cell death and disease. Cold Spring Harb Perspect Biol 5(4):a008656
42. Ellis HM, Horvitz HR (1986) Genetic control of programmed cell death in the nematode C. elegans. Cell 44(6):817–829
43. Czabotar PE, Lessene G, Strasser A, Adams JM (2014) Control of apoptosis by the BCL-2 protein family: implications for physiology and therapy. Nat Rev Mol Cell Biol 15(1):49–63
44. Gaur U, Aggarwal BB (2003) Regulation of proliferation, survival and apoptosis by members of the TNF superfamily. Biochem Pharmacol 66(8):1403–1408
45. Fulda S (2015) Targeting extrinsic apoptosis in cancer: challenges and opportunities. Semin Cell Dev Biol 39:20–25
46. Goll DE, Thompson VF, Li H, Wei W, Cong J (2003) The calpain system. Physiol Rev 83 (3):731–801
47. Hu H, Li X, Li Y, Wang L, Mehta S, Feng Q, Chen R, Peng T (2009) Calpain-1 induces apoptosis in pulmonary microvascular endothelial cells under septic conditions. Microvasc Res 78(1):33–39
48. Covington MD, Schnellmann RG (2012) Chronic high glucose downregulates mitochondrial calpain 10 and contributes to renal cell death and diabetes-induced renal injury. Kidney Int 81(4):391–400
49. Bajaj G, Sharma RK (2006) TNF-alpha-mediated cardiomyocyte apoptosis involves caspase-12 and calpain. Biochem Biophys Res Commun 345(4):1558–1564
50. de Duve C (2005) The lysosome turns fifty. Nat Cell Biol 7(9):847–849
51. Deiss LP, Galinka H, Berissi H, Cohen O, Kimchi A (1996) Cathepsin D protease mediates programmed cell death induced by interferon-gamma, Fas/APO-1 and TNF-alpha. EMBO J 15(15):3861–3870
52. Turk B, Stoka V (2007) Protease signalling in cell death: caspases versus cysteine cathepsins. FEBS Lett 581(15):2761–2767
53. Timmer JC, Salvesen GS (2007) Caspase substrates. Cell Death Differ 14(1):66–72
54. Cirman T, Oresic K, Mazovec GD, Turk V, Reed JC, Myers RM, Salvesen GS, Turk B (2004) Selective disruption of lysosomes in HeLa cells triggers apoptosis mediated by cleavage of Bid by multiple papain-like lysosomal cathepsins. J Biol Chem 279(5):3578–3587
55. Blomgran R, Zheng L, Stendahl O (2007) Cathepsin-cleaved Bid promotes apoptosis in human neutrophils via oxidative stress-induced lysosomal membrane permeabilization. J Leukoc Biol 81(5):1213–1223
56. Li H, Zhu H, Xu CJ, Yuan J (1998) Cleavage of BID by caspase 8 mediates the mitochondrial damage in the Fas pathway of apoptosis. Cell 94(4):491–501
57. Bidere N, Lorenzo HK, Carmona S, Laforge M, Harper F, Dumont C, Senik A (2003) Cathepsin D triggers Bax activation, resulting in selective apoptosis-inducing factor (AIF) relocation in T lymphocytes entering the early commitment phase to apoptosis. J Biol Chem 278(33):31401–31411
58. Droga-Mazovec G, Bojic L, Petelin A, Ivanova S, Romih R, Repnik U, Salvesen GS, Stoka V, Turk V, Turk B (2008) Cysteine cathepsins trigger caspase-dependent cell death through cleavage of bid and antiapoptotic Bcl-2 homologues. J Biol Chem 283(27):19140–19150
59. Klionsky DJ (2007) Autophagy: from phenomenology to molecular understanding in less than a decade. Nat Rev Mol Cell Biol 8(11):931–937
60. He C, Klionsky DJ (2009) Regulation mechanisms and signaling pathways of autophagy. Annu Rev Genet 43:67–93
61. Klionsky DJ, Cuervo AM, Dunn WA Jr, Levine B, van der Klei I, Seglen PO (2007) How shall I eat thee? Autophagy 3(5):413–416

62. Klionsky DJ (2005) The molecular machinery of autophagy: unanswered questions. J Cell Sci 118(Pt 1):7–18
63. Wullschleger S, Loewith R, Hall MN (2006) TOR signaling in growth and metabolism. Cell 124(3):471–484
64. Deretic V, Saitoh T, Akira S (2013) Autophagy in infection, inflammation and immunity. Nat Rev Immunol 13(10):722–737
65. Levine B, Kroemer G (2008) Autophagy in the pathogenesis of disease. Cell 132(1):27–42
66. Rubinsztein DC, Mariño G, Kroemer G (2011) Autophagy and aging. Cell 146(5):682–695
67. Lang T, Schaeffeler E, Bernreuther D, Bredschneider M, Wolf DH, Thumm M (1998) Aut2p and Aut7p, two novel microtubule-associated proteins are essential for delivery of autophagic vesicles to the vacuole. EMBO J 17(13):3597–3607
68. Marino G, Uria JA, Puente XS, Quesada V, Bordallo J, Lopez-Otin C (2003) Human autophagins, a family of cysteine proteinases potentially implicated in cell degradation by autophagy. J Biol Chem 278(6):3671–3678
69. Li M, Hou Y, Wang J, Chen X, Shao ZM, Yin XM (2011) Kinetics comparisons of mammalian Atg4 homologues indicate selective preferences toward diverse Atg8 substrates. J Biol Chem 286(9):7327–73238
70. Hemelaar J, Lelyveld VS, Kessler BM, Ploegh HL (2003) A single protease, Apg4B, is specific for the autophagy-related ubiquitin-like proteins GATE-16, MAP1-LC3, GABARAP, and Apg8L. J Biol Chem 278(51):51841–51850
71. Kaminskyy VO, Zhivotovsky B (2014) Free radicals in cross talk between autophagy and apoptosis. Antioxid Redox Signal 21(1):86–102
72. Norman JM, Cohen GM, Bampton ET (2015) The in vitro cleavage of the hAtg proteins by cell death proteases. Autophagy 6(8):1042–1056
73. Wolf J, Dewi DL, Fredebohm J, Müller-Decker K, Flechtenmacher C, Hoheisel JD, Boettcher M (2013) A mammosphere formation RNAi screen reveals that ATG4A promotes a breast cancer stem-like phenotype. Breast Cancer Res 15(6):R109
74. Marino G, Salvador-Montoliu N, Fueyo A, Knecht E, Mizushima N, López-Otin C (2007) Tissue-specific autophagy alterations and increased tumorigenesis in mice deficient in Atg4C/autophagin-3. J Biol Chem 282(25):18573–18583
75. Maiuri MC, Zalckvar E, Kimchi A, Kroemer G (2007) Self-eating and self-killing: crosstalk between autophagy and apoptosis. Nat Rev Mol Cell Biol 8(9):741–752
76. Debnath J, Baehrecke EH, Kroemer G (2005) Does autophagy contribute to cell death? Autophagy 1(2):66–74
77. Denton D, Nicolson S, Kumar S (2012) Cell death by autophagy: facts and apparent artefacts. Cell Death Differ 19(1):87–95
78. Cho DH, Jo YK, Hwang JJ, Lee YM, Roh SA, Kim JC (2009) Caspase-mediated cleavage of ATG6/Beclin-1 links apoptosis to autophagy in HeLa cells. Cancer Lett 274(1):95–100
79. Norman JM, Cohen GM, Bampton ET (2010) The in vitro cleavage of the hAtg proteins by cell death proteases. Autophagy 6(8):1042–1056
80. Boland B, Kumar A, Lee S, Platt FM, Wegiel J, Yu WH, Nixon RA (2008) Autophagy induction and autophagosome clearance in neurons: relationship to autophagic pathology in Alzheimer's disease. J Neurosci 28(27):6926–6937
81. Kroemer G, Jaattela M (2005) Lysosomes and autophagy in cell death control. Nat Rev Cancer 5(11):886–897
82. Vanlangenakker N, Vanden Berghe T, Krysko DV, Festjens N, Vandenabeele P (2008) Molecular mechanisms and pathophysiology of necrotic cell death. Curr Mol Med 8(3):207–220
83. Poon IK, Hulett MD, Parish CR (2010) Molecular mechanisms of late apoptotic/necrotic cell clearance. Cell Death Differ 17(3):381–397
84. Jacobson LS, Lima H Jr, Goldberg MF, Gocheva V, Tsiperson V, Sutterwala FS, Joyce JA, Gapp BV, Blomen VA, Chandran K, Brummelkamp TR, Diaz-Griffero F, Brojatsch J (2013)

Cathepsin-mediated necrosis controls the adaptive immune response by Th2 (T helper type 2)-associated adjuvants. J Biol Chem 288(11):7481–7491
85. Ueda N, Walker PD, Hsu SM, Shah SV (1995) Activation of a 15-kDa endonuclease in hypoxia/reoxygenation injury without morphologic features of apoptosis. Proc Natl Acad Sci U S A 92(16):7202–7206
86. Ueda N, Shah SV (2000) Tubular cell damage in acute renal failure-apoptosis, necrosis, or both. Nephrol Dial Transplant 15(3):318–323
87. Meli E, Pangallo M, Picca R, Baronti R, Moroni F, Pellegrini-Giampietro DE (2004) Differential role of poly(ADP-ribose) polymerase-1in apoptotic and necrotic neuronal death induced by mild or intense NMDA exposure in vitro. Mol Cell Neurosci 25(1):172–180
88. Wang X, Ryter SW, Dai C, Tang ZL, Watkins SC, Yin XM, Song R, Choi AM (2003) Necrotic cell death in response to oxidant stress involves the activation of the apoptogenic caspase-8/bid pathway. J Biol Chem 278(31):29184–29191
89. Lockshin RA, Zakeri Z (2002) Caspase-independent cell deaths. Curr Opin Cell Biol 14 (6):727–733
90. Newton K, Manning G (2016) Necroptosis and Inflammation. Annu Rev Biochem 85:743–763
91. Wilson NS, Dixit V, Ashkenazi A (2009) Death receptor signal transducers: nodes of coordination in immune signaling networks. Nat Immunol 10(4):348–355
92. Vanden Berghe T, Vanlangenakker N, Parthoens E, Deckers W, Devos M, Festjens N, Guerin CJ, Brunk UT, Declercq W, Vandenabeele P (2010) Necroptosis, necrosis and secondary necrosis converge on similar cellular disintegration features. Cell Death Differ 17 (6):922–930
93. Chen D, Yu J (1865) Zhang L (2016) Necroptosis: an alternative cell death program defending against cancer. Biochim Biophys Acta 2:228–236
94. Degterev A, Hitomi J, Germscheid M, Ch'en IL, Korkina O, Teng X, Abbott D, Cuny GD, Yuan C, Wagner G, Hedrick SM, Gerber SA, Lugovskoy A, Yuan J (2008) Identification of RIP1 kinase as a specific cellular target of necrostatins. Nat Chem Biol 4(5):313–321
95. Zychlinsky A, Prevost MC, Sansonetti PJ (1992) Shigella flexneri induces apoptosis in infected macrophages. Nature 358(6382):167–169
96. Cookson BT, Brennan MA (2001) Pro-inflammatory programmed cell death. Trends Microbiol 9(3):113–114
97. Fink SL, Cookson BT (2006) Caspase-1-dependent pore formation during pyroptosis leads to osmotic lysis of infected host macrophages. Cell Microbiol 8(11):1812–1825
98. Yang JR, Yao FH, Zhang JG, Ji ZY, Li KL, Zhan J, Tong YN, Lin LR, He YN (2014) Ischemia-reperfusion induces renal tubule pyroptosis via the CHOP-caspase-11 pathway. Am J Physiol Renal Physiol 306(1):F75–F84
99. Pilla DM, Hagar JA, Haldar AK, Mason AK, Degrandi D, Pfeffer K, Ernst RK, Yamamoto M, Miao EA, Coers J (2014) Guanylate binding proteins promote caspase-11-dependent pyroptosis in response to cytoplasmic LPS. Proc Natl Acad Sci U S A 111(16):6046–6051
100. Yang D, He Y, Munoz-Planillo R, Liu Q, Nunez G (2015) Caspase-11 Requires the Pannexin-1 Channel and the Purinergic P2X7 Pore to Mediate Pyroptosis and Endotoxic Shock. Immunity 43(5):923–932
101. Cerqueira DM, Pereira MS, Silva AL, Cunha LD, Zamboni DS (2015) Caspase-1 but not caspase-11 is required for NLRC4-mediated pyroptosis and restriction of infection by flagellated legionella species in mouse macrophages and in vivo. J Immunol 195(5):2303–2311
102. Bosman GJ, Willekens FL, Werre JM (2005) Erythrocyte aging: a more than superficial resemblance to apoptosis? Cell Physiol Biochem 16(1–3):1–8
103. Berg CP, Engels IH, Rothbart A, Lauber K, Renz A, Schlosser SF, Schulze-Osthoff K, Wesselborg S (2001) Human mature red blood cells express caspase-3 and caspase-8, but are devoid of mitochondrial regulators of apoptosis. Cell Death Differ 8(12):1197–1206

104. Bratosin D, Estaquier J, Petit F, Arnoult D, Quatannens B, Tissier JP, Slomianny C, Sartiaux C, Alonso C, Huart JJ, Montreuil J, Ameisen JC (2001) Programmed cell death in mature erythrocytes: a model for investigating death effector pathways operating in the absence of mitochondria. Cell Death Differ 8(12):1143–1156
105. Weil M, Jacobson MD, Raff MC (1998) Are caspases involved in the death of cells with a transcriptionally inactive nucleus? Sperm and chicken erythrocytes. J Cell Sci 111(Pt 18):2707–2715
106. Ogen-Shtern N, Ben David T, Lederkremer GZ (2016) Protein aggregation and ER stress. Brain Res pii: S0006-8993(16)30183-4
107. Naidoo N (2009) ER and aging-protein folding and the ER stress response. Ageing Res Rev 8(3):150–159
108. Kaufman RJ, Scheuner D, Schröder M, Shen X, Lee K, Liu CY, Arnold SM (2002) The unfolded protein response in nutrient sensing and differentiation. Nat Rev Mol Cell Biol 3 (6):411–421
109. Xu C, Bailly-Maitre B, Reed JC (2005) Endoplasmic reticulum stress: cell life and death decisions. J Clin Invest 115(10):2656–2664
110. Nakagawa T, Zhu H, Morishima N, Li E, Xu J, Yankner BA, Yuan J (2000) Caspase-12 mediates endoplasmic-reticulum-specific apoptosis and cytotoxicity by amyloid-beta. Nature 403(6765):98–103
111. Nakagawa T, Yuan J (2000) Cross-talk between two cysteine protease families. Activation of caspase-12 by calpain in apoptosis. J Cell Biol 150(4):887–894
112. Fischer H, Koenig U, Eckhart L, Tschachler E (2002) Human caspase 12 has acquired deleterious mutations. Biochem Biophys Res Commun 293(2):722–726
113. Saleh M, Vaillancourt JP, Graham RK, Huyck M, Srinivasula SM, Alnemri ES, Steinberg MH, Nolan V, Baldwin CT, Hotchkiss RS, Buchman TG, Zehnbauer BA, Hayden MR, Farrer LA, Roy S, Nicholson DW (2004) Differential modulation of endotoxin responsiveness by human caspase-12 polymorphisms. Nature 429(6987):75–79

Role of Proteases in Idiopathic Pulmonary Fibrosis

Soumya Chatterjee, Kaustav Chakraborty, Subhabrata Moitra and Arindam Bhattacharyya

Abstract

Idiopathic pulmonary fibrosis may be described as the debilitating condition of lung where excessive collagen-rich extracellular matrix (ECM) gets deposited. From the chemical and biological properties, it can be assumed that the activities of proteases can degrade matrix. Though, some of the proteases are anti-fibrotic, whereas most of them have profibrotic functions. Proteases perform important functions in a range of biological processes, like tissue repairing, remodeling, and providing immunity. However, the exact mechanism is yet to be known how these enzymes work during fibrosis; i.e., the proteins that the proteases target to perform a specific process and its effect on ECM turnover is still opaque. However, experimental models and clinical studies have identified some crucial steps that could help understanding the disease mechanism and also herald a ground for future therapy.

Keywords

Pulmonary fibrosis · Proteases · Lung disease · MMP

S. Chatterjee · K. Chakraborty · A. Bhattacharyya (✉)
Immunology Laboratory, Department of Zoology, University of Calcutta,
35, Ballygunge Circular Road, Kolkata, India
e-mail: arindam19@yahoo.com

S. Chatterjee
e-mail: soumyac.il15@gmail.com

K. Chakraborty
e-mail: kaustavc17@gmail.com

S. Moitra
Department of Pneumology, Allergy and Asthma Research Centre, Kolkata, India
e-mail: subhabrata2207@gmail.com

S. Chakraborti and N.S. Dhalla (eds.), *Pathophysiological Aspects of Proteases*,
DOI 10.1007/978-981-10-6141-7_22

Abbreviations

α-SMA	α A smooth muscle actin
AHR	Airway hyper reactivity
ALI	Acute lung injury
AR	Amphiregulin
ARDS	Acute respiratory distress syndrome
BALf	Broncho alveolar lavage fluid
Cat K	Cathepsin K
COPD	Chronic obstructive pulmonary disease
ECM	Extracellular matrix
EGFR	Epithelial growth factor receptor
EMT	Epithelial mesenchymal transition
IPF	Idiopathic pulmonary fibrosis
IGFBP-3	Insulin growth factor binding protein-3
MMP	Matrix metalloprotease
NE	Neutrophil elastase
PFT	Pulmonary function test
PTGS2	Prostaglandin G/H synthase 2
ROC	Receiver operating characteristic curve
TGF β	Transforming growth factor β

1 Introduction

Idiopathic pulmonary fibrosis (IPF) is the commonest form age-related diffuse lung disease that is known to be persistent with progressive collagen production in alveolar septa leading to the damage of the lungs affecting the gas-exchange mechanism [1–3]. With a prevalence of approximately 20 per 100,000 people, the burden of this disease is alarmingly increasing with an estimated rate of 4.6–16.3 cases per 100,000 population, each year [4, 5]. Patients with IPF suffer mostly from cough and dyspnea, thus lead a compromised quality of life. This disease is often misdiagnosed in the initial stage, and therefore, the prognosis becomes extremely marginal [6]. The pathologic mechanisms of IPF are not fully understood; hence, targeted treatment regime has been extremely limited. Risk factors, such as chronic or repetitive exposure to cigarette smoking, occupational and environmental pollutants, chronic gastroesophageal reflux and chronic lower airway infection, are trivial in the manifestation of the disease. The progressive and irreversible disruption of alveolar bed induces an accelerated accumulation of fibroblasts in the disrupted area where the cells are transformed into myofibroblasts and deploy collagen and other extracellular matrix molecules, the milieu which distorts pulmonary architecture leading to impaired lung function [7]. The immediate response due to tissue injury, shown during inflammation, is usually caused by stimuli like pathogens, different chemicals, etc. [8]. Lung inflammation is a kind of innate immune response that is directed in a coordinated expression of different inflammatory cytokines and immune cells aiming to clear the pathogenic agents [9]. Different proteases in lung regulate the activities of these inflammatory cytokines. Several studies have explored the role of proteases in different diseases [10]. Inflammatory cells like lymphocytes, neutrophils, macrophages, mast cells, as well as lung native cells like epithelial, endothelial, and fibroblasts act as the major protease source in lung [11–15]. Most of these proteases show profibrotic properties whereas some responses as anti-fibrotic components. The detailed properties of the proteases in regulation of IPF are described here.

2 Idiopathic Pulmonary Fibrosis (IPF): Overview of the Disease

Idiopathic pulmonary fibrosis (IPF) is a type of interstitial lung disease with progressive fibrotic development which has been studied from ancient time [16–18]. Modern-day science diagnoses this by surgical lung biopsy [19, 20], and the presence of patchy inflammatory cells, myofibroblasts, foci of proliferating fibroblasts, deposited collagen, and Pulmonary Function Test (PFT) [21, 22]. Recently, through some unique pathologic analysis, different types of pulmonary fibrosis have been described [23, 24]. Currently, IPF is one of the most common and severe forms of pulmonary fibrosis, with a very low rate of survival [5]. The two main biological processes, apoptosis of fibroblasts, proliferation and accumulation, breakdown of ECM are the cause behind the progression of this disease. Imbalance of ECM deposition, turn over and fibroblast proliferation, apoptosis may lead to pulmonary fibrosis. Reports have said that by circulating fibrocytes, monocytes, the bone marrow derived cells may produce the ECM-producing cells [25–28], or epithelial-mesenchymal transition (EMT) [29]. In pulmonary fibrosis, myofibroblast is the ultimate effector cell. In association with the lung fibroblasts, alveolar epitheliums also have been found to be involved in disease pathogenesis [30, 31]. The epithelial-mesenchymal interplay plays the pivotal role in pulmonary fibrosis, in which ECM gets accumulated due to coupling of the lung mesenchymal cells with the injured alveolar epithelium cells, and lung architecture alters.

Extensive studies have proved that among the cytokines, TGF-β plays a key role in stimulation of fibroblast proliferation and IPF progression [29]. Apart from that, the Epithelial Growth Factor Receptor (EGFR) and its ligand Amphiregulin (AR) have shown to induce the ECM deposition and myofibroblast proliferation [32]. It has been postulated that among the proteases, Neutrophil Elastase (NE) may be regulating the early stages of lung inflammation during the developmental stage of pulmonary fibrosis [33]. NE acts as a promising link between emphysema and

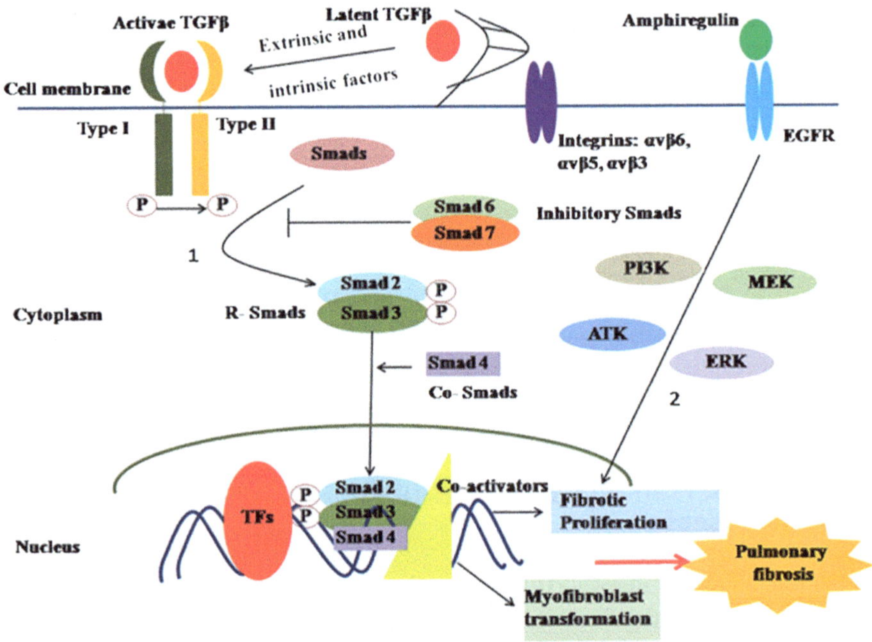

Fig. 1 Role of different extrinsic, intrinsic factors, cytokines, cell surface receptors, and different molecular pathways in regulation of pulmonary fibrosis. *1* TGFβ is the key molecule behind the fibrotic proliferation in lung that leads to IPF. TGFβ is an immunosuppressive cytokine that upon activation attaches to its receptor and activates the Smad-dependent pathway (canonical pathway) which results in activation of the profibrotic genes that later leads to pulmonary fibrosis. *2* Role of EGFR signaling on pulmonary fibrosis. Apart from TGFβ, EGFR is another important receptor that has shown impressive impact as a probable cause and therapeutic target of IPF. Amphiregulin, the EGFR ligand upon activation, phosphorylates the downstream proteins. This phosphorylation leads to activation of several Transcription factors that induce the fibrotic proliferation [32]

fibrosis [34]. But metalloproteases have established themselves as the most potent proteases that have pivotal role in the disease etiology (Fig. 1).

3 Role of Proteases in IPF

As already discussed, inflammatory cells, fibrotic foci, myofibroblasts, and collagen deposition are the notable histopathological features of IPF. These cells either secret or help secreting proteases that cause the irreversible figrogenic changes in the lungs. Tissue growth factor-β (TGF-β) stimulates fibroblast proliferation, and it has been observed that lung cells treated with (TGF-β) exhibit an increased activity of matrix metalloproteinase-2 (MMP-2) leading to progressive changes in the lungs. Another proteolytic enzyme, neutrophil elastase (NE), has also been demonstrated to be involved in the development of inflammation during pulmonary fibrosis. The

Table 1 Different types of proteases involved in Idiopathic Pulmonary Fibrosis (IPF)

Type of protease	Subtype	References
Serine protease	Neutrophil Elastase	Belaaouaj et al. [35]
Cysteine protease	Cathepsin	Lecaille et al. [77]
	ATG4B protease	Cabrera et al. [37]
Aspertic protease	Pepsin	Davis et al. [38]
	Napsin A	Samukawa et al. [39]
Metalloprotease	MMP-3	Peters et al. [41]
	MMP-7	Li et al. [46]
	MMP-8	Garcia-Verdugo et al. [48]
	MMP-9	Lee et al. [53]
	MMP-12	Kang et al. [62]
	MMP 13	McQuibban et al. [68]
	MMP 19	Yu and Stamenkovic [54]

recent understanding of the role of NE in emphysema highlights a link between emphysema and IPF and also indicates NE as a common possible inducer of these two diseases. Amphiregulin has also been found to induce IPF by regulating myofibroblast proliferation in the lungs. Proteases are now classified into seven families based on the nature of the catalytic residues. These are aspartic- cysteine-, serine-, metallo-, threonine-, glutamic-, and asparagines peptidase. Among these seven types of proteases, mainly four classes of proteases have been found in lungs that are predominantly supposed to impact in IPF (Table 1). Below we discuss the roles of different proteases in IPF.

I. Serine Protease

A. Neutrophil Elastase (NE)

Polymorphonuclear (PMN) cells have been found to contribute to the pathogenesis of IPF. NE is a neutrophil-derived serine protease that acts on a wide variety of substrates and has been found crucial in protecting against pulmonary infection [35]. NE promotes the proliferation of lung fibroblasts. In an experiment, it was found that at modest concentrations of NE fibroblast got proliferated. There are also some reports that show that NE also promotes myofibroblast differentiation.

II. Cysteine protease

A. Cathepsin

Cathepsin K (Cat K) expression gets enhanced in response to different fibrogenic particles The co-relation between Cat K response and the development of lung fibrosis was tested in an experiment. After one month of the experimental pulmonary fibrosis condition, Cathepsin K showed significant high level of elevation than the control group [36]. The expression level of Cat K was inversely associated with the fibrotic development in murine strains, suggesting the contribution of Cat K to limit lung fibrosis. Several other in vivo and in vitro studies prove the role of TGF-β in repressing the expression of Cat K in lung fibroblasts resulting in fibrosis development. The findings thus help to conclude Cat K as a potential collagenolytic mammalian protease.

B. **ATG4B protease**

ATG4B is another class of cysteine protease that has been recently investigated to modulate cellular events in pulmonary fibrosis. In a group of *Atg4b*-deficient mice treated with bleomycin, Cabrera and colleagues (2015) observed significantly higher inflammatory response, increased neutrophilic infiltration, and significant alterations in proinflammatory cytokines [37]. They also observed that *Atg4b* unsettlement resulted in increased apoptosis of the alveolar and bronchiolar epithelial cells leading to a progressive and irreversible damage to the lungs. Deregulated extracellular matrix-related gene expression and increased collagen accumulation are associated with these changes [37]. Therefore, ATG4B protease could have a protective role against bleomycin-induced fibrotic changes of the lungs.

III. **Aspertic protease**

A. **Pepsin**

Pepsin has been established as a novel biomarker for the diagnosis of IPF. Gastroesophageal reflux has been reported to jeopardize respiratory conditions. Pepsin, a proteolytic enzyme secreted by the gastric cells, has been found in BALF in patients with GERD and lung transplantation and asthma. Recently, higher BALF pepsin level has been found to associate with higher rate of exacerbation of the IPF patients [38].

B. **Napsin A**

Napsin A, aspartic protease, is mostly found in alveolar type-II cells and has been observed to impact significantly in conditions like lung cancer. In a study, it was observed that levels of napsin A in serum were higher among the IPF patients than the controls, and such elevated level also contributed to the severity of disease [39] (Fig. 2).

IV. **Metalloproteases**

A. **MMP-3**

Increased expression of MMP-1, 9, and 13 in IPF patients indicates collagenase activity of the proteins in different conditions [40]. MMP-3 (stromelysin-1) has been shown to have profibrotic activity as fibrotic change is reduced after bleomycin challenge in *Mmp*-$^{-/-}$ mice [41, 42].

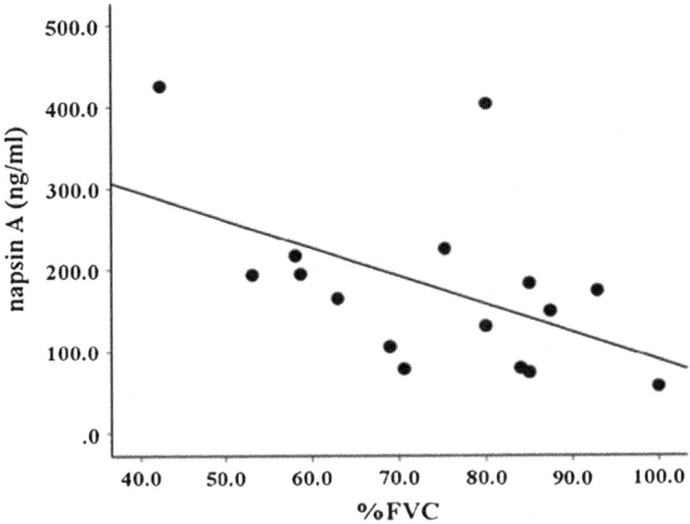

Fig. 2 Inverse correlation between napsin levels and lung function as measured by percent-predicted FVC (% FVC) ($n = 16$, $p < 0.05$) [reproduced from 39]

It is believed that the epithelial cells are pushed to an epithelial-mesenchymal transition by MMP-3, leading to the formation of myofibroblast-like cells in IPF. However, the exact role of MMP-3 has not been fully understood. Thus, the exact role of MMP-3 in promoting fibrosis through EMT remains an open question [43, 44].

B. **MMP-7**

MMP-7 or matrilysin is over-expressed in fibrosis than in healthy lung tissues [45, 46]. Alike MMP-3, MMP-7 has also been identified as profibrotic protease MMPs have been found to release chemokines [47]. According to a newly identified pathway, MMP-7 breaks down the carrier molecule of chemokine CXCL1, syndecan-1, and this cleavage in the glycosaminoglycan chain promotes the transepithelial migration of neutrophils [46]. Therefore, a possible neutrophil influx mechanism has been proposed in the development of IPF which is manifested by an up-regulated activity of MMP-7. Apart from its profibrotic nature, MMP-7 can have anti-fibrotic effects too [48].

C. **MMP-8**

MMP-8 has also been reported as profibrotic; however, its activity is opposite in liver tissues where MMP-8 serves as anti-fibrotic [49–52]. Although inactivation of interleukin IL-10 by MMP-8 was observed as a crucial step in IPF [49, 50], another contradictory report suggests that lung fibrosis is initiated by over-expression of IL-10 [53].

D. **MMP-9**

MMP-9 has also been shown to contribute to the development of IPF through a pathway involving the activation of TGFβ1 [54, 55]. In experimental models, although the occurrence of bleomycin-induced IPF was not substantially different between wild-type and MMP-9$^{-/-}$ mice [56], probability of bleomycin-induced fibrosis was lesser in transgenic mice in which MMP-9 expression in macrophages was regulated by a scavenger receptor class A promoter [57]. Still, the role of MMP-9 in fibrosis is still not clear. While MMP-9 was associated with mild fibrotic changes in one study [57], another study confirmed it as profibrotic [54], while others were unsure about its role [56, 58]. MMP-9 expression has been linked to chemokine-level modulation and leukocyte influx in allergic lung disease [59–62] indicating that more studies are warranted to elucidate its function in disease mechanism.

E. **MMP-12**

Most of the experimental fibrosis models have demonstrated MMP-12 as a profibrotic protease [63–66]; however, the role of this protein in augmenting IPF has remained contradictory. While one study reported reduced fibrotic changes in MMP-12 null mice after bleomycin challenge than the wild type [63], other study did not find any conclusive evidence of such change between the two genotypes [57, 67]. Although such contradictions can be explained by their study design and measurement time points, experimental results indicate that an MMP-12-driven effect in IPF may only be seen visible in an advanced stage. MMP-12 might have a role in the activation of fibroblast or collagen production. MMP-12 was also established as a profibrotic in another study where pulmonary fibrosis was induced by anti-Fas antibody (Jo2) [64]. The investigators observed that targeted deletion of MMP-12 significantly reduced inflammation and protected the animals against Fas-induced fibrosis. The investigators also observed that the expression of two other profibrotic genes, egr1 and cyr61, reduced in MMP-12 transgenics than the wild-type animals [64]. In another study, Madala and colleagues (2010) suggested that MMP-12 could suppress other anti-fibrotic genes, thus manifests its profibrotic activity. However, results of two studies were contradictory where the investigators reported no effect or anti-fibrotic or no effect of MMP-12 [68, 69]. Therefore, more investigation is required to confirm the role of MMP-12 in disease manifestation. Though it has been proposed as a profibrotic protease, there are contrasting findings proposing MMP-12 as anti-fibrotic protease [68]. However, a direct link is still missing between fibrotic mechanism and role of MMP-12.

F. **MMP-13**

The role of MMP-13 in fibrosis has been investigated over a long time. Flechsig and colleagues (2010) reported reduced inflammation in radiation exposed MMP-13 null mice. However, this observation was contradicted by the findings of Sen and colleagues (2010). They reported that fibrotic changes were almost similar in MMP-13 null and wild-type mice in induced-hyperoxia; however, they found an elevated inflammation in the MMP-13 null animals. MMP-13 also has roles in the metabolism of various chemokines; thus, it can induce the activity of the chemokines [70–73], and such altered activities of the chemokines could play pivotal role in the development of fibrosis and also to the severity of the disease.

G. **MMP-19**

Up-regulation of MMP-19 expression has been observed in the epithelial cells of IPF patients [55]. Fibrosis in MMP-19 null mice exhibited aggravated hydroxyproline and αSMA staining than wild-type mice which may indicate that MMP-19 could modulate disease progression. MMP-19 has been reported to exert an anti-fibrotic effect by modulating the cyclooxygenase (COX)-2 pathways that produce prostaglandin E_2 (PGE_2). PGE_2 has been found to minimize fibroblast proliferation, migration, collagen synthesis, and differentiation into myofibroblasts, thus suppress fibrosis [74–76] (Fig. 3).

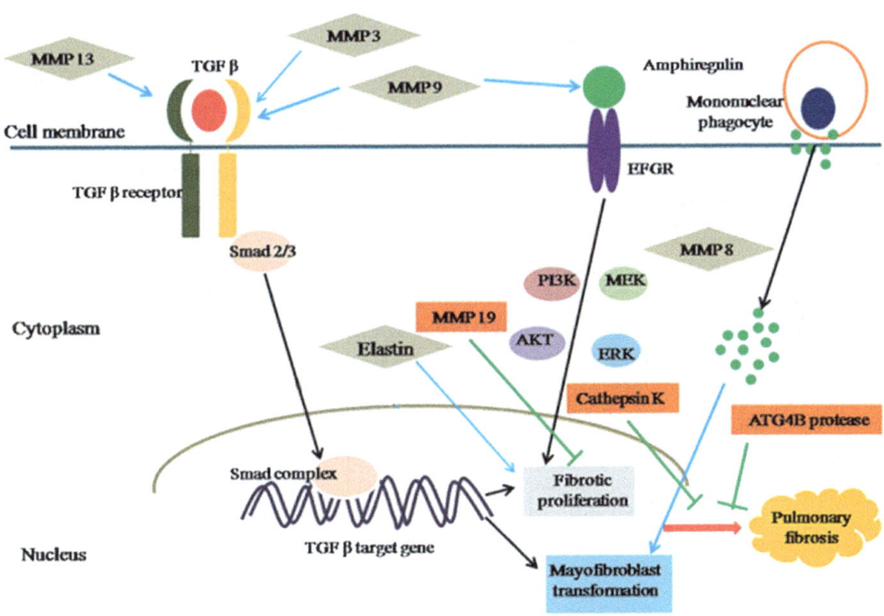

Fig. 3 Role of different proteases in IPF. MMP 9 has proved itself as a potent activator of Amphiregulin that activates the EGFR pathway whereas MMP 9 and MMP 13 and MMP 3 have shown to activate TGFβ. Mononuclear phagocyte releases MMP 8 and has positive regulatory effect on myofibroblast proliferation and ECM deposition. Increased Elastin level in lung during fibrotic condition has established itself as a novel biomarker for the disease. MMP 19 has shown to induce fibrotic proliferation of cell. Among all these profibrotic proteases, Cathepsin K and ATG4B protease negatively regulate pulmonary fibrosis

4 Protease Inhibitors: A Potent Target for Therapeutic Approach

Proteases are the potent profibrotic molecules, whereas some of the results show anti-fibrotic properties. The profibrotic proteases mostly show their effect by activating the latent TGFβ or IL-13 or by inducing the EGFR signaling cascade. They have also shown to block the anti-fibrotic protease functions resulting in imbalance of pro-/anti-fibrotic proteases in lung. Thus, targeting the profibrotic proteases may be beneficial for the attenuation of pulmonary fibrosis. There are some protease inhibitors that have been shown to exert anti-fibrotic response.

A. **ONO-5046**

As earlier discussed, NE is a potent profibrotic protease. ONO-5046 is known to be a potent antagonist of mouse and human NE. In an experiment, the later was used to see whether it could reduce pulmonary fibrosis or not. What was found is that ONO-5046 reduced lung fibrosis significantly. The results from hydroxyproline content and histological analysis proved that blocking NE attenuates experimental pulmonary fibrosis.

5 Conclusion

Pulmonary fibrosis is a diffused parenchymal lung disease with distinct radiographic, clinical, pathologic, and physiologic signs. However, till date, there is no proven and effective therapy to manage these patients. The recent discoveries of various protease-mediated pathways have been quite helpful in understanding the pathological conditions and mechanism of the disease. Modification of ECM, which is an inevitable and crucial aspect of identifying cases of IPF, has provided a ground zero as an effective treatment modality. Apart from that, molecules propagating anti-fibrotic proteases and protease inhibitors also seem to be a promising future therapy for the management of IPF and other lung diseases.

Acknowledgements The authors thank all the laboratory members for their kind support throughout the time and sharing their views and ideas during the scientific discussions. The authors also like to thank Department of Zoology, University of Calcutta, for providing the ideal environment to complete this work.

References

1. Richeldi L, Costabel U, Selman M, Kim DS, Hansell DM, Nicholson AG, Brown KK, Flaherty KR, Noble PW, Raghu G, Brun M, Gupta A, Juhel N, Kluglich M, de Bios RM (2011) Efficacy of a tyrosine kinase inhibitor in idiopathic pulmonary fibrosis. N Engl J Med 365:1079–1087

2. Hattori N, Degen JL, Sisson TH, Liu H, Moore BB, Pandrangi RG, Simon RH, Drew AF (2000) Bleomycin-induced pulmonary fibrosis in fibrinogen-null mice. J Clin Invest 106:1341–1350
3. Swaisgood CM, French EL, Noga C, Simon RH, Ploplis VA (2000) The development of bleomycin-induced pulmonary fibrosis in mice deficient for components of the fibrinolytic system. Am J Pathol 157:177–187
4. Gribbin J, Hubbard RB, Le Jeune I, Smith CJ, West J, Tata LJ (2006) Incidence and mortality of idiopathic pulmonary fibrosis and sarcoidosis in the UK. Thorax 61:980–985
5. Raghu G, Collard HR, Egan JJ, Martinez FJ, Behr J, Brown KK (2011) An official ATS/ERS/JRS/ALAT statement: idiopathic pulmonary fibrosis: evidence-based guidelines for diagnosis and management. Am J Respir Crit Care Med 183:788–824
6. Ley B, Collard HR, King TE (2011) Clinical course and prediction of survival in idiopathic pulmonary fibrosis. Am J Respir Crit Care Med 183:431–440
7. Sisson TH, Matriptase Spagnolo P (2016) Protease-activated receptor 2, and idiopathic pulmonary fibrosis. Further evidence for signaling pathway redundancy in this difficult-to-treat disease? Am J Respir Crit Care Med 193:816–817
8. Kundu S, Sengupta S, Chatterjee S (2009) Cadmium induces lung inflammation independent of lung cell proliferation: a molecular approach. J Inflamm 6:19. Exp Mol Pathol 92:287–295
9. Benabid R, Wartelle J, Malleret L (2012) Neutrophil elastase modulates cytokine expression: contribution to host defense against Pseudomonas aeruginosa-induced pneumonia. J Biol Chem 287:34883–34894
10. Cheronis JC, Repine JE (1993) Proteases, protease inhibitors, and protease-derived peptides: importance in human pathophysiology and therapeutics. Birkhauser, Basel, pp 2–25
11. Caughey GH (1994) Serine proteinases of mast cell and leukocyte granules: a league of their own. Am J Respir Crit Care Med 150:S138–S142
12. Shapiro SD (2002) Proteinases in chronic obstructive pulmonary disease. Biochem Soc Trans 30:98–102
13. Lieberman J (2003) The ABCs of granule-mediated cytotoxicity: new weapon in the arsenal. Nat Rev Immunol 3:361–370
14. Barnes PJ, Shapiro SD, Pauwels RA (2003) Chronic obstructive pulmonary disease: molecular and cellular mechanisms. Eur Respir J 22:672–688
15. Selman M, Pardo A (2003) The epithelial/fibroblastic pathway in the pathogenesis of idiopathic pulmonary fibrosis. Am J Respir Cell Mol Biol 29:S93–S97
16. Bjoraker JA, Ryu JH, Edwin MK (1998) Prognostic significance of histopathologic subsets in idiopathic pulmonary fibrosis. Am J Respir Crit Care Med 157:199–203
17. Takishima T, Shimura S (1994) Definition and classification of pulmonary fibrosis. In: Basic and clinical aspects of pulmonary fibrosis. CRC Press, Boca Raton, FL, pp 293–303
18. Orr CR, Jacobs WF (1926) Pulmonary Fibrosis. Radiology 7:318–325
19. Hamman L, Rich A (1944) Acute diffuse interstitial fibrosis of the lungs. Bull Johns Hopkins Hosp 74:177–212
20. Hamman L, Rich AR (1935) Fulminating diffuse interstitial fibrosis of the lungs. Trans Am Clin Climatol Assoc 51:154–163
21. Schwartz DA, Van Fossen DS, Davis CS (1994) Determinants of progression in idiopathic pulmonary fibrosis. Am J Respir Crit Care Med 149:444–449
22. King TEJ, Schwarz MI, Brown K (2001) Idiopathic pulmonary fibrosis. Relationship between histopathologic features and mortality. Am J Respir Crit Care Med 164:1025–1032
23. Scadding JG, Hinson KF (1967) Diffuse fibrosing alveolitis (diffuse interstitial fibrosis of the lungs). Correlation of histology at biopsy with prognosis. Thorax 22:291–304
24. Crystal RG, Fulmer JD, Roberts WC, Moss ML, Line BR, Reynolds HY (1976) Idiopathic pulmonary fibrosis. Clinical, histologic, radiographic, physiologic, scintigraphic, cytologic, and biochemical aspects. Ann Intern Med 85:769–788
25. Hashimoto N, Jin H, Liu T, Chensue SW, Phan SH (2004) Bone marrow-derived progenitor cells in pulmonary fibrosis. J Clin Invest 113:243–252

26. Bucala R, Spiegel LA, Chesney J, Hogan M, Cerami A (1994) Circulating fibrocytes define a new leukocyte subpopulation that mediates tissue repair. Mol Med 1:71–81
27. Kuwana M, Okazaki Y, Kodama H, Izumi K, Yasuoka H, Ogawa Y, Kawakami Ikeda Y (2003) Human circulating CD14+ monocytes as a source of progenitors that exhibit esenchymal cell differentiation. J Leukoc Biol 74:833–845
28. Postlethwaite AE, Shigemitsu H, Kanangat S (2004) Cellular origins of fibroblasts: possible implications for organ fibrosis in systemic sclerosis. Curr Opin Rheumatol 16:733–738
29. Willis BC, Liebler JM, Luby-Phelps K (2005) Induction of epithelial-mesenchymal transition in alveolar epithelial cells by transforming growth factor-β1: potential role in idiopathic pulmonary fibrosis. Am J Pathol 166:1321–1332
30. Coalson JJ (1982) The ultrastructure of human fibrosing alveolitis. Virchows Arch A Pathol Anat Histol 395:181–199
31. Kawanami O, Ferrans VJ, Crystal RG (1982) Structure of alveolar epithelial cells in patients with fibrotic lung disorders. Lab Invest 46:39–53
32. Lee Chang-Min, Park Jin Wook, Cho Won-Kyung, Zhou Yang, Han Boram, Yoon Pyoung Oh, Chae Jeiwook, Elias Jack A, Lee Chun Geun (2014) Modifiers of TGF-b1 effector function as novel therapeutic targets of pulmonary fibrosis. Korean J Intern Med 29:281–290
33. Taooka Y, Maeda A, Hiyama K (1997) Effects of neutrophil elastase inhibitor on bleomycin induced pulmonary fibrosis in mice. Am J Respir Crit Care Med 156:260–265
34. Lucattelli M, Cavarra E, de Santi MM (2003) Collagen phagocytosis by lung alveolar macrophages in animal models of emphysema. Eur Respir J 22:728–734
35. Belaaouaj A, McCarthy R, Baumann M (1998) Mice lacking neutrophil elastase reveal impaired host defense against gram negative bacterial sepsis. Nat Med 4:615–618
36. Lucattelli M, Bartalesi B, Cavarra E et al (2005) Is neutrophil elastase the missing link between emphysema and fibrosis? Evidence from two mouse models. Respir Res 6:83–96
37. Cabrera S, Maciel M, Herrera I, Nava T, Vergara F, Gaxiola M, López-Otín C, Selman M, Pardo A (2015) Essential role for the ATG4B protease and autophagy in bleomycin-induced pulmonary fibrosis. Autophagy 11:670–684
38. Davis CS, Mendez BM, Flint DV, Pelletiere K, Lowery E, Ramirez L, Love RB, Kovacs EJ, Fisichella PM (2013) Pepsin concentrations are elevated in the bronchoalveolar lavage fluid of patients with idiopathic pulmonary fibrosis after lung transplantation. J Surg Res 185:e101–e108
39. Samukawa T, Hamada T, Uto H, Yanagi M, Tsukuya G, Nosaki T, Maeda M, Hirano T, Tsubouchi H, Inoue H (2012) The elevation of serum napsin A in idiopathic pulmonary fibrosis, compared with KL-6, surfactant protein-A and surfactant protein-D. BMC Pulm Med 11(12):55
40. Fukada Y, Ishizaki M, Kudoh S, Kitaichi M, Yamanaka N (1998) Localization of matrix metalloproteinases-1, -2 and -9 and tissue inhibitor of metalloproteinase-2 in interstitial lung disease. Lab Invest 78:687–698
41. Peters CA, Freeman MR, Fernandez CH, Stephan J, Wiederschain DG, Moses MH (1997) Dysregulated proteolytic balance as the basis of excess extracellular matrix in fibrotic disease. Am J Physiol 272:1960–1965
42. Yamashita CM, Dolgonos L, Zemans RL, Young SK, Robertson J, Briones N, Suzuki T, Campbell MN, Gauldie J, Radisky DC et al (2011) Matrix metalloproteinase 3 is a mediator of pulmonary fibrosis. Am J Pathol 179:1733–1745
43. Gadek JE, Kelman JA, Fells G et al (1979) Collagenase in the lower respiratory tract of patients with idiopathic pulmonary fibrosis. N Engl J Med 301:737–742
44. O'Connor CM, Odlum C, van Breda A, Power C, FitzGerald MX (1988) Collagenase and fibronectin in bronchoalveolar lavage fluid in patients with sarcoidosis. Thorax 43:393–400
45. Zuo F, Kaminski N, Eugui E, Allard J, Yakhini Z, Ben-Dor A, Lollini L, Morris D, Kim Y, DeLustro B, Sheppard D, Pardo A, Selman M, Heller RA (2002) Gene expression analysis reveals matrilysin as a key regulator of pulmonary fibrosis in mice and humans. Proc Natl Acad Sci USA 99:6292–6297

46. Li Q, Park PW, Wilson CL, Parks WC (2002) Matrilysin shedding of syndecan-1 regulates chemokine mobilization and transepithelial efflux of neutrophils in acute lung injury. Cell 111:635–646
47. McQuibban GA, Gong JH, Wong JP (2002) Matrix metalloproteinase processing of monocyte chemoattractant proteins generates CC chemokine receptor antagonists with anti-inflammatory properties in vivo. Blood 100:1160–1167
48. Manicone AM, Huizar I, McGuire JK (2009) Matrilysin (Matrix Metalloproteinase-7) regulates anti-inflammatory and antifibrotic pulmonary dendritic cells that express CD103 (alpha (E)beta (7)-integrin). Am J Pathol 175:2319–2331
49. Garcia-Verdugo I, Descamps D, Chignard M (2010) Lung protease/anti-protease network and modulation of mucus production and surfactant activity. Biochimie 92:1608–1617
50. García-Prieto E, González-López A, Cabrera S, Astudillo A, Gutiérrez-Fernández A, Fanjul-Fernandez M, Batalla-Solís E, Puente XS, Fueyo A, López-Otín C, Albaiceta GM (2010) Resistance to bleomycin-induced lung fibrosis in MMP-8 deficient mice is mediated by interleukin-10. PLoS ONE 5:e13242
51. Craig VJ, Quintero PA, Fyfe SE, Patel AS, Knolle MD, Kobzik L, Owen CA (2013) Profibrotic activities for matrix metalloproteinase-8 during bleomycin-mediated lung injury. J Immunol 190:4283–4296
52. Siller-López F, Sandoval A, Salgado S, Salazar A, Bueno M, Garcia J, Vera J, Gálvez J, Hernández I, Ramos M, Aguilar-Cordova E, Armendariz-Borunda J (2004) Treatment with human metalloproteinase-8 gene delivery ameliorates experimental rat liver cirrhosis. Gastroenterology 126:1122–1133
53. Sun L, Louie MC, Vannella KM, Wilke CA, LeVine AM, Moore BB, Shanley TP (2011) New concepts of IL-10-induced lung fibrosis: fibrocyte recruitment and M2 activation in a CCL2/CCR2 axis. Am J Physiol Lung Cell Mol Physiol 300:L341–L353
54. Lee WL, Downey GP (2001) Leukocyte elastase: physiological functions and role in acute lung injury. Am J Respir Crit Care Med 164:896–904
55. Yu Q, Stamenkovic I (2000) Cell surface-localized matrix metalloproteinase-9 proteolytically activates TGF-beta and promotes tumor invasion and angiogenesis. Genes Dev 14:163–176
56. Betsuyaku T, Fukuda Y, Parks WC, Shipley JM, Senior RM (2000) Gelatinase B is required for alveolar bronchiolization after intratracheal bleomycin. Am J Pathol 157:525–535
57. Cabrera S, Gaxiola M, Arreola JL, Ramírez R, Jara P, D'Armiento J, Richards T, Selman M, Pardo A (2007) Overexpression of MMP9 in macrophages attenuates pulmonary fibrosis induced by bleomycin. Int J Biochem Cell Biol 39:2324–2338
58. Kaviratne M, Hesse M, Leusink M, Cheever AW, Davies SJ, McKerrow JH, Wakefield LM, Letterio JJ, Wynn TA (2004) IL-13 activates a mechanism of tissue fibrosis that is completely TGF-beta independent. J Immunol 173:4020–4029
59. Van Den Steen PE, Wuyts A, Husson SJ, Proost P, Van Damme J, Opdenakker G (2003) Gelatinase B/MMP-9 and neutrophil collagenase/MMP-8 process the chemokines human GCP-2/CXCL6, ENA-78/CXCL5 and mouse GCP-2/LIX and modulate their physiological activities. Eur J Biochem 270:3739–3749
60. Van den Steen PE, Proost P, Wuyts A et al (2000) Neutrophil gelatinase B potentiates interleukin-8 tenfold by aminoterminal processing, whereas it degrades CTAP-III, PF-4, and GROalpha and leaves RANTES and MCP-2 intact. Blood 96:2673–2681
61. Corry DB, Kiss A, Song LZ, Song L, Xu J, Lee SH, Werb Z, Kheradmand F (2004) Overlapping and independent contributions of MMP2 and MMP9 to lung allergic inflammatory cell egression through decreased CC chemokines. FASEB J 18:995
62. Greenlee KJ, Corry DB, Engler DA, Matsunami RK, Tessier P, Cook RG, Werb Z, Kheradmand F (2006) Proteomic identification of in vivo substrates for matrix metallopro-teinases 2 and 9 reveals a mechanism for resolution of inflammation. J Immunol 177:7312–7321
63. Kang T, Yi J, Guo A (2001) Subcellular distribution and cytokine-and chemokine regulated secretion of leukolysin/MT6-MMP/MMP-25 in neutrophils. J Biol Chem 276:21960–21968

64. Matute-Bello G, Wurfel MM, Lee JS, Park DR, Frevert CW, Madtes DK, Shapiro SD, Martin TR (2007) Essential role of MMP-12 in Fas-induced lung fibrosis. Am J Respir Cell Mol Biol 37:210–221
65. Garbacki N, Di Valentin E, Piette J, Cataldo D, Crahay C, Colige A (2009) Matrix metalloproteinase 12 silencing: a therapeutic approach to treat pathological lung tissue remodeling? Pulm Pharmacol Ther 22:267–278
66. Madala SK, Pesce JT, Ramalingam TR, Wilson MS, Minnicozzi S, Cheever AW, Thompson RW, Mentink-Kane MM, Wynn TA (2010) Matrix metalloproteinase 12-deficiency augments extracellular matrix degrading metalloproteinases and attenuates IL-13-dependent fibrosis. J Immunol 184:3955–3963
67. Manoury B, Nenan S, Guenon I, Boichot E, Planquois JM, Bertrand CP, Lagente V (2006) Macrophage metalloelastase (MMP-12) deficiency does not alter bleomycin-induced pulmonary fibrosis in mice. J Inflamm (Lond) 3:2
68. Lanone S, Zheng T, Zhu Z et al (2002) Overlapping and enzyme-specific contributions of matrix metalloproteinases-9 and -12 in IL-13-induced inflammation and remodeling. J Clin Investig 110:463–474
69. England KA, Price AP, Tram KV, Shapiro SD, Blazar BR, Panoskaltsis-Mortari A (2011) Evidence for early fibrosis and increased airway resistance in bone marrow transplant recipient mice deficient in MMP12. Am J Physiol Lung Cell Mol Physiol 301:L519–L526
70. McQuibban GA, Gong JH, Tam EM, McCulloch CA, Clark-Lewis I, Overall CM (2000) Inflammation dampened by gelatinase A cleavage of monocyte chemoattractant protein-3. Science 289:1202–1206
71. Parks WC, Wilson CL, López-Boado YS (2004) Matrix metalloproteinases as modulators of inflammation and innate immunity. Nat Rev Immunol 4:617–629
72. Van Lint P, Libert C (2007) Chemokine and cytokine processing by matrix metalloproteinases and its effect on leukocyte migration and inflammation. J Leukoc Biol 82:1375–1381
73. Gill SE, Parks WC (2008) Metalloproteinases and their inhibitors: regulators of wound healing. Int J Biochem Cell Biol 40:1334–1347
74. Kolodsick JE, Peters-Golden M, Larios J, Toews GB, Thannickal VJ, Moore BB (2003) Prostaglandin E2 inhibits fibroblast to myofibroblast transition via E. prostanoid receptor 2 signaling and cyclic adenosine monophosphate elevation. Am J Respir Cell Mol Biol 29:537–544
75. White ES, Atrasz RG, Dickie EG, Aronoff DM, Stambolic V, Mak TW, Moore BB, Peters-Golden M (2005) Prostaglandin E(2) inhibits fibroblast migration by E-prostanoid 2 receptor-mediated increase in PTEN activity. Am J Respir Cell Mol Biol 32:135
76. Huang SK, White ES, Wettlaufer SH, Grifka H, Hogaboam CM, Thannickal VJ, Horowitz JC, Peters-Golden M (2009) Prostaglandin E(2) induces fibroblast apoptosis by modulating multiple survival pathways. FASEB J 23:4317–4326
77. Lecaille F, Choe Y, Brandt W, Li Z, Craik CS, Brömme D (2002) Selective inhibition of the collagenolytic activity of human cathepsin K by altering its S2 subsite specificity. Biochemistry 41(26):8447–8454

Protease-Activated Receptor Signaling in Lung Pathology

Ganapasam Sudhandiran, Srinivasan Kalayarasan, Thomas Divya and Bakthavatchalam Velavan

Abstract

Protease-activated receptors (PARs) are self-activated G-protein-coupled receptors that have diverse roles in several disease paradigms including neurodegeneration, cancer, cardiovascular diseases, and others. Recently, extensive research on PAR family and its effect in modulating signaling pathways have gained attention. There is evidence that PARs are expressed in the airways in a variety of cell types that are relevant to inflammatory lung diseases, and activation of these receptors might be linked to significant pathological changes. Thus, PARs are exciting targets in lung disease research. This chapter mainly focuses on the role of PAR family members in several lung diseases. In this context, modulation of PAR signaling might open novel avenues in the treatment interventions in a number of respiratory conditions at least in part to reduce the burden of the diseases.

Keywords

Proteases · Protease-activated receptor · Lung
Pulmonary fibrosis · Inflammation

1 Introduction

The lungs are one of the most affected organs during infection and oxidative stress due to its large surface area and frequent exposure to oxygen. Since lungs are highly vascular organ, the functions of lungs are greatly influenced by blood coagulation

G. Sudhandiran (✉) · S. Kalayarasan · T. Divya · B. Velavan
Cell Biology Laboratory, Department of Biochemistry, University of Madras, Guindy Campus, Chennai, India
e-mail: sudhandiran@yahoo.com

S. Chakraborti and N.S. Dhalla (eds.), *Pathophysiological Aspects of Proteases*,
DOI 10.1007/978-981-10-6141-7_23

cascade. In addition to the exchange of gases between lungs and blood, they also function in filtering the venous blood which prevents blood clots and other fibers from entering the arterial circulation [1]. Proteases play an important role in the systemic coagulation cascade as the series of events during blood coagulation end up in the formation of thrombin, a serine protease. The formation of thrombin is critical during coagulation, and it catalyzes the conversion of fibrinogen to fibrin to form a clot. Hence, the serine protease thrombin is considered as the central enzyme which regulates the entire coagulation system [2]. Thrombin is generated early during acute lung injury and is thought to contribute to increased lung vascular permeability [3].

While the involvement of thrombin is crucial for the progression of lung disorders, it is interesting to know how it communicates to cells. Protease-activated receptors (PARs) provide an answer that how thrombin produces signals. PARs were first identified in the year 1991 as a class of seven trans-membrane domain G-protein-coupled receptors (GPCRs). The inactive GPCRs are activated by ligand receptor interactions which are furnished by proteolytic cleavage [4, 5]. Almost all eukaryotic cell types including epithelial cells, endothelial cells, fibroblast express PARs which regulate important physiological processes, and hence PARs are emerging as novel therapeutic target for various diseases [6–8]. Moreover, PARs are reported to play major role in airway inflammation as it is distributed throughout the airways. Endogenous or exogenous airway proteases highly regulate the activities of airway PARs by either activating or inhibiting the receptors. Elevated levels of PAR levels are associated with pathological conditions in the lung, and its role in developing lung diseases is extensively being investigated [9–11]. A thorough understanding of PAR-mediated cellular events is considered as one of the novel therapeutic approach targeted to prevent lung diseases and is the principal focus of this chapter.

2 The PAR Family Members

Till now, four different types of PARs have been identified named PAR-1 to PAR-4. Among these PAR1, PAR3, and PAR4 are thrombin receptors. PAR2 is an exceptional; serine proteases such as trypsin, tryptase as well as coagulation factors VIIa and Xa activate PAR2 [12, 13]. PAR1 is detected in fibroblasts, T cells, monocytes, hematopoietic progenitor cells, natural killer cells, smooth muscle cells, mast cells, neurons, glial cells, and epithelial cells [14]. Thrombin activates PAR1 by cleaving its extracellular amino terminus domain. Other proteases such as coagulation factor Xa, matrix metalloprotease-1 (MMP-1), and activated protein C (APC) can also activate PAR1. Out of these four proteases, PAR1 has been recognized as the major receptor that mediates pro-fibrotic and pro-inflammatory effects of thrombin [15, 16]. Initially trypsin was identified as the enzyme to activate PAR2 [17]. Since, gastrointestinal (GI) tract highly expresses PAR2 as well as trypsin; early research focus was on investigating the role of the PAR2 in diseases associated with GI tract only [18]. However, several further studies identified that mast cell tryptase and other proteases activate PAR-2, which elaborate research studies on the role of PAR2 in a wide range of other adult tissues, especially, kidney, heart, vascular endothelium, and lung [19]. The discovery of PAR3 was quite interesting. A research group has suggested the presence of further receptors for thrombin other than PAR1. They have derived platelets from PAR1-deficient mice and have analyzed its response to thrombin and it was found that the platelets were responsive to thrombin [20]. This report strongly recommends the existence of PAR3 as receptors for thrombin. The occurrence of PAR3 in mouse bone marrow, spleen, and human tissues has been reported [21]. PAR4 was identified and cloned in the year 1998 by Kahn et al. They have reported that, while PAR3 is essential for normal thrombin responses in rat platelets, still another PAR4-mediated pathway also exists for thrombin signaling [22]. Though PARs are differing in their mechanism of activation, all PAR genes are reported to have the same structure with two exons and two introns. The similar gene arrangements in PARs propose the theory that these protease receptors rose from a common ancestral gene and suggest the existence of a common gene family [23].

3 Mechanism of Activation and Signaling of PARs- General Aspects

The mechanism behind the activation of PARs was primarily recognized for PAR1 [4]. The similar paradigm was found to be associated with the activation of other PARs also. The irreversible proteolytic activation of PARs initiate by the cleavage of the amino-terminal domain of the receptor. This proteolytic cleavage produces a new amino terminus end that functions as a ligand for the receptors to initiate further trans-membrane signaling [24]. However, the mechanism of PAR3 activation differs from this model. Ishihara et al. have demonstrated the mechanism of activation of PAR3 in COS-7, kidney fibroblast cells and suggest that initial cleavage at ectodomain of PAR3 alone is sufficient to activate it [21]. Interestingly, Nakanishi et al. have reported that murine PAR3 act as a co-factor for thrombin-mediated cleavage and activation of PAR4 [25]. Studies show that when compared with PAR1 or PAR3; PAR4 requires increased thrombin level for its activation. The reason is, thrombin highly binds to the carboxyl-terminal sequence of PAR1 or PAR3 receptors which is absent in PAR4 exodomain [26]. Apart from these, the interesting fact is that activation of PARs is an irreversible mechanism and thus the regulatory mechanisms of PAR signaling are critical determinants of the protease response in cells.

4 PAR Activation and Signaling Is Tissue Specific?

An important question that could arise is whether PAR signaling is tissue specific or not. Different proteases can activate PARs under various conditions. The hypothesis behind this mechanism is that, when proteases cleave PARs at the same sites it activates same signaling cascades with common patho-physiological outcomes. However, when proteases cleave PARs at distinct sites, the consequences of PAR cleavage could be varying. In this way, the activation of PARs by proteases may differ in normal and pathological conditions depending on the site of protease cleavage. The outcome of PAR signaling or activation of PARs by same protease is also tissue specific and can vary between tissues. For example, Ku et al. have described a distinct outcome of PAR1 activation in endothelial as well as vascular smooth muscle cells. They have demonstrated that activation of PAR1 by thrombin results in relaxation of the coronary artery. However, removal of endothelium results in contraction. This could possibly due to different signaling mechanisms by endothelial versus vascular smooth muscle cells [27].

5 PAR Signaling in the Lung

All four subtypes of PARs appear to be expressed in human respiratory tract. PARs are present in alveoli, fibroblasts, smooth muscle cells, epithelial cells, endothelial cells, goblet cells, as well as nerves within the lungs [6, 9]. Any internal or external inflammatory stimuli may enhance the expression of PARs, as well as proteases that activate PARs. Among the four PARs, PARs1, and 2 are reported to have a major role in inflammatory and fibro proliferative processes in the lungs. This hypothesis is supported by a number of evidences that increased expressions of PARs1 and 2 are observed during the development of inflammatory-mediated pulmonary diseases such as asthma, bronchitis, and pulmonary fibrosis [10, 11, 28]. Various proteases that present in the airways activate PARs. For example, PAR2 is activated by mast cell tryptase; PAR1, 2, and 4 are activated by trypsin; PAR1 is specifically activated by chymase; Cathepsin G activates PAR4. Active research is currently progressing on the regulation of PAR activity which helps to define potential therapeutic approaches for the prevention of respiratory diseases.

6 'Double Edged Sword' Nature of PARs in the Airways

An increased expression of PARs 1 and 2 has been well documented in a number of inflammatory and fibrotic lung diseases. However, the argument on the bronchoprotective nature of PARs still exists. There are evidences for protective role of PAR2 as it possesses anti-inflammatory effect following activation. A research group has reported that PAR2-activating peptide protects against bronchoconstriction induced by histamine in guinea-pig model [29]. Another research group has demonstrated that activation of PAR2 provides protection against bronchoconstriction induced by 5-Hydroxytryptamine in rats [30].

On the other hand, evidences show that PAR2 stimulation results in pro-inflammatory responses, bronchial hyper-responsiveness, and contribute to allergic inflammation of the airway in mice over expressing PAR-2 [31]. Activation of PARs (PAR1, 2, and 4) in human respiratory epithelial cells is associated with Interleukin-6 (IL-6), IL-8, and prostaglandin E2 release [32]. Vliagoftis et al. have reported the important role of PAR2 in the regulation of matrix metalloproteinase-9 (MMP-9) in airway epithelial cells and demonstrated as critical elements in tissue reorganization during asthma and other inflammatory conditions in the respiratory system [33]. In addition, activation of PAR2 can directly produce airway contraction and may facilitate hyper-responsiveness in guinea pig airways [34]. PAR activation may also play a role in remodeling through the stimulated release of MMP-9 and procollagen from bronchial epithelial cells and lung fibroblasts, respectively [33]. Activation of PARs 1 and 2 in epithelial cells and fibroblasts has also been reported to stimulate the secretion of chemokines including IL-8, eosinophil survival promoting factors, granulocyte-macrophage colony-stimulating factor, as well as other pro-inflammatory and fibrotic mediators [35–37]. Taken as a whole, these observations point out that PAR activation may alternately stimulate inflammatory cell activity and play an important role in the progression of inflammatory and fibrotic lung diseases. The possible role of PARs in various lung diseases is discussed below.

7 PAR Signaling in Lung Diseases

7.1 Potential Role of PARs in Asthma

Asthma is a chronic inflammatory airway disease, characterized by inflammation of the airway epithelium which results in airway obstruction and bronchial hyper-reactivity reactions [38]. This disease has been postulated as a consequence of repeated injury to the mucosal surfaces of the lung. Both endogenous as well as exogenous proteases play major role in the development of asthma. Endogenous proteases such as thrombin, trypsin, tryptase, specific allergens, and serine proteases (Der p 3, 6, and 9), have been reported to regulate airway epithelial cell function during the progression of asthma [39–41]. Indeed, Terada et al. have found an increased level of thrombin in bronchoalveolar lavage (BAL) fluid along with other inflammatory mediators such as IL-5 and growth factors such as transforming growth factor-β (TGF-β) in asthmatic patients [42]. Since PAR1 can be activated by thrombin and increased concentration of thrombin in airway of asthmatic patients indicate that PAR1 plays a major role in the development of asthma. Similarly, increased expression of PAR2 has also been reported in bronchial epithelium of asthma patients [43]. A research group has detected the presence of PAR2 in airway epithelial cells and administration of selective PAR2 agonist through nasal cavity stimulated macrophage infiltration into BALF in ovalbumin challenged mice [31]. Moreover, it was found that inhibition of PAR2 suppresses airway hyper-responsiveness induced by ovalbumin and over expression of PAR2 worsening this conditions [31]. The possible reason could be that the allergens which express several proteases can activate PAR2 and thereby facilitate inflammatory reactions. This was supported by an interesting study in which cockroach extract used as a challenging antigen, and the investigators revealed that the allergic airway sensitization to cockroach extract induces eosinophilic airway inflammation which depends on the ability of cockroach extract to activate PAR2 [44, 45]. Yet another group has reported the expression of all the four PARs (PAR1–4) on epithelium as well as smooth muscle in human bronchial biopsy samples from patients with asthma. PAR1 and 3 were predominately present within the columnar epithelial cells, whereas PAR2 and 4 were more widely diffused throughout airways. The investigators have further concluded an increased expression of PAR2 in bronchial epithelium and airways that might play a major role in airway inflammatory response [46].

Taken together, these reports suggest that activation of PARs especially PAR2 may enhance airway hyper-responsiveness and airway inflammation in allergic reactions and asthma. Various proteases such as mast cell tryptase and proteases present in the allergens are able to cleave PAR2 and represent one of the mechanisms for the aggravation of the disease.

7.2 Role of PARs in Chronic Obstructive Pulmonary Disease

A growing attention on the role of PAR signaling exists in chronic obstructive pulmonary disease (COPD) because, like asthma, COPD also associated with airway remodeling and activation of the coagulation cascade. Cigarette smoking is certainly the predominant risk factor for COPD [47]. One interesting fact is that PAR1 signaling promotes goblet cell metaplasia and excessive mucus production which is a hallmark feature of COPD and sheds light on the involvement of PAR signaling in COPD. In addition, PAR1 activation by thrombin has been shown to endorse airway remodeling by enhancing the expression of TGF-β1 in ovalbumin-allergic rats [48]. Another group has reported that there is no difference in the expression of PAR2 in bronchial biopsy of patients with COPD [49]. In contrast, Miotto et al. have reported an increased expression of PAR2 in the respiratory tract of COPD patients [50]. BAL from COPD patients contains increased level of trypsin [51]. Though COPD patients have enhanced coagulant activity, increased thrombin levels have not been reported in BAL from patients with COPD [52]. These studies indirectly help to find out the hypothetical involvement of PAR2-activating enzymes in COPD. However, further studies are warranted to elucidate the mechanisms involved in detail.

7.3 Potential Role of PARs in Pulmonary Fibrosis

Pulmonary fibrosis (PF) is a chronic obstructive pulmonary disease characterized by the excessive accumulation of extracellular matrix components within the pulmonary interstitium [53]. Several processes such as injury to the alveolar epithelial and endothelial cells, activation and proliferation of fibroblasts, epithelial-to-mesenchymal transition, inflammatory cell influx, hypercoagulation, collagen accumulation, angiogenesis, and aberrant repair process leading to fibrogenesis of lungs [54–56]. Following tissue injury, one of the initial events occur in lungs is the activation of the coagulation cascade. Activation of coagulation cascade during the development of several fibrotic lung diseases such as systemic sclerosis [57], idiopathic pulmonary fibrosis (IPF) [58], and pneumonitis [59] have been reported. The incidence of increased thrombin levels in BALF from patients with PF associated with systemic sclerosis was illustrated nearly two decades ago itself [60]. The occurrence of high thrombin level in premature infants with chronic lung diseases was also reported [61]. The biological effect of thrombin during obstructive pulmonary diseases might be an indirect mechanism through the production of inflammatory mediators and fibroproliferative growth factors, which further activate PAR1 fibroblasts. In support of this, Mercer et al. have reported that PAR1 knockout mice are protected from the pathological consequences of bleomycin (BLM)-induced lung inflammation and fibrosis [62]. Studies involving primary human lung fibroblasts revealed that PAR1 exerts potent fibrogenic effects, including promoting platelet derived growth factor (PDGF)-mediated fibroblast proliferation and increased collagen synthesis. Furthermore, in animal models of fibrotic lung injury, the same research group has shown that BLM-induced inflammatory cell infiltration and lung collagen accumulation were reduced in PAR1 knockout mice [28]. TGF-β1 is considered as a notorious fibrotic mediator in multiple fibrotic conditions by promoting myofibroblast differentiation and extracellular matrix deposition [63]. Studies show that PAR1 contribute to fibrogenesis through the activation of TGF-β1 in fibroblasts and alveolar epithelial cells [64].

The crucial role of PAR1 in the progression of PF is well studied. However, limited reports are available on the involvement of other PARs in this disease model. In this context, we have assessed the role of PAR2 in BLM-induced experimental fibrosis model. RT-PCR, Confocal microscopic analysis, and Western blotting were performed to examine the expression of PAR2 in rat lungs administered with BLM. Normal rat exhibits little expression of PAR2. However, after BLM instillation the expression of PAR2 was markedly increased in rat lungs. Our findings suggested that PAR2 is involved in two different ways in the progression of PF. One mechanism is the MMP2/9 and cytokine-mediated fibroblast proliferation [11]. The second mechanism is PAR2-mediated epithelial cell apoptosis which involves p38 MAPK, c-JNK, and caspase-3. We showed evidences that organosulfur compound diallyl sulfide provide protection against BLM-induced lung fibrosis through the attenuation of PAR2 [11]. Further, we have shown evidence that PAR2 induces inflammation and apoptosis in lungs of rats administered with BLM [65]. Schematic representation of the PAR-mediated events taking place

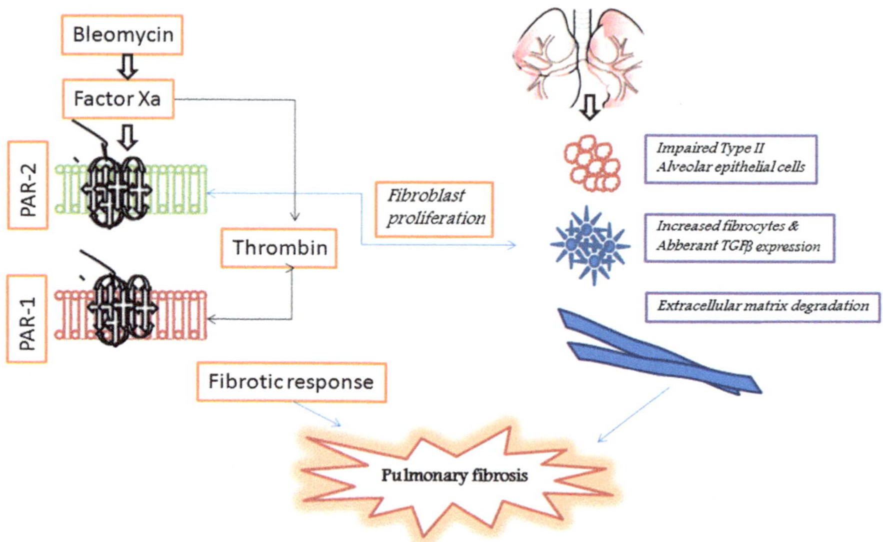

Fig. 1 PAR-mediated events in the progression of pulmonary fibrosis

in PF is depicted in Fig. 1. Available reports suggest that PAR4 may also have a fibroproliferative role. Stimulation of PAR with thrombin or a synthetic PAR4 agonist peptide induces morphological changes in primary cultured alveolar epithelial cells and A549 cells. PAR4 stimulation decreases E-cadherin expression and increases the expression of mesenchymal markers. This report indicates that PAR4 induces epithelial-to-mesenchymal transition which is a key mechanism in the progression of PF [66]. Together, PAR signaling is a promising mechanism in PF research and could be a better target for the regulation of this disease.

7.4 Role of PARs in Acute Lung Injury

Acute lung injury (ALI) is a disease condition occurs as a result of various diseases such as pneumonia, acute pancreatitis, sepsis, and trauma. The alveolar damage during ALI leads to the formation of lung edema. Injury to the alveolar cells and damage of the epithelial barrier enhance the fluid accumulation in the alveolar and interstitial space of the lungs [67]. Though the inflammatory and coagulation responses can be considered as protective mechanism to the host especially in the context of infection, the excessive coagulation and inflammation can lead to extreme tissue injury with disruption of the alveolar-endothelial capillary barrier which leads to the accumulation of protein-rich fluid in the alveolar spaces [68]. Available report suggests that PAR2 does not have a role in the development of acid-induced experimental acute lung injury in mice [69]. However, one report shows that mast cell tryptase mediates ALI through the activation of PAR2 to produce IL-8 in experimental animals [70]. In contrast, reports recommend that PAR1 signaling influences several key features of ALI including neutrophil infiltration, alveolar leakage in BLM-induced lung injury in mice through the activation of TGF-β1 [71].

7.5 Role of PARs in Bronchitis

Bronchitis is a respiratory disease in which the mucus membrane in the lungs' bronchial passages becomes inflamed. This airway inflammation is one of the three elementary components of all airway diseases, the other two are airflow obstruction and airway hyper-responsiveness [72]. The very first event of the inflammatory reaction starts upon the activation of resident cells in the damaged tissue. Among these, mast cells and macrophages play a predominant role by releasing inflammatory mediators, such as cytokines, nitric oxide (NO), prostaglandins, histamine, and serotonins. Since mast cells express PAR1 and PAR2, these two gained much attention in inflammatory lung diseases where the activation of these two receptors leads to mast cell activation and subsequent release of inflammatory mediators [73]. Several research studies have illustrated the pro-inflammatory role of PARs in the respiratory system. For instance, increased secretion of eosinophils and eotaxin content were observed in BALF from PAR2 deficient mice even after sensitization with ovalbumin, whereas wild-type mice exhibited increased infiltration of eosinophils in BALF after ovalbumin induction [31, 74]. Another report suggests a neuropeptide dependant mechanism of PAR2-mediated inflammation, which shown evidences that removal of sensory neurons significantly decreased the PAR2-mediated airway constriction, pulmonary inflammation, and edema [75]. In contrast to the pro-inflammatory role, anti-inflammatory effect of PARs has also been reported. PAR1-activating peptides (APs) cause prostaglandin E2 (PGE2) release which has a role in limiting the immune/inflammatory response especially in the lungs [76], from human bronchial epithelial cells [32] and human lung fibroblasts [77], suggest anti-inflammatory role of PARs in the lungs. In support with these findings, D'Agostino et al. have reported that PAR2-APs significantly inhibit bronchoconstriction, airway hyper-responsive reactions, and immune response induced by allergens in rabbits [78]. Together, more research on the possible role of PARs whether pro-inflammatory or anti-inflammatory, in airway inflammation and bronchitis helps to tone down the severity of the disease condition.

7.6 Possible Role of PARs in Lung Cancer

Lung cancer is one of the leading cause of cancer death worldwide and is the second most common cancer overall [79]. The important role of PARs in tumor development and metastasis has been suggested in late 1990s. Among four, PAR1 has been reported to play major role in cancer metastasis. A research group has reported that inhibition of PAR1 ligation at the site of tumor helps to regulate tumor metastasis [80].

Another group has reported the importance of thrombin receptor in cancer invasiveness [81]. Subsequently, Boire et al. have found that, MMP-1 cleaves PAR1 at the proper site to generate PAR1-dependent Ca^{2+} signals to enhance migration and tumor invasion [82]. Cisowski et al. have reported that lung cancer cell lines express PAR1 at high concentration than other group of PARs. Primary cell cultures isolated from lung cancer patients shown 40-fold higher expressions of PAR1 and exhibit increased migratory capacity than epithelial cells cultured from nonmalignant lungs. Experiments performed with cells silencing PAR1 expression revealed PAR1 as a possible target for therapy in lung cancer [83]. In continuation, another research group has reported that activation of PAR2 prevents apoptosis of lung cancer cells. They showed evidences that tryptase activates PAR2 to regulate the expression of epidermal growth factor receptor (EGFR) and to prevent apoptotic cell death of cancer cells [84]. Yet another group has identified the mechanism behind the pro-invasive property of PAR1 and has reported that the cooperation between PAR1 and integrins via its cytoplasmic tail mediates tumor invasion, therefore, PAR1-integrin complex may represent an important therapeutic target to prevent tumor invasion during lung cancer [85].

8 Conclusion

In summary, PAR family of receptors are emerging mediators of a wide range of cellular and molecular events during the development and progression of several lung diseases. The development of antagonists for PAR activation still remains to be established. Modulation of PAR signaling pathways may represent a novel therapeutic approach for the prevention of number of respiratory diseases to reduce the burden of disease condition.

Acknowledgements The authors thank Indian Council of Medical Research for supporting a grant in Protease-activated receptor signaling in pulmonary fibrosis (52/21/2008-BMS).

References

1. Fogarty AW, Lewis SA, McKeever TM, Lowe GD, Clark L, Britton J (2010) The association between blood coagulation activity and lung function: a population-based study. PLoS ONE 5 (11):e15014
2. Lane DA, Philippou H, Huntington JA (2005) Directing thrombin. Blood 106(8):2605–2612
3. Kipnis E, Guery BP, Tournoys A, Leroy X, Robriquet L, Fialdes P, Neviere R, Fourrier F (2004) Massive alveolar thrombin activation in pseudomonas aeruginosa-induced acute lung injury. Shock 21(5):444–451
4. Vu TK, Hung DT, Wheaton VI, Coughlin SR (1991) Molecular cloning of a functional thrombin receptor reveals a novel proteolytic mechanism of receptor activation. Cell 64(6):1057–1068

5. Coughlin SR (2000) Thrombin signaling and protease-activated receptors. Nature 407 (6801):258–264
6. Ramachandran R, Hollenberg MD (2008) Proteinases and signalling: pathophysiological and therapeutic implications via PARs and more. Br J Pharmacol 153(Suppl 1):S263–S282
7. Kawabata A, Nishikawa H, Saitoh H, Nakaya Y, Hiramatsu K, Kubo S, Nishida M, Kawao N, Kuroda R, Sekiguchi F, Kinoshita M, Kakehi K, Arizono N, Yamagishi H, Kawai K (2004) A protective role of protease-activated receptor 1 in rat gastric mucosa. Gastroenterology 126 (1):208–219
8. Holzhausen M, Spolidorio LC, Ellen RP, Jobin MC, Steinhoff M, Andrade-Gordon P, Vergnolle N (2006) Protease-activated receptor-2 activation: a major role in the pathogenesis of porphyromonas gingivalis infection. Am J Pathol 168(4):1189–1199
9. Sokolova E, Reiser G (2007) A novel therapeutic target in various lung diseases: airway proteases and protease-activated receptors. Pharmacol Ther 115(1):70–83
10. Lee NR, Baek SY, Gu A, da Kim H, Kim SY, Lee JS, Kim IS (2016) House dust mite allergen suppresses neutrophil apoptosis by cytokine release via PAR2 in normal and allergic lymphocytes. Immunol Res 64(1):123–132
11. Kalayarasan S, Sriram N, Soumyakrishnan S, Sudhandiran G (2013) Diallylsulfide attenuates excessive collagen production and apoptosis in a rat model of bleomycin induced pulmonary fibrosis through the involvement of protease activated receptor-2. Toxicol Appl Pharmacol 271(2):184–195
12. Molino M, Barnathan ES, Numerof R, Clark J, Dreyer M, Cumashi A, Hoxie JA, Schechter N, Woolkalis M, Brass LF (1997) Interactions of mast cell tryptase with thrombin receptors and PAR-2. J Biol Chem 272(7):4043–4049
13. Camerer E, Huang W, Coughlin SR (2000) Tissue factor- and factor X-dependent activation of protease-activated receptor 2 by factor VIIa. Proc Natl Acad Sci U S A 97(10):5255–5260
14. Macfarlane SR, Seatter MJ, Kanke T, Hunter GD, Plevin R (2001) Proteinase-activated receptors. Pharmacol Rev 53(2):245–282
15. Trejo J, Connolly AJ, Coughlin SR (1996) The cloned thrombin receptor is necessary and sufficient for activation of mitogen-activated protein kinase and mitogenesis in mouse lung fibroblasts. Loss of responses in fibroblasts from receptor knockout mice. J Biol Chem 271 (35):21536–21541
16. Cunningham MA, Rondeau E, Chen X, Coughlin SR, Holdsworth SR, Tipping PG (2000) Protease-activated receptor 1 mediates thrombin-dependent, cell-mediated renal inflammation in crescentic glomerulonephritis. J Exp Med 191(3):455–462
17. Nystedt S, Emilsson K, Wahlestedt C, Sundelin J (1994) Molecular cloning of a potential proteinase activated receptor. Proc Natl Acad Sci U S A 91(20):9208–9212
18. Vergnolle N (2000) Review article: proteinase-activated receptors—novel signals for gastrointestinal pathophysiology. Aliment Pharmacol Ther 14(3):257–266
19. D'Andrea MR, Derian CK, Leturcq D, Baker SM, Brunmark A, Ling P, Darrow AL, Santulli RJ, Brass LF, Andrade-Gordon P (1998) Characterization of protease-activated receptor-2 immunoreactivity in normal human tissues. J Histochem Cytochem 46(2):157–164
20. Connolly AJ, Ishihara H, Kahn ML, Farese RV Jr, Coughlin SR (1996) Role of the thrombin receptor in development and evidence for a second receptor. Nature 381(6582):516–519
21. Ishihara H, Connolly AJ, Zeng D, Kahn ML, Zheng YW, Timmons C, Tram T, Coughlin SR (1997) Protease-activated receptor 3 is a second thrombin receptor in humans. Nature 386 (6624):502–506
22. Kahn ML, Zheng YW, Huang W, Bigornia V, Zeng D, Moff S, Farese RV Jr, Tam C, Coughlin SR (1998) A dual thrombin receptor system for platelet activation. Nature 394 (6694):690–694
23. Bohm SK, McConalogue K, Kong W, Bunnett NW (1998) Proteinase-activated receptors: new functions for old enzymes. News Physiol Sci 13:231–240
24. Dery O, Corvera CU, Steinhoff M, Bunnett NW (1998) Proteinase-activated receptors: novel mechanisms of signaling by serine proteases. Am J Physiol 274(6 Pt 1):C1429–C1452

25. Nakanishi-Matsui M, Zheng YW, Sulciner DJ, Weiss EJ, Ludeman MJ, Coughlin SR (2000) PAR3 is a cofactor for PAR4 activation by thrombin. Nature 404(6778):609–613
26. Kahn ML, Nakanishi-Matsui M, Shapiro MJ, Ishihara H, Coughlin SR (1999) Protease-activated receptors 1 and 4 mediate activation of human platelets by thrombin. J Clin Invest 103(6):879–887
27. Ku DD, Dai J (1997) Expression of thrombin receptors in human atherosclerotic coronary arteries leads to an exaggerated vasoconstrictory response in vitro. J Cardiovasc Pharmacol 30 (5):649–657
28. Howell DC, Johns RH, Lasky JA, Shan B, Scotton CJ, Laurent GJ, Chambers RC (2005) Absence of proteinase-activated receptor-1 signaling affords protection from bleomycin-induced lung inflammation and fibrosis. Am J Pathol 166(5):1353–1365
29. Cicala C, Spina D, Keir SD, Severino B, Meli R, Page CP, Cirino G (2001) Protective effect of a PAR2-activating peptide on histamine-induced bronchoconstriction in guinea-pig. Br J Pharmacol 132(6):1229–1234
30. Cocks TM, Fong B, Chow JM, Anderson GP, Frauman AG, Goldie RG, Henry PJ, Carr MJ, Hamilton JR, Moffatt JD (1999) A protective role for protease-activated receptors in the airways. Nature 398(6723):156–160
31. Schmidlin F, Amadesi S, Dabbagh K, Lewis DE, Knott P, Bunnett NW, Gater PR, Geppetti P, Bertrand C, Stevens ME (2002) Protease-activated receptor 2 mediates eosinophil infiltration and hyperreactivity in allergic inflammation of the airway. J Immunol 169(9):5315–5321
32. Asokananthan N, Graham PT, Fink J, Knight DA, Bakker AJ, McWilliam AS, Thompson PJ, Stewart GA (2002) Activation of protease-activated receptor (PAR)-1, PAR-2, and PAR-4 stimulates IL-6, IL-8, and prostaglandin E2 release from human respiratory epithelial cells. J Immunol 168(7):3577–3585
33. Vliagoftis H, Schwingshackl A, Milne CD, Duszyk M, Hollenberg MD, Wallace JL, Befus AD, Moqbel R (2000) Proteinase-activated receptor-2-mediated matrix metalloproteinase-9 release from airway epithelial cells. J Allergy Clin Immunol 106(3):537–545
34. Ricciardolo FL, Steinhoff M, Amadesi S, Guerrini R, Tognetto M, Trevisani M, Creminon C, Bertrand C, Bunnett NW, Fabbri LM, Salvadori S, Geppetti P (2000) Presence and bronchomotor activity of protease-activated receptor-2 in guinea pig airways. Am J Respir Crit Care Med 161(5):1672–1680
35. Gordon JR, Zhang X, Stevenson K, Cosford K (2000) Thrombin induces IL-6 but not TNF alpha secretion by mouse mast cells: threshold-level thrombin receptor and very low level Fcepsilonri signaling synergistically enhance IL-6 secretion. Cell Immunol 205(2):128–135
36. Vliagoftis H, Befus AD, Hollenberg MD, Moqbel R (2001) Airway epithelial cells release eosinophil survival-promoting factors (GM-CSF) after stimulation of proteinase-activated receptor 2. J Allergy Clin Immunol 107(4):679–685
37. Sun G, Stacey MA, Schmidt M, Mori L, Mattoli S (2001) Interaction of mite allergens Der p3 and Der p9 with protease-activated receptor-2 expressed by lung epithelial cells. J Immunol 167(2):1014–1021
38. Masoli M, Fabian D, Holt S, Beasley R (2004) The global burden of asthma: executive summary of the GINA dissemination committee report. Allergy 59(5):469–478
39. Robinson C, Wan H, Winton HL (1998) Epithelial repair in asthma. Do the benefits of house dust mite avoidance result from proteinase avoidance? Clin Exp Allergy 28(5):530–533
40. Wan H, Winton HL, Soeller C, Tovey ER, Gruenert DC, Thompson PJ, Stewart GA, Taylor GW, Garrod DR, Cannell MB, Robinson C (1999) Der p 1 facilitates transepithelial allergen delivery by disruption of tight junctions. J Clin Invest 104(1):123–133
41. Winton HL, Wan H, Cannell MB, Thompson PJ, Garrod DR, Stewart GA, Robinson C (1998) Class specific inhibition of house dust mite proteinases which cleave cell adhesion, induce cell death and which increase the permeability of lung epithelium. Br J Pharmacol 124(6): 1048–1059
42. Terada M, Kelly EA, Jarjour NN (2004) Increased thrombin activity after allergen challenge: a potential link to airway remodeling? Am J Respir Crit Care Med 169(3):373–377

43. Asaduzzaman M, Nadeem A, Arizmendi N, Davidson C, Nichols HL, Abel M, Ionescu L, Puttagunta L, Thebaud B, Gordon J, DeFea K, Hollenberg MD, Vliagoftis H (2015) Functional inhibition of PAR2 alleviates allergen-induced airway hyper responsiveness and inflammation. Clin Exp Allergy 45(12):1844–1855
44. de Boer JD, Van't Veer C, Stroo I, van der Meer AJ, de Vos AF, van der Zee JS, Roelofs JJ, van der Poll T (2014) Protease-activated receptor-2 deficient mice have reduced house dust mite-evoked allergic lung inflammation. Innate Immun 20(6):618–625
45. Arizmendi NG, Abel M, Mihara K, Davidson C, Polley D, Nadeem A, El Mays T, Gilmore BF, Walker B, Gordon JR, Hollenberg MD, Vliagoftis H (2011) Mucosal allergic sensitization to cockroach allergens is dependent on proteinase activity and proteinase-activated receptor-2 activation. J Immunol 186(5):3164–3172
46. Knight DA, Lim S, Scaffidi AK, Roche N, Chung KF, Stewart GA, Thompson PJ (2001) Protease-activated receptors in human airways: upregulation of PAR-2 in respiratory epithelium from patients with asthma. J Allergy Clin Immunol 108(5):797–803
47. Trupin L, Earnest G, San Pedro M, Balmes JR, Eisner MD, Yelin E, Katz PP, Blanc PD (2003) The occupational burden of chronic obstructive pulmonary disease. Eur Respir J 22 (3):462–469
48. Gong JH, Cho IH, Shin D, Han SY, Park SH, Kang YH (2014) Inhibition of airway epithelial-to-mesenchymal transition and fibrosis by kaempferol in endotoxin-induced epithelial cells and ovalbumin-sensitized mice. Lab Invest 94(3):297–308
49. Matej R, Vasakova M, Kukal J, Sterclova M, Olejar T (2014) Higher TGF-β with lower CD124 and TSLP, but no difference in PAR-2 expression in bronchial biopsy of bronchial asthma patients in comparison with COPD patients. Appl Immunohistochem Mol Morphol 22 (7):543–549
50. Miotto D, Hollenberg MD, Bunnett NW, Papi A, Braccioni F, Boschetto P, Rea F, Zuin A, Geppetti P, Saetta M, Maestrelli P, Fabbri LM, Mapp CE (2002) Expression of protease activated receptor-2 (PAR-2) in central airways of smokers and non-smokers. Thorax 57 (2):146–151
51. Prikk K, Maisi P, Sepper R, Stenman UH, Salo T, Sorsa T (2001) Association of trypsin-2 with activation of gelatinase B and collagenase-2 in human bronchoalveolar lavage fluid in vivo. Ann Med 33(6):437–444
52. Ashitani J, Mukae H, Arimura Y, Matsukura S (2002) Elevated plasma procoagulant and fibrinolytic markers in patients with chronic obstructive pulmonary disease. Intern Med 41 (3):181–185
53. Wynn TA (2011) Integrating mechanisms of pulmonary fibrosis. J Exp Med 208(7):1339–1350
54. Cheresh P, Kim SJ, Tulasiram S, Kamp DW (2013) Oxidative stress and pulmonary fibrosis. Biochim Biophys Acta 1832(7):1028–1040
55. Tzouvelekis A, Herazo-Maya J, Sakamoto K, Bouros D (2016) Biomarkers in the evaluation and management of idiopathic pulmonary fibrosis. Curr Top Med Chem 16(14):1587–1598
56. Jones MG, Fletcher S, Richeldi L (2013) Idiopathic pulmonary fibrosis: recent trials and current drug therapy. Respiration 86(5):353–363
57. Imokawa S, Sato A, Hayakawa H, Kotani M, Urano T, Takada A (1997) Tissue factor expression and fibrin deposition in the lungs of patients with idiopathic pulmonary fibrosis and systemic sclerosis. Am J Respir Crit Care Med 156(2 Pt 1):631–636
58. Chambers RC (2003) Role of coagulation cascade proteases in lung repair and fibrosis. Eur Respir J Suppl 44:33s–35s
59. Michelin E, Snijders D, Conte S, Dalla Via P, Tagliaferro T, Da Dalt L, Monciotti CM, Simioni P, Stefanutti G, Ghirardo V, Gamba P, Barbato A (2008) Procoagulant activity in children with community acquired pneumonia, pleural effusion and empyema. Pediatr Pulmonol 43(5):472–475

60. Hernandez-Rodríguez NA, Cambrey AD, Harrison NK, Chambers RC, Gray AJ, Southcott AM, duBois RM, Black CM, Scully MF, McAnulty RJ (1995) Role of thrombin in pulmonary fibrosis. Lancet 346(8982):1071–1073
61. Dik WA, Zimmermann LJ, Naber BA, Janssen DJ, van Kaam AH, Versnel MA (2003) Thrombin contributes to bronchoalveolar lavage fluid mitogenicity in lung disease of the premature infant. Pediatr Pulmonol 35(1):34–41
62. Mercer PF, Johns RH, Scotton CJ, Krupiczojc MA, Königshoff M, Howell DC, McAnulty RJ, Das A, Thorley AJ, Tetley TD, Eickelberg O, Chambers RC (2009) Pulmonary epithelium is a prominent source of proteinase-activated receptor-1-inducible CCL2 in pulmonary fibrosis. Am J Respir Crit Care Med 179(5):414–425
63. Ask K, Bonniaud P, Maass K, Eickelberg O, Margetts PJ, Warburton D, Groffen J, Gauldie J, Kolb M (2008) Progressive pulmonary fibrosis is mediated by TGF-beta isoform 1 but not TGF-beta3. Int J Biochem Cell Biol 40(3):484–495
64. Jenkins RG, Su X, Su G, Scotton CJ, Camerer E, Laurent GJ, Davis GE, Chambers RC, Matthay MA, Sheppard D (2006) Ligation of protease-activated receptor 1 enhances alpha (v)beta6 integrin-dependent TGF-beta activation and promotes acute lung injury. J Clin Invest 116(6):1606–1614
65. Soumyakrishnan S, Divya T, Kalayarasan S, Sriram N, Sudhandiran G (2014) Daidzein exhibits anti-fibrotic effect by reducing the expressions of proteinase activated receptor 2 and TGFβ1/smad mediated inflammation and apoptosis in Bleomycin-induced experimental pulmonary fibrosis. Biochimie 103:23–36
66. Ando S, Otani H, Yagi Y, Kawai K, Araki H, Fukuhara S, Inagaki C (2007) Proteinase-activated receptor 4 stimulation-induced epithelial-mesenchymal transition in alveolar epithelial cells. Respir Res 8:31
67. Mokra D, Kosutova P (2015) Biomark Acute Lung Inj. Respir Physiol Neurobiol 209:52–58
68. Günther A, Mosavi P, Heinemann S, Ruppert C, Muth H, Markart P, Grimminger F, Walmrath D, Temmesfeld-Wollbrück B, Seeger W (2000) Alveolar fibrin formation caused by enhanced procoagulant and depressed fibrinolytic capacities in severe pneumonia. Comparison with the acute respiratory distress syndrome. Am J Respir Crit Care Med 161(2 Pt 1):454–462
69. Su X, Matthay MA (2009) Role of protease activated receptor 2 in experimental acute lung injury and lung fibrosis. Anat Rec (Hoboken) 292(4):580–586
70. Gan X, Liu D, Huang P, Gao W, Chen X, Hei Z (2012) Mast-cell-releasing tryptase triggers acute lung injury induced by small intestinal ischemia-reperfusion by activating PAR-2 in rats. Inflammation 35(3):1144–1153
71. Scotton CJ, Krupiczojc MA, Königshoff M, Mercer PF, Lee YC, Kaminski N, Morser J, Post JM, Maher TM, Nicholson AG, Moffatt JD, Laurent GJ, Derian CK, Eickelberg O, Chambers RC (2009) Increased local expression of coagulation factor X contributes to the fibrotic response in human and murine lung injury. J Clin Invest 119(9):2550–2563
72. Dasgupta A, Neighbour H, Nair P (2013) Targeted therapy of bronchitis in obstructive airway diseases. Pharmacol Ther 140(3):213–222
73. Naldini A, Sower L, Bocci V, Meyers B, Carney DH (1998) Thrombin receptor expression and responsiveness of human monocytic cells to thrombin is linked to interferon-induced cellular differentiation. J Cell Physiol 177(1):76–84
74. Takizawa T, Tamiya M, Hara T, Matsumoto J, Saito N, Kanke T, Kawagoe J, Hattori Y (2005) Abrogation of bronchial eosinophilic inflammation and attenuated eotaxin content in protease-activated receptor 2-deficient mice. J Pharmacol Sci 98(1):99–102
75. Su X, Camerer E, Hamilton JR, Coughlin SR, Matthay MA (2005) Protease-activated receptor-2 activation induces acute lung inflammation by neuropeptide-dependent mechanisms. J Immunol 175(4):2598–2605
76. Vancheri C, Mastruzzo C, Sortino MA, Crimi N (2004) The lung as a privileged site for the beneficial actions of PGE2. Trends Immunol 25(1):40–46

77. Sokolova E, Hartig R, Reiser G (2008) Downregulation of protease-activated receptor-1 in human lung fibroblasts is specifically mediated by the prostaglandin E receptor EP2 through cAMP elevation and protein kinase A. FEBS J 275(14):3669–3679
78. D'Agostino B, Roviezzo F, De Palma R, Terracciano S, De Nardo M, Gallelli L, Abbate GF, D'Aiuto E, Russo M, Cirino G, Rossi F (2007) Activation of protease-activated receptor-2 reduces airways inflammation in experimental allergic asthma. Clin Exp Allergy 37 (10):1436–1443
79. Jemal A, Siegel R, Ward E, Murray T, Xu J, Thun MJ (2007) Cancer statistics, 2007. CA Cancer J Clin 57(1):43–66
80. Nierodzik ML, Chen K, Takeshita K, Li JJ, Huang YQ, Feng XS, D'Andrea MR, Andrade-Gordon P, Karpatkin S (1998) Protease-activated receptor 1 (PAR-1) is required and rate-limiting for thrombin-enhanced experimental pulmonary metastasis. Blood 92(10):3694–3700
81. Henrikson KP, Salazar SL, Fenton JW 2nd, Pentecost BT (1999) Role of thrombin receptor in breast cancer invasiveness. Br J Cancer 79(3–4):401–406
82. Boire A, Covic L, Agarwal A, Jacques S, Sherifi S, Kuliopulos A (2005) PAR1 is a matrix metalloprotease-1 receptor that promotes invasion and tumorigenesis of breast cancer cells. Cell 120(3):303–313
83. Cisowski J, O'Callaghan K, Kuliopulos A, Yang J, Nguyen N, Deng Q, Yang E, Fogel M, Tressel S, Foley C, Agarwal A, Hunt SW 3rd, McMurry T, Brinckerhoff L, Covic L (2011) Targeting protease-activated receptor-1 with cell-penetrating pepducins in lung cancer. Am J Pathol 179(1):513–523
84. Huang SH, Li Y, Chen HG, Rong J, Ye S (2013) Activation of proteinase-activated receptor 2 prevents apoptosis of lung cancer cells. Cancer Invest 31(9):578–581
85. Zhu L, Wang X, Wu J, Mao D, Xu Z, He Z, Yu A (2012) Cooperation of protease-activated receptor 1 and integrin $\alpha v\beta 5$ in thrombin-mediated lung cancer cell invasion. Oncol Rep 28 (2):553–560

Unfolding the Mechanism of Proteases in Pathophysiology of Gastrointestinal Diseases

Sharmistha Banerjee, Sumit Ghosh, Krishnendu Sinha and Parames C. Sil

Abstract

The intestinal epithelial biology is controlled by the presence of the gastrointestinal protease pool and their inhibitors. Enteric bacterial proteases and their signalling mechanisms contribute to the pathogenesis of IBD. Proteases derived from *Helicobacter pylori* (*H. pylori*) are mainly responsible for ulceration, low stomach acid and gastritis. *H. pylori* infection initiates the inflammatory and apoptotic pathways. The apoptotic pathways involve the proteolytic activities of caspases, i.e. cysteine aspartate proteases. However, some types of matrix metalloproteinases restrain such infection while other types act as the causal agent for gastric ulceration. Serine proteases and PAR-2 have been found to elicit signalling pathways related to induction of pain in patients suffering from irritable bowel syndrome. The prognostic role of tumour-associated proteases in colorectal cancer has also been investigated. The increased levels of cathepsin in Gaucher storage cells point towards their role in pathogenesis of the disease. On the other hand, excessive trypsin and chymotrypsin activity has been observed in case of chronic pancreatitis. This chapter focusses on the role of proteases on the pathology of various diseases.

Keywords

Gastritis · Gastropathy · Matrix metalloprotease
Inflammatory bowel disease · Tumour necrosis factor-α

S. Banerjee · S. Ghosh · P.C. Sil (✉)
Division of Molecular Medicine, Bose Institute, P-1/12, CIT Scheme VII M, Kolkata 700054, West Bengal, India
e-mail: parames@jcbose.ac.in; parames_95@yahoo.co.in

K. Sinha
Jhargram Raj College, Government of West Bengal, Jhargram, India

S. Chakraborti and N.S. Dhalla (eds.), *Pathophysiological Aspects of Proteases*,
DOI 10.1007/978-981-10-6141-7_24

1 Introduction

Proteases are proteolytic enzymes that cleave the peptide bonds formed between the amino and carboxyl groups of adjacent amino acid residues of a polypeptide chain for the purpose of protein catabolism. About, 550 different types of proteases have been identified from the human genome, and more than 100 have been predicted. Proteases account for about 5% of the genome of infectious organisms such as bacteria and viruses, thereby establishing themselves as potential drug targets.

Proteases can be classified into seven groups: serine proteases, aspartate proteases, cysteine proteases, glutamic acid proteases, threonine proteases, metalloproteases and asparagine peptide lyases [1]. Of these, the mechanism of action of serine proteases has been extensively studied till date. Aspartic, glutamic and metalloproteases activate water molecules for nucleophilic attack on the peptide bonds for their hydrolysis. On the other hand, the serine, threonine and cysteine proteases use a catalytic triad for nucleophilic attack. A catalytic triad refers to three amino acids which function together at the active centre of transferase or hydrolase enzymes. Such triads point towards the convergent evolution in the proteolytic activity of such proteases [2].

The broad spectrum of protease activity is attributable to their wide range of applications as important biological tools in industrial and therapeutic research. Nowadays, the gastrointestinal diseases are found to be quite common in both developing and developed countries. Lifestyle, personal preferences for food, rampant exposure to pollutants, etc. are important causes for the prevalence of gastric diseases. The role of gastric proteases in the progression of these diseases has been elaborated in this chapter. Diseases like gastritis, sepsis, Gaucher disease, colorectal cancer, inflammatory bowel disease, acute and chronic gastropathy. Figure 1 has been discussed in this chapter, mainly focussing on the role of

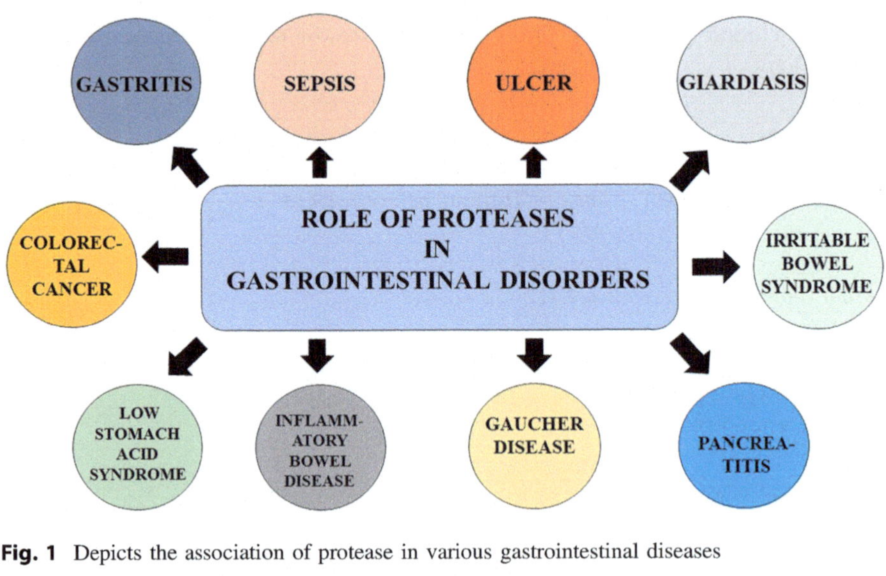

Fig. 1 Depicts the association of protease in various gastrointestinal diseases

proteases and the associated molecular mechanisms responsible for the regulation of concerned pathogenicity.

The cosmopolitan occurrence of proteases in lower to higher groups of organisms impart immense significance to these enzymes as they are being the important sources for the study of evolutionary convergence or divergence in structure–activity relationships. Their ability to adapt to the mechanisms of 'limited' to 'unlimited' proteolysis and vice versa (depending on the substrate or other micro-environmental factors) has increased their domain of functional activities.

Modern molecular reductionism has enlightened us on the role of enzymatic activities in the onset and maintenance of different diseases. The major theme of this chapter is centred on the holistic overview of the potential role of proteases in different gastrointestinal disorders. This approach has been adopted with the intention of providing an essence of the various mechanisms associated with the protease-mediated pathogenicity of gut-associated diseases.

2 Mechanisms of Gastrointestinal Protease Activity: A Brief Overview

Though proteases are holistically considered as enzymes for degradation, a much wider aspect of their functions is chiefly attributable to the emergence of the concept of 'limited proteolysis'. Proteolytic cleavage is considered to be an ideal multicomponent irreversible signalling event. Such events are regulated by different enzymes, growth factors, surface receptors, etc. On the other hand, 'limited proteolysis', a term coined by Lang and Linderstrom, refers to non-specific protein degradation. It plays a vital role in maintaining the function of intestinal epithelial cells (IEC).

The gastric functions are dependent on pericellular proteolysis. Intestinal proteases induce the pathogenesis of gastrointestinal inflammatory diseases [3]. Most of these proteases are synthesized as concealed zymogens which undergo proteolytic cleavage for activation [4, 5]. Multiple families of proteases function together to form networks [6]. Matrix metalloproteinase (MMP) zymogens which degrade the ECM are cleaved to active enzymatic forms by serine proteases associated with the plasminogen activator system and [7] also by furin-like prohormone convertases [8]. The functional specificity of enterocyte proteases to suitable substrates is controlled through spatial and chronological compartmentalization. Mast cell proteases are catalogued within intracellular granules and undergo intestinal inflammation-mediated retortion [9]. Proteases are localized in the pericellular environment through various mechanisms. For example, matriptase, a type II epithelial transmembrane serine protease, is reported to get fastened directly to the plasma membrane by its membrane-spanning domain [10, 11]. Termination of protease activity is accomplished by several mechanisms. Serpins, a family of inhibitors of serine protease, target different stages of the serine protease signalling cascades by functioning as irreversible 'suicide substrates' [12]. Proteolysis-mediated activation of PARs is reported to exhibit gastrointestinal and physiological activities [13]. Serine proteases of the intestinal lumen activate PARs to exert their functions [14]. Proteases are essential for wound repair and inflammation [15]. Components of the ECM, such as proteoglycans, fibronectin, collagen undergo MMP-10-mediated degradation while different protease inhibitors are regulated by cytokines released in response to repair of wound [16].

3 Gastritis

Helicobacter pylori (*H. pylori*), a gram-negative microaerophilic bacteria colonizing the human gastric mucosa, is the causative agent of gastritis and is approximated to affect 50% of the world population. Gastritis is associated with inflammation of the gastric mucosa. In 1984, Warren and Marshall discovered the role of *H. pylori* in the pathogenesis of gastritis and consequently received the Nobel Prize in 2005. Different types of matrix metalloproteases (MMPs), tissue inhibitors of MMPs (TIMPs) and a disintegrin and metalloproteases (ADAMs) are involved in progression of the disease [17].

H. pylori infection is acquired mainly through the faecal–oral route during infancy and such bacterial colonisations may persist throughout lifetime thereby offering a range of clinical manifestations [18]. The severity and progression of the disease depend on interactions of different factors related to both the host (genetic make-up, immunological state, etc.) and the bacteria (adaptation capacity, genomic plasticity, generation of virulence factors like vacuolating cytotoxin (VacA) and cytotoxin-associated antigen A (CagA)). CagA and VacA are coded by a pathogenicity island, called cag PAI and are present in the genome of *H. pylori* [19].

The extracellular matrix metalloproteases (MMPs) are a family of zinc-dependent proteases which vitiate the machinery of the extracellular matrix (ECM) and also participate in various physiological and pathological processes like morphogenesis, tumoural angiogenesis, metastatic invasions [20].

MMP-1 degrades interstitial collagen types I, II and III and thereby participates in the pathogenesis of gastritis [21]. MMP-2 and MMP-9 (i.e., Gelatinase A and B, respectively) degrade the major structural unit of the basement membrane, i.e. collagen type IV. In gastric tissues of gastritis patients, IL-21-mediated increase in expression of MMP-9 and MMP-2 compared to that in normal tissue has been reported [22]. Increased expression of MMP-8 has also been reported in case of gastritis patients [23].

MMP-7 can degrade fibronectin, proteoglygans, elastin, etc. and is reported to exhibit a vital function in the progression of the disease. MMP-7 is associated with induction of epithelial to mesenchymal transition of human gastric cells and is considered to play an important role in the development of gastric cancer [24]. It induces IGF-II (insulin-like growth factor binding protein)-mediated metastatic invasion of neoplastic cells from affected gastric tissues to other parts of the body [25, 26]. MMP-7-181G allele induces the expression of MMP-7, thereby stimulating the degradation of the extracellular matrix in gastric tissues of gastritis patients [27].

The cag PAI-mediated expression of MMP-7 is regulated by the phosphorylation and intracellular redistribution of p120-catenin. On being translocated to the nucleus of gastric cells, p120-catenin releases a repressor of MMP7, thereby controlling its expression [28].

The MMPs are inhibited non-covalently by the tissue inhibitors of MMPs (TIMPs). Increased mRNA levels of TIMP-1, -2, -3 and -4 are observed in case of *H. pylori* infected patients [29].

The A disintegrin and metalloproteinase (ADAM) family of proteases is associated with the activation of signalling pathways related to cellular adhesion and fusion and the release of growth factors and cytokines [30]. High levels of ADAM10 and ADAM17 in the gastric antrum of *H. pylori* infected individuals activate an array of signalling molecules such as tumour necrosis factor, c-Met, E-cadherin, thereby leading to the progression of gastric cancer [31, 32].

Neprilysin, a zinc-dependent metalloprotease, is also reported to play an important role in the prevention of gastritis. *H. pylori* infection reduces the level of expression of substance P, vasoactive intestinal peptide (VIP), neprilysin, etc. [33].

A subfamily of G-protein-coupled receptors, i.e. the protease-activated receptors (PARs), regulates cellular functions through the proteolysis of a receptor fastened peptide domain. PAR-1 and PAR-2 are ubiquitously present along the entire gastrointestinal (GI) tract. Thrombin and trypsin-like proteases activate PAR-2 and PAR-1 on the neurones of the vagus nerve, thereby increasing the intracellular levels of calcium via activation of phospholipase C and inositol triphosphate (IP3) signalling pathway. Increased expression of PAR-2 and PAR-1 transcripts is observed with the progression of inflammation. Such deviations from the normal physiological conditions are also reported to cause changes in gastric motility and induction of gastritis or inflammatory bowel disease. The activation of vagal efferent DMV neurons, displaying functional PAR-2 and PAR-1, leads to robust gastric stasis [34].

4 Gaucher Disease

Gaucher's disease or Gaucher disease (GD) is characterized by the accumulation of glucocerebroside (a sphingolipid) in white blood cells (mainly tissue resident macrophages i.e. Gaucher cells) and different organs like the kidneys, lungs, brain, bone marrow, etc. due to deficiency of glucocerebrosidase activity. The symptoms include anaemia, low platelet count and enlargement of the spleen and liver.

Gaucher disease is of three clinical subtypes:

GD type I (non-neuropathic) is the most common form of the disease and is associated with hepatosplenomegaly, anaemia, leukopenia and thrombocytopenia.

GD type II (acute neuropathic) is characterized by progressive damage of the brain, disorders of optic movements, limb rigidity and a poor ability to suck and swallow in infants.

GD type III (chronic neuropathic) is characterized by skeletal defects, anaemia, respiratory problems, etc.

In GD type I, massive accretion of glucosylceramide occurs in the lysosomes of tissue resident macrophages, called Gaucher cells. These cells are mainly found in the spleen, liver and bone marrow and are associated with hepatosplenomegaly, avascular necrosis, pancytopenia, etc. [35]. Substrate reduction therapy and enzyme replacement therapy are associated with treatment of the disease. The enzyme replacement therapy is based on the oral administration of N-butyl-deoxynojirimycin, an inhibitor of glucosylceramide biosynthesis [36]. Substrate reduction therapy is based on the oral administration of miglustat, an inhibitor of glucosylceramide synthase, the key enzyme in glucosylceramide biosynthesis [37]. The high cost of such therapies led to the quest for detecting different plasma biomarkers related to Gaucher disease. It is a cost-effective approach for personalized treatment. Two specific Gaucher cell markers that have been identified include chitotriosidase and CCL18, both of whose levels get elevated in Gaucher patients [38, 39]. Investigation on the plasma protein expression patterns in Gaucher patients using the conventional two-dimensional gel electrophoresis (2DGE) within the acidic range led to the observation, that in the case of plasma samples of patients, there was lack of common high molecular weight proteins while novel low molecular weight proteins (degradation products) were present. Such a unique observation indicated the involvement of proteases functioning preferably in an acidic environment. The responsible protease(s) in Gaucher plasma have not been identified yet. The involvement of MMPs and Cathepsin B, D, K, S has been elucidated through extensive research. Increased concentrations of different cathepsins [36] and angiotensin-converting enzyme (ACE) [25] have been observed in the plasma of symptomatic Gaucher patients.

5 Sepsis

Sepsis is a serious pathophysiological condition, generated when the body's internal response to infection harms its own tissues and organs. The common symptoms include fever, increased breathing rate, increased heart rate and confusion [40].

PARs are activated through proteolysis, which is triggered by extracellular proteases. PARs can convert events of extracellular proteolytic cleavage into an intracellular signal. These receptors carry their own fastened ligands, which remain ambiguous until unveiled by receptor N-terminal cleavage. An intramolecular rearrangement follows which then allows the receptor moieties and the ligand to interact with each other [41].

PAR-1 plays an important role in coordinating the interaction between inflammation and coagulation, and thus, extensive research in this field may lead to documentation of novel therapeutic targets for the treatment of sepsis [42]. PAR-1 is associated with thrombin-mediated aggregation of platelets in humans. It can trigger various cellular responses through different signalling mechanisms by switching from one conformational state to another. PAR-1 activated by thrombin increases endothelial permeability [43]. It is responsible for *Streptococcus pneumonia*-induced sepsis through an interaction with platelet-activating factor receptor (PAFR). It also induces the influenza virus infection-associated pathogenicity and aggravates pulmonary oedema [44]. PAR-1 is reported to amplify PAFR dependent pneumococcal dissemination thereby inducing severe pneumonia. It opposes the progression of sepsis, and so, PAR-1 activation or blockade proves to be alternatively favourable at different stages of various sepsis models [45].

The proteasome degrades damaged proteins by proteolysis, a chemical reaction catalysed by proteases. Experiments on rabbit (as the working model) have shown that Toll-like receptor 4, the IκB kinase and the proteasome can be potential therapeutic targets in the treatment of lipopolysaccharide (LPS)-induced intestinal sepsis. It is reported that systemic blockade of Toll-like receptor 4, IκB kinase or the proteasome ameliorates the effect of LPS on the mRNA expressions of acetylcholine, substance P, prostaglandin E2, IL-8 and IL-6 mRNA expression in the rabbit intestine [46].

6 Giardiasis

Giardia duodenalis is the major aetiological agent of waterborne diarrhoeal diseases throughout the world. It is estimated to taint about 280 million individuals annually. The parasite has been included in the neglected diseases initiative of the World Health Organization (WHO) since 2006 [47]. Giardiasis is characterized by stunted growth in children, anion hypersecretion, shortening of microvilli of the intestinal brush border, enterocyte apoptosis, etc. [48]. When the infection reaches its peak, the trophozoites of *G. duodenalis* induce various pathological processes within the epithelial cells of the intestine thereby leading to the development of diarrhoeal disease.

Cathepsin cysteine proteases of *G. duodenalis* play a vital role in the induction of Giardiasis. The trophozoites produce cathepsin B and L in various different strains. The cathepsin cysteine proteases help in trophozoite encystation and excystation [49]. These proteases contain cysteine and histidine residue bearing active site [50]. The genome of *G. duodenalis* contains genes for different catL- and catB-like cysteine proteases [51]. The catB-like cysteine proteases degrade CXCL8, thereby attenuating neutrophil chemotaxis [52]. Giardia contributes to diarrhoea by disrupting the organization of the epithelial brush border microvilli. Giardia is also reported to induce epithelial apoptosis [53]. Villin is a pro-survival factor whose overexpression delays epithelial apoptosis [54]. It also helps in cellular migration and wound healing [55]. Thus, villin maintains homeostasis at the cellular level. The Giardial cathepsin cysteine proteases are reported to degrade host intestinal villin via activation of the myosin light chain kinase (MLCK). This helps in progression of the disease. *G. duodenalis* uses various mechanisms for the disruption of villin and the intestinal epithelial brush border.

G. duodenalis cathepsin cysteine proteases act as potential parasitic virulence factors having the capacity to induce different pathological responses within the host intestinal epithelial cells [56].

7 Inflammatory Bowel Disease

Inflammatory bowel disease (IBD) is an inflammatory disorder which affects the gastrointestinal tract and colon mucosa. IBD is primarily manifested as ulcerative colitis (UC) and Crohn's disease (CD). One of the characteristic features of this disease is that relapse and remissions occur frequently. Environmental factors, immune response-mediated tissue injury, and genetic factor play major role in the pathophysiology of IBD [57]. Cytokines play a significant role in the regulation of this disease [58]. Gastrointestinal tract (GI) of human contains more than 500 species of bacteria. These bacteria secrete various compounds which have beneficial as well as adverse effects. These bacteria colonizing the GI tract helps to maintain gut homeostasis. The intestinal immune system has evolved to protect against luminal bacteria. Under normal condition, gastrointestinal injury and tissue repair are tightly regulated by interaction and interplay between array of defensive and aggressive factors in the lumen and wall of intestine. Inflammosomes are multiprotein complexes that regulate various mechanisms during homeostasis and inflammation. These complexes consist of different NLR proteins like NLRP1, 2, 3, 6 and others. NLR6 has been reported to play crucial roles in protection against infection and inflammation [59, 60]. This molecule performs crucial function in the maintenance of intestinal homeostasis and healthy intestinal microbiota. NLRP6 is also required for mucosal self-renewal and proliferation. Secretion of mucus by goblet cell of large intestine serves for a protective mechanism at the interface between eukaryotic and prokaryotic cells of the intestinal ecosystem.

NLRP6 deficiency leads to microbial dysbiosis by favouring the development of colitogenic microbiota. This microbial population is associated at the base of the crypt and stimulates the release of pro-inflammatory cytokines ultimately leading to chronic inflammation paving a road for inflammatory bowel disease [61]. It can be concluded that microbial dysbiosis in the intestine is one of the major causes of IBD [62].

The complex system of intestinal homeostasis is also modulated by environmental stress. In animal models, stress has been found to reactivate colitis. The central nervous system responds to environmental stress and links with the intestine through dorsal root nuclei, spinal cord and intestinal neurons on one hand and via neuroendocrine system on the other hand. This coordination leads to activation or inhibition of specific neurons in the intestine and stimulates the release of certain molecules which help to aggravate the disease [63, 64].

Proteases are associated with the progression of the disease. Metalloprotease (MMP) is one family of protease. MMPs are zinc-dependent endoproteases which help in degradation and remodelling of extracellular matrix (ECM). These proteases are secreted as inactive precursor known as zymogens, and they are converted to active enzyme in the extracellular space. Regulation of MMPs occurs in four ways: at the level of their transcription from gene, through interaction with specific ECM components, at the level of activation from precursor enzyme and by TIMP inhibition [65, 66].

MMPs can proteolytically cleave and activate various non-matrix substances like cytokines, chemokines and growth factors.

MMP13 is a matrix metalloprotease which belongs to the class collagenase. Literature suggests that MMP13 is involved in ulcerative colitis and Crohn's disease. mRNA expression of this MMP was significantly increased in IBD biopsy specimens [67]. There was a positive association between MMP13 expression and histological inflammation scores in mucosal samples taken from IBD patients [68]. MMP13 deficiency protected mice from LPS-induced inflammation, intestinal permeability, mucus depletion and endoplasmic reticulum stress (ER stress). This protease converted pro-TNF-α into mature bioactive TNF which aided in augmenting inflammation and mucus depletion MMP13 deficiency also resulted in reduced symptoms of clinical colitis in DSS treated mice [69] (Fig. 2). Increased mRNA expression of MMP-13, MMP-7, and MMP-2 mRNA were observed in biopsy specimens of Crohn's disease and ulcerative colitis. Expression of MMP-7 and MMP-2 was enhanced also at protein level [70]. Serum level of MMP-3 and MMP-9 has been co-related with the disease symptoms and activity in children.

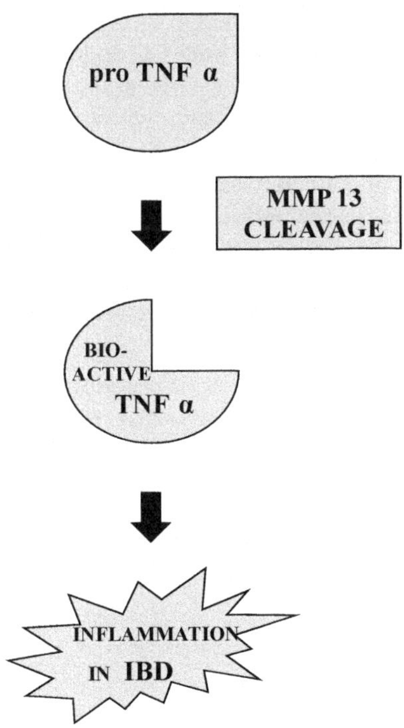

Fig. 2 Depicts the activation of TNF-α by MMP13

In a study involving, 31 children aged between 3 and 18 years with ulcerative colitis had higher level of serum MMP-3 and 9 as compared with children without ulcerative colitis. Elevation of these two proteins suggests the possibility of their use as clinical biomarkers of IBD [71]. A positive correlation was found between elevated expression of MMP-13 and histological evidence of inflammation in patients suffering from Crohn's disease (CD) and ulcerative colitis (UC). Expression of this protease was absent in non-inflamed colonic mucosa [72]. Enhanced expression of interstitial collagenase (MMP-1) and stromelysin-1 (MMP-3) transcripts was observed in granular tissue around ulceration site in patients with Crohn's disease or ulcerative colitis (UC). Elevated expression of MMP-13 and TIMP-3 is observed in fibroblasts of colonic ulcer [73]. MMP-3 activation has been implicated in colonic inflammation. Stromal myofibroblasts play a role in directing extracellular matrix components through metalloproteinases secretion. Exposure of the human colonic myofibroblast cell line 18 Co to TNF-α and bradykinin stimulated expression of MMP-3 mRNA. This TNF-α and bradykinin-induced MMP-3 expression are regulated through signalling mechanism involving PKC, PKD1 and MEK [74]. MMP-9, MMP-8 and prolyl endopeptidase (PEP) degrade collagen to form proline–glycine–proline (PGP). This degradation product of collagen has chemotactic effects on neutrophil. This proteolytic cascade has been found in intestine of patients with IBD [75]. Neutrophil infiltration into ulcer site can be partially prevented by reducing collagen degradation.

Macrophage metalloelastase MMP-12 and membrane type 1-MMP (MMP-14) were over expressed in lamina propria and macrophages of gut [76]. It is frequently observed that the gut is massively infiltrated with plasma cells and B cells in the pathogenesis of both UC and CD. Human B cells secrete granzyme B (GrB) in the presence of IL-21. GrB producing CD19+ and IgA+cell numbers were significantly higher in intestinal mucosa of both CD and UC than that of normal mucosa. Pro-inflammatory cytokine, IL-21 enhanced the expression of GrB in CD19+B cells and ultimately lead to enhanced cytotoxic activities. IBD related pathophysiology is marked by the accumulation of GrB-expressing CD19+ and IgA+cells [77]. Recent evidence suggests that patients suffering from IBD fail to respond and are not operative for the treatment of Crohn's disease. Activated matrix metalloproteinase 3 (MMP3) and MMP12 whose levels are elevated in inflamed mucosa of patients with IBD, have a wide range of substrates, including IgG1. Proteolytic degradation helps to contribute to non-responsiveness to anti-TNF agents in IBD patients [78]. MMPs are also associated with *Helicobacter pylori*-associated gastritis, and gastric and duodenal ulcer. Gastrointestinal mucosal inflammatory responses to *Helicobacter pylori* are characterized by increased production of MMP-9 and induction of mitogen-activated protein kinase (MAPK) and Rac1 activation. This H. pylori-induced MMP-9 also requires the ERK-mediated phosphorylation of cPLA2 on serine 505 and is necessary for membrane translocation of Rac1.

8 Pancreatitis

Pancreatitis is an inflammatory disease which affects pancreas. The main events which lead to the initiation of acute pancreatitis are protease activation and the inhibition of zymogen secretion [79]. In cerulein-induced acute pancreatitis, zymogens are missorted and co-localized with lysosomal cathepsin B in large cytoplasmic vacuoles, the sites of trypsinogen activation [80]. Premature activation of trypsinogen in pancreas triggers the activation of the cascade of all pancreatic enzymes. Chymotrypsin C, a protease is reported to limit proteolytic activities of trypsinogen and trypsin. Hereditary pancreatitis is concomitant with CRTC mutants (p.A73T and p.G61R) which precipitate in ER, resulting in ER stress. These mutants, on getting expressed, reduce the secretion of amylase from the pancreatic acinar cells AR42 J of carbachol-stimulated rat and the pancreatic acini isolated from mice. This also resulted in acetylated tubulin level through increased phosphorylation of deacetylase, SIRT2. Inhibition of SIRT2 restored amylase secretion as well as tubulin acetylation in isolated acini and pancreatic acinar cells [81]. Cationic trypsinogen (PRSS1) and anionic trypsinogen (PRSS2) are the major trypsinogen isoforms found in human pancreas. Mutation in PRSS1 causes hereditary pancreatitis. This mutation alters cleavage of regulatory nick sites by chymotrypsin C (CTRC) and resulted in reduced degradation of trypsinogen leading to its increased autoactivation. CTRC aided to promote degradation and suppressed autoactivation of human anionic trypsinogen. This increased sensitivity to CTRC-mediated degradation was the result of an additional cleavage site at Leu148 in the autolysis loop and the absence of the conserved Cys139–Cys206 disulphide bond. Single mutations of either Leu81 or Arg122 of cationic trypsinogen resulted in total resistance to chymotrypsin-mediated degradation. This shows that CTRC suppressed autoactivation of human anionic trypsinogen more effectively than the cationic trypsinogen which explains the lack of association of anionic trypsinogen with hereditary pancreatitis [82].

Cathepsin D (CD) is an aspartic protease found in lysosomes. This protease participates in various physiological functions like regulation of apoptosis, aids in activation of enzyme precursor and macroautophagy. CD remains localized to lysosomes as a 52 kDa precursor [83] and is converted to intermediate form by cleavage of 44 amino acids from amino terminus thus producing a 48 kDa single chain. This proteolytic degradation is executed by lysosomal cysteine proteases or by autolysis of CD [84, 85]. Further, cleavage of this intermediate 48 kDa into double chain enzyme composed of heavy (34 kDa) and light (14 kDa) chains by lysosomal cysteine protease yields mature CD [86]. CD-deficient mice exhibited autophagic activity and normal pancreatic development. Accumulation of LC3II is observed in both physiological and pancreatitis conditions. CD deficiency also leads to the accretion of matured cathepsin B (CB) and cathepsin L (CL) [87]. Acute renal failure is associated with severe acute pancreatitis (SAP).

Levels of creatinine, serum amylase and blood urea nitrogen were higher in rats of SAP group after 36 h. Expression of MMP-9 and vasodilator-stimulated phosphoprotein was significantly increased in the kidney of SAP animals. Capillary endothelial cells in the renal interstitium were weakened. This sheds some light on the association of MMP-9 in renal damage in SAP [88]. In early stages of SAP, the balance between TIMP-1 and MMP-9 get disturbed. This indicates that endogenous TIMP-1 is unable to prevent excess activation and release of MMP-9. This matrix metalloprotease may be regarded as marker of severity of pancreatitis in patients [89]. Gene conversion between pseudogene trypsinogen 6 (PRSS3P2) and cationic trypsinogen (PRSS1) is observed in patients suffering from chronic pancreatitis [90].

9 Colorectal Cancer

It is reported that colorectal cancer (CRC) is the second most diagnostic cancer in females and third most in males worldwide [91]. The most prognostic factors are tumour size, grade and stage. These factors do not present the clinical outcome of a patient completely. So, it is essential to have a biomarker to predict the outcome of the patients. A meta-analysis comprising of 2985 patients from 17 studies concluded that MMP-7 can be used prognostic indicator and target for treatment of CRC [92]. Capthepsin B is a lysosomal cysteine protease which is responsible for progression of cancer. The expression of this protease is upregulated in adenomas and carcinomas with respect to normal colorectal mucosa. Expression of capthepsin B was moderate in stroma and variable in epithelium in normal mucosa. Stroma of adenoma and carcinoma expresses moderate-to-high level of the protein. The activity of the protease is very high in areas of angiogenesis, inflammation and necrosis. The activity of capthepsin B is found in granular form of the epithelium in adenoma and carcinoma. Activity of this protease is associated with invasion of cancer cells, inflammatory cells, endothelial cells and cell death in human colon [93]. MMP-9 and MMP-2 are intensely associated with metastatic progression of CRC in animal model as well as in human patients [94]. MMP-9 inhibition has been found to inhibit progression of CRC [95].

It is well established that tumour derived trypsin plays a vital role in the growth and invasion of cancer cells. In colon cancer cell, trypsin acts as a growth factor through activation of the G-protein-coupled receptor protease-activated receptor-2 (PAR-2). The activation leads to matrix metalloprotease-dependent release of transforming growth factor β (TGF-β), and this, TGF-β activates epidermal growth factor receptor (EGFR) and ERK1/2 [96] (Fig. 3). Proteases play a pivotal role in the breakdown and reconstitution of basement membrane and extracellular matrix components (ECM), angiogenesis and metastasis of CRC [97]. The disruption of fine balance between MMP and TIMP is a crucial event in CRC. TIMP-1 inhibits MMP-9 facilitated release of vascular endothelial growth factor (VEGF) from matrix thereby affecting angiogenesis [98]. The spatial and temporal dysregulation of protease activity play a vital role in the progression of malignant disease. Regular

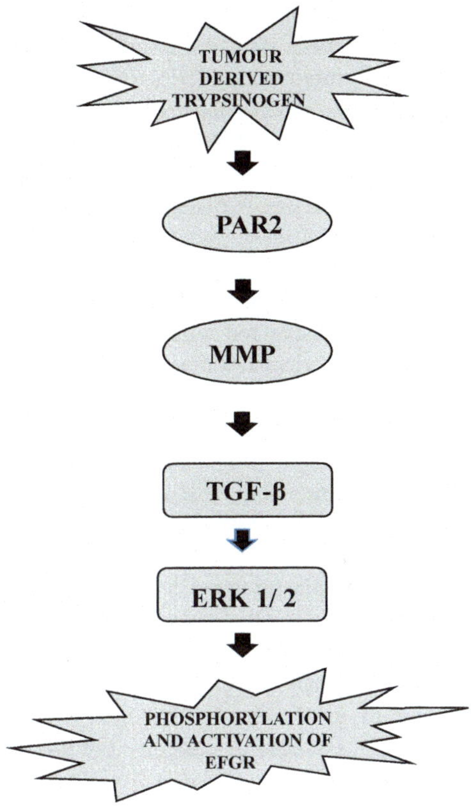

Fig. 3 Depicts the mechanism of tumour derived trypsinogen in activation of EGFR

monitoring of tumour-associated protease activity in blood sample of patients might help to improve the diagnosis of cancer and prediction of various stages. Reporter peptide is spiking and the quantification of proteolytic fragments with LC–MS showed good imitation in case of colorectal cancer patients. Functional protease profiling will improve the prognosis of colorectal cancer in future [99].

10 Gastropathy

One of the major causative agents of different gastropathies is the spiral-shaped, gram-negative, microaerophilic bacterium *Helicobactor pylori* (*H. pylori*). It has the distinctive characteristic of inhabiting in the gastric epithelium. Several studies have established that *H. pylori* are the causative agent of chronic gastritis and peptic ulcer disease (PUD). It is also highly correlated with gastric carcinoma and mucosa-associated lymphatic tissue (MALT) lymphoma [100, 101]. Its routes of transmission of this cosmopolitan pathogen may include oral–oral and faecal–oral route of transmission [100, 102]. Scientists confirmed a number of virulence factors which are related to *H. pylori* with respect to its special ecological niche and its aetiology. These are comprised of cytotoxins, several adhesins, neutrophil-activating protein and expression of laminin, collagen, extracellular matrix proteins, type IV and vitronectin. *H. pylori* strains usually express binding of soluble plasminogen whose binding was optimal at pH 7.0 [100]. Tissue plasminogen activator converted the plasminogen to plasmin. But in case of *H. pylori* infection, the activity of tissue plasminogen activator is diminished. But in contrast, the activity of urokinase increased. For enhance tissue penetration by *H. pylori*, plasminogen binding and conversion to plasmin are very important. It is the only proteolytic activity of plasminogen involved in the aetiology of the pathogen. This activity also associated with carcinogenesis by *H. pylori*, whereas *H. pylori* eradication therapy reversed these effects. Though *H. pylori* can survive routing through the gastric juice's acid milieu, it quickly drifts through the mucus layer towards the epithelial surface which has a higher pH value. Plasminogen helps this migration [100]. Conversion of plasminogen to plasmin on the cell surface potentiate the bacterium with enhance tissue breakdown ability and thus assist the invasion of *H. pylori* through the gastric epithelium [100]. The role of plasminogen in healing of epithelial wounds still remains unclear but Bugge et al. reported that homozygous deficiency of plasminogen led to multiple abnormalities including gastrointestinal tract ulceration [103]. Studies showed that, plasminogen is also involved in invasion of tumour cells by proteolytic degradation of ECM and tissue remodelling or by matrix metalloproteases activation [100, 104]. There is also a possibility in relation to gastric carcinoma and MALT-lymphoma that plasminogen helps in progression of these diseases by binding and activation [100].

Another protease, the metalloproteinase-2 (MMP-2) has controversial roles in the treatment of gastric cancer [105]. Shen et al. systematically reviewed the indication for assessment of MMP-2 expression in gastric cancer to clarify this unclear issue [105]. They found that MMP-2 overexpression was significantly correlated with the overall decrease in the survival of gastric cancer patients [105]. They also showed, by subgroup analysis, that MMP-2 overexpression had a disparaging impact on overall survival in Asian and European countries, and that the MMP-2 overexpression was expressively related with the depth of invasion, lymph node metastasis and distant metastasis [105]. Overall, the study by Shen et al. showed that MMP-2 overexpression might be a predictive factor for poor prognosis for gastric cancer [105].

Now, coming to an important intracellular proteases, the calpains (CAPN), which play a very important role in gastric health. It constitutes a varied group of intracellular cysteine proteases that are found in nearly all eukaryotes. Calpains are cytosolic proteases, which at a neutral pH, exhibit Ca^{2+}-dependent proteolytic activity. Its activity is under strict cellular regulations. One interesting thing to know about calpain is that, unlike proteasomes and lysosomal proteases, calpain proteolyse their substrates and thus are involved in proteolytic processing, instead of for degradation. Calpain-mediated proteolysis converts the substrate into new functional states, and thus differentially affects various cellular functions. Hence, it is also referred as 'modulator protease'. The importance of the calpains in gastropathy is exhibited by a variety of defects caused by compromised calpain function. CAPN8 and CAPN9 are gastrointestinal tract-specific calpains. They are being expressed mostly by the pit cells (surface mucus-secreting cells) in the stomach. Researchers have shown that endogenous CAPN8 and CAPN9 form a co-dependent heterodimer [CAPN8/9], G-calpain, in the stomach of mouse. Hata et al. showed that, though Capn8$^{-/-}$ and Capn9$^{-/-}$ mice appear healthy under normal circumstances, they are very much prone to ethanol-induced gastric ulcers [106]. They also showed that, CAPN8 protease-inactive knock-in mutant shows stress-induced gastropathy. These studies indicate that, CAPN8 and CAPN9 can be regarded as key components of gastric mucosal defence [106]. A single nucleotide polymorphism (SNP) database search has been adopted to studying the physiological functions of G-calpain. This alternative approach showed that human CAPN8 and CAPN9 contain several SNPs which remain responsible for resulting in amino acid substitutions. Besides, a study of in vitro expression suggested that the G-calpain variants, created because of these SNPs, have compromised proteolytic activity. Combining, it can be said that expression of certain CAPN8 and CAPN9 variants in individuals might cause gastrointestinal dysfunction. Now the question comes, 'How does calpain regulate gastric mucosal health?' Basically, there are two suggested pathways. Gastric mucosal defence is a complex process involving, mucus secretion, migration and differentiation of pit cells from their stem cell progenitors. Here, it is interesting to note that a subunit of the COPI concatamer complex involved in ER-Golgi retrograde transportation (major cellular phenomenon in the mucous secretion), the b-COP, is a substrate for CAPN8 [107, 108]. This suggests the probable link between G-calpain and the regulation of mucus secretion. Cousin et al. also showed that the expression of CAPN8 in cranial neural crest cells is involved in the cellular motility. This study in turn suggests that the G-calpain may be involved in the pit cell migration. Another class of ubiquitously present cysteine proteases, caspases plays significant roles in the gastropathy. In our study, we have shown that caspase-3-mediated cell death plays a pivotal role in indomethacin-induced gastropathy [109]. In that study, we have proposed a feedback loop triggered by the interplay between inflammation and oxidative stress which exaggerated the indomethacin-induced gastric damage executed by caspase-3 [109].

11 Conclusion

Digestive enzymes constitute a major portion of the protease pool within the human body. Therefore, the course of evolution of these proteases, in turn, reflects the changes in the feeding habits and the evolution of the digestive system, as a whole, from lower to higher groups of organisms. Within the digestive tract, the proteases are mainly concerned with metabolic processes. Acid proteases are found in the stomach and serine proteases in the duodenum.

The gastrointestinal tract is the largest reservoir of exogenous and endogenous proteases. Thus, proteases play a significant role in the progression of disorders of the digestive tract. Enteric microbes and parasites can seize the proteolytic pathways and this is referred to as ‘pathogen host mimicry’. Therefore, the protease balance in the enteric milieu is necessary for the physiological maintenance of the gut and disruption of this balance leads to pathological outcomes [110].

Proteases play a pivotal role in the commencement and advancement of various diseases. Transcriptomic and proteomic profiling of proteases in blood samples can help in opening new avenues for detecting possible biomarkers associated with various gastrointestinal diseases.

Thus, this chapter provides a holistic approach towards the potent role of different proteases in the pathogenesis of a wide array of gastrointestinal disorders.

References

1. Salvesen GS, Hempel A, Coll NS (2015) Protease signaling in animal and plant regulated cell death. FEBS J 22:21–25
2. Dodson G, Wlodawer A (1998) Catalytictriads and their relatives. Trends Biochem Sci 23 (9):347–352
3. Medina C, Radomski MW (2006) Role of matrix metalloproteinases in intestinal inflammation. J Pharmacol Exp Ther 318:933–938
4. Neurath H, Walsh KA (1976) Role of proteolytic enzymes in biological regulation (a review). Proc Natl Acad Sci USA 73:3825–3832
5. Eggermont E et al (1971) Distribution of enterokinase activity in the human intestine. Acta Gastroenterol Belg 34:655–662
6. Turk B (2006) Targeting proteases: successes, failures and future prospects. Nat Rev Drug Discov 5:785–799
7. Netzel-Arnett S et al (2002) Collagen dissolution by keratinocytes requires cell surface plasminogen activation and matrix metalloproteinase activity. J BiolChem 277:45154–45161
8. Yana I, Weiss SJ (2000) Regulation of membrane type-1 matrix metalloproteinase activation by proprotein convertases. MolBiol Cell 11:2387–2401
9. Jacob C et al (2005) Mast cell tryptase controls paracellular permeability of the intestine. Role of protease-activated receptor 2 and beta-arrestins. J Biol Chem 280:31936–31948
10. Hooper JD et al (2001) Type II transmembrane serine proteases. Insights into an emerging class of cell surface proteolytic enzymes. J Biol Chem 276:857–860
11. Netzel-Arnett S et al (2003) Membrane anchored serine proteases: a rapidly expanding group of cell surface proteolytic enzymes with potential roles in cancer. Cancer Metastasis Rev 22:237–258
12. Antalis TM, Lawrence DA (2004) Serpin mutagenesis. Methods 32:130–140
13. Kawabata A (2003) Gastrointestinal functions of proteinase-activated receptors. Life Sci 74:247–254
14. MacNaughton WK (2005) Epithelial effects of proteinase-activated receptors in the gastrointestinal tract. Mem Inst Oswaldo Cruz 100(Suppl 1):211–215
15. Broughton G et al (2006) The basic science of wound healing. Plast Reconstr Surg 117: 12S–34S
16. Salmela MT et al (2004) Collagenase-1 (MMP-1), matrilysin-1 (MMP-7), and stromelysin-2 (MMP-10) are expressed by migrating enterocytes during intestinal wound healing. Scand J Gastroenterol 39:1095–1104

17. Bartnik W (2008) Clinical aspects of *Helicobacter pylori* infection. Pol Arch Med Wewn 118:426–430
18. Vogiatzi P, Cassone M, Luzzi I et al (2007) *Helicobacter pylori* as a class I carcinogen: physiopathology and management strategies. J Cell Biochem 102:264–273
19. Hatakeyama M, Higashi H (2005) *Helicobacter pylori* CagA: a new paradigm for bacterial carcinogenesis. Cancer Sci 96:835–843
20. Clark IM, Swingler TE, Sampieri CL et al (2008) The regulation of matrix metalloproteinases and their inhibitors. Int J Biochem Cell Biol 40:1362–1378
21. Krueger S, Hundertmark T, Kalinski T et al (2006) *Helicobacter pylori* encoding the pathogenicity island activates matrix metalloproteinase 1 in gastric epithelial cells via JNK and ERK. J BiolChem 281:2868–2875
22. Bergin PJ, Anders E, Sicheng W et al (2004) Increased production of matrix metalloproteinases in *Helicobacter pylori*-associated human gastritis. Helicobacter 9:201–210
23. Rautelin HI, Oksanen AM, Veijola LI et al (2009) Enhanced systemic matrix metalloproteinase response in *Helicobacter pylori* gastritis. Ann Med 41:208–215
24. Yin Y, Grabowska AM, Clarke PA et al (2010) *Helicobacter pylori* potentiates epithelial: mesenchymal transition in gastric cancer: links to soluble HB-EGF, gastrin and matrix metalloproteinase-7. Gut 59:1037–1045
25. Wroblewski LE, Noble PJ, Pagliocca A et al (2003) Stimulation of MMP-7 (matrilysin) by *Helicobacter pylori* in human gastric epithelial cells: role in epithelial cell migration. J Cell Sci 116:3017–3026
26. McCaig C, Duval C, Hemers E et al (2003) The role of matrix metalloproteinase-7 in redefining the gastric microenvironment in response to *Helicobacter pylori*. Gastroenterology 130:1754–1763
27. Achyut BR, Ghoshal UC, Moorchung N et al (2003) Transforming growth factor-B1 and matrix metalloproteinase-7 promoter variants induce risk for *Helicobacter pylori*-associated gastric precancerous lesions. DNA Cell Biol 28:295–301
28. Ogden SR, Wroblewski LE, Weydig C et al (2008) p120 and Kaiso regulate *Helicobacter pylori*-induced expression of matrix metalloproteinase-7. Mol Biol Cell 19:4110–4121
29. Bodger K, Ahmed S, Pazmany L et al (2006) Altered gastric corpus expression of tissue inhibitors of metalloproteinases in human and murine helicobacter infection. J Clin Pathol 61:72–78
30. Edwards DR, Handsley MM, Pennington CJ (2008) The ADAM metalloproteinases. Mol Aspects Med 29:258–289
31. Schirrmeister W, Gnad T, Wex T et al (2009) Ectodomain shedding of E-cadherin and c-Met is induced by *Helicobacter pylori* infection. Exp Cell Res 315:3500–3508
32. Sampieri Clara L (2013) *Helicobacter pylori* and gastritis: the role of extracellular matrix metalloproteases, their inhibitors, and the disintegrins and metalloproteases—a systematic literature review. Dig Dis Sci 58:2777–2783
33. Erin N, Türker S, Elpek Ö, Yıldırım B (2012) Differential changes in substance P, VIP as well as neprilysin levels in patients with gastritis or ulcer. Peptides 35:218–224
34. Browning KN (2010) Protease-activated receptors: novel central role in modulation of gastric functions. Neuro Gastroenterol Motil 22:361–365
35. Beutler E, Grabowski G, Scriver CR, Beaudet AL, Sly WS, Valle D, editors (1995) Gaucher disease In: The metabolic basis of inherited disease. McGraw-Hill, New York 2641–2670
36. Moran MT, Schofield JP, Hayman AR, Shi G-P, Young E, Cox TM (2000) Pathologic gene expression in gaucher disease: up-regulation of cysteine proteinases including osteoclastic cathepsin K. Blood 96:1969–1978
37. Moyses C (2003) Substrate reduction therapy: clinical evaluation in type 1 gaucherdisease. Philos Trans R Soc Lond B Biol Sci 358(1433):955–960
38. Villanueva J, Shaffer DR, Philip J et al (2006) Differential exoprotease activities confer tumor-specific serum peptidome patterns. J Clin Invest 116:271–284

39. van Breemen MJ, Aerts JMFG, Sprenger RR, Speijer D (2006) Potential artefacts in proteome analysis of plasma of gaucher patients due to protease abnormalities. Clinica Chimica Acta 396:26–32
40. Aerts JMFG, Hollak CEM (1997) Plasma and metabolic abnormalities in gaucher's disease. Baillieres Clin Haematol 10:691–709
41. Asehnoune K, Moine P (2013) Protease activated receptor-1: key player in the sepsis coagulation—inflammation crosstalk. Crit Care 17:119
42. Adams MN, Ramachandran R, Yau M-K, Suen JY, Fairlie DP, Hollenberg MD, Hooper JD (2013) Structure, function and pathophysiology of protease activated receptors. Pharmacol Ther 130:248–282
43. van der Poll T, Levi M (2012) Crosstalk between inflammation and coagulation: the lessons of sepsis. CurrVasc Pharmacol 10:632–638
44. Petäjä J (2013) Inflammation and coagulation. an overview. Thromb Res 127(Suppl 2): S34–S37
45. Jenkins RG, Su X, Su G, Scotton CJ, Camerer E, Laurent GJ, Davis GE, Chambers RC, Matthay MA, Sheppard D (2006) Ligation of protease-activated receptor 1 enhances alpha (v)beta6 integrin-dependent TGF-beta activation and promotes acute lung injury. J Clin Invest 116:1606–1614
46. Gonzalo S, Valero MS, de Salinas FM, Vergara C, Arruebo MP, Plaza MA, Murillo MD, Grasa L (2015) Roles of toll-like receptor 4, IjB kinase, and the proteasome in the intestinal alterations caused by sepsis. Dig Dis Sci 60:1223–1231
47. Savioli L, Smith H, Thompson A (2006) Giardia and cryptosporidium join the 'neglected diseases initiative'. Trends Parasitol 22(5):203–208
48. Cotton JA, Beatty JK, Buret AG (2011) Host parasite interactions and pathophysiology in giardia infections. Int J Parasitol 41(9):925–933
49. DuBois KN, Abodeely M, Sakanari J, Craik CS, Lee M, McKerrow JH et al (2008) Identification of the major cysteine protease of giardia and its role in encystation. J Biol Chem 283(26):18024–18031
50. Kissoon-Singh V, Mortimer L, Chadee K (2011) Entamoebahistolyticacathepsin-like enzymes: interactions with the host gut. Adv Exp Med Biol 712:62–83
51. Aurrecoechea C, Brestelli J, Brunk BP, Carlton JM, Dommer J, Fischer S et al (2009) GiardiaDB and TrichDB: integrated genomic resources for the eukaryotic protist pathogens giardia lamblia and trichomonas vaginalis. Nucleic acids Res 37(Database issue): D526–D530
52. Cotton JA, Bhargava A, FerrazJG Yates RM, Beck PL, Buret AG (2014) Giardia duodenaliscathepsin B proteases degrade intestinal epithelial interleukin-8 and attenuate interleukin-8-induced neutrophil chemotaxis. Infect Immun 45:53–67
53. Scott KG, Meddings JB, Kirk DR, Lees-Miller SP, Buret AG (2002) Intestinal infection with Giardia spp. reduces epithelial barrier function in a myosin light chain kinase-dependent fashion. Gastroenterology 123(4):1179–1190
54. Teoh DA, Kamieniecki D, Pang G, Buret AG (2000) Giardia lamblia rearranges F-actin and alpha-actinin in human colonic and duodenal monolayers and reduces transepithelial electrical resistance. J Parasitol 86(4):800–806
55. Wang Y, Srinivasan K, Siddiqui MR, George SP, Tomar A, Khurana S (2008) A novel role for villin in intestinal epithelial cell survival and homeostasis. J Biol Chem 283(14):9454–9464
56. Bhargaval A, Cotton JA, Dixon BR, Gedamu L, Yates RM, Buret AG (2010) Giardia duodenalis surface cysteine proteases induce cleavage of the intestinal epithelial cytoskeletal protein villin via myosin light chain kinase. PLOS One 1110–1117
57. Kaser A, Zeissig S, Blumberg RS (2010) Genes and environment: how will our concepts on the pathophysiology of IBD develop in the future? Dig Dis 28:395–405

58. Bouguen G, Chevaux JB, Peyrin-Biroulet L (2011) Recent advances in cytokines: therapeutic implications for inflammatory bowel diseases. World J Gastroenterol 17:547–556
59. Anand PK, Malireddi RK, Lukens JR, Vogel P et al (2012) NLRP6 negatively regulates innate immunity and host defence against bacterial pathogens. Nature 488:389–393
60. Chen GY, Liu M, Wang F, Bertin J et al (2011) Afunctional role for Nlrp6 in intestinal inflammation and tumorigenesis. J. Immunol. 186:7187–7194
61. Elinav E, Strowig T, Kau AL, Thaiss CA et al (2011) NLRP6 inflammasome regulates colonic microbial ecology and risk for colitis. Cell 145:745–757
62. Qie BS, Vallance BA, Blennerhassett PA, Collins SM (1999) The role of CD4+lymphocytes in thesusceptibility of mice to stress-induced reactivation of experimental colitis. Nat Med 5:1–5
63. Mayer EA (2000) The neurobiology of stress and gastrointestinal disease. Gut 47:861–869
64. Hollander D (2003) Inflammatory bowel diseases and brain-gut axis. J Physiol Pharmacol 54:183–190
65. Gomez DE, Alonso DF, Yoshiji H, Thorgeirsson UP (1997) Tissue inhibitors of metalloproteinases: structure, regulation and biological functions. Eur J Cell Biol 74:111–122
66. Brew K, Dinakarpandian D, Nagase H (2000) Tissue inhibitors of metalloproteinases: evolution, structure and function. Biochim Biophys Acta 1477:267–283
67. Rath T, Roderfeld M, Graf J, Wagner S et al (2006) Enhanced expression of MMP-7 and MMP-13 in inflammatory bowel disease: a precancerous potential? Inflamm Bowel Dis 12:1025–1035
68. Vizoso FJ, Gonzalez LO, Corte MD, Corte MG, Bongera M, Martinez A, Martin A, Andicoechea A, Gava RR (2006) Collagenase-3 (MMP-13) expression by inflamed mucosa in inflammatory bowel disease. Scand J Gastroenterol 41:1050–1055
69. Roosmarijn VE, Hauwermeiren FV et al (2013) Matrix metalloproteinase 13 modulates intestinal epithelial barrier integrity in inflammatory diseases by activating TNF. EMBO Mol Med 5: 932–948
70. Rath T, Roderfeld M, Graf J, Wagner S, Vehr AK et al (2006) Enhanced expression of MMP-7 and MMP-13 in inflammatory bowel disease: a precancerous potential? Inflamm Bowel Dis 12:1025–1035
71. Dłubacz AK,, Matusiewicz M, Krzesiek E et al (2014) Metalloproteinase-3 and -9 as novel markers in the evaluation of ulcerative colitis activity in children. Adv Clin Exp Med 23: 103–110
72. Vizoso FJ, González LO, Corte MD, Corte MG, Bongera M et al (2006) Collagenase-3 (MMP-3) expression by inflamed mucosa in inflammatory bowel disease. Scand J Gastroenterol 41:1050–1055
73. Kirkegaard T, Hansen A, Bruun E, Brynskov J (2004) Expression and localisation of matrix metalloproteinases and their natural inhibitors in fistulae of patients with Crohn's disease. Gut 53:701–709
74. Yoo J, Perez CER, Nie W, Smith JS et al (2011) Protein Kinase D1 mediates synergistic MMP-3 expression induced by TNF-α and bradykinin in human colonic myofibroblasts. Biochem Biophys Res Commun 413:30–35
75. Saskia Braber S, Mary E Morgan ME, A J HenricksA J, Roda MA et al (2014) Collagen degradation and neutrophilic infiltration: a vicious circle in inflammatory bowel disease. Gut 63:578–587
76. Stallmach A, Chan CC, Ecker KW, Feifel G, Herbst H et al (2000) Comparable expression of matrix metalloproteinases 1 and 2 in pouchitis and ulcerative colitis. Gut 47:415–422
77. Cupi ML, Sarra M, Marafini I, Monteleone I, Ortenzi FEA (2014) Plasma cells in the mucosa of patients with inflammatory bowel disease produce granzyme b and possess cytotoxic activities. J Immunol 192:6083–6091

78. Biancheri P, Brezski J, Di Sabatino, Greenplate R et al (2015) Proteolytic cleavage and loss of function of biologic agentsthat neutralize tumor necrosis factor in the mucosa of patients with inflammatory bowel disease. Gastroenterology 149:1564–1574
79. Gaisano HY, Gorelick FS (2009) New insights into the mechanisms of pancreatitis. Gastroenterology 136(7) 2040–2044
80. Willemer S, Bialek R, Adler G (1990) Localization of lysosomal and digestive enzymes in cytoplasmic vacuoles in caerulein-pancreatitis. Histochem 94:161–170
81. Binker MG, Daniel Richards D, Gaisano HY et al (2015) ER stress-associated CTRC mutants decrease stimulated pancreatic zymogen secretion through SIRT2-mediated microtubule Dysregulation. Biochem Biophys Res Commun 463:329–335
82. Jancsó Z, Sahin-Tóth M (2016) Tighter control by chymotrypsin C (CTRC) explains lack of association between human anionic trypsinogen and hereditary pancreatitis. J Biol Chem 291 (25) pii: jbc.M116.725374
83. von Figura K, Hasilik A (1986) Lysosomal enzymes and their receptors. Annu Rev Biochem 55:167–193
84. Hentze M, Hasilik A, von Figura K (1984) Enhanced degradation of cathepsin D synthesized in the presence of the threonine analog beta-hydroxynorvaline. Arch Biochem Biophys 230:375–382
85. Samarel AM, Ferguson AG, Decker RS, Lesch M (1989) Effects of cysteine protease inhibitors on rabbit cathepsin D maturation. Am J Physiol 257:C1069–C1079
86. Gieselmann V, Von Figura AK (1985) Processing of human cathepsin D inlysosomes in vitro. J Biol Chem 260:3215–3220
87. Mehanna S, Suzuki C, Shibata M, Sunabori T et al (2016) Cathepsin D in pancreatic acinar cells is implicated in cathepsin B and L degradation, but not in autophagic activity. Biochem Biophys Res Commun 469:405–411
88. Li H, Liu J, Wang W, Zhang Z et al (2015) Matrix metalloproteinase 9 and vasodilator-stimulated phosphoprotein related to acute kidney injury in severe acute pancreatitis rats. Dig Dis Sci 60:3647–3655
89. Wereszczynska-Siemiatkowska U, Siemiatkowski A, Swidnicka-Siergiejko A, Mroczko B, Dabrowski A (2015) The imbalance between matrix metalloproteinase 9 and tissue inhibitor of metalloproteinase 1 in acutepancreatitis. Z Gastroenterol. 53:199–204
90. Rygiel AM, Beer S, Simon P, Tysarowska KW (2015) Gene conversion between cationic trypsinogen (*PRSS1*) and the pseudogene trypsinogen 6 (*PRSS3P2*) in patients with chronic pancreatitis. Hum Mutat 36:350–356
91. Emal A, Bray F, Center MM, Ferlay J, Ward E, Forman D (2011) Global cancer statistics. CA Cancer J Clin 61:69–90
92. Sun D, Zhang Y, Qi Y, Xing-tong Zhou X et al (2015) Prognostic significance of MMP-7 expression in colorectal cancer: a meta-analysis. Cancer Epidemiol 39(2):135–142.
93. Lonny GM, Fonnet H, Bleeker E, Lauritzen B, Bahns S et al (2000) Comparative localization of cathepsin B protein and activity in colorectal cancer. J Histochem Cytochem 48:1421–1430
94. Mook ORF, Frederiks WM, Van Noorden CJF (2004) The role of gelatinases in colorectal cancer progression and metastasis. Biochem Biophy Acta 1705:69–89
95. Ndinguri MW, Bhowmick M, Tokmina-Roszyk D, Robichaud TK (2012) Peptide-based selective inhibitors of matrix metalloproteinase mediated activities. Molecules 30:14230–14248
96. Darmoul D, Gratio V, Devaud H (2004) Protease-activated Receptor 2 in Colon Cancer. J Biol Chem 279:20927–20934
97. Herszényi L, Lakatos G, Hritz I, Varga MZ et al (2012) The role of inflammation and proteinases intumor progression. Dig Dis 30:249–254
98. Lambert E, Dassé E, Haye B, Petitfrère E (2004) TIMPs as multifacial proteins. Crit Rev Oncol Hematol 49:187–198

99. Yepes D, Costina V, Lothar R et al (2014) Multiplex profiling of tumor-associated proteolytic activity in serum of colorectal cancer patients. Proteomics Clin Appl 8: 308–316
100. Ljungh A (2000) *Helicobacter pylori* interactions with plasminogen methods 21:151–157
101. Correa P, Fox J, Fontham E (1993) *Helicobacter pylori* and gastric carcinoma: serum antibody prevalence in populations with contrasting cancer risks. Cancer 66:2569–2574
102. Makristathis A, Pasching E, Schütze K, Wimmer M, Rotter (1998) Detection of *Helicobacter pylori* in stool specimens by PCR and antigen enzyme immunoassay J Clin Microbiol 36:2772–2774
103. Bugge TH, Kombrinck KW, Flick MJ, Daugherty CC (1996) Loss of fibrinogen rescues mice from the pleiotropic effects of plasminogen deficiency. Cell 87:709–716
104. Romer J, Bugge TH, Pyke C, Lund LR et al (1996) Impaired wound healing in mice with a disrupted plasminogen gene. Nat Med 2:287–292
105. Shen W, Xi H, Wei B, Chen Lin (2014) The prognostic role of matrix metalloproteinase 2 in gastric cancer: a systematic review with meta–analysis. J Cancer Res Clin Oncol 140: 1003–1009
106. Hata S, Abe M, Suzuki H, Kitamura F et al (2010) Calpain 8/nCL-2 and Calpain 9/nCL-4 constitute an active protease complex, G-calpain, involved in gastric mucosal defense. PLoS Genet 6:e1001040
107. Cousin H, Abbruzzese G, Kerdavid E, Gaultier AA (2011) Translocation of the cytoplasmic domain of ADAM13 to the nucleus is essential for calpain8-a expression and cranial neural crest cell migration. Dev Cell 20:256–263
108. Sorimachi H, Hata S, Ono Y (2011) Impact of genetic insights into calpain biology. J Biochem 150:23–37
109. Sinha K, Sadhukhan P, Saha S, Pal PB, SilPC (2015) Morin protects gastric mucosa from nonsteroidal anti-inflammatory drug, indomethacin induced inflammatory damage and apoptosis by odulating NF-κB pathway. Biochim Biophys Acta 1850:769–783
110. Antalis TM, Donohue TS, Stefanie N Vogel et al (2007) Mechanisms of disease: protease functions in intestinal mucosal pathobiology. Nat Clin Pract Gastroenterol Hepatol 4: 393–402

Matrix Metalloproteases: Potential Role in Type 2 Diabetic Nephropathy

Gantala Srilatha Reddy and Hanumanth Surekha Rani

Abstract

Type 2 diabetes mellitus is the most common form and constitutes a major diabetic population in all countries. The complications of diabetes mellitus (DM) include nephropathy, neuropathy, retinopathy, and cardiovascular disease. Type 2 diabetic nephropathy (DN) is a devastating complication of DM and a main cause of end-stage renal failure. Evidences show that susceptibility to Type 2 DN has a significant genetic component in addition to environmental factors. In Type 2 DN, hyperglycemia-induced changes include extracellular matrix (ECM) deposition, basement membrane (BM) thickening, as well as vascular smooth muscle and mesangial cell growth. ECM proteins are degraded by zinc-dependent endopeptidases called matrix metalloproteases (MMPs) which in turn are regulated by tissue inhibitors of metalloproteases (TIMPs). The proteases (MMPs) and antiproteases (TIMP) offer the opportunity to identify the determinants of the disease that are very likely to be causative and might lead to new therapeutics with strong molecular underpinning.

Keywords

Diabetes mellitus · Type 2 diabetic nephropathy · Glomerular basement membrane · Matrix metalloproteases · Tissue inhibitors of metalloproteases Polymorphism

G. Srilatha Reddy · H. Surekha Rani (✉)
Department of Genetics & Biotechnology, Osmania University, Hyderabad 500007, Telangana, India
e-mail: surekharanih@gmail.com

S. Chakraborti and N.S. Dhalla (eds.), *Pathophysiological Aspects of Proteases*,
DOI 10.1007/978-981-10-6141-7_25

1 Introduction

1.1 Diabetes Mellitus

Diabetes Mellitus (DM) is a chronic multifactorial disorder associated with a relative or absolute deficiency of insulin or its function and presently reconciles as the twenty-first century's most common chronic disease worldwide. Globally the incidence of DM is increasing in all age-groups and the prevalence was estimated to be 2.8% (175 million people) in 2000 while recent trends show that it might increase to 4.4% (366 million people) in 2030 [1, 2].

DM is the most incessant disease and is the cause of death throughout the world and now ranks the fifth, succeeding communicable diseases, cancer, cardiovascular disease, and chronic respiratory diseases [3]. The long-term effects of DM include gradual development of specific complications of retinopathy (potential blindness), nephropathy (leading to kidney failure), and/or neuropathy, etc. Renal failure in diabetes, particularly in Type 2 DM, has become "a medical catastrophe of worldwide dimension."

1.2 Diabetic Nephropathy

One of the extreme complications of DM is diabetic nephropathy (DN) and develops in 25–40% of patients with Type 1 or Type 2 DM and an essential cause of increased morbidity and mortality in these patients [4]. It is the common cause of end-stage renal disease (ESRD) requiring dialysis. The classical definition of Type 2 DN is a progressive rise in urine albumin excretion, copulates with increasing blood pressure, and leads to declining glomerular filtration and eventually ESRD.

The risk factors for Type 2 DN include reduced glycemic control, prolonged duration of diabetes, insulin resistance, high blood pressure (BP), advanced age, smoking, race, genetic propensity, etc.

The existence of microalbuminuria is the earliest clinical evidence of nephropathy. Microalbuminuria progresses result in proteinuria. Once overt proteinuria develops, renal function progressively declines and ESRD attains [5]. Both environmental and genetic factors accord to the development and outcome of Type 2 DN.

The functional unit of the kidney is the glomerulus and glomerular mesangial cells together with resident monocytes/macrophages, and surrounding matrix material constitutes the mesangium. It is involved in regulation of glomerular circulation and filtration [6]. In addition to regulating glomerular filtration, it is also involved in basement membrane (BM) remodeling. Hyperglycemia induces hemodynamic and metabolic stimuli mediators for kidney injury. These activate ischemic, pro-oxidant, fibrotic, and inflammatory pathways leading to mesangial matrix accumulation.

Prolonged duration of hyperglycemia in Type 2 DN leads to noticeable changes like: diffuse glomerulosclerosis, thickened glomerular basement membrane (GBM), nodular glomerulosclerosis, podocyte loss, exudative lesions in the Bowman's capsule, glomerular hyperperfusion, and hyperfiltration. Many factors have been divulged to be involved in this defective regulation, including vascular endothelial growth factor (VEGF), nitric oxide, and the renin-angiotensin system, especially angiotensin II [7].

Alteration in the morphology occurs in mesangial cells due to continuous cycle of stretch/relaxation and leads to enhanced proliferation and increased production of extra cellular matrix (ECM) components. The effect of stretch occurs partially as a result of increase in expression of ECM components, fibronectin, collagen I, III, and IV, and laminin. The hemodynamic changes increase the production of mesangial cell matrix, promotes the leakage of albumin from the glomerular capillaries, expands the GBM, and causes injury to podocytes. It can also induce the release of localized growth factors and cytokines which in turn activates some of the matrix metalloproteases (MMPs) which further degrade the ECM present on GBM.

2 Matrix-Metalloproteases: General Aspects

MMPs are tightly regulated enzymes having the ability to degrade the ECM and BM constituents. They are a family of endopeptidases that share structural domains but diverge in cellular sources, substrate specificity, and inducibility. All MMPs have several prevailing characteristics which are as follows:

1. MMP family members within their organization have a conserved domain pattern.
2. They contain Zn^{2+} at their active site and require calcium a cofactor for their stability.
3. They are secreted as inactive zymogens and require activation for further ECM degradation.
4. The proteins which build up the BM and the ECM are the common substrates for all MMPs.
5. The MMP's enzymatic activity is optimal at physiological pH.
6. The MMP's proteolytic activity is inhibited by TIMPs.

The MMPs have been classified into six subgroups, based on the sequence homology and substrate specificity, Collagenases (MMP-1, 8, 13 and 18), Gelatinases (MMP-2 and 9), Stromelysins (MMP-3, 10 and 11), Matrilysins (MMP-7 and 26), Membrane-type matrix metalloproteases (MT-MMP-14, 15, 16, 17, 24 and 25), and other MMPs (MMP-12, 19, 20, 21, 23, 27 and 28). Except for the MT-MMPs, most of the MMPs are secreted out and have extracellular distribution; however, recent evidence suggests that some MMPs like MMP-1, 2 and 11 have intracellular expression where they may merge with cytosolic proteins to modulate various biological processes [8].

3 The Biological Roles of the MMPs

Numerous physiological and developmental events are regulated by ECM. MMPs degrade ECM, and the main function is assumed to be the remodeling of the ECM. Morphogenesis and tissue growth is a critical process of remodeling of ECM and MMPs control angiogenesis by releasing pro-angiogenic factors such as basic fibroblast growth factor (bFGF) or VEGF [9]. MMPs uphold cell proliferation by augmenting the release of insulin-like growth factor (IGF) and the transforming growth factor α (TGF-α). Regulation of apoptosis is the main biological role of MMPs as an increase in apoptotic cell death is substantiated by the over expression of certain MMPs like MMP-3, 7, 9, and 11.

4 Regulation of MMPs

MMPs are neutral proteases maintain the equity between synthesis and degradation of matrix proteins. The MMP proteolytic actions are controlled at three levels involving, proenzyme activation, transcription, and inhibition by the TIMPs. The MMPs are synthesized as pro-MMPs (latent zymogens), and their enzymatic activation requires prodomain removal. The serine proteases or the other active MMPs also extracellularly activate most MMPs, for example, MMP-2 requires active MT1-MMP and TIMP-2 which binds to C-terminus of pro-MMP-2 which further undergoes a complex activation pathway.

The MMP transcriptional regulation mechanism in kidney disease is quite complex and in turn induced by various signals, such as cytokines, oncogene products, growth factors and also the activation of many other signal transduction pathways [10]. Metabolic pathways are the major arbitrators of Type 2 DN involving the activation of the immune system and chronic inflammation. Several studies suggest that the meager raise in monocytes/macrophages noticed in glomeruli contributes significantly to the development of Type 2 DN. Inherent renal cells, in conjunction with mesangial, glomerular endothelial, dendritic, and renal tubular cells, are able to upregulate the inflammatory factors and cytokines, mainly interleukin 1 (IL-1), IL-6, and IL-18, VEGF, tumor necrosis factor a (TNF-a), and TGF-β, which all have been implicated in transcription regulation of MMPs in Type 2 DN progression.

Lastly, the MMP's activity is controlled tightly by the action of endogenous inhibitors (TIMPs). They have an N-terminal and C-terminal domain of 125 and 65 amino acids with each comprising three conserved disulfide bonds. The N-terminal domain folds as a separate unit and is capable of inhibiting most of the MMPs. In addition to TIMPs, another important inhibitor of MMPs is α-2 macroglobulin which binds to MMP receptors generating an MMP-macroglobulin complex which is inactive.

5 High Glucose (HG) and MMPs

There are several mechanisms by which dysregulation in renal MMPs and TIMPs ratio or activity in kidney could commit to the development of progressive Type 2 DN. As previously mentioned, MMP's expression is firmly controlled by diverse mechanisms that include transcription and posttranscription. Evidence illustrates that increase in glucose levels may also regulate MMP gene expression via varied transcription factors NF-κB and AP-1 or relies upon growth factors like connective tissue growth factor (CTGF) and transforming growth factor (TGF-β) [11, 12].

6 AGEs and MMPs

Glycol-oxidation end products or advanced glycation end products are known as AGEs are one of the major aberrantly synthesized molecules that can influence the progression of Type 2 DN. Hyperglycemia leads to the generation of AGEs which is primarily as a consequence of a condensation of free amino group and sugar with the development of a labile Schiff base which undergoes complicated intramolecular modification to generate complex toxic AGEs. Such derivatization can occur between sugar and lipids as well, and their production can be initiated in both intra and extracellular compartments [13, 14].

Several forms of AGE derivatives have been described in renal injury related to diabetes, and morphologic changes that are often related with it encompass glomerular and tubular BM thickening, capillary aneurismal delay, mesangial expansion with production of Kimmelstiel-Wilson nodules, arteriolar thickening, and hyalinosis. AGE formation can modify the function of certain important ECM molecules, such as collagen Type 1, 3, 4, fibronectin, laminin, etc. Furthermore, AGEs can modulate the intracellular signaling pathways, gene and protein expression by interacting with their receptors, i.e., RAGE.

The AGEs via AGE: RAGE synergy can stimulate PKC, MAPK, and NF-kB, which can, in turn, modulate the expression of TGF-β and subsequently MMPs. Such ligand: receptor synergy can also generate reactive oxygen species (ROS), which then can regulate the MMPs expression via articulation of various transcription factors [15].

7 MMPs in Type 2 Diabetic Nephropathy

Type 2 diabetic nephropathy is a feature of renal and interstitial fibrosis, which ultimately progresses toward ESRD. Normal growth and development is the main physiologic feature of ECM remodeling. At several levels of ECM turnover MMPs are involved, playing a crucial role in glomerular ECM synthesis and degradation [16]. MMPs are redox sensitive, as high levels of oxidative stress and inflammatory markers activate them from their inactive latent form to active form. The amount and durational activity of MMPs determine the extent of glomerular structural damage as demonstrated by higher MMP levels in various glomerulonephritis forms [17].

The observations of Johnson et al. [18] could relate redox stress, inflammation, and MMP in glomerular failure. The presence of elevated levels of MMPs within inflammatory glomeruloscelerosis has led to the indication that increased inflammatory response coupled with increased MMP levels may cause the glomerular damage in Type 2 DN [19].

With a robust activation of MMPs, the BM and supporting ECM could be degraded allowing a complete detachment of the renal cells (mesangial cell, podocyte, and endothelial cell) within the glomerulus and the proximal renal tubule ensuing in ECM remodeling, apoptosis, atrophy, dysfunction, and cytoskeleton rearrangement. MMPs produced in various cell types in the kidney suggest their involvement in ECM degradation in glomerulus, renal morphogenesis, and remodeling [20, 21]. The dysregulated remodeling and accumulation of ECM in renal fibrosis affect all main compartments of the kidney being termed glomerulosclerosis in the glomeruli, tubulointerstitial fibrosis in the tubulointerstitium, and arterio- and arteriolosclerosis in the vasculature, where all renal cells are involved in fibrosis [22].

8 Polymorphic Studies of MMPs in Type 2 DN

Single nucleotide polymorphism (SNPs) can be used to identify the DNA sequence variation, susceptibility to disease and to understand an individual response to certain drugs.

8.1 Collagenases

Collagenases are enzymes that cleave the peptide bonds of collagen, the main component of ECM, includes collagenase-1, 8 and 13. MMP-1 (Interstitial Collagenase) which is localized on chromosome 11q22 is the most ubiquitously considered interstitial collagenase and it's over expression is associated with several diseases. MMP-1 having collagen as the main substrate is a key determinant of

ECM degradation, degrades the interstitial collagens, collagen Type 1, 2, and 3 [23]. A SNP at −1607 bp promoter region in human MMP-1 gene has been found to be associated with elevated risk of various diseases including Type 2 DN. The MMP-1 promoter region contains consensus sequences for DNA-binding proteins such as Ets/PEA-3, AP-1, and AP-2 [24].

The transcription is augmented by creating an Ets binding site at SNP (−1607) promoter region. The transcriptional start site of the MMP-1 gene at −1607 position is proportionate by two alleles (1G) or (2G) guanine nucleotides [25]. A 20 fold higher transcriptional activity seen in 2G allelic promoter of MMP-1 gene than 1G allele and is associated with higher MMP-1 levels.

MMP-8 (Matrix metalloproteinase-8) plays a major role in degradation of ECM components, collagens Type 1, 2 and 3, modifies the immune responses and regulates the cytokine activity. The gene coding for MMP-8 is located on chromosome 11q22.3. The three polymorphisms at −381A/G, −799 C > T and +17C/G in the MMP-8 gene are known to modify the gene transcription [26].

Human collagenase-3 (MMP-13), a protease, plays a role in ECM degradation and cleaves collagen Type 1, 2, 3, 4, 14 and Type 10 and also activates or degrades key regulatory proteins such as CTGF and TGFB1. The gene coding for MMP-13 is located on chromosome 11q22.2. A SNP at 77A → G in the promoter region of the MMP-13 gene has been detected. The SNP at this position leads to alterations in the gene expression and plays a modulatory role on transcriptional activity in the pathogenesis of various diseases [27].

8.2 Gelatinases

Gelatinases are proteolytic enzymes that hydrolyze gelatin into polypeptides, peptides, and amino acids. In humans, gelatinases are MMP-2 and MMP-9. MMP-2 is notably interesting because of its multiple functions and ubiquitous expression [28]. MMP-2, also known as gelatinase A, is a 72 kDa that degrades Type 4 collagen and is located on chromosome 16q12.2. It is a fundamental component of the BM. Increased expression of MMP-2 may accelerate the degeneration of gap junction protein and Type 4 collagen, leading to the vascular complications of diabetes and Type 2 DN.

Several SNPs have been described in the promoter region of MMP-2 and of these two SNPs C-1306T and C-735T prevailing upstream from the transcriptional start site effect MMP-2 transcriptional activity. MMP-2 is likely to be regulated by transcription factors; among these controlling elements, Sp1 binding site is important.

Price et al. [29] suggested that the MMP-2 transition at promoter region C-1306T could remarkably alter the promoter activity, due to disruption of the Sp1 binding site (CCACC box). MMP-2 is expressed in mesangial cells, and as a result of proinflammatory signaling, the dramatical expression of MMP-2 is elevated in various glomerulopathies [30].

MMP-9 (gelatinase B) degrades the ECM components exclusively Type 4 collagen, gelatin, and laminin, and the gene coding for MMP-9 is located on chromosome 20q11.2-q13.1 [31]. A number of SNPs found in the promoter region particularly an SNP at −1562C > T has distinct implication and has an allele-specific effect on MMP-9 transcription suggesting that MMP-9 may play a critical role in the turnover of the mesangial matrix in Type 2 DN [32]. This promoter contains the 9-bp sequence (GCGCAC/TGCC) an important gene expression regulatory element. The increased levels of MMP-9 results are due to the loss of DNA–protein interaction due to a C to T substitution.

In addition, a recent study has shown that Type 2 DN patients has elevated levels of MMP-9 and had a substantial association with age, hyperglycemia, blood pressure, body mass index, glycated hemoglobin (HbA1c), and progression of diabetes [33].

8.3 Stromelysins

Stromelysin, the enzymes of MMP family, plays a major role in the degradation of proteoglycans, gelatin, and other ECM constituents. It includes MMP-3 (stromelysin-1), MMP-10 (stromelysin-2) and MMP-11 (stromelysin-3). MMP-3 (stromelysin-1), a member of MMP-family degrades collagen Types 3, 4, 9 and 10, proteoglycans, elastin, fibronectin and laminin, and also involved in the activation of other MMPs (e.g. pro-MMP-1, -8, -9 and -13), and auto activation of pro-MMP-3. The gene is located on the chromosome 11q22.2-22.3, and the expression level of this gene was found to be altered by SNPs.

At the upstream of the MMP-3 transcription start site, the SNP identified in the promoter region at −1171 bp has one allele having five adenosines (5A) and other having run of six adenosines (6A) and was found the allele having five adenosines has higher transcriptional activity [34]. MMP-3 elevated levels have been observed in sera from patients with a number of diseases like active lupus nephritis, mesangial proliferative glomerulonephritis, IgA nephropathy, etc.

Stromelysin-2 (MMP-10) is another important Zn-dependent endopeptidase cleaves laminin, fibronectin, proteoglycan core protein, elastin, gelatins of Type 1, 3, 4, and 5. MMP-10 also activates pro-MMP-1, -7, -8 and -9, which degrades extracellular collagen in different pathological conditions along with other MMPs. The gene is part of a cluster of MMP genes and located on chromosome 11q22.3. The two polymorphisms rs17435959 (G > C) at position 102780582 and rs17293607 (C > T) at position 102779658 are identified. These two polymorphisms lead to the substitution of specific amino acids and may have functional effects. MMP-10 serum levels have a close relationship with some risk factors for ischemic stroke, such as carotid intima-media thickness, presence of carotid plaques, inflammatory markers and smoking and its role in the pathogenesis of Type 2 DN are still unexplored.

Stromelysin-3 (MMP-11), the enzyme encoded by this gene is intracellularly activated by furin, in contrast to other MMP's, and cleaves alpha 1-proteinase inhibitor but weakly degrades structural proteins of the ECM. The gene for MMP-11 is located on chromosome 22q11.23. MMP-11 associated diseases mostly include ophthalmomyiasis and colorectal cancer.

8.4 Tissue Inhibitors Metalloproteases (TIMPs)

MMPs are inhibited by TIMPs consisting of 184–194 amino acids. Four structurally related TIMP family members include TIMP-1, -2, -3 and -4. Inhibition is accomplished by their ability to interact with the zinc-binding site within the catalytic domain of active MMPs [35]. TIMPs have a certain degree of specificity toward MMP family members. To maintain the sustainability of healthy tissues, the balance between MMPs and TIMPs is important. The balance disrupted by the polymorphic variants within TIMP gene is associated with development of various diseases.

The biological activities of TIMPs involve anti-angiogenesis, cell migration, effects on cell growth and differentiation, synaptic plasticity, anti- and pro-apoptosis. Type 2 DN certainly associate with altered activity of the MMP and TIMP which in turn leads to glomerulosclerosis [36].

8.4.1 Metallopeptidase Inhibitor 1/TIMP-1

The best-identified gene is the TIMP-1, present on X chromosome (Xp11.3-p11.23). TIMPs bind with active MMPs with high affinity and have complicated roles in pathological and physiological tissue remodeling. TIMPs inhibit all MMPs, but TIMP-1 is an indigent inhibitor of MT1-MMP, MT3-MMP, MT5-MMP, MMP-19 and, ADAM-10. At exon 6 of TIMP-1 gene located 536C/T polymorphism capable of binding and preventing the activation of most MMPs [37].

8.4.2 Metallopeptidase Inhibitor 2/TIMP-2

TIMP-2 is a secretory protein located at 17q25, inhibits the proteolytic activity of MMP-2 involved in the ECM degradation. Additionally, Apoptosis and cell growth are regulated by TIMPs [38]. An SNP identified at G > C −418 position in the promoter region of TIMP-2 may influence the binding of Sp1 transcription factor and downregulates the TIMP-2 transcriptional activity [39].

8.4.3 Metallopeptidase Inhibitor 3/TIMP-3

TIMP-3, a member of the TIMP family, inhibits the action of MMPs which are involved in ECM degradation. It has also been shown to have inhibitory effects on angiogenesis and tumor growth [12, 13]. ADAM-10, 12, 17 and ADAMTS-1, 4 and 5 are inhibited by TIMP-3, and the gene is located on chromosome 22q12.3. Recently, three novel polymorphisms in the promoter region of TIMP-3 gene (−899T/A, −915A/G and −1296T/C) have been identified. TIMP-3 polymorphism has an impact on various complex diseases such as spontaneous abortion, cancer, diabetic nephropathy, macular degeneration, and hypertension [40].

8.4.4 Metallopeptidase Inhibitor 4/TIMP-4

TIMP-4 belongs to the TIMP gene family and is expressed predominantly in the heart, which inhibits the action of MMP-2 by binding strongly to MMP-2 carboxyl hemopexin domain. The gene is located on chromosome 3p25.2. Two SNPs (rs3755724, −55C/T, promoter; rs17035945, 3′-untranslated region) were genotyped in TIMP-4 gene [41].

9 Conclusion

Type 2 end-stage renal disease is a devastating condition, MMPs are the major regulators of ECM degradation in Type 2 DN pathology and progression. The proteases (MMPs) and antiproteases (TIMP) offer the opportunity to identify the determinants of the disease that are very likely to be causative and may also lead to new therapeutics with strong molecular underpinning.

References

1. Amos AF, McCarty DJ, Zimmet P (1997) The rising global burden of diabetes and its complications: estimates and projections to the year 2010. Diabet Med 19:S1–S85
2. Chan JC, Malik V, Jia W et al (2009) Diabetes in Asia: epidemiology, risk factors, and pathophysiology. JAMA 301:2129–2140
3. King GL (2008) The role of inflammatory cytokines in diabetes and its complications. J Periodontol 79:1527–1534
4. Murphy M, Crean J, Brazil DP, Sadlier D, Martin F, Godson C (2008) Regulation and consequences of differential gene expression in diabetic kidney disease. Biochem Soc Trans 36:941–1005
5. Hemanth KN, Prashanth S, VidyaSagar J (2011) Diabetic nephropathy pathogenesis and newer targets in treatment. Int J Pharm Sci Rev Res 6:91–101
6. Kimberly R, Hyun Mi K, Thomas H, Katalin S (2014) Molecular mechanisms of diabetic kidney disease. J Clin Investig 124(6):2333–2340
7. Ziyadeh FN, Wolf G (2008) Pathogenesis of the podocytopathy and proteinuria in diabetic glomerulopathy. Curr Diab Rev 4:39–45
8. Xu X, Xiao L, Xiao P et al (2014) A glimpse of matrix metalloproteinases in diabetic nephropathy. Curr Med Chem 21(28):3244–3260
9. Bergers G, Brekken R, McMahon G et al (2000) Matrix metalloproteinase-9 triggers the angiogenic switch during carcinogenesis. Nat Cell Biol 2:737–744
10. Egeblad M, Werb Z (2002) New functions for the matrix metalloproteinase in cancer progression. Nature 2:161–174
11. Hsieh HL, Chi PL, Lin CC, Yang CC, Yang CM (2014) Up-regulation of ROS-dependent matrix metalloproteinase-9 from high-glucose-challenged astrocytes contributes to the neuronal apoptosis. Mol Neurobiol 50:520–533
12. Kanwar YS, Sun L, Xie P, Liu FY, Chen S (2011) A glimpse of various pathogenetic mechanisms of diabetic nephropathy. Annu Rev Pathol 6:395–423
13. Busch M, Franke S, Ruster C, Wolf G (2010) Advanced glycation end-products and the kidney. Eur J Clin Invest 40(8):742–755

14. Schrijvers BF, De Vriese AS, Flyvbjerg A (2004) From hyperglycemia to diabetic kidney disease: the role of metabolic, hemodynamic, intracellular factors and growth factors/cytokines. Endocr Rev 25(6):971–1010
15. Wendt T, Tanji N, Guo J et al (2003) Glucose, glycation, and RAGE: implications for amplification of cellular dysfunction in diabetic nephropathy. J Am Soc Nephrol 14 (5):1383–1395
16. Engelmyer E, van Goor H, Edwards DR, Diamond JR (1995) Differential mRNA expression of renal cortical tissue inhibitor of metalloproteinase-1, -2, and -3 in experimental hydronephrosis. J Am Soc Nephrol 5(9):1675–1683
17. Akiyama K, Shikata K, Sugimoto H et al (1997) Changes in serum concentrations of matrix metalloproteinases, tissue inhibitors of metalloproteinases and type IV collagen in patients with various types of glomerulonephritis. Res Commun Mol Pathol Pharmacol 95:115–128
18. Johnson PJ, Tyagi SC, Katwa LC et al (1998) Activation of extracellular matrix metalloproteinases in equine laminitis. Vet Rec 142:392–396
19. Catania JM, Chen G, Parrish AR (2007) Role of matrix metalloproteinases in renal pathophysiologies. Am J Physiol Renal Physiol 292:F905–F911
20. Dimas G, Iliadis F, Grekas D (2013) Matrix metalloproteinases, atherosclerosis, proteinuria and kidney disease: link-age–based approaches. Hippokratia 17:292–297
21. Keeling J, Herrera GA (2008) Human matrix metalloproteinases: characteristics and pathologic role in altering mesangial homeostasis. Microsc Res Tech 71:371–379
22. Matthew G, William CP (2014) Diverse functions of matrix metalloproteinases during fibrosis. Dis Model Mech 7:193–203
23. Saffarian S, Collier IE, Marmer BL, Elson EL (2004) Goldberg G. Interstitial collagenase is a brownian ratchet driven by proteolysis of collagen. Science 306:108–111
24. Rutter JL, Mitchell TI, Buttice G et al (1998) A single nucleotide polymorphism in the matrix metalloproteinase-1 promoter creates an Ets binding site and augments transcription. Cancer Res 58:5321–5325
25. Arakaki PA, Marques MR, Santos MCLG (2009) MMP-1 polymorphism and its relationship to pathological processes. J Biosci 34:313–320
26. Izakovicova HL, Hrdlickova B, Vokurka J, Fassmann A (2012) Matrix metalloproteinase 8 (MMP8) gene polymorphisms in chronic periodontitis. Arch Oral Biol 57:188–196
27. Leeman MF, Curran S, Murray GI (2003) The structure, regulation, and function of human matrix metalloproteinase-13. Crit Rev Biochem Mol Biol 37(3):149–166
28. Aimes RT, Quigley JP (1995) Matrix metalloproteinase-2 is an interstitial collagenase. Inhibitor-free enzyme catalyzes the cleavage of collagen fibrils and soluble native type I collagen generating the specific 3/4- and 1/4-length fragments. J Biol Chem 270:5872–5876
29. Price SJ, Greaves DR, Watkins H (2000) Identification of novel, functional genetic variants in the human matrix metalloproteinase-2 gene: role of Sp1 in allele-specific transcriptional regulation. J Biol Chem 276(10):7549–7558
30. Pan MR, Hung WC (2002) Nonsteroidal anti-inflammatory drugs inhibit matrix metalloproteinase-2 via suppression of the ERK/Sp1-mediated transcription. J Biol Chem 277(36):32775–32780
31. Eberhardt W, Huwiler A, Beck KF, Walpen S, Pfeilschifter J (2000) Amplification of IL-1ß-induced matrix metalloproteinase-9 expression by superoxide in rat glomerular mesangial cells is mediated by increased activities of NF-kB and activating protein-1 and involves activation of the mitogen-activated protein kinase pathways. J Immunol 165:5788–5797
32. Zhang B, Ye S, Herrmann SM et al (1999) Functional polymorphism in the regulatory region of gelatinase B gene in relation to severity of coronary atherosclerosis. Circulation 99:1788–1794
33. Tadahide K, Yutaka Y, Junichi KPK et al (2004) Expression of MMP-9 in mesangial cells and its changes in anti-GBM glomerulonephritis in WKY rats. Clin Exp Nephrol 8:206–215

34. Humphries S, Bauters C, Meirhaeghe A, Luong L, Bertrand M, Amouyel P (2002) The 5A6A polymorphism in the promoter of the stromelysin-1 (MMP3) gene as a risk factor for restenosis. Eur Heart J 23:721–725
35. Keith B, Hideaki N (2010) The tissue inhibitors of metalloproteinases (TIMPs): an ancient family with structural and functional diversity. Biochim Biophys Acta 1803:55–71
36. Kanauchi M, Nishioka H, Nakashima Y, Hashimoto T, Dohi K (1996) Role of tissue inhibitors of metalloproteinase in diabetic nephropathy. Nihon Jinzo Gakkai Shi 38:124–128
37. Lambert E, Boudot C, Kadri Z et al (2003) Tissue inhibitor of metalloproteinases-1 signalling pathway leading to erythroid cell survival. Biochem J 372:767–774
38. Hoegy SE, Oh HR, Corcoran ML, Stetler-Stevenson WG (2001) Tissue inhibitor of metalloproteinases-2 (TIMP-2) suppresses TKR-growth factor signaling independent of metalloproteinase inhibition. J Biol Chem 276:3203–3214
39. Jaworski DM, Perez-Martinez L (2006) Tissue inhibitor of metalloproteinase-2 (TIMP-2) expression is regulated by multiple neural differentiation signals. J Neuro chem 98:234–247
40. Beranek M et al (2003) Three novel polymorphisms in the promoter region of the TIMP-3 gene are not associated with proliferative diabetic retinopathy in Type 2 diabetes mellitus. Curr Eye Res 27(2):91–93
41. Olson TM, Hirohata S, Ye J, Leco K, Seldin MF, Apte SS (1998) Cloning of the human tissue inhibitor of metalloproteinase-4 gene (TIMP4) and localization of the TIMP4 and Timp4 genes to human chromosome 3p25 and mouse chromosome 6, respectively. Genomics 51 (1):148–151

Antihypertensive Role of Kidney: Focus on Tissue Kallikreins

Amritlal Mandal, Tapati Chakraborti and Sajal Chakraborti

Abstract

Regulation of blood pressure (BP) depends mostly on genetic and environmental factors. Among different physiological mechanisms responsible for a discernible increase in BP (hypertension), two major mechanisms are important: (a) matrix remodeling associated with arterial wall thickening with apparent reduction in blood flow and (b) activation of renin–angiotensin system (RAS). To counteract the BP rise beyond physiological limit and to ameliorate BP-rise-associated complications, nature has designed an important endogenous regulatory mechanism operative especially in kidney, the kallikrein–kinin system (KKS). In cortical collecting duct of the kidney, KKS plays a significant role in BP regulation under salt excess condition. This protective phenomenon potentially acts as a hypotensive mechanism to regulate abnormal BP increase. The functional components of the KKS are the kallikrein, a serine protease, and kinin. Proteolytic action of kallikrein on precursor kininogen forms vasoactive peptide kinin by enzymatic cleavage. Upon release, kinin binds to the B2 receptor on cell surface and exerts antihypertensive effect. Though KKS operates in both cardiovascular and renal systems, for the sake of simplicity and to have a focused but detailed understanding, only the regulatory mechanisms of renal KKS on BP homeostasis has been considered for discussion in this review.

Keywords

Hypertension · Kallikrein–kinin system · Renin–angiotensin system
Epithelial sodium channel · Kallikrein gene delivery

A. Mandal (✉)
Department of Physiology, University of Arizona, Tucson, AZ 85724, USA
e-mail: mandal@email.arizona.edu

T. Chakraborti · S. Chakraborti
Department of Biochemistry and Biophysics, University of Kalyani,
Kalyani 741235, West Bengal, India

S. Chakraborti and N.S. Dhalla (eds.), *Pathophysiological Aspects of Proteases*,
DOI 10.1007/978-981-10-6141-7_26

1 Introduction

High blood pressure (BP), commonly known as hypertension, is a manifestation of physiological conditions due to many complex factors including genetic, drug, and environmental components. Excess salt intake is a potential threat in developing BP rise [1]. Recent development of new information elucidates the basic mechanisms of BP regulation in salt excess condition.

Kallikrein, a family of serine proteases, has been implicated in BP regulation. Kallikrein-mediated blood pressure regulation is controlled by fine-tuning of the kallikrein–kinin system (KKS) in the kidney. The KKS consists of tissue kallikrein, kininogen, kinin, kininases and kinin receptor B2. Though endogenous tissue kallikrein (TK) inhibitor, kallistatin, is not considered to be an integral part of the KKS, kallistatin plays an important role in regulating kallikrein's action. These components are mainly expressed along the collecting duct of the distal nephron. The KKS maintains blood pressure homeostasis in salt excess condition by virtue of controlled electrolyte and salt absorption mechanisms in kidney. Thus, an impaired or faulty KKS could potentially generate a threat of blood pressure increase and hypertension.

The recent advances in understanding physiological and functional role of kallikrein as one of the significant protective mechanisms in hypertension and BP regulation in mineralocorticoid and salt excess conditions have established kallikrein as a potential therapeutic target to control hypertension and BP regulation. Though renin–angiotensin system (RAS), potentially a cardiovascular mechanism, is responsible for hypertensive regulation of BP, it works as an opposing mechanism of renal KKS-mediated hypotensive effect in kidney. This chapter is aimed at understanding the physiological functions of kallikrein in regulating hypertension.

In an earlier studies, participation of KKS in the intra-hepatic blood flow of patients with secondary pulmonary hypertension has been reported [2]. A separate endotoxin-mediated study in experimental ovine model demonstrated a marked decrease in PK level upon chronic exposure to endotoxin [3]. The model demonstrated a link between elevated pulmonary microvascular permeability and a tendency toward hyperdynamic circulation [4]. Urinary kallikrein excretion in salt-sensitive hypertensive subjects is found less in comparison with the salt-resistant hypertensive counterparts [5, 6]. These earlier studies have provided a strong foundation for a separate area of research on kallikreins and its role in hypertension.

2 Proteases Involved in Blood Pressure Regulation

Two main classes of neutral proteases based on their active sites are widely studied in the context of BP homesostasis. These are as follows: (i) serine proteases, including endogenous vascular elastase (EVE), the plasminogen activator/plasmin system, kallikreins, and (ii) the matrix metalloproteinases (MMPs) [7].

Kallikreins are divided into two groups: (1) plasma kallikreins (PKs) and (2) tissue kallikreins (TKs). They both differ by molecular weights, biological functions, physiochemical, and immunological properties and on their distribution in the body. Tissue kallikrein 1 (KLK1), kallikrein-related peptidases 2-15 (KLK2-15), and plasma kallikreins are a family of secreted serine protease that forms kinins. Tissue kallikreins, KLK 1–15 genes, are located on chromosomes 19q13.3-q13.4. Plasma kallikrein, KLKB1 gene, is located on chromosome 4q35.

3 Kallikrein–Kinin System (KKS)

The kallikrein/kinin system (KKS) is a very important mechanism in a wide range of physiological functions and in the development of many pathological conditions including blood pressure (BP) regulation [8]. The plasma KKS forms a cell-bound protease activation cascade with an ultimate releases of the inflammatory peptide bradykinin (BK) from high molecular weight kinininogen (H-kininogen) [9]. This highly orchestrated sequential event occurs by binding a complex of H-kininogen and plasma pre-kallikrein (PPK) to a multiprotein receptor on the cell surface requiring an optimal Zn^{2+} concentration [10–12]. Formation of HK-PPK is critical, which relies on a stoichiometric interaction between PPK and prolylcarboxypeptidase (PRCP) [13] or heat shock protein 90 (HSP90) [14]. Factor XII-independent activation of PK has also been reported by Røjkjær and Schmaier [15]. Important steps involved in plasma and tissue KKSs, including B1- and B2-receptor-mediated initiation of downstream signaling pathways, are schematically shown in Fig. 1. The KKS consists of several components that work as a cascade of fine-tuned mechanisms for blood pressure regulation. In isolated mouse-microperfused cortical collecting ducts, luminal TK was found to activate epithelial sodium channels (ENaC) [16].

4 Plasma Kallikreins (PKs)

PKs (EC 3.4.21.34) are present in circulation as inactive pre-kallikreins or Fletcher factor. Plasma pre-kallikrein is a precursor of PK, produced in liver and circulates in blood, bound with high molecular weight kininogen (HMW kininogen). Activation of pre-kallikreins and consequent formation of kallikreins have been shown to occur by the action of Hageman factor (HF) [17]. Upon activation, PK acts on endothelial cell surface during intrinsic coagulation steps and cleaves HMW kininogen to release bradykinin (BK) [18, 19].

PK directly regulates cardiovascular system by activating different proteolytic cascades which are intrinsic to BP regulation. These include KKS, the fibrinolytic system, and the alternate complement activation pathways. PK is cardioprotective in normal physiological condition.

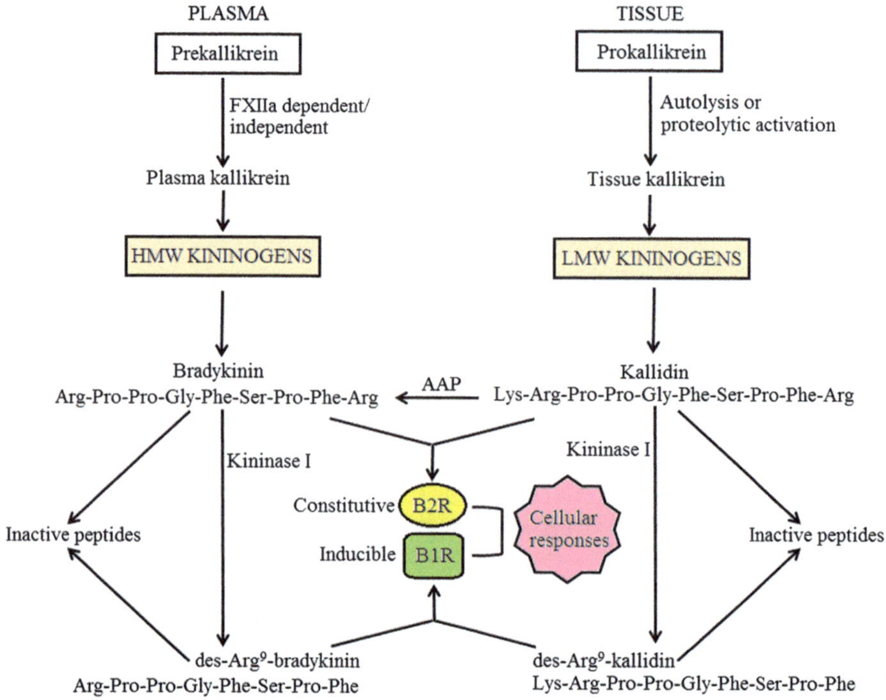

Fig. 1 Schematic diagram showing major events of the kallikrein–kinin system (KKS). Plasma pre-kallikrein and tissue pro-kallikreins secreted as zymogens and become activated by proteolytic cleavage to plasma kallikrein (PK) and tissue kallikrein (TK), respectively. PK acts on HMW kininogen, and TK acts on LMW kininogen to generate bradykinin (BK) and kallidin, which act through the constitutive B2 receptor (B2R). BK is also generated from kallidin by the action of an arginine aminopeptidase (AAP). These peptides are further processed to their des-Arg^9 (removal of Arg at position 9) metabolites by the kininases carboxypeptidases (kininases I). The Des-Arg kinins bind to the inducible B1 receptor (B1R). Kinins are rapidly degraded to the inactive peptides mainly by angiotensin-converting enzyme (ACE, kininase II). Kinin-mediated B receptor stimulation is involved in diverse physiological processes including blood pressure regulation, electrolyte balance, and inflammation

Kallikrein B1 (KLKB1), another member of kallikrein, is one of the intrinsic components of blood caoagulation cascade. KLKB1 is produced as precursor of plasma kallikrein (Pre-KLKB1) by KLKB1 gene. KLKB1 is expressed in liver, kidney, and pancreas. Low KLKB1 expression level has also been detected in heart, brain, spleen, thymus, intestine, and testis [20]. Impaired or faulty KLKB1 activity has been observed to be responsible for developing arterial dysfunction and in renal tubular defects [21, 22]. KLKB1 acts on two major substrates: coagulation factor XII and HMW kininogen [23].

5 Tissue Kallikreins (TKs)

Major source of tissue kallikrein (EC 3.4.21.35) is exocrine glands, such as the kidney, pancreas, intestine, salivary glands, bronchoalveolar lavage fluid of asthmatic patients and in synovial tissues [24–29]. TKs responsible for BP regulation are mostly synthesized in arteries and distal renal tubule, and TK is the main target of aldosterone [30]. Existence of KKS has also been reported in the vascular wall, where the presence of secreted kallikrein-like kininogenase activity has been observed [31, 32]. TKs share similar physiochemical properties and are found to be immunologically identical in the same species [33].

Expression of mRNA for TK and kininogen has been demonstrated in vascular tissue [34] and endothelial cells. TK protein expression has also been detected in human endothelial cells [35, 36]. Human umbilical vein endothelial cells (HUVEC) have been reported to be a site of synthesis and storage of TK [37].

In kidney, major quantity of TK is synthesized by connecting tubule cells. Distal convoluted tubules and the cortical collecting ducts are the minor contributing cell types for tissue kallikrein synthesis [38]. The release of TK into the tubular fluid and the peritubular interstitium eventually forms kinins. Locally available endogenous kininogens are cleaved by TK leading to the formation of kallidin (lysyl-bradykinin). The cleavage of kallidin by N-aminopeptidase produces bradykinin, which in turn activates bradykinin type 2 (B2) receptor [39]. Antihypertensive effects of ACE1 not only inhibit formation of angiotensin II, but also inhibit kinin degradation. In rodent and human, an inverse relationship exists between urinary tissue kallikrein excretion and blood pressure regulation [39].

6 Kinins

Kinins are the vasoactive polypeptides, released in the circulating bloodstream from precursor kininogens by the proteolytic action of kallikreins. Kininogens are the substrate for kallikreins. High molecular weight (HMW, ~76 kDa) and low molecular weight (LMW, ~48 kDa) kininogens have been isolated from bovine plasma [40]. Bradykinin (BK), a nonapeptide, is released in the circulation by the activation of HMW kininogen. Kallidin (KD), a decapeptide, is similarly released from LMW kinininogen by tissue kallikrein.

Kinins resembles bradykinin (BK) (Arg^1-Pro^2-Pro^3-Gly^4-Phe^5-Ser^6-Pro^7-Phe^8-Arg^9) in structure and in pharmacological action [41]. Many of the effects of kinin are attributed to eicosanoid [42], nitric oxide (NO) [42], endothelium-derived hyperpolarizing factor (EDHF) [43, 44], or through tissue plasminogen factor (tPA) [45]. Kinins play protective role in cardiac ischemia preconditioning and in cardiovascular diseases.

Kinin's renal protective function has been a well-studied area. Renoprotective effects of kinin have been attributed due to either angiotensin-converting enzyme (ACE) inhibition [46, 47] or by angiotensin type I receptor blocker (ARB) [45, 48].

In a small family-based study, enhanced urinary kallikrein excretion has been found to be linked with a decrease of risk for essential hypertension [45]. Interestingly, human, rat, or mice with established deficiency of KKS or with a missing component of KKS did not show any discernible hypertension. In kidney, papillary blood flow largely depends on availability of kinin. Kinins play important regulatory roles in sodium and water excretion through kidney [45, 49, 50]. Indeed, B2 knockout mice show enhanced hypertension due to excess salt intake, which further established a role of kinin as antihypertensive molecule [51].

7 Kininases

Kininases act on substrate kinins in blood or tissue homogenates to rapidly degrade kinins into inactive peptides. Kininase I (carboxypeptidase N) is present in human and in animal plasma. Kininanses were first detected in kidney cortex and later in the plasma [52], which inactivates BK by cleaving C-terminal amino acids including Arg^9 [53]. Later, this enzyme has been renamed as angiotensin 1-converting enzyme (ACE). ACE inhibitors are established cardioprotective agents, which improves cardiac performance after heart failure, helps recovering from hypertensive cardiac hypertrophy, reduces myocardial infarct size, and prevents reperfusion injury [54–58]. ACE inhibitors are important clinical tools for treating hypertension and albuminuria.

8 Kallikrein Activation Mechanisms

PK in endothelial cells is activated by a membrane-associated cysteine protease. In general, autoactivation of factor XII (FXII) to factor XIIa (FXIIa) is the initial step of PK activation. FXIIa acts on inactive plasma pre-kallikrein to produce active kallikrein which in turn reciprocally activates FXII by an autocatalytic mechanism in the surface of endothelial cells. The mechanisms of plasma KKS activation in endothelial cell surface received inadequate attention due to abundant localization of FXII on the endothelial cell surface [59, 60]. This has been suggested for plasma kallekrein activation in endothelial cells. In many studies, PK activation in endothelial cells has been observed to be associated with intrinsic availability of FXII on the endothelial cell surface for autoactivation to FXIIa [61–63]. Motta et al. [64] and Rojkjaer et al. [65] have demonstrated FXII-independent activation of plasma kallikreins in cultured HUVEC cells. These studies have restricted FXII for consideration as an essential and required component of PK activation in endothelial cells. However, FXII is not an absolute requirement for endothelial PK activation and accelerated synthesis of factor XII by kallikrein [15].

Recent evidence showed that TK employs kinin-independent effects on different renal transporters [66]. The examples include epithelial Na^+ channel, transient receptor potential channel vanilloid subtype 5 (TRPV5), and the colonic H^+, K^+-ATPase.

Studies with experimental animal models demonstrated kinin-induced increase in renal blood flow, diuresis and natriuresis [67, 68]. Increased electrolyte and water excretion have been observed to be linked to the renal hemodynamic effects of kinins. Bradykinin (BK) is a very well-studied and potent inhibitor of NaCl absorption in isolated, perfused cortical collecting ducts (CCDs) [69, 70]. The BK-induced inhibition of Na^+ absorption in CCDs is possibly exerted through the activation of B2 receptors [69, 70]. Impaired B2 receptor function has been positively correlated with increased NaCl absorption, leading to increased blood volume and blood pressure increase [71].

Knockdown of TK gene in experimental mouse model has been shown to decreased kinin production [72, 73]. However, on the other hand, B2 receptor knockout mice and TK-deficient mice failed to show any changes in altered Na^+ level and BP [16], which suggests that TK has kinin-independent effects in BP regulation.

Duka et al. [74] have demonstrated that under normal physiological condition, vasoregulatory function of bradykinin occurs through B2 receptor activation. When B2 receptor expression is either selectively knocked out, B1 receptor is upregulated [75] and comes into play by overtaking the responsible functions of B2 receptors. Earlier, arachidonc acid activation cascade has been observed due to B1 receptor activation [75, 76], which directly linked to insulin-dependent metabolic controls, indicating that the B1 receptor's role as a vasoregulatory component is a novel finding.

Inactive tissue kallikreins are synthesized as pre-proserine proteases [77]. 16-30 amino acid presequence at the N-terminal in the inactive precursor is enzymatically cleaved prior to the secretion of the active kallikriens [78]. Pro-sequence containing 37 amino acids in kallikrein 5 (KLK5) and 4-9 amino acids in other kallikreins is cleaved upon activation [78]. Cleavage of arginine or lysine in the pro-domain by trypsin is believed to be the key steps in the tissue kallikrein activation. The presence of histidine and aspartic acid was found to be conserved among mature and active tissue kallikreins. Activation and physiological roles of tissue kallikreins have been reviewed by Emami et al. [79].

9 Genetic Models of TK Deficiency

Due to unavailability of potent and selective pharmacological inhibitor of TK, a need for genetically engineered animal model was felt necessary. In an effort to have this tool available, scientists have developed two important TK-deficient models. The first one is TK-deficient murine model, where the mice TK gene has been selectively inactivated by genetic engineering [16, 73]. The second is human

TK-deficient phenotype, which has been achieved by a major loss of function due to polymorphism of the TK gene [80]. Physiological studies with both models reveal its important role in arterial function, flow-induced vasodilation, and adaptation to shear stress [72, 81].

Studies on kallikrein-deficient mice provided data showing that kallikrein directly influences BP in mineralocorticoid excess and salt retention but not in normotensive animals in high renin hypertension. This critical observation sheds light on KLK as an antihypertensive agent, where Na^+ retention is the main cause of hypertension and blood pressure elevation [82].

10 TK and Blood Pressure Regulation

10.1 TK Deficiency and Impaired Blood Pressure Control

Reduced excretory level of kallikrein in urine samples has been recorded in clinical and experimental hypertension [83–86]. TK deficiency and blood pressure regulation have been extensively reviewed by Sharma et al. [87]. Role of KKS as a potent BP lowering mechanism is very well known [88]. Kinins have been observed to be effective in lowering BP, arterial vasodilatation, natriuresis, and an increase in renal blood flow with the decrease in peripheral resistance [83, 89–92]. Additionally, experimental evidence also proved a faulty or inhibited KKS could potentially cause Na^+ retention, arterial vasoconstriction, raised peripheral resistance, increased vascular or plasma volume and hypertension. A decrease of blood kinin level has been thought to be directly linked to the development of hypertension.

Arterial dysfunction is commonly observed in both capacitative and resistance vessels of TK-deficient mice [72, 93]. Non-TK-deficient wild-type mice did not show any alteration in baseline blood pressure recordings [73, 94]. The missense mutation (R53H) in human TK gene (KLK1) reduces the kinin-forming activity of TK [80, 81]. In another study, Yu et al. [94] demonstrated that multiallelic promoter variation of the TK gene is associated with hypertension. No association of this promoter variation with TK activity level has, however, been observed. Genetic model of TK deficiency has been extensively reviewed by Potier et al. [82]. KKS has been shown to play protective roles in mitigating BP increase and hypertension [95, 96].

Experimental data have provided evidence for the development of tubulointerstitial injury in kallekrein-deficient experimental models. These subjects elicit salt-sensitive hypertension. Reduced kallikrein expression has also been shown to result a reduction in kinin formation, favoring vasoconstrictive sodium retention. An impaired or faulty KKS has been shown to lead to development of renal fibrosis, which could be reverted by using pharmacological stimulus of KKS, thereby preventing progression of salt-sensitive hypertension and kidney diseases [97].

10.2 TK Deficiency in Experimental Hypertension

Blood pressure in mice with renovascular hypertension remains unaltered in TK deficiency [98]. This model of hypertension is renin dependent, and Na^+ balance has been found to be either neutral or negative, depending upon the origin and cause of hypertension. Urinary TK excretion was found to be unaltered in WT mice, where renal artery had been clipped to induce TK deficiency [98]. This study clearly shows that TK does not have any counter-regulatory hypertensive effect in renin–angiotensin model of hypertension. However, in another experimental model, where hypertension has been induced by aldosterone and salt administration, a prominent hypertension was observed [30]. This observation was consistently found to be reproducible across different mouse genetic strains.

Slight or no increase of BP has been reported in mice of different genetic background. In wild-type (WT) animals, aldosterone/salt treatment for one month elicited sustained hypertension. TK gene inactivation in experimental mice and subsequent BP control in shorter and longer time periods for both in active and inactive phases has been studied by Waeckel et al. [30]. The study showed that chronic aldosterone excess had no effect on BP in wild-type mice, whereas in TK inactivated group, severe hypertension has been observed. This animal model of hypertension is based on Na^+ retention in suppressed renin secretion. The model largely relies on renal excretion of urinary TK by aldosterone administration. Thus, a counter-regulatory mechanism exists that is dependent on increased TK synthesis in the setting of mineralocorticoid excess and prevents hypertension. Further studies in TK-deficient mice explained the antihypertensive effect of TK at the level of renal tubule by potentially inhibiting ENaC synthesis and with a resultant Na^+ retention [30, 99, 100]. No convincing mechanisms has, however, been available to link kinin-mediated effect of TK [101] to other proteolytic action of TK or both [16, 30, 102]. Azizi et al. [81] have also provided evidence supporting antihypertensive effect of TK in human under mineralocorticoid and salt excess condition. In this regard, proper evaluation of urinary and plasma TK levels could be interesting in patients with adrenal adenoma or adrenal hyperplasia to assess the degree of disease pathology in relation to hypertension.

Kinins elicit renal protective effects in diabetes [59, 103]. Incipient diabetic nephropathy has shown no significant blood pressure alteration in patients with TK deficiency. TK deficiency is usually associated with renal dysfunction and albuminurea [104]. Hypertension in association with diabetic nephropathy and the role of TK deficiency in this scenario need further investigation to understand the complex disease mechanism.

Two opposing mechanisms, the renin–angiotensin system (RAS) and KKS (the fluid homeostasis mechanism) in kidney, are crucial in maintaining normal BP regulation. KKS-mediated cardioprotective mechanisms by counteracting RAS, a potential hypertensive mechanisms, have been reviewed by Regoli et al. [105]. Concerted balancing effects of RAS-induced BP elevation in certain instances get normalized by the hypotensive effect of KKS. Renin (inactive) becomes activated by plasma or tissue kallikreins [106, 107]. Plasma KKS directly interacts with the

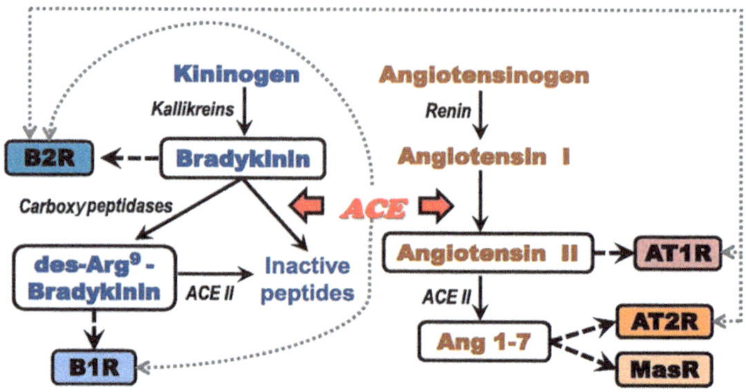

Fig. 2 Interactions between the KKS and the renin–angiotensin system (RAS). ACE represents a central physiological bridge connecting the KKS (*left side*) and the RAS (classical RAS on the right side). ACE regulates the levels of endogenous Ang II and kinins by degrading BK and converting Ang I to Ang II. A cross-talk between ACE, B1R, and/or B2R is also postulated (*broken gray arrow*) in some systems and increases the complexity of these interactions. Similarly, cross-talks between angiotensin II receptor type 1 (AT1R) and angiotensin II receptor type 2 (AT2R) with B2R have also been described (*broken black arrow*). Mas receptor (MasR), a G protein-coupled receptor, is associated with cardiac, renal, and cerebral protective responses. Taken from Ref. [129] with permission

RAS. The initial step of PK activation in the endothelial cells takes place by prolylcarboxypeptidase (PRCP) and that was recognized as the major angiotensin II-converting peptide to act as an activator of PK [106]. Increased urinary kallikrein expression has been noted in dogs upon intra-arterial infusion of angiotensin II [108]. Interaction between KKS and RAS is schematically represented in Fig. 2.

Physiological Na^+ homeostasis and BP regulation are important for healthy kidney and its capacity to appropriately reabsorb salt and water. This is achieved by proper activities of the ion transporters and channels located in the nephron [109]. The epithelial sodium channel (ENaC), located in the kidney, plays a pivotal role in extracellular volume and BP homeostasis. Cortical collecting tubule of nephron is the major site of ENaC expression which plays an important role in Na^+ homeostasis. When ENaC function is gained by mutation in β subunit, hypertension results (Liddle's syndrome) [110]. On the other hand, loss of function due to mutation leads to renal salt wasting (pseudohypoaldosterism) [111]. Recent studies demonstrated proteolytic cleavage of γ-subunit of ENaC by kallikrein and possible protection mechanisms against hypertension [112].

RAS is a well-established regulator of ENaC [109, 113]. Angiotensin I is formed from the precursor angiotensinogen by the renin-mediated proteolytic cleavage. Angiotensin-converting enzyme (ACE) acts on angiotensin I to form angiotensin II, the main regulatory component of RAS [113]. Angiotensin II causes vasoconstriction and consequent rise in BP [114]. Regulation of ENaC by bradykinin in the distal nephron in low and high salt conditions is schematically represented in Fig. 3.

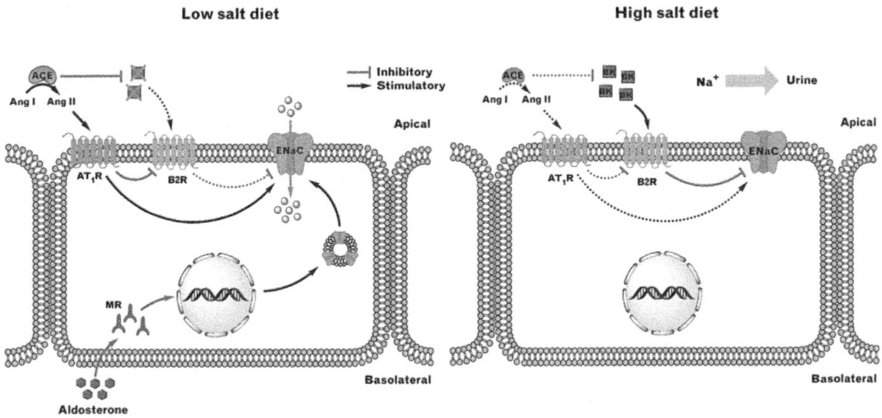

Fig. 3 Principal scheme of interaction between the kallikrein–kinin and renin–angiotensin systems in the distal nephron during different salt intakes. *ACE* angiotensin-converting enzyme; *Ang I* angiotensin I; *Ang II* angiotensin II; *AT1R* angiotensin type 1 receptor; *B2R* bradykinin type 2 receptor; *ENaC* epithelial Na^+ channel; *MR* mineralocorticoid receptor. *Dotted arrows* represent inactive mechanisms of ENaC regulation under particular sodium intake. Taken from Ref. [50] with permission

11 Kallikrein Gene Delivery

Ischemia/reperfusion (I/R) injury-induced myocardial damage is commonly observed during thrombolytic therapy for myocardial infarction and in several restorative treatment options available to the cardioplegic arrest and cardiovascular surgery. Kallikrein gene delivery using adenovirus vector to the experimental rat model of icatibant (B2 receptor antagonist)-induced hypertension was found to be effective in reducing myocardial I/R injury [115].

In a study with Dahl-salt-sensitive (Dahl-SS) hypertensive rats, human tissue kallikrein gene delivery using cytomegalovirus (CMV) vector has been found to be effective in reducing BP rise from day 2 of post-injection [116]. Adenovirus-mediated tissue kallikrein gene delivery was also found to be effective in reducing left ventricular mass, cardiomyocyte hypertrophy, attenuation of glomerular sclerotic lesions, tubular dilation, and proximal tubule brush-bordered cell damage in the Goldblatt hypertensive rat model [117] and in Dahl-SS rat model [118] of hypertension.

These findings indicated potentials of tissue kallikrein gene delivery in cardiac and renal functions. The shortcoming of adenoviral tissue kallikrein gene delivery is the eventual degradation of the DNA due to lack of viral integration factor. The host immune system plays a significant role in rapidly destroying the injected gene [119]. Modified adenovirus vectors have been shown useful for long-term transgene expression in experimental systems [120] to enable better protection against hypertension.

12 Kallistatin, the Endogenous Inhibitor of TK

Bioavailability of kallikreins depend on factors that either affect kallikrein biosynthesis or its clearance mechanisms. Posttranslational regulation of kallikreins is well studied [121]. Post-translational modification and inactivation of kallikreins by inhibitors also play regulatory roles of kallikrein's activity. Kallistatin, an endogenous serine protease inhibitor of kallikrein, has been very well investigated [122–125]. Liver synthesizes kallistatin predominantly, albeit pancreas and kidney elicit little kallistatin expression. Plasma and urine levels of kallistatin have been measured in normal and several groups of patients with a wide range of disease conditions [126, 127]. A reduced plasma kallikrein level has been observed in patients with liver diseases and sepsis [126]. The presence of higher urinary kallistatin level has been documented in urines of patients with pregnancy-related hypertension [127]. The anti-hypotensive property of endogenous kallikreins was reviewed by Chao et al. [128].

13 Conclusions and Future Direction

Kallikreins have been established as one of the intrinsic components of KKS, which plays antihypertensive role in mammalian cardiovascular and renal systems. Experimental animal models including pharmacological inhibition and conditional gene knockout for kallikreins have made it possible to study the physiological roles of the enzyme in diverse pathological conditions. Kidney tissue kallikrein plays pivotal role in renovascular blood flow regulation and epithelial salt and electrolyte excretion by employing ENaC channels located at the distal loop of the cortical collecting duct. Thus, TK is an important regulatory component in physiological salt retention and blood volume regulation. In a small family-based study, dominant kallikrein expression has been positively correlated to be cardioprotective and reduces occurrence of hypertension during the condition of excess salt intake. The subjects have also shown to increase excretion of urinary kallikreins, which was known to be renoprotective. Both plasma and tissue kallikrein-mediated complex physiological fine regulation of blood pressure usually occurs in concerted mechanisms. The activation of plasma and tissue kallikreins differs widely. Plasma kallikreins are activated on the endothelial cell surface by the action of factor XIIa. But evidence of factor XIIa independent plasma kallikrein activation has failed to prove that factor XIIa is essential for the plasma kallikrein activation. Tissue kallikrein activation has been observed to occur through proteolytic cleavage of the pre- and pro-peptide regions of the precursor molecule as posttranslational modification steps. Components responsible to limit kinin formation in kidney act toward the development of salt-induced hypertension.

In addition to the availability of endogenous kallikrein, development of kallikrein-like factors could be a potential area, where significant research effort needs to be targeted for therapeutic management of hypertension during salt excess condition.

Gene therapy is a demanding area in clinical trials. Effective design of viral vectors for a safe and successful targeting of the kallikrein gene in a tissue-specific manner could prove useful as a therapeutic approach to treat hypertension and associated complications.

The apparent decrease in excretory kallikrein levels in hypertensive subject was found to be effectively reversed by antihypertensive drug or diuretic treatment. Further studies are needed to understand the mechanism of action of hypotensive drugs and the increase in urinary kallikrein excretion.

Acknowledgements Thanks are due to DST-SERB (Govt. of India) and the University of Kalyani, Kalyani 741235, West Bengal, India, for financial assistance.

References

1. Majid DSA, Prieto MC, Gabriel Navar L (2015) Salt-sensitive hypertension: perspectives on intrarenal mechanisms. Curr Hypertens Rev 11:38–48
2. Tatarkina ND, Avdeeva EV, Orlovskaia TG et al (1991) Central, pulmonary and hepatic hemodynamics at the developmental stages of cardiopulmonary insufficiency. Sov Med 4:5–8
3. Traber DL, Flynn JT, Herndon DN et al (1989) Comparison of the cardiopulmonary responses to single bolus and continuous infusion of endotoxin in an ovine model. Circ Shock 27:123–138
4. Godsoe A, Kimura R, Herndon DN et al (1988) Cardiopulmonary changes with intermittent endotoxin administration in sheep. Circ Shock 25:61–74
5. Ferri C, Bellini C, Carlomagno A et al (1996) Active kallikrein response to changes in sodium-chloride intake in essential hypertensive patients. J Am Soc Nephrol 7:443–453
6. Ferri C, Bellini C, Carlomagno A et al (1994) Urinary kallikrein and salt sensitivity in essential hypertensive males. Kidney Int 46:780–788
7. d'Ortho M-P (2008) MMPs, inflammation and pulmonary arterial hypertension. In: Lagente V, Boichot E (eds) Matrix metalloproteinases in tissue remodelling and inflammation. Birkhäuser Verlag, Basel/Switzerland, pp 81–97
8. Yarovaya GA, Neshkova EA (2015) Kallikrein-kinin system. Long history and present. (to 90th anniversary of discovery of the system). Bioorg Khim 41:275–291
9. Kerbiriou DM, Griffin JH (1979) Human high molecular weight kininogen. Studies of structure-function relationships and of proteolysis of the molecule occurring during contact activation of plasma. J Biol Chem 254:12020–12027
10. Gustafson EJ, Schmaier AH, Wachtfogel YT et al (1989) Human neutrophils contain and bind high molecular weight kininogen. J Clin Invest 84:28–35
11. Gustafson EJ, Schutsky D, Knight LC et al (1986) High molecular weight kininogen binds to unstimulated platelets. J Clin Invest 78:310–318
12. Schmaier AH, Kuo A, Lundberg D et al (1988) The expression of high molecular weight kininogen on human umbilical vein endothelial cells. J Biol Chem 263:16327–16333
13. Shariat-Madar Z, Mahdi F, Schmaier AH (2002) Identification and characterization of prolylcarboxypeptidase as an endothelial cell prekallikrein activator. J Biol Chem 277:17962–17969

14. Joseph K, Tholanikunnel BG, Kaplan AP (2002) Heat shock protein 90 catalyzes activation of the prekallikrein-kininogen complex in the absence of factor XII. Proc Natl Acad Sci 99:896–900
15. Røjkjær R, Schmaier AH (1999) Activation of the plasma kallikrein/kinin system on endothelial cells. Proc Assoc Am Phys 111:220–227
16. Picard N, Eladari D, El Moghrabi S et al (2008) Defective ENaC processing and function in tissue kallikrein-deficient mice. J Biol Chem 283:4602–4611
17. Kaplan AP, Austen KF (1971) A prealbumin activator of prekallikrein: II. Derivation of activators of prekallikrein from active Hageman factor by digestion with plasmin. J Exp Med 133:696–712
18. Kaplan AP, Joseph K, Shibayama Y et al (2001) Activation of the plasma kinin forming cascade along cell surfaces. Int Arch Allerg Immunol 124:339–342
19. Motta G, Shariat-Madar Z, Mahdi F et al (2001) Assembly of high molecular weight kininogen and activation of prekallikrein on cell matrix. Thromb Haemost 86:840–847
20. Fink E, Bhoola KD, Snyman C et al (2007) Cellular expression of plasma prekallikrein in human tissues. Biol Chem 388:957–963
21. Bianchi G, Ferrari P, Cusi D et al (1986) Cell membrane abnormalities and genetic hypertension. J Clin Hypertens 2:114–119
22. Bianchi G, Ferrari P, Salvati P et al (1986) A renal abnormality in the Milan hypertensive strain of rats and in humans predisposed to essential hypertension. J Hypertens Suppl 4:S33–S36
23. Marcondes S, Antunes E (2005) The plasma and tissue kininogen-kallikrein-kinin system: Role in the cardiovascular system. Curr Med Chem Cardiovasc Hematol Agents 3:33–44
24. Bhoola KD, Morley J, Schachter M et al (1965) Vasodilatation in the submaxillary gland of the cat. J Physiol 179:172–184
25. Christiansen SC, Proud D, Cochrane CG (1987) Detection of tissue kallikrein in the bronchoalveolar lavage fluid of asthmatic subjects. J Clin Invest 79:188–197
26. Krantz S, Lober M, Fiedler H (1970) Isoelectric focusing of fibrinogens on polyacrylamide gels. FEBS Lett 11:100–102
27. Nustad K, Vaaje K, Pierce JV (1975) Synthesis of kallikreins by rat kidney slices. Br J Pharmacol 53:229–234
28. Sharma JN, Zeitlin IJ, Deodhar SD, Buchanan WW (1983) Detection of kallikrein-like activity in inflamed synovial tissue. Arch Int Pharmacodyn Ther 262:279–286
29. Zeitlin IJ (1972) Rat intestinal kallikrein. In: Back N, Sicuteri F (eds) Vasopeptides: chemistry, pharmacology, and pathophysiology. Springer, Boston, pp 289–296
30. Waeckel L, Potier L, Chollet C et al (2012) Antihypertensive role of tissue kallikrein in hyperaldosteronism in the mouse. Endocrinology 153:3886–3896
31. Nolly H, Scicli AG, Scicli G, Carretero OA (1985) Characterization of a kininogenase from rat vascular tissue resembling tissue kallikrein. Circ Res 56:816–821
32. Oza NB, Schwartz JH, Goud HD (1990) Rat aortic smooth muscle cells in culture express kallikrein, kininogen, and bradykininase activity. J Clin Invest 85:597–600
33. Fiedler F (1979) Enzymology of Glandular Kallikreins. In: Erdös EG (ed) Bradykinin, kallidin and kallikrein: Supplement. Springer, Berlin, pp 103–161
34. Saed GM, Carretero OA, MacDonald RJ et al (1990) Kallikrein messenger RNA in rat arteries and veins. Circ Res 67:510–516
35. Graf K, Gräfe M, Auch-Schwelk W et al (1994) Tissue kallikrein activity and kinin release in human endothelial cells. Eur J Clin Chem Clin Biochem 32:495–500
36. Wolf WC, Harley RA, Sluce D et al (1999) Localization and expression of tissue kallikrein and kallistatin in human blood vessels. J Histochem Cytochem 47:221–228
37. Dedio J, Wiemer G, Rütten H et al (2001) Tissue kallikrein KLK1 is expressed de novo in endothelial cells and mediates relaxation of human umbilical veins. Biol Chem 382: 1483–1490

38. Figueroa CD, MacIver AG, Mackenzie JC et al (1988) Localisation of immunoreactive kininogen and tissue kallikrein in the human nephron. Histochem 89:437–442
39. Bhoola KD, Figueroa CD, Worthy K (1992) Bioregulation of kinins: kallikreins, kininogens, and kininases. Pharmacol Rev 44:1–80
40. Komiya M, Kato H, Suzuki T (1974) Bovine plasma kininogens: III. Structural comparison of high molecular weight and low molecular weight kininogens. J Biochem 76:833–845
41. Sharma JN, Buchanan WW (1979) Kinin system in clinical and experimental rheumatoid inflammation: a short review. Curr Med Res Opin 6:314–321
42. Waeckel L, Potier L, Richer C et al (2013) Pathophysiology of genetic deficiency in tissue kallikrein activity in mouse and man. Thromb Haemost 110:476–483
43. Mombouli JV, Bissiriou I, Agboton V et al (1996) Endothelium-derived hyperpolarizing factor: a key mediator of the vasodilator action of bradykinin. Immunopharmacology 33: 46–50
44. Mombouli JV, Vanhoutte PM (1995) Endothelium-derived hyperpolarizing factor(s) and the potentiation of kinins by converting enzyme inhibitors. Am J Hypertens 8:19S–27S
45. Rhaleb N-E, Yang X-P, Carretero OA (2011) The kallikrein-kinin system as a regulator of cardiovascular and renal function. Comp Physiol 1:971–993
46. Mombouli JV (1997) ACE inhibition, endothelial function and coronary artery lesions. Role of kinins and nitric oxide. Drugs 54:12–22
47. Kaplan AP (2014) Bradykinin-mediated diseases. Chem Immunol Allergy 100:140–147
48. Billings FT, Balaguer JM, Yu C et al (2012) Comparative effects of angiotensin receptor blockade and ACE Inhibition on the fibrinolytic and inflammatory responses to cardiopulmonary bypass. Clin Pharmacol Therap 91:1065–1073
49. Mamenko M, Zaika O, Boukelmoune N et al (2015) Control of ENaC-mediated sodium reabsorption in the distal nephron by bradykinin. Vitam Horm 98:137–154
50. Mamenko M, Zaika O, Pochynyuk O (2014) Direct regulation of ENaC by bradykinin in the distal nephron. Implications for renal sodium handling. Curr Opin Nephrol Hypertens 23:122–129
51. Sabatini RA, Guimarães PB, Fernandes L et al (2008) ACE activity is modulated by kinin B2 receptor. Hypertension 51:689–695
52. Yang HYT, Erdos EG (1967) Second kininase in human blood plasma. Nature 215: 1402–1403
53. Erdos EG, Sloane EM (1962) An enzyme in human blood plasma that inactivates bradykinin and kallidins. Biochem Pharmacol 11:585–592
54. Ertl G, Kloner RA, Alexander RW et al (1982) Limitation of experimental infarct size by an angiotensin-converting enzyme inhibitor. Circulation 65:40–48
55. Garg R, Yusuf S, Bussmann WD et al (1995) Overview of randomized trials of angiotensin-converting enzyme inhibitors on mortality and morbidity in patients with heart failure. J Am Med Assoc 273:1450–1456
56. Hock CE, Ribeiro LGT, Lefer AM (1985) Preservation of ischemic myocardium by a new converting enzyme inhibitor, enalaprilic acid, in acute myocardial infarction. Am Heart J 109:222–228
57. Linz W, Schölkens BA, Han YF (1986) Beneficial effects of the converting enzyme inhibitor, ramipril, in ischemic rat hearts. J Cardiovasc Pharmacol 8:S91–S99
58. Schmieder RE, Martus P, Klingbeil A (1996) Reversal of left ventricular hypertrophy in essential hypertension: A meta-analysis of randomized double-blind studies. J Am Med Assoc 275:1507–1513
59. Kakoki M, Smithies O (2009) The kallikrein-kinin system in health and in diseases of the kidney. Kidney Int 75:1019–1030
60. Reddigari SR, Shibayama Y, Brunnée T et al (1993) Human Hageman factor (factor XII) and high molecular weight kininogen compete for the same binding site on human umbilical vein endothelial cells. J Biol Chem 268:11982–11987

61. Gurewich V, Johnstone M, Loza J-P et al (1993) Pro-urokinase and prekallikrein are both associated with platelets. FEBS Lett 318:317–321
62. Lenich C, Pannell R, Gurewich V (1995) Assembly and activation of the intrinsic fibrinolytic pathway on the surface of human endothelial cells in culture. Thromb Haemost 74:698–703
63. Loza JP, Gurewich V, Johnstone M et al (1994) Platelet-bound prekallikrein promotes pro-urokinase-induced clot lysis: a mechanism for targeting the factor XII dependent intrinsic pathway of fibrinolysis. Thromb Haemost 71:347–352
64. Motta G, Rojkjaer R, Hasan AAK et al (1998) High molecular weight kininogen regulates prekallikrein assembly and activation on endothelial cells: a novel mechanism for contact activation. Blood 91:516–528
65. Røjkjær R, Hasan AAK, Motta G et al (1998) Factor XII does not initiate prekallikrein activation on endothelial cells. Thromb Haemost 80:74–81
66. Chambrey R, Picard N (2011) Role of tissue kallikrein in regulation of tubule function. Curr Opin Nephrol Hypertens 20:523–528
67. Olsen UB (1978) Kidney volume expansion and prostaglandin release by bradykinin. The effect of indomethacin pretreatment. Acta Physiol Scand 102:129–136
68. Wilson TW (1992) Renal prostaglandin synthesis and angiotensin-converting enzyme inhibition. J Cardiovasc Pharmacol 19:S39–S44
69. Tomita K, Pisano JJ, Burg MB et al (1986) Effects of vasopressin and bradykinin on anion transport by the rat cortical collecting duct. Evidence for an electroneutral sodium chloride transport pathway. J Clinl Invest 77:136–141
70. Tomita K, Pisano JJ, Knepper MA (1985) Control of sodium and potassium transport in the cortical collecting duct of the rat. Effects of bradykinin, vasopressin, and deoxycorticosterone. J Clin Invest 76:132–136
71. Charbonneau H, Buléon M, Minville V et al (2016) Acute bradykinin receptor blockade during hemorrhagic shock in mice prevents the worsening hypotensive effect of angiotensin-converting enzyme inhibitor. Crit Care Med [Epub ahead of print]
72. Bergaya S, Meneton P, Bloch-Faure M et al (2001) Decreased flow-dependent dilation in carotid arteries of tissue kallikrein-knockout mice. Circ Res 88:593–599
73. Meneton P, Bloch-Faure M, Hagege AA et al (2001) Cardiovascular abnormalities with normal blood pressure in tissue kallikrein-deficient mice. Proc Natl Acad Sci 98:2634–2639
74. Duka I, Shenouda S, Johns C et al (2001) Role of the B2 receptor of bradykinin in insulin sensitivity. Hypertension 38:1355–1360
75. Duka I, Kintsurashvili E, Gavras I et al (2001) Vasoactive potential of the B1 bradykinin receptor in normotension and hypertension. Circ Res 88:275–281
76. Kohlman O, de Assis Rocha Neves F, Ginoza M et al (1995) Role of bradykinin in insulin sensitivity and blood pressure regulation during hyperinsulinemia. Hypertension 25: 1003–1007
77. Obiezu CV, Diamandis EP (2005) Human tissue kallikrein gene family: applications in cancer. Cancer Lett 224:1–22
78. George MY, Eleftherios PD (2001) The new human tissue kallikrein gene family: Structure, function, and association to disease. Endocr Rev 22:184–204
79. Emami N, Diamandis EP (2007) Human tissue kallikreins: a road under construction. Clin Chim Acta 381:78–84
80. Slim R, Torremocha F, Moreau T et al (2002) Loss-of-function polymorphism of the human kallikrein gene with reduced urinary kallikrein activity. J Am Soc Nephrol 13:968–976
81. Azizi M, Emanueli C, Peyrard S et al (2008) Genetic and dietary control of plasma tissue kallikrein secretion and urinary kinins exretion in man. J Hypertens 26:714–720
82. Potier L, Waeckel L, Richer C et al (2013) Tissue kallikrein, blood pressure regulation, and hypertension: insight from genetic kallikrein deficiency. Biol Chem 394:329–333
83. Adetuyibi A, Mills IH (1972) Relation between urinary kallikrein and renal function, hypertension, and excretion of sodium and water in man. Lancet 2:203–207

84. Croxatto HR, Martin MS (1970) Kallikrein-like activity in the urine of renal hypertensive rats. Experientia 26:1216–1217. doi:10.1007/bf01897974
85. Lechi A, Covi G, Lechi C et al (1978) Urinary kallikrein excretion and plasma renin activity in patients with essential hypertension and primary aldosteronism. Clin Sci Mol Med 55: 51–55
86. Margolius H, Pisano J, Geller R et al (1971) Altered urinary kallekrein excretion in human hypertension. Lancet 298:1063–1065
87. Sharma JN, Narayanan P (2014) The kallikrein-kinin pathways in hypertension and diabetes. Prog Drug Res 69:15–36
88. Frey EK (1926) Zusammenhänge zwischen Herzarbeit und Nierentätigkeit. Arch Klin Chir 142:663–669
89. De Freitas FM, Faraco EZ, De Azevedo DF (1964) General circulatory alterations induced by intravenous infusion of synthetic bradykinin in man. Circulation 29:66–70
90. Mills IH (1982) The renal kallikrein-kinin system and sodium excretion. Q J Exp Physiol 67:393–399
91. Nasjletti A, McGiff JC, Colina-Chourio J (1978) Interrelations of the renal kallikrein-kinin system and renal prostaglandins in the conscious rat influence of mineralocorticoids. Circ Res 43:799–807
92. Willis LR, Ludens JH, Hook JB, Williamson HE (1969) Mechanism of natriuretic action of bradykinin. Am J Physiol 217:1–5
93. Bergaya S, Hilgers RHP, Meneton P et al (2004) Flow-dependent dilation mediated by endogenous kinins requires angiotensin AT2 receptors. Circ Res 94:1623–1629
94. Trabold F, Pons S, Hagege AA et al (2002) Cardiovascular phenotypes of kinin B2 Receptor- and tissue kallikrein-deficient mice. Hypertension 40:90–95
95. Carretero OA, Carbini LA, Scicli AG (1993) The molecular biology of the kallikrein-kinin system: I. General description, nomenclature and the mouse gene family. J Hypertens 11:693–697
96. Carretero OA, Scicli AG (1978) The renal kallikrein-kinin system in human and in experimental hypertension. Klin Wochenschr 56:113–125
97. Ardiles LG, Figueroa CD, Mezzano SA (2003) Renal kallikrein-kinin system damage and salt sensitivity: insights from experimental models. Kidney Int Suppl 86:S2–S8
98. Griol-Charhbili V, Sabbah L, Colucci J et al (2009) Tissue kallikrein deficiency and renovascular hypertension in the mouse. Am J Physiol Regul Integr Comp Physiol 296: R1385–R1391
99. Marchetti J, Imbert-Teboul M, Alhenc-Gelas F et al (1984) Kallikrein along the rabbit microdissected nephron: a micromethod for its measurement. Pflugers Arch 401:27–33
100. Pratt JH (2005) Central role for ENaC in development of hypertension. J Am Soc Nephrol 16:3154–3159
101. Alfie ME, Yang X-P, Hess F et al (1996) Salt-sensitive hypertension in bradykinin B2 receptor knockout mice. Biochem Biophys Res Commun 224:625–630
102. Hecquet C, Tan F, Marcic BM et al (2000) Human bradykinin B2 receptor is activated by kallikrein and other serine proteases. Mol Pharmacol 58:828–836
103. Huang W, Gallois Y, Bouby N et al (2001) Genetically increased angiotensin I-converting enzyme level and renal complications in the diabetic mouse. Proc Natl Acad Sci 98:13330–13334
104. Bodin S, Chollet C, Goncalves-Mendes N et al (2009) Kallikrein protects against microalbuminuria in experimental type I diabetes. Kidney Int 76:395–403
105. Regoli D, Plante GE, Gobeil jr F (2012) Impact of kinins in the treatment of cardiovascular diseases. Pharmacol Therap 135:94–111
106. Derkx FHM, Tan-Tjiong HL, Man In'T V et al (1979) Activation of inactive plasma renin by tissue kallikreins. J Clin Endocrinol Metab 49:765–769
107. Sealey JE, Atlas SA, Laragh JH et al (1978) Human urinary kallikrein converts inactive to active renin and is a possible physiological activator of renin. Nature 275:144–145

108. Macfarlane NA, Adetuyibi A, Mills IH (1974) Proceedings: changes in kallikrein excretion during arterial infusion of angiotensin. J Endocrinol 61:LXXII
109. Lifton RP, Gharavi AG, Geller DS (2001) Molecular mechanisms of human hypertension. Cell 104:545–556
110. Shimkets RA, Warnock DG, Bositis CM et al (1994) Liddle's syndrome: heritable human hypertension caused by mutations in the β subunit of the epithelial sodium channel. Cell 79:407–414
111. Rossier BC, Pradervand S, Schild L et al (2002) Epithelial sodium channel and the control of sodium balance: Interaction between genetic and environmental factors. Annu Rev Physiol 64:877–897
112. Patel AB, Chao J, Palmer LG (2012) Tissue kallikrein activation of the epithelial Na channel. Am J Physiol Renal Physiol 303:F540–F550
113. White PC (1994) Disorders of aldosterone biosynthesis and action. New Engl J Med 331:250–258
114. Connell JMC, MacKenzie SM, Freel EM et al (2008) A lifetime of aldosterone excess: long-term consequences of altered regulation of aldosterone production for cardiovascular function. Endocr Rev 29:133–154
115. Yoshida H, Zhang JJ, Chao L et al (2000) Kallikrein gene delivery attenuates myocardial infarction and apoptosis after myocardial ischemia and reperfusion. Hypertension 35:25–31
116. Chao J, Zhang JJ, Lin K-F et al (1998) Human kallikrein gene delivery attenuates hypertension, cardiac hypertrophy, and renal injury in dahl salt-sensitive rats. Hum Gene Ther 9:21–31
117. Yayama K, Wang C, Chao L et al (1998) Kallikrein gene delivery attenuates hypertension and cardiac hypertrophy and enhances renal function in goldblatt hypertensive rats. Hypertension 31:1104–1110
118. Bledsoe G, Shen B, Yao Y et al (2006) Reversal of renal fibrosis, inflammation, and glomerular hypertrophy by kallikrein gene delivery. Huma Gene Ther 17:545–555
119. Wolf WC, Yoshida H, Agata J et al (2000) Human tissue kallikrein gene delivery attenuates hypertension, renal injury, and cardiac remodeling in chronic renal failure. Kidney Int 58:730–739
120. Ilan Y, Droguett G, Chowdhury NR et al (1997) Insertion of the adenoviral E3 region into a recombinant viral vector prevents antiviral humoral and cellular immune responses and permits long-term gene expression. Proc Natl Acad Sci 94:2587–2592
121. Lawrence MG, Lai J, Clements JA (2010) Kallikreins on steroids: Structure, function, and hormonal regulation of prostate-specific antigen and the extended lallikrein locus. Endocr Rev 31:407–446
122. Chao J, Miao Robert Q, Chen V et al (2001) Novel roles of kallistatin, a specific tissue kallikrein inhibitor, in vascular remodeling. Biol Chem 382:15–21
123. Chen L-M, Song Q, Chao L, Chao J (1995) Cellular localization of tissue kallikrein and kallistatin mRNAs in human kidney. Kidney Int 48:690–697
124. Wang MY, Day J, Chao L et al (1989) Human kallistatin, a new tissue kallikrein-binding protein: purification and characterization. Adv Exp Med Biol 247B:1–8
125. Zhou GX, Chao L, Chao J (1992) Kallistatin: a novel human tissue kallikrein inhibitor. Purification, characterization, and reactive center sequence. J Biol Chem 267:25873–25880
126. Chao J, Schmaier A, Chen L-M, Yang Z et al (1996) Kallistatin, a novel human tissue kallikrein inhibitor: Levels in body fluids, blood cells, and tissues in health and disease. J Lab Clin Med 127:612–620
127. Madeddu P, Parpaglia PP, Demontis MP et al (1995) Effects of kinin blockade on the blood pressure of salt-loaded pregnant rats. Hypertension 25:823–827
128. Chao J, Chao L (1995) Biochemistry, regulation and potential function of kallistatin. Biol Chem Hoppe Seyler 376:705–713
129. da Costa PLN, Sirois P, Tannock IF et al (2014) The role of kinin receptors in cancer and therapeutic opportunities. Cancer Lett 345:27–38

Proteases—The Sharp Scissors in Human Diseases

Sandipan Chakraborty and Soumalee Basu

1 Introduction

Proteases or more generally peptidases [1] are a type of enzyme that perform peptide bond hydrolysis in proteins and are found ubiquitously in eukaryotes, prokaryotes, and viruses. They cover about 2% of the total number of proteins present in an organism. All proteases share in common the general mechanism of a nucleophilic attack on the carbonyl carbon of an amide bond [2]. This results in a general acid-base hydrolytic process that disrupts the covalent bond. Different proteases utilize different strategies to generate the nucleophile and to bring together the nucleophile with the targeted bond. These distinctions form the basis for the classification of this class of enzymes. Resting on the presence of important chemical groups in their active sites, proteases were typically categorized into four major classes, namely serine, cysteine, metallo- and aspartate proteases till the end of the last century. Metallo- and aspartate proteases utilize heavy metals and aspartate residues, respectively, to immobilize and polarize a water molecule so that the oxygen atom in water becomes the nucleophile [3]. The central feature of aspartyl proteases is the presence of a pair of aspartic acid residue in different protonation states that together facilitate the attack on the peptide bond by water molecule. The deprotonated aspartic acid residue activates the water molecule by assuring its deprotonation, whereas the protonated residue polarizes the peptide carbonyl thus making it susceptible to the attack. The active site of metallo-proteases, on the other

S. Chakraborty
Department of Physical Chemistry, Indian Association for the Cultivation of Science, Jadavpur, Kolkata 700032, West Bengal, India

S. Basu (✉)
Department of Microbiology, University of Calcutta, 35 Ballygunge Circular Road, Kolkata 700019, West Bengal, India
e-mail: soumalee@gmail.com

S. Chakraborti and N.S. Dhalla (eds.), *Pathophysiological Aspects of Proteases*,
DOI 10.1007/978-981-10-6141-7_27

hand, is constituted of a bound metal ion, mostly zinc that activates a water molecule to attack the peptide carbonyl group as a nucleophile. Serine and cysteine proteases utilize their HO and HS side chains, respectively, directly as nucleophiles. Although not identical, the catalytic mechanisms of serine and cysteine proteases are remarkably similar. The formation of an oxyanion or thiolate anion (the nucleophile) in serine and cysteine proteases, respectively, is essential for catalysis. Thus, the presence or absence of a covalent acyl-enzyme intermediate on the reaction pathway demarcates the mechanism of catalysis by serine and cysteine proteases from aspartic and metallo-proteases. In general, these cysteine and serine proteases are folded into two relatively large globular domains surrounding a cleft containing the active site residues. Substrate entry into the cleft is a prerequisite for cleavage, and efficient entry is dictated by the structural fit between the potential substrate and the topology of the cleft, a major determinant of enzyme specificity. The formation of a spatial fit between a targeted bond of the substrate and the active site nucleophile is obviously also a critical determinant of substrate specificity.

In 1995 and 2004, threonine and glutamate proteases were included in the classification. Threonine proteases use threonine as the nucleophile, while glutamate proteases use water as nucleophiles. Inclusion of a seventh group of protease named as asparagine peptide lyase may be debatable as the proteolytic mechanism involves elimination reaction instead of hydrolysis. Proteases are also classified into acid, neutral, and basic/alkaline proteases depending on the optimal pH at which they remain active. Exopeptidases, namely aminopeptidase and carboxypeptidase A as well as endopeptidases, are also used as classes of proteases.

Protease activity can be highly specific such that hydrolysis would occur only at a particular site. It can be highly promiscuous too such that a wide range of substrates may be hydrolyzed. Specificity is dictated not only by desired interactions at the active site but more importantly by the influence of remote binding interactions on the orientation of the substrate toward the catalytic site for desired interactions to occur [4]. Nonspecific hydrolysis on the other hand occurs after the protease binds to a particular residue of the substrate and hydrolyses relatively nonspecifically at positions with preferences for certain amino acids at the P1, P2, P1′, or P2′ positions.

The human genome codes for about 560 proteases and their homologs that are distributed into 186 metallo-, 178 serine, 21 aspartic, 148 cysteine and 28 threonine proteases [5]. According to the most recent consensus, among all the known proteases, 1.97% is from archaeal origin and 67.8 and 30.23% constitute the bacterial and eukaryotic proteases (Fig. 1a). Interestingly, the percentages of proteases in whole genome sequences from different super-kingdom greatly vary. A list of percentage of the number of proteases in whole genome for the most and least abundant organism from three super-kingdoms is shown in Fig. 1b.

(a)

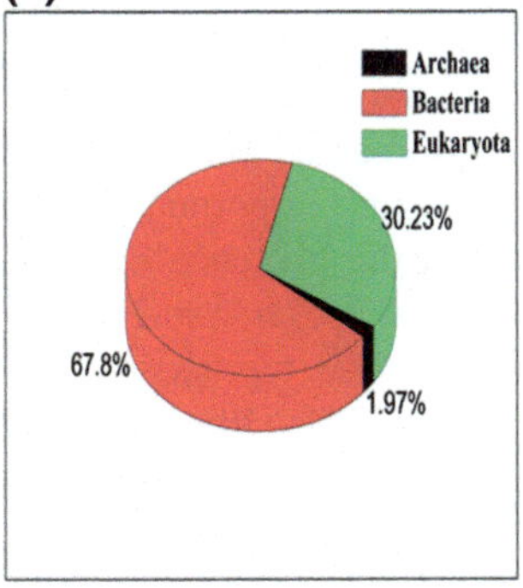

(b)

Super kingdom	Name of the organism	% of peptidases in whole genome sequences
Archaea	*Pyrococcus abyssi strain GE5*	4.61
	Natrinema pellirubrum strain DSM15624	1.12
Bacteria	*Bacillus anthracis strain A2012*	53.54
	Vibrio sp. RC586	0
Eukaryota	*Eimeria tenella*	58.61
	Passalora fulva	0

Fig. 1 **a** A pie chart representation of all known proteases among different super-kingdom. **b** A list of percentage of the number of proteases in whole genome for most (*top*) and least (*below*) abundant organism from three super-kingdoms (http://merops.sanger.ac.uk/)

Proteases play a very prominent role in various aspects of physiological processes such as food digestion, blood coagulation, wound healing, apoptosis, signal transduction, and immune responses to name a few. A delicate control on the activation and inactivation of protease cascades at different levels of protease gene transcription, mRNA translation, activation of zymogens, substrate specificity, enzyme kinetics, and even by means of enzyme-inhibitors needs to be maintained. A number of studies illustrate that apart from proteolysis, proteases possess a variety of regulatory functions that are mediated through intracellular signaling pathways, caspase-like enzyme activity, and/or regulation of specific cytokines and signaling receptors. Proteases are, therefore, viewed as multifunctional, hormone-like signaling molecules that are cardinal to various physiological and pathological processes [6]. Aside from their roles in the physiology of organisms, they also play crucial roles in the pathogenesis of a number of diseases, such as several parasitic infection, cardiopulmonary disease, emphysema, gingivitis, several inflammatory infections, tumor invasion, traumatic brain injury, ischemic brain injury, Alzheimer's disease, the destruction of the extracellular matrix of articular cartilage and bone in arthritic joints. Out of these, parasitic diseases represent major global health problems of immense proportion. Parasite-derived proteases have emerged as the major among the virulence factors identified. Well-characterized roles in pathogenesis played by proteases include their involvement in invasion of the host by parasite migration through tissue barriers, degradation of hemoglobin and other blood proteins, immune evasion, activation of inflammation, egression. Therefore, many of these enzymes that originate from the human genome or from the genome of the disease-causing organism form potential targets.

2 Disease-Causing Serine Proteases—*Intrinsic and Extrinsic*

Serine proteases that constitute one-third of peptidases are so named because of the presence of a nucleophilic serine residue at the active site playing important roles in mediating protein hydrolysis. The class is distinguished by the presence of novel catalytic triads and dyads with the classic catalytic His-Asp-Ser/Asp-His-Ser/Ser-Asp-His/Ser-His-Asp triad spanning over four different clans typified by chymotrypsin, subtilisin, carboxypeptidase Y, and Clp protease [4]. Other catalytic triads and dyads include Ser-His-Glu, His-Ser-His, Ser-Lys/His, and N-terminal serine [7]. Chymotrypsin, the first three-dimensional structure of protease to be determined by Blow [8], depicted the presence of catalytic, substrate recognition, and zymogen activation domain components, later found to be common to all chymotrypsin-like proteases.

Serine proteases are involved in a wide range of biological processes, such as intracellular and extracellular protein metabolism, digestion, fertilization, blood coagulation, and regulation of development. Human serine proteases like elastase, cathepsin G, PR-3, thrombin, kallikreins, tryptase, and chymase all play a role in cardiovascular disease [9]. Cystic fibrosis (CF) lung disease is referred to as chronic infection of the lung characterized by inflammation of the airway. Elevated levels of neutrophils and subsequently that of the three neutrophil serine proteases, namely neutrophil elastase, proteinase 3 (PR3), and cathepsin G, have been observed in sputum of CF patients. Excess protease activity results in the detrimental cycle of inflammation [10]. Celiac disease (CD) is triggered by the ingestion of gluten in genetically susceptible individuals and is yet another example of a disease where serine protease seems to play a significant role in the development of the disease as elafin which is a serine protease inhibitor has been found to be expressed less in CD patients. This is true for patients suffering from inflammatory bowel disease (IBD) too.

Quite a few number of serine proteases have been identified in protozoan and parasitic helminths that have putative roles in parasite development and nutrition, host tissues and cell invasion, anticoagulation, and immune evasion. Humans are constantly threatened by a wide variety of infectious diseases caused by these pathogens. The rhomboid family of serine protease, first identified in Drosophila, is conserved in prokaryotes and eukaryotes. In toxoplasma and *P. falciparum*, rhomboid proteases were identified to function as shedding of adhesion proteins during invasion [11]. The serine proteases that have been identified in parasitic helminths include nematodes or roundworms (*Trichinella spiralis*, *T. pseudospiralis*, *Onchocerca volvulus*, *O. lienalis*, *Brugia malayi*, *Ancylostoma caninum*), cestodes or tapeworm (*Spirometra mansoni, Echinococcus granulosus*) and trematodes or flatworm (*Fasciola hepatica*, *F. gigantica*, and *Schistosoma mansoni*). TspSP-1 from *T. spiralis* plays an important role in degrading cytoplasmic or intercellular proteins, thereby facilitating the movement of larvae [12]. Ts23-2 is another serine protease that is expressed after cyst formation. TspSP-1.2 has been isolated as another member of this family that contributes to the larval invasion into the host epithelial cells and is being considered as a potential vaccine candidate against *T. spiralis* infection [13]. TsSerP is yet another member that is involved in molting process of the parasite and digestive functions. *Onchocerca volvulus* is an important filarial nematode that causes subcutaneous filariasis of humans and affects the eyes and skin. A neutral elastase of the nematode is known to degrade the elastic fibers of the host tissue, and a blisterase belonging to the subtilisin-like serine protease plays an important role in cuticle production and maintenance, neural signaling, and nematode development [14] wherefore it forms a potential drug target for controlling parasite infection. Another serine protease from *Brugia malayi* microfilariae is believed to be involved in suppressing the immune system by cleaving C5a of the complement system of the host [15]. Infection by a hematophagous nematode of the genus *Ancylostoma*, living in the small intestine of hosts, causes hookworm disease. *Ancylostoma caninum* uses an elastolytic enzyme to feed on the villous capillaries by preventing the blood from clotting. This protease is thus a potential target for chemotherapeutic or immunological intervention [16]. Serine proteases from cestodes have been found to be involved in host tissue invasion and immune evasion and have been characterized from plerocercoid larvae of *Spirometra mansoni* or from *Echinococcus granulosus* or *Taenia solium cysticercus* [17]. Serine proteases from the liver fluke disease-causing parasites *Fasciola hepatica* and *F. Gigantica* have also gained attention [18]. Schistosomiasis caused by the parasite blood fluke or *Schistosoma* spp. is a chronic disease [19] in which its serine protease "cercarial elastase" facilitates skin invasion by the infective schistosome larvae to permissive hosts including humans. The stage-specific expression of the proteases contributes to successful parasitism lasting for decades [20].

The malarial proteases are a group of molecules that are essential for parasite life cycle stages out of which serine proteases are of special interest due to their involvement in parasite-specific egress and invasion [21].

3 Disease-Causing Cysteine Proteases—*Intrinsic and Extrinsic*

Cysteine proteases are grouped into two superfamilies—the family of enzymes related to interleukin 1β converting enzyme (ICE), and the papain superfamily of cysteine proteases [22]. Other than the presence of the active site cysteine, the ICE superfamily does not share much homology with the papain family [23]. The papain superfamily is composed of the three groups, namely calpain, bleomycin hydrolase, and the papain family itself. Cathepsin from humans also belongs to this group.

Apoptosis, the physiological death of cells, is implemented by caspases. Once activated, caspases act as cysteine proteases. Enzymatic activation of initiator caspases leads to proteolytic activation of downstream executioner (effector) caspases leading to the cleavage of a number of vital proteins, resulting in the orderly removal of the cell [24]. Thus, both reduced and increased apoptosis lead to pathology and the associated diseases include cancer, neurological disorders, cardiovascular disorders, and autoimmune diseases.

In humans, cathepsin cysteine proteases consist of a family of 11 members, namely cathepsins B, C, F, H, K, L, O, S, V, W, and Z [25]. Many cellular events in the development of atherosclerosis-based cardiovascular disease depend on the cathepsin-mediated degradation of intracellular and extracellular proteins, including cell adhesion, transmigration, differentiation, proliferation, apoptosis, neovascularization, and antigen presentation. Even though cathepsins are abundant in human, the exact role each specific cathepsin plays in heart disease development and the mechanism and significance behind their function are unknown. Cathepsins S, K, B, and L have been shown to be involved in osteoporosis and cancer. Calpain cathepsin and caspases are involved in traumatic brain injury (TBI), ischemic brain injury, and many neurodegenerative diseases. In Alzheimer's disease, cathepsin B is found in neurites and dendrites, whereas in the normal condition its activity is localized in lysosomes [26]. Neurofibrillary tangles (NFT) in brains suffering from AD contain increased levels of calpain 2 and cathepsin [27], while calpain 1 has been found to be hyperactive in brains. Hsp70, which has a dual role as a chaperone in damaged protein and also as an important factor in the maintenance of lysosomal integrity, is cleaved by calpain after modification by oxidative stress resulting in the impairment of lysosomal autophagy [28]. Cathepsin B has been shown to have multiple roles in cancer with an increased activity in lung tumor and lymph node metastases [29]. Cysteine proteases are essential for the induction and development of both innate and adaptive immune responses in humans. Their role ranges from antigen and pathogen recognition and elimination, signal processing and cell homeostasis.

There are pathogens that secrete cysteine proteases that often target the same proteins as the mammalian proteases and thereby modulate host immunity [30]. Cysteine proteases that are involved in parasitic diseases include falcipains 2, 2′, and 3 from *P. falciparum*. They are referred to as malarial hemoglobinase as they degrade hemoglobin after it has been cleaved into peptide fragments [31]. *Trypanosoma brucei* is a protozoan parasite and the causative agent of human African trypanosomiasis, a fatal disease that is transmitted by the bite of the tsetse fly. Two clan CA cysteine proteases have been identified in *T. brucei*, rhodesain, which is cathepsin L-like and most abundant and tbcatB, a cathepsin B-like enzyme. Studies show that despite being less abundant, tbcat is essential to *T. brucei* as it is implicated in iron acquisition whereby transferrin may be a natural substrate [32]. *T. brucei* cathepsin B is a logical target for the development of new anti-trypanosomal chemotherapy as *T. brucei* cathepsin B-like protease is a key enzyme in host protein turnover. It may be mentioned that the cysteine proteases staphopain A and B secreted by the human pathogen *Staphylococcus aureus* could not be inhibited by any of the seven human cystatins [33]. Cysteine proteases are also used by the worms of schistosome to digest the blood meal.

4 Disease-Causing Metallo-Proteases—*Intrinsic and Extrinsic*

Well-timed degradation of extracellular matrix (ECM) is an important feature of development, morphogenesis, tissue repair, and remodeling. Upon disruption of the precise regulation pertaining under normal physiological conditions, diseases like arthritis, nephritis, cancer, encephalomyelitis, chronic ulcers, fibrosis occur. Cardiovascular disorders such as atherosclerosis, stenosis, left ventricular hypertrophy, heart failure and aneurysm [34] result from uncontrolled ECM remodeling of the myocardium and vasculature. Major enzymes responsible for ECM degradation are the matrix metallo-proteinases (MMPs) also called matrixins. They form a large family of calcium-dependent zinc-containing endopeptidases. Humans have about 30 matrixin genes with a duplicate MMP-23 gene. Expression of matrixins is transcriptionally controlled by inflammatory cytokines, growth factors, hormones, cell–cell and cell–matrix interaction, while its activity is regulated by activation of the precursor zymogens and inhibition by endogenous inhibitors, tissue inhibitors of metallo-proteinases (TIMPs) [35]. In pathological conditions, this equilibrium is shifted toward increased MMP activity leading to tissue degradation. On the basis of the specificity of MMPs, they are classified into collagenase (MMP-1, -8, and -13), gelatinase (MMP-2 and -9), stromelysin (MMP-3, -10, -11), and matrilysin (MMP-7 and -12). Another subclass is the membrane-type MMPs (MT-MMP). Most of the matrix metallo-proteinases consist of four distinct domains, which are N-terminal pro-domain, catalytic domain, hinge region, and C-terminal hemopexin-like domain. This may be responsible for the macromolecular substrate recognition and/or interaction with TIMPs. The membrane-type MMPs (MT-MMPs) contain an additional transmembrane domain that anchors them in the cell surface [36]. MMPs in the CNS are secreted by microglia, astrocytes and neurons. Under normal conditions they are either absent or present at undetectable levels in the brain and deregulation of their activity shifts the balance and induces perpetuation of chronic inflammation. This has been shown in different peripheral chronic diseases, such as atherosclerosis and rheumatoid arthritis [37, 38] and neuroinflammatory diseases, such as cerebral ischemia, stroke, and bacterial meningitis [39]. MMP-1 in synovial fluid of rheumatoid arthritis patients has shown promise as a marker of the disease, MMP-8 may be useful as a marker for plaque instability in atherosclerosis, MMP-13 has a specialized role in bone development and hence it plays important role in osteoarthritis, MMP-2, MMP-3 and MMP-9 are also important [40]. During infection of glial cells with human T-lymphotropic virus type I (HTLV-I), the causative agent of a progressive chronic myelopathy, MMP-9, increases [41]. It is worth mentioning that the role of MMPs has become well appreciated in neurodegenerative diseases with the listed diseases being Alzheimer's disease, Parkinson's disease, amyotrophic lateral sclerosis and multiple sclerosis.

Metallo-protease falcilysin from *P. falciparum* is also involved in degradation of hemoglobin after it has been cleaved into peptide fragments [42]. The metallo-protease toxolysin 4 from *Toxoplasma gondii* localizes to the micronemes and its secretion coincides with discharge of micronemal contents [43], thus implying its function in invasion. *H. pylori* infection increases secretion of both MMP-1 and MMP-3 by gastric epithelial cells. *M. tuberculosis* infection increases MMP-1 from human airway epithelial cells [44]. An efficient way for a pathogen to cause pathology is by operating at the top of a proteolytic cascade. An example of this is the protease of the thermolysin family secreted by *Pseudomonas aeruginosa* and *Vibrio cholera* that activate pro-MMP-1, -8, and -9 [45], while proteases from the oral pathogen *Porphyromonas gingivalis* activate MMP-1, -3, and -9 [46].

5 Disease-Causing Aspartyl Proteases—*Intrinsic and Extrinsic*

Aspartyl proteases form the smallest group of proteases in humans that has been shown to play important roles in physiological and pathological processes. Aspartic proteases such as cathepsin D, gastricin, pepsin, renin, HIV protease and others have been the subject of research in the last years. The involvement of renin in hypertension, cathepsin D in metastasis of breast cancer, BACE 1 (beta site-activated APP cleaving enzyme) in Alzheimer's disease and the protease of human immunodeficiency virus (HIV) in acquired immune deficiency syndrome (AIDS) is being vigorously studied. Multicellular organisms maintain endogenous proteins that function as protease inhibitors to regulate proteolytic activity. Most of these inhibitory proteins are directed against serine proteases, although some are known to target cysteine, aspartyl, or metallo-proteases. Indeed, inhibitors of serine, cysteine and metallo-proteases are distributed ubiquitously throughout the biological world. In sharp contrast, however, naturally occurring inhibitors of aspartic proteases are relatively uncommon and are found in only certain specialized locations.

The extended family of plasmepsin aspartyl proteases in *P. falciparum* has been the focus of intensive research mainly due to the parasite-specific nature of these enzymes. Plasmepsins are made up of four main members, two of which are thought to initiate hemoglobin degradation, while the other two function in the degradation of peptides generated by the action of upstream members of the family [47]. Plasmepsins have therefore evolved as very popular antimalarial agents and are of immense therapeutic importance. Aspartic proteases are important to the worms of schistosome as they use the protease to digest the blood meal for survival in the host and hence cause the disease.

6 Drugs as Protease Inhibitors

Deviation from the physiological role of intrinsic/endogenous proteases in humans or a spoilt balance in their amount results in severe diseases. Additionally, usage of intrinsic and/or extrinsic proteases by the human pathogens during invasion and/or egression for subsequent survival has led to the idea of employment of protease inhibitors in therapeutics. Several important drugs are protease inhibitors. For example, captopril, an inhibitor of the metallo-protease angiotensin-converting enzyme (ACE), has been used to regulate blood pressure. Crixivan, an inhibitor of the HIV protease, is used in the treatment of AIDS. This protease cleaves multidomain viral proteins into their active forms; complete blocking of this process prevents the virus from being infectious. To prevent unwanted side effects, protease inhibitors used as drugs must be specific for one enzyme without inhibiting other proteins within the body.

7 Important Databases

The MEROPS, an online database at http://merops.sanger.ac.uk, is a rich source of information on peptidases and their inhibitors. CutDB is a proteolytic event database at cutdb.burnham.org that focuses on the annotation of individual proteolytic events both actual and predicted.

The mammalian degradome database at http://degradome.uniovi.es gives complete information of the degradome of an organism whose whole genome is available. Degradome is defined as the complete set of proteases present in an organism. This Web site also provides detailed information about genetic diseases of proteolysis. Finally, the user can find additional information about protease structures, protease inhibitors, ancillary domains of proteases and differences between mammalian degradomes.

References

1. Barett AJ, McDonald JK (1986) Nomenclature: protease, proteinase and peptidase. Biochem J 237:935
2. Polgar L (ed) (1989) Metalloproteases. In: Mechanisms of protease action, pp. 208–210. CRC Press, Boca Ratan, FL
3. Menard R, Storer A (1992) Oxyanion hole interactions in serine and cysteine proteases. Hoppe-seyler's Z Biol Chem 373:393–400
4. Hedstrom L (2002) Serine protease mechansm and specificity. Chem Rev 102:4501–4523
5. Puente XS, Lopez-Otin C (2004) A genomic analysis of rat proteases and protease inhibitors. Genome Res 14:609–622
6. Ramachandran R, Hollenberg MD (2008) Proteinases and signalling: pathophysiological and therapeutic implications via PARs and more. Br J Pharmacol 153(Suppl 1):S263–S282
7. Dodson G, Wlodawr A (1998) Catalytic triads and their relatives. Trends Biochem Sci 23:347–352

8. Blow DM (1971) The enzymes, 3rd edn, vol 3. Boyer PD (ed) Academic Press, Boca Raton
9. Sharony R, Yu P-J, Park J, Galloway AC, Mignatti P, Pintucci G (2010) Protein targets of inflammatory serine proteases and cardiovascular disease. J Inflam 7:45
10. Twigg MS, Brockbank S, Lowry P, FitzGerald SP, Taggart C, Weldon S (2015) The role of serine proteases and antiproteases in cystic fibrosis. Mediators Inflamm. Article ID: 293053
11. Zhou XW, Blackman MJ, Howell SA, Carruthers VB (2004) Proteomic analysis of cleavage events reveals a dynamic two-step mechanism for proteolysis of a key parasite adhesive complex. Mol Cell Proteomics 3:565–576
12. Romaris F, North SJ, Gagliardo LF, Butcher BA, Ghosh K, Beiting DP, Panico M, Arasu P, Dell A, Morris HR, Appleton JA (2002) A putative serine protease among the excretory-secretory glycoproteins of L1 *Trichinella spiralis*. Mol Biochem Parasitol 122:149–160
13. Wang B, Wang ZQ, Jin J, Ren HJ, Liu LN, Cui J (2013) Cloning, expression and characterization of a *Trichinella spiralis* serine protease gene encoding a 35.5 kDa protein. Exp Parasitol 134:148–154
14. Poole CB, Jin J, McReynolds LA (2003) Cloning and biochemical characterization of blisterase, a subtilisin-like convertase from the filarial parasite, *Onchocerca volvulus*. J Biol Chem 278:36183–36190
15. Rees-Roberts D, Mullen LM, Gounaris K, Selkirk ME (2010) Inactivation of the complement anaphylatoxin C5a by secreted products of parasitic nematodes. Int J Parasitol 40:527–532
16. Hotez PZ, Cerami A (1983) Secretion of a proteolytic anticoagulant by Ancylostoma hookworms. J Exp Med 157:1594–1603
17. Kong Y, Chung YB, Cho SY, Choi SH, Kang SY (1994) Characterization of three neutral proteases of *Spirometra mansoni* plerocercoid. Parasitology 108:359–368
18. Mohamed SA, Fahmy AS, Mohamed TM, Hamdy SM (2005) Proteases in egg, miracidium and adult of *Fasciola gigantica*. Characterization of serine and cysteine proteases from adult. Comp Biochem Physiol B Biochem Mol Biol 142:192–200
19. Steinmann P, Keiser J, Bos R, Tanner M, Utzinger J (2006) Schistosomiasis and water resources development: systematic review, meta-analysis, and estimates of people at risk. Lancet Infect Dis 6:411–425
20. Horn M, Fajtová P, Rojo Arreola L, Ulrychová L, Bartošová-Sojková P et al (2014) Trypsin- and chymotrypsin-like serine proteases in schistosoma mansoni—'the undiscovered country'. PLoS Negl Trop Dis 8(3):e2766. doi:10.1371/journal.pntd.0002766
21. Alam A (2014) Serine proteases of malaria parasite *Plasmodium falciparum*: potential as antimalarial drug targets. Interdisc Perspect Infect Dis. doi:10.1155/2014/453186
22. Berti PJ, Storer AC (1995) Alignment/phylogeny of the papain superfamily of cysteine proteases. J Mol Biol 246:273–283
23. Thornberry N, Bull HG, Calaycay JR, Chapman KT, Howard AD et al (1992) A novel heterodimeric cysteine protease is required for inteleukin-1 beta processing in monocytes. Nature 356:768–774
24. Nuñez G, Benedict MA, Hu Y, Inohara N (1998) Caspases: the proteases of the apoptotic pathway. Oncogene 17:3237–3245
25. Turk V, Turk B, Turk D (2001) Lysosomal cysteine proteases: facts and opportunities. EMBO J 20:4629–4633
26. Cataldo AM, Hamilton DJ, Nixon RA (1994) Lysosomal abnormalities in degenerating neurons link neuronal compromise to senile plaque development in Alzheimer disease. Brain Res 640:68–80
27. Grynspan F, Griffin WR, Cataldo A, Katayama S, Nixon RA (1997) Active site-directed antibodies identify calpain II as an early-appearing and pervasive component of neurofibrillary pathology in Alzheimer's disease. Brain Res 763:145–158
28. Yamashima T (2012) Hsp70.1 and related lysosomal factors for necrotic neuronal death. J Neurochem 120:477–494

29. Werle B, Kraft C, Lah TT, Kos J, Schanzenbächer U, Kayser K et al (2000) Cathepsin B in infiltrated lymph nodes is of prognostic significance for patients with nonsmall cell lung carcinoma. Cancer 89:2282–2291
30. Donnelly S, Dalton JP, Robinson MW (2011) How pathogen-derived cysteine proteases modulate host immune responses. Adv Exp Med Biol 712:192–207
31. Salas F, Fichmann J, Lee GK, Scott MD, Rosenthal PJ (1995) Functional expression of falcipain, a *Plasmodium falciparum* cysteine proteinase, supports its role as a malarial hemoglobinase. Infect Immun 63(6):2120–2125
32. O'Brien TC, Mackey ZB, Fetter RD, Choe Y, O'Donoghue AJ, Zhou M, Craik CS, Caffrey CR, McKerrow JH (2008) A parasite cysteine protease is key to host protein degradation and iron acquisition 24; 283(43): 28934–28943
33. Vincents B, Onnerfjord P, Gruca M, Potempa J, Abrahamson M (2007) Down-regulation of human extracellular cysteine protease inhibitors by the secreted staphylococcal cysteine proteases, staphopain A and B. Biol Chem 388:437–446
34. Shah PK (1997) Inflammation, metalloproteinases, and increased proteolysis—an emerging pathophysiological paradigm in aortic aneurysm. Circulation 96:2115–2117
35. Nagase H, Visse R, Murphy G (2006) Structure and function of matrix metalloproteinases and TIMPs. Cardiovasc Res 69:562–573
36. Skiles JW, Monovich LG, Jeng AY (2000) Matrix metalloproteinase inhibitor in the treatment of cancer. Annu Rep Med Chem 35:167–176
37. Newby AC (2005) Dual role of matrix metalloproteinases (matrixins) in intimal thickening and atherosclerotic plaque rupture. Physiol Rev 85:1–31
38. Arend WP, Dayer J-M (1995) Inhibition of the production and effects of interleukin-1 and tumor necrosis factor alpha in rheumatoid arthritis. Arthritis Rheum 38:151–160
39. Candelario-Jalil E, Yang Y, Rosenberg GA (2009) Diverse roles of matrix metalloproteinases and tissue inhibitors of metalloproteinases in neuroinflammation and cerebral ischemia. Neuroscience 158:983–994
40. Klein T, Bischoff R (2011) Physiology and pathophysiology of matrix metalloproteases. Amino Acids 41:271–290
41. Giraudon P, Buart S, Bernard A, Thomasset N, Belin MF (1996) Extracellular matrix-remodeling metalloproteinases and infection of the central nervous system with retrovirus human T-lymphotropic virus type I (HTLV-I). Prog Neurobiol 49:169–184
42. Eggleson KK, Duffin KL, Goldberg DE (1999) Identification and characterization of falcilysin, a metallopeptidase involved in hemoglobin catabolism within the malaria parasite *Plasmodium falciparum*. J Biol Chem 274:32411–32417
43. Laliberté J, Carruthers VB (2011) Toxoplasma gondii toxolysin 4 is an extensively processed putative metalloproteinase secreted from micronemes. Mol Biochem Parasitol 177:49–56
44. Elkington PTG, Emerson JE, Lopez-Pascua LDC et al (2005) Mycobacterium tuberculosis up-regulates matrix metalloproteinase-1 secretion from human airway epithelial cells via a p38 MAPK switch. J Immunol 175:5333–5340
45. Okamoto T, Akaike T, Suga M et al (1997) Activation of human matrix metalloproteinases by various bacterial proteinases. J Biol Chem 272:6059–6066
46. DeCarlo AA Jr, Windsor LJ, Bodden MK, Harber GJ, Birkedal-Hansen B, Birkedal-Hansen H (1997) Activation and novel processing of matrix metalloproteinases by a thiol-proteinase from the oral anaerobe *Porphyromonas gingivalis*. J Dent Res 76:1260–1270
47. Banerjee R, Liu J, Beatty W, Pelosof L, Klemba M, Goldberg DE (2002) Four plasmepsins are active in the *Plasmodium falciparum* food vacuole, including a protease with an active-site histidine. Proc Natl Acad Sci U S A 99:990–995

Pathophysiological Aspects of *Aspergillus* sp.

Tapan Kumar Das

1 Introduction

Aspergillosis was firstly identified as fungal infection developed in the lungs of a jackdaw and animals [1] which was firstly focused in human in 1953 [2], and since then, the same infection increases gradually throughout European countries. Bennett examined sputum of the patients suffering from tuberculosis and identified aspergillosis in the cavities of lungs [3]. Spreading of invasive aspergillosis was reported to enhance several times day by day as described by Cawley in [4] (and by Hinson et al. in [5]).

The genus *Aspergillus* has been identified as a class of Euascomycetes of the Phylum Ascomycota [6]. This is asexual species which is supposed to be present as telomorphic forms in the ancestor of Trichocomaceae. Under the heading of the genus, *Aspergillus*, seven subgroups with containing numerous number of species are already recognized. 40 number of *Aspergillus* spp. out of approximately 184 species are found to show human or animal infections, *and* it is reported that it can multiply very rapidly with different colors of conidia like black, white, green, gray, etc. [7].

Aspergillus species living on dead or decomposing ingredients considered as saprophytic fungi are available abundantly everywhere in the world. About 1000 species have been identified in the genus *Aspergillus*. *Aspergillus flavus* and *Aspergillus fumigatus* among other species of *Aspergillus* are recognized commonly as infectious fungi to infect human severely which are about to more than 90% and about 10% are identified are invasive diseases [7]. *Aspergillus clavatus*, *A. cameus*, *A. niger*, *A. terreus*, *A. parasiticus*, *A. oryzae*, *A. wentii*, *A. versicolor*, and

T.K. Das (✉)
Department of Biochemistry and Biophysics, University of Kalyani,
Kalyani 741235, West Bengal, India
e-mail: tapndas@hotmail.com

S. Chakraborti and N.S. Dhalla (eds.), *Pathophysiological Aspects of Proteases*,
DOI 10.1007/978-981-10-6141-7_28

some other *Aspergillus* species could show infectivity character toward human and animals.

After infecting human by *A. flavus*, it causes swelling of the inner membrane in any of the hollow areas of the skull around the nose developing sinuses [8]; Otomycosis is majorly caused by *A. niger* infection to human, and in the same infection, mycelia of *A. niger* are exposed in cerumen and desquamated cells in the external auditory canal, and symptoms are found primarily as scaling, itching, and pain, but *A. niger* could not attack the tissues, and overall tissues remain intact [9]. The most dangerous pathogenic species are *A. flavus* and *A. fumigatus* which could produce alfatoxin showing as toxcin as well as carcinogen properties. Infection of *Aspergillus* causes development of a group of diseases in human and animal called aspergillosis reflecting symptoms of fever, cough, chest pain, or problem in respiration, and it may disorder the immune systems, and multiple infections may be developed in lungs.

Aspergilli also have a less reputable side in the agricultural industries; it could not favor *Aspergilli* as a few species of Aspergillus, like *A. flavus* and *A. parasiticus*, [7] are reported to pollute daily needs crops with aflatoxin secreted by the same *Aspergillus* spp. which are already examined as very toxic and carcinogen, and it can show immunosuppressive properties. In the present scenario of the developing countries in the world, after intake of such type of polluted grains products as daily food, general people are found to be infective with such type toxic materials and feeling severe life-threatening sickness, and they would be on the verge of death.

2 Aspergillosis Infection and Its Symptoms

Approximately, 40 of several number of species of *Aspergillus* are concerned in the development of infection in human or animal systems, and it can be commonly expressed as aspergillosis. *A. fumigates* and *A. flavus* are identified as a severe infectious *Aspergillus* sp. Infecting human to develop a group disease called aspergillosis in major cases and other species like *A. niger* are considered as second line of pathogenic fungi causing fungal infection to human and animal in the world. Infection of *Aspergillus* sp. could develop clinical allergies by *Aspergillus* sp. include clinical allergies, external, and confined infections, and some infections are found to develop in human degenerating tissues. In most cases, *A. fumigatus* can attack human and animals severely to develop insidious aspergillosis. Characteristics of infection are dependent on *Aspergillus* sp. and mainly on immunity of individual persons [10]. As reported, *A. flavus* along with other species of *Aspergillus* could cause severe infections in human and animals. *Aspergillus* species can be attracted at the site of sino-pulmonary tract with lung in human to initiate its infection with the symptom like high fever, pain in head, and face headache, inflammations in face, bleeding of infected nose as it is affluent in blood vessels, projection of one or both eyeball, disorders of smell, vision, eyes, taste, and positional vertigo, ulceration and necrosis of the soft palate, and corrosion of bone. In addition to the commonly observed symptoms, like fever, cold, and cough and breathing disorder, shortness of breath (serious disease at heart and lungs, undefined symptoms are also found in vascular incursion and pleural chest pain while infected persons have been suffering from pulmonary aspergillosis for long time without treatment and extensive rate of caseating visceral tuberculosis occurs by hematogenous dissemination in any organ, and in fine, central nervous system would be arrested intensively succeeding strokes.

Most of the species of *Aspergillus* can grow on departed or putrefying substances available in the surroundings. At the first stage of the life cycle, *Aspergillus* is to generate spore which can easily spread at any places of environment through natural air circulation. The spore carried by air can enter the lungs or other organs initially through inhalation during respiration and occupy the position of alveolar spaces which would not be detached from the deposited places of alveoli by mucociliary escalator due to creation of obstruction by epithelial cells, and there is chance of killing of *Aspergillus* spores initially due to presence of alveolar macrophages by phagocytosis.

3 Aspergillosis

The reason of the infection, aspergillosis in human, is due to inhale of conidia of *Aspergillus* from the environment, and it causes lung diseases as well as immune system could be destabilized. The infection aspergillosis is classified on the basis of localization of *Aspergillus* infection. Problem in respiratory tract is found to develop after infecting with *Aspergillus* [11]. Symptoms of the various type of aspergillosis are being mentioned as below [12].

3.1 Allergic Bronchopulmonary Aspergillosis

Shortly, the same infection is called ABPA. The fungus, *Aspergillus* predominately *A. fumigatus* infects mainly at lungs after inhaling the conidia of the same fungus during respiration [13]. It could initiate the infection through vigorous response of the immune system which may be considered as one type of hypersensitivity response) to *A. fumigatus*. This is also found in a person suffering from cystic fibrosis creating very thick salty mucus and sweat; ultimately, lungs would be affected severely creating problem in respiration with allergic symptoms like generation of coughing and breathless, and at last, severe and permanent bronchitis problem are found to develop. Conidia of *A. fumigatus* would damage tissues of lungs severely if any patients suffer long time with negligence of medical treatment.

3.1.1 Symptom

(i) Reflection of allergy to coughing and wheezing
(ii) Asthma

3.2 Allergic *Aspergillus* Sinusitis

The abbreviation of allergic *Aspergillus* sinusitis is AFS. The AFS is almost similar to ABPA. The same infectious disease may be developed through the various interconnected factors and proceedings as proposed by Manning and Holman [14]. Initially, conidia of *A. fumigatus* could enter by inhalation through nasal respiration, and after entering, the host of human having inclination toward growing up an allergy in hereditary is disclosed to the fungi, *Aspergillus* showing the primary antigenic motivation, and this provocative reaction follows primarily as a consequence reaction between Gell and Coombs type I considered as an IgE mediated and type III considered as immune complex-mediated resulting tissue edema. Finally, the sinus ostium is obstructed; as a result, an accumulation of fluid occurs in the sinus which environments encourage the multiplication of *Aspergillus* creating severe allergic symptoms. Under this condition, it would produce thick, persistent, and highly dense *Aspergillus* mucin with the change of color from chocolate to shady bottle green color [15].

3.2.1 Symptoms

(i) Tissue edema due to an inflammatory reaction
(ii) Obstruction of sinus
(iii) Allergies and nasal polyps
(iv) Asthma

3.3 Aspergilloma

Aspergilloma is also acknowledged as mycetoma/fungal ball developed after assembling of fungi, *Aspergillus* sp. at hollow space of human body as at nasal cavity and also at hollow space of lungs.

In Aspergilloma, organ, lung is commonly affected in human. The patients suffering from the common aviary lung infection like tuberculosis, cystic fibrosis, bronchiectasis, any immunodeficiency, etc., can be affected generally by Aspergilloma [16]. In same, infection *A. fumigatus* is found to be common fungus to infect lung of human after inhaling the same fungal conidia through nasal route, and then, conidia is found to deposit in a hole of lung for multiplying the same fungus as the hole is apart for any intervention since decisive components of the immune coordination are not capable of infiltrating into the hole. The fungal ball is found to constitute after assembling the growing fungus in the hole, and the same ball can associate with dead tissue around the lung, mucus, and other garbage The infection, Aspergilloma, can grow in hard organ as abscesses like in brain and kidney, generally of immune negotiated persons, and this can also be found in cavities of body in human.

3.3.1 Symptom

(i) Feeling of pain mainly in chest and bones
(ii) Visual problem
(iii) Excretion of urine with blood
(iv) Excrete decreased amount of urine
(v) Feeling of pain in head
(vi) Feeling cold
(vii) Breathing problem
(viii) Skin aching

3.4 Chronic Pulmonary Aspergillosis

This infection is also called as CPA, and it belongs to the infection aspergillosis which is considered as a long-period infection of aspergillosis found in human, while *A. fumigatus* would infect the lung of human after inhalation of conidia of the same fungus during respiration [17]. This infection is found to be very rare creating severe problem in lung resulting turmoil in respiration, but CPA is reflected generally as a widespread form considered as chronic cavitary pulmonary aspergillosis and abbreviation is CCPA. Chronic fibrosis pulmonary aspergillosis can be developed while a person has been suffering from CCPA for long time or a patient suffering from CPA remain medically untreated condition. Patients suffering from CPA may show well immune system that cures the infection created by this fungus under this condition. Accordingly, patients can not be attacked speedily by the infection; however, it would supervise to reside in the body region at which it can grip. The fungi can choose the appropriate region through which it would get access. Accordingly, fungal conidia can pass through the air route of lungs in environment, and then, it is necessary to avoid the immune system for creation of inhabitant in a cavities which are far away from the immune system like the destructed tissues which are found after tuberculosis or such type of infectious diseases, and the damaged tissues supply the garbage available after normal infectious diseases to fungus as per requirement for the footing of fungus, and the fungus initiates its growth slowly in a restricted region nearby the tissues at which the immune system of the persons suffered by the infection can hit it which are present around the cavities. So, it is found to remain unseen for several periods with the reflection of a few symptoms of the infection; however, it can initiate to corrode the around tissues; as a result, swelling of tissues occurs in contact with the fungus. Finally, there is a formation of fungal ball in the same places of cavities with the coughing up of blood. This infection may make misconception about tuberculosis.

3.4.1 Symptoms

(i) Feeling fatigue
(ii) Loss of weight
(iii) Severe respiratory troubles
(iv) Increase amount of cough productivity
(v) Excretion of cough with blood
(vi) Presence of fungal ball/nodules in pulmonary cavity

3.5 Invasive Aspergillosis

Invasive aspergillosis is found to develop in cancer/leukemia/acquired immune syndrome disease suffered people whenever their immune system becomes destabilized, while chemotherapy and other treatment for aids are going on, and the impaired immune system can not struggle for fighting against different types infection [18]; as a result, the same aspergillosis can enter forcibly the lung tissues as well as becoming widen to the kidneys or brain, and at last, it is reported to develop severe health hazardous infectious pneumonia, while the patients suffering from invasive aspergillosis with conciliated immune organization remain untreated medically for long time [19]. In addition to these aspergillosis infections, *A. fumigatus* can affect at skin of outer surface in human to create infection called Cutaneous aspergillosis. In this case, conidia of *A. fumigatus* may be allowed to pass through the body skin around the surgery or burn wound, and the same fungal spores can make infection which may be considered as invasive aspergillosis. Cutaneous aspergillosis may also be exposed in the body skin after severely affecting lungs by invasive infection [20].

3.5.1 Symptoms

(i) Cough occasionally with blood
(ii) Chest Pain
(iii) Decreasing of breathing time
(iv) High body temperature

4 Molecular Mechanism of Aspergillosis Infection Made *Aspergillus* sp.

Aspergillus species are capable of growing on departed or decomposing materials available in the environment. Life cycle of *Aspergillus* starts with the formation of spores which can spread all over air of the environment and can move everywhere through air in combination of moisture and dust, and the same spores can easily be inhaled by human during respiration from air which is considered as a principal entrance of the *Aspergillus,* and then, spores are assembled in alveolar regions to start the infectious diseases. Macrophages found to be present in alveolar can destruct the deposited *Aspergillus* spores through phagocytosis in a well-established immune system [21]. These fungal spores can overcome the phagocytosis killing and can multiply easily with the aim of defending neutrophils. As a result, formation of invasive aspergillosis will be developing with loss of capability of host defense and allow *A. fumigatus* for survival with its growth in pulmonary surroundings. The *A. fumigatus* can grow in very fast at physiological system compared to others *Aspergillus* sp. as growth rate of *A. fumigatus* is found to go faster about 40% in physiological and pharmacological concentrations of hydrocortisone. So, growth rate is one of the important switches for reflection to the progress of the diseases, that is, of the pathogenicity. Under this condition, *A. fumigatus* can generate very small size of spores which are allowed to penetrate intensively in the lungs. The spores are able to remain survival under extreme conditions of the surroundings as the same spores are covered with special type of layer consisting of hydrophobic proteins and lipids, and these special type of spores stuck with laminin and fibrinogen are capable of defending the same spore from the attack of the host. During germination of spores of *A. fumigatus* as well as life cycle of *A. fumigatus*, different toxic materials like several proteases, ribotoxin phospholipases, a hemolysin, gliotoxin, aflatoxin, phthioic acid, etc., are reported to excrete during infection in lungs and other organs [22]. Epithelial cells are found to become loosen from pulmonary region with induction of the excreted protease with the release of pro-provocative cytokine. Host cell membrane can be damaged by phospholipases excreted by the *A. fumigatus* as well as destruction of tissues occurs. Gliotoxin excreted by the fungi is capable of reducing macrophage and neutrophils causing phagocytosis of inhaled spores of *Aspergillus* sp., and so, all of the spores remain viable. Direct role of Gliotoxin in pathogenecity is not clear till date. As reported, a rare number of *A. fumigatus* are capable producing phthioic acid. Role of phthioic acid in pathogenicity is reported to form granuloma which is also found in tuberculosis. In addition to these, supper oxide dismutase and catalases can be synthesized by *A. fumigatus* by which fungus *A. fumigatus* cannot be damaged by singlet oxygen, hydrogen peroxide, and other free radicals generated by phagocytes as well as fungal cell can attack the tissues of lungs and other organs developing aspergillosis.

5 Conclusion

Among *Aspergillus* spp., *A. fumigatus* and *A. flavus* have been recognized as general human pathogen causing development of aspergillosis in human system. Human with immune negotiated is suffering from diseases related to hematological malignancy, and transplanted patients are infected severely with *Aspergillus* sp mainly with *A. fumigatus* and *A. flavus* IA and cause of development of invasive aspergillosis. Special characteristics of such type of fungi are reflected to the formation of spores which are very small in size with the presence of a few pathogenic factors created during its germination and multiplication which are associated very deeply with the infection aspergillosis as the same fungi can grow very well at body temperature with the easy entrance of the airborne spores in the way of interior respiratory tract during respiration. Spores of *A. fumigatus* become accustomed very well to the tissues of lungs of immune-suppressed patients as well as spores can start the colony formation in affected tissues of lungs with the avoid of attack of macrophage and split up the attached tissues for required nutrients for the uncontrolled growth of same fungus as a result extreme irritation occurs with the combinations of several pathogenic factors.

References

1. Mayer AC (1815) Deutsch Arch Physiol
2. Rankin NE (1953) Disseminated aspergillosis and moniliasis associated with neutropenia as hemorrhage and other operative complications agranulocytosis and antibiotic therapy. Br Med J 183:918–919
3. Bennett JH (1842) On the parasitic vegetable structures found growing in living may be critical, but the temporal impact is small. Granulocyte animals. Trans R Soc Edinb 15 (Colony-stimulating factor, granulocyte-macrophage colony-277)
4. Cawley EP (1947) Aspergillosis and the aspergilli. Arch Intern Med 80:423–434
5. Hinson KFW, Moon AJ, Plummer NS (1952) Broncho-pulmonary aspergillosis: a review and a report of eight new cases. Thorax 7:317–333
6. Segal BH (2009) Aspergillosis. N Engl J Med 360(18):1870–1884
7. Verweij PE, Brandt ME (2007) Aspergillus, fusarium and other opportunistic moniliaceous fungi. In: Murray PR (ed), 9th edn. ASM Press, Washington D.C., pp 1802–1838
8. Machida M, Gomi K (2010) Aspergillus: molecular biology and genomics. Horizon Scientific Press, p 157. ISBN 978-1-904455-53-0
9. Schuster E, Dunn-Coleman N, Frisvad JC, Van Dijck PW (2002) On the safety of Aspergillus niger—a review. Appl Microbiol Biotechnol 59(4–5):426–435
10. Denning DW, Pleuvry A, Cole DC (2013) Global burden of chronic pulmonary aspergillosis complicating sarcoidosis. Eur Respir J 41(3):621–626
11. Knutsen AP, Bush RK, Demain JG (2012) Fungi and allergic lower respiratory tract diseases. J Allergy Clin Immunol 129(2):280–291
12. Ederies A, Chen J, Aviv RI (2010) Aspergillosis of the petrous apex and Meckel's cave. Skull Base 20(3):189–192
13. Agarwal R (2009) Allergic bronchopulmonary aspergillosis. Chest 135(3):805–826
14. Manning SC, Holman M (1998) Further evidence for allergic pathophysiology in allergic fungal sinusitis. Laryngoscope 108(10):1485–1496

15. Glass D, Amedee RG (2011) Allergic fungal rhinosinusitis: a review. Ochsner J 11(3):271–275
16. Addrizzon Harris DJ, Harkin TJ, McGuinness G, Naidich DP, Rom WN (1997) Pulmonary aspergilloma and AIDS: a comparison of HIV-infected and HIV-negative individuals. Chest 111:612–618
17. Soubani AO, Chandrasekar PH (2002) The clinical spectrum of pulmonary aspergillosis. Chest 121(6):1988–1989
18. Peral-Cagigal B, Redondo-Gonzalez LM, Verrier-Hernandez A (2014) Invasive maxillary sinus aspergillosis: a case report successfully treated with voriconazole and surgical debridement. J Clin Exp Dent 1;6(4):e448–e451
19. Chen J, Yang Q, Huang J, Li L (2013) Risk factors for invasive pulmonary aspergillosis and hospital mortality in acute-on-chronic liver failure patients: a retrospective cohort study. Int J Med Sci 10(12):1625–1631
20. Van Burik JA, Colven R, Spach DH (1998) Cutaneous aspergillosis. J Clin Microbiol 36 (11):3115–3121
21. Brakhage A, Bruns S, Thywissen A, Zipfel PF, Behnsen J (2010) Interaction of phagocytes with filamentous fungi. Curr Opin Microbiol 13:409–415
22. Amitani R, Taylor G, Elezis EN, Llewellyn-Jones C, Mitchell J, Kuze F, Cole PJ, Wilson R (1995) Purification and characterization of factors produced by *Aspergillus fumigatus* which affect human ciliated respiratory epithelium. Infect Immun 63:3266–3271

Cysteine Proteases of Parasitic Helminths

Alok K. Dixit, Pooja Dixit and Ramji L. Sharma

Abstract

Cysteine protease and/or thiol peptidase/sulphydryl peptidase/cysteinyl peptidase invariably contain cysteine at their active site and hence collectively termed as cathepsins. This proteolytic enzyme, secreted and/or regurgitated/excreted by the in situ invasive forms of metazoan parasites of men and domestic livestock, often involve in the host tissue penetration, digestion of nutrients, and their availability to the parasite for its growth and development, evasion of host defense, and in of late documented their significant role in detection of in situ parasites at an early developmental stage(s) [Immunodiagnosis] beside an important entity to develop immunoprotection strategies against the livestock diseases, so much so, to contain and curb the colossal recurring losses, incidental to widely prevalent parasitic disease, to the fast developing livestock products based food industry in the developing world. Herein, the authors discuss the historical background, molecular structure, and classification of cathepsin proteases and update advancements in exploiting the enzyme for early diagnosis and planning immunoprotection strategies against metazoan parasitic diseases with special reference to tropical fasciolosis.

A.K. Dixit (✉)
Department of Veterinary Parasitology, College of Veterinary Science & A. H., Nanaji Deshmukh Veterinary Science University, South Civil Lines, Jabalpur 482001, Madhya Pradesh, India
e-mail: alokdixit7@yahoo.com

P. Dixit
Department of Veterinary Medicine, College of Veterinary Science & A. H., Nanaji Deshmukh Veterinary Science University, South Civil Lines, Jabalpur 482001, Madhya Pradesh, India

R.L. Sharma
843-44, Ranisati Nagar, P.O. Shyam Nagar, Ajmer Road, Jaipur 302019, Rajasthan, India

S. Chakraborti and N.S. Dhalla (eds.), *Pathophysiological Aspects of Proteases*,
DOI 10.1007/978-981-10-6141-7_29

Keywords
Cysteine protease · Immunodiagnosis · Vaccine development
Fasciolosis · Trematodes · Nematodes

1 Introduction

Parasites invariably infect their host(s), in situ survive, and sustain metabolic activities for development and growth besides propagation. Peptidases are protein/polypeptide catabolizing enzymes which act specifically on peptide bonds [1, 2]. Proteases, proteinases, and peptide hydrolases are the synonyms of these enzymes. Peptidases are responsible for various biological activities in parasitic organisms. Immunochemical investigations revealed tissue localization to the epithelia and microvilli of the gut of the metazoans [3–5]. These enzymes help the parasite to penetrate through various barriers in the host, in addition to digestion of host proteins for their nutrition [6, 7]. Parasite peptidases also help the parasite to escape the immune response of the host [6]. In addition, these are involved in secondary protein modifications and are used in immunodiagnosis of infection [8, 9]. These enzymes also have role in triggering allergic responses in hosts [10]. Evidently, these are actively involved in penetration of tissues, parasite nutrition, and in situ protection from host immune response.

2 Classification

Depending upon the site to be cleaved by the proteases, these can be broadly divided into two groups: exopeptidases which act on terminal carboxyl or amino group and endopeptidases which act on peptide bonds. Proteases are of four types on the basis of reaction mechanism and nature of active site residue in the mechanism. Serine and cysteine proteases directly act as nucleophile, and aspartyl and metalloprotease use activated oxygen of water molecule as nucleophile. As cysteine proteases contain cysteine in their active site, these are called cysteine cathepsins, while other cathepsins may contain serine or aspartate in their active site.

2.1 Papain Family of Cathepsin Proteases

Cysteine proteases are present in almost all life-forms, even in viruses and are one of the most important enzyme groups. In 1879, the first cysteine protease was purified and characterized from *Carica papaya*, the papaya fruit, and was thus named papain. In parasites, cysteine proteases have various roles like immune evasion, enzyme activation, cell/tissue invasion, excystment, hatching, molting, and gene encoding which help in better parasite survival. As these enzymes contribute 2% of the genomic content of the parasite and are highly immunogenic, it is used as an important vaccine candidate. Cysteine proteases elicit the immune response of the host towards either Th1 or Th2 response whichever may be insufficient for expulsion of parasite. On the basis of sequence similarity, possession of inserted peptide loops, and biochemical specificity to small peptide substrates, cysteine proteases are divided into many clans like: CA, CB, CC and CD. The main cysteine proteases are included in clan CA which is further subdivided into two families C1 and C2. Important parasitic cysteine proteases included in C1 family are cathepsin B and L like.

Though the parasites possess many types of proteases, the proteases which are released by the parasites are of sole importance in host parasite interactions. In helminths, the main source of these proteases is their gut. As these proteases are digestive proteases that too of invertebrates, these are mainly cysteine and aspartic proteases in contrast to the serine proteases secreted in intestinal digestion of vertebrates [11, 12].

2.2 The Trematode Cathepsins

Peptidases act as essential biocatalysts in various basic biological processes in trematodes too. As these proteases have their role in host parasite interactions, these can be targeted for vaccines and chemotherapeutic drug development.

Various stages of *Fasciola* express cathepsin L and B which were originally described as immunoglobulin-cleaving enzyme or hemoglobinase [13, 14]. In *Fasciola* excretory secretary (ES) cysteine proteinase group, the major component is of 26–30 kDa in size. In amino acid sequence and substrate specificity, these proteases resemble mammalian liver cathepsin L protease [15, 16]. Nowadays, these are called as the *Fasciola* cathepsin L cysteine proteinases. As these are secreted by the gastrodermis of juvenile and adult flukes, these can be in vitro isolated/purified from fluke regurgitants.

From *Fasciola hepatica* ES products, first parasitic helminth cathepsin L proteinase was described as a single chain form at 27 kDa (FhCL-1) on SDS-PAGE, and it had multiple band complexes ranging between 60 and 90 kDa in non-reducing conditions [17]. Later on a 29.5 kDa, cathepsin L-like cysteine proteinase (FhCL-2) was found in ES products of *Fasciola*, and it was also reported that cathepsin L-like cysteine proteinase alone constitutes more than 80% of ES products of the fluke [18]. From adult *Fasciola gigantica* also, cathepsin L-like cysteine proteinase of 27 kDa FgCL-1 [19], 27–28 kDa FgCL-2 [20], and 28 kDa FgCL-3 [21, 22] were isolated. In *F. hepatica* and *F. gigantica,* many cathepsin L genes were identified which help in in situ development of these parasites. The recombinant cathepsin L-like proteases like RFhCL-1 [23], RFhCL-2 [24], RFhCL-3 [25], RFhCL-5 [26, 27], RFgCL-1 [28, 29], RFgCL-1D [30], and RFgCL-2 [31] were also produced from adult *Fasciola*.

Cathepsins secreted from newly excysted juvenile (NEJ) and immature intraperitoneal flukes were different from that of adults [32–34]. The most characterized proteases in *Fasciola* spp. belong to the cathepsin L group of cysteine proteases, while cathepsin B activity is present in juvenile stages or immature flukes [35, 36]. The most abundant proteases produced by NEJs and immature *Fasciola* belong to a family of cathepsin B-like proteases [33, 35–38].

Recent analyses of the secretome of *F. hepatica* and *F. gigantica* juvenile and immature flukes have identified six types of cathepsin Bs in the ES products and even in metacercariae stage which is the infective form of the parasite [37–40]. From NEJs to 5-week-old in situ developing immature parasites FhCatB1 (CB2), one isotype of cathepsin B, was found in both somatic and ES products [33, 35, 36]. Three stage-specific isotypes of cathepsin B proteases of *F. gigantica* FgCatB1, FgCatB2, and FgCatB3 cloned from cDNA libraries were found to be orthologous to isotypes (CB1, CB2, and CB3) of *F. hepatica*. RT–PCR analysis showed that FgCatB2 and FgCatB3 were expressed in NEJs and metacercariae. FgCatB1 is mainly expressed in adult parasites though found in all stages [37, 38].

Proteomics and transcriptome analysis revealed seven additional cathepsin B sequences in *F. hepatica* [39]. The expression of these proteases is regulated by the stage of the parasite whether juvenile, immature, or adult. Six cathepsin Bs are released by invasive stages of *Fasciola* and help in the establishment of infection. FhCB2 protease is released in the initial migratory stage when the infection is asymptomatic, and this is recognized by the host immune system also. In sheep, antibody titers to FhCB2 were increased 2–5 weeks post-infection and between 8 and 10 weeks post-infection. By adult stages, FhCB6–FhCB10 are secreted which are found to be cross reactive to FhCB2. FhCB4 of adult stages is an inactive protein as it has no cysteine residue at active site. Only FhCB6 and FgCB1 are secreted by both the adult and migratoty stages of *Fasciola;* therefore, these two cathepsins are thought to play a major role in fluke biology.

2.3 The Nematode Cathepsins

In nematodes also, papain-like enzymes are common. In *Haemonchus contortus*, many cathepsin Bs have been cloned [41–44]. Cysteine protease of *H. contortus* extracts and ES products of *Necator americanus* have shown the capability of digesting haemoglobin (Hb), fibrinogen, and antibodies [45, 46].

In *Ancylostoma caninum,* cathepsin Bs have a significant role in hematophagus activity. Two cathepsin B-encoding mRNAs have been cloned from adult *A. caninum* [47]. AcCP1 cathepsin is expressed only in secretory glands and esophagus [47], having only in vitro Hb-digesting activity, while AcCP2 is produced in intestine leading to in vivo Hb-digesting activity. Optimal hemoglobinolytic activity was seen after initial cleavage of Hb by aspartic protease.

In parasitic and non-parasitic nematodes, cathepsin B-like protease genes constitute large multigene families of which the best known is from intestinal tissues of *H. contortus* [44, 48, 49]. These play a role in digestion of Hb and other nutrients.

In blood-feeder nematodes with the help of a specific sequence alignment program, hemoglobinase motif was identified in cathepsin B-like proteases [50]. As the use of gene knockout and RNA-silencing techniques are limited in helminths, therefore, it is difficult to exactly determine the role of these proteases in lysis of HB. In nematodes, tissue localization activity of recombinant enzymes, and over expression in female worms suggests a significant role of these proteases in degradation of substrate for nutrition. Cathepsin Ls are less abundant in nematodes and have comparatively more role in embryogenesis and molting than digestion of nutrients contrary to that of hematophagous trematodes [51–53].

3 Vital Role in Host's Immune Evasion

Chapman and Mitchell [13], initially postulated the role of proteinases in immune evasion. These enzymes prevent the migrating flukes from host immune attack by activating effector functions in them. In in vitro assays, it has been found that the presence of proteinases coincides with the inability of the eosinophils to maintain antibody-mediated adherence to newly excysted/migrating juveniles. Proteinases secreted by the parasites confer protection against host immune responses as proteinases cause turnover in parasite tegument, hindering the recognition of the parasite by the host immune mechanisms [32, 54].

Fasciola spp. causes a chronic liver disease in most mammalian hosts. This parasite diverts the immune response to Th2 response which helps in its longer survival and provides resistance from host immune attacks. Th2 response predominance is shown by IL4, IL5, and IL10 production but very little IFN-γ [55]. Experimentally also, it has been found that these responses are of Th2 type with progression of infection. In cattle, it was observed that in early stages of infection, mixed Th1 and Th2 responses are there which change toward Th2 response predominance as the infection progresses, but in mice, Th1 response was significantly suppressed [55, 56]. In some studies, it was found that cathepsin L1 suppresses IFN-γ leading to decrease in Th1 response and increased Th2 response [17]. This suppressive effect of FhCL-1 was further confirmed in IL4 defective mice [57].

When *Fasciola* infection occurs, all stages of parasite liberate cathepsin L which quickly subverts the immune response of the host [58]. Prowse et al. [59] reported in vitro proliferation of sheep T lymphocyte and reduction of expression of CD4 in both ovine and human cells by *F. hepatica* cathepsin L (RFhCL-5). This was further confirmed by addition of E64 (a cysteine proteinase inhibitor). CD4 is a surface marker of 64 kDa found on T cells, and it acts as a co-receptor along with TCR: CD3 complex in antigen recognition by MHC-II. This is necessary for T helper cell activation. The co-receptor functions of CD4 increase the T cell signaling activity up to 10–100 folds so that the number of molecules required for T cell activation decreases. CD4 is required for T cell activation, and cathepsin cleaves the CD4 which helps in suppression of lymphocyte proliferation. In sheep, cathepsin L of *Fasciola* induces cleavage of CD4 to suppress lymphocyte proliferation for evasion of host immune system. In *Paragonimus westermanni*, the cysteine proteases secreted from the gut are known to degrade Hb and collagen besides suppressing immune system and inducing immune tolerance.

4 Facilitatory Role in Parasite in situ Migration and Feeding Activities

The feeding and migration of the parasite are interdependent as the degraded host tissues during migration of the parasite act as substrate for parasite feeding. Similar to mammalian host, cathepsin B enzymes have optimal activity at pH range of 5–7. This pH range indicates that the enzyme has activity inside the gut of fluke as well as host tissue pH i.e., slight acidic to neutral pH. These enzymes show their activity over a wide pH range (4–8.5) and are stable at neutral pH. It is thought that those parasitic enzymes, which show optimal activity between pH 3–4.5, does digestion in the acidic environment of the gut of the parasite, while proteases which are active at physiological pH are responsible for the digestion of the blood cells/proteins, etc., and thus help in penetration. FhCL-2 cleaving of fibrinogen produces fibrin clot so as to prevent the access of immune effecter cells to the fluke surface [60]. Role of FhCL-1 and FhCL-2 in degrading extra cellular matrix and basement membrane to facilitate in situ migration of the parasite was demonstrated by Berasain et al. [61].

As cathepsin B-like proteases are expressed in the early stages of the liver fluke, these may have important role in excystation and invasion of the host tissues [35, 36, 62]. Cathepsin B enzymes also have a role in parasite homeostasis, and this was supported by a study using RNAi to knockdown FhCB-2 expression in immature flukes [63]. In the study, RNAi treatment of NEJs was done, and this led to 78% reduction in expression of FhCB-2, thus 50% reduction in ability to penetrate rat gut in vitro. Further, prolonged incubation of these flukes led to death of the flukes too [63]. These findings are suggestive of the role of FhCB-2 in gut penetration, tissue invasion and parasite homeostasis.

5 Cysteine Protease and Fecundity of Flukes

The Mehlis glands of *F. hepatica* secrete substances, steering the formation of the egg cells [64]. The secretion contain cathepsin L, influencing in situ maturation of egg shell components and other reproductive activities of the parasite [24]. Based on an experimental study in buffaloes immunized with the cysteine proteinase, it was postulated that cathepsin L induced immunity, targeted the mehlis glands and interfered with the egg shell synthesis. Obviously, the inhibitory effects of the enzyme influences fluke fecundity and posses epizootiological impact on disease transmission from molluscan to definative mammalian hosts.

6 Immunodiagnostic Target

For control of fasciolosis, immunodiagnosis is of utmost importance as the infection can be detected in early prepatent period i.e., before the pasture contamination. This early diagnosis helps timely treatment of the infection leading to prevention of contamination of water bodies. Immunodiagnostic tests of varying sensitivity and specificity developed earlier, had restricted application incidental to cross reactivity with other in situ metazoans. This necessitated the isolation, purification and identification of specific immunodominant molecules from the complex ES antigens of the fluke [66]. For early diagnosis of *Fasciola* infections, cathepsin L cysteine proteinase has been used to develop reliable test. A single epitope was used as an antigen in peptide-based indirect ELISA to detect antibodies against *F. hepatica* in cattle. In experimental studies, *F. hepatica* specific antibodies were first detected between third and fourth week post-infection in cattle and persisted till 183rd day post-infection in peptide ELISA. However, FhCL-3 ELISA was found superior to peptide ELISA in both natural as well as experimental studies. A diagnostic ELISA with RFhCL-3 was developed to detect antibodies against *F. gigantica* in sheep [67]. The 28 kDa recombinant protein was shown to be cross reactive and detected *F. gigantica* antibodies fifth week post-infection in sheep, while 28 kDa FgCL-3 detected antibodies as early as 4 weeks post-infection (wpi) by indirect ELISA, dipstick ELISA, and western blot [25]. Dixit et al. [21] detected *F. gigantica*-specific antibodies as early as 2–4 wpi i.e., earlier than the test using RFhCL3. This study revealed the possibilities and advantages of using homologous antigen i.e., FgCL3 in the diagnosis of experimental *F. gigantica* infection. Some experimental studies have also shown 100% sensitivity of FgCL3 ELISA in bubaline fasciolosis diagnosis [68], but in natural infections, this sensitivity declined to 97% with 100% specificity [30]. Early studies with FgCL3 using serum of goats and buffaloes experimentally infected with *Paramphistomum epiclitum* evidenced no cross reactivity and encouraged its use in specific diagnosis of fasciolosis [69].

Synthetic peptides similar to *F. hepatica* cathepsin L were developed and used in immunodiagnosis of tropical ovine fasciolosis. ELISA could not detect antibodies against different peptides up to 4 weeks of infection, but these peptide constructs were of use to detect low and high degree of infection of *F. gigantica* in sheep [70].

Purified 28 kDa FgCL3 (by ion exchange chromatography) was used for immunodiagnosis of experimental *F. gigantica* infected sheep and buffaloes to further confirm earlier reports [21, 68]. The same has also been evaluated for early prepatent detection of *F. gigantica* infection (4 wpi) in experimentally infected calves. In field studies also, FgCL3 was used to detect *F. gigantica* infection using ELISA [71].

6.1 Immunoprotection

6.1.1 Fasciolosis

Immunological control of fasciolosis is preferred over chemotherapy as it is comparatively cheaper, more efficient, and provides long-term cure. A number of molecules like cathepsin L, Glutathione-S-transferase (GST), Leucine aminopeptidase (LAP), and Fatty acid binding proteins (FABP) were used for inducing immunity against *Fasciola* spp. in animals. Among all, cysteine proteinase family cathepsin L enzymes were mainly tested and given good protection if used as vaccines. In one trial, only 10 mg of cathepsin L given on three occasions induced protective response [72]. Subsequently, cathepsin L at the dose of 200 mg was also protective [73]. Cattle vaccinated with *F. gigantica* cathepsin L induced very high antibody titer, but there was no effect on in situ migration and fecundity of the parasite. There was variation in the response depending upon the adjuvant used, and the best response was found when Fruend's complete adjuvant was used with cathepsin L, while with heme-binding protein antigen, only 73% protection could be achieved [20, 72].

Vaccination trials with a combination of FhCL1 and FhCL2 in cattle and sheep produced 53 and 30–60% protection, respectively, but when these two cathepsins were combined with *F. hepatica* LAP, the protection level increased to 79% in sheep [73, 74]. The morphology of the flukes recovered from vaccinated sheep was normal indicating that vaccination has no effect on development of fluke [64], but some workers observed a reduction in the size as well as in the fecundity of recovered flukes from buffaloes [65, 72]. Evidently, the immune response depends upon the immunogen used.

Despite all these efforts, a commercialized vaccine has not yet been developed which can give sufficient protection against fasciolosis. On the basis of these studies, it was concluded that the immune response to a vaccine depends upon route of administration, immunogen type, and adjuvant used. Besides this, the commercial vaccine production should be cost-effective to compete with the chemotherapy. Various studies involving different antigens and different routes of administration induced varying levels of protection. In rats, oral inclusion bodies containing RFhCL-1-like cysteine proteinase resulted in 70–80% protection, and the acquired immunity conferred by intramuscular injection of cDNA of the enzyme, was equally comparable [75, 76]. Intranasal vaccination with the cDNA of enzyme or intramuscular injection of the recombinant enzyme with FCA, resulted in lesser protection [77]. An antigen given without adjuvant in the form of inclusion bodies, was suitable for oral vaccination, as it is protected from being digested in the gut. The immunity conferred by the intranasal vaccination with inclusion bodies of RFhCL-1 was 54 and 56.5% in calves and lambs, respectively. Besides, the flukes developed in vaccinated calves had lower reproductive potential [78].

The recombinant FhCB2 proteinas vaccine in sheep resulted in high antibody titers up to 10^6 substantiating that FhCB2 is a leading vaccine candidate [35, 79] However, it is not very clear whether the antibody titers in sheep are efficient enough to interrupt the development of the in situ flukes. Jayaraj et al. [80] have recently validated FhCB2 as vaccine for fasciolosis in rat model. Recombinant vaccines involving FhCB2 alone as well as multivalent vaccine with recombinant proteins of proteases expressed in adult as well as in metacercariae of liver fluke were evaluated. The strategy behind this trial is multivalent vaccination (using many proteases expressed at different stages of life cycle of the parasite) will protect the host throughout the migration period as well as the patent period of the parasite thus giving long-term protection. Vaccination with FhCB2, a juvenile protease, alone was found to be highly immunogenic. It reduced 60% fluke burden and 63% reduction in size of the recovered flukes, indicating that it inhibits the development of surviving flukes by affecting the feeding behavior or migration of the parasite. Significant reduction in liver damage (61%) by the challenge dose of flukes post FhCB2 vaccination indicated that killing of parasite has occurred [80].

Multivalent vaccination with FhCB2 in combination with FhCatL5 or FhCatL1g or with all three proteins resulted in 83% protection and 76% reduction in parasite size as well as 73% reduction in liver damage. Further, FhCB2 combined with FhCatL5 had given best results indicating that if vaccination is done targeting juvenile and adult proteases, it will give comparatively better protection. As juvenile and immature *F. hepatica* express six cathepsin B cDNA sequences (FhCB1–FhCB6), there is possibility that vaccination targeting these cathepsins can add better protection [38, 39].

Only a few DNA vaccines were targeted on cysteine protease of viral and bacterial infections. DNA vaccines encoding cathepsin L-like proteases of *F. hepatica* resulted in fluke load reduction in rats [58]. Likewise, in FhCB2 DNA vaccine, the immune response was enhanced as this vaccine construct produces FhCB2 protein. In addition, these vaccines will also produce a strong memory response and antibody titers in sheep. Immune responses following FhCB2 vaccine can be augmented by the use of CpG augmentation and CTLA4-mediated targeting [79].

6.1.2 Other Metazoan Diseases

Due to problem of anthelmintic resistance and drug residues, there is a need for effective anti-nematode vaccines [81]. Vaccine against cryptic antigen of blood-feeding helminths can be a good alternative as these antigens are usually not recognized by the host in natural infections. The molecules are being employed in the vaccines against ectoparasites of domestic animals [82, 83] and *Haemonchus* in sheep [84]. Some gut membrane protease like H-gal-GP [85] are hidden from host immune system during natural infection, but host antibodies can reach these hidden antigens upon ingestion of the host blood by the parasite. Thus, these antigens can be utilized for vaccination against obligate blood-feeder parasites. In sheep, protective immunity was developed following administration of H-gal-GP antigen, and the antibodies produced could be used for passive immunization [86]. *Whereas,* vaccination studies on haemonchosis revealed that studies on individual component of the complex antigen in recombinant form are essentially required. This would be the first recombinant protein vaccine against nematode infection(s) prevalent in field.

In *Ancylostoma* parasite of dogs, immunization with recombinant Ac-APR-1 vaccine has shown a reduction in adult parasite burden [87]. In the study, the adult worm burden in terminal part of intestine increased indicating that antibodies against Ac-APR-1 forced the adult *Ancylostoma* to shift to colon from small intestine. The possible inference drawn from the study was that the female worms being more voracious blood suckers shifted in large number to colon as these were more exposed to anti Ac-APR-1 immunoglobulin [87].

Experimentally, immunization in rats with *Clonorchis sinensis* cysteine protease (CsCp) produced 31.5% protection [88]. Cathepsin B-like cysteine proteases were also found immunogenic in *H. contortus* in sheep [43]. In another study on goats, these cysteine proreases of *Haemonchus* also provided protection [89]. Conclusively, proteases are of utmost importance in parasitic vaccine development strategies.

7 Conclusion and Future Perspective

Cysteine proteases are present in almost all life-forms including viruses. Parasitic cysteine protease enzymes are diverse, and being immunogenic in nature, play important role in the in situ survival of the parasites. These enzymes are being extensively exploited for developing efficient diagnostics and vaccines against the parasitic diseases of economic significance. The cysteine proteases modulate either Th1 or Th2 responses to protect the parasites inside the host. At the same time, being potent immunogens, induce acquired immunity in the host and confer effective protection against future infections. Undoubtedly, cysteine proteinases as vaccines are claimed to be a preferred choice over chemotherapeutic control of the parasitic disease in the field.

Due to the ubiquitous presence of the cysteine proteases in helminth parasites, these molecules represent intriguing targets for development of antiparasitic strategies for field application. Homologous family members occur in nearly every major parasitic organism, and in light of the fact that the catalytic mechanism is identical in each of these homologue, it is reasonable to infer that a relatively large inhibitor library would contain potential leads for a number of different parasitic diseases. This would provide a very cost-effective approach to new antiparasitic chemotherapy – a necessity while dealing with diseases that are endemic in economically poor regions of the world including the Indian subcontinent. Although cysteine proteinases proved as diagnostic markers in fluke infections especially fasciolosis, however further research is required to fully exploit these molecules as potential vaccine, as an integrated strategy for ongoing antiparasitic drug development programme worldover.

References

1. McKerrow JH, Caffrey C, Kelly B, Loke PN, Sajid M (2006) Proteases in parasitic diseases. Annu Rev Pathol 1:497–536
2. Sajid M, Blackman MJ, Doyl P, He C, Land KM, Lobo C, Mackey Z, Ndao M, Reed SL, Shiels B, Swenerton R (2009) Proteases of parasitic protozoa—current status and validation. In: Selzer P (ed) Drug discovery in infectious diseases: from drug targets to drug candidates. Wiley–VCH Verlag, pp 177–208
3. Chapell CL, Dresden MH (1986) Characterization of a cysteinyl protease from human parasite, *Schistosoma mansoni*. In: Turk V (ed) Cysteine protease and their inhibitors. Walter de Gruyter, New York, pp 199–208
4. Chapell CL, Dresden MH (1987) *Schistosoma mansoni* adult worm protease: tissue localization in the parasites and antibody response in infected animals. Molecular Strategies of Parasitic Invasion. A.R. Liss Inc, New York, pp 399–408
5. Ray C, McKerrow JH (1992) Gut specific and development expression of a *Caenorhabditis elegans* cysteine protease gene. Mol Biochem Parasitol 51:239–250
6. Dixit AK, Pooja Dixit, Sharma RL (2008) Immunodiagnostic/protective role of cathepsin L cysteine proteinases secreted by *Fasciola* species. Vet Parasitol 154:177–184
7. Goldberg DE (2005) Hemoglobin degradation. Curr Top Microbiol Immunol 295:275–291
8. Ponder EL, Bogyo M (2007) Ubiquitin-like modifiers and their deconjugating enzymes in medically important parasitic protozoa. Eukaryot Cell 6:1943–1952
9. Zacks MA, Garg N (2006) Recent developments in the molecular, biochemical and functional characterization of GPI8 and the GPI anchoring mechanism. Mol Membr Biol 23:209–225
10. Sajid M, McKerrow JH (2002) Cysteine proteases of parasitic organisms. Mol Biochem Parasitol 120:1–21
11. Rawlings ND, Morton FR, Barrett AJ (2006) MEROPS: the peptidase database. Nucleic Acids Res 34:D270–D272
12. Delcroix M, Sajid M, Caffrey CR, Lim KC, Dvorak J, Hsieh I, Bahgat M, Dissous C, McKerrow JH (2006) A multienzyme network functions in intestinal protein digestion by a platyhelminth parasite. J Biol Chem 281:39316–39329
13. Chapman CB, Mitchell GF (1982) Proteolytic cleavage of immunoglobin by enzymes released by *Fasciola hepatica*. Vet Parasitol 11:165–178
14. Coles GC, Rubano D (1988) Antigenicity of a proteolytic enzyme of *Fasciola hepatica*. J Helminthol 62:257–260
15. Brady MT, O'Neill SM, Dalton JP, Mills KHG (1999) *Fasciola hepatica* a protective Th1 response against *Bordetella pertussis*. Infect Immun 67:5372–5378
16. Tort J, Brindley PJ, Knox D, Wolfe KH, Dalton JP (1999) Proteinases and associated genes of parasitic helminths. Adv Parasitol 43:161–266
17. Smith AM, Dowd AJ, McGonigle S, Keegan PS, Brennan G, Trudgett A, Dalton JP (1993) Purification of a cathepsin L like proteinase secreted by adult *Fasciola hepatica*. Mol Biochem Parasitol 62:1–8
18. Dowd AJ, Smith AM, McGonigle S, Dalton JP (1994) Purification and characterisation of a second cathepsin L proteinase secreted by the parasitic trematode *Fasciola hepatica*. Eur J Biochem 223:91–98
19. Maleewong W, Wongkham C, Intapan PM, Pipitgool V (1999) *Fasciola gigantica*-specific antigens: purification by a continuous elution method and its evaluation for the diagnosis of human fascioliasis. Am J Trop Med Hyg 61:648–651
20. Estuningsih SE, Smooker PM, Wiedosari E, Widjajanti S, Vaiano S, Partoutomo S, Spithill TW (1997) Evaluation of antigens of *Fasciola gigantica* as vaccines against tropical fasciolosis in cattle. Int J Parasitol 27:1419–1428
21. Dixit AK, Yadav SC, Sharma RL (2002) 28 kDa *Fasciola gigantica* cysteine proteinase in the diagnosis of prepatent ovine fasciolosis. Vet Parasitol 109:233–247

22. Yadav SC, Saini M, Raina OK, Nambi PA, Jadav K, Sriveny D (2005) *Fasciola gigantica* Cathepsin L cysteine proteinase in the detection of early experimental fasciolosis in ruminants. Parasitol Res 97:527–534
23. Roche L, Dowd AJ, Tort J, McGonigle S, McSweeney A, Curley GP, Ryan T, Dalton JP (1997) Functional expression of *Fasciola hepatica* cathepsin L1 in *Saccharomyces cerevisiae*. Eur J Biochem 245:373–380
24. Dowd AJ, Tort J, Roche L, Ryan T, Dalton JP (1997) Isolation of a cDNA encoding *Fasciola hepatica* cathepsin L and functional expression in *Saccharomyces cerevisiae*. Mol Biochem Parasitol 88:163–174
25. Cornelissen JBWJ, Gaasenbeek CPH, Borgesteede FHM, Holland WG, Harmsen MM, Boersma WJA (2001) Early immunodiagnosis of fasciolosis in ruminants using recombinant *Fasciola hepatica* cathepsin L-like protease. Int J Parasitol 31:728–737
26. Spithill TW, Piedrafita D, Smooker PM (1997) Immunological approaches for the control of fasciolosis. Int J Parasitol 27:1221–1235
27. Smooker PM, Whisstock J, Irving J, Siyaguna S, Spithill TW, Pike RN (2000) A single amino acid substitution affects substrate specificity in cysteine proteinases from *Fasciola hepatica*. Protein Sci 9:2567–2572
28. Grams R, Vichasri-Grams S, Sobhon P, Upatham ES, Viyanant V (2001) Molecular cloning and characterisation of cathepsin L encoding genes from *Fasciola gigantica*. Parasitol Int 50:105–114
29. Tantrawatpan C, Maleewong W, Wongkham C, Wongkham S, Intapan PM, Nakashima K (2005) Serodiagnosis of human fascioliasis by a cystatin capture enzyme-linked immunosorbent assay with recombinant *Fasciola gigantica* cathepsin L antigen. Am J Trop Med Hyg 72:82–86
30. Raina OK, Yadav SC, Sriveny D, Gupta SC (2006) Immunodiagnosis of bubaline fasciolosis with *Fasciola gigantica* Cathepsin L and recombinant Cathepsin L 1-D proteases. Acta Trop 98:145–151
31. Yamasaki H, Mineki R, Murayama F, Ito A, Aoki T (2002) Characterisation and expression of the *Fasciola gigantica* Cathepsin L gene. Int J Parasitol 32:1031–1042
32. Carmona C, Dowd AJ, Smith AM, Dalton JP (1993) Cathepsin L proteinase secreted by *Fasciola hepatica in vitro* prevents antibody mediated eosinophil attachment to newly excysted juveniles. Mol Biochem Parasitol 62:9–18
33. Tkalcevic J, Ashman K, Meeusen E (1995) *Fasciola hepatica*: rapid identification of newly excysted juvenile proteins. Biochem Biophys Res Commun 213:169–174
34. Tkalcevic J, Brandon MR, Meeusen EN (1996) *Fasciola hepatica*: rapid switching of stage-specific antigen expression after infection. Parasite Immunol 18:139–147
35. Law RH, Smooker PM, Irving JA, Piedrafita D, Ponting R, Kennedy NJ, Whisstock JC, Pike RN, Spithill TW (2003) Cloning and expression of the major secreted cathepsin B-like protein from juvenile *Fasciola hepatica* and analysis of immunogenicity following liver fluke infection. Infect Immun 71:6921–6932
36. Wilson LR, Good RT, Panaccio M, Wijffels GL, Sandeman RM, Spithill TW (1998) *Fasciola hepatica*: characterization and cloning of the major cathepsin B protease secreted by newly excysted juvenile liver fluke. Exp Parasitol 88:85–94
37. Meemon K, Grams R, Vichasri-Grams S, Hofmann A, Korge G, Viyanant V, Upathum ES, Habe S, Sobhon P (2004) Molecular cloning and analysis of stage and tissue-specific expression of cathepsin B encoding genes from *Fasciola gigantica*. Mol Biochem Parasitol 136:1–10
38. Cancela M, Acosta D, Rinaldi G, Silva E, Durán R, Roche L, Zaha A, Carmona C, Tort JF (2008) A distinctive repertoire of cathepsins is expressed by juvenile invasive *Fasciola hepatica*. Biochimie 90:1461–1475
39. Robinson MW, Menon R, Donnelly SM, Dalton JP, Ranganathan S (2009) An integrated transcriptomics and proteomics analysis of the secretome of the helminth pathogen *Fasciola*

hepatica: proteins associated with invasion and infection of the mammalian host. Mol Cell Proteomics 8:1891–1907

40. Robinson MW, Dalton JP, Donnelly S (2008) Helminth pathogen cathepsin proteases: it's a family affair. Trends Biochem Sci 33:601–608
41. Pratt D, Cox GN, Milhausen MJ, Boisvenue RJ (1990) A developmentally regulated cysteine protease gene family in *Haemonchus contortus*. Mol Biochem Parasitol 43:181–191
42. Pratt D, Armes LG, Hageman R, Reynolds V, Boisvenue RJ, Cox GN (1992) Cloning and sequence comparisons of four distinct cysteine proteases expressed by *Haemonchus contortus* adult worms. Mol Biochem Parasitol 51:209–218
43. Skuce PJ, Redmond DL, Liddell S, Stewart EM, Newlands GF, Smith WD, Knox DP (1999) Molecular cloning and characterization of gut-derived cysteine proteinases associated with a host protective extract from *Haemonchus contortus*. Parasitology 119:405–412
44. Rehman A, Jasmer DP (1999) Defined characteristics of cathepsin B-like proteins from nematodes: inferred functional diversity and phylogenetic relationships. Mol Biochem Parasitol 102:297–310
45. Knox DP, Redmond DL, Jones DG (1993) Characterization of proteinases in extracts of adult *Haemonchus contortus*, the ovine abomasal nematode. Parasitology 106:395–404
46. Kumar S, Pritchard DI (1993) New observations on proteases of the human hookworm *Necator americanus*. Biochem Mol Biol Int 30:13–19
47. Harrop SA, Sawangjaroen N, Prociv P, Brindley PJ (1995) Characterization and localization of cathepsin B proteinases expressed by adult *Ancylostoma caninum* hookworms. Mol Biochem Parasitol 71:163–171
48. Jasmer DP, Roth J, Myler PJ (2001) Cathepsin B-like cysteine proteases and *Caenorhabditis elegans* homologues dominate gene products expressed in adult *Haemonchus contortus* intestine. Mol Biochem Parasitol 116:159–169
49. Shompole S, Jasmer DP (2001) Cathepsin B-like cysteine proteases confer intestinal cysteine protease activity in *Haemonchus contortus*. J Biol Chem 276:2928–2934
50. Baig S, Damian RT, Peterson DS (2002) A novel cathepsin B active site motif is shared by helminth blood feeders. Exp Parasitol 101:83–89
51. Loukas A, Selzer PM, Maizels RM (1998) Characterisation of Tc-cpl-1, a cathepsin L-like cysteine protease from *Toxocara canis* infective larvae. Mol Biochem Parasitol 92:275–289
52. Hashmi S, Britton C, Liu J, Guiliano DB, Oksov Y, Lustigman S (2002) Cathepsin L is essential for embryogenesis and development of *Caenorhabditis elegans*. J Biol Chem 277:3477–3486
53. Britton C, Murray L (2002) A cathepsin L protease essential for *Caenorhabditis elegans* embryogenesis is functionally conserved in parasitic nematodes. Mol Biochem Parasitol 122:21–33
54. Howard RJ, Chapman CB, Mitchell GF (1980) A difference in surface proteins of *Fasciola hepatica* larvae from intact and nude mice. Aus J Exp Bio Med Sci 58:201–205
55. O'Neill SM, Brady MT, Callanan JJ, Mulcahy G, Joyce P, Mills KHG, Dalton JP (2000) *Fasciola hepatica* infection down-regulates Th1 responses in mice. Parasite Immunol 22:147–155
56. Mulcahy G, Joyce P, Dalton JP (1999) Immunology of *Fasciola hepatica* infection. In: Dalton JP (ed) Fasciolosis. CAB International, Wallingford, pp 341–376
57. O'Neill SM, Mills KHG, Dalton JP (2001) *Fasciola hepatica* cathepsin L cysteine proteinase suppresses *Bordetella pertussis* specific interferon-gamma production in vivo. Parasite Immunol 23:541–547
58. Harmsen MM, Cornelissen JBWJ, Buijs HE, Boersma WJ, Jeurissen SH, van Milligen FJ (2004) Identification of a novel *Fasciola hepatica* cathepsin L protease containing protective epitopes within the propeptide. Int J Parasitol 34:675–682
59. Prowse RK, Chaplin P, Robinson HC, Spithill TW (2002) *Fasciola hepatica* cathepsin L suppresses sheep lymphocyte proliferation in vitro and modulates surface CD4 expression on human and ovine T cell. Parasite Immunol 24:57–66

60. Dowd AJ, McGonigle S, Dalton JP (1995) *Fasciola hepatica* cathepsin L cleaves fibrinogen and produces a novel type of fibrinogen clot. Eur J Biochem 232:241–246
61. Berasain P, Goni F, McGonigle S, Dowd AJ, Dalton JP, Frangione B, Carmona C (1997) Proteinases secreted by *Fasciola hepatica* degrade extracellular matrix and basement membrane components. J Parasitol 83:1–5
62. Beckham SA, Piedrafita D, Phillips CI, Samarawickrema N, Law RH, Smooker PM, Quinsey NS, Irving JA, Greenwood D, Verhelst SH, Bogyo M (2009) A major cathepsin B protease from the liver fluke *Fasciola hepatica* has atypical active site features and a potential role in the digestive tract of newly excysted juvenile parasites. Int J Biochem Cell Biol 41:1601–1612
63. McGonigle L, Mousley A, Marks NJ, Brennab GP, Dalton JP, Spithill TW, Day TA, Maule AG (2008) The silencing of cysteine proteases in *Fasciola hepatica* newly excysted juveniles using RNA interference reduces gut penetration. Int J Parasitol 38:149–155
64. Wijffels GL, Salvatore L, Dosen M, Waddington J, Wilson L, Thompson C, Campbell N, Sexton J, Wicker J, Bowen F, Friedel T, Spithill TW (1994) Vaccination of sheep with purified cysteine proteinases of *Fasciola hepatica* decreases worm fecundity. Exp Parasitol 78:132–148
65. Edith R (2004) Comparative immunoprotection profile of infection specific and excretory secretory proteins of *Fasciola gigantica* against tropical fasciolosis in buffaloes. MVSc thesis, IVRI Deemed University, Izatnagar, pp 1–96
66. Yadav SC, Mandal S, Sharma RL (1999) A comparative evaluation of *Fasciola gigantica* antigens in immunodiagnosis of bubaline fasciolosis. Rivista Di Parasitologia 16:73–81
67. Cornelissen JBWJ, Gaasenbeek CPH, Boersma W, Borgsteede FHM, van Milligen FJ (1999) Use of a preselected epitope of cathepsin L1 in a highly specific peptide based immunoassay for the diagnosis of *Fasciola hepatica* infections in cattle. Int J Parasitol 29:685–696
68. Dixit AK, Yadav SC, Sharma RL (2004) Kinetics of antibody response using 28 kDa *Fasciola gigantica* cysteine proteinase as antigen. Trop Anim Health Prod 36:49–54
69. Dixit AK, Yadav SC, Saini M, Sharma RL (2003) Purification and characterization of 28 kDa cysteine proteinase for immunodiagnosis of tropical fasciolosis. J Vet Parasitol 17:5–9
70. Jezek J, Ridi RE, Salah M, Wagih A, Aziz HW, Tallima H, Shafie MH, Khalek TA, Ammou FF, Strongylis C, Moussis V, Tsikaris V (2007) *Fasciola gigantica* cathepsin L proteinase based synthetic peptide for immunodiagnosis and prevention of sheep fasciolosis. Peptide Sci 90(3):349–357
71. Sriveny D, Raina OK, Yadav SC, Chandra D, Jayaraw AK, Singh M, Velusamy R, Singh BP (2006) Cathepsin L cysteine proteinase in the diagnosis of bovine *Fasciola gigantica* infection. Vet Parasitol 135:25–31
72. Dalton JP, McGonigle S, Rolph TP, Andrews SJ (1996) Induction of protective immunity in cattle against infection with *Fasciola hepatica* by vaccination with cathepsin L proteinases and with hemoglobin. Inf Immun 64:5066–5074
73. Mulcahy G, O'Corner F, McGonigle S, Dowd AJ, Clery D, Andrews SJ, Dalton JP (1998) Correlation of specific antibody titre and avidity with protection in cattle immunized against *Fasciola hepatica*. Vaccine 16:932–939
74. Piacenza L, Acosta D, Basmadjian I, Dalton JP, Carmona C (1999) Vaccination with cathepsin L-proteinases and with leucine amino peptidase induces high levels of protection against fascioliasis in sheep. Inf Immun 67:1954–1961
75. Kesik M, Jedlina-Panasiuk L, Kozak-Cieszczyk M, Plucienniczak A, Wedrychowicz H (2007) Enteral vaccination of rats against *Fasciola hepatica* using recombinant cysteine proteinase (cathepsin L1). Vaccine 25:3619–3628
76. Kofta W, Mieszczanek J, Plucienniczak G, Wedrychowicz H (2000) Successful DNA immunisation of rats against fasciolosis. Vaccine 18:2985–2990
77. Wedrychowicz H, Lamparska M, Kesik M, Kotomski G, Mieszczanek J, Jedlina-Panasiuk L, Jaros S, Plucienniczak A (2003) The immune response of rats to vaccination with the cDNA

or protein forms of the cysteine proteinase of *Fasciola hepatica*. Vet Immunol Immunopathol 94:83–93
78. Wedrychowicz H, Kesik M, Kaliniak M, Kozak-Cieszczyk M, Jedlina-Panasiuk L, Jaros S, Plucienniczak A (2007) Vaccine potential of inclusion bodies containing cysteine proteinase of *Fasciola hepatica* in calves and lambs experimentally challenged with metacercariae of the fluke. Vet Parasitol 147:77–88
79. Kennedy NJ, Spithill TW, Tennent J, Wood PR, Piedrafita D (2006) DNA vaccines in sheep: CTLA- 4 mediated targeting and CpG motifs enhance immunogenicity in a DNA prime/protein boost strategy. Vaccine 24:970–979
80. Jayaraj R, Piedrafita D, Dynon K, Grams R, Spithill TW, Smooker PM (2009) Vaccination against fasciolosis by a multivalent vaccine of stage-specific antigens. Vet Parasitol 160:230–236
81. Knox DP, Skuce PJ, Newlands GF, Redmond DL (2001) Nematode gut peptidases, proteins and vaccination. In: Kennedy MW, Harnett W (eds) Parasitic nematodes: molecular biology, biochemistry and immunology. CAB International, pp 247–268
82. Willadsen P, Riding GA, McKenna RV, Kemp DH, Tellam RL, Nielsen JN, Lahnstein J, Cobon GS, Gough JM (1989) Immunologic control of a parasitic arthropod. Identification of a protective antigen from *Boophilus microplus*. J Immunol 143:1346–1351
83. Opdebeeck JP (1994) Vaccines against blood-sucking arthropods. Vet Parasitol 54:205–222
84. Knox DP, Smith WD (2001) Vaccination against gastrointestinal nematode parasites of ruminants using gut-expressed antigens. Vet Parasitol 100:21–32
85. Smith WD (1999) Prospects for vaccines of helminth parasites of grazing ruminants. Int J Parasitol 29:17–24
86. Smith WD (1993) Protection in lambs immunised with *Haemonchus contortus* gut membrane proteins. Res Vet Sci 54:94–101
87. Hotez PJ, Ashcom J, Bin Z, Bethony J, Williamson A, Hawdon JM, Jianjun F, Dobardzic A, Rizo I, Bolden J, Jin Q (2002) Effects of vaccinations with recombinant fusion proteins on the *Ancylstoma caninum* habitat selection in the canine intestine. J Parasitol 88:684–690
88. Lee JS, Kim IS, Sohn WM, Lee J, Yong TS (2006) Vaccination with DNA encoding cysteine proteinase confers protective immune response to rats infected with *Clonorchis sinensis*. Vaccine 24(13):2358–2366
89. Muleke CI, Yan R, Sun Y, Zhao G, Xu L, Li X (2007) Vaccination of goats against *Haemonchus contortus* with a recombinant cysteine protease. Small Rumin Res 73:95–102

Printed by Printforce, the Netherlands